human physiology

seventh edition

The Mechanisms of Body Function

Arthur Vander
Professor Emeritus of Physiology
University of Michigan

James Sherman
Associate Professor of Physiology
University of Michigan

Dorothy Luciano
Formerly of the Department of Physiology
University of Michigan

WCB
McGraw-Hill

Boston, Massachusetts   Burr Ridge, Illinois   Dubuque, Iowa
Madison, Wisconsin   New York, New York   San Francisco, California   St. Louis, Missouri

WCB/McGraw-Hill

A Division of The **McGraw·Hill** Companies

 This book is printed on recycled, acid-free paper containing 10% postconsumer waste.

2 3 4 5 6 7 8 9 0   VNH VNH 9 0 9 8

The credits section for this book is on page 789 and is considered an extension of the copyright page.

Cover credit: Corel-CD *Amateur Sports*

ISBN 0-07-067065-X

Vice president, editorial director: *Kevin T. Kane*
Publisher: *James M. Smith*
Sponsoring editor: *Kristine Noel Tibbetts*
Developmental editor: *Brittany J. Rossman*
Marketing manager: *Keri L. Witman*
Project manager: *Peggy J. Selle*
Production supervisor: *Sandra Hahn*
Designer: *K. Wayne Harms*
Cover designer: *Jeff Storm*
Compositor: *York Graphics*
Typeface: *10/12 New Caledonia*
Printer: *Von Hoffman Press, Inc.*

INTERNATIONAL EDITION

2 3 4 5 6 7 8 9 0   VNH VNH 9 0 9 8

When ordering this title, use ISBN 0-07-115624-0

Vander, Arthur J., 1933–
    Human physiology : the mechanisms of body function / Arthur Vander,
James Sherman, Dorothy Luciano. — 7th ed.
       p.   cm.
    Includes bibliographical references and index.
    ISBN 0-07-067065-X
    1. Human physiology.   I. Sherman, James H., 1936–
II. Luciano, Dorothy S.   III. Title.
    [DNLM: 1. Physiology.   QT 104 V228h 1998]
QP34.5.V36   1998
612—dc21
DNLM/DLC
for Library of Congress   97-28929
                                                    CIP

www.mhhe.com

*To our parents,
and to Judy, Peggy, and Joe
without whose understanding
it would have been
impossible*

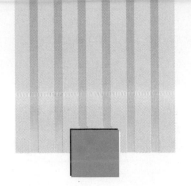

REFERENCES FOR FIGURE ADAPTATIONS

Berne, R. M., and N. M. Levy: "Cardio-vascular Physiology," 6th ed., Mosby, St. Louis, 1992.

Bloom, F. E., A. Lazerson, and L. Hofstadter: "Brain, Mind, and Behaviour," Freeman, New York, 1985.

Carlson, B. M.: "Patten's Foundations of Embryology," 5th ed., McGraw-Hill, New York, 1988.

Chaffee, E. E., and I. M. Lytle: "Basic Physiology and Anatomy," 4th ed., Lippincott, Philadelphia, 1980.

Chapman, C. B., and J. H. Mitchell: *Scientific American*, May 1965.

Comroe, J. H.: "Physiology of Respiration," Year Book, Chicago, 1965.

Crosby, W. H.: *Hospital Practice*, February 1987.

Davis, H., and H. R. Silverman: "Hearing and Deafness," Holt, Rinehart, and Winston, New York, 1970.

Dowling, J. E. and B. B. Boycott: *Proceedings of the Royal Society, London* B, **166:**80 (1966).

Egan, S. E., and R. A. Weinberg: *Nature*, **365:**281, (1993).

Elias, H., J. E. Pauly, and E. R. Burns: "Histology and Human Microana-tomy," 4th ed., Wiley, New York, 1978.

Erickson, G. F., D. A. Magoffin, C. A. Dyer, and C. Hofeditz: *Endocrine Reviews*, Summer 1985.

Felig, P., and J. Wahren: *New England Journal of Medicine*, **293:**1078 (1975).

Ganong, W. F.: "Review of Medical Physiology," 4th ed., Lange, Los Altos, Calif., 1969.

Gardner, E. "Fundamentals of Neurology," 5th ed., Saunders, Philadelphia, 1968.

Golde, D. W., and J. C. Gasson: *Scientific American*, July 1988.

Goodman, H. M.: "Basic Medical Endocrinology," Raven, New York, 1988.

Gray, H. M., A. Sette, and S. Buus (drawing by G. B. Kelvin): *Scientific American*, November 1989.

Guyton, A. C.: "Functions of the Human Body," 3d ed., Saunders, Philadelphia, 1969.

Hedge, G. A., H. D. Colby, and R. L. Goodman: "Clinical Endocrine Physio-logy," Saunders, Philadelphia, 1987.

Hoffman, B. F., and P. E. Cranefield: "Electrophysiology of the Heart," McGraw-Hill, New York, 1960.

Hubel, D. H., and T. N. Wiesel: *Journal of Physiology*, **154:**572 (1960).

Hudspeth, A. J.: *Scientific American*, January 1983.

Kandel, E. R., and J. H. Schwartz: "Principles of Neural Science," 2d ed., Elsevier/North-Holland, New York, 1985.

Kappas, A., and A. P. Alvares: *Scientific American*, June 1975.

Kuffler, W. W., J. G. Nicholls, and A. R. Martin: "From Neuron to Brain," 2d ed., Sinauer Associates, Sunderland, MA, 1984.

Lambersten, C. J.: in P. Bard (ed.), "Medical Physiological Psychology," 11th ed., Mosby, St. Louis, 1961.

Lehninger, A. L.: "Biochemistry," Worth, New York, 1970.

Lentz, T. L.: "Cell Fine Structure," Saunders, Philadelphia, 1971.

Little, R. C.: "Physiology of the Heart and Circulation," 2d ed., Year Book, Chicago, 1981.

Maxwell, D. J., and M. Rethely: *Trends in Neurosciences*, **10:**117 (1987).

Nauta, W. J. H., and M. Fiertag: "Funda-mental Neuroanatomy," Freeman, New York, 1986.

Olds, J.: *Scientific American*, October 1956.

Rasmujssen, A. T.: "Outlines of Neuro-anatomy," 2d ed., Wm. C. Brown, Dubuque, IA, 1943.

Rhoades, R., and R. Pflanzer: "Human Physiology," Saunders, Philadelphia, 1989.

Roberts, H. R., and J. N. Lozier (drawing by Ilil Arbel): *Hospital Practice*, January 1992.

Rushmer, R. F.: "Cardiovascular Dynamics," 2d ed., Saunders, Philadelphia, 1961.

Scales, W. E., A. J. Vander, M. B. Brown, and M. J. Kluger: *American Journal of Physiology*, **65:**1840 (1988).

Snyder, S. H.: "Drugs and the Brain," Freeman, New York, 1986.

Tung, K.: *Hospital Practice*, June 1988.

Young, M. P.: *Proc. R. Soc. Lond.* B, **252:**13 (1993).

von Bekesy, G.: *Scientific American*, August 1957.

Wang, C. C., and M. I. Grossman: *American Journal of Physiology*, **164:**527 (1951).

Wersall, J., L. Gleisner, and P. G. Lundquist: in A. V. S. de Reuck and J. Knight (eds.), "Myostatic, Kinesthetic, and Vestibular Mechanisms," Ciba Foundation Symposium, Little Brown, Boston, 1967.

CONTENTS IN BRIEF

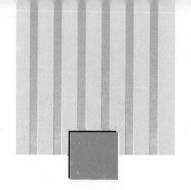

CONTENTS

9   THE SENSORY SYSTEMS   223

12   CONTROL OF BODY MOVEMENT   330

13   CONSCIOUSNESS AND BEHAVIOR   351

PART THREE: COORDINATED BODY FUNCTIONS

14 CIRCULATION 372

15 RESPIRATION   461

**18 REGULATION OF ORGANIC METABOLISM,
GROWTH, AND ENERGY BALANCE**   **590**

**SECTION A CONTROL AND INTEGRATION
OF CARBOHYDRATE, PROTEIN, AND
FAT METABOLISM**   **591**

19 REPRODUCTION 635

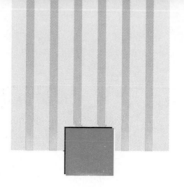

Preface

GOALS AND ORIENTATION

The purpose of this book remains what it was in the first six editions: to present the fundamental principles and facts of human physiology in a format that is suitable for undergraduate students, regardless of academic backgrounds or fields of study: liberal arts, biology, nursing, pharmacy, or other allied health professions. The book is also suitable for dental students, and many medical students have also used previous editions to lay the foundation for the more detailed coverage they receive in their courses.

The most significant feature of this book is its clear, up-to-date, accurate explanations of **mechanisms**, rather than the mere description of facts and events. Because there are no limits to what can be covered in an introductory text, it is essential to reinforce over and over, through clear explanations, that physiology can be understood in terms of basic themes and principles. As evidenced by the very large number of flow diagrams employed, the book emphasizes understanding based on the ability to think in **clearly defined chains of causal links.** This approach is particularly evident in our emphasis of the dominant theme of human physiology and in this book—**homeostasis** as achieved through the coordinated function of **homeostatic control systems.**

To repeat, we have attempted to explain, integrate, and synthesize information rather than simply to describe, so that students will achieve a working knowledge of physiology, not just a memory bank of physiological facts. Since our aim has been to tell a coherent story, rather than to write an encyclopedia, we have been willing to devote considerable space to the logical development of difficult but essential concepts; examples are second messengers (chapter 7), membrane potentials (chapter 8), and the role of intrapleural pressure in breathing (chapter 15).

In keeping with our goals, the book progresses from the cell to the total body (Figure 1-1), utilizing information and principles developed previously at each level of complexity. One example of this approach is as follows: the characteristics that account for this protein specificity are presented in Part One (Chapter 4), and this concept is used in that section to explain the "recognition" process exhibited by enzymes. It is then used again in Part Two (Chapter 7) for membrane receptors, and again in Part Three (Chapter 20) for antibodies. In this manner, the student is helped to see the basic foundations upon which more complex functions such as homeostatic neuroendocrine and immune responses are built.

Another example: Rather than presenting, in a single chapter, a gland-by-gland description of all the hormones, we give a description of the basic principles of endocrinology in Chapter 10, but then save the details of the individual hormones for later chapters. This permits the student to focus on the functions of the hormones in the context of the homeostatic control systems in which they participate.

ALTERNATIVE SEQUENCES

Given the inevitable restrictions of time, our organization permits a variety of sequences and approaches to be adopted. Chapter 1 should definitely be read first as it introduces the basic themes that dominate the book. Depending on the time available, the instructor's goals, and the students' backgrounds in physical science and cellular biology, the chapters of Part One can be either worked through systematically at the outset or be used more selectively as background reading in the contexts of Parts Two and Three.

In Part Two, the absolutely essential chapters are, in order, Chapters 7, 8, 10, and 11, for they present the basic concepts and facts relevant to homeostasis, intercellular communication, signal transduction, and nervous and endocrine systems and muscle. The material in Chapters 9, 12, and 13 is not as critical for an understanding of later chapters.

It would be best to begin the coordinated body functions of Part Three with circulation (Chapter 14), but otherwise the chapters of Part Three, as well as Chapters 9, 12, and 13 of Part Two, can be rearranged and used or not used to suit individual instructor's preferences and time availability.

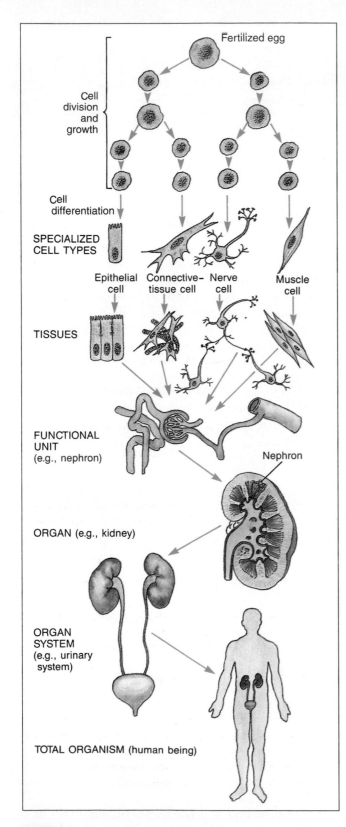

FIGURE 1-1
Levels of cellular organization.

1. Our goal for this revision was to completely update all material and assure the greatest accuracy possible. The following is a partial list of topics that have either been significantly altered or added for the first time:

Chapter 1
Structure and functions of extracellular matrix

Chapters 2 and 4
Free radicals

Chapter 5
Human Genome Project
mRNA splicing
Chaperones
Transcription factors (see figure, p.xxiii)
Telomeres and DNA replication
Nucleosomes
Growth factors
DNA repair mechanisms
Genetic diseases
Genetic engineering: DNA fingerprinting, transgenic organisms, DNA cloning, knockout organisms
Cancer and p53 mutations

Chapter 6
Aquaporins
Endosomes

Chapter 7
Darwinian medicine
Learning and feedforward regulation
Genetic mutations and premature aging
Receptor structure
Receptors and cytoplasmic tyrosine kinases
Phosphatases in signal transduction
Plasma membrane receptors and gene transduction
Primary response genes
"Cross-talk" in signal transduction pathways

Chapter 8
Multiple effects of a single neurotransmitter

Chapter 9
Signal transduction in retinal receptors
Pain
Hair cells
Devices used to assist hearing
Vertigo and motion sickness
Taste and smell

Chapter 10
Nongenomic effects of steroid hormones
Roles of hypophysiotropic hormones

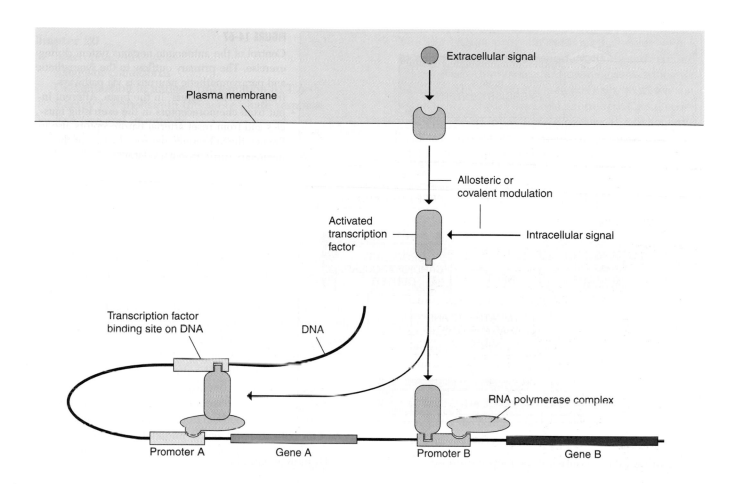

FIGURE 5-9

Transcription of gene B is modulated by the binding of an activated transcription factor directly to the promoter region. In contrast, transcription of gene A is modulated by the same transcription factor which, in this case, binds to a region of DNA considerably distant from the promoter region.

One cannot meaningfully analyze the complex activities of the human body without a framework upon which to build, a set of viewpoints to guide one's thinking. It is the purpose of this chapter to provide such an orientation to the subject of **human physiology**—the mechanisms by which the body functions.

MECHANISM AND CAUSALITY

The **mechanist view** of life, the view taken by physiologists, holds that all phenomena, no matter how complex, can ultimately be described in terms of physical and chemical laws. In contrast, **vitalism** is the view that some "vital force" beyond physics and chemistry is required to explain life. The mechanist view has predominated in the twentieth century because virtually all information gathered from observation and experiment has agreed with it.

Physiologists should not be misunderstood when they sometimes say that "the whole is greater than the sum of its parts." This statement in no way implies a vital force but rather recognizes that *integration* of an enormous number of individual physical and chemical events occurring at all levels of organization is required for biological systems to function. Thus the scope of physiology is extremely broad. At one end of the spectrum, it includes the study of individual molecules—for example, how a particular protein's shape and electrical properties allow it to function as a channel for sodium ions to move into or out of a cell. At the other end, it is concerned with complex processes that depend on the interplay of many widely separated organs in the body—for example, regulation of the amount of sodium in the entire body.

A common denominator of physiological processes is their contribution to survival. Unfortunately, it is easy to misunderstand the nature of this relationship. Consider, for example, the statement, "During exercise a person sweats because the body *needs* to get rid of the excess heat generated." This type of statement is an example of **teleology** (pronounced teal-ee-OL-oh-gee), the explanation of events in terms of purpose, but it is not an explanation at all in the scientific sense of the word. It is somewhat like saying, "The furnace is on because the house needs to be heated." Clearly, the furnace is on not because it senses in some mystical manner the house's "needs," but because the temperature has fallen below the thermostat's set point and the electric current in the connecting wires has turned on the heater.

Is it not true that sweating serves a useful purpose during exercise because the excess heat, if not eliminated, might cause sickness or even death? Yes, it is, but this is totally different from stating that a need to avoid injury causes the sweating. The cause of the sweating is a sequence of events initiated by the increased heat generation: increased heat generation → increased blood temperature → increased activity of specific nerve cells in the brain → increased activity of a series of nerve cells → increased production of sweat by the sweat-gland cells. Each step occurs by means of physicochemical changes in the cells involved. In science, to explain a phenomenon is to reduce it to a causally linked sequence of physicochemical events. This is the scientific meaning of causality, of the word "because."

This is a good place to emphasize that causal chains can be not only long, as in the example just cited, but also multiple. In other words, one should not assume the simple relationship of one cause, one effect. We shall see that multiple factors often must interact to elicit a response. To take an example from medicine, cigarette smoking can cause lung cancer, but the likelihood of cancer developing in a smoker depends on a variety of other factors, including the way that person's body processes the chemicals in cigarette smoke, the rate at which damaged molecules are repaired, and so on.

That a phenomenon is beneficial to a person, while not explaining the *mechanism* of the phenomenon, is of obvious interest and importance. Evolution is the key to understanding why most body activities do indeed appear to be purposeful, since responses that have survival value undergo natural selection. Throughout this book we emphasize how a particular process contributes to survival, but the reader must never confuse the survival value of a process with the explanation of the mechanisms by which the process occurs.

A SOCIETY OF CELLS

Cells: The Basic Units of Living Organisms

The simplest structural units into which a complex multicellular organism can be divided and still retain the functions characteristic of life are called **cells.** One of the unifying generalizations of biology is that certain fundamental activities are common to almost all cells and represent the minimal requirements for maintaining cell integrity and life. Thus, a human liver cell and an amoeba are remarkably similar in their means of exchanging materials with their immediate environments, of obtaining energy from organic nutrients, of synthesizing complex molecules, and of duplicating themselves.

Each human organism begins as a single cell, a fertilized egg, which divides to create two cells, each of which divides in turn, resulting in four cells, and so on. If cell multiplication were the only event occurring, the end result would be a spherical mass of indentical cells. During development, however, each cell becomes specialized for the performance of a particular function, such as producing force and movement (muscle cells) or generating electric signals (nerve cells). The process of transforming an unspecialized cell into a specialized cell is known as **cell differentiation.**

In addition to differentiating, cells migrate to new locations during development and form selective adhesions with other cells to produce multicellular structures. In this manner, the cells of the body are arranged in various combinations to form a hierarchy of organized structures. Differentiated cells with similar properties aggregate to form **tissues** (nerve tissue, muscle tissue, and so on), which combine with other types of tissues to form **organs** (the heart, lungs, kidneys, and so on), which are linked together to form **organ systems** (Figure 1-1).

About 200 distinct kinds of cells can be identified in the body in terms of differences in structure and function. When cells are classified according to the broad types of function they perform, however, four categories emerge: (1) muscle cells, (2) nerve cells, (3) epithelial cells, and (4) connective-tissue cells. In each of these functional categories, there are several cell types that perform variations of the specialized function. For example, there are three types of muscle cells—skeletal, cardiac, and smooth—which differ from each other in shape, in the mechanisms controlling their contractile activity, and in their location in the various organs of the body.

Muscle cells are specialized to generate the mechanical forces that produce force and movement. They may be attached to bones and produce movements of the limbs or trunk. They may be attached to skin, as for example, the muscles producing facial expressions. They may enclose hollow cavities so that their contraction expels the contents of the cavity, as in the pumping of the heart. Muscle cells also surround many of the tubes in the body—blood vessels, for example—and their contraction changes the diameter of these tubes.

Nerve cells are specialized to initiate and conduct electric signals, often over long distances. A signal may initiate new electric signals in other nerve cells, or it may stimulate secretion by a gland cell or contraction of a muscle

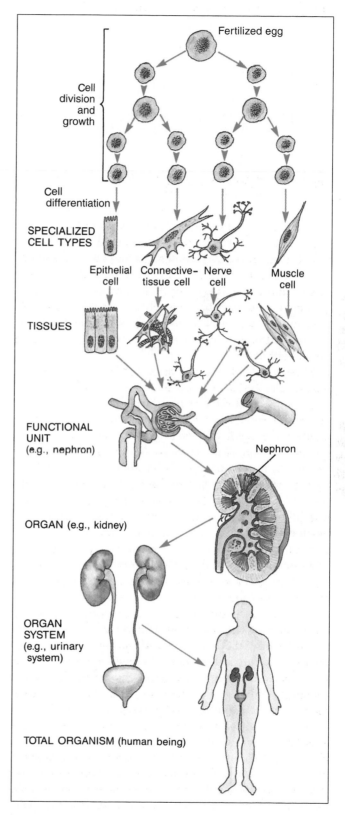

FIGURE 1-1

Levels of cellular organization.

cell. Thus, nerve cells provide a major means of controlling the activities of other cells. Their activity also underlies such phenomena as consciousness and perception.

Epithelial cells are specialized for the selective secretion and absorption of ions and organic molecules. They are located mainly at the surfaces that either cover the body or individual organs or else line the walls of various tubular and hollow structures within the body. Epithelial cells, which rest on a homogeneous extracellular protein layer called the **basement membrane,** form the boundaries between compartments and function as selective barriers regulating the exchange of molecules across them. For example, the epithelial cells at the surface of the skin form a barrier that prevents most substances in the external environment from entering the body through the skin. Epithelial cells are also found in glands that form from the invagination of epithelial surfaces.

Connective-tissue cells, as their name implies, have as their major function connecting, anchoring, and supporting the structures of the body. These cells typically have a large amount of extracellular material between them. Some connective-tissue cells are found in the loose meshwork of cells and fibers underlying most epithelial layers; other types include fat-storing cells, bone cells, and red blood cells and white blood cells.

We will emphasize later in this chapter that the immediate environment of each individual cell in the body is the extracellular fluid. Actually this fluid is interspersed within a complex **extracellular matrix** consisting of a mixture of protein molecules (and, in some cases, minerals) specific for any given tissue. The matrix serves two general functions: (1) It provides a scaffold for cellular attachments, and (2) it transmits to the cells information, in the form of chemical messengers, that helps regulate their migration, growth, and differentiation.

The proteins of the extracellular matrix consist of **fibers**—ropelike **collagen fibers** and rubberband-like **elastin fibers**—and a mixture of other proteins that contain chains of complex sugars (carbohydrates). In some ways, the structure of the extracellular matrix is analogous to that of reinforced concrete. The fibers of the matrix, particularly collagen, which constitutes one-third of all bodily proteins, are like the reinforcing iron mesh or rods in the concrete, and the carbohydrate-containing protein molecules are the surrounding cement. However, these latter molecules are not merely inert "packing material," as in concrete, but function as adhesion/recognition molecules between cells and as important links in the communication chains between extracellular messenger molecules and cells.

Tissues

Most specialized cells are associated with other cells of a similar kind to form tissues. Corresponding to the four general categories of differentiated cells, there are four general classes of tissues: (1) **muscle tissue,** (2) **nerve tissue,** (3) **epithelial tissue,** and (4) **connective tissue.** It should be noted that the term "tissue" is used in different ways. It is formally defined as an aggregate of a single type of specialized cell. However, it is also commonly used to denote the general cellular fabric of any organ or structure, for example, kidney tissue or lung tissue, each of which in fact usually contains all four classes of tissue.

Organs and Organ Systems

Organs are composed of the four kinds of tissues arranged in various proportions and patterns: sheets, tubes, layers, bundles, strips, and so on. For example, the kidneys consist of (1) a series of small tubes, each composed of a single layer of epithelial cells; (2) blood vessels, whose walls contain varying quantities of smooth muscle and connective tissue; (3) nerve-cell extensions that end near the muscle and epithelial cells; and (4) a loose network of connective-tissue elements that are interspersed throughout the kidneys and also form enclosing capsules.

Many organs are organized into small, similar subunits often referred to as **functional units,** each performing the function of the organ. For example, the kidneys' 2 million functional units are termed nephrons (which contain the small tubes mentioned in the previous paragraph), and the total production of urine by the kidneys is the sum of the amounts formed by the individual nephrons.

Finally we have the organ system, a collection of organs that together perform an overall function. For example, the kidneys, the urinary bladder, the tubes leading from the kidneys to the bladder, and the tube leading from the bladder to the exterior constitute the urinary system. There are 10 organ systems in the body. Their components and functions are given in Table 1-1.

To sum up, the human body can be viewed as a complex society of differentiated cells structurally and functionally combined and interrelated to carry out the functions essential to the survival of the organism. The individual cells constitute the basic units of this society, and almost all of these cells individually exhibit the fundamental activities common to all forms of life. Indeed, many of the cells can be removed and maintained in test tubes as free-living organisms (this is termed *"in vitro,"* literally "in glass," as opposed to *"in vivo,"* meaning within the body).

There is a paradox in this analysis: How is it that the functions of the organ systems are essential to the survival of the organism when each individual cell seems capable of performing its own fundamental activities? The resolution of this paradox is found in the isolation of most of the cells of a multicellular organism from the external environment—the environment surrounding the organism—and the presence of an internal environment.

TABLE 1-1 ORGAN SYSTEMS OF THE BODY

System	Major organs or tissues	Primary functions
Circulatory	Heart, blood vessels, blood (Some classifications also include lymphatic vessels and lymph in this system.)	Transport of blood throughout the body's tissues
Respiratory	Nose, pharynx, larynx, trachea, bronchi, lungs	Exchange of carbon dioxide and oxygen; regulation of hydrogen-ion concentration
Digestive	Mouth, pharynx, esophagus, stomach, intestines, salivary glands, pancreas, liver, gallbladder	Digestion and absorption of organic nutrients, salts, and water
Urinary	Kidneys, ureters, bladder, urethra	Regulation of plasma composition through controlled excretion of salts, water, and organic wastes
Musculoskeletal	Cartilage, bone, ligaments, tendons, joints, skeletal muscle	Support, protection, and movement of the body; production of blood cells
Immune	White blood cells, lymph vessels and nodes, spleen, thymus, and other lymphoid tissues	Defense against foreign invaders; return of extracellular fluid to blood; formation of white blood cells
Nervous	Brain, spinal cord, peripheral nerves and ganglia, special sense organs	Regulation and coordination of many activities in the body; detection of changes in the internal and external environments; states of consciousness; learning; cognition
Endocrine	All glands secreting hormones: Pancreas, testes, ovaries, hypothalamus, kidneys, pituitary, thyroid, parathyroid, adrenal, intestinal, thymus, heart, and pineal	Regulation and coordination of many activities in the body
Reproductive	Male: Testes, penis, and associated ducts and glands	Production of sperm; transfer of sperm to female
	Female: Ovaries, uterine tubes, uterus, vagina, mammary glands	Production of eggs; provision of a nutritive environment for the developing embryo and fetus; nutrition of the infant
Integumentary	Skin	Protection against injury and dehydration; defense against foreign invaders; regulation of temperature

THE INTERNAL ENVIRONMENT AND HOMEOSTASIS

An amoeba and a human liver cell both obtain their energy by breaking down certain organic nutrients. The chemical reactions involved in this intracellular process are remarkably similar in the two types of cells, and involve the utilization of oxygen and the production of carbon dioxide. The amoeba picks up oxygen directly from the fluid surrounding it (its external environment) and eliminates carbon dioxide into the same fluid. But how can the liver cell and all other internal parts of the body obtain oxygen and eliminate carbon dioxide when, unlike the amoeba, they are not in direct contact with the external environment—the air surrounding the body?

Figure 1-2 summarizes the exchanges of matter that occur in a person. Supplying oxygen is the function both of the respiratory system, which takes up oxygen from the external environment, and of the circulatory system, which distributes the oxygen to all parts of the body. In addition, the circulatory system carries the carbon dioxide generated by all the cells of the body to the lungs, which eliminate it to the exterior. Similarly, the digestive and circulatory systems working together make nutrients from the external environment available to all the body's cells. Wastes other than carbon dioxide are carried by the

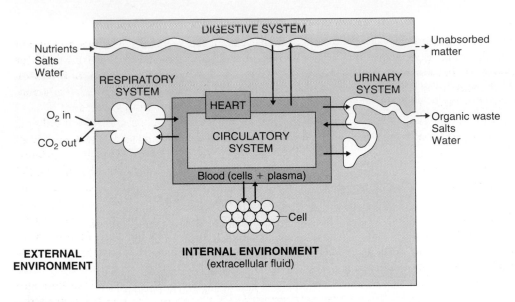

FIGURE 1-2

Exchanges of matter occur between the external environment and the circulatory system via the digestive, respiratory, and urinary systems. Extracellular fluid (plasma and interstitial fluid) is the internal environment of the body. The external environment is the air surrounding the body.

circulatory system from the cells that produced them to the kidneys and liver, which excrete them from the body. The kidneys also regulate the amounts of water and many essential minerals in the body. The nervous and hormonal systems coordinate and control the activities of all the other organ systems.

Thus the overall effect of the activities of organ systems is to create within the body an environment in which all cells can survive and function. This fluid environment surrounding each cell is called the **internal environment.** The internal environment is not merely a theoretical physiological concept. It can be identified quite specifically in anatomical terms. The body's internal environment is the **extracellular fluid** (literally, fluid outside the cells), which bathes each cell (Figure 1-2).

In other words, the environment in which each cell lives is not the external environment surrounding the entire body but the local extracellular fluid surrounding that cell. It is from this fluid that the cells receive oxygen and nutrients and into which they excrete wastes. A multicellular organism can survive only as long as it is able to maintain the composition of its internal environment in a state compatible with the survival of its individual cells. In 1857, the French physiologist Claude Bernard clearly described the central importance of the extracellular fluid: *"It is the fixity of the internal environment that is the condition of free and independent life. . . . All the vital mechanisms, however varied they may be, have only one object,* *that of preserving constant the conditions of life in the internal environment."*

The relative constancy of the internal environment is known as **homeostasis** (pronounced home-ee-oh-STAY-sis). Changes do occur but the magnitudes of these changes are small and are kept within narrow limits. As emphasized by the twentieth-century American physiologist, Walter B. Cannon, such stability can be achieved only through the operation of carefully coordinated physiological processes. The activities of the cells, tissues, and organs must be regulated and integrated with each other in such a way that any change in the extracellular fluid initiates a reaction to minimize the change. A collection of body components that functions to keep a physical or chemical property of the internal environment relatively constant is termed a **homeostatic control system.** As will be described in detail in Chapter 7, such a system must be capable of detecting changes in the magnitude of the property, relay this information to an appropriate site for integration with other incoming information, and elicit a "command" to particular cells to alter their rates of function in such a way as to restore the property toward its original value.

The description at the beginning of this chapter of how sweating is brought about in response to increased heat generation during exercise is an example of a homeostatic control system in operation; the sweating (more precisely, the evaporation of the sweat) removes heat from the body and keeps the body temperature relatively constant even though more heat is being produced by the exercising muscles. Here is another example: A mountaineer who ascends to high altitude suffers a decrease in the concentration of oxygen in his or her blood because of the decrease in the amount of oxygen in inspired air; the nervous system detects this change in the blood and increases its signals to the skeletal muscles responsible for breathing. The result is that the mountaineer breathes more rapidly and deeply, and the increase in the amount of air inspired helps keep the blood oxygen concentration from falling as much as it otherwise would.

To summarize, the total activities of every individual cell in the body fall into two categories: (1) Each cell performs for itself all those fundamental basic cellular processes—movement of materials across its membrane, extraction of energy, protein synthesis, and so on—that represent the minimal requirements for maintaining its individual integrity and life; and (2) each cell simultaneously performs one or more specialized activities that, in concert with the activities performed by the other cells of its tissue or organ system, contribute to the survival of the organism by helping maintain the stable internal environment required by all cells.

BODY-FLUID COMPARTMENTS

Although the internal environment can be equated with the extracellular fluid, it was not stated earlier that extracellular fluid exists in two locations—surrounding cells and inside blood vessels. Approximately 80 percent of the extracellular fluid surrounds all the body's cells except the blood cells. Because it lies between cells, this 80 percent of the extracellular fluid is known as **interstitial fluid** (or intercellular fluid). The remaining 20 percent of the extracellular fluid is the fluid portion of the blood, the **plasma,** in which the various blood cells are suspended.

As the blood (plasma plus suspended blood cells) is continuously circulated by the action of the heart to all parts of the body, the plasma exchanges oxygen, nutrients, wastes, and other metabolic products with the interstitial fluid surrounding the smallest of the blood vessels. Because of the exchanges between plasma and interstitial fluid, concentrations of dissolved substances are virtually identical in the two fluids, except for protein concentration. With this major exception—higher protein concentration in plasma than in interstitial fluid—the entire extracellular fluid may be considered to have a homogeneous composition. In contrast, the composition of the extracellular fluid is very different from that of the **intracellular fluid,** the fluid inside the cells (the actual differences will be presented in Chapter 6, Table 6-1).

In essence, the fluids in the body are enclosed in "compartments." The volumes of the body-fluid compartments are summarized in Figure 1-3 in terms of water, since water is by far the major component of the fluids in each compartment. Water accounts for about 60 percent of normal body weight. Two-thirds of this water (28 L in a typical normal 70-kg person) is intracellular fluid. The remaining one-third (14 L) is extracellular, and as described above, 80 percent of this extracellular fluid is interstitial fluid (11 L) and 20 percent (3 L) is plasma.

FIGURE 1-3.

Fluid compartments of the body. Volumes are for an average 70-kg (154-lb) man. TBW = Total body water; ECF = Extracellular fluid.

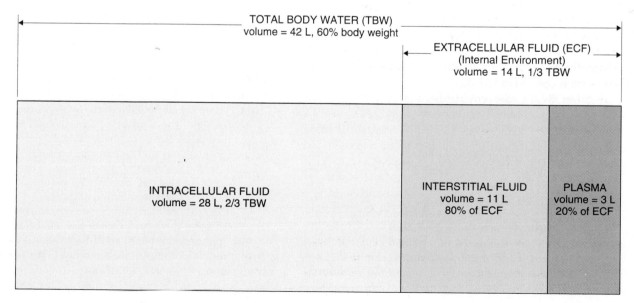

Compartmentalization is an important general principle in physiology (we shall see in Chapter 3 that the inside of cells also is divided into compartments). Compartmentalization is achieved by barriers between the compartments, and the properties of the barriers determine which substances can move between contiguous compartments. These movements in turn account for the differences in composition of the different compartments. In the case of the body-fluid compartments, the intracellular fluid is separated from the extracellular fluid by membranes that surround the cells; the properties of these membranes and how they account for the profound differences between intracellular and extracellular fluid are described in Chapter 6. In contrast, the two components of extracellular fluid—the interstitial fluid and the blood plasma—are separated by the cellular wall of the smallest blood vessels, the capillaries. How this barrier normally keeps 80 percent of the extracellular fluid in the interstitial compartment and restricts proteins mainly to the plasma is described in Chapter 14.

This completes our introductory framework. With it in mind, the overall organization and approach of this book should easily be understood. Because the fundamental features of cell function are shared by virtually all cells and because these features constitute the foundation upon which specialization develops, we devote Part 1 of the book to an analysis of basic cell physiology.

Part 2 provides the principles and information required to bridge the gap between the functions of individual cells and the integrated systems of the body. Chapter 7 describes the basic characteristics of homeostatic control systems and the required cellular communications. The other chapters of Part 2 deal with the specific components of the body's control systems: nerve cells, muscle cells, and gland cells.

Part 3 describes the coordinated functions (circulation, respiration, and so on) of the body, emphasizing how they result from the precisely controlled and integrated activities of specialized cells grouped together in tissues and organs. The theme of these descriptions is that each function, with the obvious exception of reproduction, serves to keep some important aspect of the body's internal environment relatively constant. Thus, homeostasis, achieved by homeostatic control systems, is the single most important unifying idea to be kept in mind in Part 3.

SUMMARY

MECHANISM AND CAUSALITY

I. The mechanist view of life, the view taken by physiologists, holds that all phenomena can be described in terms of physical and chemical laws.

II. Vitalism holds that some additional force is required to explain the function of living organisms.

A SOCIETY OF CELLS

I. Cells are the simplest structural units into which a complex multicellular organism can be divided and still retain the functions characteristic of life.

II. Cell differentiation results in the formation of four categories of specialized cells.
    A. Muscle cells generate the mechanical activities that produce force and movement.
    B. Nerve cells initiate and conduct electric signals.
    C. Epithelial cells selectively secrete and absorb ions and organic molecules.
    D. Connective-tissue cells connect, anchor, and support the structures of the body.

III. Specialized cells associate with similar cells to form tissues: muscle tissue, nerve tissue, epithelial tissue, and connective tissue.

IV. Organs are composed of the four kinds of tissues arranged in various proportions and patterns; many organs contain multiple small, similar functional units.

V. An organ system is a collection of organs that together perform an overall function.

THE INTERNAL ENVIRONMENT AND HOMEOSTASIS

I. The extracellular fluid surrounding cells is the body's internal environment.

II. The function of organ systems is to maintain the internal environment relatively constant—homeostasis. This is achieved by homeostatic control systems.

III. Each cell performs the basic cellular processes required to maintain its own integrity plus specialized activities that help achieve homeostasis.

BODY-FLUID COMPARTMENTS

I. The body fluids are enclosed in compartments.
    A. The extracellular fluid is composed of the interstitial fluid (the fluid between cells) and the blood plasma. Of the extracellular fluid, 80 percent is interstitial fluid, and 20 percent is plasma.
    B. Interstitial fluid and plasma have essentially the same composition except that plasma contains a much higher concentration of protein.
    C. Extracellular fluid differs markedly in composition from the fluid inside cells—the intracellular fluid.
    D. Approximately one-third of body water is in the extracellular compartment, and two-thirds is intracellular.

II. The differing compositions of the compartments reflect the activities of the barriers separating them.

extracellular fluid   interstitial fluid
homeostasis   plasma
homeostatic control system   intracellular fluid

KEY TERMS

human physiology	basement membrane
mechanist view	connective-tissue cell
vitalism	extracellular matrix
teleology	fiber
cell	collagen fiber
cell differentiation	elastin fiber
tissue	muscle tissue
organ	nerve tissue
organ system	epithelial tissue
muscle cell	connective tissue
nerve cell	functional unit
epithelial cell	internal environment

REVIEW QUESTIONS

1. Describe the levels of cellular organization and state the four types of specialized cells and tissues.

2. List the 10 organ systems of the body and give one-sentence descriptions of their functions.

3. Contrast the two categories of functions performed by every cell.

4. Name two fluids that constitute the extracellular fluid. What are their relative proportions in the body, and how do they differ from each other in composition?

5. State the relative volumes of water in the body-fluid compartments.

CHAPTER 2

Chemical Composition of the Body

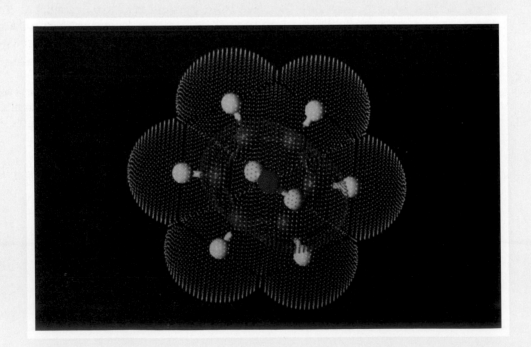

toms and molecules are the chemical units of cell structure and function. In this chapter we describe the distinguishing characteristics of the major chemicals in the human body. The specific roles of these substances will be discussed in subsequent chapters. This chapter is, in essence, an expanded glossary of chemical terms and structures, and like a glossary, it should be consulted as needed.

ATOMS

The units of matter that form all chemical substances are called **atoms.** The smallest atom, hydrogen, is approxi- mately 2.7 billionths of an inch in diameter. Each type of atom—carbon, hydrogen, oxygen, and so on—is called a **chemical element.** A one- or two-letter symbol is used as a shorthand identification for each element. Although slightly more than 100 elements exist in the universe, only 24 (Table 2-1) are known to be essential for the structure and function of the human body.

The chemical properties of atoms can be described in terms of three subatomic particles—**protons, neutrons,** and **electrons** (Table 2-2). The protons and neutrons are confined to a very small volume at the center of an atom, the **atomic nucleus,** whereas the electrons revolve in or- bits at various distances from the nucleus. This miniature- solar-system model of an atom is an oversimplification, but it is sufficient to provide a conceptual framework for un- derstanding the chemical and physical interactions of atoms.

Each of the subatomic particles has a different electric charge: Protons have one unit of positive charge, electrons have one unit of negative charge, and neutrons are elec- trically neutral and have no electric charge. Since the pro- tons are located in the atomic nucleus, the nucleus has a net positive charge equal to the number of protons it con- tains. The entire atom has no net electric charge, however, because the number of negatively charged electrons or- biting the nucleus is equal to the number of positively charged protons in the nucleus.

Protons and neutrons are approximately equal in mass and are about 1800 times heavier than an electron (Table 2-2). Thus, almost all the mass of an atom is located in its nucleus.

TABLE 2-1   ESSENTIAL ELEMENTS IN THE BODY

Element	Symbol
Major elements: 99.3% of total atoms	
Hydrogen	H (63%)
Oxygen	O (26%)
Carbon	C (9%)
Nitrogen	N (1%)
Mineral elements: 0.7% of total atoms	
Calcium	Ca
Phosphorus	P
Potassium	K (Latin *kalium*)
Sulfur	S
Sodium	Na (Latin *natrium*)
Chlorine	Cl
Magnesium	Mg
Trace elements: less than 0.01% of total atoms	
Iron	Fe (Latin *ferrum*)
Iodine	I
Copper	Cu (Latin *cuprum*)
Zinc	Zn
Manganese	Mn
Cobalt	Co
Chromium	Cr
Selenium	Se
Molybdenum	Mo
Fluorine	F
Tin	Sn (Latin *stannum*)
Silicon	Si
Vanadium	V

TABLE 2-2   CHARACTERISTICS OF MAJOR SUBATOMIC PARTICLES

Particle	Mass relative to electron mass	Electric charge	Location in atom
Electron	1	−1	Orbiting the nucleus
Proton	1836	+1	Nucleus
Neutron	1839	0	Nucleus

Atomic Number

Every atom of each chemical element contains a specific number of protons, and it is this number that distinguishes one type of atom from another. This number is known as the **atomic number.** For example, hydrogen, the simplest atom, has an atomic number of 1, corresponding to its single proton; calcium has an atomic number of 20, corresponding to its 20 protons. Since an atom is electrically neutral, the atomic number is also equal to the number of electrons in the atom.

Atomic Weight

Atoms have very little mass. A single hydrogen atom, for example, has a mass of only 1.67×10^{-24} g. The **atomic weight** scale indicates an atom's mass relative to the mass of other atoms. This scale is based upon assigning the carbon atom a mass of 12. On this scale a hydrogen atom has an atomic weight of approximately 1, indicating that it has

one-twelfth the mass of a carbon atom; a magnesium atom, with an atomic weight of 24, has twice the mass of a carbon atom.

Since the atomic weight scale is a *ratio* of atomic masses, it has no units. The unit of atomic mass is known as a dalton, and 1 dalton (d) equals one-twelfth the mass of a single carbon atom. Thus, carbon has an atomic weight of 12, and a single carbon atom has an atomic mass of 12 daltons.

Although the number of neutrons in the nucleus of an atom is often equal to the number of protons, many chemical elements can exist in multiple forms, called **isotopes,** which differ in the number of neutrons they contain. For example, the most abundant form of the carbon atom, ^{12}C, contains 6 protons and 6 neutrons, and thus has an atomic number of 6 and an atomic weight of 12, whereas the radioactive carbon isotope ^{14}C contains 6 protons and 8 neutrons giving it an atomic number of 6 but an atomic weight of 14.

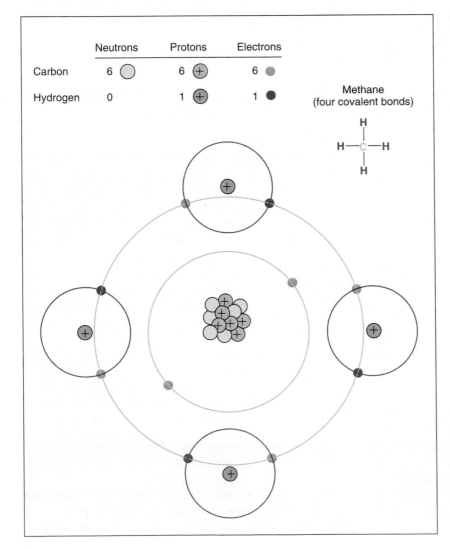

FIGURE 2-1

Each of the four hydrogen atoms in a molecule of methane (CH_4) forms a covalent bond with an atom of carbon by sharing its one electron with one of the electrons in the carbon atom. Each shared pair of electrons forms a covalent bond.

One **gram atomic mass** of a chemical element is the amount of the element in grams that is equal to the numerical value of its atomic weight. Thus, 12 g of carbon is 1 gram atomic mass of carbon, and 1 g of hydrogen is 1 gram atomic mass of hydrogen. *One gram atomic mass of any element contains the same number of atoms.* For example, 1 g of hydrogen contains 6×10^{23} atoms, and 12 g of carbon, whose atoms have 12 times the mass of a hydrogen atom, also has 6×10^{23} atoms.

Atomic Composition of the Body

Just four of the body's essential elements (Table 2-1)— hydrogen, oxygen, carbon, and nitrogen—account for over 99 percent of the atoms in the body.

The seven essential mineral elements are the most abundant substances dissolved in the extracellular and intracellular fluids. Most of the body's calcium and phosphorus atoms, however, make up the solid matrix of bone tissue.

The 13 essential **trace elements** are present in extremely small quantities, but they are nonetheless essential for normal growth and function. For example, iron plays a critical role in the transport of oxygen by the blood. Additional trace elements will likely be added to this list as the chemistry of the body becomes better understood.

Many other elements, in addition to the 24 listed in Table 2-1, can be detected in the body. These elements enter in the foods we eat and the air we breathe but are not essential for normal body function and may even interfere with normal body chemistry. For example, ingested arsenic has poisonous effects.

MOLECULES

Two or more atoms bonded together make up a **molecule.** For example, a molecule of water contains two hydrogen atoms and one oxygen atom, which can be represented by H_2O. The atomic composition of glucose, a sugar, is $C_6H_{12}O_6$, indicating that the molecule contains 6 carbon atoms, 12 hydrogen atoms, and 6 oxygen atoms. Such formulas, however, do not indicate how the atoms are linked together in the molecule.

Covalent Chemical Bonds

The atoms in molecules are held together by chemical bonds, which are formed when electrons are transferred from one atom to another or are shared between two atoms. The strongest chemical bond between two atoms, a **covalent bond,** is formed when one electron in the outer electron orbit of each atom is shared between the two atoms. The atoms in most molecules found in the body are linked by covalent bonds.

The atoms of some elements can form more than one covalent bond and thus become linked simultaneously to two or more other atoms (Figure 2-1). Each type of atom forms a characteristic number of covalent bonds which depends on the number of electrons in its outermost electron orbit. The number of chemical bonds formed by the four most abundant atoms in the body are hydrogen, one; oxygen, two; nitrogen, three; and carbon, four. When the structure of a molecule is diagramed, each covalent bond is represented by a line indicating a pair of shared electrons. The covalent bonds of the four elements mentioned above can be represented as

$$H- \qquad -O- \qquad -\overset{\displaystyle |}{N}- \qquad -\overset{\displaystyle |}{\underset{\displaystyle |}{C}}-$$

A molecule of water H_2O can be diagramed as

$$H-O-H$$

In some cases, two covalent bonds—a double bond—are formed between two atoms by the sharing of two electrons from each atom. Carbon dioxide (CO_2) contains two double bonds:

$$O=C=O$$

Note that in this molecule the carbon atom still forms four covalent bonds and each oxygen atom only two.

Molecular Shape

When atoms are linked together, molecules with various shapes can be formed. Although we draw diagrammatic structures of molecules on flat sheets of paper, these molecules actually have a three-dimensional shape. When more than one covalent bond is formed with a given atom, the bonds are distributed about the atom in a pattern that may or may not be symmetrical (Figure 2-2).

Molecules are not rigid, inflexible structures. Within certain limits, the shape of a molecule can be changed without breaking the covalent bonds linking its atoms together. A covalent bond is like an axle around which the joined atoms can rotate. As illustrated in Figure 2-3, a sequence of six carbon atoms can assume a number of shapes as a result of rotations around various covalent bonds. As we shall see, the three-dimensional shape of molecules is one of the major factors governing molecular interactions.

IONS

A single atom is electrically neutral since it contains equal numbers of negative electrons and positive protons. If, however, an atom gains or loses one or more electrons, it

TABLE 2-3 MOST FREQUENTLY ENCOUNTERED IONIC FORMS OF ELEMENTS

Atom	Chemical symbol	Ion	Chemical symbol	Electrons gained or lost
Hydrogen	H	Hydrogen ion	H^+	1 lost
Sodium	Na	Sodium ion	Na^+	1 lost
Potassium	K	Potassium ion	K^+	1 lost
Chlorine	Cl	Chloride ion	Cl^-	1 gained
Magnesium	Mg	Magnesium ion	Mg^{2+}	2 lost
Calcium	Ca	Calcium ion	Ca^{2+}	2 lost

acquires a net electric charge and becomes an **ion.** For example, when a sodium atom (Na), which has 11 electrons, loses 1 electron, it becomes a sodium ion (Na^+) with a net positive charge; it still has 11 protons, but it now has only 10 electrons. On the other hand, a chlorine atom (Cl), which has 17 electrons, can gain an electron and become a chloride ion (Cl^-) with a net negative charge—it now has 18 electrons but only 17 protons. Some atoms can gain or lose more than 1 electron to become ions with two or even three units of net electric charge, for example, calcium, Ca^{2+}. Because of their ability to conduct electricity when dissolved in water, ions are collectively referred to as **electrolytes.**

Hydrogen atoms and most mineral and trace element atoms readily form ions. The number of electrons an atom may gain or lose in becoming an ion is a specific characteristic of each type of atom. Table 2-3 lists the ionic forms of some of these elements. Ions that have a net positive

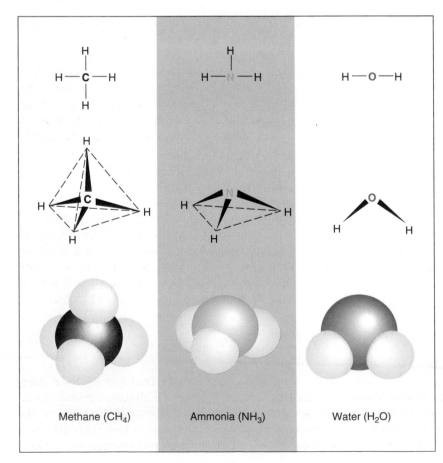

Methane (CH_4)   Ammonia (NH_3)   Water (H_2O)

FIGURE 2-2

Geometric configuration of covalent bonds around the carbon, nitrogen, and oxygen atoms bonded to hydrogen atoms.

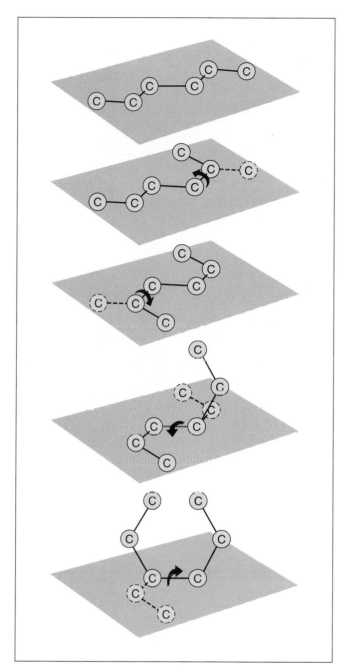

FIGURE 2-3

Changes in molecular shape occur as portions of a molecule rotate around different carbon-to-carbon bonds, transforming this molecule's shape, for example, from a relatively straight chain into a ring.

charge are called **cations,** while those that have a net negative charge are called **anions.**

The process of ion formation, known as ionization, can occur in single atoms or in atoms that are covalently linked in molecules. Within molecules two commonly encountered groups of atoms that undergo ionization are the

carboxyl group (—COOH) and the **amino group** (—NH$_2$). The formulas for molecules containing such groups are written as R—COOH or R—NH$_2$, where R signifies the remaining portion of the molecule. The carboxyl group ionizes when the oxygen atom linked to the hydrogen atom captures the hydrogen atom's only electron to form a carboxyl ion (R—COO$^-$) and releases a hydrogen ion (H$^+$):

$$R\text{—COOH} \rightleftharpoons R\text{—COO}^- + H^+$$

The amino group can bind a hydrogen ion to form an ionized amino group (R — NH$_3^+$):

$$R\text{—NH}_2 + H^+ \rightleftharpoons R\text{—NH}_3^+$$

The ionization of each of the above groups can be reversed, as indicated by the double arrows; the ionized carboxyl group can combine with a hydrogen ion to form an un-ionized carboxyl group, and the ionized amino group can lose a hydrogen ion and become an un-ionized amino group.

FREE RADICALS

The electrons that revolve around the nucleus of an atom occupy regions known as orbitals, each of which can be occupied by two electrons. An atom is most stable when each orbital is occupied by two electrons of opposite spin, one clockwise and the other anticlockwise. An atom containing a single electron in its outermost orbital is known as a **free radical,** as are molecules containing such atoms. Most free radicals react rapidly with other atoms, thereby filling the unpaired orbital; thus free radicals normally exist for only brief periods of time before combining with other atoms.

Free radicals are diagramed with a dot next to the atomic symbol. Examples of biologically important free radicals are the superoxide anion, $O_2 \cdot^-$; hydroxyl radical, $OH \cdot$; and nitric oxide, $NO \cdot$. Note that a free radical configuration can occur in either an ionized or an un-ionized atom. A number of free radicals play important roles in the normal and abnormal functioning of the body.

POLAR MOLECULES

As we have seen, when the electrons of two atoms interact, the two atoms may share the electrons equally, forming an electrically neutral covalent bond, or one of the atoms may completely capture an electron from the other, forming two ions. Between these two extremes are covalent

TABLE 2-4 EXAMPLES OF NONPOLAR AND POLAR BONDS, AND IONIZED CHEMICAL GROUPS

Nonpolar bonds	$-\overset{\displaystyle \|}{\underset{\displaystyle \|}{C}}-H$	Carbon-hydrogen bond
	$-\overset{\displaystyle \|}{\underset{\displaystyle \|}{C}}-\overset{\displaystyle \|}{\underset{\displaystyle \|}{C}}-$	Carbon-carbon bond
Polar bonds	$\overset{(-)\ (+)}{R-O-H}$	Hydroxyl group (R—OH)
	$\overset{(-)\ (+)}{R-S-H}$	Sulfhydryl group (R—SH)
	$R-\overset{\displaystyle H\ (+)}{\underset{\displaystyle \|}{N}}-R$ $^{(-)}$	Nitrogen-hydrogen bond
Ionized groups	$R-\overset{\displaystyle O}{\underset{\displaystyle \|}{C}}-O^{\ominus}$	Carboxyl group (R—COO⁻)
	$R-\overset{\displaystyle H}{\underset{\displaystyle H}{\overset{\displaystyle \|}{\underset{\displaystyle \|}{N}}}}{}^{\oplus}-H$	Amino group ($R-NH_3^+$)
	$R-O-\overset{\displaystyle O}{\underset{\displaystyle O^{\ominus}}{\overset{\displaystyle \|}{\underset{\displaystyle \|}{P}}}}-O^{\ominus}$	Phosphate group ($R-PO_4^{2-}$)

bonds in which the electrons are not shared equally between the two atoms, but instead reside closer to one atom of the pair. This atom thus acquires a slight negative charge, while the other atom, having partly lost an electron, becomes slightly positive. Such bonds are known as **polar covalent bonds** (or, simply, polar bonds) since the atoms at each end of the bond have an opposite electric charge. For example, the bond between hydrogen and oxygen in a **hydroxyl group** (—OH) is a polar covalent bond in which the oxygen is slightly negative and the hydrogen slightly positive:

$$\overset{(-)\ (+)}{R-O-H}$$

The electric charge associated with the ends of a polar bond is considerably less than the charge on a fully ionized atom (for example, the oxygen in the polarized hydroxyl group has only about 13 percent of the negative charge associated with the oxygen in an ionized carboxyl group, R—COO⁻). Polar bonds do not have a *net* electric charge, as do ions, since they contain equal amounts of negative and positive charge.

Atoms of oxygen and nitrogen, which have a relatively strong attraction for electrons, form polar bonds with hydrogen atoms; in contrast, bonds between carbon and hydrogen atoms and between two carbon atoms are electrically neutral (Table 2-4).

Different regions of a single molecule may contain nonpolar, polar, and ionized bonds. Molecules containing significant numbers of polar bonds or ionized groups are known as **polar molecules,** whereas molecules composed predominantly of electrically neutral bonds are known as **nonpolar molecules.** As we shall see, the physical characteristics of these two classes of molecules, especially their solubility in water, are quite different.

Hydrogen Bonds

The electrical attraction between the hydrogen atom in a polar bond in one molecule and an oxygen or nitrogen atom in a polar bond of another molecule—or within the same molecule if the bonds are sufficiently separated from

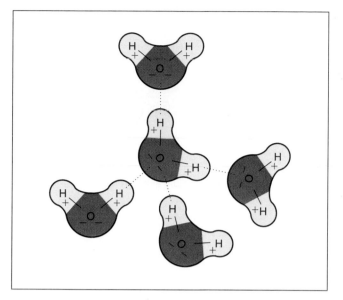

FIGURE 2-4

Five water molecules. Note that polarized covalent bonds link the hydrogen and oxygen atoms within each water molecule and that hydrogen bonds occur between adjacent water molecules. Hydrogen bonds are represented in diagrams by dashed or dotted lines, and covalent bonds by solid lines.

each other—forms a **hydrogen bond.** This type of bond is very weak, having only about 4 percent of the strength of the polar bonds linking the hydrogen and oxygen within a water molecule (H_2O). Hydrogen bonds are represented in diagrams by dashed or dotted lines to distinguish them from stronger covalent bonds (Figure 2-4). Hydrogen bonds between and within molecules play an important role in molecular interactions and in determining the shape of large molecules.

Water

Hydrogen is the most numerous atom in the body, and water is the most numerous molecule. Out of every 100 molecules, 99 are water. The covalent bonds linking the two hydrogen atoms to the oxygen atom in a water molecule are polar. Therefore, the oxygen in water has a slight negative charge, and each hydrogen atom has a slight positive charge. The positively polarized regions near the hydrogen atoms of one water molecule are electrically attracted to the negatively polarized regions of the oxygen atoms in adjacent water molecules by hydrogen bonds (Figure 2-4).

At body temperature, water exists as a liquid because the weak hydrogen bonds between water molecules are continuously being formed and broken. If the temperature is increased, the hydrogen bonds are broken more readily and molecules of water escape into the gaseous state, whereas if the temperature is lowered, hydrogen bonds are broken less frequently so that larger and larger

clusters of water molecules are formed until at 0° C water freezes into a continuous crystalline matrix—ice.

Water molecules take part in many chemical reactions of the general type:

$$R_1 - R_2 \ + \ H - O - H \ \longrightarrow \ R_1 - OH \ + \ H - R_2$$

In this reaction the covalent bond between R_1 and R_2 and the bond between a hydrogen atom and oxygen in water are broken, and the hydroxyl group and hydrogen atom are transferred to R_1 and R_2, respectively. Reactions of this type are known as hydrolytic reactions, or **hydrolysis.** Many large molecules in the body are broken down into smaller molecular units by hydrolysis.

SOLUTIONS

Substances dissolved in a liquid are known as **solutes,** and the liquid in which they are dissolved is the **solvent.** Solutes dissolve in a solvent to form a **solution.** Water is the most abundant solvent in the body, accounting for 60 percent of the total body weight. A majority of the chemical reactions that occur in the body involve molecules that are dissolved in water, either in the intracellular or extracellular fluid. However, not all molecules dissolve in water.

Molecular Solubility

In order to dissolve in water, a substance must be electrically attracted to water molecules. For example, table salt (NaCl) is a solid crystalline substance because of the strong electrical attraction between positive sodium ions and negative chloride ions. This strong attraction between two oppositely charged ions is known as an **ionic bond.** When a crystal of sodium chloride is placed in water, the polar water molecules are attracted to the charged sodium and chloride ions (Figure 2-5). The ions become surrounded by clusters of water molecules, allowing the sodium and chloride ions to separate from the salt crystal and enter the water, that is, to dissolve.

Molecules having a number of polar bonds and/or ionized groups will dissolve in water. Such molecules are said to be **hydrophilic,** or "water-loving." Thus, the presence in a molecule of ionized groups, such as carboxyl and amino groups, or of polar groups, such as hydroxyl groups, promotes solubility in water. In contrast, molecules composed predominantly of carbon and hydrogen are insoluble in water since their electrically neutral covalent bonds are not attracted to water molecules. These molecules are **hydrophobic,** or "water-fearing."

When nonpolar molecules are mixed with water, two phases are formed, as occurs when oil is mixed with water. The strong attraction between polar molecules

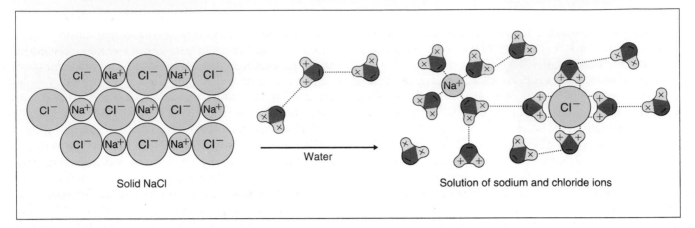

FIGURE 2-5

The ability of water to dissolve sodium chloride crystals depends upon the electrical attraction between the polar water molecules and the charged sodium and chloride ions.

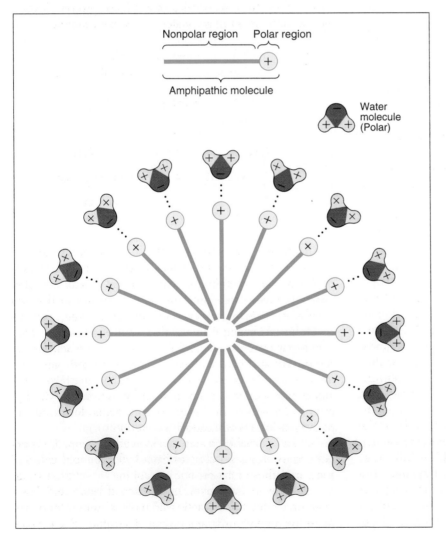

FIGURE 2-6

In water, amphipathic molecules aggregate into spherical clusters. Their polar regions form hydrogen bonds with water molecules at the surface of the cluster.

"squeezes" the nonpolar molecules out of the water phase. Such a separation is never 100 percent complete, however, and very small amounts of nonpolar solutes remain dissolved in the water phase.

Molecules that have a polar or ionized region at one end and a nonpolar region at the opposite end are called **amphipathic**—consisting of two parts. When mixed with water, amphipathic molecules form clusters, with their polar (hydrophilic) regions at the surface of the cluster where they are attracted to the surrounding water molecules. The nonpolar (hydrophobic) ends are oriented toward the interior of the cluster (Figure 2-6). Such an arrangement provides the maximal interaction between water molecules and the polar ends of the amphipathic molecules. Nonpolar molecules can dissolve in the central nonpolar regions of these clusters and thus exist in aqueous solutions in far higher amounts than would otherwise be possible based on their low solubility in water. As we shall see, the orientation of amphipathic molecules plays an important role in the structure of cell membranes and in both the absorption of nonpolar molecules from the gastrointestinal tract and their transport in the blood.

Concentration

Solute **concentration** is defined as the amount of the solute present in a unit volume of solution. One measure of the amount of a substance is its mass given in grams. The unit of volume in the metric system is a liter (L). (One liter equals 1.06 quarts. See Appendix C for metric and English units.) Smaller units are the milliliter (ml, or 0.001 liter) and the microliter (μl, or 0.001 ml). The concentration of a solute in a solution can then be expressed as the number of grams of the substance present in one liter of solution (g/L).

A comparison of the concentrations of two different substances on the basis of the number of *grams* per liter of solution does not directly indicate how many *molecules* of each substance are present. For example, 10 g of compound X, whose molecules are heavier than those of compound Y, will contain fewer molecules than 10 g of compound Y. Concentrations in units of grams per liter are most often used when the chemical structure of the solute is unknown. When the structure of a molecule is known, concentrations are usually expressed as moles per liter, which provides a unit of concentration based upon the number of solute molecules in solution, as described next.

The **molecular weight** of a molecule is equal to the sum of the atomic weights of all the atoms in the molecule. For example, glucose ($C_6H_{12}O_6$) has a molecular weight of 180 ($6 \times 12 + 12 \times 1 + 6 \times 16 = 180$). One **mole** (abbreviated mol) of a compound is the amount of the compound in grams equal to its molecular weight. A solution containing 180 g of glucose (1 mol) in 1 L of solution is a 1 molar solution of glucose (1 mol/L). If 90 g of glucose are dissolved in enough water to produce 1 L of solution, the solution will have a concentration of 0.5 mol/L. Just as 1 gram atomic mass of any element contains the same number of atoms, 1 mol (1 gram molecular mass) of any molecule will contain the same number of molecules—6×10^{23}. Thus, a 1 mol/L solution of glucose contains the same number of solute molecules per liter as a 1 mol/L solution of urea or any other substance.

The concentrations of solutes dissolved in the body fluids are much less than 1 mol/L. Many have concentrations in the range of millimoles per liter (1 mmol/L = .001 mol/L), while others are present in even smaller concentrations—micromoles per liter (1 μmol/L = 0.000001 mol/L) or nanomoles per liter (1 nmol/L = 0.000000001 mol/L).

Hydrogen Ions and Acidity

As mentioned earlier, a hydrogen atom has a single proton in its nucleus orbited by a single electron. A hydrogen ion (H^+), formed by the loss of the electron, is thus a single free proton. Hydrogen ions are formed when the proton of a hydrogen atom in a molecule is released, leaving behind its electron. Molecules that release protons (hydrogen ions) in solution are called **acids,** for example:

$$HCl \longrightarrow H^+ + Cl^-$$
hydrochloric acid $\qquad$ chloride

$$H_2CO_3 \rightleftharpoons H^+ + HCO_3^-$$
carbonic acid $\qquad$ bicarbonate

$$CH_3-\underset{\underset{H}{|}}{\overset{\overset{OH}{|}}{C}}-COOH \rightleftharpoons H^+ + CH_3-\underset{\underset{H}{|}}{\overset{\overset{OH}{|}}{C}}-COO^-$$
lactic acid $\qquad\qquad$ lactate

Conversely, any substance that can accept a hydrogen ion (proton) is termed a **base.** In the reactions above, bicarbonate and lactate are bases since they can combine with hydrogen ions (note the double arrows in the two reactions). It is important to distinguish between the un-ionized and ionized forms of these molecules and to note that separate terms are used for the un-ionized acid forms, lactic acid and carbonic acid, and the ionized bases derived from the acids, lactate and bicarbonate. Bases remove hydrogen ions from solution by combining with hydrogen ions to form the un-ionized acids, lowering the hydrogen-ion concentration of a solution.

When hydrochloric acid is dissolved in water, 100 percent of its atoms separate to form hydrogen and chloride ions, and these ions do not recombine in solution (note the one-way arrow above). In the case of lactic acid, however, only a fraction of lactic acid molecules in solution release hydrogen ions at any instant. Therefore, if a 1 mol/L solution of hydrochloric acid is compared with a 1 mol/L solution of lactic acid, the hydrogen-ion concentration will be lower in the lactic acid solution than in the hydrochloric acid solution. Hydrochloric acid and other acids that are 100 percent ionized in solution are known as **strong acids,** whereas carbonic and lactic acids and other acids that do not completely ionize in solution are **weak acids.** The same principles apply to bases.

It must be understood that the hydrogen-ion concentration of a solution refers only to the hydrogen ions that are free in solution and not to those that may be bound, for example, to amino groups ($R—NH_3^+$). The **acidity** of a solution refers to the *free* (unbound) hydrogen-ion concentration in the solution; the higher the hydrogen-ion concentration, the greater the acidity. The hydrogen-ion concentration is frequently expressed in terms of the **pH** of a solution, which is defined as the negative logarithm to the base 10 of the hydrogen-ion concentration (the brackets around the symbol for the hydrogen ion in the formula below indicate concentration):

$$pH = -\log [H^+]$$

Thus, a solution with a hydrogen-ion concentration of 10^{-7} mol/L has a pH of 7, whereas a more acidic solution with a concentration of 10^{-6} mol/L has a pH of 6. Note that as the acidity *increases,* the pH *decreases;* and a change in pH from 7 to 6 represents a tenfold increase in the hydrogen-ion concentration.

Pure water, due to the ionization of some of the molecules into H^+ and OH^-, has a hydrogen-ion concentration of 10^{-7} mol/L (pH = 7.0) and is termed a **neutral solution. Alkaline solutions** have a lower hydrogen-ion concentration (a pH higher than 7.0), while those with a higher hydrogen-ion concentration (a pH lower than 7.0) are **acidic solutions.** The extracellular fluid of the body has a hydrogen-ion concentration of about 4×10^{-8} mol/L (pH = 7.4), with a normal range of about pH 7.35 to 7.45, and is thus slightly alkaline. Most intracellular fluids have a slightly higher hydrogen-ion concentration (pH 7.0 to 7.2) than extracellular fluids.

As we saw earlier, the ionization of carboxyl and amino groups involves the release and uptake, respectively, of hydrogen ions. These groups behave as weak acids and bases. Changes in the acidity of solutions containing molecules with carboxyl and amino groups alters the net electric charge on these molecules by shifting the ionization reaction to the right or left, as illustrated for the carboxyl group:

$$R—COOH \rightleftharpoons R—COO^- + H^+$$

If the electric charge on a molecule is altered, its interaction with other molecules or with other regions within the same molecule is altered, and thus its functional characteristics are altered. In the extracellular fluid, hydrogen-ion concentrations beyond the tenfold pH range of 7.8 to 6.8 are incompatible with life if maintained for more than a brief period of time. Even small changes in the hydrogen-ion concentration can produce large changes in molecular interactions, as we shall see.

CLASSES OF ORGANIC MOLECULES

Because most naturally occurring carbon-containing molecules are found in living organisms, the study of these compounds became known as organic chemistry. (Inorganic chemistry is the study of non-carbon-containing molecules.) However, the chemistry of living organisms, **biochemistry,** now forms only a portion of the broad field of organic chemistry.

One of the properties of the carbon atom that makes life possible is its ability to form four covalent bonds with other atoms, in particular with other carbon atoms. Since carbon atoms can also combine with hydrogen, oxygen, nitrogen, and sulfur atoms, a vast number of compounds can be formed with relatively few chemical elements. Some of these molecules are extremely large (**macromolecules**), being composed of thousands of atoms. Such large molecules are formed by linking together hundreds of smaller molecules (subunits) and are thus known as **polymers** (many small parts). The structure of macromolecules depends upon the structure of the subunits, the number of subunits linked together, and the position along the chain of each type of subunit.

Most of the organic molecules in the body can be classified into one of four groups: carbohydrates, lipids, proteins, and nucleic acids (Table 2-5).

Carbohydrates

Although carbohydrates account for only about 1 percent of the body weight, they play a central role in the chemical reactions that provide cells with energy. Carbohydrates are composed of carbon, hydrogen, and oxygen atoms in the proportions represented by the general formula $Cn(H_2O)n$, where *n* is any whole number. It is from this formula that the class of molecules gets its name, **carbohydrate**—water-containing (hydrated) carbon atoms. Linked to most of the carbon atoms in a carbohydrate are a hydrogen atom and a hydroxyl group:

$$H—\overset{|}{\underset{|}{C}}—OH$$

Chemical groups containing nitrogen and phosphorus atoms may also be linked to this basic carbohydrate structure. The presence of numerous polar hydroxyl groups makes carbohydrates readily soluble in water.

Most carbohydrates taste sweet, and it is among the carbohydrates that we find the substances known as sug-

TABLE 2-5 MAJOR CATEGORIES OF ORGANIC MOLECULES IN THE BODY

Category	Percent of body weight	Majority of atoms	Subclass	Subunits
Carbohydrates	1	C, H, O	Monosaccharides (sugars)	
			Polysaccharides	Monosaccharides
Lipids	15	C, H	Triacylglycerols	3 fatty acids + glycerol
			Phospholipids	2 fatty acids + glycerol + phosphate + small charged nitrogen molecule
			Steroids	
Proteins	17	C, H, O, N	Peptides	Amino acids
			Proteins	Amino acids
Nucleic acids	2	C, H, O, N	DNA	Nucleotides containing the bases adenine, cytosine, guanine, thymine, the sugar deoxyribose, and phosphate
			RNA	Nucleotides containing the bases adenine, cytosine, guanine, uracil, the sugar ribose, and phosphate

ars. The simplest sugars are the **monosaccharides** (single-sweet), the most abundant of which is **glucose,** a six-carbon molecule ($C_6H_{12}O_6$) often called "blood sugar" because it is the major monosaccharide found in the blood.

There are two ways of representing the linkage between the atoms of a monosaccharide, as illustrated in Figure 2-7. The first is the conventional way of drawing the structure of organic molecules, but the second gives a better representation of the three-dimensional shape of monosaccharides. Five carbon atoms and an oxygen atom form a ring that lies in an essentially flat plane; the hydrogen and

hydroxyl groups on each carbon lie above and below the plane of this ring. If one of the hydroxyl groups below the ring is shifted to a position above the ring, as shown in Figure 2-8, a different monosaccharide is produced.

Most monosaccharides in the body contain five or six carbon atoms and are called **pentoses** and **hexoses,** respectively. Larger carbohydrate molecules can be formed by linking a number of monosaccharides together. Carbohydrates composed of two monosaccharides are known as **disaccharides. Sucrose,** or table sugar (Figure 2-9), is composed of two monosaccharides, glucose and

FIGURE 2-7

Two ways of diagraming the structure of the monosaccharide glucose.

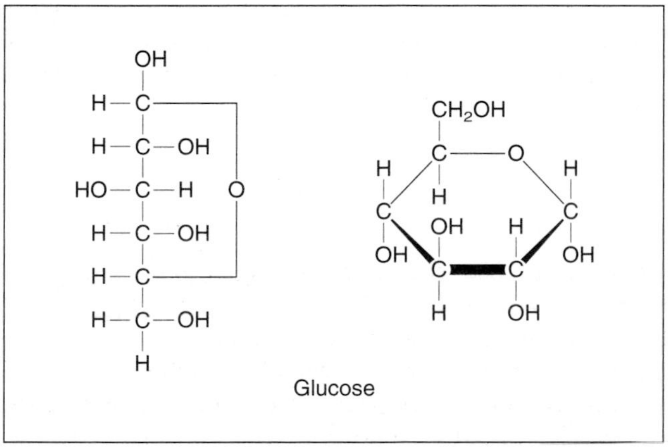

FIGURE 2-8

The structural difference between the monosaccharides glucose and galactose has to do with whether the hydroxyl group at the position indicated lies below or above the plane of the ring.

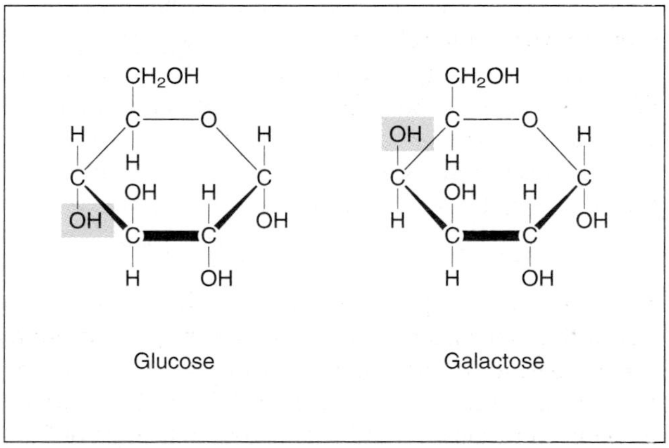

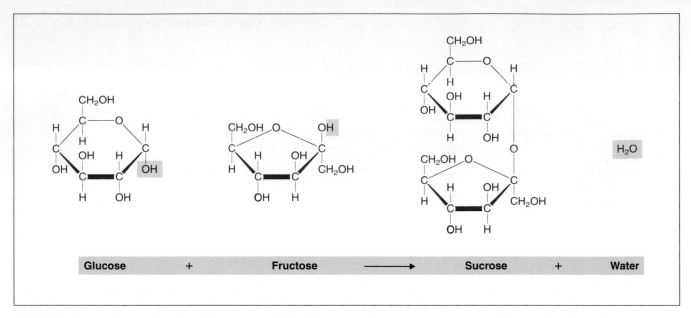

| Glucose | + | Fructose | $\longrightarrow$ | Sucrose | + | Water |

FIGURE 2-9

Sucrose (table sugar) is a disaccharide formed by the linking together of two monosaccharides, glucose and fructose.

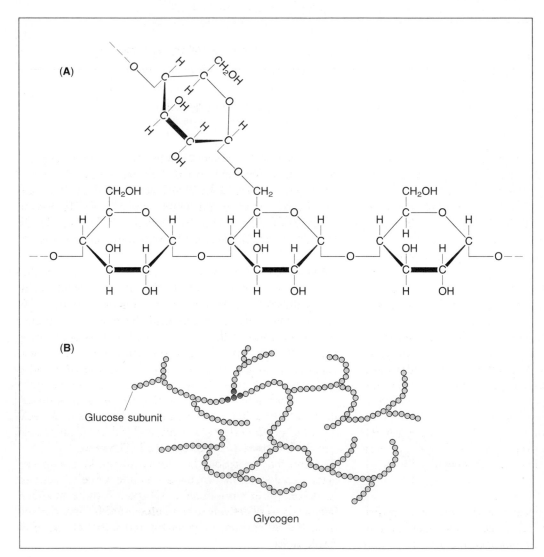

(A)

(B)

Glucose subunit

Glycogen

FIGURE 2-10

Many molecules of glucose linked end-to-end and at branch points form the branched-chain polysaccharide glycogen (A), shown in diagrammatic form in (B). The four red subunits in (B) correspond to the four glucose subunits in (A).

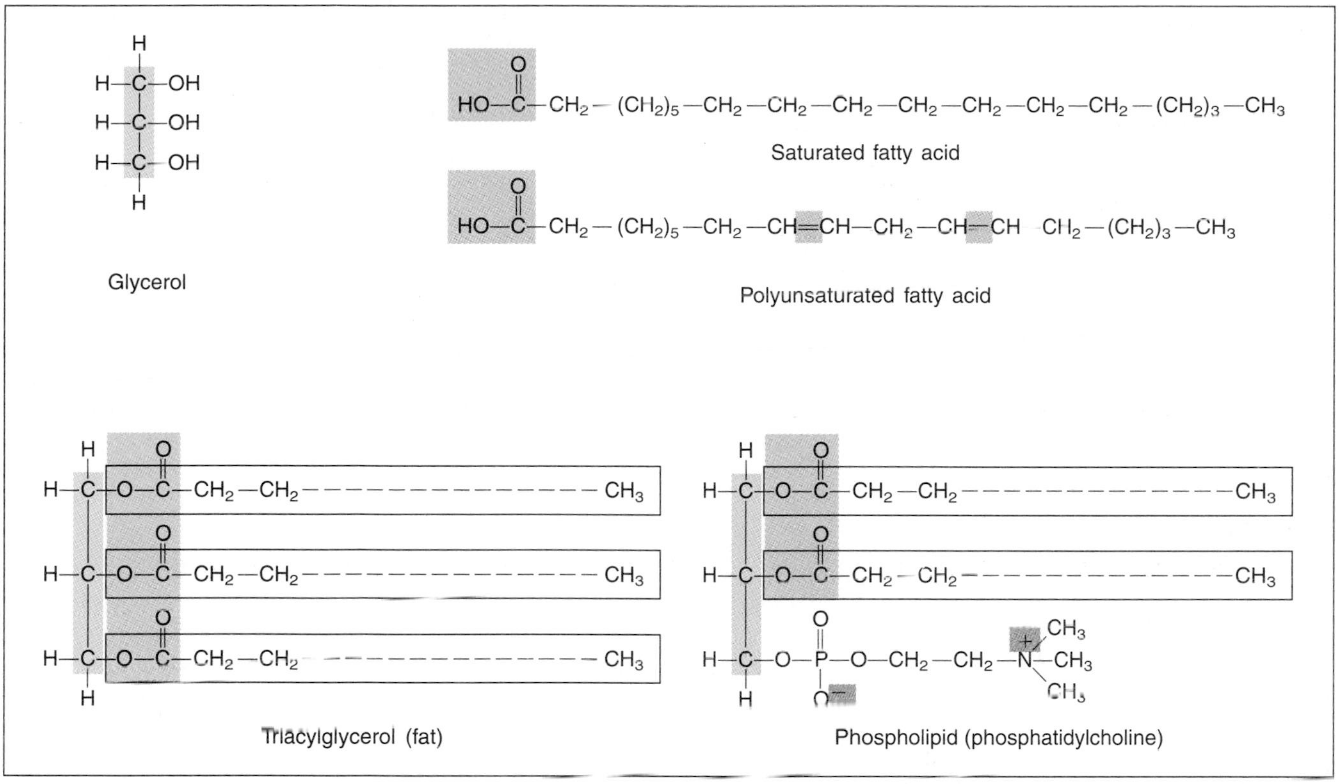

FIGURE 2-11
Glycerol and fatty acids are the major subunits that combine to form triacylglycerols and phospholipids.

fructose. The linking together of most monosaccharides involves the removal of a hydroxyl group from one monosaccharide and a hydrogen atom from the other, giving rise to a molecule of water and linking the two sugars together through an oxygen atom. Conversely, hydrolysis of the disaccharide breaks this linkage by adding back the water and thus uncoupling the two monosaccharides. Additional disaccharides frequently encountered are maltose (glucose-glucose), formed during the digestion of large carbohydrate molecules in the intestinal tract, and lactose (glucose-galactose), present in milk.

When many monosaccharides are linked together to form large polymers, the molecules are known as **polysaccharides.** Starch, found in plant cells, and **glycogen** (Figure 2-10), present in animal cells and often called "animal starch," are examples of polysaccharides. Both of these polysaccharides are composed of thousands of glucose molecules linked together in long chains. They differ only in the degree of branching along the chain. Hydrolysis of these polysaccharides leads to release of the glucose subunits.

Lipids

Lipids are molecules composed predominantly of hydrogen and carbon atoms. Since these atoms are linked by neutral covalent bonds, lipids are nonpolar molecules and thus

have a very low solubility in water. It is the physical property of insolubility in water that characterizes this class of organic molecules. Lipids, which account for about 40 percent of the organic matter in the average body (15 percent of the body weight), can be divided into four subclasses: fatty acids, triacylglycerols, phospholipids, and steroids.

Fatty Acids. A **fatty acid** consists of a chain of carbon atoms with a carboxyl group at one end. Because fatty acids are synthesized in the body by the linking together of 2-carbon fragments, most fatty acids have an even number of carbon atoms, with 16- and 18-carbon fatty acids being the most common. When all the carbons in a fatty acid chain are linked by single covalent bonds, the fatty acid is said to be a **saturated fatty acid.** Some fatty acids contain one or more double bonds, and these are known as **unsaturated fatty acids.** If one double bond is present, the acid is said to be **monounsaturated,** and if there is more than one double bond, **polyunsaturated** (Figure 2-11).

Some fatty acids can be altered to produce a special class of molecules that regulate a number of cell functions. As described in more detail in Chapter 7, these modified fatty acids—collectively termed **eicosanoids**—are derived from the 20-carbon, polyunsaturated fatty acid **arachidonic acid.**

Triacylglycerols. The **triacylglycerols** (also known as triglycerides) constitute the majority of the lipids in the body, and it is these molecules that are generally referred to simply as "fat." Triacylglycerols are formed by the linking together of **glycerol,** a three-carbon carbohydrate, with three fatty acids (Figure 2-11). Each of the three hydroxyl groups in glycerol is linked to the carboxyl group of a fatty acid by the removal of a molecule of water.

The three fatty acids in a molecule of triacylglycerol need not be identical; therefore, a variety of fats can be formed with fatty acids of different chain lengths and degrees of saturation. Animal fats generally contain a high proportion of saturated fatty acids, whereas vegetable fats contain more unsaturated fatty acids. Hydrolysis of triacylglycerols releases the fatty acids from glycerol, and these products can then be metabolized to provide energy for cell functions.

Phospholipids. **Phospholipids** are similar in overall structure to triacylglycerols, with one important difference. The third hydroxyl group of glycerol, rather than being attached to a fatty acid, is linked to phosphate. In addition, a small polar or ionized nitrogen-containing molecule is usually attached to this phosphate (Figure 2-11). These groups constitute a polar (hydrophilic) region at one end of the phospholipid molecule, whereas the fatty acid chains provide a nonpolar (hydrophobic) region at the opposite end. Therefore, phospholipids are amphipathic. In water, they become organized into clusters, with their polar ends being attracted to the water molecules.

Steroids. **Steroids** have a distinctly different structure from that of the other subclasses of lipid molecules. Four interconnected rings of carbon atoms form the skeleton of all steroids (Figure 2-12). A few polar hydroxyl groups may be attached to this ring structure, but they are not numerous enough to make a steroid water-soluble. Examples of steroids are cholesterol, cortisol from the adrenal glands, and female (estrogen) and male (testosterone) sex hormones secreted by the gonads.

Proteins

The term **"protein"** comes from the Greek *proteios* (of the first rank), which aptly describes their importance. These molecules, which account for about 50 percent of the organic material in the body (17 percent of the body weight), play critical roles in almost every physiological process. Proteins are composed of atoms of carbon, hydrogen, oxygen, and nitrogen, and small amounts of other elements, notably sulfur. They are macromolecules, often containing thousands of atoms, and like most large molecules, they are formed by the linking together of a large number of small subunits to form long chains.

Amino Acid Subunits. The subunit of protein structure is an **amino acid;** thus, proteins are polymers of amino acids. Every amino acid except proline has an amino ($-NH_2$) and a carboxyl ($-COOH$) group linked to the terminal carbon atom in the molecule

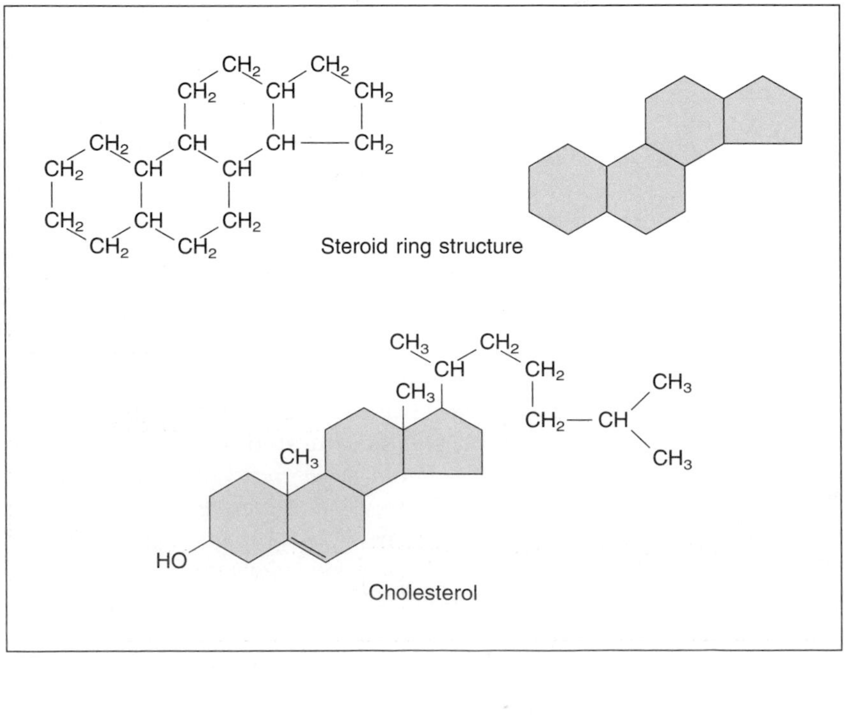

FIGURE 2-12

Steroid ring structure, shown with all the carbon and hydrogen atoms in the rings and again without these atoms to emphasize the overall ring structure of this class of lipids. Different steroids have different types and numbers of chemical groups attached at various locations on the steroid ring, as shown by the structure of cholesterol.

$$R\!-\!\underset{\underset{\displaystyle NH_2}{|}}{\overset{\overset{\displaystyle H}{|}}{C}}\!-\!COOH$$

The third bond of this terminal carbon atom is linked to a hydrogen atom and the fourth to the remainder of the molecule, which is known as the **amino acid side chain** (R in the formula). These side chains are relatively small, ranging from a single hydrogen atom to 9 carbon atoms.

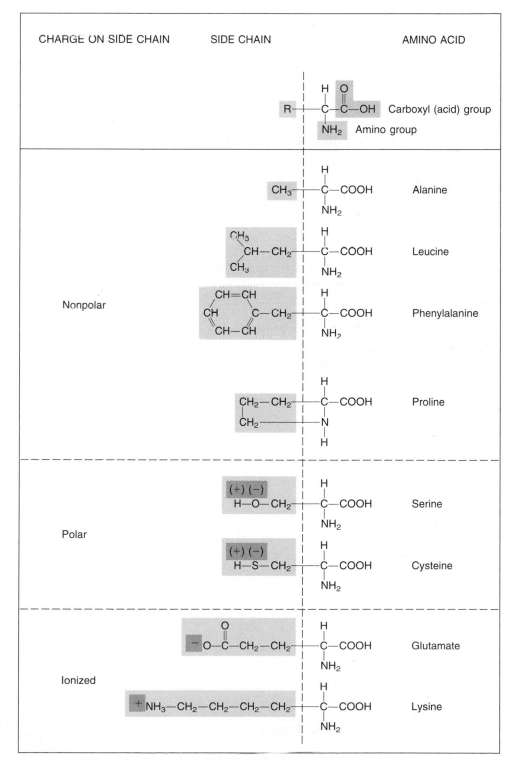

FIGURE 2-13

Structures of 8 of the 20 amino acids found in proteins. Note that proline does not have a free amino group, but it can still form a peptide bond.

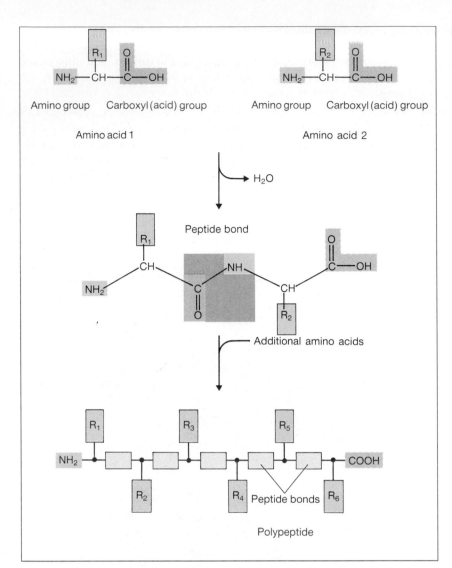

FIGURE 2-14
Linkage of amino acids by peptide bonds to form a polypeptide.

The proteins of all living organisms are composed of the same set of 20 different amino acids, corresponding to 20 different side chains. The side chains may be nonpolar (8 amino acids), polar (7 amino acids), or ionized (5 amino acids) (Figure 2-13).

Polypeptides.   Amino acids are joined together by linking the carboxyl group of one amino acid to the amino group of another. In the process, a molecule of water is formed (Figure 2-14). The bond formed between the amino and carboxyl groups is called a **peptide bond,** and it is a polar covalent bond. Note that when two amino acids are linked together, one end of the resulting molecule has a free amino group and the other has a free carboxyl group. Additional amino acids can be linked by peptide bonds to these free ends. A sequence of amino acids linked by peptide bonds is known as a **polypeptide.** The peptide bonds form the backbone of the polypeptide, and

the side chain of each amino acid sticks out from the side of the chain. If the number of amino acids in a polypeptide is 50 or less, the molecule is known as a **peptide;** if the sequence is more than 50 amino acid units long, the polypeptide is known as a **protein.** The number 50 is arbitrary but has become the convention for distinguishing between large and small polypeptides.

After certain proteins have been synthesized, one or more monosaccharides are covalently attached to the side chains of specific amino acids (serine and threonine) to form a class of proteins known as **glycoproteins.**

Primary Protein Structure.   Two variables determine the primary structure of a polypeptide: (1) the number of amino acids in the chain, and (2) the specific type of amino acid at each position along the chain (Figure 2-15). Each position along a polypeptide chain can be occupied by any one of the 20 different amino acids. Let us consider the number of different peptides that can be formed that have a sequence of three amino acids. Any one of the 20 different amino acids may occupy the first position in the sequence, any one of the 20 the second position, and any one of the 20 the third position, for a total of $20 \times 20 \times 20 = 20^3 = 8000$ possible sequences of three amino acids. If the peptide is 6 amino acids in length, $20^6 = 64,000,000$ possible combinations can be formed. Peptides that are only 6 amino acids long are still very small molecules compared to proteins, which may have sequences of 1000 or more amino acids. Thus, with 20 different amino acids an almost unlimited variety of polypeptides can be formed by altering both the amino acid sequence and the total number of amino acids in the chain.

Protein Conformation. The structure of a polypeptide is analogous to a string of beads, each bead representing a single amino acid (Figure 2-15). Moreover, since amino acids can rotate around their peptide bonds, a polypeptide chain is flexible and can be bent into a number of shapes,

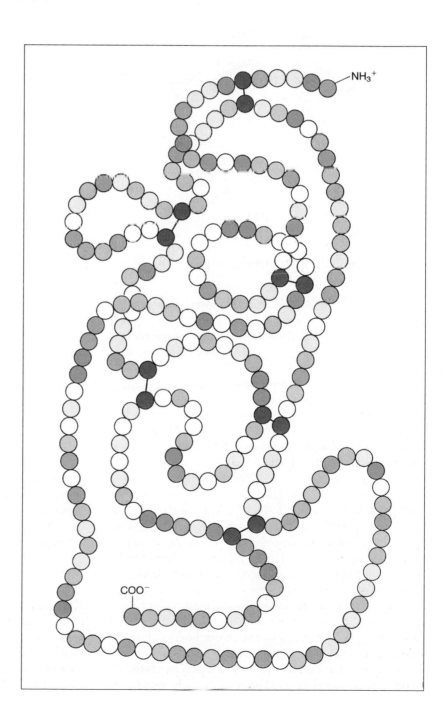

FIGURE 2-15
The position of each type of amino acid in a polypeptide chain and the total number of amino acids in the chain distinguish one polypeptide from another. The polypeptide illustrated contains 223 amino acids; different types of amino acids are represented by different-colored circles. The bonds between different regions of the chain (red to red) represent covalent disulfide bonds between cysteine side chains.

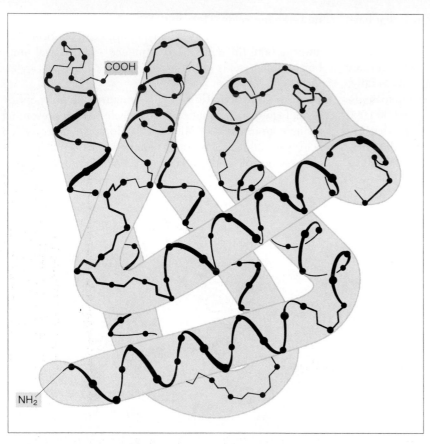

FIGURE 2-16
Conformation (shape) of a protein molecule (myoglobin). *(Adapted from Albert L. Lehninger.)*

just as a string of beads can be twisted into many configurations. The three-dimensional shape of a molecule is known as its **conformation** (Figure 2-16). The conformations of peptides and proteins play a major role in the functioning of these molecules, as we shall see in Chapter 4.

Four factors determine the conformation of a polypeptide chain once the amino acid sequence has been formed:

(1) hydrogen bonds between portions of the chain or with surrounding water molecules; (2) ionic bonds between polar and ionized regions along the polypeptide chain; (3) **van der Waals forces,** which are very weak forces of attraction between nonpolar (hydrophobic) regions in close proximity to each other; and (4) covalent bonds linking the side chains of two amino acids (Figure 2-17).

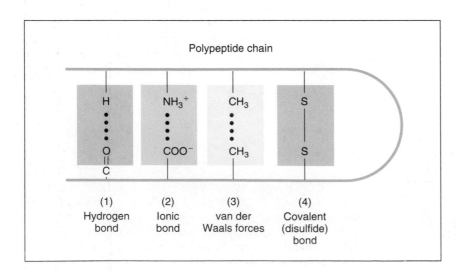

FIGURE 2-17
Factors that contribute to the folding of polypeptide chains and thus to their conformation are: (1) hydrogen bonds between side chains or with surrounding water molecules, (2) ionic bonds between polar or ionized amino acid side chains, (3) weak van der Waals forces between nonpolar side chains, and (4) covalent bonds between side chains.

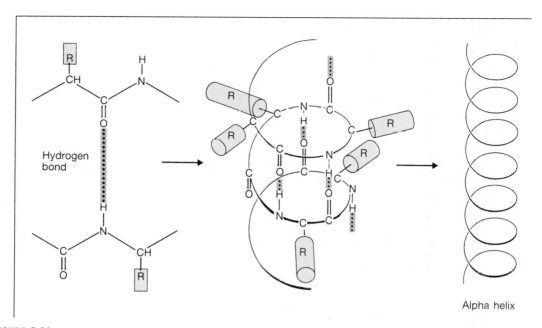

FIGURE 2-18

Hydrogen bonds between regularly spaced peptide bonds can produce a helical configuration in a polypeptide chain.

An example of the attractions between various regions along a polypeptide chain is the hydrogen bond that can occur between the hydrogen linked to the nitrogen atom in one peptide bond and the double-bonded oxygen in another peptide bond (Figure 2-18). Since peptide bonds occur at regular intervals along a polypeptide chain, the hydrogen bonds between them tend to force the polypeptide chain into a coiled configuration known as an **alpha helix.** Hydrogen bonds can also be formed between peptide bonds when extended regions of a polypeptide chain run approximately parallel to each other, forming a relatively straight, extended region known as a **beta sheet** (Figure 2-19). However, for several reasons, a given region of a polypeptide chain may not assume either a helical or beta sheet conformation. For example, the sizes of the side chains and ionic bonds between oppositely charged ionized side chains can interfere with the repetitive hydrogen bonding required to produce these shapes. These irregular regions are known as loop conformations and occur in regions linking the more regular helical and beta sheet patterns (Figure 2-19).

Covalent bonds between certain side chains can also distort the regular folding patterns; in addition, these bonds can link two polypeptides together. The side chain

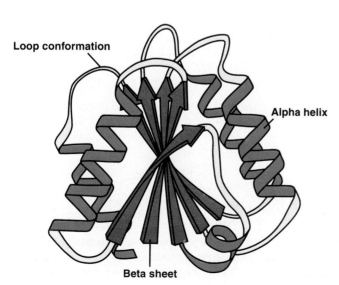

FIGURE 2-19

A ribbon diagram illustrating the pathway followed by the backbone of a polypeptide chain. Helical regions (blue) are coiled, beta sheets (red) of parallel chains are shown as relatively straight arrows, and loop conformations (yellow) connect the various helical and beta sheet regions. Beginning at the end of the chain labeled beta sheet, trace the path of the chain as it passes through various conformations.

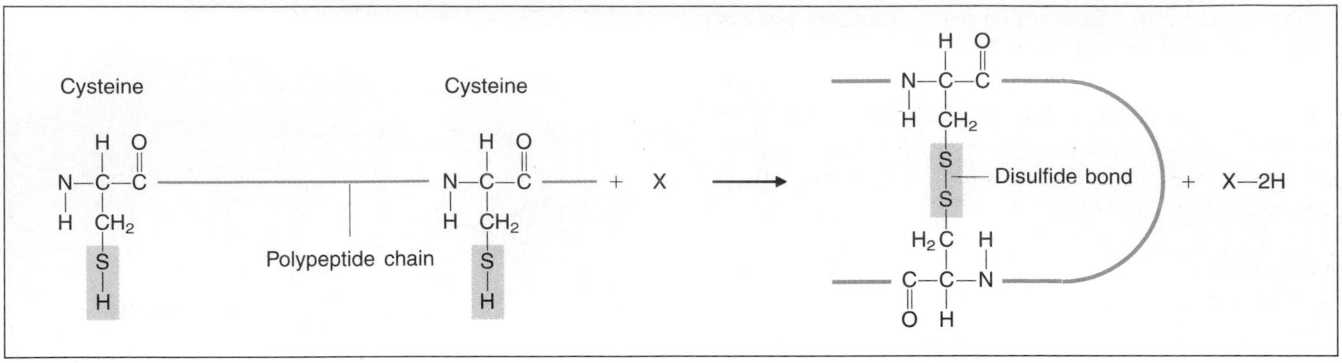

FIGURE 2-20
Formation of a disulfide bond between the side chains of two cysteine amino acids links two regions of the polypeptide together. The hydrogen atoms on the sulfhydryl groups of the cysteines are transferred to another molecule, X, during the formation of the disulfide bond.

of the amino acid cysteine contains a sulfhydryl group (R—SH), which can react with a sulfhydryl group in another cysteine side chain to produce a **disulfide bond** (R—S—S—R), which links the two amino acid side chains together (Figure 2-20). Disulfide bonds form strong covalent bonds between portions of a polypeptide chain, in contrast to the weaker hydrogen and ionic bonds, which are more easily broken. Table 2-6 provides a summary of the types of bonding forces that contribute to the conformation of polypeptide chains. These same bonds are also involved in other intermolecular interactions, which will be described in later chapters.

A number of proteins contain more than one polypeptide chain. The same factors that influence the conformation of a single polypeptide also determine the interactions between the polypeptides in a multichain protein. Thus, the separate chains can be held together by interactions between various ionized, polar, and nonpolar side chains, as well as by disulfide covalent bonds between the chains. The polypeptide chains in a multichain protein may be identical or different. For example, hemoglobin, the protein that transports oxygen in the blood, is composed of four polypeptide chains, two of one kind and two of another.

TABLE 2-6   BONDING FORCES BETWEEN ATOMS AND MOLECULES

Bond	Strength	Characteristics	Examples
Hydrogen	Weak	Electrical attraction between polarized bonds, usually hydrogen and oxygen	Attractions between peptide bonds forming the alpha helix structure of proteins and between polar amino acid side chains contributing to protein conformation; attractions between water molecules
Ionic	Strong	Electrical attraction between oppositely charged ionized groups	Attractions between ionized groups in amino acid side chains contributing to protein conformation; attractions between ions in a salt
van der Waals	Very weak	Attraction between nonpolar molecules and groups when very close to each other	Attractions between nonpolar amino acids in proteins contributing to protein conformation; attractions between lipid molecules
Covalent	Strong	Shared electrons between atoms	Most bonds linking atoms together to form molecules

The primary structures (amino acid sequences) of a large number of proteins are known, but three-dimensional conformations have been determined for only a few. Because of the multiple factors that can influence the folding of a polypeptide chain, it is not yet possible to predict accurately the conformation of a protein from its primary amino acid sequence.

Nucleic Acids

Nucleic acids account for only 2 percent of the body's weight, yet these molecules are extremely important because they are responsible for the storage, expression, and transmission of genetic information. It is the expression of genetic information that determines whether one is a human being or a mouse, or whether a cell is a muscle cell or a nerve cell.

There are two classes of nucleic acids, **deoxyribonucleic acid (DNA)** and **ribonucleic acid (RNA).** DNA molecules store genetic information coded in the sequence of their structural subunits, whereas RNA molecules are involved in the decoding of this information into instructions for linking together a specific sequence of amino acids to form a specific polypeptide chain. The mechanisms of gene expression and protein synthesis will be described in Chapter 5. Both types of nucleic acids are polymers composed of linear sequences of repeating subunits. Each subunit, known as a **nucleotide,** has three components: a phosphate group, a sugar, and a ring of carbon and nitrogen atoms known as a base because it can accept hydrogen ions (Figure 2-21). The phosphate group of one nucleotide is linked to the sugar of the adjacent nucleotide to form a chain, with the bases sticking out from the side of the phosphate-sugar backbone (Figure 2-22).

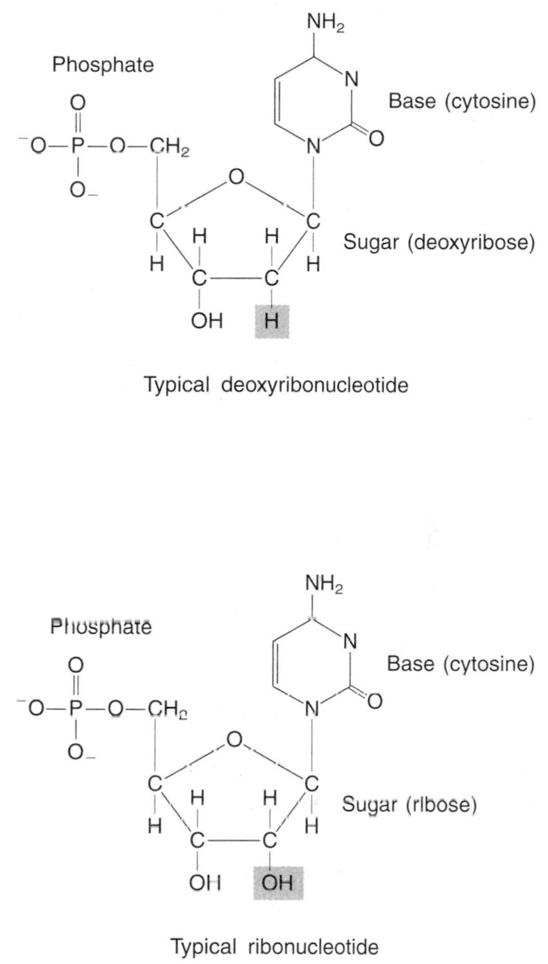

DNA. The nucleotides in DNA contain the five-carbon sugar **deoxyribose;** hence the name "deoxyribonucleic acid." Four different nucleotides are present in DNA, corresponding to the four different bases that can be linked to deoxyribose. These bases are divided into two classes: (1) the **purine** bases, **adenine** (A) and **guanine** (G), which have double (fused) rings of nitrogen and carbon atoms, and (2) the **pyrimidine** bases, **cytosine** (C) and **thymine** (T), which have only a single ring (Figure 2-22).

A DNA molecule consists of not one but two chains of nucleotides coiled around each other in the form of a double helix (Figure 2-23). The two chains are held together by hydrogen bonds between purine and pyrimidine bases. This base pairing maintains a constant distance between the sugar-phosphate backbones of the two chains as they coil around each other. Specificity is imposed on the base pairings by the location of the hydrogen-bonding groups in the four bases (Figure 2-23). Three hydrogen bonds are formed between the purine guanine and the pyrimidine cytosine (G—C pairing), while only

FIGURE 2-21

Nucleotide subunits of DNA and RNA. Nucleotides are composed of a sugar, a base and phosphate. Deoxyribonucleotides present in DNA contain the sugar deoxyribose. The sugar in ribonucleotides, present in RNA, is ribose, which has an OH at the position that lacks this group in deoxyribose.

two hydrogen bonds can be formed between the purine adenine and the pyrimidine thymine (A—T pairing). As a result, G is always paired with C, and A with T. In Chapter 5 we shall see how this specificity in base pairing provides the mechanism for duplicating and transferring genetic information.

RNA. The structure of RNA molecules differs in only a few respects from that of DNA (Table 2-7): (1) RNA consists of a single (rather than a double) chain of nucleotides;

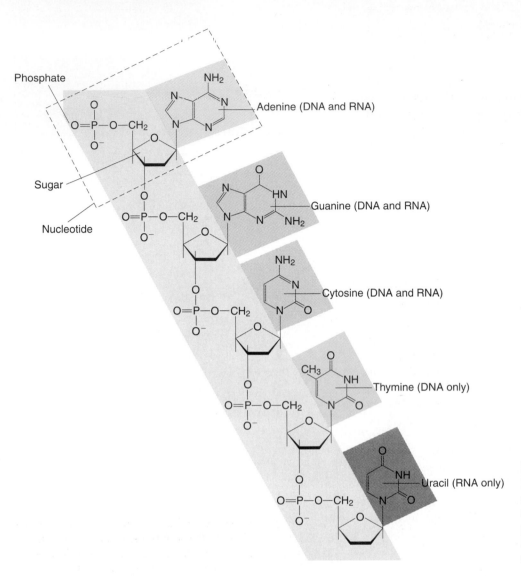

Phosphate

Sugar

Nucleotide

Adenine (DNA and RNA)

Guanine (DNA and RNA)

Cytosine (DNA and RNA)

Thymine (DNA only)

Uracil (RNA only)

FIGURE 2-22

Phosphate-sugar bonds link nucleotides in sequence to form nucleic acids. Note that the pyrimidine base thymine is only found in DNA, and uracil is only present in RNA.

TABLE 2-7	COMPARISON OF DNA AND RNA COMPOSITION	
	DNA	**RNA**
Nucleotide sugar	Deoxyribose	Ribose
Nucleotide bases		
Purines	Adenine	Adenine
	Guanine	Guanine
Pyrimidines	Cytosine	Cytosine
	Thymine	Uracil
Number of chains	Two	One

(2) in RNA, the sugar in each nucleotide is **ribose** rather than deoxyribose; and (3) the pyrimidine base thymine in DNA is replaced in RNA by the pyrimidine base **uracil** (U) (Figure 2-21), which can base-pair with the purine adenine (A—U pairing). The other three bases, adenine, guanine and cytosine, are the same in both DNA and RNA. Although RNA contains only a single chain of nucleotides, portions of this chain can bend back upon itself and undergo base pairing with nucleotides in the same chain or with nucleotides in other molecules of DNA or RNA.

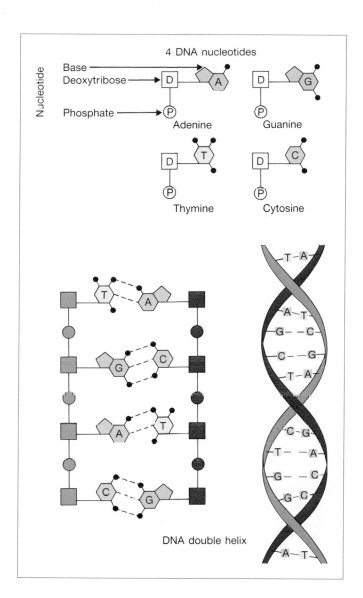

FIGURE 2-23

Base pairings between two nucleotide chains form the double-helical structure of DNA. D and P represent deoxyribose and phosphate, which form the backbone of each polynucleotide chain. The four bases are indicated by A, T, C and G.

SUMMARY

ATOMS

I. Atoms are composed of three subatomic particles: positive protons and neutral neutrons, both located in the nucleus, and negative electrons revolving around the nucleus.

II. The atomic number is the number of protons in an atom, and because atoms are electrically neutral, it is also equal to the number of electrons.

III. The atomic weight of an atom is the ratio of the atom's mass relative to that of a carbon-12 atom.

IV. One gram atomic mass is the number of grams of an element equal to its atomic weight. One gram atomic mass of any element contains the same number of atoms—6×10^{23}.

V. The 24 elements essential for normal body function are listed in Table 2-1.

MOLECULES

I. Molecules are formed by linking atoms together.

II. A covalent bond is formed when two atoms share a pair of electrons. Each type of atom can form a characteristic number of covalent bonds: hydrogen forms one; oxygen, two; nitrogen, three; and carbon, four.

III. Molecules have characteristic shapes, which can be altered within limits by the rotation of their atoms around covalent bonds.

IONS

When an atom gains or loses one or more electrons, it acquires a net electric charge and becomes an ion.

FREE RADICALS

Free radicals are atoms or molecules that contain atoms that have an unpaired electron in their outer electron orbital.

POLAR MOLECULES

I.   In polar covalent bonds, one atom attracts the bonding electrons more than the other atom does.

II. The electrical attraction between hydrogen and the oxygen or nitrogen atom in a separate molecule or different region of the same molecule forms a hydrogen bond.

III. Water, a polar molecule, is attracted to other water molecules by hydrogen bonds.

SOLUTIONS

I.   Substances dissolved in a liquid are solutes, and the liquid in which they are dissolved is the solvent. Water is the most abundant solvent in the body.

II. Substances that have polar or ionized groups dissolve in water by being electrically attracted to the polar water molecules.

III. In water, amphipathic molecules form clusters with the polar regions at the surface and the nonpolar regions in the interior of the cluster.

IV. The molecular weight of a molecule is the sum of the atomic weights of all its atoms. One mole of any substance is its molecular weight in grams and contains 6×10^{23} molecules.

V.   Substances that release a hydrogen ion in solution are called acids. Those that accept a hydrogen ion are bases.

   A. The acidity of a solution is determined by its free hydrogen-ion concentration; the greater the hydrogen-ion concentration, the greater the acidity.

   B. The pH of a solution is the negative logarithm of the hydrogen-ion concentration. As the acidity of a solution increases, the pH decreases. Acid solutions have a pH less than 7.0 whereas alkaline solutions have a pH greater than 7.0.

CLASSES OF ORGANIC MOLECULES

I.   Carbohydrates are composed of carbon, hydrogen, and oxygen in the proportions $Cn(H_2O)n$.

   A. The presence of the polar hydroxyl groups makes carbohydrates soluble in water.

   B. The most abundant monosaccharide in the body is glucose ($C_6H_{12}O_6$), which is stored in cells in the form of the polysaccharide glycogen.

II. Lipids lack polar and ionized groups, a characteristic that makes them insoluble in water.

   A. Triacylglycerols (fats) are formed when fatty acids are linked to each of the three hydroxyl groups in glycerol.

   B. Phospholipids contain two fatty acids linked to two of the hydroxyl groups in glycerol, with the third hydroxyl linked to phosphate, which in turn is linked to a small charged or polar compound. The polar and ionized groups at one end of phospholipids make these molecules amphipathic.

   C. Steroids are composed of four interconnected rings, often containing a few polar hydroxyl groups.

III. Proteins, macromolecules composed primarily of carbon, hydrogen, oxygen, and nitrogen, are polymers of 20 different amino acids.

   A. Amino acids have an amino ($-NH_2$) and a carboxyl ($-COOH$) group linked to their terminal carbon atom.

   B. Amino acids are linked together by peptide bonds between the carboxyl group of one amino acid and the amino group of the next.

   C. The primary structure of a polypeptide chain is determined by (1) the number of amino acids in sequence, and (2) the type of amino acid at each position.

   D. The factors that determine the conformation of a polypeptide chain are summarized in Figure 2-17.

   E. Hydrogen bonds between peptide bonds along a polypeptide chain force much of the chain into the form of an alpha helix.

   F. Covalent disulfide bonds can form between the sulfhydryl groups of cysteine side chains to hold regions of a polypeptide chain close to each other.

   G. Some proteins have multiple polypeptide chains.

IV. Nucleic acids are responsible for the storage, expression, and transmission of genetic information.

   A. Deoxyribonucleic acid (DNA) stores genetic information.

   B. Ribonucleic acid (RNA) is involved in the decoding of the information coded in DNA into instructions for linking amino acids together to form proteins.

   C. Both types of nucleic acids are polymers of nucleotides, each containing a phosphate group, a sugar, and a base of carbon, hydrogen, oxygen, and nitrogen atoms.

   D. DNA contains the sugar deoxyribose and consists of two chains of nucleotides coiled around each other in a double helix. The chains are held together by hydrogen bonds between purine and pyrimidine bases in the two chains.

   E. Base pairings in DNA always occur between guanine and cytosine and between adenine and thymine.

   F. RNA consists of a single chain of nucleotides, containing the sugar ribose and three of the four bases found in DNA. The fourth base in RNA is the pyrimidine uracil rather than thymine. Uracil base-pairs with adenine.

KEY TERMS

atom
chemical element
proton
neutron
electron
atomic nucleus
atomic number
atomic weight
isotope
gram atomic mass
trace element
molecule
covalent bond
ion
electrolyte
cation
anion
carboxyl group
amino group
free radical
polar covalent bond
hydroxyl group
polar molecule
nonpolar molecule
hydrogen bond
hydrolysis
solute
solvent
solution
ionic bond
hydrophilic
hydrophobic
amphipathic
concentration
molecular weight
mole
acid
base
strong acid
weak acid
acidity
pH
neutral solution
alkaline solution
acidic solution
biochemistry
macromolecule
polymer

carbohydrate
monosaccharide
glucose
pentose
hexose
disaccharide
sucrose
polysaccharide
glycogen
lipid
fatty acid
saturated fatty acid
unsaturated fatty acid
monounsaturated fatty acid
polyunsaturated fatty acid
eicosanoid
arachidonic acid
triacylglycerol
glycerol
phospholipid
steroid
protein
amino acid
amino acid side chain
peptide bond
polypeptide
peptide
glycoprotein
conformation
van der Waals forces
alpha helix
beta sheet
disulfide bond
nucleic acid
deoxyribonucleic acid (DNA)
ribonucleic acid (RNA)
nucleotide
deoxyribose
purine
adenine
guanine
pyrimidine
cytosine
thymine
ribose
uracil

REVIEW QUESTIONS

1. Describe the electric charge, mass, and location of the three major subatomic particles in an atom.
2. Which four kinds of atoms are most abundant in the body?
3. Describe the distinguishing characteristics of the three classes of essential chemical elements found in the body.
4. How many covalent bonds can be formed by atoms of carbon, nitrogen, oxygen, and hydrogen?
5. What property of molecules allows them to change their three-dimensional shape?
6. Describe how an ion is formed.
7. Draw the structures of an ionized carboxyl group and an ionized amino group.
8. Define a free radical.
9. Describe the polar characteristics of a water molecule.
10. What determines a molecule's solubility or lack of solubility in water?
11. Describe the organization of amphipathic molecules in water.
12. What is the molar concentration of 80 g of glucose dissolved in sufficient water to make 2 L of solution?
13. What distinguishes a weak acid from a strong acid?
14. What effect does increasing the pH of a solution have upon the ionization of a carboxyl group? An amino group?
15. Name the four classes of organic molecules in the body.
16. Describe the three subclasses of carbohydrate molecules.
17. To which subclass of carbohydrates does each of the following molecules belong: glucose, sucrose, and glycogen?
18. What properties are characteristic of lipids?
19. Describe the subclasses of lipids.
20. Describe the linkage between amino acids to form a polypeptide chain.
21. What is the difference between a peptide and a protein?
22. What two factors determine the primary structure of a polypeptide chain?
23. Describe the types of interactions that determine the conformation of a polypeptide chain.
24. Describe the structure of DNA and RNA.
25. Describe the characteristics of base pairings between nucleotide bases.

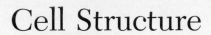

Cell Structure

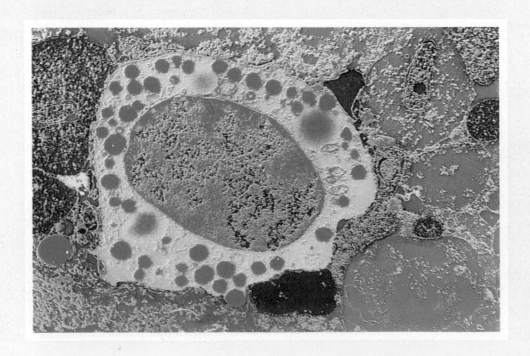

As we learned in Chapter 1, cells are the structural and functional units of all living organisms. The word "cell" means "a small chamber" (like a jail cell). The human body is composed of trillions of cells, each a microscopic compartment (Figure 3-1). In this chapter, we describe the structures present in most of the body's cells and state their functions. Subsequent chapters describe the mechanisms by which these structures perform their functions.

The cells of a mouse, a human being, and an elephant are all approximately the same size. An elephant is large because it has more cells, not because it has larger cells. A majority of the cells in a human being have diameters in the range of 10 to 20 μm, although cells as small as 2 μm and as large as 120 μm are present. A cell 10 μm in diameter is about one-tenth the size of the smallest object that can be seen with the naked eye; a microscope must therefore be used to observe cells and their internal structure.

MICROSCOPIC OBSERVATIONS OF CELLS

The smallest object that can be resolved with a microscope depends upon the wavelength of the radiation used to illuminate the specimen—the shorter the wavelength, the smaller the object that can be seen. With a **light microscope,** objects as small as 0.2 μm in diameter can be resolved, whereas an **electron microscope,** which uses electron beams instead of light rays, can resolve structures as small as 0.002 μm. A greater resolution is achieved with an electron microscope because electrons behave as waves

FIGURE 3-1
Cellular organization of tissues, as illustrated by a portion of spleen. Oval, clear spaces in the micrograph are blood vessels. (*From Johannes A. G. Rhodin, "Histology, A Text & Atlas," Oxford University Press, New York, 1974.*)

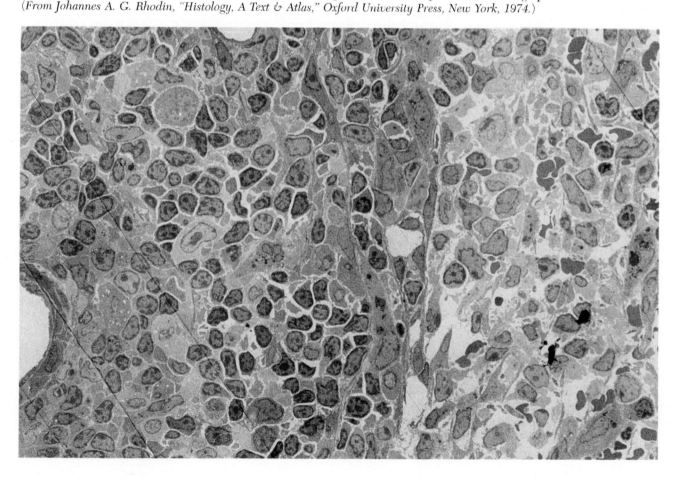

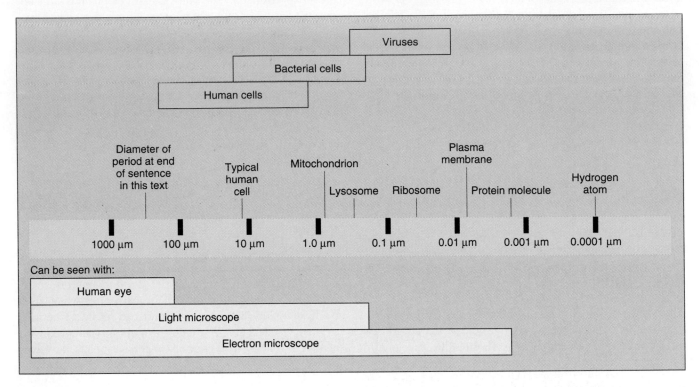

FIGURE 3-2

Size range of cell structures, plotted on a logarithmic scale. Typical sizes are given. Each component has a range of sizes around this typical value.

with much shorter wavelengths than those of visible light. Typical sizes of cells and cellular components are illustrated in Figure 3-2.

Although *living* cells can be observed with a light microscope, this is not possible with an electron microscope. To form an image with an electron beam, most of the electrons must pass through the specimen, just as light passes through a specimen in a light microscope. However, electrons can penetrate only a short distance through matter; therefore, the observed specimen must be very thin. Cells to be observed with an electron microscope must be cut into thin sections, like slices of salami. Each section is on the order of 0.1 μm thick, which is about one-hundredth of the thickness of a typical cell.

Because electron micrographs (such as Figure 3-3) are images of very thin sections of a cell, they can often be misleading. Structures that appear as separate objects in the electron micrograph may actually be continuous structures that are connected through a region lying outside the plane of the section. As an analogy, a thin section through a ball of string would appear as a collection of separate lines and disconnected dots even though the piece of string was originally continuous.

There are three major classes of biological units: **prokaryotic cells, eukaryotic cells,** and **viruses.** The cells of the human body, as well as those of other multicellular animals and plants, are eukaryotic (true-nucleus) cells. These cells are distinguished by a nuclear membrane surrounding the cell nucleus and numerous other membrane-bound structures within the cell. Prokaryotic cells lack these membranous structures; bacteria make up the largest class of prokaryotic cells. Viruses have the simplest structure, consisting of only a nucleic acid molecule surrounded by a protein shell. This chapter describes the structure of eukaryotic cells only.

CELL COMPARTMENTS

Compare an electron micrograph of a section through a cell (Figure 3-3) with a diagrammatic illustration of a typical human cell (Figure 3-4). What is immediately obvious from both the diagram and the electron micrograph is the extensive structure inside the cell. The cell is surrounded by a limiting barrier, the **plasma membrane,** which covers the cell surface. The cell interior is divided into a number of compartments also surrounded by membranes. These membrane-bound compartments, along with some particles and filaments, are known as **cell organelles** (little organs). Each cell organelle performs specific functions that contribute to the cell's survival.

The interior of a cell is divided into two regions: (1) the **nucleus,** a spherical or oval structure usually near the center of the cell, and (2) the **cytoplasm,** the region out-side the nucleus (Figure 3-5). The cytoplasm contains two components: (1) cell organelles and (2) the fluid known as the **cytosol** (cytoplasmic solution) surrounding these organelles. The term **intracellular fluid** refers to *all* the fluid inside a cell—in other words, cytosol plus the fluid inside all the organelles, including the nucleus. The chemical compositions of the fluids in these cell organelles differ from that of the cytosol. The cytosol is by far the largest intracellular fluid compartment.

FIGURE 3-3

Electron micrograph of a thin section through a portion of a rat liver cell. [*From K. R. Porter in T. W. Goodwin and O. Lindberg (eds.), "Biological Structure and Function," vol. I, Academic Press, Inc., New York, 1961.*]

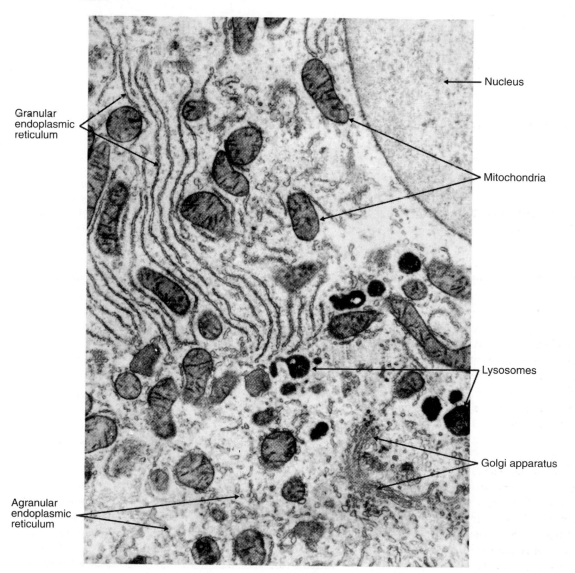

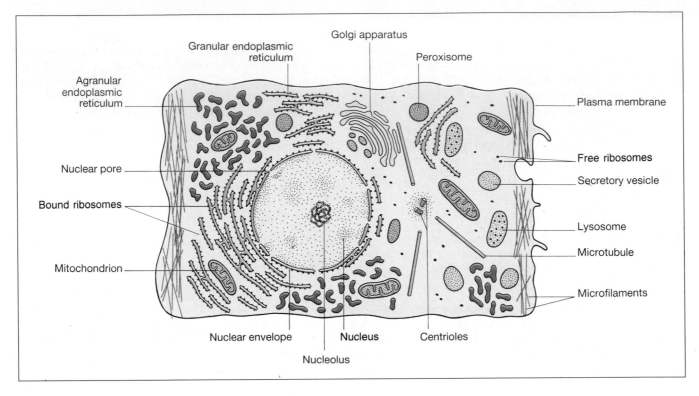

FIGURE 3-4

Structures found in most human cells.

Membranes

Membranes form a major structural element in human cells. They are found surrounding the entire cell (the plasma membrane) and enclosing most of the cell organelles. Although membranes perform a variety of functions, their most universal role is to act as a selective bar-rier to the passage of molecules, allowing some molecules to cross while excluding others. The plasma membrane regulates the passage of substances into and out of the cell, whereas the membranes surrounding cell organelles allow selective movement of substances between the organelles and the cytosol. One of the advantages of restricting the movements of molecules across membranes is confining the products of chemical reactions to specific cell organelles. As we shall see in Chapter 6, the hindrance offered by a membrane to the passage of substances can be altered to allow increased or decreased flow of molecules or ions across the membrane.

The plasma membrane, in addition to acting as a selective barrier, plays an important role in detecting chemical signals from other cells, anchoring cells to adjacent cells, and providing sites for the attachment of protein filaments associated with the generation and transmission of force (Table 3-1).

Membrane Structure.   All membranes consist of a double layer of lipid molecules in which proteins are embedded (Figure 3-6). The major membrane lipids are **phospholipids.** As described in Chapter 2, these are amphipathic molecules: one end has a charged region, and the remainder of the molecule, which consists of two long

FIGURE 3-5

Comparison of cytoplasm and cytosol. (A) Cytoplasm (colored area) is the region of the cell outside the nucleus. (B) Cytosol (colored area) is the fluid portion of the cytoplasm outside the cell organelles.

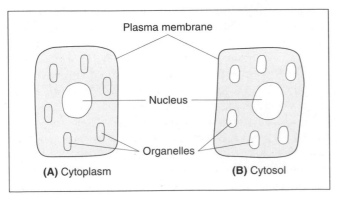

Plasma membrane

Nucleus

Organelles

(A) Cytoplasm   **(B)** Cytosol

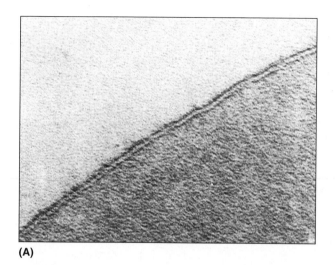

(A)

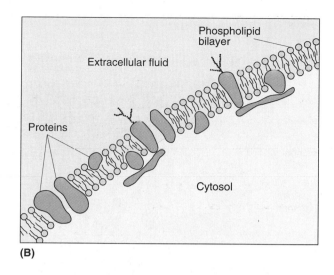

(B)

FIGURE 3-6

(A) Electron micrograph of a human red-cell plasma membrane. Cell membranes are 6 to 10 nm thick, too thin to be seen without the aid of an electron microscope. In an electron micrograph a membrane appears as two dark lines separated by a light interspace. The dark lines correspond to the polar regions of the proteins and lipids, whereas the light interspace corresponds to the nonpolar regions of these molecules. [*From J. D. Robertson in Michael Locke (ed.), "Cell Membranes in Development," Academic Press, Inc., New York, 1964.*] (B) Arrangement of the proteins and lipids in a membrane.

fatty acid chains, is nonpolar. The phospholipids in cell membranes are organized into a bimolecular layer with the nonpolar fatty acid chains in the middle. The polar regions of the phospholipids are oriented toward the surfaces of the membrane as a result of their attraction to the polar water molecules in the extracellular fluid and cytosol.

There are no chemical bonds linking the phospholipids to each other or to the membrane proteins. Therefore, each molecule is free to move independently of the others. This results in considerable random lateral movement parallel to the surfaces of the bilayer. In addition, the long fatty acid chains can bend and wiggle back and forth. Thus, the lipid bilayer has the characteristics of a fluid, much like a thin layer of oil on a water surface, and this makes the membrane quite flexible. Like a piece of cloth, membranes can be bent and folded but cannot be stretched without being torn. This flexibility, along with

the fact that cells are filled with fluid, allows cells to undergo considerable changes in shape without disruption of their structural integrity.

The plasma membrane also contains **cholesterol** (about one molecule of cholesterol for each molecule of phospholipid), whereas intracellular membranes contain very little cholesterol. Cholesterol, a steroid, is slightly amphipathic because of a single polar hydroxyl group (Figure 2-12) on its nonpolar ring structure. Therefore, cholesterol, like the amphipathic phospholipids, is inserted into the lipid bilayer with its polar region at a bilayer surface and its nonpolar rings in the interior in association with the fatty acid chains. The cholesterol in the plasma membrane plays an important role in maintaining the fluidity of the lipid bilayer, preventing close association of the long fatty acid chains, which are attracted to each other by van der Waals forces (Chapter 2). In the absence of cholesterol, these forces would tend to solidify the lipid bilayer.

There are two classes of membrane proteins: integral and peripheral. **Integral membrane proteins** are closely associated with the membrane lipids and cannot be extracted from the membrane without disrupting the lipid bilayer. Like the phospholipids, the integral proteins are amphipathic, having polar amino acid side chains in one region of the molecule and nonpolar side chains clustered together in a separate region. Because they are amphipathic, integral proteins are arranged in the membrane with the same orientation as amphipathic lipids—the polar regions are at the surfaces in association with polar water molecules, and the nonpolar regions are in the interior

---

TABLE 3-1   FUNCTIONS OF PLASMA MEMBRANES

1. Regulate the passage of substances into and out of cells
2. Detect chemical messengers arriving at the cell surface
3. Link adjacent cells together by membrane junctions
4. Anchor a variety of proteins, including intracellular and extracellular protein filaments involved in the generation and transmission of force

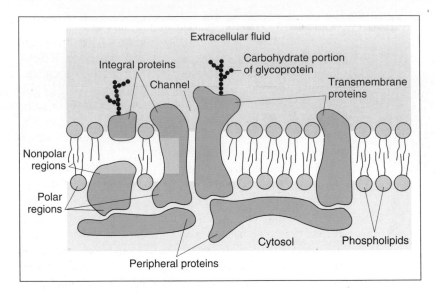

FIGURE 3-7

Arrangement of integral and peripheral membrane proteins in association with bimolecular layer of phospholipids. Dark blue areas indicate the polar regions of integral membrane proteins.

in association with nonpolar fatty acid chains (Figure 3-7). Like the membrane lipids, many of the integral proteins can move laterally in the plane of the membrane, but others are immobilized because they are linked to a network of peripheral proteins located primarily at the cytosolic surface of the membrane.

Most integral proteins span the entire membrane and are referred to as **transmembrane proteins.** These proteins have two polar regions connected by a nonpolar segment that associates with the nonpolar regions of the lipids in the interior of the membrane. The polar regions of transmembrane proteins may extend far beyond the surfaces of the lipid bilayer. Some transmembrane proteins form channels through which ions or water can move across the membrane, whereas others are associated with the transmission of chemical signals across the membrane or the anchoring of extracellular and intracellular filaments to the plasma membrane.

Peripheral membrane proteins are not amphipathic and do not associate with the nonpolar regions of the lipids in the interior of the membrane. They are located at the membrane surface where they are bound to the polar regions of the integral membrane proteins (Figure 3-7). Most of the peripheral proteins are on the cytosolic surface of the plasma membrane where they are associated with cytoskeletal elements that influence cell shape and motility.

The extracellular surface of the plasma membrane also contains small amounts of carbohydrate covalently linked to some of the membrane lipids and proteins. These carbohydrates consist of short, branched chains of monosac-

charides that extend from the cell surface into the extracellular fluid where they form a fuzzy, "sugar-coated" layer known as the **glycocalyx.** These surface carbohydrates play important roles in enabling cells to identify and interact with each other.

The lipids in the outer half of the bilayer differ somewhat in kind and amount from those in the inner half, and, as we have seen, the proteins or portions of proteins on the outer surface differ from those on the inner surface. Many membrane functions are related to these asymmetries in chemical composition between the two surfaces of a membrane.

All membranes have the general structure described above, which has come to be known as the **fluid-mosaic model** (Figure 3-8). However, the proteins and, to a lesser extent, the lipids (the distribution of cholesterol, for example) in the plasma membrane are different from those in organelle membranes. Thus, the special functions of membranes, which depend primarily on the membrane proteins, may differ in the various membrane-bound organelles and in the plasma membranes of different types of cells.

Membrane Junctions.   In addition to providing a barrier to the movements of molecules between the intracellular and extracellular fluids, plasma membranes are involved in interactions between cells to form tissues. Some cells, particularly those of the blood, do not associate with other cells but remain as independent cells suspended in a fluid—the blood plasma in the case of blood cells. Most cells, however, are packaged into tissues and are not free to move around the body. But even the cells in a tissue are

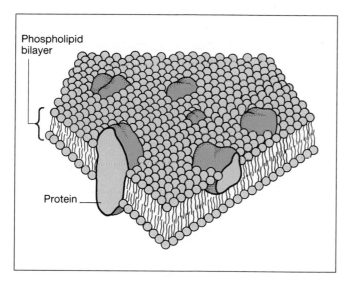

FIGURE 3-8

Fluid-mosaic model of cell membrane structure. (*Redrawn from S. J. Singer and G. L. Nicholson, Science, 175:723. Copyright 1972 by the American Association for the Advancement of Science.*)

not packed so tightly that the adjacent cell surfaces are in direct contact with each other, except at certain specialized junctions. There is usually a space between the plasma membranes of adjacent cells; this space is filled with extracellular fluid and provides the pathway for substances to pass between cells on their way to and from the blood.

The forces that organize cells into tissues and organs are poorly understood, but they depend, at least in part, on the ability of certain transmembrane proteins in the plasma membrane, known as **integrins,** that bind to specific proteins in the extracellular matrix and to membrane proteins on adjacent cells. In addition, many cells are physically joined at discrete locations along their membranes by specialized types of junctions known as desmosomes, tight junctions, and gap junctions.

Desmosomes (Figure 3-9A) consist of a region between two adjacent cells where the apposed plasma membranes are separated by about 20 nm and have a dense accumulation of protein at the cytoplasmic surface of each membrane and in the space between the two membranes. In addition, fibers extend from the cytoplasmic surface into the cell and are linked to other desmosomes on the opposite side of the cell. The function of the desmosome is to hold adjacent cells firmly together in areas that are subject to considerable stretching, such as in the skin. The specialized area of the membrane in the region of a desmosome is usually disk-shaped, and these membrane junctions could be likened to rivets or spot-welds.

A second type of membrane junction, the **tight junction,** (Figure 3-9B) is formed when the extracellular surfaces of two adjacent plasma membranes are joined together so that there is no extracellular space between them. Unlike the desmosome, which is limited to a disk-shaped area of the membrane, the tight junction occurs in a band around the entire circumference of the cell. Tight junctions greatly restrict the movement of most organic molecules through the extracellular space between the joined cells, but they have a variable leakiness to small ions and water.

Most epithelial cells are joined by tight junctions. For example, epithelial cells cover the inner surface of the intestinal tract, where they come in contact with the digestion products in the cavity of the tract. During absorption, the products of digestion move across the epithelium and enter the blood. This transfer could take place theoretically by movement either through the extracellular space between the epithelial cells or through the epithelial cells themselves. For many substances, however, movement through the extracellular space is blocked by the tight junctions that encircle the cells near their luminal border. Thus, organic nutrients are required to pass through cells, rather than between them. In this way, the selective barrier properties of the plasma membrane can control the types and amounts of absorbed substances. Figure 3-9D shows both a tight junction and a desmosome near the luminal border between two epithelial cells.

A third type of junction, the **gap junction,** consists of protein channels linking the cytosols of adjacent cells (Figure 3-9C). In the region of the gap junction, the two opposing plasma membranes come within 2 to 4 nm of each other, allowing specific proteins from the two membranes to join, forming small, protein-lined channels linking the two cells. The smaller diameter of these channels (about 1.5 nm) limits what can pass between the connected cells to small molecules and ions, such as sodium and potassium, and excludes the exchange of large protein molecules. A variety of cell types possess gap junctions, including the muscle cells of the heart and smooth-muscle cells. As we shall see in Chapter 11, gap junctions play a very important role in the transmission of electrical activity between certain types of muscle cells. In other cases, gap junctions coordinate the activities of adjacent cells by allowing chemical messengers to move from one cell to another.

CELL ORGANELLES

Nucleus

Almost all cells contain a single nucleus. A few specialized cells, for example, skeletal-muscle cells, contain multiple nuclei, while the mature red blood cell has none. The primary function of the nucleus is the storage and the

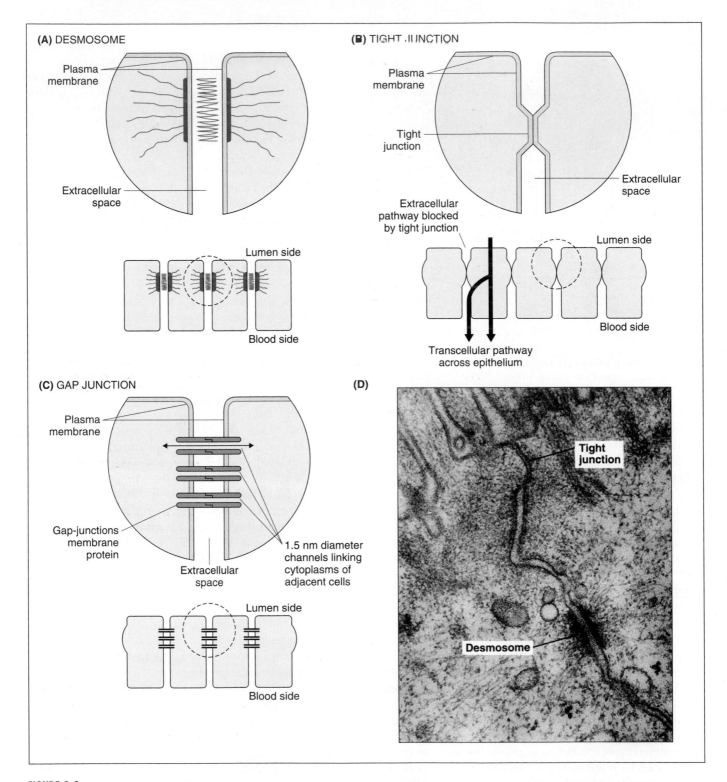

FIGURE 3-9
Three types of specialized membrane junctions: (A) desmosome, (B) tight junction, and (C) gap junction. (D) Electron micrograph of two intestinal epithelial cells joined by a tight junction near the luminal surface and a desmosome below the tight junction. [*Electron micrograph from M. Farquhar and G. E. Palade, J. Cell. Biol., 17:375–412 (1963).*]

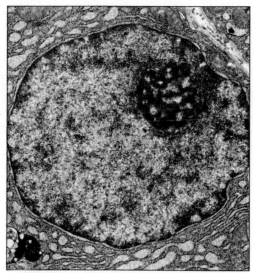

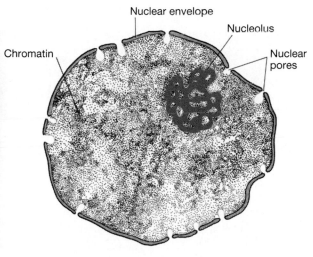

NUCLEUS

Structure: Largest organelle. Round or oval body located near cell center. Surrounded by nuclear envelope composed of two membranes. Envelope contains nuclear pores through which messenger molecules pass between the nucleus and the cytoplasm. No membrane-bound organelles are present in nucleus, which contains coiled strands of DNA known as chromatin, which condense to form chromosomes at the time of cell division.

Function: Stores and transmits genetic information in the form of DNA. Genetic information passes from the nucleus to the cytoplasm, where amino acids are assembled into proteins.

NUCLEOLUS

Structure: Densely stained filamentous structure within nucleus. Consists of proteins associated with DNA in regions where information concerning ribosomal proteins is being expressed.

Function: Site of ribosomal RNA synthesis. Assembles RNA and protein components of ribosomal subunits, which then move to the cytoplasm through nuclear pores.

FIGURE 3-10

Nucleus. (*Electron micrograph courtesy of K. R. Porter.*)

transmission to the next generation of cells, of genetic information. This information, in the form of DNA, is used to synthesize the proteins that determine the structure and function of the cell (Chapter 5).

Surrounding the nucleus is a barrier, the **nuclear envelope,** composed of two membranes. At regular intervals along the surface of the nuclear envelope, the two membranes are joined to each other, forming the rims of circular openings known as **nuclear pores** (Figure 3-10). Molecules of RNA that regulate the expression of genetic information move between the nucleus and cytoplasm through these nuclear pores. The movement of very large molecules, such as RNA and proteins, is selective, that is, restricted to specific macromolecules. An energy-dependent process that alters

the diameter of the pore in response to specific signals is involved in the transfer process.

Within the nucleus, DNA, in association with proteins, forms a fine network of threads known as **chromatin;** the threads are coiled to a greater or lesser degree, producing the variations in the density of the nuclear contents seen in electron micrographs (Figure 3-10). At the time of cell division the chromatin threads become tightly condensed, forming rodlike bodies known as **chromosomes.**

The most prominent organelle in the nucleus is the **nucleolus,** a highly coiled structure associated with numerous particles, but not surrounded by a membrane. The nucleolus is composed of RNA and proteins that are associated with specific regions of DNA. This DNA

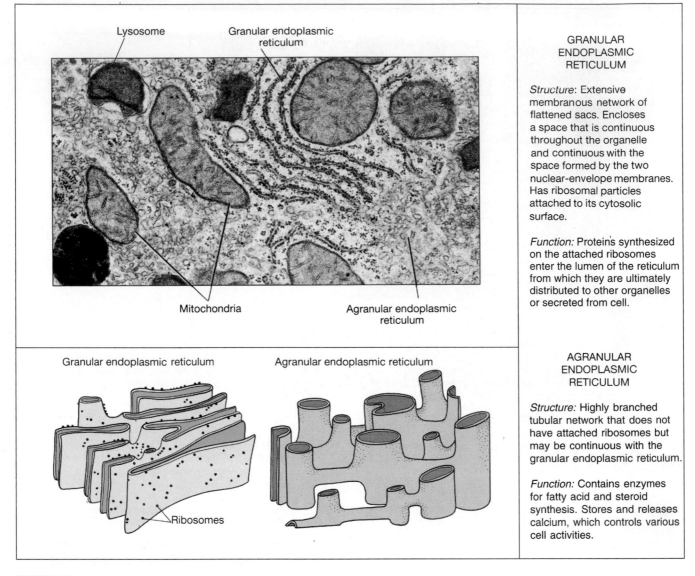

Lysosome   Granular endoplasmic reticulum

Mitochondria   Agranular endoplasmic reticulum

Granular endoplasmic reticulum   Agranular endoplasmic reticulum

Ribosomes

GRANULAR ENDOPLASMIC RETICULUM

Structure: Extensive membranous network of flattened sacs. Encloses a space that is continuous throughout the organelle and continuous with the space formed by the two nuclear-envelope membranes. Has ribosomal particles attached to its cytosolic surface.

Function: Proteins synthesized on the attached ribosomes enter the lumen of the reticulum from which they are ultimately distributed to other organelles or secreted from cell.

AGRANULAR ENDOPLASMIC RETICULUM

Structure: Highly branched tubular network that does not have attached ribosomes but may be continuous with the granular endoplasmic reticulum.

Function: Contains enzymes for fatty acid and steroid synthesis. Stores and releases calcium, which controls various cell activities.

FIGURE 3-11

Endoplasmic reticulum. (*Electron micrograph from D. W. Fawcett, "The Cell, An Atlas of Fine Structure," W. B. Saunders Company, Philadelphia, 1966.*)

contains the genes for forming the RNA found in particular cytoplasmic organelles called ribosomes (see below). It is in the nucleolus that the RNA and protein components of ribosomal subunits are assembled; these subunits are then transferred through the nuclear pores to the cytoplasm, where they combine to form functional ribosomes.

Ribosomes

Ribosomes are the protein factories of a cell. On ribosomes protein molecules are synthesized from amino acids, using genetic information carried by RNA messenger molecules from DNA in the nucleus. Ribosomes are large particles, about 20 nm in diameter, composed of about 70 proteins and several RNA molecules (Chapter 5).

Ribosomes are either bound to the organelle called granular endoplasmic reticulum (described next) or are found free in the cytoplasm.

The proteins synthesized on the free ribosomes are released into the cytosol, where they perform their functions. The proteins synthesized by ribosomes attached to the granular endoplasmic reticulum pass into the lumen of the reticulum and are then transferred to yet another organelle, the Golgi apparatus (see next page). They are ultimately secreted from the cell or distributed to other organelles.

Endoplasmic Reticulum

The most extensive cytoplasmic organelle is the network of membranes that forms the **endoplasmic reticulum**

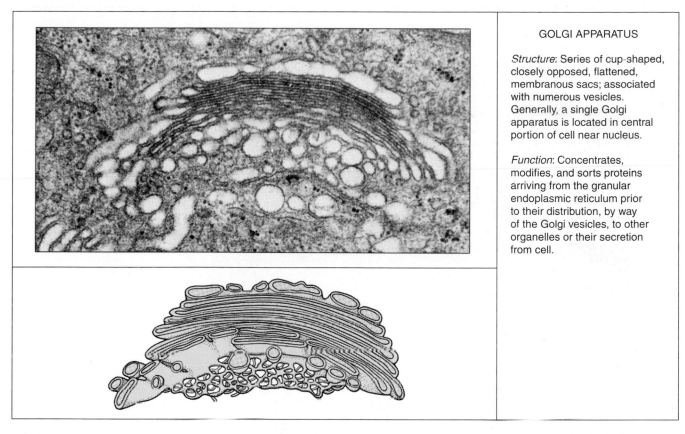

GOLGI APPARATUS

Structure: Series of cup-shaped, closely opposed, flattened, membranous sacs; associated with numerous vesicles. Generally, a single Golgi apparatus is located in central portion of cell near nucleus.

Function: Concentrates, modifies, and sorts proteins arriving from the granular endoplasmic reticulum prior to their distribution, by way of the Golgi vesicles, to other organelles or their secretion from cell.

FIGURE 3-12

Golgi apparatus. (*Electron micrograph from W. Bloom and D. W. Fawcett, "Textbook of Histology," 9th ed. W. B. Saunders Company, Philadelphia, 1968.*)

(Figure 3-11). The membranes enclose a space that is continuous throughout the network. (The continuity of the endoplasmic reticulum is not obvious when examining a single electron micrograph because only a portion of the network is present in any one section.)

Two forms of endoplasmic reticulum can be distinguished: **granular** (rough-surfaced) and **agranular** (smooth-surfaced). As noted above, the granular endoplasmic reticulum has ribosomes bound to its cytosolic surface, and it has a flattened-sac appearance. The outer membrane of the nuclear envelope also has ribosomes on its surface, and the space between the two nuclear-envelope membranes is continuous with that of the granular endoplasmic reticulum. The agranular endoplasmic reticulum has no ribosomal particles on its surface and has a branched, tubular structure.

Both granular and agranular endoplasmic reticulum exist in the same cell, but the relative amounts of the two types vary in different cells and even within the same cell during different periods of cell activity. As noted above, granular endoplasmic reticulum is involved in the packaging of proteins that, after processing in the Golgi apparatus, are to be secreted by cells or distributed to other cell organelles (Chapter 6). The agranular endoplasmic reticulum is the site at which lipid molecules are synthesized (Chapter 4), and it also stores and releases calcium ions involved in controlling various cell activities (Chapter 7).

Golgi Apparatus

The **Golgi apparatus** is a series of closely apposed, flattened membranous sacs that are slightly curved, forming a cup-shaped structure (Figure 3-12). Most cells have a single Golgi apparatus located near the nucleus, although some cells may have several. Associated with this organelle, particularly near its concave surface, are a number of approximately spherical, membrane-enclosed vesicles.

Proteins arriving at the Golgi apparatus from the granular endoplasmic reticulum undergo a sequential series of modifications as they pass from one Golgi compartment to the next. For example, oligosaccharides are linked to proteins to form glycoproteins, and the length of the protein is often shorted by removing a terminal portion of the

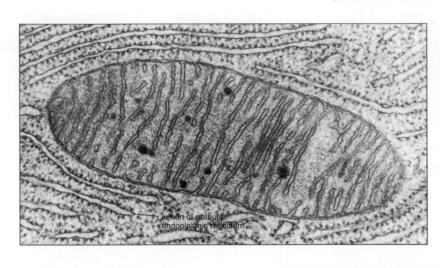

Lumen of granular
endoplasmic reticulum

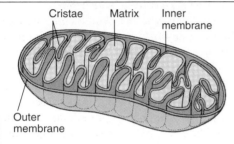

Cristae   Matrix   Inner
                    membrane

Outer
membrane

MITOCHONDRION

Structure: Rod- or oval-shaped body
surrounded by two membranes. Inner
membrane folds into matrix of the
mitochondrion, forming cristae.

Function: Major site of ATP production,
O_2 utilization, and CO_2 formation.
Contains enzymes of Krebs cycle and
oxidative phosphorylation.

FIGURE 3-13

Mitochondrion. (*Electron micrograph courtesy of K. R. Porter.*)

polypeptide chain. The Golgi apparatus sorts the modified proteins into discrete classes of transport vesicles that will be delivered to various parts of the cell, including the plasma membrane, where the protein contents of the vesicle are released to the outside of the cell. Vesicles containing proteins to be secreted from the cell are known as **secretory vesicles.**

Mitochondria

Mitochondria (singular, *mitochondrion*) are primarily concerned with the chemical processes by which energy is made available to cells in the form of adenosine triphosphate (ATP) molecules (Chapter 4). Most of the ATP used by cells is formed in the mitochondria by a process that consumes oxygen and produces carbon dioxide.

Mitochondria are spherical or elongated, rodlike structures surrounded by an inner and an outer membrane (Figure 3-13). The outer membrane is smooth, whereas the inner membrane is folded into sheets or tubules known as **cristae,** which extend into the inner mitochondrial compartment, the **matrix.** Mitochondria are found throughout the cytoplasm. Large numbers of them, as many as 1000, are present in cells that utilize large amounts of energy, whereas less active cells contain fewer.

Mitochondria have small amounts of DNA that contain the genes for the synthesis of some of the mitochondrial proteins. This DNA, which uses a slightly different genetic code than nuclear DNA (Chapter 5), was acquired millions of years ago when a bacteria-like organism was engulfed by another cell and, rather than being destroyed, its metabolic functions became integrated with those of the host cell.

Lysosomes

Lysosomes are spherical or oval organelles surrounded by a single membrane (Figure 3-4). A typical cell may contain several hundred lysosomes. The fluid within a lysosome is highly acidic and contains a variety of digestive enzymes. Lysosomes act as "cellular stomachs," breaking down bacteria and the debris from dead cells that have been engulfed by a cell. They may also break down cell organelles that have been damaged and no longer function normally. They play an especially important role in the various cells that make up the defense systems of the body (Chapter 20).

Peroxisomes

The structure of **peroxisomes** is similar to that of lysosomes, that is, oval bodies enclosed by a single membrane, but their chemical composition is quite different. Like mitochondria, peroxisomes consume oxygen, although in much smaller amounts, but this oxygen is not used in chemical reactions associated with ATP formation. Rather, peroxisomes destroy certain products formed from oxygen, notably hydrogen peroxide, which can be quite toxic to cells—thus the organelles' name. In addition, the peroxisomes in liver and kidney cells actually make hydrogen peroxide and use it to detoxify various ingested molecules.

Filaments

In addition to the membrane-enclosed organelles, the cytoplasm of most cells contains a variety of protein filaments. This filamentous network is referred to as the cell's **cytoskeleton** (Figure 3-14), and, like the bony skeleton of the body, it is associated with processes that maintain and change cell shape and produce cell movements.

There are four classes of cytoskeletal filaments, based on their diameter and the types of protein they contain (Figure 3-15). In order of size, starting with the thinnest, they are (1) microfilaments, (2) intermediate filaments, (3) muscle thick filaments, and (4) microtubules. Microfilaments and microtubules can be assembled and disassembled rapidly, allowing a cell to change its cytoskeletal framework according to changing requirements. In contrast, intermediate filaments and muscle thick filaments, once assembled, are less readily disassembled.

Microfilaments, which are composed of the contractile protein **actin,** make up a major portion of the cytoskeleton in all cells. **Intermediate filaments** are most

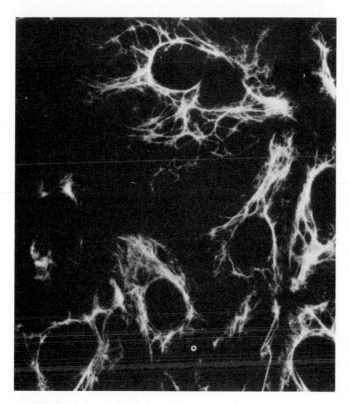

FIGURE 3-14

Cells stained to show the intermediate filament components of the cytoskeleton. (*From Roy A. Quinlan, et. al., Annals of the New York Academy of Sciences, vol. 455, New York, 1985.*)

FIGURE 3-15

Cytoskeletal filaments associated with cell shape and motility.

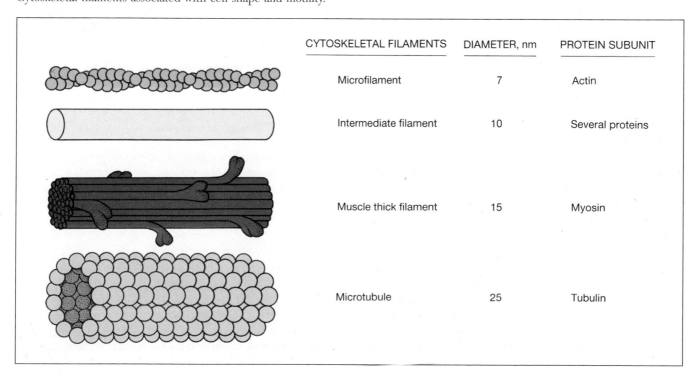

CYTOSKELETAL FILAMENTS	DIAMETER, nm	PROTEIN SUBUNIT
Microfilament	7	Actin
Intermediate filament	10	Several proteins
Muscle thick filament	15	Myosin
Microtubule	25	Tubulin

extensively developed in regions of cells that are subject to mechanical stress. **Myosin thick filaments,** composed of bundles of the contractile protein **myosin,** are found only in muscle cells. Individual molecules of myosin, however, are present in most nonmuscle cells where they are involved in producing forces associated with various cell movements.

Microtubles are hollow tubes about 25 nm in diameter, whose subunits are composed of the protein **tubulin.** They are the most rigid of the cytoskeletal filaments and are present in the long processes of nerve cells, where they provide the framework that maintains the processes' cylindrical shape. Microtubules radiate from a region of the cell known as the **centrosome,** which consists of two small cylindrical bodies, **centrioles,** composed of nine sets of fused microtubules. Surrounding the centrioles is a cloud of amorphous material that regulates the formation and elongation of microtubules. During cell division the centrosome generates the microtubular spindle fibers used in chromosome separation. Microtubules and microfilaments have also been implicated in the movements of organelles within the cytoplasm. These fibrous elements form the tracks along which organelles are propelled by contractile proteins attached to the surface of the organelles.

Cilia, the hair-like motile extensions on the surfaces of some epithelial cells, have a central core of microtubules organized in a pattern similar to that found in the centrioles; these microtubules in combination with a contractile protein produce the movements of the cilia. In hollow organs that are lined with ciliated epithelium, the cilia wave back and forth, propelling the luminal contents along the surface of the epithelium.

SUMMARY

MICROSCOPIC OBSERVATIONS OF CELLS

I. All living matter is composed of cells.

II. There are three classes of functional biological units: viruses, prokaryotic cells (bacteria) and eukaryotic cells (plant and animal cells).

CELL COMPARTMENTS

I. Every cell is surrounded by a plasma membrane.

II. Within each cell are numerous membrane-bound compartments, nonmembranous particles, and filaments, known collectively as cell organelles.

III. A cell is divided into two regions, the nucleus and the cytoplasm, the latter composed of the cytosol and cell organelles other than the nucleus.

IV. The membranes that surround the cell and cell organelles regulate the movements of molecules and ions into and out of the cell and its compartments.

   A. Membranes consist of a bimolecular lipid layer, composed of phospholipids and cholesterol, in which proteins are embedded.

   B. Integral membrane proteins are amphipathic proteins that often span the membrane, whereas peripheral membrane proteins are confined to the surfaces of the membrane.

V. Three types of membrane junctions link adjacent cells.

   A. Desmosomes link cells that are subject to considerable stretching.

   B. Tight junctions, found primarily in epithelial cells, limit the passage of molecules through the extracellular space between the cells.

   C. Gap junctions form channels between the cytosols of adjacent cells.

CELL ORGANELLES

I. The nucleus transmits and expresses genetic information.

   A. Threads of chromatin, composed of DNA and protein, condense to form chromosomes when a cell divides.

   B. The nucleolus is the site at which ribosomal subunits are formed.

II. Ribosomes, composed of RNA and protein, are the sites of protein synthesis.

III. The endoplasmic reticulum is a network of flattened sacs and tubules in the cytoplasm.

   A. Granular endoplasmic reticulum has attached ribosomes and is primarily involved in the packaging of proteins that are to be secreted by the cell or distributed to other organelles.

   B. Agranular endoplasmic reticulum is tubular, lacks ribosomes, and is the site of lipid synthesis and calcium accumulation and release.

IV. The Golgi apparatus modifies and sorts the proteins that are synthesized on the rough endoplasmic reticulum and packages them into secretory vesicles.

V. Mitochondria are the major cell sites that consume oxygen and produce carbon dioxide in chemical processes that transfer energy to ATP, which can then provide energy for energy-requiring cell functions.

VI. Lysosomes digest particulate matter that enters the cell.

VII. Peroxisomes break down certain toxic products formed from oxygen.

VIII. The cytoplasm contains a network of four types of filaments that form the cytoskeleton: (1) microfilaments, (2) intermediate filaments, (3) muscle thick filaments, and (4) microtubules.

KEY TERMS

light microscope
electron microscope
prokaryotic cell
eukaryotic cell
virus
plasma membrane
cell organelle
nucleus
cytoplasm
cytosol
intracellular fluid
phospholipid
cholesterol
integral membrane protein
transmembrane protein
peripheral membrane
   protein
glycocalyx
fluid-mosaic model
integrin
desmosome
tight junction
gap junction
nuclear envelope
nuclear pore
chromatin

chromosome
nucleolus
ribosome
endoplasmic reticulum
granular endoplasmic
   reticulum
agranular endoplasmic
   reticulum
Golgi apparatus
secretory vesicle
mitochondria
mitochondrial cristae
mitochondrial matrix
lysosome
peroxisome
cytoskeleton
microfilament
actin
intermediate filament
myosin thick filament
myosin
microtubule
tubulin
centrosome
centriole
cilia

REVIEW QUESTIONS

1. In terms of the size and number of cells, what makes an elephant larger than a mouse?

2. Identify the location of cytoplasm, cytosol, and intracellular fluid within a cell.

3. Identify the classes of organic molecules found in cell membranes.

4. Describe the orientation of the phospholipid molecules in a membrane.

5. Which plasma membrane components are responsible for membrane fluidity?

6. Describe the location and characteristics of integral and peripheral membrane proteins.

7. Describe the structure and function of the three types of junctions found between cells.

8. What function is performed by the nucleolus?

9. Describe the location and function of ribosomes.

10. Contrast the structure and functions of the granular and agranular endoplasmic reticulum.

11. What function is performed by the Golgi apparatus?

12. Describe the structure and primary function of mitochondria

13. What functions are performed by lysosomes and peroxisomes?

14. List the four types of filaments associated with the cytoskeleton. Identify the structures in cells that are composed of microtubules.

In adults the rates at which organic molecules are continuously synthesized (anabolism) and broken down (catabolism) are approximately equal.

CHEMICAL REACTIONS

I. The difference in the energy content of reactants and products is the amount of energy (measured in calories) that is released or added during a reaction.

II. The energy released during a chemical reaction either is released as heat or is transferred to other molecules.

III. The four factors that can alter the rate of a chemical reaction are listed in Table 4-1.

IV. The activation energy required to initiate the breaking of chemical bonds in a reaction is usually acquired through collisions with other molecules.

V. Catalysts increase the rate of a reaction by lowering the activation energy.

VI. The characteristics of reversible and irreversible reactions are listed in Table 4-2.

VII. The net direction in which a reaction proceeds can be altered, according to the law of mass action, by increases or decreases in the concentrations of reactants or products.

ENZYMES

I. Nearly all chemical reactions in the body are catalyzed by enzymes, the characteristics of which are summarized in Table 4-3.

II. Some enzymes require small concentrations of cofactors for activity.

    A. The binding of trace metal cofactors maintains the conformation of the enzyme's binding site so that it is able to bind substrate.

    B. Coenzymes, derived from vitamins, transfer small groups of atoms from one substrate to another. The coenzyme is regenerated in the course of these reactions and can be used over and over again.

REGULATION OF ENZYME-MEDIATED REACTIONS

The rates of enzyme-mediated reactions can be altered by changes in temperature, substrate concentration, enzyme concentration, and enzyme activity. Enzyme activity is altered by allosteric or covalent modulation.

MULTIENZYME METABOLIC PATHWAYS

I. The rate of product formation in a metabolic pathway can be controlled by allosteric or covalent modulation of the enzyme mediating the rate-limiting reaction in the pathway. The end product often acts as a modulator molecule, inhibiting the rate-limiting enzyme's activity.

II. An "irreversible" step in a metabolic pathway can be reversed by the use of two enzymes, one for the forward reaction and one for the reverse direction via another, energy-yielding reaction.

ATP

In all cells, energy from the catabolism of fuel molecules is transferred to ATP. The hydrolysis of ATP to ADP and P_i then transfers this energy to cell functions.

metabolism	active site
anabolism	cofactor
catabolism	coenzyme
calorie	vitamin
kilocalorie	NAD^+
activation energy	FAD
catalyst	enzyme activity
reversible reaction	metabolic pathway
chemical equilibrium	rate-limiting reaction
irreversible reaction	end-product inhibition
law of mass action	adenosine triphosphate
enzyme	(ATP)
substrate	

1. How do molecules acquire the activation energy required for a chemical reaction?

2. List the four factors that influence the rate of a chemical reaction and state whether increasing the factor will increase or decrease the rate of the reaction.

3. What characteristics of a chemical reaction make it reversible or irreversible?

4. List five characteristics of enzymes.

5. What is the difference between a cofactor and a coenzyme?

6. From what class of nutrients are coenzymes derived?

7. Why are small concentrations of coenzymes sufficient to maintain enzyme activity?

8. List three ways in which the rate of an enzyme-mediated reaction can be altered.

9. How can an irreversible step in a metabolic pathway be reversed?

10. What is the function of ATP in metabolism?

11. Approximately how much of the energy released from the catabolism of fuel molecules is transferred to ATP? What happens to the rest?

SECTION C
METABOLIC PATHWAYS

Three distinct but linked metabolic pathways are used by cells to transfer the energy released from the breakdown of fuel molecules to ATP. They are known as glycolysis, the Krebs cycle, and oxidative phosphorylation (Figure 4-18). In the following section we will describe the major characteristics of these three pathways in terms of the location of the pathway enzymes in a cell, the relative contribution of each pathway to ATP production, the sites of carbon dioxide formation and oxygen utilization, and the key molecules that enter and leave each of the pathways. One important generalization to keep in mind is that glycolysis can occur in either the presence or absence of oxygen, whereas the other two pathways require oxygen.

CELLULAR ENERGY TRANSFER

Glycolysis

Glycolysis (from the Greek *glycos,* sugar, and *lysis,* breakdown) is a pathway that partially catabolizes carbohydrates, primarily glucose. It consists of 10 enzymatic reactions that convert a six-carbon molecule of glucose into two three-carbon molecules of **pyruvate,** the ionized form of pyruvic acid (Figure 4-19). The reactions produce a net gain of two molecules of ATP and transfer four atoms of hydrogen to yield $2NADH + 2H^+$:

$$Glucose + 2\,ADP + 2\,P_i + 2\,NAD^+ \longrightarrow \quad (4\text{-}1)$$
$$2\,Pyruvate + 2\,ATP + 2\,NADH + 2\,H^+ + 2\,H_2O$$

These 10 reactions, *none of which utilizes oxygen,* take place in the cytosol. Note (Figure 4-19) that all the intermediates between glucose and the end product pyruvate contain one or more ionized phosphate groups. As we shall learn in Chapter 6, plasma membranes are impermeable to such highly ionized molecules, and thus these molecules remain trapped within the cell.

Note next that the early steps in glycolysis (reactions 1 and 3) each *use,* rather than produce, one molecule of ATP, to form phosphorylated intermediates. Note in addition

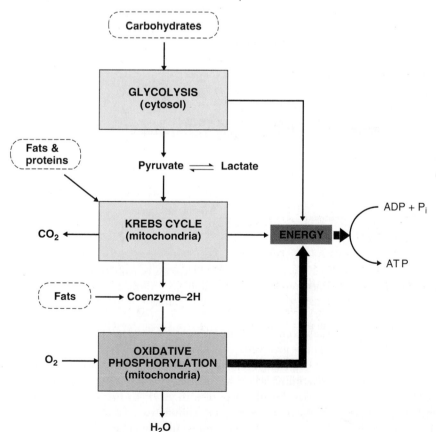

FIGURE 4-18

Pathways linking the energy released from the catabolism of fuel molecules to the formation of ATP. For simplicity, this figure does not show that glycolysis can also provide some coenzyme-2H molecules to oxidative phosphorylation.

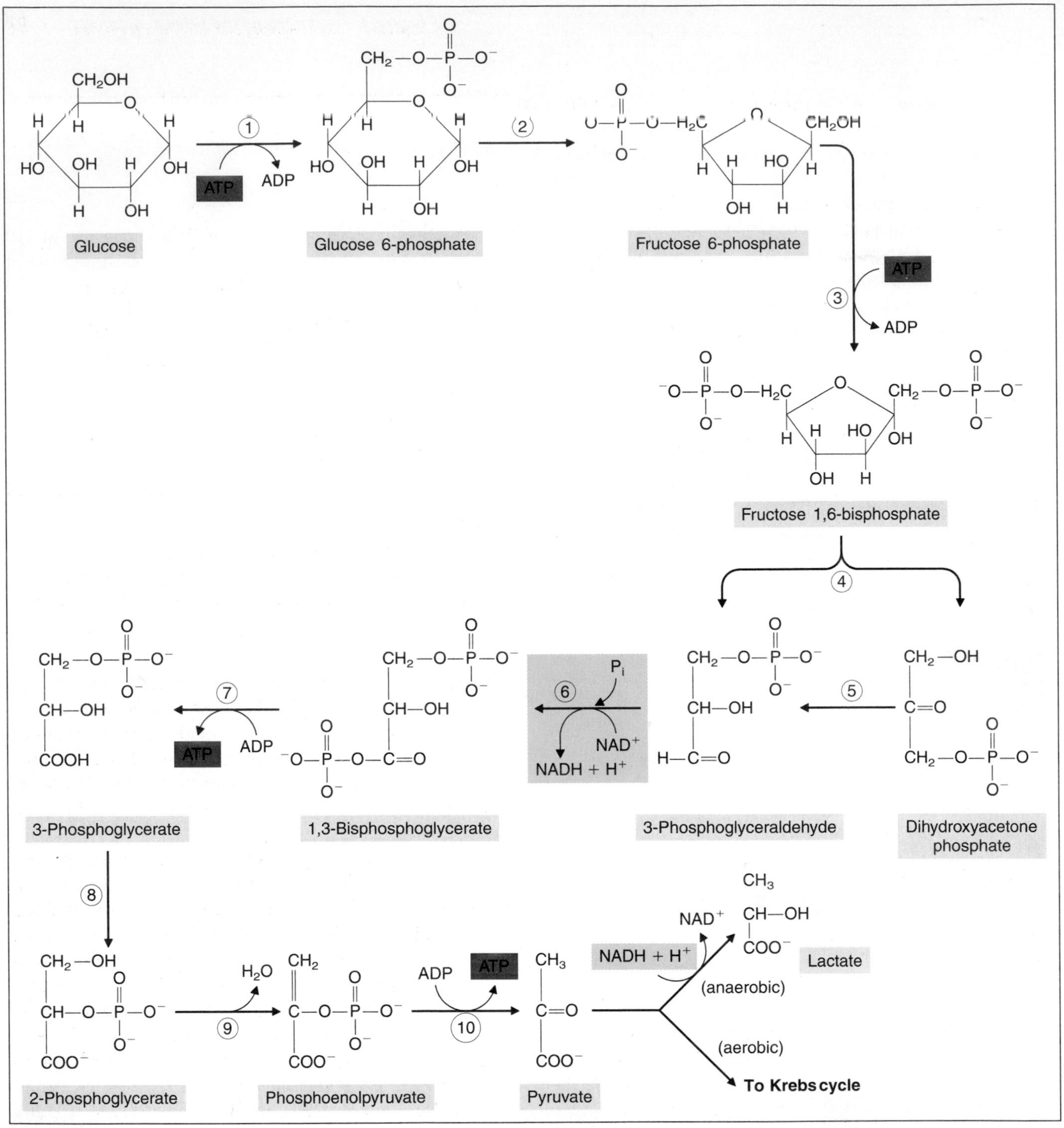

FIGURE 4-19

Glycolytic pathway. Under anaerobic conditions there is a net synthesis of two molecules of ATP for every molecule of glucose that enters the pathway. Note that at the pH existing in the body, the products produced by the various glycolytic steps exist in the ionized, anionic form (pyruvate for example). They are actually produced as acids (pyruvic acid, for example) that then ionize. ▭▭

that reaction 4 splits a six-carbon intermediate into two three-carbon molecules—dihydroxyacetone phosphate and 3-phosphoglyceraldehyde—and that reaction 5 converts dihydroxyacetone phosphate into another molecule of 3-phosphoglyceraldehyde. Thus, at this point in the pathway we have two molecules of 3-phosphoglyceraldehyde derived from one molecule of glucose. Keep in mind, then,

that from this point on, *two* molecules of each intermediate are involved.

The first *formation* of ATP in glycolysis occurs during reaction 7 when a phosphate group is transferred from 1,3-bisphosphoglycerate to ADP to form ATP. Since, as stressed above, two intermediates exist at this point, reaction 7 produces two molecules of ATP, one from each of them. In this reaction the mechanism of forming ATP is known as **substrate-level phosphorylation** since the phosphate group is transferred from a substrate molecule to ADP. As we shall see, this mechanism is quite different from that used during oxidative phosphorylation in which *free* inorganic phosphate is coupled to ADP to form ATP.

A similar substrate-level phosphorylation of ADP occurs during reaction 10, where again two molecules of ATP are formed. Thus, reactions 7 and 10 generate a total of four molecules of ATP for every molecule of glucose entering the pathway. There is a net gain, however, of only two molecules of ATP during glycolysis because two molecules of ATP were used in reactions 1 and 3.

The end product of glycolysis, pyruvate, can proceed in one of two directions, depending on the availability of molecular oxygen, which, as we stressed earlier, is *not* utilized in any of the glycolytic reactions themselves. If oxygen is present, that is, if **aerobic** conditions exist, pyruvate enters the Krebs cycle and is broken down into carbon dioxide, as described in the next section. In contrast, in the absence of oxygen (**anaerobic** conditions), pyruvate is converted to **lactate** (the ionized form of lactic acid) by a single enzyme-mediated reaction. In this reaction (Figure 4-20) two hydrogen atoms derived from $NADH + H^+$ are transferred to pyruvate to form lactate, and NAD^+ is regenerated. These hydrogens had originally been transferred to NAD^+ during reaction 6 of glycolysis, so the coenzyme NAD^+ shuttles hydrogen between the two reactions during anaerobic glycolysis. The *overall* reaction for anaerobic glycolysis is

$$\text{Glucose} + 2\,\text{ADP} + 2\,P_i \longrightarrow$$
$$2\,\text{Lactate} + 2\,\text{ATP} + 2\,H_2O \quad (4\text{-}2)$$

Recall that under aerobic conditions pyruvate is not converted to lactate but rather enters the Krebs cycle. Therefore, the mechanism just described for regenerating NAD^+ from $NADH + H^+$ by forming lactate no longer occurs. (Compare Equations 4-1 and 4-2.) Instead, as we shall see, $NADH + H^+$ are converted back to NAD^+ by the transfer of the hydrogens to oxygen during oxidative phosphorylation.

In most cells, the amount of ATP produced by glycolysis is much smaller than the amount formed by the other two ATP-generating pathways—the Krebs cycle and oxidative phosphorylation. There are special cases, however, in which glycolysis supplies most, or even all, of a cell's ATP. For example, erythrocytes contain the enzymes for glycolysis but have no mitochondria, which, as we shall see, are required for the other pathways. All of their ATP production occurs, therefore, by glycolysis. Also, certain types of skeletal muscles contain considerable amounts of glycolytic enzymes but have few mitochondria. During intense muscle activity, glycolysis provides most of the ATP

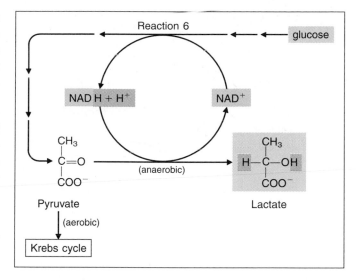

FIGURE 4-20

Under anaerobic conditions the coenzyme NAD^+ utilized in the glycolytic reaction 6 (Figure 4-19) is regenerated when it transfers its hydrogen atoms to pyruvate during the formation of lactate.

in these cells and is associated with the production of large amounts of lactate. Despite these exceptions most cells do not have sufficient concentrations of glycolytic enzymes or enough glucose to provide, by glycolysis alone, the high rates of ATP production necessary to meet their energy requirements and thus are unable to function for long under anaerobic conditions.

Our discussion of glycolysis has focused upon glucose as the major carbohydrate entering the glycolytic pathway. However, other carbohydrates such as fructose, derived from the disaccharide sucrose (table sugar), and galactose, from the disaccharide lactose (milk sugar), can also be catabolized by glycolysis since these carbohydrates are converted into several of the intermediates that participate in the early portion of the glycolytic pathway.

In some microorganisms, yeast cells for example, pyruvate is converted under anaerobic conditions to carbon dioxide and alcohol (CH_3CH_2OH) rather than to lactate. This process is known as fermentation and forms the basis for the production of alcohol from cereal grains rich in carbohydrates.

Table 4-4 summarizes the major characteristics of glycolysis.

Krebs Cycle

The **Krebs cycle,** named in honor of Hans Krebs, who worked out the intermediate steps in this pathway (also known as the **citric acid cycle** or **tricarboxylic acid cycle**), is the second of the three pathways involved in fuel catabolism and ATP production. It utilizes molecular fragments formed during carbohydrate, protein, and fat

TABLE 4-4 CHARACTERISTICS OF GLYCOLYSIS

Entering substrates	Glucose and other monosaccharides
Enzyme location	Cytosol
Net ATP production	2 ATP formed directly per molecule of glucose entering pathway
	Can be produced in the absence of oxygen (anaerobically)
Coenzyme production	2 NADH + 2 H$^+$ formed under aerobic conditions
Final products	Pyruvate—under aerobic conditions
	Lactate—under anaerobic conditions
Net reaction	
Aerobic:	Glucose + 2 ADP + 2 P$_i$ + 2 NAD$^+$ $\longrightarrow$
	2 pyruvate + 2 ATP + 2 NADH + 2 H$^+$ + 2 H$_2$O
Anaerobic:	Glucose + 2 ADP + 2 P$_i$ $\longrightarrow$ 2 lactate + 2 ATP + 2 H$_2$O

breakdown, and it produces carbon dioxide, hydrogen atoms bound to coenzymes, and small amounts of ATP. The enzymes for this pathway are located in the mitochondria, mostly in the inner mitochondrial compartment, the matrix.

The primary molecule entering at the beginning of the Krebs cycle is **acetyl coenzyme A (acetyl CoA):**

$$CH_3-\overset{\overset{\displaystyle O}{\|}}{C}-S-CoA$$

Coenzyme A (CoA) is derived from the B vitamin pantothenic acid and functions primarily to transfer two-carbon acetyl groups from one molecule to another. These acetyl groups come either from pyruvate, which, as we have just seen, is the end product of aerobic glycolysis, or from the breakdown of fatty acids and some amino acids, as we shall see in a later section.

Pyruvate, upon entering mitochondria from the cytosol, undergoes the following reaction to produce acetyl CoA:

$$\text{Pyruvate} + \text{CoA} + \text{NAD}^+ \longrightarrow \underset{+ \text{CO}_2 + \text{NADH} + \text{H}^+}{\text{Acetyl CoA}} \quad (4\text{-}3)$$

Note that this reaction produces the first molecule of CO$_2$ formed thus far in the pathways of fuel catabolism, and that hydrogen atoms have been transferred to NAD$^+$.

The Krebs cycle begins with the transfer of the acetyl group of acetyl CoA to the four-carbon molecule, oxaloacetate, to form the six-carbon molecule, citrate (Figure 4-21). At the third step in the cycle a molecule of CO$_2$ is produced, and again at the fourth step. Thus, two carbon atoms entered the cycle in the form of the acetyl group attached to CoA, and two carbons (although not the same ones) have left in the form of CO$_2$. Note also that the oxygen that appears in the CO$_2$ is not derived from molecular oxygen but from the carboxyl groups of Krebs-cycle intermediates.

In the remainder of the cycle, the four-carbon molecule formed in reaction 4 is modified through a series of reactions to produce the four-carbon molecule oxaloacetate, which becomes available to accept another acetyl group and repeat the cycle.

Now we come to a crucial fact: In addition to producing carbon dioxide, intermediates in the Krebs cycle donate hydrogen atoms to the coenzymes NAD$^+$ and FAD. Two hydrogen atoms are transferred to NAD$^+$ (to form NADH + H$^+$) in each of steps 3, 4, and 8, and to FAD in reaction 6. These hydrogens will be transferred from the coenzymes to oxygen in the next stage of fuel metabolism to be described, oxidative phosphorylation. Since oxidative phosphorylation is necessary for regeneration of the hydrogen-free form of these coenzymes, *the Krebs cycle can operate only under aerobic conditions.* There is no pathway in the mitochondria that can remove the hydrogen from these coenzymes under anaerobic conditions.

So far we have said nothing of how the Krebs cycle contributes to the formation of ATP. In fact, the Krebs cycle directly produces only one high-energy nucleotide triphosphate. This occurs during reaction 5 in which inorganic phosphate is transferred to guanosine diphosphate (GDP) to form guanosine triphosphate (GTP). The hydrolysis of GTP, like that of ATP, can provide energy for some energy-requiring reactions. In addition, the energy in GTP can be transferred to ATP by the reaction

$$\text{GTP} + \text{ADP} \rightleftharpoons \text{GDP} + \text{ATP}$$

This reaction is reversible, and the energy in ATP can be used to form GTP from GDP when additional GTP is required for protein synthesis (Chapter 5) and signal transduction (Chapter 7).

To reiterate, the formation of ATP from GTP is the only mechanism by which ATP is formed within the Krebs cycle. Why, then, is the Krebs cycle so important? Because the hydrogen atoms transferred to coenzymes during the cycle are used in the next pathway, oxidative phosphorylation, to form large amounts of ATP.

The net result of the catabolism of one acetyl group from acetyl CoA by way of the Krebs cycle can be written:

$$\begin{aligned} \text{Acetyl CoA} + 3\text{NAD}^+ + \text{FAD} + \text{GDP} + \text{P}_i \\ + 2\,\text{H}_2\text{O} \longrightarrow 2\,\text{CO}_2 + \text{CoA} \\ + 3\,\text{NADH} + 3\,\text{H}^+ + \text{FADH}_2 + \text{GTP} \end{aligned} \quad (4\text{-}4)$$

One more point should be noted: Although the major function of the Krebs cycle is the provision of hydrogen atoms to the oxidative-phosphorylation pathway, some of the intermediates in the cycle can be used to synthesize organic molecules, especially several types of amino acids, required by cells. Oxaloacetate is one of the intermediates used in this manner. When a molecule of oxaloacetate is removed from the Krebs cycle in the process of forming amino acids, however, it is not available to combine with the acetate fragment of acetyl CoA at the beginning of the cycle. Thus, there must be a way of replacing the oxaloacetate

FIGURE 4-21

The Krebs-cycle pathway. Note that the carbon atoms in the two molecules of CO_2 produced by a turn of the cycle are not the same two carbon atoms that entered the cycle as an acetyl group (identified by the dashed box throughout the figure).

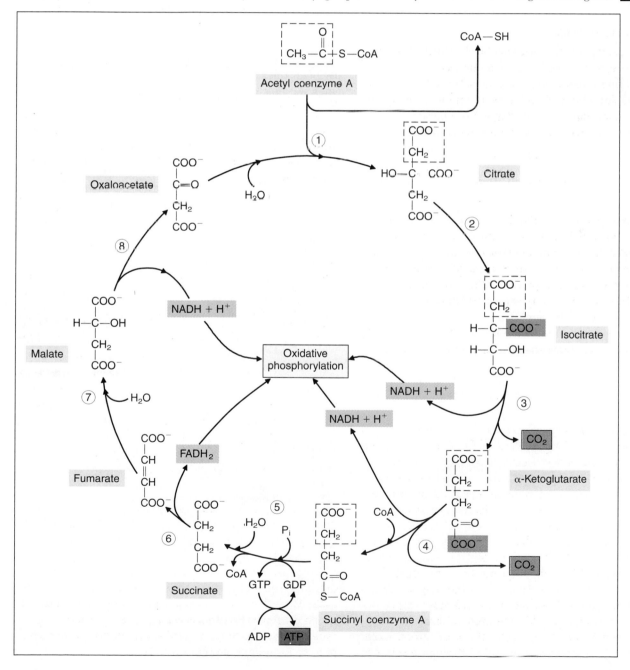

and other Krebs-cycle intermediates that are consumed in synthetic pathways. Carbohydrates provide one source of oxaloacetate replacement by the following reaction, which converts pyruvate into oxaloacetate (contrast this reaction to Equation 4-3):

$$\text{Pyruvate} + CO_2 + ATP \longrightarrow$$
$$\text{Oxaloacetate} + ADP + P_i \qquad (4\text{-}5)$$

Certain amino acid derivatives, as we shall see, can also be used to form oxaloacetate and other Krebs-cycle intermediates.

Table 4-5 summarizes the characteristics of the Krebs cycle reactions.

Oxidative Phosphorylation

Oxidative phosphorylation provides the third, and quantitatively most important, mechanism by which energy derived from fuel molecules can be transferred to ATP. The basic principle behind this pathway is simple: The energy transferred to ATP is derived from the energy released when hydrogen atoms combine with molecular oxygen to form water. The source of the hydrogen is the NADH + H^+ and $FADH_2$ coenzymes generated by the Krebs cycle and, to a much lesser extent, by glycolysis. The net reaction involving NADH + H^+ is

$$\tfrac{1}{2}O_2 + \text{NADH} + H^+ \longrightarrow H_2O + \text{NAD}^+ + 53 \text{ kcal/mol}$$

The proteins that mediate oxidative phosphorylation are embedded in the inner mitochondrial membrane unlike the enzymes of the Krebs cycle, which are soluble enzymes in the mitochondrial matrix. The proteins for oxidative phosphorylation can be divided into two groups: (1) those that mediate the series of reactions by which hydrogen atoms are transferred to molecular oxygen, and

(2) those that couple the energy released by these reactions to the synthesis of ATP.

Most of the first group of proteins contain iron and copper cofactors, forming proteins known as **cytochromes** (because in pure form they are brightly colored). Their structure resembles the red iron-containing hemoglobin molecule, which binds oxygen in red blood cells. The cytochromes form the components of the **electron transport chain,** in which two electrons from the hydrogen atoms are initially transferred either from NADH + H^+ or $FADH_2$ to one of the elements in this chain. These electrons are then successively transferred to other compounds in the chain, often to or from an iron or copper ion, until the electrons are finally transferred to molecular oxygen, which then combines with hydrogen ions (protons) to form water. These hydrogen ions, like the electrons, come from the hydrogen-bearing coenzymes, having been released from them early in the transport chain when the electrons from the hydrogen atoms were transferred to the cytochromes.

Importantly, in addition to transferring the coenzyme hydrogens to water, this process regenerates the hydrogen-free form of the coenzymes, which then become available to accept two more hydrogens from intermediates in the Krebs cycle. Thus, the electron transport chain provides the *aerobic* mechanism for regenerating the hydrogen-free form of the coenzymes, whereas, as described earlier, the *anaerobic* mechanism is coupled to the formation of lactate.

At each step along the electron transport chain small amounts of energy are released, which in total account for the full 53 kcal/mol released from a direct reaction between hydrogen and oxygen. Because this energy is released in small steps, it can be linked to the synthesis of several molecules of ATP, each of which requires only 7 kcal/mol.

TABLE 4-5	**CHARACTERISTICS OF THE KREBS CYCLE**
Entering substrate	Acetyl coenzyme A—acetyl groups derived from pyruvate and fatty acids
	Some intermediates derived from amino acids
Enzyme location	Inner compartment of mitochondria (the mitochondrial matrix)
ATP production	1 GTP formed directly, which can be converted into ATP
	Operates only under aerobic conditions even though molecular oxygen is not used directly in this pathway
Coenzyme production	3 NADH + 3 H^+ and 1 $FADH_2$
Final products	2 CO_2 for each molecule of acetyl coenzyme A entering pathway
	Some intermediates used to synthesize amino acids and other organic molecules required for special cell functions
Net reaction	Acetyl CoA + 3 NAD^+ + FAD + GDP + P_i + 2 H_2O $\longrightarrow$ 2 CO_2 + CoA + 3 NADH + 3 H^+ + $FADH_2$ + GTP

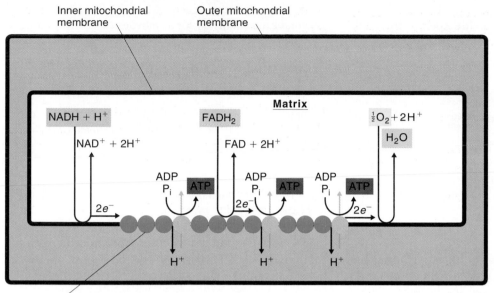

Inner mitochondrial membrane

Outer mitochondrial membrane

Matrix

NADH + H$^+$

NAD$^+$ + 2H$^+$

FADH$_2$

FAD + 2H$^+$

$\frac{1}{2}$O$_2$+2H$^+$

H$_2$O

ADP P$_i$ ATP

ADP P$_i$ ATP

ADP P$_i$ ATP

2e$^-$

2e$^-$

2e$^-$

H$^+$

H$^+$

H$^+$

Cytochromes in electron transport chain

FIGURE 4-22

ATP is formed during oxidative phosphorylation by the flow of hydrogen ions across the inner mitochondrial membrane. Two or three molecules of ATP are produced per pair of electrons donated, depending on the point at which a particular coenzyme enters the electron transport chain.

ATP is formed at three points along the electron transport chain. The mechanism by which this occurs is known as the **chemiosmotic hypothesis.** As electrons are transferred from one cytochrome to another along the electron transport chain, the energy released is used to move hydrogen ions (protons) from the matrix into the compartment between the inner and outer mitochondrial membranes (Figure 4-22), thus producing a source of potential energy in the form of a hydrogen-ion gradient across the membrane. At three points along the chain a protein complex forms a channel through which the hydrogen ions can flow back to the matrix side and in the process transfer their energy to the formation of ATP from ADP and P$_i$. FADH$_2$, which is formed during step 6 in the Krebs cycle, has a slightly lower chemical energy content than does NADH + H$^+$ and enters the electron transport chain at a point beyond the first site of ATP generation (Figure 4-22). Thus, the transfer of its electrons to oxygen produces only two ATP rather than the three formed from NADH + H$^+$.

To repeat, the majority of the ATP formed in the body is produced during oxidative phosphorylation as a result of the processing of hydrogen atoms originating largely from the Krebs cycle, during the breakdown of carbohydrates, fats, and proteins. The mitochondria, where the oxidative phosphorylation and the Krebs cycle reactions occur, are thus considered the powerhouses of the cell. In addition, as we have just seen, it is within these organelles that the majority of the oxygen we breathe is consumed and the carbon dioxide we expire is produced.

Table 4-6 summarizes the key features of oxidative phosphorylation.

Free-Radical Formation

As we have just seen, the formation of ATP by oxidative phosphorylation involves the successive transfer of four electrons along with four hydrogen ions to molecular oxygen to form two molecules of water. However, several

TABLE 4-6   CHARACTERISTICS OF OXIDATIVE PHOSPHORYLATION

Entering substrates	Hydrogen atoms obtained from NADH + H$^+$ and FADH$_2$ formed (1) during glycolysis, (2) by the Krebs cycle during the breakdown of pyruvate and amino acids, and (3) during the breakdown of fatty acids
	Molecular oxygen
Enzyme location	Inner mitochondrial membrane
ATP production	3 ATP formed from each NADH + H$^+$
	2 ATP formed from each FADH$_2$
Final products	H$_2$O—one molecule for each pair of hydrogens entering pathway.
Net reaction with NADH + H$^+$	$\frac{1}{2}$O$_2$ + NADH + H$^+$ + 3 ADP + 3 P$_i$ $\longrightarrow$ H$_2$O + NAD$^+$ + 3 ATP

highly reactive transient oxygen derivatives are formed during this step—**hydrogen peroxide** and the free radicals **superoxide anion** and **hydroxyl radical.**

$$O_2 \xrightarrow{e^-} O_2^- \cdot \xrightarrow[2H^+]{e^-} H_2O_2 \xrightarrow{e^-}$$

Superoxide anion   Hydrogen peroxide

$$OH^- + OH \cdot \xrightarrow{e^-} 2OH^- \xrightarrow[2H^+]{} 2 H_2O$$

Hydroxyl radical

Although most of the electrons transferred along the electron transport chain go into the formation of water, small amounts of reactive oxygen species do escape. These species can react with and damage proteins, membrane phospholipids, and nucleic acids. Such damage has been implicated in the aging process and in inflammatory reactions to tissue injury. Some cells use these reactive molecules to kill invading bacteria, as described in Chapter 20.

Reactive oxygen molecules are also formed by the action of ionizing radiation on oxygen and by reactions of oxygen with heavy metals such as iron. Cells contain several enzymatic mechanisms for removing these free radicals.

CARBOHYDRATE, FAT, AND PROTEIN METABOLISM

Having described the three pathways by which energy is transferred to ATP, we now consider how each of the three classes of fuel molecules enters the ATP-generating pathways. We also consider the synthesis of these fuel molecules and the pathways and restrictions governing their conversion from one class to another. These anabolic pathways are also used to synthesize molecules that have functions other than the storage and release of energy. For example, with the addition of a few enzymes, the pathway for fat synthesis is also used for synthesis of the phospholipids found in membranes.

Carbohydrate Metabolism

Carbohydrate Catabolism. In the previous sections we described the major pathways of carbohydrate catabolism: the breakdown of glucose to pyruvate or lactate by way of the glycolytic pathway, and the metabolism of pyruvate to carbon dioxide and water by way of the Krebs cycle and oxidative phosphorylation.

FIGURE 4-23

Pathways of aerobic glucose catabolism and their linkage to ATP formation.

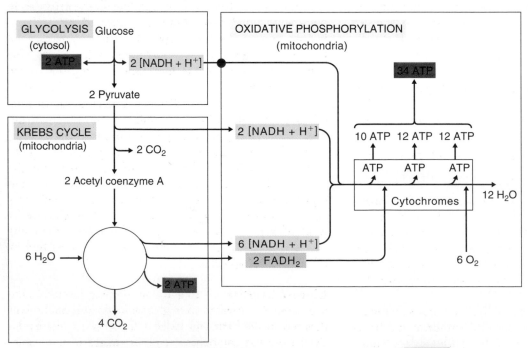

The amount of energy released during the catabolism of glucose to carbon dioxide and water is 686 kcal/mol of glucose:

$$C_6H_{12}O_6 + 6\ O_2 \longrightarrow 6\ H_2O + 6\ CO_2 + 686\ \text{kcal/mol}$$

As noted earlier, about 40 percent of this energy is transferred to ATP. Figure 4-23 illustrates the points at which ATP is formed during glucose catabolism. As we have seen, a net gain of two ATP molecules occurs by substrate-level phosphorylation during glycolysis, and two more are formed during the Krebs cycle from GTP, one from each of the two molecules of pyruvate entering the cycle. The major portion of ATP molecules produced in glucose catabolism—34 ATP per molecule—is formed during oxidative phosphorylation from the hydrogens generated at various steps during glucose breakdown.

To reiterate, in the absence of oxygen, only 2 molecules of ATP can be formed by the breakdown of glucose to lactate. This yield represents only 2 percent of the energy stored in glucose. Thus, the evolution of aerobic metabolic pathways greatly increased the amount of energy available to a cell from glucose catabolism. For example, if a muscle consumed 38 molecules of ATP during a contraction, this amount of ATP could be supplied by the breakdown of 1 molecule of glucose in the presence of oxygen or 19 molecules of glucose under anaerobic conditions.

It is important to note, however, that although only 2 molecules of ATP are formed per molecule of glucose under anaerobic conditions, large amounts of ATP can still be supplied by the glycolytic pathway if large amounts of glucose are broken down to lactate. This is not an efficient utilization of fuel energy, but it does permit continued ATP production under anaerobic conditions, such as occur during intense exercise (Chapter 11).

Glycogen Storage.   A small amount of glucose can be stored in the body to provide a reserve supply for use during those periods when glucose is not being absorbed into the blood from the intestinal tract. It is stored in the form of the polysaccharide **glycogen,** mostly in skeletal muscles and the liver.

Glycogen is synthesized from glucose by the pathway illustrated in Figure 4-24. The enzymes for both glycogen synthesis and glycogen breakdown are located in the cytosol. The first step, the transfer of phosphate from a molecule of ATP to glucose, forming glucose 6-phosphate, is the same as the first step in glycolysis. Thus, glucose 6-phosphate can either be broken down to pyruvate or used to form glycogen.

Note that, as indicated by the bowed arrows in Figure 4-24, different enzymes are used to synthesize and break down glycogen. The existence of two pathways containing enzymes that are subject to both covalent and allosteric

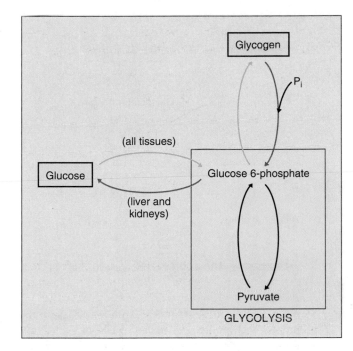

FIGURE 4-24

Pathways for glycogen synthesis and breakdown. Each bowed arrow indicates an irreversible reaction that requires one enzyme to catalyze the reaction in the forward direction and a separate enzyme (or enzymes) for the reverse reaction.

modulation provides a mechanism for regulating the flow of glucose to and from glycogen. When an excess of glucose is available to a liver or muscle cell, the enzymes in the glycogen synthesis pathway are activated by the chemical signals described in Chapter 18, and the enzyme that breaks down glycogen is simultaneously inhibited. This combination leads to the net storage of glucose in the form of glycogen.

When less glucose is available, the reverse combination of enzyme stimulation and inhibition occurs, and net breakdown of glycogen to glucose 6-phosphate occurs. Two paths are available to this glucose 6-phosphate: (1) In most cells, including skeletal muscle, it enters the glycolytic pathway where it is catabolized to provide the energy for ATP formation; (2) in liver and kidney cells, glucose 6-phosphate can be converted to free glucose by removal of the phosphate group, and glucose is then able to pass out of the cell into the blood, for use as fuel by other cells (Chapter 18).

Glucose Synthesis.   In addition to being formed in the liver from the breakdown of glycogen, glucose can be synthesized in the liver and kidneys from intermediates derived from the catabolism of other classes of fuel molecules. This process of generating new molecules of glucose

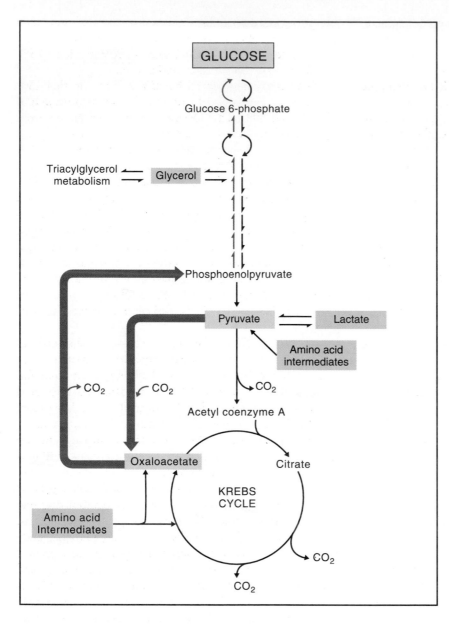

FIGURE 4-25

Gluconeogenic pathway by which pyruvate, lactate, glycerol, and various amino acid intermediates can be converted into glucose in the liver. Note the points at which each of these precursors, supplied by the blood, enters the pathway.

is known as **gluconeogenesis.** The major substrate in gluconeogenesis is pyruvate, formed from lactate and from several amino acids during protein breakdown. In addition, glycerol derived from the hydrolysis of triacylglycerols can be converted into glucose via a pathway that does not involve pyruvate.

The pathway for gluconeogenesis in the liver and kidneys (Figure 4-25) makes use of many but not all of the enzymes used in glycolysis because most of these reactions are reversible. However, reactions 3 and 10 (Figure 4-19) are irreversible, and additional enzymes are required, therefore, to form glucose from pyruvate. Pyruvate is converted to phosphoenolpyruvate by a series of mitochondrial reactions in which CO_2 is added to pyruvate to form the four-carbon Krebs-cycle intermediate oxaloacetate.

[In addition to being an important intermediary step in gluconeogenesis, this reaction (Equation 4-5) provides a pathway for replacing Krebs-cycle intermediates, as described earlier.] An additional series of reactions leads to the transfer of a four-carbon intermediate derived from oxaloacetate out of the mitochondria and its conversion to phosphoenolpyruvate in the cytosol. Phosphoenolpyruvate then reverses the steps of glycolysis back to the level of reaction 3, in which a different enzyme from that used in glycolysis is required to convert fructose 1,6-bisphosphate to fructose 6-phosphate. From this point on the reactions are again reversible, leading to glucose 6-phosphate, which can be converted to glucose in the liver and kidneys or stored as glycogen. Since energy is released during the glycolytic breakdown of glucose to pyruvate in the form of heat and ATP generation, energy must be added to reverse this pathway. A total of six ATP are consumed in the reactions of gluconeogenesis.

Many of the same enzymes are used in glycolysis and gluconeogenesis, so the question arises: What controls the direction of the reactions in these pathways? What conditions determine whether glucose is broken down to pyruvate or whether pyruvate is converted into glucose? The answer lies in the concentrations of glucose or pyruvate in a cell and in the control of the enzymes involved in the irreversible steps in the pathway, a control exerted via various hormones that alter the concentration and activities of these key enzymes (Chapter 18).

Fat Metabolism

Fat Catabolism. Triacylglycerol (fat) consists of three fatty acids linked to glycerol (Chapter 2). Fat accounts for the major portion (approximately 80 percent) of the energy stored in the body (Table 4-7). Under resting conditions, approximately half the energy used by such tissues as muscle, liver, and kidneys is derived from the catabolism of fatty acids.

Although most cells store small amounts of fat, the majority of the body's fat is stored in specialized cells known as **adipocytes.** Almost the entire cytoplasm of these cells is filled with a single large fat droplet. Clusters of adipocytes form **adipose tissue,** most of which is in deposits underlying the skin. The function of adipocytes is to synthesize and store triacylglycerols during periods of food uptake and then, when food is not being absorbed from the intestinal tract, to release fatty acids and glycerol into the blood for uptake and use by other cells to provide the energy for ATP formation. The factors controlling fat storage and release from adipocytes will be described in Chapter 18. Here we will emphasize the pathway by which fatty acids are catabolized by most cells to provide the energy for ATP synthesis, and the pathway for the synthesis of fatty acids from other fuel molecules.

Figure 4-26 shows the pathway for fatty acid catabolism, which is achieved by enzymes present in the mitochondrial matrix. The breakdown of a fatty acid is initiated by linking a molecule of coenzyme A to the carboxyl end of the fatty acid. This initial step is accompanied by the *breakdown* of ATP to AMP and two P_i.

The coenzyme-A derivative of the fatty acid then proceeds through a series of reactions, known as **beta oxidation,** which split off a molecule of acetyl coenzyme A from the end of the fatty acid and transfer two pairs of hydrogen atoms to coenzymes (one pair to FAD and the other to NAD^+). The hydrogen atoms from the coenzymes then enter the oxidative-phosphorylation pathway to form ATP.

When an acetyl coenzyme A is split from the end of a fatty acid, another coenzyme A is added (ATP is not required for this step), and the sequence is repeated. Each passage through this sequence shortens the fatty acid chain by two carbon atoms until all the carbon atoms have been transferred to coenzyme A. As we saw, the acetyl coenzyme A molecules then enter the Krebs cycle to produce CO_2 and ATP via the Krebs cycle and oxidative phosphorylation.

How much ATP is formed as a result of the total catabolism of a fatty acid? Most fatty acids in the body contain 14 to 22 carbons, 16 and 18 being most common. The catabolism of one 18-carbon saturated fatty acid yields 146 ATP molecules. In contrast, as we have seen, the catabolism of one glucose molecule yields a maximum of 38 ATP molecules. Thus, taking into account the difference in molecular weight of the fatty acid and glucose, it turns out that the amount of ATP formed from the catabolism of a gram of fat is about $2\frac{1}{2}$ times greater than the amount of ATP produced by catabolizing one gram of carbohydrate. If an average person stored most of his or her fuel as carbohydrate rather than fat, body weight would have to be approximately 30 percent greater in order to store the same amount of usable energy, and the person would consume more energy moving this extra weight around. Thus, a major step in fuel economy occurred when animals evolved the ability to store fuel as fat. In contrast, plants store almost all their fuel as carbohydrate (starch).

Fat Synthesis. The synthesis of fatty acids occurs by reactions that are almost the reverse of those that degrade them. However, the enzymes in the synthetic pathway are in the cytosol, whereas (as we have just seen) the enzymes catalyzing fatty acid breakdown are in the mitochondria. Fatty acid synthesis begins with cytoplasmic acetyl coenzyme A. Through a series of reactions requiring ATP and $NADPH + H^+$ (a phosphorylated form of $NADH + H^+$), the acetyl coenzyme A transfers its acetyl group to another molecule of acetyl coenzyme A to form a four-carbon chain. By repetition of this process, long-chain fatty acids are built up two carbons at a time, which accounts for the fact that all the fatty acids synthesized in the body contain

TABLE 4-7 FUEL CONTENT OF A 70-KG PERSON

	Total-body content, kg	Energy content, kcal/g	Total-body energy content	
			kcal	%
Triacylglycerols	15.6	9	140,000	78
Proteins	9.5	4	38,000	21
Carbohydrates	0.5	4	2,000	1

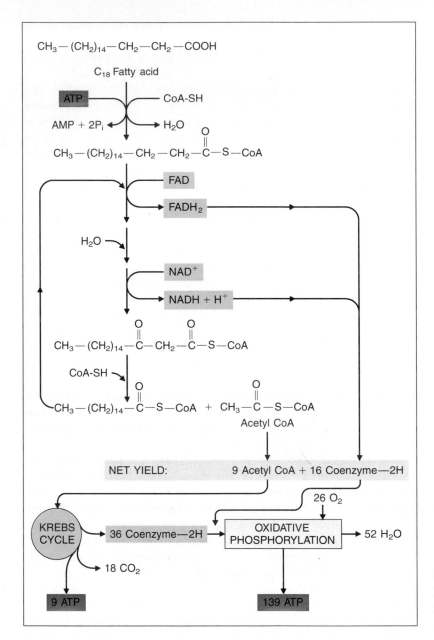

FIGURE 4-26

Pathway of fatty acid catabolism, which takes place in the mitochondria. The energy equivalent of two ATP is consumed at the start of the pathway. ▶️

an even number of carbon atoms. The overall reaction for the synthesis of an 18-carbon fatty acid can be written:

$$9 \overset{\overset{\text{O}}{\|}}{CH_3C}-S-CoA + 8\ ATP + 16\ NADPH + 16\ H^+ \longrightarrow$$
$$CH_3(CH_2)_{16}COOH + 8\ ADP + 8\ P_i + 16\ NADP^+$$
$$+ 7\ H_2O + 9\ CoA-SH$$

Once the fatty acids are formed, triacylglycerol can be synthesized by linking fatty acids to each of the three hydroxyl groups in glycerol, more specifically, to a phosphorylated form of glycerol called **α-glycerol phosphate.** The synthesis of triacylglycerol is carried out by enzymes asso-

ciated with the membranes of the smooth endoplasmic reticulum.

Compare the molecules produced by glucose catabolism with those required for synthesis of both fatty acids and α-glycerol phosphate. First, acetyl coenzyme A, the starting material for fatty acid synthesis, can be formed from pyruvate, the end product of glycolysis. Second, the other ingredients required for fatty acid synthesis— hydrogen-bound coenzymes and ATP—are produced during carbohydrate catabolism. Third, α-glycerol phosphate can be formed from a glucose intermediate. It should not be surprising, therefore, that much of the carbohydrate in food is converted into fat and stored in adipose tissue shortly after its absorption from the gastrointestinal tract.

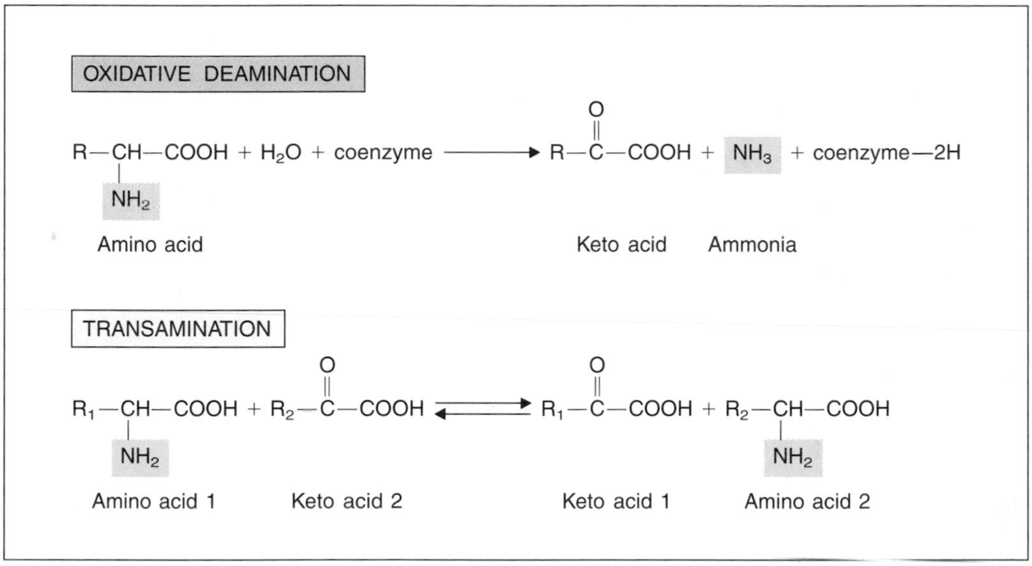

OXIDATIVE DEAMINATION

$$R-\underset{\underset{NH_2}{|}}{CH}-COOH + H_2O + coenzyme \longrightarrow R-\underset{\underset{O}{\parallel}}{C}-COOH + NH_3 + coenzyme-2H$$

Amino acid   Keto acid   Ammonia

TRANSAMINATION

$$R_1-\underset{\underset{NH_2}{|}}{CH}-COOH + R_2-\underset{\underset{O}{\parallel}}{C}-COOH \rightleftharpoons R_1-\underset{\underset{O}{\parallel}}{C}-COOH + R_2-\underset{\underset{NH_2}{|}}{CH}-COOH$$

Amino acid 1   Keto acid 2   Keto acid 1   Amino acid 2

FIGURE 4-27
Oxidative deamination and transamination of amino acids.

Mass action resulting from the increased concentration of glucose intermediates, as well as the specific hormonal regulation of key enzymes, promotes this conversion, as will be described in Chapter 18.

It is very important to note that fatty acids, or more specifically the acetyl coenzyme A derived from fatty acid breakdown, cannot be used to synthesize new molecules of glucose. The reasons for this can be seen by examining the pathways for glucose synthesis (Figure 4-25). First, because the reaction in which pyruvate is broken down to acetyl coenzyme A and carbon dioxide is irreversible, acetyl coenzyme A cannot be converted into pyruvate, a molecule that could lead to the production of glucose. Second, the equivalent of the two carbon atoms in acetyl coenzyme A are converted into two molecules of carbon dioxide during their passage through the Krebs cycle before reaching oxaloacetate, another takeoff point for glucose synthesis, and therefore cannot be used to synthesize *net* amounts of oxaloacetate.

Thus, glucose can readily be converted into fat, but the fatty acid portion of fat cannot be converted to glucose. However, the three-carbon glycerol backbone of fat can be converted into an intermediate in the gluconeogenic pathway and thus give rise to glucose, as mentioned earlier.

Protein and Amino Acid Metabolism

In contrast to the complexities of protein synthesis, described in Chapter 5, protein catabolism requires only a few enzymes, termed **proteases,** to break the peptide bonds between amino acids. Some of these enzymes split off one amino acid at a time from the ends of the protein chain, whereas others break peptide bonds between specific amino acids within the chain, forming peptides rather than free amino acids.

Amino acids can be catabolized to provide energy for ATP synthesis, and they can also provide intermediates for the synthesis of a number of molecules other than proteins. Since there are 20 different amino acids, a large number of intermediates can be formed, and there are many pathways for processing them. A few basic types of reactions common to most of these pathways can provide an overview of amino acid catabolism.

Unlike most carbohydrates and fats, amino acids contain nitrogen atoms (in their amino groups) in addition to carbon, hydrogen, and oxygen atoms. Once the nitrogen containing amino group is removed, the remainder of most amino acids can be metabolized to intermediates capable of entering either the glycolytic pathway or the Krebs cycle.

The two types of reactions by which the amino group is removed are illustrated in Figure 4-27. In the first reaction, **oxidative deamination,** the amino group gives rise to a molecule of ammonia (NH_3) and is replaced by an oxygen atom derived from water to form a **keto acid,** a categorical name rather than the name of a specific molecule. The second means of removing an amino group is known as **transamination** and involves transfer of the amino group from an amino acid to a keto acid. Note that the keto acid to which the amino group is transferred becomes an amino acid. The nitrogen derived from amino groups can also be used by cells to synthesize other important nitrogen-

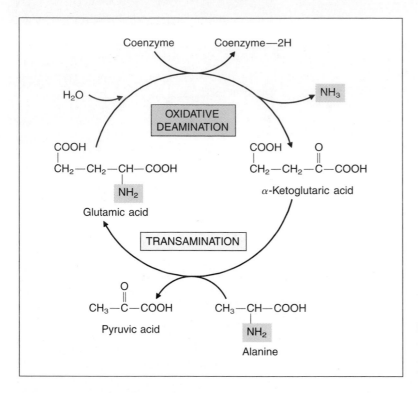

FIGURE 4-28

Oxidative deamination and transamination of the amino acids glutamate and alanine lead to keto acids that can enter the carbohydrate pathways.

containing molecules, such as the purine and pyrimidine bases found in nucleic acids.

Figure 4-28 illustrates the oxidative deamination of the amino acid glutamate and the transamination of the amino acid alanine. Note that the keto acids formed are intermediates either in the Krebs cycle (α-ketoglutarate) or glycolytic pathway (pyruvate). Once formed, these keto acids can be metabolized to produce carbon dioxide and form ATP, or they can be used as intermediates in the synthetic pathway leading to the formation of glucose. As a third alternative, they can be used to synthesize fatty acids after their conversion to acetyl coenzyme A by way of pyruvic acid. Thus, amino acids can be used as a source of energy, and some can be converted into carbohydrate and fat.

As we have seen, the oxidative deamination of amino acids yields ammonia. This substance, which is highly toxic to cells if allowed to accumulate, readily passes through cell membranes and enters the blood, which carries it to the liver (Figure 4-29). The liver contains enzymes that can link two molecules of ammonia with carbon dioxide to form **urea.** Thus, urea, which is relatively nontoxic, is the major nitrogenous waste product of protein catabolism. It leaves the liver and is excreted by the kidneys into the urine. Two of the 20 amino acids also contain atoms of sulfur, which can be converted to sulfate, SO_4^{2-}, and excreted in the urine.

Thus far, we have discussed mainly amino acid *catabolism.* The keto acids pyruvic acid and α-ketoglutaric acid can be derived from the breakdown of glucose; they can then be transaminated, as described above, to form glutamate and alanine. Thus glucose can be used to produce certain amino acids, provided other amino acids are available in the diet to supply amino groups for transamination. However, only 11 of the 20 amino acids can be formed by this process because 9 of the specific keto acids cannot be synthesized from other intermediates. The 9 amino acids corresponding to these keto acids must be obtained from the food we eat and are known as **essential amino acids.**

Figure 4-30 provides a summary of the multiple routes by which amino acids are handled by the body. The amino acid pools, which consist of the body's total free amino acids, are derived from (1) ingested protein, which is degraded to amino acids during digestion in the intestinal tract, (2) the synthesis of nonessential amino acids from the keto acids derived from carbohydrates and fat, and (3) the continuous breakdown of body proteins. These pools are the source of amino acids for the resynthesis of body protein and a host of specialized amino acid derivatives, as well as for conversion to carbohydrate and fat. A very small quantity of amino acids and protein is lost from the body via the urine, skin, hair, fingernails, and in women, the menstrual fluid. The major route for the loss of amino acids is not their excretion but rather their deamination, with ultimate excretion of the nitrogen atoms as urea in the urine. The terms **negative nitrogen balance** and **positive nitrogen balance** refer to whether there is a net loss or gain, respectively, of amino acids in the body over any period of time.

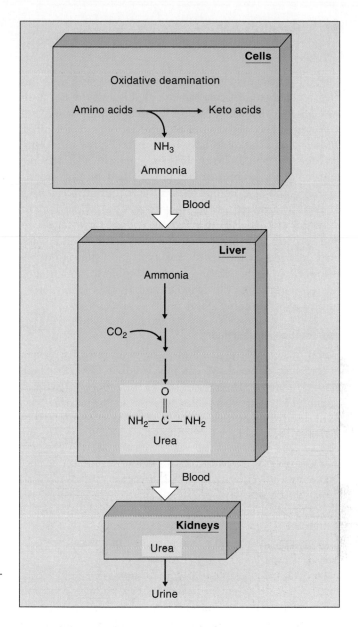

FIGURE 4-29

Formation and excretion of urea, the major nitrogenous waste product of protein catabolism.

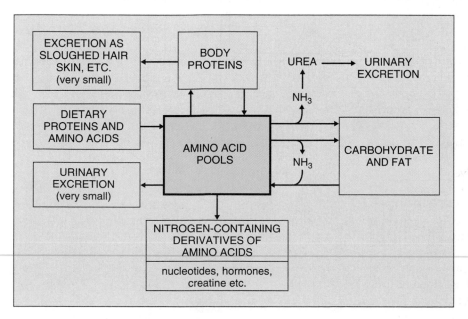

FIGURE 4-30

Pathways of amino acid metabolism.

If any of the essential amino acids are missing from the diet, a negative nitrogen balance, that is, loss greater than gain, always results. The proteins that require a missing essential amino acid cannot be synthesized, and the other amino acids that would have been incorporated into these proteins are metabolized. This explains why a dietary requirement for protein cannot be specified without regard to the amino acid composition of that protein. Protein is graded in terms of how closely its relative proportions of essential amino acids approximate those in the average body protein. The highest quality proteins are found in animal products, whereas the quality of most plant proteins is lower. Nevertheless, it is quite possible to obtain adequate quantities of all essential amino acids from a mixture of plant proteins alone.

Fuel Metabolism Summary

Having discussed the metabolism of the three major classes of organic molecules, we can now briefly review how each class is related to the others and to the process of synthesizing ATP. Figure 4-31 illustrates the major pathways we have discussed and the relations of the common intermediates. All three classes of molecules can enter the Krebs cycle through some intermediate, and thus all three can be used as a source of energy for the synthesis of ATP. Glucose can be converted into fat or into some amino acids by way of common intermediates such as pyruvate, oxaloacetate, and acetyl coenzyme A. Similarly, some amino acids can be converted into glucose and fat. Fatty acids cannot be converted into glucose because of the irreversibility of the reaction converting pyruvate to acetyl coenzyme A, but the glycerol portion of triacylglycerols can be converted into glucose. Fatty acids can be used to synthesize portions of the keto acids used to form some amino acids. Metabolism is thus a highly integrated process in which all classes of molecules can be used, if necessary, to provide energy, and in which each class of molecule can provide the raw materials required to synthesize most but not all members of other classes.

ESSENTIAL NUTRIENTS

About 50 substances required for normal or optimal body function cannot be synthesized by the body or are synthesized in amounts inadequate to keep pace with the rates

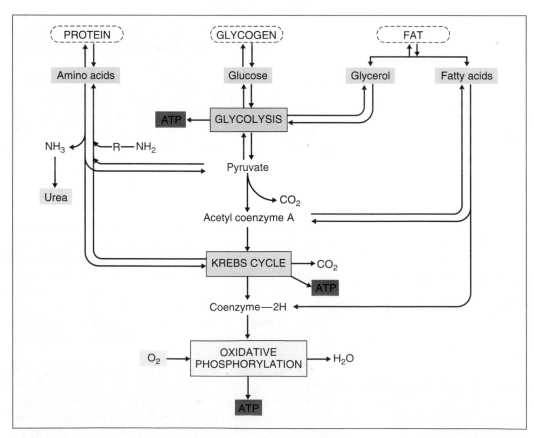

FIGURE 4-31
Interrelations between the pathways for the metabolism of carbohydrate, fat, and protein.

at which they are broken down or excreted. Such substances are known as **essential nutrients** (Table 4-8). Because they are all removed from the body at some finite rate, they must be continually supplied in the foods we eat.

TABLE 4-8   ESSENTIAL NUTRIENTS

Water

Mineral elements
   7 major mineral elements (Table 2-2)
   13 trace elements (Table 2-2)

Essential amino acids
   Isoleucine
   Leucine
   Lysine
   Methionine
   Phenylalanine
   Threonine
   Tryptophan
   Tyrosine
   Valine

Essential fatty acids
   Linoleic
   Linolenic

Vitamins
   Water-soluble vitamins
     B_1: thiamine
     B_2: riboflavin
     B_6: pyridoxine
     B_{12}: cobalamine
     Niacin
     Pantothenic acid
     Folic acid
     Biotin
     Lipoic acid
     Vitamin C
   Fat-soluble vitamins
     Vitamin A
     Vitamin D
     Vitamin E
     Vitamin K

Other essential nutrients
   Inositol
   Choline
   Carnitine

It must be emphasized that the term "essential nutrient" is reserved for substances that fulfill *two* criteria: (1) they must be essential for health, and (2) they must not be synthesized by the body in adequate amounts. Thus, glucose, although "essential" for normal metabolism, is not classified as an essential nutrient because the body normally can synthesize all it needs, from amino acids, for example. Furthermore, the quantity of an essential nutrient that must be present in the diet in order to maintain health is not a criterion for determining if the substance is essential. Approximately 1500 g of water, 2 g of the amino acid methionine, but only about 1 mg of the vitamin thiamine are required per day.

Water is an essential nutrient because far more of it is lost in the urine and from the skin and respiratory tract than can be synthesized by the body. (Recall that water is formed as an end product of oxidative phosphorylation as well as from several other metabolic reactions.) Therefore, to maintain water balance, water intake is essential.

The mineral elements provide an example of substances that cannot be synthesized or broken down but are continually lost from the body in the urine, feces, and various secretions. The major minerals must be supplied in fairly large amounts, whereas only small quantities of the trace elements are required.

We have already noted that 9 of the 20 amino acids are essential. Two fatty acids, linoleic and linolenic acid, which contain a number of double bonds and serve important roles in chemical messenger systems, are also essential nutrients. Three additional essential nutrients—inositol, choline, and carnitine—have functions that will be described in later chapters but do not fall into any common category other than being essential nutrients. Finally, the class of essential nutrients known as vitamins deserves special attention.

Vitamins

Fourteen vitamins are essential for normal growth and health in humans. The exact chemical structures of the first vitamins to be discovered were unknown, and they were simply identified by letters of the alphabet. Vitamin B turned out to be composed of eight substances now known as the vitamin B complex. Plants and bacteria have the enzymes necessary for vitamin synthesis, and it is by eating either plants or meat from animals that have eaten plants that we get our vitamins.

The vitamins as a class have no particular chemical structure in common, but they can be divided into the **water-soluble vitamins** and the **fat-soluble vitamins.** The water-soluble vitamins form portions of coenzymes such as NAD^+, FAD, and coenzyme A. The fat-soluble vitamins (A, D, E, and K) in general do not function as coenzymes. For example, vitamin A (retinol) is used to form the light-sensitive pigment in the eye, and lack of this vitamin leads to night blindness. The specific functions of each of the fat-soluble vitamins will be described in later chapters.

The catabolism of vitamins does not provide chemical energy, although they may participate as coenzymes in chemical reactions that release energy from other molecules. Increasing the amount of vitamins in the diet beyond a certain minimum does not necessarily increase the activity of those enzymes for which the vitamin functions as a coenzyme. Only very small quantities of coenzymes are necessary to saturate the coenzyme binding sites on enzyme molecules, and increasing the concentration above this level does not increase the enzyme's activity.

The fate of large quantities of ingested vitamins varies depending upon whether the vitamin is water-soluble or fat-soluble. As the amount of water-soluble vitamins in the diet is increased, so is the amount excreted in the urine; thus the accumulation of these vitamins in the body is limited. On the other hand, fat-soluble vitamins can accumulate in the body because they are poorly excreted by the kidneys and because they dissolve in the fat stores in adipose tissue. The intake of very large quantities of fat-soluble vitamins can produce toxic effects.

SECTION C SUMMARY

CELLULAR ENERGY TRANSFER

I. The end products of glycolysis under aerobic conditions are ATP and pyruvate, whereas ATP and lactate are the end products under anaerobic conditions.

   A. Carbohydrates are the only fuel molecules that can enter the glycolytic pathway, enzymes for which are located in the cytosol.

   B. During anaerobic glycolysis, hydrogen atoms are transferred to NAD^+, which then transfers them to pyruvate to form lactate, thus regenerating the original coenzyme molecule.

   C. During aerobic glycolysis, $NADH + H^+$ transfers its hydrogen atoms to the oxidative-phosphorylation pathway.

   D. The formation of ATP in glycolysis is by substrate-level phosphorylation, a process in which a phosphate group is transferred from a phosphorylated metabolic intermediate directly to ADP.

II. The Krebs cycle, the enzymes of which are in the matrix of the mitochondria, catabolizes molecular fragments derived from fuel molecules and produces carbon dioxide, hydrogen atoms, and ATP.

   A. Acetyl coenzyme A, the acetyl portion of which is derived from all three types of fuel molecules, is the major substrate entering the Krebs cycle. Amino acids can also enter at several sites in the cycle by being converted to cycle intermediates.

   B. During one rotation of the Krebs cycle two molecules of carbon dioxide are produced, and four pairs of hydrogen atoms are transferred to coenzymes. Substrate-level phosphorylation produces one molecule of GTP, which can be converted to ATP.

III. Oxidative phosphorylation forms ATP from ADP and P_i, using the energy released when molecular oxygen ultimately combines with hydrogen atoms to form water.

   A. The enzymes for oxidative phosphorylation are located on the inner membrane of mitochondria.

   B. Hydrogen atoms derived from glycolysis, the Krebs cycle, and the breakdown of fatty acids are delivered to the electron transport chain, which regenerates the hydrogen-free forms of the coenzymes NAD^+ and FAD by transferring the hydrogens to molecular oxygen to form water.

   C. The reactions of the electron transport chain produce a hydrogen-ion gradient across the inner mitochondrial membrane. The flow of hydrogen ions back across the membrane provides the energy for ATP synthesis.

   D. Small amounts of reactive oxygen species, which can damage proteins, lipids and nucleic acids, are formed during electron transport.

CARBOHYDRATE, FAT, AND PROTEIN METABOLISM

I. The aerobic catabolism of carbohydrates proceeds through the glycolytic pathway to pyruvate, which enters the Krebs cycle and is broken down to carbon dioxide and to hydrogens, which are transferred to coenzymes.

   A. About 40 percent of the chemical energy in glucose can be transferred to ATP under aerobic conditions, the rest is released as heat.

   B. Under aerobic conditions, 38 molecules of ATP can be formed from 1 molecule of glucose: 34 from oxidative phosphorylation, 2 from glycolysis, and 2 from the Krebs cycle.

   C. Under anaerobic conditions, 2 molecules of ATP are formed from 1 molecule of glucose during glycolysis.

II. Carbohydrates are stored as glycogen, primarily in the liver and skeletal muscles.

   A. Two different enzymes are used to synthesize and break down glycogen. The control of these enzymes regulates the flow of glucose to and from glycogen.

   B. In most cells glucose 6-phosphate is formed by glycogen breakdown and is catabolized to produce ATP. In liver and kidney cells, glucose can be derived from glycogen and released from the cells into the blood.

III. New glucose can be synthesized (gluconeogenesis) from some amino acids, lactate, and glycerol via the enzymes that catalyze reversible reactions in the glycolytic pathway. Fatty acids cannot be used to synthesize glucose.

IV. Fat, stored primarily in adipose tissue, provides about 80 percent of the stored energy in the body.

    A. Fatty acids are broken down, two carbon atoms at a time, in the mitochondrial inner compartment by beta oxidation, to form acetyl coenzyme A and hydrogen atoms, which combine with coenzymes.

    B. The acetyl portion of acetyl coenzyme A is catabolized to carbon dioxide in the Krebs cycle, and the hydrogen atoms generated there, plus those generated during beta oxidation, enter the oxidative-phosphorylation pathway to form ATP.

    C. The amount of ATP formed by the catabolism of 1 g of fat is about $2\frac{1}{2}$ times greater than the amount formed from 1 g of carbohydrate.

    D. Fatty acids are synthesized from acetyl coenzyme A by enzymes in the cytosol and are linked to α-glycerol phosphate, produced from carbohydrates, to form triacylglycerols by enzymes in the smooth endoplasmic reticulum.

V. Proteins are broken down to free amino acids by proteases.

    A. The removal of amino groups from amino acids leaves keto acids, which can either be catabolized via the Krebs cycle to provide energy for the synthesis of ATP or be converted into glucose and fatty acids.

    B. Amino groups are removed by (1) oxidative deamination, which gives rise to ammonia, or by (2) transamination, in which the amino group is transferred to a keto acid to form a new amino acid.

    C. The ammonia formed from the oxidative deamination of amino acids is converted to urea by enzymes in the liver and then excreted in the urine by the kidneys.

VI. Some amino acids can be synthesized from keto acids derived from glucose, whereas others cannot be synthesized by the body and must be provided in the diet.

ESSENTIAL NUTRIENTS

I. Approximately 50 essential nutrients, listed in Table 4-8, are necessary for health but cannot be synthesized in adequate amounts by the body and must therefore be provided in the diet.

II. A large intake of water-soluble vitamins leads to their rapid excretion in the urine, whereas large intakes of fat-soluble vitamins lead to their accumulation in adipose tissue and may produce toxic effects.

SECTION C KEY TERMS

glycolysis	aerobic
pyruvate	anaerobic
substrate-level	lactate
phosphorylation	Krebs cycle
citric acid cycle	adipose tissue
tricarboxylic acid cycle	beta oxidation
acetyl coenzyme A	α-glycerol phosphate
(acetyl CoA)	protease
oxidative phosphorylation	oxidative deamination
cytochrome	keto acid
electron transport chain	transamination
chemiosmotic hypothesis	urea
hydrogen peroxide	essential amino acid
superoxide anion	negative nitrogen balance
hydroxyl radical	positive nitrogen balance
glycogen	essential nutrient
gluconeogenesis	water-soluble vitamin
adipocyte	fat-soluble vitamin

SECTION C REVIEW QUESTIONS

1. What are the end products of glycolysis under aerobic and anaerobic conditions?

2. To which molecule are the hydrogen atoms in $NADH + H^+$ transferred during anaerobic glycolysis? During aerobic glycolysis?

3. What are the major substrates entering the Krebs cycle, and what are the products formed?

4. Why does the Krebs cycle operate only under aerobic conditions even though molecular oxygen is not used in any of its reactions?

5. Identify the molecules that enter the oxidative-phosphorylation pathway and the products that are formed.

6. Where are the enzymes for the Krebs cycle located? The enzymes for oxidative phosphorylation? The enzymes for glycolysis?

7. How many molecules of ATP can be formed from the breakdown of one molecule of glucose under aerobic conditions? Under anaerobic conditions?

8. Describe the origin and effects of reactive oxygen molecules.

9. Describe the pathways by which glycogen is synthesized and broken down by cells.

10. What molecules can be used to synthesize glucose?

11. Why can't fatty acids be used to synthesize glucose?

12. Describe the pathways used to catabolize fatty acids to carbon dioxide.

13. Why is it more efficient to store fuel as fat than as glycogen?

14. Describe the pathway by which glucose is converted into fat.

15. Describe the two processes by which amino groups are removed from amino acids.

16. What can keto acids be converted into?

17. What is the source of the nitrogen atoms in urea, and in what organ is urea synthesized?

18. Why is water considered an essential nutrient whereas glucose is not?

19. What is the consequence of ingesting large quantities of water-soluble vitamins? Fat-soluble vitamins?

CHAPTER 4 THOUGHT QUESTIONS

(Answers are given in Appendix A.)

1. A variety of chemical messengers that normally regulate acid secretion in the stomach bind to proteins in the plasma membranes of the acid-secreting cells. Some of these binding reactions lead to increased acid secretion, and others to decreased secretion. In what ways might a drug that causes decreased acid secretion be acting on these cells?

2. In one type of diabetes the plasma concentration of the hormone insulin is normal, but the response of the cells to which insulin usually binds is markedly decreased. Suggest a reason for this in terms of the properties of protein binding sites.

3. Given the following substances in a cell and their effects on each other, predict the change in compound H that will result from an increase in compound A, and diagram this sequence of changes.
Compound A is a modulator molecule that allosterically activates protein B.
Protein B is a protein kinase enzyme that activates protein C.
Protein C is an enzyme that converts substrate D to product E.
Compound E is a modulator molecule that allosterically inhibits protein F.
Protein F is an enzyme that converts substrate G to product H.

4. Shown below is the relation between the amount of acid secreted and the concentration of compound X, which stimulates acid secretion in the stomach by binding to a membrane protein.

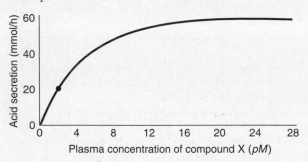

At a plasma concentration of 2 pM, compound X produces an acid secretion of 20 mmol/h.

    A. Specify two ways in which acid secretion by compound X could be increased to 40 mmol/h.

    B. Why will increasing the concentration of compound X to 28 pM not produce more acid secretion than increasing the concentration of X to 18 pM.

5. How would protein regulation be affected by a mutation that causes the loss of phosphoprotein phosphatase from cells?

6. How much energy is added to or released from a reaction in which reactants A and B are converted to products C and D if the energy content, in kilocalories per mole, of the participating molecules is: A = 55, B = 93, C = 62, and D = 87? Is this reaction reversible or irreversible? Explain.

7. In the following metabolic pathway, what is the rate of formation of the end product E if substrate A is present at a saturating concentration? The maximal rates (products formed per second) of the individual steps are indicated.

$$A \xrightarrow{30} B \xrightarrow{5} C \xrightarrow{20} D \xrightarrow{40} E$$

8. If the concentration of oxygen in the blood delivered to a muscle is increased, what effect will this have on the rate of ATP production by the muscle?

9. During prolonged starvation, when glucose is not being absorbed from the gastrointestinal tract, what molecules can be used to synthesize new glucose?

10. Why does the catabolism of fatty acids occur only under aerobic conditions?

11. Why do certain forms of liver disease produce an increase in the blood levels of ammonia?

CHAPTER 5

Genetic Information and Protein Synthesis

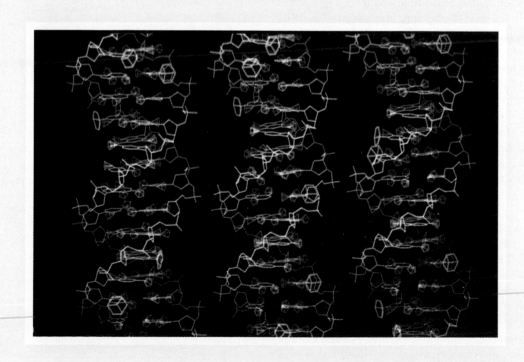

The outstanding accomplishment of twentieth-century biology has been the discovery of the chemical basis of heredity and its relationship to protein synthesis. Whether an organism is a human being or a mouse, has blue eyes or black, has light skin or dark, is determined by its proteins. Moreover, within an individual organism, the properties of muscle cells differ from those of nerve cells and epithelial cells because of the types of proteins present in each cell type and the functions performed by these proteins.

The hereditary material in each cell contains the instructions for synthesizing the cell's proteins. These instructions are coded into DNA molecules. Given that different cell types have different proteins and that the specifications for these proteins are coded in DNA, one might be led to conclude that different cell types contain different DNA molecules. However, this is not the case. All cells in the body, with the exception of sperm or egg cells, receive the same genetic information when DNA molecules are duplicated and passed on to daughter cells at the time of cell division. Cells differ in structure and function because only a portion of the total genetic information common to all cells is used by any given cell to synthesize proteins.

This chapter describes how genetic information is used to synthesize proteins, some of the factors that govern the selective expression of genetic information, the process by which DNA molecules are replicated and their genetic information passed on to daughter cells during cell division, and how altering the genetic message—mutation—can lead to the class of diseases known as inherited disorders as well as to cancers.

GENETIC CODE

Molecules of DNA contain instructions, coded in the sequence of nucleotides, for the synthesis of proteins. A sequence of DNA nucleotides containing the information that specifies the amino acid sequence of a single polypeptide chain is known as a **gene.** A gene is thus a unit of hereditary information. A single molecule of DNA contains many genes.

Taken together, all the nucleotide sequences that specify an organism's structure and function are known as its **genome.** The human genome contains between 50,000 and 100,000 genes, the instructions for producing 50,000 to 100,000 different proteins. Currently scientists from around the world are collaborating in the Human Genome Project to determine, by around the turn of the century, all the nucleotide sequences of the human genome. This will involve locating the position of each nucleotide in the sequences of approximately 3 billion nucleotides that make up the human genome.

Although DNA contains the information specifying the amino acid sequences in proteins, it does not itself participate *directly* in the assembly of protein molecules. Most of a cell's DNA is in the nucleus (a small amount is in the mitochondria), whereas most protein synthesis occurs in the cytoplasm. The transfer of information from DNA to the site of protein synthesis is the function of RNA molecules, whose synthesis is governed by the information coded in DNA. Genetic information flows from DNA to RNA and then to protein (Figure 5-1). The process of transferring genetic information from DNA to RNA in the nucleus is known as **transcription;** the process that uses the coded information in RNA to assemble a protein in the cytoplasm is known as **translation.**

$$\text{DNA} \xrightarrow{\text{transcription}} \text{RNA} \xrightarrow{\text{translation}} \text{protein}$$

FIGURE 5-1

The expression of genetic information in a cell occurs through the transcription of coded information from DNA to RNA in the nucleus, followed by the translation of the RNA information into protein synthesis in the cytoplasm.

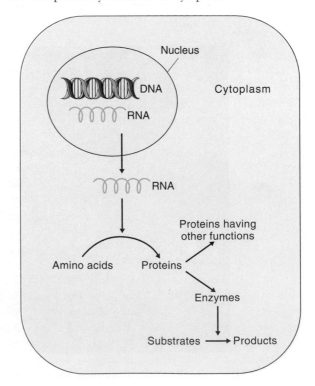

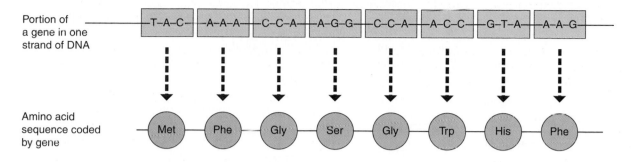

Portion of a gene in one strand of DNA

T-A-C A-A-A C-C-A A-G-G C-C-A A-C-C G-T-A A-A-G

Amino acid sequence coded by gene

Met Phe Gly Ser Gly Trp His Phe

FIGURE 5-2

The sequence of triplet nucleotides in a gene determines the sequence of amino acids in a polypeptide chain. The names of the amino acids are abbreviated. Note that more than one triplet code sequence can indicate the same amino acid, for example, the amino acid phenylalanine (Phe) is coded by two triplet codes, AAA and AAG.

As described in Chapter 2, a molecule of DNA consists of two chains of nucleotides coiled around each other to form a double helix (see Figure 2-23). Each DNA nucleotide contains one of four bases—adenine (A), guanine (G), cytosine (C), or thymine (T)—and each of these bases is specifically paired by hydrogen bonds with a base on the opposite chain of the double helix. In this base pairing, A and T bond together and G and C bond together. Thus, both nucleotide chains contain a specifically ordered sequence of bases, one chain being complementary to the other.

The genetic language is similar in principle to a written language, which consists of a set of symbols, such as α, β, δ, γ, that form an alphabet. The letters are arranged in specific sequences to form words, and the words are arranged in linear sequences to form sentences. The genetic language contains only four letters, corresponding to the bases A, G, C, and T. The words are three-base sequences that specify particular amino acids, that is, each word in the genetic language is only three letters long. This is termed a triplet code. The sequence of triplets along a gene in a single strand of DNA designates the sequence of amino acids in a polypeptide chain (Figure 5-2). Thus, a gene is equivalent to a sentence, and the total genetic information contained in human DNA is equivalent to a book—the human genome. Using a single letter (A, T, C, G) to specify each of the four bases in the DNA nucleotides, it will require about 550,000 pages equivalent to this text page to print the nucleotide sequence of the human genome.

The four bases in the DNA alphabet can be arranged in 64 different three-letter combinations ($4 \times 4 \times 4 = 64$). Thus, a triplet code actually provides more than enough words to code for the 20 different amino acids that can be found in proteins. It turns out, however, that not just 20, but 61 of the 64 possible triplets are used to specify amino acids. This means that a given amino acid is usually specified by more than one code word. For example, the four DNA triplets C-C-A, C-C-G, C-C-T, and C-C-C all specify the amino acid glycine. The three triplets that do not specify amino acids are known as **termination code words.** They perform the same function as does a period at the end of a sentence—they indicate that the end of a genetic message has been reached.

The genetic code is a universal language used by all living cells. For example, the code words for the amino acid tryptophan are the same in the DNA of a bacterium, an amoeba, a plant, and a human being. Although the same code words are used by all living cells, the messages they spell out—the sequences of code words which determine the amino acid sequences in proteins—are different in each organism. The universal nature of the genetic code supports the concept that all forms of life on earth evolved from a common ancestor.

Before we turn to the specific mechanisms by which the DNA code is used in protein synthesis an important clarification and qualification is required. As noted earlier, the information coded in genes is always first transcribed into RNA. As we shall see in the next section there are three major classes of RNA—messenger RNA, ribosomal RNA, and transfer RNA. Only messenger RNA directly codes for the amino acid sequences of proteins even though the other RNA classes participate in the overall process of protein synthesis. For this reason, the customary definition of a gene as the sequence of DNA nucleotides that specifies the amino acid sequence of a protein is true only for those genes that are transcribed into messenger RNA. The vast majority of genes are of this type, but it should at least be noted that the genes that code for the other classes of RNA do not technically fit this definition.

PROTEIN SYNTHESIS

The first step in using the genetic information in DNA to synthesize a protein is called *transcription,* and it involves the synthesis of an RNA molecule containing coded information that corresponds to the information in a single gene. As noted above, several classes of RNA molecules take part in protein synthesis; the class of RNA molecules that specifies the amino acid sequence of a protein and carries this message from DNA to the cytoplasm is known as **messenger RNA (mRNA).**

Transcription: mRNA Synthesis

As described in Chapter 2, ribonucleic acids are single-chain polynucleotides whose nucleotides differ from DNA in that they contain the sugar ribose (rather than deoxyribose) and the base uracil (rather than thymine). The other three bases—adenine, guanine, and cytosine—occur in both DNA and RNA. The pool of subunits used to synthesize mRNA are free (uncombined) ribonucleotide triphosphates: ATP, GTP, CTP, and UTP.

As mentioned earlier, the two polynucleotide chains in DNA are linked together by hydrogen bonds between specific pairs of bases—A and T and G and C. Transcription begins with the breakage of these hydrogen bonds so that a portion of the two chains of the DNA double-helix separates (Figure 5-3). The bases in the exposed DNA nucleotides are then able to pair with the bases in free ribonucleotide triphosphates. Free ribonucleotides containing A bases pair with the exposed T bases in DNA, and likewise, free ribonucleotides containing G, C, or U pair with the exposed DNA bases C, G, and A, respectively. Note that uracil, which is present in RNA but not DNA, pairs with the base adenine in DNA. In this way, the nucleotide sequence in one strand of DNA acts as a template that determines the sequence of nucleotides in mRNA.

The aligned ribonucleotides are joined together by the enzyme **RNA polymerase,** which hydrolyses the nucleotide triphosphates, releasing two of the terminal phosphate groups, and joining the remaining phosphate in covalent linkage to the ribose of the adjacent nucleotide.

Since DNA consists of two strands of polynucleotides, both of which are exposed during transcription, it should theoretically be possible to form two different RNA molecules, one from each strand. However, only one of the two potential RNAs is ever formed. Which of the two DNA strands is used as the template for RNA synthesis for a particular gene is determined by a specific sequence of DNA nucleotides called the **promoter,** which is located near the beginning of the gene (Figure 5-3). It is to this promoter region that RNA polymerase binds. For any given gene, only one strand is used, and that is the strand with the promoter region at the beginning of the gene. However, a transcribed gene may be located on either of the two strands of the DNA double helix.

Beginning at the promoter end of a gene, RNA polymerase moves along the template strand, joining one ribonucleotide at a time, at a rate of about 30 nucleotides per second, to the growing RNA chain. Shortly after transcription begins, the nucleotide at the beginning of the RNA chain is modified, which serves to decrease the degradation of RNA by digestive enzymes. Upon reaching a specific nucleotide sequence specifying the end of the gene the RNA polymerase releases the newly formed primary RNA transcript. After the RNA transcript is released, a series of 100 to 200 adenine nucleotides is added to its end, forming a poly A tail that acts as a signal allowing

FIGURE 5-3

Transcription of a gene from the template strand of DNA to a primary RNA transcript.

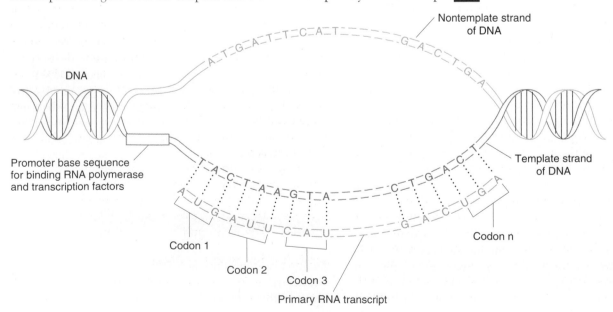

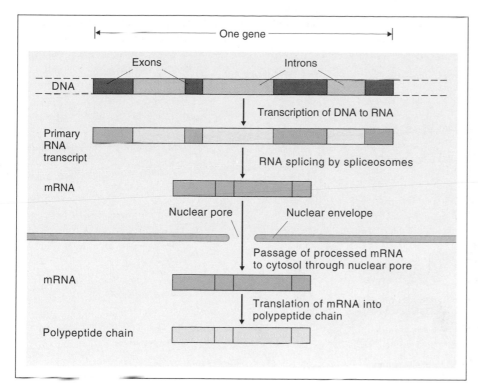

FIGURE 5-4

Spliceosomes remove the noncoding intron-derived segments from a primary RNA transcript and link the exon-derived segments together to form the mRNA molecule that passes through the nuclear pores to the cytosol. The lengths of the intron- and exon-derived segments represent the relative lengths of the base sequences in these regions.

RNA to move out of the nucleus. It also aids in the next step, the binding of mRNA to ribosomes in the cytoplasm.

In a given cell, the information in only a few of the 50,000–100,000 genes present in DNA is transcribed into mRNA at any given time. Genes are transcribed only when RNA polymerase can bind to their promoter sites. Various mechanisms, described later in this chapter, are used by cells either to block or to make accessible the promoter region of any particular gene to RNA polymerase. Such regulation of gene transcription provides a means of controlling the synthesis of specific proteins and thereby the activities of cells.

It must be emphasized that the base sequence in the primary RNA transcript is *not identical* to that in the template strand of DNA, since the RNA's formation depends on the pairing between complementary, not identical, bases (Figure 5-3). A three-base sequence in RNA that specifies one amino acid is called a **codon.** Each codon is *complementary* to a three-base sequence in DNA. For example, the base sequence T-A-C in the template strand of DNA corresponds to the codon A-U-G in transcribed RNA.

Although the entire sequence of nucleotides in the template strand of a gene is transcribed into a complementary sequence of nucleotides in RNA, only certain segments of the gene actually code for sequences of amino acids. These regions of the gene, known as **exons** (expression regions), are separated by noncoding sequences of nucleotides known as **introns** (intervening sequences). Before passing to the cytoplasm, a newly formed primary

RNA transcript must undergo splicing (Figure 5-4) to remove the sequences that correspond to the DNA introns and thereby form the messenger RNA that will be translated into protein (only after this process is the RNA termed messenger RNA).

Splicing occurs in the nucleus and is performed by a **spliceosome,** a complex of proteins and small nuclear RNAs. The spliceosome identifies specific nucleotide sequences at the beginning and end of each intron-derived sequence, removes the sequence from the newly formed primary transcript of RNA, and splices the end of one exon-derived sequence to the beginning of another to form mRNA with a continuous coding sequence. Moreover, in some cases, during the splicing process the exon-derived sequences from a single gene can be spliced together in different ways or some exon-derived sequences can be deleted entirely; these processes result in the formation of different mRNA sequences from the same gene and give rise, in turn, to proteins with slightly different amino acid sequences.

The mRNAs formed as a result of splicing are 75 to 90 percent shorter than the primary RNA transcript, meaning that 75 to 90 percent of the nucleotide sequences in DNA are introns. What role, if any, such large amounts of "nonsense" DNA may perform is unclear.

Translation: Polypeptide Synthesis

Once formed by transcription and splicing, the mRNA moves through the pores in the nuclear envelope into the

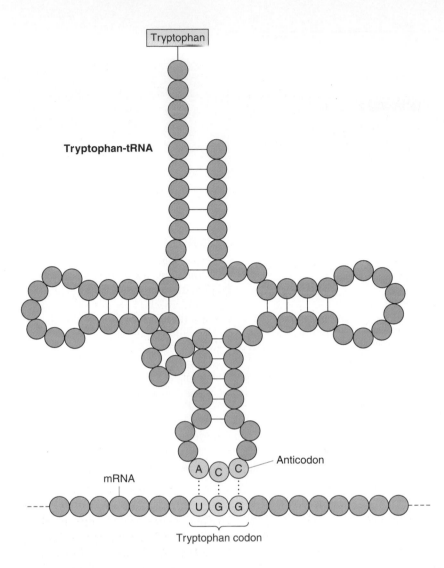

FIGURE 5-5

Base-pairing between the anticodon region of a tRNA molecule and the corresponding codon region of a mRNA molecule.

cytoplasm. Although the nuclear pores allow the diffusion of small molecules and ions between the nucleus and cytoplasm, they have specific mechanisms for the selective transport of large molecules such as proteins and RNA. These mechanisms are energy-dependent, requiring the hydrolysis of ATP and the binding of the molecules to be transported to specific recognition sites in the nuclear pore complex. Only those molecules recognized by the sites can move between the nucleus and cytoplasm. The modified ends of the mRNA molecules are necessary for the interaction with the nuclear pore transport proteins, which enables the mRNA to enter the cytoplasm.

In the cytoplasm, mRNA binds to a ribosome, the cell organelle that contains the enzymes and other components required for the translation of mRNA's coded message into protein. Before describing this assembly process, we will examine the structure of a ribosome and the characteristics of the two additional major classes of RNA involved in protein synthesis.

Ribosomes and rRNA.   As described in Chapter 3, ribosomes are small granules located in the cytoplasm, either suspended in the cytosol (free ribosomes) or attached to the surface of the endoplasmic reticulum (bound ribosomes).

A ribosome is a complex particle composed of about 80 different proteins in association with a class of RNA molecules known as **ribosomal RNA (rRNA).** The genes for rRNA are transcribed from DNA in a process similar to that for mRNA except that a different RNA polymerase is used. Ribosomal RNA transcription occurs in the region of the nucleus known as the nucleolus. Ribosomal proteins, like other proteins, are synthesized in the cytoplasm from the mRNAs specific for them. These proteins then move back through nuclear pores to the nucleolus where they combine with rRNA to form two ribosomal subunits. After the two types of subunits are formed (Chapter 2), they are individually transported to the cytoplasm where they finally combine to form a functional ribosome during protein translation.

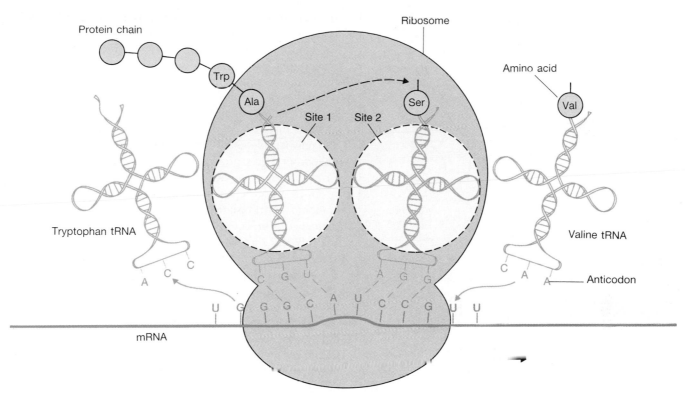

FIGURE 5-6

Sequence of events during the synthesis of a protein by a ribosome.

Transfer RNA. How do individual amino acids identify the appropriate codons in mRNA during the process of translation? By themselves, free amino acids do not have the ability to bind to the bases in mRNA codons. This process of identification involves the third major class of RNA, known as **transfer RNA (tRNA).** Transfer RNA molecules are the smallest (about 80 nucleotides long) of the major classes of RNA. The single chain of tRNA loops back upon itself, forming a structure resembling a cloverleaf with three loops (Figure 5-5).

Like mRNA and rRNA, tRNA molecules are synthesized in the nucleus by base-pairing with DNA nucleotides at specific tRNA genes and then move to the cytoplasm. The key to tRNA's role in protein synthesis is its ability to combine with both a specific amino acid and a codon in ribosome-bound mRNA specific for that amino acid. This permits tRNA to act as the link between an amino acid and the mRNA codon for that amino acid.

A tRNA molecule is covalently linked to an amino acid by an enzyme known as aminoacyl-tRNA synthetase. There are at least 20 different aminoacyl-tRNA synthetases, each of which catalyzes the linkage of a specific amino acid to a particular type of tRNA. The next step is to link the tRNA, bearing its attached amino acid, to the mRNA codon for that amino acid. As one might predict, this is achieved by base-pairing between tRNA and mRNA. A

three-nucleotide sequence at the end of one of the loops of tRNA can base-pair with a complementary codon in mRNA. This tRNA triplet sequence is appropriately termed an **anticodon.** Figure 5-5 illustrates the binding between mRNA and a tRNA specific for the amino acid tryptophan. Note that tryptophan is covalently linked to one end of tRNA and does not bind to either the anticodon region of tRNA or the codon region of mRNA.

Protein Assembly. The process of assembling a polypeptide chain based on an mRNA message involves three stages—initiation, elongation, and termination. Synthesis is initiated by the binding of a mRNA molecule to a ribosome. A number of additional proteins known as **initiation factors** are required to establish an initiation complex, which positions the tRNA opposite the mRNA codon that signals the start site at which assembly is to begin. This initiation phase is the slowest step in protein assembly, and the rate of protein synthesis can be regulated by factors that influence the activity of initiation factors.

Following the initiation process, the protein chain is elongated by the successive addition of amino acids (Figure 5-6). A ribosome has two binding sites for tRNA. One holds the tRNA attached to the most recently added amino acid, and the other holds the tRNA containing the next amino acid to be added to the chain. Ribosomal enzymes

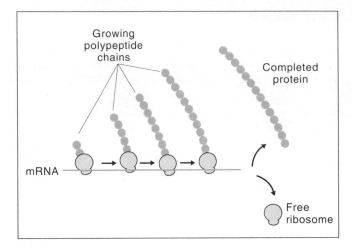

FIGURE 5-7

Several ribosomes can simultaneously move along a strand of mRNA, producing the same protein in different states of assembly.

catalyze the formation of a peptide bond between these two amino acids. Following the formation of the peptide bond, the tRNA at the first binding site is released from the ribosome, and the tRNA at the second site—now linked to the peptide chain—is transferred to the first binding site. The ribosome moves down one codon along the mRNA, making room for the binding of the next amino acid-tRNA molecule. This process is repeated over and over as amino acids are added to the growing peptide chain at an average rate of two to three per second. When the ribosome reaches a termination codon in mRNA specifying the end of the protein, the link between the polypeptide chain and the last tRNA is broken, and the completed protein is released from the ribosome.

RNA molecules are not destroyed during protein synthesis, so they may be used to synthesize many protein molecules. Moreover, while one ribosome is moving along a particular strand of mRNA, a second ribosome may become attached to the start site on that same mRNA and begin the synthesis of a second identical protein molecule. Thus, a number of ribosomes, as many as 70, may be moving along a single strand of mRNA, each at a different stage of the translation process (Figure 5-7).

Molecules of mRNA do not, however, remain in the cytoplasm indefinitely. Eventually they are broken down into nucleotides by cytoplasmic enzymes. Therefore, if a gene corresponding to a particular protein ceases to be transcribed into mRNA, the protein will no longer be formed after its cytoplasmic mRNA molecules are broken down.

For small proteins the folding that gives the protein its characteristic three-dimensional shape occurs spontaneously as the polypeptide chain emerges from the ribosome. Large proteins have a folding problem because their

final conformation may depend upon interactions with portions of the molecule that have not yet emerged from the ribosome. In addition, a large segment of unfolded protein tends to aggregate with other proteins, which inhibits its proper folding. These problems are overcome by a complex of proteins known as **chaperones,** which form a small, hollow chamber into which the emerging protein chain is inserted. Within the confines of the chaperone, the polypeptide chain is able to complete its folding. The chaperones thus provide an isolated environment where protein folding can occur without interference.

Once a polypeptide chain has been assembled it may undergo posttranslational modifications to its amino acid sequence. For example, the amino acid that was used to identify the start site of the assembly process is cleaved from the end of most proteins. In some cases specific peptide bonds within the polypeptide chain are broken, producing a number of smaller peptides, each of which may perform a different function. For example, as illustrated in Figure 5-8, five proteins can be derived from the same mRNA as a result of posttranslational cleavage. The same initial polypeptide may be split at different points in different cells depending on the specificity of the hydrolyzing enzymes present.

Carbohydrates and lipid derivatives are often covalently linked to particular amino acid side chains. These additions act as signals to direct the protein to those locations in the cell where it is to function. The addition of a fatty acid to a protein, for example, can lead to the anchoring of the protein to a membrane as the nonpolar portion of the fatty acid becomes inserted into the lipid bilayer.

The steps leading from DNA to a functional protein are summarized in Table 5-1.

FIGURE 5-8

Posttranslational splitting of a protein can result in several smaller proteins, each of which may perform a different function. All these proteins are derived from the same gene.

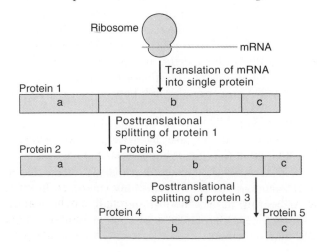

TABLE 5-1 EVENTS LEADING FROM DNA TO PROTEIN SYNTHESIS

Transcription

1. RNA polymerase binds to the promoter region of a gene and separates the two strands of the DNA double helix in the region of the gene to be transcribed.

2. Free ribonucleotide triphosphates base-pair with the deoxynucleotides in DNA.

3. The ribonucleotides paired with one strand of DNA are linked by RNA polymerase to form mRNA containing a sequence of bases complementary to one strand of the DNA base sequence.

4. RNA splicing removes the intron-derived regions of RNA, which contain noncoding sequences, and splices together the exon-derived regions, which code for specific amino acids.

Translation

5. Processed mRNA passes from the nucleus to the cytoplasm, where one end of the mRNA binds to a ribosome.

6. Free amino acids are linked to their corresponding tRNAs by aminoacyl-tRNA synthetase.

7. The three-base anticodon in an amino acid–tRNA complex pairs with the corresponding codon in the region of the mRNA bound to the ribosome.

8. The portion of the peptide that has already been synthesized (and is still attached to a tRNA bound to the ribosome) is now linked by a peptide bond to the amino acid on the tRNA next to it, thereby adding one more amino acid to the chain.

9. The tRNA that has been freed of the peptide chain is released from the ribosome.

10. The ribosome moves one codon step along mRNA.

11. Steps 7 to 10 are repeated over and over until the end of the mRNA message is reached.

12. The completed protein chain is released from the ribosome when the termination codon in mRNA is reached.

13. Chaperone proteins guide the folding of some proteins into their proper conformation.

14. In some cases, the protein undergoes posttranslational processing in which various chemical groups are attached to specific side chains and/or the protein is split into several smaller peptide chains.

Regulation of Protein Synthesis

As noted earlier, in any given cell only a small fraction of the 50,000 to 100,000 genes in the human genome are ever transcribed into mRNA and translated into proteins. Of this fraction, a small number of genes are *continuously* being transcribed into mRNA, but the transcription of most genes is regulated and can be turned on or off in response either to signals generated within the cell or to external signals received by the cell. In order for a gene to be transcribed, RNA polymerase must be able to bind to the promoter region of the gene and be in an activated configuration.

Transcription of most genes is regulated by a class of proteins known as **transcription factors,** which act as gene switches, interacting in a variety of ways to activate or repress the initiation process that takes place at the promoter region of a particular gene. The influence of a transcription factor on transcription is not necessarily all or none, on or off; it may have the effect of slowing or speeding up the initiation of the transcription process. Many of these regulatory proteins have a conformation that allows them to bind to the promoter region of a specific gene, while others bind to RNA polymerase or other proteins that interact in the initiation of the transcription process. Some transcription factors bind to regions of DNA that are far removed from the promoter region of the gene whose transcription they regulate. In this case the DNA containing the bound transcription factor forms a loop that brings the transcription factor into contact with the promoter region where it may activate or repress transcription (Figure 5-9).

There are numerous transcription factors and numerous mechanisms by which a transcription factor can influence gene transcription. For example, by binding to the promoter region a transcription factor may block the binding of RNA polymerase, it may prevent the polymerase from assuming an active conformation, or it may promote polymerase binding and activation.

Many genes contain regulatory sites that can be influenced by a common transcription factor; thus there does not need to be a different transcription factor for every gene. In addition, more than one transcription factor may interact in controlling the transcription of a given gene.

Since transcription factors are proteins, the activity of a particular transcription factor, that is, its ability to bind to DNA or to other regulatory proteins, can be turned on or off by allosteric or covalent modulation in response to signals either received by a cell or generated within a cell. Thus, specific genes can be regulated in response to specific signals. These signaling mechanisms will be discussed in Chapter 7.

To summarize, the rate of a protein's synthesis can be regulated at various points: (1) gene transcription into mRNA; (2) the initiation of protein assembly on a ribosome; and (3) mRNA degradation in the cytoplasm.

PROTEIN DEGRADATION

We have thus far emphasized protein synthesis, but an important fact is that the concentration of a particular protein in a cell at a particular time depends not only upon its rate of synthesis but upon its rates of degradation and secretion, if any, from the cell.

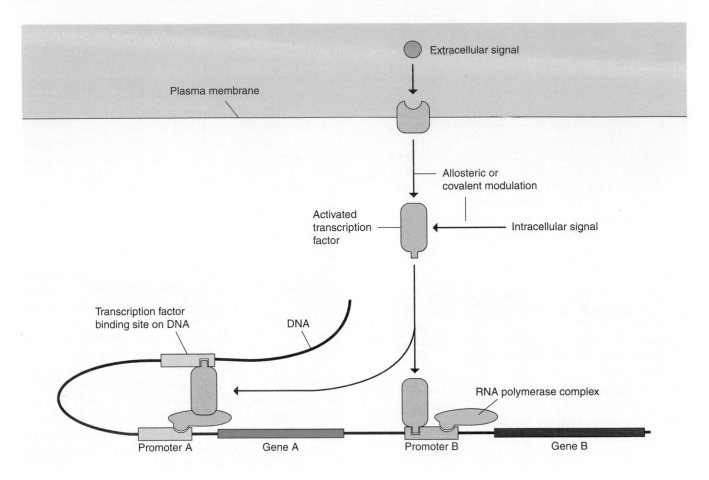

FIGURE 5-9

Transcription of gene B is modulated by the binding of an activated transcription factor directly to the promoter region. In contrast, transcription of gene A is modulated by the same transcription factor which, in this case, binds to a region of DNA considerably distant from the promoter region. ▶

Different proteins are degraded at different rates. In part this depends on the structure of the protein, with some proteins having a higher affinity for certain proteolytic enzymes than others. A denatured (unfolded) protein is more readily digested than a protein with an intact conformation. Proteins can be targeted for degradation by the attachment of a small peptide, **ubiquitin,** to the protein. This peptide directs the protein to a protein complex known as a **proteosome,** which unfolds the protein and breaks it down into small peptides. In addition, the number and types of proteolytic enzymes in a cell can be regulated by controlling the expression of their genes and their activities by allosteric and covalent modulation in response to various signals.

PROTEIN SECRETION

Most proteins synthesized by a cell remain in the cell, providing structure and function for the cell's survival. Some proteins, however, are secreted into the extracellular fluid, where they act as signals to other cells or provide material for forming the extracellular matrix to which tissue cells are anchored. Since proteins are large, charged molecules that cannot diffuse through cell membranes (as will be described in more detail in Chapter 6), special mechanisms are required to insert them into or move them through membranes.

Proteins destined to be secreted from a cell or become integral membrane proteins are recognized during the early stages of protein synthesis. For such proteins, the first 15 to 30 amino acids at the carboxyl end that emerge from the surface of the ribosome act as a recognition signal, known as the **signal sequence,** or signal peptide. The signal sequence directs the protein along a pathway leading to secretion.

The signal sequence binds to a complex of proteins known as a signal recognition particle, which temporarily inhibits further growth of the polypeptide chain. The signal recognition particle, bound to the signal sequence,

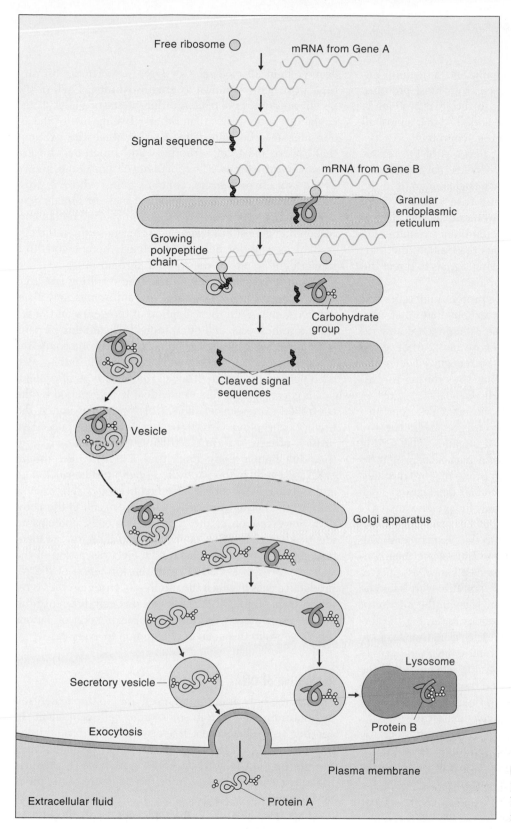

Free ribosome

mRNA from Gene A

Signal sequence

mRNA from Gene B

Granular
endoplasmic
reticulum

Growing
polypeptide
chain

Carbohydrate
group

Cleaved signal
sequences

Vesicle

Golgi apparatus

Lysosome

Secretory vesicle

Protein B

Exocytosis

Plasma membrane

Extracellular fluid

Protein A

FIGURE 5-10
Pathway of proteins destined to
be secreted by cells or transferred to lysosomes.

then binds to a specific membrane protein on the surface of the granular endoplasmic reticulum. This binding restarts the process of protein assembly, and the growing polypeptide chain is fed through a protein complex in the membrane into the lumen of the reticulum (Figure 5-10).

Thus, upon completion of protein assembly, proteins that are to be secreted find themselves in the lumen of the granular endoplasmic reticulum. (Proteins that are destined to function as integral membrane proteins remain embedded in the reticulum membrane.)

Within the lumen of the endoplasmic reticulum, enzymes remove the signal sequence from most proteins, and so this portion is not present in the final protein. In addition, carbohydrate groups are added so that almost all proteins secreted from the cell are glycoproteins.

Following these modifications, portions of the reticulum membrane bud off, forming vesicles that contain the newly synthesized proteins. These vesicles migrate to the Golgi apparatus (Figure 5-10) and fuse with the Golgi membranes. Vesicle budding, movement through the cytosol, and fusion with the Golgi membranes requires the interaction of a number of proteins that initiate the budding process and provide the docking signals to direct the vesicles to the appropriate membrane.

Within the Golgi apparatus the protein undergoes still further modification. Some of the carbohydrates that were added in the granular endoplasmic reticulum are now removed and new groups added. These new carbohydrate groups function as labels that can be recognized when the protein encounters various binding sites during the remainder of its trip through the cell.

While in the Golgi apparatus, the many different proteins that have been funneled into this organelle become sorted out according to their final destination. This sorting involves the binding of regions of a particular protein to specific proteins in the Golgi membrane that are destined to form vesicles targeted to a particular destination.

Following modification and sorting, the proteins are packaged into vesicles that bud off the surface of the Golgi membrane. Some of the vesicles travel to the plasma membrane where they fuse with the membrane and release their contents to the extracellular fluid, a process known as exocytosis (Chapter 6). Other vesicles dock and fuse with lysosome membranes, delivering digestive enzymes to the interior of this organelle. The specific interactions governing the formation and distribution of these vesicles from the Golgi apparatus are similar in mechanism to those involved in vesicular shuttling between the endoplasmic reticulum and the Golgi apparatus. Specific proteins on the surface of a vesicle are recognized by specific docking proteins on the surface of the membranes with which the vesicle finally fuses.

In contrast to this entire story, if a protein does not have a signal sequence, synthesis continues on a free ribosome until the completed protein is released into the cytosol, where it will carry out its function.

REPLICATION AND EXPRESSION OF GENETIC INFORMATION

The set of genes present in each cell of an individual is inherited from the father and mother at the time of fertilization of an egg by a sperm. The egg and sperm cell

each contain 23 molecules of DNA associated with proteins in structures known as **chromosomes.** Each of the 23 chromosomes contains a different set of genes, some containing more genes than others. Twenty-two of the 23 chromosomes contain genes that produce the proteins that govern most cell structures and functions and are known as **autosomes.** The remaining chromosome, known as the **sex chromosome,** contains genes whose expression determines the development of male or female gender. The 22 autosomes in the egg and sperm contain corresponding genes. For example, a chromosome in the egg contains a gene for hemoglobin that is homologous to a similar gene in one of the sperm's chromosomes.

When the egg and sperm unite, the resulting fertilized egg contains 46 chromosomes—44 autosomes and 2 sex chromosomes. With the exception of the genes on the sex chromosomes, each cell of an individual contains 22 pairs of homologous genes. Of each pair, one chromosome was inherited from the mother and one from the father, with each potentially able to produce the same type of protein.

The development of an individual is determined by the controlled expression of the set of genes inherited at the time of conception. Growth occurs through the successive division of cells to form the trillions of cells that make up the adult human body. Each time a cell divides, the 46 DNA molecules in the 46 chromosomes must be replicated, and identical DNA copies passed on to each of the two new cells, termed **daughter cells.** Thus every cell in the body, with the exception of the reproductive cells, contains an identical set of 46 DNA molecules, and therefore an identical set of genes. (See Chapter 19 for a discussion of the special processes associated with the formation of the reproductive cells in which the number of chromosomes is reduced from 46 to 23.) What makes one cell different from another depends on the differential expression of various sets of genes in this gene pool common to every cell.

Replication of DNA

DNA is the only molecule in a cell able to duplicate itself without information from some other cell component. In contrast, as we have seen, RNA can only be formed using the information present in DNA, protein formation uses the information in mRNA, and all other molecules use protein enzymes that determine the structure of the products formed.

DNA replication is, in principle, similar to the process whereby RNA is synthesized. During DNA replication (Figure 5-11), the two strands of the double helix separate and each exposed strand acts as a template to which free *deoxyribonucleotide* triphosphates can base-pair, A with T and G with C. An enzyme, **DNA polymerase,** then links the free nucleotides together at a rate of about 50 nucleotides per second as it moves along the strand, forming

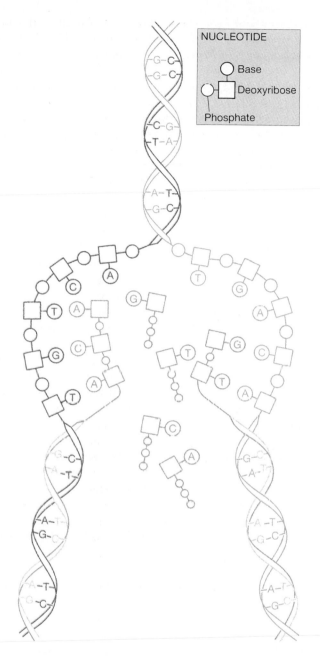

FIGURE 5-11

Replication of DNA involves the pairing of free deoxyribonucleotides with the bases of the separated DNA strands, giving rise to two new identical DNA molecules, each containing one old and one new polynucleotide strand.

a new strand complementary to the template strand of DNA. The end result is two identical molecules of DNA, each called a "copy." In each copy, one strand of nucleotides, the template strand, was present in the original DNA molecule and one strand has been newly synthesized.

This description of DNA synthesis provides an overview of the basic elements of the process, but the individual

steps are considerably more complex. A number of proteins in addition to DNA polymerase are required. Some of these proteins determine where along the DNA strand replication will begin, others open the DNA helix so that it can be copied, while still others prevent the tangling that can occur as the helix unwinds and rewinds.

A special problem arises as the replication process approaches the end of the DNA molecule being replicated. The complex of proteins that carry out the replication sequence are in part anchored to a portion of the DNA molecule that lies ahead of the site at which two strands separate during replication. If a DNA molecule ended at the very end of the last gene, this gene could not be copied during DNA replication because there are no more downstream sites to anchor the replication complex.

This problem is overcome by an additional set of proteins that add to the ends of DNA a chain of nucleotides composed of several hundred to several thousand repeats of the six-nucleotide sequence TTAGGG. This terminal repetitive segment is known as a **telomere,** and the key enzyme that catalyzes the formation of a telomere is **telomerase.** In the absence of telomerase, each replication of DNA results in a shorter molecule because of failure to replicate the ends of DNA.

Cells that continue to divide throughout the life of an organism contain telomerase, as do the cells that give rise to sperm and egg cells. The presence of telomerase prevents shortening of their DNA. However, many cells do not express telomerase, and each replication of DNA leads to a loss of coded genetic information. It is hypothesized that the telomeres serve as a biological clock that sets the number of divisions a cell can undergo and still remain viable.

In order to form the approximately 40 trillion cells of the adult human body, a minimum of 40 trillion individual cell divisions must occur. Thus, the DNA in the original fertilized egg must be replicated at least 40 trillion times. Actually, many more than 40 trillion divisions occur during the growth of a fertilized egg into an adult human being since many cells die during development and are replaced by the division of existing cells.

If a secretary were to type the same manuscript 40 trillion times, one would expect to find some typing errors. Therefore, it is not surprising to find that during the duplication of DNA, errors occur that result in an altered sequence of bases and a change in the genetic message. What is amazing is that DNA can be duplicated so many times with relatively few errors.

A mechanism called **proofreading** corrects errors in the base sequence as it is being duplicated and is largely responsible for the low error rate observed during DNA replication. If an incorrect free nucleotide has become temporarily paired with a base in the template strand of DNA (for example, C pairing with A rather than its appropriate

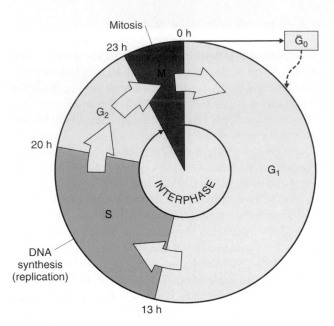

FIGURE 5-12

Phases of the cell cycle with approximate elapsed times in a cell that divides every 24 h. A cell may leave the cell cycle and enter the G_0 phase where division ceases unless the cell receives a specific signal to reenter the cycle.

partner G), the DNA polymerase somehow "recognizes" this abnormal pairing and will not proceed in the linking of nucleotides until the abnormal pairing has been replaced. Note that in performing this proofreading, the DNA polymerase needs to identify only two configurations, the normal A-T and G-C pairing; any other combination halts polymerase activity. In this manner each nucleotide, as it is inserted into the new DNA chain, is checked for its appropriate complementarity to the base in the old chain.

Cell Division

Starting with a single fertilized egg, the first cell division produces 2 cells. When these daughter cells divide, they each produce 2 cells, giving a total of 4. These 4 cells produce a total of 8, and so on. Thus, starting from a single cell, 3 division cycles will produce 8 cells (2^3), 10 division cycles will produce $2^{10} = 1024$ cells, and 20 division cycles will produce $2^{20} = 1,048,576$ cells. If the development of the human body involved only cell division and growth without any cell death, only about 46 division cycles would be needed to produce all the cells in the adult body. However, large numbers of cells die during the course of development, and even in the adult many cells survive only a few days and are continually replaced by the division of existing cells.

The time between cell divisions varies considerably in different types of cells, with the most rapidly growing cells

dividing about once every 24 h. During most of this period, there is no visible evidence that the cell will divide. For example, in a 24-h division cycle, changes in cell structure begin to appear only 1 h before division. The period between the end of one division and the appearance of the structural changes that indicate the beginning of the next division is known as **interphase.** Since the physical process of dividing one cell into two cells takes only about 1 h, the cell spends most of its time in interphase, and most of the cell properties described in this book are properties of interphase cells.

One very important event related to subsequent cell division does occur during interphase, namely, the replication of DNA, which begins about 10 h before the first visible signs of division and lasts about 7 h. This period of the cell cycle is known as the S phase (synthesis) (Figure 5-12). Following the end of DNA synthesis there is a brief interval, G_2 (second gap), before the signs of cell division begin. The period from the end of cell division to the beginning of the S phase is the G_1 (first gap) phase of the cell cycle.

In terms of the capacity to undergo cell division, there are two classes of cells in the adult body. Some cells proceed continuously through one cell cycle after another, while others seldom or never divide once they have been formed. The first group consists of the stem cells, which provide a continuous supply of cells that differentiate into specialized cells to replace those (such as blood cells, skin cells, and the cells lining the intestinal tract) that are continuously lost. The second class includes a number of differentiated, specialized cell types, such as nerve and striated muscle cells, that never divide once they have differentiated. Also included in this second class are cells that leave the cell cycle and enter a phase known as G_0 (Figure 5-12) in which the process that initiates DNA replication is blocked. A cell in the G_0 phase, upon receiving an appropriate signal, can reenter the cell cycle, begin replicating DNA, and proceed to divide.

Cell division involves two processes: nuclear division, or **mitosis,** and cytoplasmic division, or **cytokinesis.** Although mitosis and cytokinesis are separate events, the term mitosis is often used in a broad sense to include the subsequent cytokinesis, and so the two events constitute the M phase (mitosis) of the cell cycle. Mitosis that is not followed by cytokinesis produces multinucleated cells found in the liver, placenta, and some embryonic cells and cancer cells.

As was stated earlier, the 3 billion nucleotides in DNA that make up the human genome are distributed among 23 pairs of chromosomes in the cell nucleus. These very long DNA molecules, having lengths a thousand times greater than the diameter of the nucleus, are packaged into the small confines of the nucleus by coiling them around clusters of DNA-associated proteins known as histones. This complex of DNA and protein, called **chromatin,** is organized in a pattern that resembles beads on a string. Each bead, called a **nucleosome,** consists of a region of DNA coiled around a histone complex.

When each of the 46 DNA molecules replicates, the result is two identical threads termed **sister chromatids,** which initially are joined together at a single point called the **centromere** (Figure 5-13). As a cell begins to divide, each chromatid pair becomes highly coiled and condensed, forming visible, rod-shaped bodies that can be stained by the dyes used to visualize structures under a microscope. These structures are called chromosomes (colored bodies), a term that also refers to a DNA molecule that contains a specific set of genes. In the condensed state prior to division, each of the 46 chromosomes, consisting of 2 chromatids, can be identified microscopically by its length and the position of its centromere.

As the duplicated chromatin threads condense, the nuclear membrane breaks down, and the chromosomes become linked in the region of their centromeres to spindle fibers (Figure 5-13C). The **spindle fibers,** composed of microtubules, are formed in the region of the cell known as the centrosome. The centrosome, which contains the centrioles (described in Chapter 2), is where microtubular assembly begins. When a cell enters the mitotic phase of the cell cycle, the centriole divides, and the two centrioles migrate to opposite sides of the cell, thus establishing the axis of cell division. One centrosome will pass to each of the daughter cells during cytokinesis. Some of the spindle fibers extend between the two centrioles, while others

connect the centrioles to the chromosomes. The spindle fibers and centrosome constitute the **mitotic apparatus.**

As mitosis proceeds, the sister chromatids of each chromosome separate at the centromere and move toward opposite centrioles (Figure 5-13D). Cytokinesis begins as the sister chromatids separate. The cell begins to constrict along a plane perpendicular to the axis of the mitotic apparatus, and constriction continues until the cell has been pinched in half, forming the two daughter cells (Figure 5-13E), each having half the volume of the parent cell. Following cytokinesis, in each daughter cell, the spindle fibers dissolve, a nuclear envelope forms, and the chromatids uncoil to form chromatin threads once again.

The forces producing the movements associated with mitosis and cytokinesis are generated by a family of contractile proteins similar to those producing the forces generated by muscle cells (described in Chapter 11) and the chemical kinetics associated with the elongation and shrinkage of microtubular filaments.

There are two critical check points in the cell cycle, at which special events must occur in order for a cell to progress to the next phase. One is at the boundary between G_1 and S, and the other between G_2 and M.

Two classes of proteins are the major players in regulating progression through these checkpoints —**cell division cycle kinases** (cdc kinases) and **cyclins.** Cyclins act as modulator molecules to activate the cdc kinases. The concentration of cyclins progressively increases during interphase and then rapidly falls during mitosis. Once activated, the kinase enzymes phosphorylate, and thus activate or

FIGURE 5-13

Mitosis and cytokinesis. (Only 4 of the 46 chromosomes in a human cell are illustrated.) (A) During interphase, chromatin exists in the nucleus as long, extended threads. The threads are partially coiled around clusters of histone proteins, producing a beaded appearance. (B) Prior to the onset of mitosis, DNA replicates, forming two sister chromatids that are joined at the centromere. A second centriole is also formed at this time. (C) As mitosis begins, the chromatids become highly condensed and become attached to spindle fibers. (D) The two chromatids of each chromosome separate and move toward opposite poles of the cell (E) as the cell divides (cytokinesis) into two daughter cells. ▮◼▮

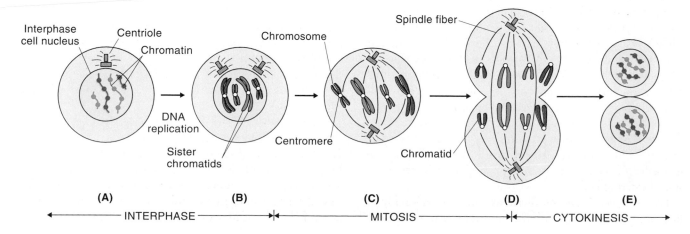

inhibit, a variety of proteins involved in the division process, including an enzyme that digests cyclin and thus prepares the cell to begin the next division cycle.

As we have noted, different types of cells progress through the cell cycle at different rates, some remaining for long periods of time in interphase. In order to progress to DNA replication, most cells must receive an external signal delivered by a group of proteins known as **growth factors.** Growth factors act at the cell membrane to generate intracellular signals; these signals activate various transcription factors that control the synthesis of key proteins involved in the division process and the checkpoint mechanisms. At least 50 growth factors have been identified. Many are secreted by one cell and stimulate other cells to divide; others stimulate division in the cell that secretes them. Growth factors also influence various aspects of metabolism and cell differentiation. In the absence of the appropriate growth factor, most cells will not divide.

Mutation

Any alteration in the nucleotide sequence that spells out a genetic message in DNA is known as a *mutation.* Certain chemicals and various forms of ionizing radiation, such as x-rays, cosmic rays, and atomic radiation, can break the chemical bonds between DNA nucleotides. This can result in the loss of segments of DNA or the incorporation of the wrong base when the broken bonds are reformed. Environmental factors that increase the rate of mutation are known as *mutagens.* Even in the absence of environmental mutagens, the mutation rate is never zero. In spite of proofreading, some errors are made during the replication of DNA, and some of the normal compounds present in cells, particularly oxygen free radicals, can damage DNA, leading to mutations.

Types of Mutations.   The simplest type of mutation, known as a point mutation, occurs when a single base is replaced by a different one. For example, the base sequence C-G-T is the DNA code word for the amino acid alanine. If guanine (G) is replaced by adenine (A), the sequence becomes C-A-T, which is the code for valine. If, however, cytosine (C) replaces thymine (T), the sequence becomes C-G-C, which is also a code for alanine, and the amino acid sequence transcribed from the mutated gene would not be altered. If an amino acid code is mutated to one of the three termination code words, the translation of the mRNA message will cease when this code word is reached, resulting in the synthesis of a shortened, typically nonfunctional protein.

Assume that a mutation has altered a single code word in a gene so that it now codes for a different amino acid, for example, alanine C-G-T changed to valine C-A-T. What effect does this mutation have upon the cell? The answer depends upon where in the gene the mutation has occurred. Although proteins are composed of many amino acids, the properties of a protein often depend upon only a very small region of the total molecule, such as the binding site of an enzyme. If the mutation does not alter the conformation of the binding site, there may be little or no change in the protein's properties. On the other hand, if the mutation alters the binding site, a marked change in the protein's properties may occur. Thus, if the protein is an enzyme, a mutation may change its affinity for a substrate or render the enzyme totally inactive. If the mutation occurs within an intron segment of a gene, it will have no effect upon the amino acid sequence coded by the exon segments (unless it alters the ability of the intron segment to undergo normal splicing from the primary RNA transcript).

In a second type of mutation, sections of DNA are deleted or single bases are added or deleted. Such mutations may result in the loss of an entire gene or group of genes or may cause the misreading of a sequence of bases. Figure 5-14 shows the effect of removing a single base on the reading of the genetic code. Since the code is read in three-base sequences, the removal of one base not only alters the code word containing that base, but also causes a misreading of all subsequent bases by shifting the reading sequence. Addition of an extra base causes a similar misreading of all subsequent triplet code words, which often

FIGURE 5-14

A deletion mutation caused by the loss of a single base G in one of the two DNA strands causes a misreading of all code words beyond the point of the mutation.

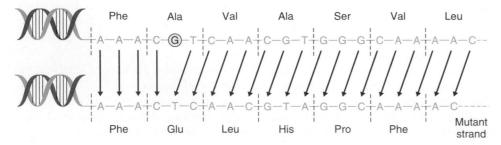

results in a protein having an amino acid sequence that does not correspond to any functional protein.

What effects do these various types of mutation have upon the functioning of a cell? If a mutated, nonfunctional enzyme is in a pathway supplying most of a cell's chemical energy, the loss of the enzyme's function *could* lead to the death of the cell. The story is more complex, however, since the cell contains a second gene for this enzyme on its homologous chromosome, one which has not been mutated and is able to form an active enzyme. Thus, little or no change in cell function would result from this mutation. If both genes had mutations that rendered their products inactive, then no functional enzyme would be formed and the cell would die. In contrast, if the active enzyme were involved in the synthesis of a particular amino acid, and if the cell could obtain that amino acid from the extracellular fluid, the cell function would not be impaired by the absence of the enzyme.

To generalize, a mutation may have any one of three effects upon a cell: (1) It may cause no noticeable change in cell function; (2) it may modify cell function, but still be compatible with cell growth and replication; or (3) it may lead to cell death.

With one exception—cancer, to be described below—the malfunction of a single cell, other than a sperm or egg, as a result of mutation usually has no significant effect because there are many cells performing the same function in the individual. Unfortunately, the story is different when the mutation occurs in a sperm or egg. In this case, the mutation will be passed on to *all* the cells in the body of the new individual. Thus, mutations in a sperm or egg cell that unite at conception, do not affect the individual in which they occur but do affect, often catastrophically, the child produced by these cells. Moreover, these mutations may be passed on to some individuals in future generations descended from the individual carrying the mutant gene.

DNA Repair Mechanisms. Cells possess a number of enzymatic mechanisms for repairing DNA that has been altered. These repair mechanisms all depend on the damage occurring in only one of the two DNA strands, so that the undamaged strand can provide the correct code for rebuilding the damaged strand. A repair enzyme identifies an abnormal region in one of the DNA strands and cuts out the damaged segment. DNA polymerase then rebuilds the segment after base-pairing with the undamaged strand just as it did during DNA replication. If adjacent regions in both strands of DNA are damaged, a permanent mutation is created that cannot be repaired by these mechanisms.

This repair mechanism is particularly important for long-lived cells, such as nerve and muscle cells, that do not divide and therefore do not replicate their DNA. This means that the same molecule of DNA must continue to function and maintain the stability of its genetic information for as long as the cell lives, which could be as long as 100 years. One aspect of aging may be related to the accumulation of unrepaired mutations in these long-lived cells.

Mutations and Evolution. Mutations contribute to the evolution of organisms. Although most mutations result in either no change or an impairment of cell function, a very small number may alter the activity of a protein in such a way that it is more, rather than less, active, or they may introduce an entirely new type of protein activity into a cell. If an organism carrying such a mutant gene is able to perform some function more effectively than an organism lacking the mutant gene, it has a better chance of surviving and passing the mutant gene on to its descendants. On the other hand, if the mutation produces an organism that functions less effectively than organisms lacking the mutation, the organism is less likely to survive and pass on the mutant gene. This is the principle of **natural selection.** Although any one mutation, if it is able to survive in the population, may cause only a very slight alteration in the properties of a cell, given enough time, a large number of small changes can accumulate to produce very large changes in the structure and function of an organism.

The Gene Pool. Given the fact that there are billions of people living on the surface of the earth, all carrying genes encoded in DNA and subject to mutation, any given gene is likely to have a slightly different sequence in some individuals as a result of these ongoing mutations. These variants of the same gene are known as **alleles,** and the number of different alleles for a particular gene in the population is known as the **gene pool.** At conception, one allele of each gene from the father and one allele from the mother are present in the fertilized egg. If both alleles of the gene have the same sequence, the individual is said to be **homozygous** for that particular gene, but if the two alleles differ in sequence the individual is **heterozygous.**

The particular alleles present in an individual is referred to as the individual's **genotype.** With the exception of the genes in the sex chromosomes, both of the homologous genes inherited by an individual can be transcribed and translated into proteins, given the appropriate signals to the transcription factors controlling expression of the given gene. The expression of this genotype produces a specific structural or functional activity that is recognized as a particular trait in the individual known as the person's **phenotype.** For example, blue eyes and black eyes represent the phenotypes of the genes involved in the formation of eye pigments.

The expression of a particular phenotype is said to be **dominant** when only one of the two inherited alleles is required to express the trait, and **recessive** when both inherited alleles must be the same, that is, the individual must be homozygous, for the trait to be expressed. For example, black eye color is inherited as a dominant trait, while blue eyes are a recessive trait. If an individual receives an allele of the gene controlling black eye pigment from either parent, the individual will have black eyes. A single copy of the allele for black eye color is sufficient to express the proteins forming black eye pigment. In contrast, the expression of the blue-eyed phenotype occurs only when both alleles in the individual code for a protein able to form the blue-eyed pigment. Although *genes* are often described as dominant or recessive, it is the activity or lack of activity of the *proteins* expressed by the genes that determines the phenotypic characteristics observed.

Genetic Disease. A growing number of diseases are being recognized as genetic, that is due to abnormal structure or function resulting from the inheritance of mutant genes, rather than the result of microbial infections, toxic agents, or improper nutrition. Over 4000 diseases have been linked to genetic abnormalities, and these diseases are currently a major cause of infant mortality. Genetic diseases can be inherited as either a dominant or recessive trait.

Familial hypercholesterolemia is an autosomal dominant disease affecting 1 in 500 individuals. These individuals have elevated blood levels of cholesterol because of a defect in a protein involved in cholesterol removal from the blood (cholesterol will be discussed in Chapter 18) and are at increased risk of developing heart disease. Inheritance of only a single mutant allele from either the mother or father is sufficient to produce this condition.

Cystic fibrosis, an autosomal recessive disease, is the most common lethal congenital disease among Caucasians, with a prevalence of about 1 in 2000 births. Because of a defective mechanism for the transfer of fluid across epithelial membranes (to be discussed in Chapter 6), various ducts in the lungs, intestines, and reproductive tracts become obstructed, with the most serious complications generally developing in the lungs and leading to death from respiratory failure. An individual must inherit a mutant allele from both parents in order for this recessive disease to be expressed. Individuals who are heterozygous, having only one copy of the mutant allele, do not show the symptoms of the disease but are carriers who are able to transmit the gene to their offspring. A single copy of the normal allele is sufficient to maintain epithelial fluid transport.

Familial hypercholesterolemia and cystic fibrosis are examples of **single gene diseases,** as are sickle cell anemia, hemophilia, and muscular dystrophy. Two other recognized classes of genetic disease are **chromosomal** and **polygenic diseases,** both of which require the expression or lack of expression of multiple genes to produce the phenotypic trait. Chromosomal diseases are the result of the addition or deletion of chromosomes or portions of chromosomes during the process of reducing the 46 chromosomes to 23 during the formation of egg and sperm

cells (to be discussed in Chapter 19). The classic example of a chromosomal disease is **Down's syndrome** in which the fertilized egg has an extra copy, or translocation, of chromosome 21. This abnormality occurs in approximately 1 of every 800 births and is characterized by retardation of growth and mental function. Other forms of chromosomal abnormalities are the major cause of spontaneous abortions or miscarriages.

Polygenic diseases result from the interaction of multiple mutant genes, any one of which by itself produces little or no effect, but when present with other mutant genes produces disease. This category of genetic disease is involved in most forms of the major diseases in our modern society, such as diabetes and hypertension.

CANCER

Like the inherited genetic diseases described above, cancer results from gene mutations. However, with a few exceptions, cancer is not an inherited genetic disease that depends on mutations in the reproductive cells. Rather, most cancers arise from mutations that can occur in any cell at anytime in life. As noted earlier, most mutations in a single nonreproductive cell have no effect upon the overall functioning of an organism, even if they lead to the death of that particular cell. If, however, mutations result in the failure of the control systems that regulate cell division, a cell with a capacity for uncontrolled growth, a **cancer cell,** may be formed.

Cancer is the second leading cause of death in America after heart disease, with approximately 25 percent of all deaths due to cancer. Fifty percent of cancers occur in three organs—lung (28 percent), colon (13 percent), and breast (9 percent). About 90 percent of cancers develop in epithelial cells and are known as **carcinomas.** Those derived from connective tissue and muscle are called **sarcomas,** and those from blood cells are **leukemias** and **lymphomas.**

The uncontrolled replication of cells forms a growing mass of tissue known as a **tumor.** If the cells remain localized and do not invade surrounding tissues, the tumor is said to be a **benign tumor.** If, however, the tumor cells grow into the surrounding tissues, disrupting their functions, or spread to other regions of the body via the circulation, a process known as **metastasis,** the tumor is said to be a **malignant tumor** (that is, cancer) and may lead to the death of the individual.

The transformation of a normal cell into a cancer cell is a multistep process that involves altering not only the mechanisms that regulate cell replication but also those that control the invasiveness of the cell and its ability to subvert the body's defense mechanisms. (As will be discussed in Chapter 20, the body's defense system is normally able to detect and destroy most cancerous cells when they first appear.) A cancer cell does not arise in its fully malignant form from a single mutation, but progresses through various stages as a result of successive mutations. The incidence of cancer increases with age as a result of the accumulation of these mutations. Some of the early stages of transformation result in changes in the cell's morphology, known as *dysplasia,* a precancerous state that can be detected by microscopic examination. At this stage, the cell has not yet acquired a capacity for unlimited growth or an ability to invade surrounding tissues.

As mentioned earlier, a number of agents in the environment can damage DNA, increasing the mutation rate. Agents that increase the probability of a cancerous transformation in a cell are known as **carcinogens;** examples of carcinogens are tobacco smoke, radiation, certain microbes, and synthetic chemicals in our food, water and air. Some of these carcinogens act directly on DNA while others are converted in the body into compounds that damage DNA. It is estimated that approximately 90 percent of all cancers are induced by environmental factors, some of which have been added to the environment by our modern lifestyle.

Over the last 20 years a growing number of genes have been identified that contribute to the cancerous state when they mutate. These cancer-related genes fall into two classes: dominant and recessive. The dominant cancer-producing genes are called **oncogenes** (Greek *onkos,* mass, tumor—the branch of medicine that deals with cancer is known as oncology). Oncogenes arise as mutations of normal genes known as **proto-oncogenes.** For example, some oncogenes code for abnormal cell surface receptors that bind growth factors, producing a state in which the altered receptor produces a continuous growth signal in the absence of bound growth factor. The oncogenes are considered dominant since only one of the two homologous proto-oncogenes needs to be mutated for the mutation to contribute to the cancerous state.

The second class of genes involved in cancer are genes known as **tumor suppressor genes.** In their unmutated state these genes prevent the cancerous transformations produced by oncogenes. Mutation of one of the pair of alleles of tumor suppressor genes inactivates its function, but leaves a normal gene on the homologous chromosome that can still suppress tumor development. It is only when both alleles have been mutated that a cell may become cancerous. Thus, this type of cancer phenotype is recessive.

One of the most frequently encountered mutations in cancer cells is of a tumor suppressor gene that codes for a phosphoprotein known as **p53** (because it has a molecular mass of 53,000 daltons). Normally, p53 functions as a transcription factor that stimulates transcription of a gene that codes for a protein that inhibits the cdc kinase re-

quired for progression of a cell from the G_1 to the S phase of the mitotic cycle. The concentration of normal p53 is increased in cells that have suffered damage to their DNA and acts to prevent the replication of these damaged cells, including cells that have undergone cancerous mutations at other gene sites. Mutation of both homologous copies of p53 results in the loss of a cell's ability to inhibit the proliferation of damaged cells and thus provides one step in the progression to a fully malignant cancer cell. Mutation of only one of the two homologous p53 genes does not cause loss of suppressor activity because there is still one normal p53 gene functioning in the cell. Of course, cells carrying one copy of a mutated p53 are at increased risk of progressing to a cancerous state if the remaining normal gene becomes mutated. Although most cancers are not directly inherited, the risk of developing cancer can be increased if, for example, one mutant p53 gene is inherited and is therefore present in all cells of the body. Because cells contain multiple control systems to regulate various stages of cell proliferation, disruption of one system, although it may produce a pre-cancerous state, is not sufficient to form a fully malignant cell.

If a cancer is detected in the early stages of its growth, before it has metastasized, the tumor may be removed by surgery. Once it has metastasized to many organs, surgery is no longer possible. Drugs and radiation can be used to inhibit cell multiplication and destroy malignant cells, both before and after metastasis, although these treatments unfortunately also damage the growth of normal cells. Some cancer cells retain the ability to respond to normal growth signals, such as the growth of breast tissue in response to the hormone estrogen. Blocking the action of the hormones on hormone-dependent tumor cells can inhibit their growth. The development of more selective drugs and the mechanisms for targeting them to cancer cells is one of the benefits that may arise from the field of genetic engineering.

GENETIC ENGINEERING

Since the discovery of the structure of DNA in the early 1950s techniques have been developed that enable scientists not only to determine the base sequence of a particular DNA molecule but to modify that sequence by the addition or deletion of specific bases, altering in a controlled manner the message encoded by the DNA. Through the use of these techniques it may become possible to successfully replace mutated genes in specific cells with normal genes. We end this chapter with a discussion of some of the ways in which DNA can be studied and manipulated.

In order to manipulate a gene, it must first be identified among the thousands of genes in the genome, isolated in

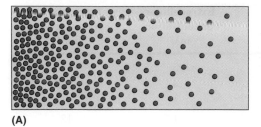 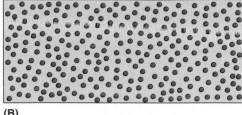

(A)   (B)

FIGURE 6-1

Molecules initially concentrated in one region of a solution (A) will, due to their random thermal motion, undergo a net movement from the region of higher to the region of lower concentration—that is, they will *diffuse* until they become uniformly distributed throughout the solution (B).

flux of glucose from compartment 1 to compartment 2 depends on the concentration of glucose in compartment 1. If the number of molecules in a unit of volume is doubled, the flux of molecules across each surface of the unit will also be doubled, since twice as many molecules will be moving in any direction at a given time.

After a short time, some of the glucose molecules that have entered compartment 2 will randomly move back into compartment 1 (Figure 6-2 Time B). The magnitude of the glucose flux from compartment 2 to compartment 1 depends upon the concentration of glucose in compartment 2 at any time. The **net flux** of glucose between the two compartments at any instant is the difference between the two one-way fluxes. It is the net flux that determines the net gain of molecules by compartment 2 and the net loss from compartment 1.

Eventually the concentrations of glucose in the two compartments become equal at 10 mmol/L. The two one-way fluxes are then equal in magnitude but opposite in direction, and the net flux of glucose is zero (Figure 6-2 Time C). The system has now reached **diffusion equilibrium,** and no further change in the glucose concentration of the two compartments will occur, since thermal motion will continue to move equal numbers of glucose molecules in both directions between the two compartments.

Several important properties of diffusion can be reemphasized using this example. Three fluxes can be identified at any surface—the two one-way fluxes occurring in opposite directions from one compartment to the other, and the net flux, which is the difference between them (Figure 6-3). The net flux is the most important component in diffusion since it is the net amount of material transferred from one location to another. Although the movement of individual molecules is random, *the net flux always proceeds from regions of higher concentration to regions of lower concentration.* For this reason, we often say that substances move "downhill" by diffusion. The greater the difference in concentration between any two

regions, the greater the magnitude of the net flux. Thus, both the direction and the magnitude of the net flux are determined by the concentration difference.

At any concentration difference, however, the magnitude of the net flux depends on several additional factors: (1) temperature—the higher the temperature, the greater the speed of molecular movement and the greater the net flux; (2) mass of the molecule—large molecules (for example, proteins) have a greater mass and lower speed than smaller molecules (for example, glucose) and thus have a smaller net flux; (3) surface area—the greater the surface area between two regions, the greater the space available for diffusion and thus the greater the net flux; and (4) medium through which the molecules are moving—molecules diffuse more rapidly in a gas phase than in water because collisions are less frequent in a gas phase, and when a membrane is involved, its chemical composition influences diffusion rates.

Diffusion Rate versus Distance

The distance over which molecules diffuse is an important factor in determining the rate at which they can reach a cell from the blood or move throughout the interior of a cell after crossing the plasma membrane. Although individual molecules travel at high speeds, the number of collisions they undergo prevents them from traveling very far in a straight line. Diffusion times increase in proportion to the *square* of the distance over which the molecules diffuse. It is for this reason, for example, that it takes glucose approximately 3.5 s to reach 90 percent of diffusion equilibrium at a point 10 μm away from a source of glucose, such as the blood, but it would take over 11 years to reach the same concentration at a point 10 cm away from the source.

Thus, although diffusion equilibrium can be reached rapidly over distances of cellular dimensions, it takes a very long time when distances of a few centimeters or more are involved. For an organism as large as a human being, the diffusion of oxygen and nutrients from the body surface to tissues located only a few centimeters below the

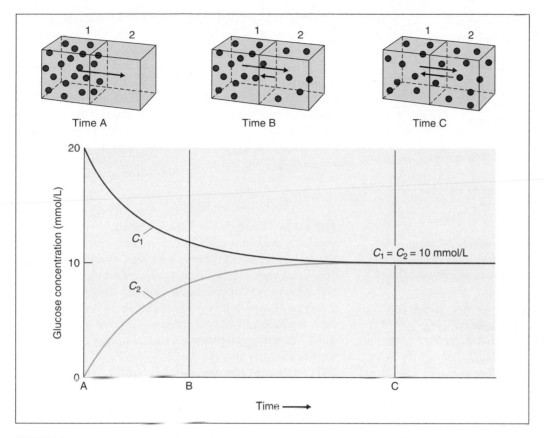

FIGURE 6-2

Diffusion of glucose between two compartments of equal volume separated by a permeable barrier. At time A, compartment 1 contains glucose at a concentration of 20 mmol/L, and no glucose is present in compartment 2. At time B, some glucose molecules have moved into compartment 2, and some are moving back into compartment 1. The length of the arrows represents the magnitude of the one-way movement. At time C, diffusion equilibrium has been reached, the concentration of glucose is equal in the two compartments (10 mmol/L) and the net movement is zero.

In the graph at the bottom of the figure, the red line represents the concentration in compartment 1 (C_1), and the blue line the concentration in compartment 2 (C_2).

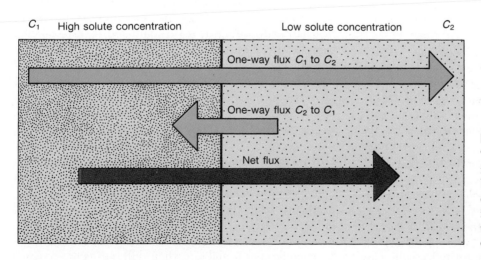

FIGURE 6-3

The two one-way fluxes occurring during the diffusion of solute across a boundary and the net flux, which is the difference between the two one-way fluxes. The net flux always occurs in the direction from higher to lower concentration.

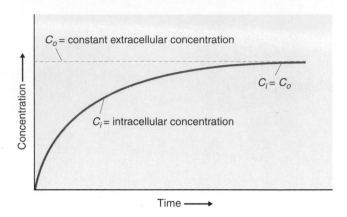

FIGURE 6-4

The increase in intracellular concentration as a substance diffuses from a constant extracellular concentration until diffusion equilibrium is reached across the plasma membrane of a cell.

surface would be far too slow to provide adequate nourishment. Accordingly, the circulatory system provides the mechanism for rapidly moving materials over large distances (by blood flow using a mechanical pump, the heart), with diffusion providing movement between the blood and tissue cells.

The rate at which diffusion is able to move molecules *within* a cell is one of the reasons that cells must be small. A cell would not have to be very large before diffusion failed to provide sufficient nutrients to its central regions. For example, the center of a 20-μm diameter cell reaches diffusion equilibrium with extracellular oxygen in about 15 ms, but it would take 265 days to reach equilibrium at the center of a cell the size of a basketball.

Diffusion Through Membranes

The rate at which a substance diffuses across a plasma membrane can be measured by monitoring the rate at which its intracellular concentration approaches diffusion equilibrium with its concentration in the extracellular fluid. Let us assume that since the volume of extracellular fluid is large, its solute concentration will remain essentially constant as the substance diffuses into the small intracellular volume (Figure 6-4). As with all diffusion processes, the net flux F of material across the membrane is from the region of higher concentration (the extracellular solution in this case) to the region of lower concentration (the intracellular fluid). The magnitude of the net flux is directly proportional to the difference in concentration across the membrane $(C_o - C_i)$, the surface area of the membrane A, and the membrane **permeability constant k_p**:

$$F = k_p A(C_o - C_i)$$

The numerical value of the permeability constant k_p is an experimentally determined number for a given type of molecule at a given temperature, and it reflects the ease

with which the molecule is able to move through a given membrane. In other words, the greater the permeability constant, the larger the net flux across the membrane for any given concentration difference and membrane area.

The rates at which molecules diffuse across membranes, as measured by their permeability constants, are a thousand to a million times smaller than the diffusion rates of the same molecules through a water layer of equal thickness. In other words, a membrane acts as a barrier that considerably slows the diffusion of molecules across its surface. The major factor limiting diffusion across a membrane is its lipid bilayer.

Diffusion Through the Lipid Bilayer. When the permeability constants of different organic molecules are examined in relation to their molecular structures, a correlation emerges. Whereas most polar molecules diffuse into cells very slowly or not at all, nonpolar molecules diffuse much more rapidly across plasma membranes, that is, they have large permeability constants. The reason is that nonpolar molecules can dissolve in the nonpolar regions of the membrane—regions occupied by the fatty acid chains of the membrane phospholipids. In contrast, polar molecules have a much lower solubility in the membrane lipids. Increasing the lipid solubility of a substance (decreasing the number of polar or ionized groups it contains) will increase the number of molecules dissolved in the membrane lipids and thus increase its flux across the membrane. Oxygen, carbon dioxide, fatty acids, and steroid hormones are examples of nonpolar molecules that diffuse rapidly through the lipid portions of membranes. Most of the organic molecules that make up the intermediate stages of the various metabolic pathways (Chapter 4) are ionized-polar molecules, often containing an ionized phosphate group, and thus having a low solubility in the lipid bilayer. Most of these substances are retained within cells because they cannot diffuse across the lipid barrier of the plasma membrane.

That it is the lipid-bilayer portion of the membrane, and not the membrane proteins, that provides the selective barrier to diffusion is substantiated by experiments using artificial membranes consisting of a bimolecular layer of lipids but containing no protein. For most substances, the rates of diffusion across these artificial lipid bilayers and the selectivity of these artificial bilayers are very similar to those of plasma membranes. An exception to this generalization, however, occurs in the case of ions.

Diffusion of Ions Through Protein Channels. Ions such as Na^+, K^+, Cl^-, and Ca^{2+} diffuse across plasma membranes at rates that are much faster than would be predicted from their very low solubility in membrane lipids. Moreover, different cells have quite different permeabilities to these ions, whereas nonpolar substances have similar permeabilities when different cells are com-

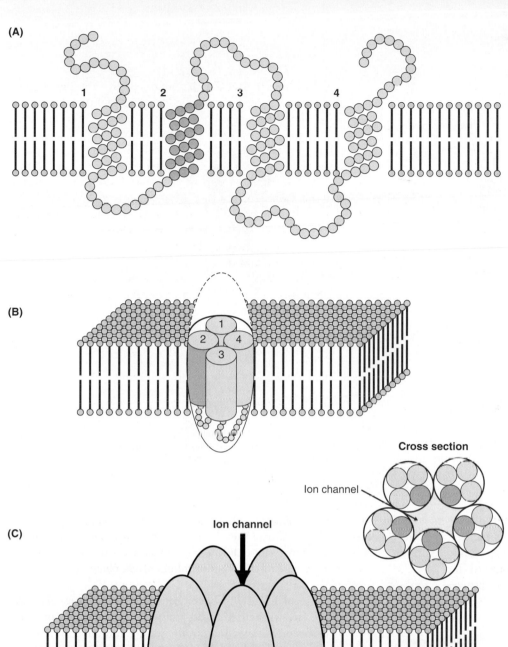

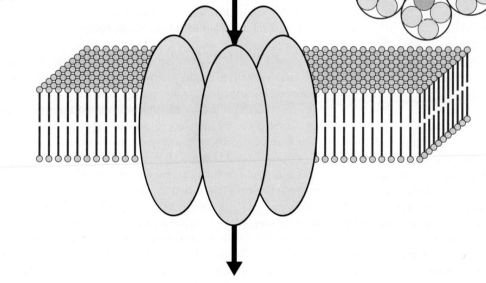

FIGURE 6-5

Model of an ion channel composed of five polypeptide subunits. (A) A channel subunit consisting of an integral membrane protein containing four transmembrane segments (1, 2, 3 and 4), each with an alpha-helix configuration within the membrane. Although this model has only 4 transmembrane segments, some channel proteins have as many as 12. (B) The same subunit as in A shown in three-dimensions within the membrane with the four transmembrane helices aggregated together. (C) The ion channel consists of five of the subunits illustrated in B, which form the sides of the channel. As shown in cross section, the helical transmembrane segment (A,2), shown in green, of each subunit forms the sides of the channel opening; the presence of ionized amino acid side chains along this region determines the selectivity of the channel to ions. ▭

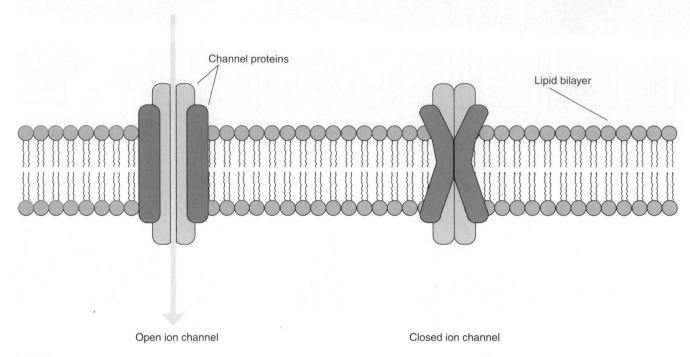

FIGURE 6-6

An ion channel may be open, allowing ions to diffuse across the membrane, or closed, as a result of conformational changes in the membrane proteins forming the channel.

pared. The fact that artificial lipid bilayers containing no protein are practically impermeable to these ions suggests that it is the protein component of the membrane that is responsible for these permeability differences.

As we have seen (Chapter 3), integral membrane proteins can span the lipid bilayer. Some of these proteins form **channels** through which ions can diffuse across the membrane. A single protein may have a conformation similar to that of a doughnut, with the hole in the middle providing the channel for ion movement. More often, several proteins aggregate to form the walls of a channel (Figure 6-5). The diameters of protein channels are very small, only slightly larger than those of the ions that pass through them. The small size of the channels prevents larger, polar, organic molecules from entering the channel.

Very importantly, channels can exist in an open or closed state (Figure 6-6). The greater the number of open channels, the greater the ion flux across the membrane for any given ion concentration difference. Furthermore, ion channels show a selectivity for the type of ions that can pass through them. This selectivity is based partially on the channel diameter and partially on the charged and polar surfaces of the proteins that form the channel walls and electrically attract or repel the ions. For example, some channels (K channels) allow only potassium ions to pass, others are specific for sodium (Na channels), and still others allow both sodium and potassium ions to pass (Na,K channels). Thus,

two membranes that have the same permeability to potassium because they have the same number of K channels may have quite different permeabilities to sodium because they contain different numbers of Na channels.

Role of Electric Forces on Ion Movement. Thus far we have described the direction and magnitude of solute diffusion across a membrane with a certain area; that is, a diffusion determined by the solute's concentration difference across the membrane, its solubility in the membrane lipids, and the presence of membrane ion channels. When describing the diffusion of ions, since they are charged, one additional factor must be considered: the presence of electric forces acting upon the ions.

There exists a separation of electric charge across plasma membranes, known as a **membrane potential** (Figure 6-7), the origin of which will be described in Chapter 8. The membrane potential provides an electric force that influences the movement of ions across the membrane. Electric charges of the same sign, both positive or both negative, repel each other, while opposite charges attract. For example, if the inside of a cell has a net negative charge with respect to the outside, as it does in most cells, there will be an electric force attracting positive ions into the cell and repelling negative ions. Even if there were no difference in ion concentration across the membrane, there would still be a net movement of posi-

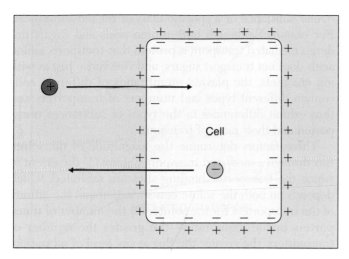

FIGURE 6-7

The separation of electric charge across a plasma membrane (the membrane potential) provides the electric force that drives positive ions into a cell and negative ions out.

tive ions into and negative ions out of the cell because of the membrane potential. Thus, the direction and magnitude of *ion* fluxes across membranes depend on both the concentration difference and the electrical difference (the membrane potential). These two driving forces are collectively known as the **electrochemical gradient,** also termed the electrochemical difference across a membrane.

It is worth noting that the two forces that make up the electrochemical difference may oppose each other. Thus, the *membrane potential* may be driving potassium ions, for example, in one direction across the membrane, while the *concentration difference* for potassium is driving these ions in the opposite direction. The net movement of potassium in this case would be determined by the magnitudes of the two opposing forces, that is, by the electrochemical difference across the membrane.

Regulation of Diffusion Through Ion Channels. Changes in a membrane's permeability to ions can occur rapidly as a result of the opening or closing of ion channels. The process of opening and closing ion channels is known as **channel gating,** like the opening and closing of a gate in a fence. A single ion channel may open and close many times in a single second, suggesting that the channel protein fluctuates between two (or more) conformations. Over an extended period of time, at any given electrochemical gradient, the total number of ions that pass through a single channel depends on how frequently the channel opens and how long it stays open.

Three factors can alter the stability of the channel protein conformations, producing changes in the opening frequency or in the duration of the opening: (1) As described in Chapter 7 the binding of specific molecules to channel proteins may directly or indirectly produce either an allosteric or covalent change in the shape of the channel protein—**ligand-sensitive channels.** (2) Changes in the membrane potential can cause movement of the charged regions on a channel protein, altering its shape—**voltage-sensitive channels.** (3) Stretching the membrane may affect the conformation of some channel proteins — **mechanosensitive channels.** A single channel may be affected by more than one of these factors.

A particular type of ion may pass through several different types of channels. For example, a membrane may contain ligand-sensitive potassium channels, voltage-sensitive potassium channels, and mechanosensitive potassium channels. Moreover, the same membrane may have several types of voltage-sensitive potassium channels, each responding to a different range of membrane voltage, or several types of ligand-sensitive potassium channels, each responding to a different chemical messenger. The roles of these gated channels in cell communication and electrical activity will be discussed in Chapters 7 through 9.

MEDIATED-TRANSPORT SYSTEMS

Although diffusion accounts for some of the transmembrane movement of ions, it does not account for all. Moreover, there are a number of other molecules, including amino acids and glucose, that are able to cross membranes yet are too polar to diffuse through the lipid bilayer and too large to diffuse through protein channels. The passage of these molecules and the nondiffusional movements of ions through membranes are mediated by still other integral membrane proteins known as **transporters** (or carriers). Movement of substances through a membrane by these **mediated-transport** systems is dependent on the conformational changes in the transporters.

The transported solute must first bind to a specific site on a transporter (Figure 6-8), a site that is exposed to the solute on one surface of the membrane. A portion of the transporter then undergoes a change in shape, exposing this same binding site to the solution on the opposite side of the membrane. The dissociation of the substance from the transporter binding site completes the process of moving the material through the membrane. Using this mechanism, molecules can move in either direction across the membrane, getting on the transporter on one side and off at the other.

The diagram of the transporter in Figure 6-8 is only a model, since we have little information concerning the specific conformational changes of any transport protein. It is assumed that the changes in the shape of transporters are analogous to those undergone by channel proteins that open and close. These oscillations in conformation are

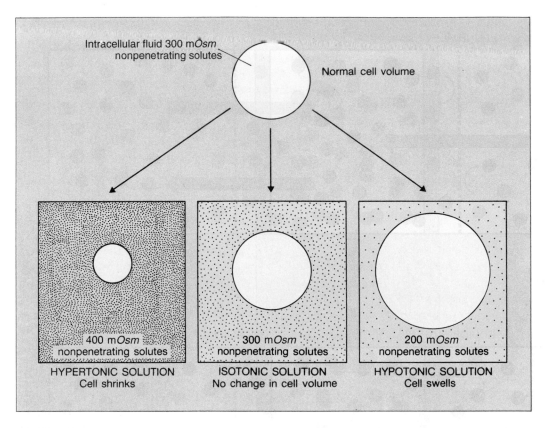

Intracellular fluid 300 m*Osm* nonpenetrating solutes

Normal cell volume

400 m*Osm* nonpenetrating solutes

HYPERTONIC SOLUTION
Cell shrinks

300 m*Osm* nonpenetrating solutes

ISOTONIC SOLUTION
No change in cell volume

200 m*Osm* nonpenetrating solutes

HYPOTONIC SOLUTION
Cell swells

FIGURE 6-20

Changes in cell volume produced by hypertonic, isotonic, and hypotonic solutions.

across plasma membranes. In contrast, if the walls of compartment 2 could not expand, the movement of water into compartment 2 would raise the pressure in compartment 2, which would oppose further net water entry. By the same token, the movement of water into compartment 2 could have been completely prevented by the application of a pressure to compartment 2. This leads to a crucial definition: When a *solution* containing nonpenetrating solutes is separated from pure water by a membrane, the pressure that must be applied to the *solution* to prevent the net flow of water into the *solution* is termed the **osmotic pressure** of the *solution*. The greater the osmolarity of a solution, the greater its osmotic pressure. It is important to recognize that the osmotic pressure of a solution does not push water molecules into the solution. Rather it is the amount of pressure that must be applied to the solution to *prevent* the net flow of water into the solution. Like osmolarity, osmotic pressure is a property of a solution that depends on the solution's water concentration—the lower the water concentration, the higher the osmotic pressure.

Extracellular Osmolarity and Cell Volume

We can now apply the principles learned about osmosis to cells, which all meet the criteria necessary to produce an osmotic flow of water across a membrane. Both the intracellular and extracellular fluids contain water, and cells are surrounded by a membrane that is very permeable to water but impermeable to many substances (**nonpenetrating solutes**).

About 85 percent of the extracellular solute particles are sodium and chloride ions, which can diffuse through protein channels in the plasma membrane or enter the cell during secondary active transport. As we have seen, however, the plasma membrane contains Na,K-ATPase pumps that actively move sodium ions out of the cell. Thus sodium moves into cells and is pumped back out, behaving as if it never entered in the first place; that is sodium behaves like a nonpenetrating solute confined to the extracellular fluid. Also, secondary active-transport pumps and the membrane potential move chloride ions out of cells as rapidly as they enter, with the result that chloride ions also behave as if they were nonpenetrating solutes.

Inside the cell, the major solute particles are potassium ions and a number of organic solutes. Most of the latter are large polar molecules unable to diffuse through the plasma membrane. Although potassium ions can diffuse out of a cell through potassium channels, they are actively transported back by the Na,K-ATPase pump. The net effect, as with sodium, is that potassium behaves as if it were a nonpenetrating solute, but in this case one confined to the intracellular fluid. Thus, sodium and chloride outside the cell and potassium and organic solutes inside the cell behave as nonpenetrating solutes on the two sides of the plasma membrane.

The osmolarity of the extracellular fluid is normally about 300 m*Osm*. Since water can diffuse across plasma membranes, the water in the intracellular fluid will come to diffusion equilibrium with the water in the extracellular fluid. At equilibrium, therefore, the osmolarity of the intracellular fluid is the same—300 m*Osm*—as the osmolarity of the extracellular fluid. Changes in extracellular osmolarity can cause cells to shrink or swell as a result of the movements of water across the plasma membrane.

If cells are placed in a solution of nonpenetrating solutes having an osmolarity of 300 m*Osm*, they will neither swell nor shrink since the water concentrations in the intra- and extracellular fluid are the same, and the solutes cannot leave or enter. Such solutions are said to be **isotonic** (Figure 6-20), defined as having the same concentration of *nonpenetrating* solutes as normal extracellular fluid. Solutions containing less than 300 m*Osm* of nonpenetrating solutes (**hypotonic** solutions) cause cells to swell because water diffuses into the cell from its higher concentration in the extracellular fluid. Solutions containing greater than 300 m*Osm* of nonpenetrating solutes (**hypertonic** solutions) cause cells to shrink as water diffuses out of the cell into the fluid with the lower water concentration. Note that the concentration of *nonpenetrating* solutes in a solution, not the total osmolar-

ity, determines its tonicity—hypotonic, isotonic, or hypertonic. Any penetrating solutes simply don't contribute to the tonicity of a solution.

In contrast, another set of terms—**isoosmotic, hyperosmotic,** and **hypoosmotic**—denotes simply the osmolarity of a solution relative to that of normal extracellular fluid without regard to whether the solute is penetrating or nonpenetrating. The two sets of terms are therefore not synonymous. For example, a 1-L solution containing 300 mOsmol of nonpenetrating NaCl and 100 mOsmol of urea, which can cross plasma membranes, would have a total osmolarity of 400 m*Osm* and would be hyperosmotic. It would, however, also be an isotonic solution, producing no change in the equilibrium volume of cells immersed in it. The reason is that urea will diffuse into the cells and reach the same concentration as the urea in the extracellular solution, and thus both the intracellular and extracellular solutions will have the same osmolarity (400 m*Osm*). Therefore, there will be no difference in the water concentration across the membrane and thus no change in cell volume.

Table 6-3 provides a comparison of the various terms used to describe the osmolarity of solutions. All hypoosmotic solutions are also hypotonic, whereas a hyperosmotic solution can be hypertonic, isotonic, or hypotonic.

As we shall see in Chapter 16, one of the major functions of the kidneys is to regulate the excretion of water in

TABLE 6-3	TERMS REFERRING TO THE SOLUTE OSMOLARITY AND TONICITY OF SOLUTIONS
Isotonic	A solution containing 300 mOsmol/L of nonpenetrating solutes, regardless of the concentration of membrane-penetrating solutes that may be present
Hypertonic	A solution containing greater than 300 mOsmol/L of nonpenetrating solutes, regardless of the concentration of membrane-penetrating solutes that may be present
Hypotonic	A solution containing less than 300 mOsmol/L of nonpenetrating solutes, regardless of the concentration of membrane-penetrating solutes that may be present
Isoosmotic	A solution containing 300 mOsmol/L of solute, regardless of its composition of membrane-penetrating and nonpenetrating solutes
Hyperosmotic	A solution containing greater than 300 mOsmol/L of solutes, regardless of the composition of membrane-penetrating and nonpenetrating solutes
Hypoosmotic	A solution containing less than 300 mOsmol/L of solutes, regardless of the composition of membrane-penetrating and nonpenetrating solutes

19. Under what conditions will a hyperosmotic solution be isotonic?

20. Endocytotic vesicles deliver their contents to which parts of a cell?

21. How do the mechanisms for actively transporting glucose and sodium across an epithelium differ?

22. By what mechanism does the active transport of sodium lead to the osmotic flow of water across an epithelium?

23. What is the difference between an endocrine gland and an exocrine gland?

THOUGHT QUESTIONS

(Answers are given in Appendix A.)

1. In two cases (A and B), the concentrations of solute X in two 1-L compartments separated by a membrane through which X can diffuse are

| | Concentration of X, mM | |
Case	Compartment 1	Compartment 2
A	3	5
B	32	30

A. In what direction will the net flux take place in case A and in case B?

B. When diffusion equilibrium is reached, what will be the concentration of solute in each compartment in case A and in case B?

C. Will A reach diffusion equilibrium faster, slower, or at the same rate as B?

2. When the extracellular concentration of the amino acid alanine is increased, the net flux of the amino acid leucine into a cell is decreased. How might this observation be explained?

3. If a transporter that mediates active transport of a substance has a lower affinity for the transported substance on the extracellular surface of the plasma membrane than on the intracellular surface, in what direction will there be a net transport of the substance across the membrane? (Assume

that the rate of transporter conformational change is the same in both directions.)

4. Why will inhibition of ATP synthesis by a cell lead eventually to a decrease and ultimately, cessation in secondary active transport?

5. Given the following solutions, which has the lowest water concentration? Which two have the same osmolarity?

| | Concentration, mM | | | |
Solution	Glucose	Urea	NaCl	CaCl$_2$
A	20	30	150	10
B	10	100	20	50
C	100	200	10	20
D	30	10	60	100

6. Assume that a membrane separating two compartments is permeable to urea but not permeable to NaCl. If compartment 1 contains 200 mmol/L of NaCl and 100 mmol/L of urea and compartment 2 contains 100 mmol/L of NaCl and 300 mmol/L of urea, which compartment will have increased in volume when osmotic equilibrium is reached?

7. What will happen to cell volume if a cell is placed in each of the following solutions?

| | Concentration, mM | |
Solution	NaCl (nonpenetrating)	Urea (penetrating)
A	150	100
B	100	150
C	200	100
D	100	50

8. Characterize each of the solutions in question 7 as to whether it is isotonic, hypotonic, hypertonic, hypoosmotic, or hyperosmotic.

9. By what mechanism might an increase in intracellular sodium concentration lead to an increase in exocytosis?

CHAPTER 7

Homeostatic Mechanisms and Cellular Communication

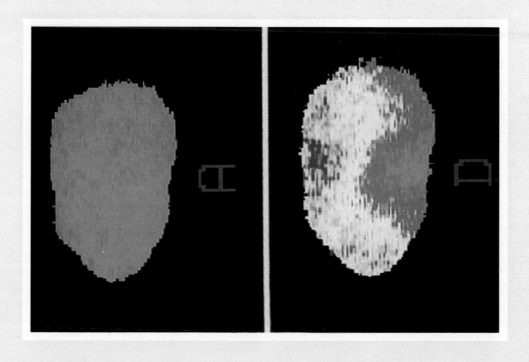

Part 2 of this book provides the information needed to bridge the gap between the cell physiology of Part 1 and the coordinated body functions of Part 3. Section A of this chapter begins that process by amplifying the concept of homeostasis first presented in Chapter 1 and describing the general characteristics and components of homeostatic control systems, as well as several processes that are related to and influence homeostasis. In these discussions many specific examples will be used purely for the purpose of illustration—for example, certain features of temperature regulation. The reader should recognize that such information will be presented again in its more specific context in later chapters.

The operation of control systems requires that cells be able to communicate with each other, often over long distances. Much of this intercellular communication is mediated by chemical messengers. Section B describes how these messengers interact with their target cells to elicit the cells' response.

SECTION A
HOMEOSTATIC CONTROL SYSTEMS

GENERAL CHARACTERISTICS

As described in Chapter 1, the activities of cells, tissues, and organs must be regulated and integrated with each other in such a way that any change in the extracellular fluid initiates a reaction to minimize the change. **Homeostasis** denotes the relatively stable conditions of the internal environment that result from these compensating regulatory responses performed by **homeostatic control systems.**

Consider the regulation of body temperature. Our subject is a resting, lightly clad man in a room having a temperature of 20°C and moderate humidity. His internal body temperature is 37°C, and he is losing heat to the external environment because it is at a lower temperature. However, the chemical reactions occurring within the cells of his body are producing heat at a rate equal to the rate of heat loss. Under these conditions, the body undergoes no net gain or loss of heat and the body temperature remains constant. The system is said to be in a **steady state,** defined as a system in which a particular variable (temperature, in this case) is not changing, but energy (in this case, heat) must be added continuously to maintain this variable in a constant state. (Steady states differ from equilibria, in which a particular variable is not changing but no input of energy is required to maintain the constancy.) The steady-state temperature in our example is known as the **operating point** (also termed the set point) of the thermoregulatory system.

This example illustrates a crucial generalization about homeostasis: Stability of an internal environmental variable is achieved by the balancing of inputs and outputs. In this case, the variable (body temperature) remains constant because metabolic heat production (input) equals heat loss from the body (output).

Now we lower the temperature of the room rapidly, say to 5°C, and keep it there. This immediately increases the loss of heat from our subject's warm skin, upsetting the dynamic balance between heat gain and loss. The body temperature therefore starts to fall. Very rapidly, however, a variety of homeostatic responses occur to limit the fall. These are summarized in Figure 7-1. *The reader is urged to study Figure 7-1 and its legend carefully because the figure is typical of those used throughout the remainder of the book to illustrate homeostatic systems, and the legend emphasizes several conventions common to such figures.*

The first homeostatic response is that blood vessels to the skin narrow, reducing the amount of warm blood flowing through the skin and thus reducing heat loss. At a room temperature of 5°C, however, blood vessel constriction cannot completely eliminate the extra heat loss from the skin. Our subject curls up in order to reduce the surface area of skin available for heat loss. This helps a bit, but excessive heat loss still continues, and body temperature keeps falling, although at a slower rate. He has a strong desire to put on more clothing— "voluntary" behavioral responses are often crucial events in homeostasis—but no clothing is available. Clearly, then, if excessive heat loss (output) cannot be prevented, the only way of restoring the balance between heat input and output is to increase input, and this is precisely what occurs. He begins to shiver, and the chemical reactions responsible for the skeletal muscular contractions that constitute shivering produce large quantities of heat.

Indeed, heat production may transiently exceed heat loss so that body temperature begins to go back toward the value existing before the room temperature was lowered (Figure 7-2). It eventually stabilizes at a temperature *a bit below this original value;* at this new steady state, heat input and heat output are both higher than their original values but are once again equal to each other.

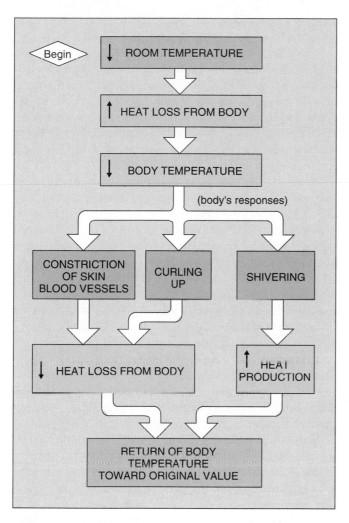

FIGURE 7-1

The homeostatic control system that maintains a relatively constant body temperature when room temperature decreases. This diagram is typical of those used throughout the remainder of this book to illustrate homeostatic systems, and several conventions should be noted. The "begin" sign indicates where to start. The arrows next to each term denote increases or decreases. The arrows connecting any two terms in the figure denote cause and effect; that is, an arrow can be read as "causes" or "leads to" (for example, decreased room temperature "leads to" increased heat loss from the body). In general, one should add the words "tends to" in thinking about these cause-and-effect relationships. For example, decreased room temperature tends to cause an increase in heat loss from the body and curling up tends to cause a decrease in heat loss from the body. Qualifying the relationship in this way is necessary because variables like heat production and heat loss are under the influence of many factors, some of which oppose each other.

The thermoregulatory system just described is an example of a **negative-feedback** system, in which an increase or decrease in the variable being regulated brings about responses that tend to move the variable in the di-

rection opposite ("negative" to) the direction of the original change. Thus, in our example, the *decrease* in body temperature led to responses that tended to *increase* the body temperature, that is, move it toward its original value.

Negative-feedback control systems are the most common homeostatic mechanisms in the body, but there is another type of feedback known as **positive feedback** in which an initial disturbance in a system sets off a train of events that increase the disturbance even further. Thus positive feedback does not favor stability and often abruptly displaces a system away from its steady-state operating point. As we shall see, several important positive-feedback relationships occur in the body, contractions of the uterus during labor being one example.

Note that in our thermoregulatory example the negative-feedback system did not bring the person's temperature back completely to its original value. This illustrates another important generalization about homeostasis: Homeostatic control systems do not maintain *complete* constancy of the internal environment in the face of continued change in the external environment, but can only minimize changes. This is the reason we have said that homeostatic systems maintain the internal environment *relatively* constant. The explanation is that as long as the initiating event (exposure to cold, in our example) continues, some change in the regulated variable (the decrease in body temperature, in our example) must persist to serve as a signal to maintain the homeostatic responses. (This last statement will be qualified

FIGURE 7-2

Changes in internal body temperature during exposure to a low external environmental temperature. As long as the environmental perturbation persists, the homeostatic responses do not return the regulated variable completely to its original value. The deviation from the original value is called the error signal.

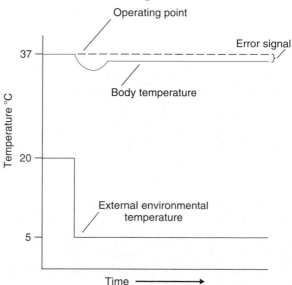

below.) Such a persisting signal is termed an **error signal** (Figure 7-2). It is crucial to recognize that this situation applies only when the initiating event continues; thus, in our example if the external temperature eventually goes back up to its original value, the homeostatic systems will be able to restore body temperature completely back to its original value.

Inherent in the concept of error signals is another generalization about homeostasis: Even in reference to one individual, thus ignoring variation among persons, any regulated variable in the body cannot be assigned a single "normal" value but has a more-or-less narrow range of normal values, depending on the magnitude of the changes in the external conditions and the sensitivity of the responding homeostatic system. The more precise the mechanisms for regulating a variable are, that is, the smaller the error signal need be to drive the system, the narrower is the range. For example, the temperature-regulating systems of the body are extremely sensitive so that body temperature normally varies by only about 1°C even in the face of marked changes in the external environment or heat production during exercise.

As we have seen, perturbations in the external environment can displace a variable from its preexisting operating point. In addition, we now recognize that the operating points for many regulated variables can be physiologically altered or *reset;* that is, the values that the homeostatic control systems are "trying" to keep relatively constant can be altered. A common example is fever, the increase in body temperature that occurs in response to infection and that is analogous to raising the setting of your house's thermostat. The homeostatic control systems regulating body temperature are still functioning during a fever, but they maintain the temperature at a higher value. We shall see in Chapter 20 that this regulated rise in body temperature is adaptive for fighting the infection.

The fact that operating points can be reset adaptively, as in the case of fever, raises important challenges for medicine, as another example illustrates. Plasma iron concentration decreases significantly during many infections. Until recently it was assumed that this decrease is a symptom caused by the infectious organism and that it should be treated with iron supplements. In fact, just the opposite is true: As described in Chapter 20 the decrease in iron is brought about by the body's defense mechanisms and serves to deprive the infectious organisms of the iron they require to replicate. In several controlled studies it has been shown that iron replacement can make the illness much worse. Clearly it is crucial to distinguish between those deviations of homeostatically controlled variables that are truly part of a disease and those that, through resetting, are part of the body's defenses against the disease. Considerations of the evolutionary significance of any particular phenomenon manifested during disease are part of a fascinating new approach termed "Darwinian medicine" (see Suggested Readings).

The examples of fever and plasma iron concentration may have left the impression that operating points are reset only in response to external stimuli, such as the presence of bacteria, but this is not the case. Indeed, as described in the next section, the operating points for many regulated variables change on a rhythmical basis every day; for example, the operating point for body temperature is higher during the day than at night.

Although the resetting of a set point is adaptive in some cases, in others it simply reflects the clashing demands of different regulatory systems. This brings us to one more generalization: It is not possible for everything to be maintained relatively constant by homeostatic control systems. In our example, body temperature was kept relatively constant, but only because large changes in skin blood flow and skeletal muscle contraction were brought about by the homeostatic control system. Moreover, because so many properties of the internal environment are closely interrelated, it is often possible to keep one property relatively constant only by moving others farther from their usual operating point. This is what we meant by "clashing demands."

The generalizations we have given concerning homeostatic control systems are summarized in Table 7-1. One additional point is that, as is illustrated by the regulation of body temperature, *multiple* systems frequently control a *single* parameter. The adaptive value of such redundancy is that it provides much greater fine-tuning and also permits regulation to occur even when one of the systems is not functioning properly because of disease.

Feedforward Regulation

Another type of regulatory process frequently used in conjunction with negative-feedback systems is feedforward. Let us give an example of feedforward and then define it. The temperature-sensitive nerve cells that trigger negative-feedback regulation of body temperature when body temperature begins to fall are located *inside* the body. In addition, there are temperature-sensitive nerve cells in the skin, and these cells, in effect, monitor *outside* temperature. When outside temperature falls, as in our example, these nerve cells immediately detect the change and relay this information to the brain, which then sends out signals to the blood vessels and muscles, resulting in heat conservation and increased heat production. In this manner compensatory thermoregulatory responses are activated *before* the colder outside temperature can cause the internal body temperature to fall. Thus, **feedforward** regulation anticipates changes in a regulated variable such as internal body temperature, improves the speed of the body's homeostatic responses, and minimizes fluctuations in the level of the variable being regulated—that is, it reduces the amount of deviation from the set point.

TABLE 7-1   SOME IMPORTANT GENERALIZATIONS ABOUT HOMEOSTATIC CONTROL SYSTEMS

1. Stability of an internal-environmental variable is achieved by balancing inputs and outputs. It is not the absolute magnitudes of the inputs and outputs that matter but the balance between them.

2. In negative-feedback systems, a change in the variable being regulated brings about responses that tend to move the variable in the direction opposite the original change, that is, back toward the initial value.

3. Homeostatic control systems cannot maintain complete constancy of any given feature of the internal environment. Therefore, any regulated variable will have a more-or-less narrow range of normal values depending on the external-environmental conditions.

4. The operating point of some variables regulated by homeostatic control systems can be reset, that is, physiologically raised or lowered.

5. It is not possible for everything to be maintained relatively constant by homeostatic control systems. There is a hierarchy of importance, such that the constancy of certain variables may be altered markedly to maintain others at relatively constant levels.

In our example feedforward control utilizes a set of "external-environmental" detectors. It is likely, however, that most feedforward control is the result of a different phenomenon—learning. The first times they occur, early in life, perturbations in the external environment probably cause relatively large changes in regulated internal-environmental factors, and in responding to these changes the central nervous system learns to anticipate them and resist them more effectively. A familiar form of this is learning to ride a bicycle with minimal swaying. Learning of this type probably explains many situations in which the error signals are extremely small or even undetectable despite profound perturbations in the system.

COMPONENTS OF HOMEOSTATIC CONTROL SYSTEMS

Reflexes

The thermoregulatory system we used as an example in the previous section, and many of the body's other homeostatic control systems, belong to the general category of stimulus-response sequences known as reflexes. Although in some reflexes we are aware of the stimulus and/or the response, many reflexes regulating the internal environment occur without any conscious awareness.

In the most narrow sense of the word, a **reflex** is an involuntary, unpremeditated, unlearned "built-in" response to a stimulus. Examples of such reflexes include pulling one's hand away from a hot object or shutting one's eyes as an object rapidly approaches the face. There are also many responses, however, that appear to be automatic and stereotyped but are actually the result of learning and practice. For example, an experienced driver performs many complicated acts in operating a car. To the driver these motions are, in large part, automatic, stereotyped, and unpremeditated, but they occur only because a great deal of conscious effort was spent learning them. We term such reflexes **learned,** or **acquired.** In general, most reflexes, no matter how basic they may appear to be, are subject to alteration by learning; that is, there is often no clear distinction between a basic reflex and one with a learned component.

The pathway mediating a reflex is known as the **reflex arc,** and its components are shown in Figure 7-3.

A **stimulus** is defined as a detectable change in the internal or external environment, such as a change in temperature, plasma potassium concentration, or blood pressure. A **receptor** detects the environmental change; we referred to the receptor as a "detector" earlier. A stimulus acts upon a receptor to produce a signal that is relayed to an integrating center. The pathway traveled by the signal

FIGURE 7-3

General components of a reflex arc that functions as a negative-feedback control system. The response of the system has the effect of counteracting or eliminating the stimulus. This phenomenon of negative feedback is emphasized by the minus sign in the feedback loop.

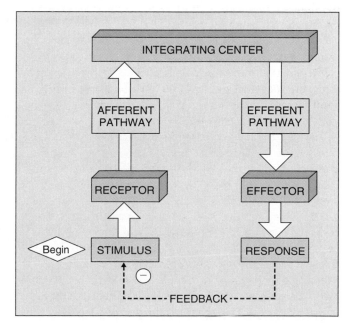

between the receptor and the integrating center is known as the **afferent pathway** (the general term afferent means "to carry to," in this case, to the integrating center).

An **integrating center** often receives signals from many receptors, some of which may be responding to quite different types of stimuli. Thus, the output of an integrating center reflects the net effect of the total afferent input; that is, it represents an integration of numerous bits of information.

The output of an integrating center is sent to the last component of the system, a device whose change in activity constitutes the overall response of the system. This component is known as an **effector.** The information going from an integrating center to an effector is like a command directing the effector to alter its activity. The pathway along which this information travels is known as the **efferent pathway** (the general term efferent means "to carry away from," in this case, away from the integrating center).

Thus far we have described the reflex arc as the sequence of events linking a stimulus to a response. If the response produced by the effector causes a decrease in the magnitude of the stimulus that triggered the sequence of events, then the reflex leads to negative feedback and we have a typical homeostatic control system. Not all reflexes are associated with such feedback. For example, the smell of food stimulates the secretion of a hormone by the stomach, but this hormone does not eliminate the smell of food (the stimulus).

To illustrate the components of a negative-feedback homeostatic reflex arc, let us use Figure 7-4 to apply these terms to thermoregulation. The temperature receptors are the endings of certain nerve cells in various parts of the body. They generate electric signals in the nerve cells at a rate determined by the temperature. These electric signals are conducted by the nerve fibers—the afferent pathway—to a specific part of the brain—the integrating center for temperature regulation. The integrating center, in turn, determines the signals sent out along those nerve cells that cause skeletal muscles and the muscles in skin blood vessels to contract. The nerve fibers to the muscles are the efferent pathway, and the muscles are the effectors. The dashed arrow and the (−) indicate the negative-feedback nature of the reflex. It is essential to recognize that normally a reflex response does not overcompensate, that is, drive the system beyond the normal set point to create another physiological imbalance.

Almost all body cells can act as effectors in homeostatic reflexes. There are, however, two specialized classes of tissues—muscle and gland—that are the major effectors of biological control systems. The physiology of glands is described in Chapter 6, that of muscle in Chapter 11.

FIGURE 7-4

Reflex for minimizing the decrease in body temperature that occurs on exposure to a reduced external environmental temperature. This figure provides the internal components for the reflex shown in Figure 7-1. The dashed arrow and the (−) indicate the negative-feedback nature of the reflex, denoting that the reflex responses cause the decreased body temperature to return toward normal.

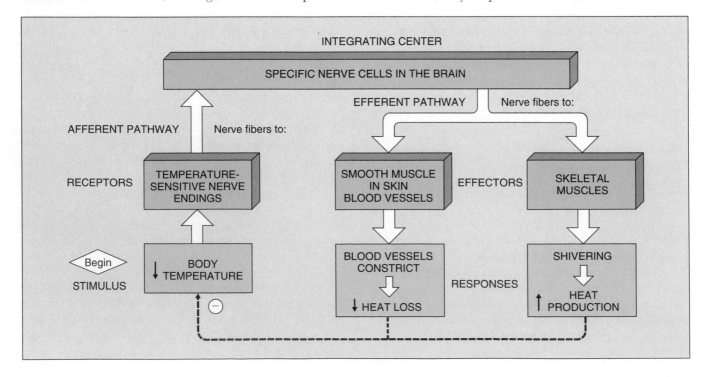

Traditionally, the term "reflex" was restricted to situations in which the receptors, afferent pathway, integrating center, and efferent pathway are all parts of the nervous system, as in the thermoregulatory reflex. Present usage is not so restrictive, however, and recognizes that the principles are essentially the same when a blood-borne chemical messenger known as a **hormone,** rather than a nerve fiber, serves as the efferent pathway, or when a hormone-secreting gland (termed an **endocrine gland**) serves as the integrating center. Thus, in the thermoregulation example, the integrating center in the brain not only sends signals by way of nerve fibers, as shown in Figure 7-4, but also causes the release of several hormones that travel via the blood to many cells, where they produce an increase in the amount of heat produced by these cells. These hormones therefore also serve as an efferent pathway in thermoregulatory reflexes.

Accordingly, in our use of the term "reflex," we include hormones as reflex components. Moreover, depending on the specific nature of the reflex, the integrating center may reside either in the nervous system or in an endocrine gland.

In conclusion, many reflexes function in a homeostatic manner to keep a physical or chemical variable of the body relatively constant. One can analyze any such system by answering the questions listed in Table 7-2.

Local Homeostatic Responses

In addition to reflexes, another group of biological responses is of great importance for homeostasis. We shall call them **local homeostatic responses.** They are initiated by a change in the external or internal environment, that is, a stimulus, and they induce an alteration of cell activity with the net effect of counteracting the stimulus. Like a reflex, therefore, a local response is the result of a sequence of events proceeding from a stimulus. Unlike a reflex, however, the entire sequence occurs only in the area of the stimulus. For example, damage to an area of skin causes cells in the damaged area to release certain chemicals that help the local defense against further injury. The significance of local responses is that they provide individual areas of the body with mechanisms for local self-regulation.

Intercellular Chemical Messengers

Essential to reflexes and local homeostatic responses, and therefore to homeostasis, is the ability of cells to communicate with one another. In most cases, this communication *between* cells—*inter*cellular communication—is performed by chemical messengers. Thus, a hormone functions as a chemical messenger that enables the hormone-secreting cell to communicate with the cell acted upon by the hormone—its **target cell,** with the blood acting as the delivery service. Also, most nerve cells communicate with each other or with effector cells by means of chemical messengers called **neurotransmitters.** Thus, one nerve cell alters the activity of another by releasing from its ending a neurotransmitter that diffuses through the extracellular fluid separating the two nerve cells and acts upon the second cell (Figure 7-5). Similarly, neurotransmitters released from nerve cells onto effector cells constitute the immediate signal, or input, to the effector cells.

As described more fully in Chapter 10, the chemical messengers released by certain nerve cells do not act on adjacent nerve cells or effector cells but rather enter the bloodstream to act on target cells elsewhere in the body. These messengers therefore are properly termed hormones (or neurohormones), not neurotransmitters.

Chemical messengers participate not only in reflexes but also in local responses. The class of chemical messengers involved in local responses are known as **paracrine agents.** They are synthesized by cells and released, once given the appropriate stimulus, into the extracellular fluid. They then diffuse to neighboring cells, some of which are their target cells. (Note that, given this broad definition, neurotransmitters could be classified as a subgroup of paracrine agents, but by convention they are not.) Paracrine agents are generally inactivated rapidly by locally existing enzymes so that they do not enter the bloodstream in large quantities.

There is one category of local chemical messengers that are not *inter*cellular messengers, that is, they do not communicate *between* cells. Rather, the chemical is secreted by a cell into the extracellular fluid and then acts upon the very cell that secreted it. Such messengers are termed **autocrine agents.** Frequently a messenger may serve both paracrine and autocrine functions simultaneously, that is, the messenger molecules released by a cell may act locally on adjacent cells as well as on the cell that released them.

TABLE 7-2   QUESTIONS TO BE ASKED ABOUT ANY HOMEOSTATIC REFLEX

1. What is the variable (for example, plasma potassium concentration, body temperature, blood pressure) that is maintained relatively constant in the face of changing conditions?

2. Where are the receptors that detect changes in the state of this variable?

3. Where is the integrating center to which these receptors send information and from which information is sent out to the effectors, and what is the nature of these afferent and efferent pathways?

4. What are the effectors, and how do they alter their activities so as to maintain the regulated variable near the operating point of the system?

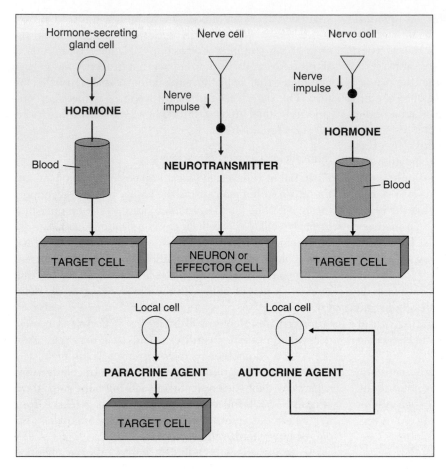

FIGURE 7-5

Categories of chemical messengers. Note that chemical messengers that are secreted by nerve cells and act on adjacent nerve cells or effector cells are termed neurotransmitters, whereas those that enter the blood and act on distant effector cells are classified as hormones (also termed neurohormones). With the exception of autocrine agents, all messengers act between cells, that is, *inter*cellularly.

Stimuli for the release of paracrine/autocrine agents are extremely varied and include neurotransmitters and hormones. In these latter cases the paracrine/autocrine agent often serves to modulate any other effects induced simultaneously in the same locale by the neurotransmitter or hormone.

A point of great importance must be emphasized here to avoid later confusion: A nerve cell, endocrine gland cell, or other cell type may all secrete the same chemical messenger. Thus, a particular messenger may sometimes be referred to as a neurotransmitter, as a hormone, or as a paracrine/autocrine agent, depending both on the type of cell doing the secreting and on the location of the messenger's target cells.

All the types of intercellular communication described so far in this section involve secretion of a chemical messenger into the extracellular fluid. However, there are two important types of chemical communication between cells that do not require such secretion. In one, which occurs via gap junctions (Chapter 3), chemicals move from one cell to an adjacent cell without ever entering the extracellular fluid. In the other, the chemical messenger is not actually released from the cell producing it but rather is located in the plasma membrane of that cell; when the cell encounters another cell type capable of responding to the messenger, the two cells link up via the membrane-bound messenger. This type of signaling (sometimes termed "juxtacrine") is of particular importance in the growth and differentiation of tissues as well as in the functioning of cells that protect the body against microbes and other foreign cells (Chapter 20).

Eicosanoids

The general approach of this text is to describe specific chemical messengers in the context of the functions they influence. However, one set of local chemical messengers exerts such a wide variety of effects in virtually every tissue and organ system that it is best described separately at this point. These are the **eicosanoids,** a family of substances produced from the polyunsaturated fatty acid **arachidonic acid,** which is present in plasma-membrane phospholipids. The eicosanoids include the **cyclic endoperoxides,** the **prostaglandins,** the **thromboxanes,** and the **leukotrienes** (Figure 7-6).

The synthesis of eicosanoids begins when an appropriate stimulus—hormone, neurotransmitter, paracrine agent, drug, or toxic agent—activates an enzyme, **phospholipase A$_2$,** in the plasma membrane of the stimulated

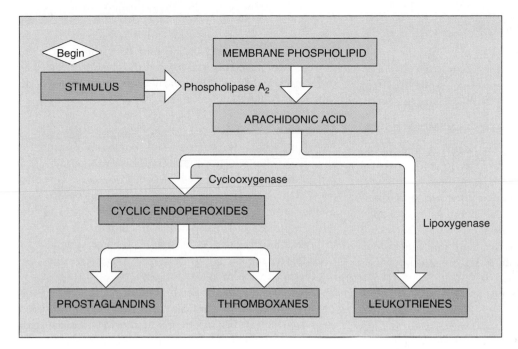

FIGURE 7-6

Pathways for the synthesis of eicosanoids. Phospholipase A_2 is the one enzyme common to the formation of all the eicosanoids; it is the site at which stimuli act. Anti-inflammatory steroids inhibit phospholipase A_2. The step mediated by cyclooxygenase is inhibited by aspirin and other nonsteroidal anti-inflammatory drugs (termed NSAIDs).

cell. As shown in Figure 7-6, this enzyme splits off arachidonic acid from the membrane phospholipids, and the arachidonic acid can then be metabolized by two pathways. One pathway is initiated by the enzyme **cyclooxygenase** and leads ultimately to formation of the cyclic endoperoxides, prostaglandins, and thromboxanes. The other pathway is initiated by the enzyme **lipoxygenase** and leads to formation of the leukotrienes. Within both of these pathways, synthesis of the various specific eicosanoids is enzyme-mediated. Accordingly, beyond phospholipase A_2, the eicosanoid-pathway enzymes found in a particular cell determine which eicosanoids the cell synthesizes in response to a stimulus.

Each of the major eicosanoid subdivisions contains more than one member, as indicated by the use of the plural in referring to them (prostaglandins, for example). On the basis of structural differences, the different molecules within each subdivision are designated by a letter—for example, PGA and PGE for prostaglandins of the A and E types, which then may be further subdivided—for example, PGE_2 (Figure 7-7).

Once synthesized in response to a stimulus, the eicosanoids are not stored to any extent but are released immediately and act locally; accordingly, the eicosanoids are categorized as paracrine and autocrine agents. After they act, they are quickly metabolized by local enzymes to inactive forms. The eicosanoids exert a bewildering array of effects, many of which we will describe in future chapters.

Finally, a word about drugs that influence the eicosanoid pathways since these are perhaps the most commonly used drugs in the world today. At the top of the list must come **aspirin,** which inhibits cyclooxygenase and, therefore, blocks the synthesis of the endoperoxides, prostaglandins, and thromboxanes. It and the newer drugs that also block cyclooxygenase are collectively termed **nonsteroidal anti-inflammatory drugs (NSAIDs).** Their major uses are to reduce pain, fever, and inflammation. The term "nonsteroidal" distinguishes them from the **adrenal steroids** (Chapters 10 and 20) that are used in large doses as anti-inflammatory drugs; these steroids inhibit phospholipase A_2 and thus block the production of all eicosanoids.

FIGURE 7-7

Structure of arachidonic acid and PGE_2. A subscript (as in PGE_2) denotes the number of double bonds in the fatty acid portion of an eicosanoid molecule.

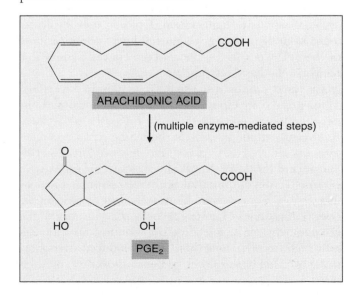

PROCESSES RELATED TO HOMEOSTASIS

A variety of seemingly unrelated processes, such as biological rhythms and aging, have important implications for homeostasis and are discussed here to emphasize this point.

Acclimatization

The term **adaptation** denotes a characteristic that favors survival in specific environments. Homeostatic control systems are inherited biological adaptations. An individual's ability to respond to a particular environmental stress is not fixed, however, but can be enhanced, with no change in genetic endowment, by prolonged exposure to that stress. This type of adaptation—the improved functioning of an already existing homeostatic system—is known as **acclimatization.**

Let us take sweating in response to heat exposure as an example and perform a simple experiment. On day 1 we expose a person for 30 min to a high temperature and ask her to do a standardized exercise test. Body temperature rises, and sweating begins after a certain period of time. The sweating provides a mechanism for increasing heat loss from the body and thus tends to minimize the rise in body temperature in a hot environment. The volume of sweat produced under these conditions is measured. Then, for a week, our subject enters the heat chamber for 1 or 2 h per day and exercises. On day 8, her body temperature and sweating rate are again measured during the same exercise test performed on day 1; the striking finding is that the subject begins to sweat earlier and much more profusely than she did on day 1. Accordingly, her body temperature does not rise to nearly the same degree. The subject has become acclimatized to the heat; that is, she has undergone an adaptive change induced by exposure to the heat and is now better able to respond to heat exposure.

The precise anatomical and physiological changes that bring about increased capacity to withstand change during acclimatization are highly varied. Typically, they involve an increase in the number, size, or sensitivity of one or more of the cell types in the homeostatic control system that mediates the basic response.

Acclimatizations are usually completely reversible. Thus, if the daily exposures to heat are discontinued, the sweating rate of our subject will revert to the preacclimatized value within a relatively short time. If an acclimatization is induced very early in life, however, at the **critical period** for development of a structure or response, it is termed a **developmental acclimatization** and may be irreversible. For example, the barrel-shaped chests of natives of the Andes Mountains represent not a genetic difference between them and their lowland compatriots but rather an irreversible acclimatization induced during the first few years of their lives by their exposure to the low-oxygen environment of high altitude. The altered chest size remains even though the individual moves to a lowland environment later in life and stays there. Lowland persons who have suffered oxygen deprivation from heart or lung disease during their early years show precisely the same chest shape.

Biological Rhythms

A striking characteristic of many body functions is the rhythmical changes they manifest. The most common type is the **circadian rhythm,** which cycles approximately once every 24 h. Waking and sleeping, body temperature, hormone concentrations in the blood, the excretion of ions into the urine, and many other functions undergo circadian variation (Figure 7-8). Other cycles have much longer periods, the menstrual cycle (approximately 28 days) being the most well known.

What have biological rhythms to do with homeostasis? They add yet another "anticipatory" component to homeostatic control systems, in effect a feedforward system oper-

FIGURE 7-8

Circadian rhythms of several physiological variables in a human subject with room lights on (open bars at top) for 16 h and off (black bars at top) for 8 h. As is usual in dealing with rhythms, we have used a 24-h clock in which both 0 and 24 designate midnight and 12 designates noon. Cortisol is a hormone secreted by the adrenal glands. (*Adapted from Moore-Ede and Sulzman*)

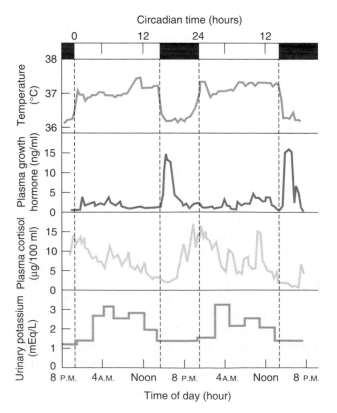

ating without detectors. The negative-feedback homeostatic responses we described earlier in this chapter are *corrective* responses, in that they are initiated *after* the steady state of the individual has been perturbed. In contrast, biological rhythms enable homeostatic mechanisms to be utilized immediately and automatically by activating them at times when a challenge is *likely* to occur but before it actually does occur. For example, there is a rhythm in the urinary excretion of potassium such that excretion is high during the day and low at night; this makes sense since we ingest potassium in our food during the day, not at night when we are asleep. Therefore, the total amount of potassium in the body fluctuates less than if the rhythm did not exist.

A crucial point concerning most body rhythms is that they are *internally* driven. Environmental factors do not drive the rhythm but rather provide the timing cues important for **entrainment** (that is, setting of the actual hours) of the rhythm. A classic experiment will clarify this distinction.

Subjects were put in experimental chambers that completely isolated them from their usual external environment. For the first few days, they were exposed to a 24 h rest-activity cycle in which the room lights were turned on and off at the same time each day. Under these conditions, their sleep-wake cycles were 24 h long. Then, all environmental time cues were eliminated, and the individuals were allowed to control the lights themselves. Immediately, their sleep-wake patterns began to change. On the average, bedtime began about 30 min later *each day* and so did wake-up time. Thus a sleep-wake cycle persisted in the complete absence of environmental cues, and such a rhythm is called a **free-running rhythm.** In this case it was approximately 25 h rather than 24; that is, cues are required to entrain a circadian rhythm to 24 h.

One more point should be mentioned: By altering the duration of the light-dark cycles, sleep-wake cycles can be entrained to any value between 23 and 27 h, but shorter or longer durations cannot be entrained; instead, the rhythm continues to free-run. Because of this, people whose work causes them to adopt sleep-wake cycles longer than 27 h are never able to make the proper adjustments and achieve stable rhythms. The result is symptoms similar to those of jet lag, to be described later.

The light-dark cycle is the most important environmental time cue in our lives but not the only one. Others include external environmental temperature, meal timing, and many social cues. Thus, if several people were undergoing the experiment just described in isolation from each other, their free-runs would be different, but if they were all in the same room, social cues would entrain all of them to the same rhythm.

Environmental time cues also function to **phase-shift** rhythms, in other words, to reset the internal clock. Thus if one jets west or east to a different time zone, the sleep-wake cycle and other circadian rhythms slowly shift to the new light-dark cycle. These shifts take time, however, and the disparity between external time and internal time is one of the causes of the symptoms of **jet lag**—disruption of sleep, gastrointestinal disturbances, decreased vigilance and attention span, and a general feeling of malaise.

Similar symptoms occur in workers on permanent or rotating night shifts. Such individuals generally do not adapt to these schedules even after several years because they are exposed to the usual outdoor light-dark cycle (normal indoor lighting is too dim to function as a good entrainer). In recent experiments, night-shift workers were exposed to extremely bright indoor lighting while they worked and 8 h of total darkness during the day while they slept. This schedule produced total adaptation to the night-shift work within 5 days.

What is the neural basis of body rhythms? In the part of the brain called the hypothalamus is a specific collection of nerve cells (the suprachiasmatic nucleus) that function as the principal **pacemaker** (time clock) for circadian rhythms. How it "keeps time" independent of any external environmental cues is not really understood. The input the pacemaker receives from the eyes and many other parts of the nervous system mediates the entrainment effects exerted by the external environment.

It should not be surprising that rhythms have effects on the body's resistance to various stresses and responses to different drugs. Also, certain diseases have characteristic rhythms. For example, **heart attacks** are almost twice as common in the first hours after waking, and **asthma** frequently flares at night. Insights about these rhythms have already been incorporated into therapy; for example, once-a-day timed release pills for asthma are taken at night and deliver a high dose of medication between midnight and 6 A.M.

Regulated Cell Death: Apoptosis

It is obvious that the proliferation and differentiation of cells are important for the development and maintenance of homeostasis in multicellular organisms. Only recently, however, have physiologists come to appreciate the contribution of another characteristic shared by virtually all cells—the ability to self-destruct by activation of an intrinsic cell suicide program. This type of cell death, termed **apoptosis,** plays important roles in the sculpting of a developing organism and in the elimination of undesirable cells (for example, cells that have become cancerous), but it is particularly crucial for regulating the number of cells in a tissue or organ. Thus, the control of cell number within each cell lineage is normally determined by a balance between cell proliferation and cell death, both of which are regulated processes. For example, white blood cells called neutrophils are programmed to die by apoptosis 24 hours after they are produced in the bone marrow.

Apoptosis occurs by controlled autodigestion of the cell contents. Within a cell, endogenous enzymes are activated that break down the cell nucleus and its DNA, as well as other cell organelles. Importantly, the plasma membrane is maintained as the cell dies so that the cell contents are not dispersed. Instead the apoptotic cell sends out chemical messengers that attract neighboring phagocytic cells (cells that "eat" matter or other cells), which engulf and digest the dying or dead cell. In this way the leakage of breakdown products, many of which are toxic, from apoptotic cells is prevented. Apoptosis is, therefore, very different from the death of a cell due to externally imposed injury; in that case (termed necrosis) the plasma membrane is disrupted, and the cell swells and releases its cytoplasmic material, inducing an inflammatory response, as described in Chapter 20.

The fact that virtually all normal cells contain the enzymes capable of carrying out apoptosis means that these enzymes must normally remain inactive if the cell is to survive. In most tissues this inactivity is maintained by the constant supply to the cell of a large number of chemical "survival signals" provided by neighboring cells, hormones, and the extracellular matrix. In other words, most cells are programmed to commit suicide if survival signals are not received from the internal environment. For example, prostate-gland cells undergo apoptosis when the influence on them of testosterone, the male sex hormone, is removed. In addition, there are other chemical signals, some exogenous to the organism (for example, certain viruses and bacterial toxins) and some endogenous (for example, certain messengers released by nerve cells and white blood cells) that can inhibit or override survival signals and induce the cell to undergo apoptosis.

It is very likely that abnormal inhibition of appropriate apoptosis may contribute to diseases, like cancer, characterized by excessive numbers of cells. At the other end of the spectrum too high a rate of apoptosis probably contributes to degenerative diseases, such as that of bone in the disease called *osteoporosis.* The hope is that therapies designed to enhance or decrease apoptosis, depending on the situation, would ameliorate these diseases.

Aging

The physiological manifestations of aging are a gradual deterioration in the function of virtually all tissues and organ systems and in the capacity of the body's homeostatic control systems to respond to environmental stresses.

Aging represents the operation of a distinct process that is distinguishable from those diseases, such as heart disease, frequently associated with aging. Aging is associated with a decrease in the number of cells in the body, due to some combination of decreased cell division and increased cell death, and to malfunction of many of the cells that remain. The immediate cause of these changes is probably an interference in the function of the cells' macromolecules—DNA, RNA, and cell proteins—and in the flow of information between these macromolecules.

The crucial question, unanswered at present, concerns what causes those changes.

One broad theory places the blame on progressive accumulation of damage to the macromolecules caused by other molecules produced in the normal course of living, including oxygen free radicals. Another hypothesis is that cellular senescence is actually programmed in our genes, and the genes responsible for the aging process are activated or inhibited in a predetermined manner as one grows older. That genes can be critically involved in aging is demonstrated by the fact that a rare inherited disease (*Werner's syndrome*), which is characterized by premature aging, is caused by mutation of a single gene. The protein encoded by this gene is an enzyme that unwinds paired DNA strands as a prelude to repair, replication, or expression of genes.

It is difficult to sort out the extent to which any particular age-related change in physiological function is due to aging itself and the extent to which it is secondary to disease and lifestyle changes. For example, until recently it was believed that the functioning of the nervous system markedly deteriorates as a result of aging per se, but this view is incorrect. It was based on studies of the performance of individuals with age-related diseases. Studies of people without such diseases do document changes, including loss of memory, increased difficulty in learning new tasks, decrease in the speed of processing by the brain, and loss of brain mass, but these changes are modest. Most brain functions considered to underlie intelligence seem to remain relatively intact.

In contrast, the 30- to 40-percent decrease (somewhat less in women) in the mass and strength of limb muscles that occurs in men between 30 and 80 years of age is due to aging changes per se. Physically active individuals have greater strength at any given age, compared to inactive persons, but the rate of decline with age is similar.

Can the aging process be inhibited or at least slowed down? One factor—physical exercise—has been shown to prolong life in human beings although it is not clear that it does so by altering the aging process itself. In various chapters we shall mention the beneficial effects of exercise in various diseases (coronary heart disease and diabetes mellitus, for example), and the explanations for these benefits are usually logical in terms of the pathophysiology of the diseases. But the remarkable finding has been that persons who participate in aerobic sports activities of moderate intensity have a significantly lower risk of dying *from any cause.* Moreover, such persons are much less likely to develop life-disturbing disabilities. It is this nonspecific aspect of exercise's benefits that raises the question as to whether being physically fit somehow alters the aging process itself rather than simply the diseases associated with aging.

A second approach—marked restriction of calories, but with enough protein, fat, vitamins, and minerals provided to prevent malnutrition—is the only intervention that has consistently been shown to prolong life in experimental animals, mainly rodents. This type of caloric restriction increases not only the *average* life span of the animals but also the maximum span—that is, the lifetime of the longest-surviving members of the group. It is this increase in maximum life span that indicates that caloric restriction is influencing the basic aging process, not simply postponing the major diseases that are common late in life (caloric restriction does that, too). How it delays aging and the relevance of these findings, if any, to human beings are still unclear.

Balance in the Homeostasis of Chemicals

Many homeostatic systems are concerned with the balance between the addition to and removal from the body of a chemical substance. Figure 7-9 is a generalized schema of the possible pathways involved in such balance. The **pool** occupies a position of central importance in the balance sheet. It is the body's readily available quantity of the particular substance and is frequently identical to the amount present in the extracellular fluid. The pool receives substances from and contributes them to all the pathways.

The pathways on the left of the figure are sources of net gain to the body. A substance may enter the body through the gastrointestinal (GI) tract or the lungs. Alternatively, a substance may be synthesized within the body from other materials.

The pathways on the right of the figure are causes of net loss from the body. A substance may be lost in the urine, feces, expired air, or menstrual fluid, as well as from the surface of the body as skin, hair, nails, sweat, and tears. The substance may also be chemically altered and thus removed by metabolism.

The central portion of the figure illustrates the distribution of the substance within the body. The substance may be taken from the pool and accumulated in storage depots, for example, the accumulation of fat in adipose tissue. Conversely, it may leave the storage depots to reenter the pool. Finally, the substance may be incorporated reversibly into some other molecular structure, such as fatty acids into membranes or iodine into thyroxine. Incorporation is reversible in that the substance is liberated again whenever the more complex structure is broken down. This pathway is distinguished from storage in that the incorporation of the substance into other molecules produces new molecules with specific functions.

It should be recognized that not every pathway of this generalized schema is applicable to every substance. For example, mineral electrolytes such as sodium cannot be synthesized, do not normally enter through the lungs, and cannot be removed by metabolism.

The orientation of Figure 7-9 illustrates two important generalizations concerning the balance concept: (1) Total-body balance depends upon the relative rates of net gain and net loss to the body; and (2) the pool concentration depends not only upon total body balance but also upon exchanges of the substance within the body.

For any chemical, three states of total-body balance are possible: (1) Loss exceeds gain, so that the total amount of the substance in the body is decreasing and the person is said to be in **negative balance;** (2) gain exceeds loss, so that the total amount of the substance in the body is increasing and the person is said to be in **positive balance;** and (3) gain equals loss, and the person is in **stable balance.**

Clearly a stable balance can be upset by alteration of the amount being gained or lost in any single pathway in the schema; for example, severe negative water balance can be caused by increased sweating. Conversely, stable balance can be restored by homeostatic control of water intake and output.

Let us take sodium balance as another example. The control systems for sodium balance have as their targets the kidneys, and the systems operate by inducing the kidneys to excrete into the urine an amount of sodium approximately equal to the amount ingested daily. In this

FIGURE 7-9

Balance diagram for a chemical substance.

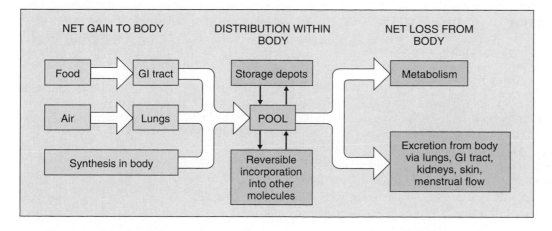

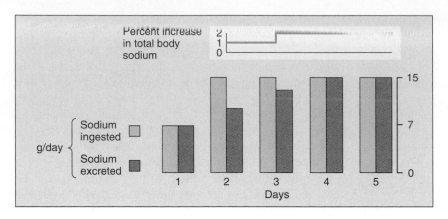

FIGURE 7-10

Effects of a continued change in the amount of sodium ingested on sodium excretion and total-body sodium balance. Stable sodium balance is reattained by day 4, but with some gain of total-body sodium.

example, we assume for simplicity that all sodium loss from the body occurs via the urine. Now imagine a person with a daily intake and excretion of 7 g of sodium—a moderate intake for most Americans—and a stable amount of sodium in her body (Figure 7-10). On day 2 of our experiment, the subject changes her diet so that her daily sodium consumption rises to 15 g—a fairly large but commonly observed intake—*and remains there indefinitely.* On this same day, the kidneys excrete into the urine somewhat more than 7 g of sodium, but not all the ingested 15 g. The result is that some excess sodium is retained in the body on that day, that is, the person is in positive sodium balance. The kidneys do somewhat better on day 3, but it is probably not until day 4 or 5 that they are excreting 15 g. From this time on, output from the body once again equals input and sodium balance is once again stable. (The delay of several days before stability is reached is quite typical for the kidneys' handling of sodium, but should not

be assumed to apply to other homeostatic responses, most of which are much more rapid.)

But, and this is an important point, although again in stable balance, the woman has perhaps 2 percent more sodium in her body than was the case when she was in stable balance ingesting 7 g. It is this 2 percent extra body sodium that constitutes the continuous error signal to the control systems driving the kidneys to excrete 15 g/day rather than 7 g/day. [Recall the generalization (Table 7-1, no. 3) that homeostatic control systems cannot maintain complete constancy of the internal environment *in the face of continued change in the perturbing event* since some change in the regulated variable (body sodium content in our example) must persist to serve as a signal to maintain the compensating responses.] An increase of 2 percent does not seem large, but it has been hypothesized that this small gain might facilitate the development of high blood pressure (**hypertension**) in some persons.

SECTION A SUMMARY

GENERAL CHARACTERISTICS OF HOMEOSTATIC CONTROL SYSTEMS

I.   Homeostasis denotes the stable conditions of the internal environment that result from the operation of compensatory homeostatic control systems.

   A. In a negative-feedback control system, a change in the variable being regulated brings about responses that tend to push the variable in the direction opposite to the original change. Negative feedback minimizes changes from the operating point of the system, leading to stability.

   B. In a positive-feedback system, an initial disturbance

in the system sets off a train of events that increases the disturbance even further.

   C. Homeostatic control systems minimize changes in the internal environment but cannot maintain complete constancy.

   D. Feedforward regulation anticipates changes in a regulated variable, improves the speed of the body's homeostatic responses, and minimizes fluctuations in the level of the variable being regulated.

COMPONENTS OF HOMEOSTATIC CONTROL SYSTEMS

I.   The components of a reflex arc are receptor, afferent pathway, integrating center, efferent pathway, and effector. The pathways may be neural or hormonal.

II. Local homeostatic responses are also stimulus-response sequences, but they occur only in the area of the stimulus, neither nerves nor hormones being involved.

III. Intercellular communication is essential to reflexes and local responses and is achieved by neurotransmitters, hormones, and paracrine agents. Less common is intercellular communication through either gap junctions or cell-bound messengers.

IV. The eicosanoids are a widespread family of messenger molecules derived from arachidonic acid. They function mainly as paracrine and autocrine agents in local responses.
  A. The first step in production of the eicosanoids is the splitting-off of arachidonic acid from plasma membrane phospholipids by the action of phospholipase A_2.
  B. There are two pathways from arachidonic acid, one mediated by cyclooxygenase and leading to the formation of prostaglandins and thromboxanes, and the other mediated by lipoxygenase and leading to the formation of leukotrienes.

PROCESSES RELATED TO HOMEOSTASIS

I. Acclimatization is an improved ability to respond to an environmental stress.
  A. The improvement is induced by prolonged exposure to the stress with no change in genetic endowment.
  B. If acclimatization occurs early in life, it may be irreversible and is known as a developmental acclimatization.

II. Biological rhythms provide a feedforward component to homeostatic control systems.
  A. The rhythms are internally driven by brain pacemakers, but are entrained by environmental cues, such as light, which also serve to phase-shift (reset) the rhythms when necessary.
  B. In the absence of cues, rhythms free-run.

III. Apoptosis, regulated cell death, plays an important role in homeostasis by helping to regulate cell numbers and eliminating undesirable cells.

IV. Aging is associated with a decrease in the number of cells in the body and with a disordered functioning of many of the cells that remain.
  A. It is a process distinct from the diseases associated with aging.
  B. Its physiological manifestations are a deterioration in organ-system function and in the capacity to respond homeostatically to environmental stresses.

V. The balance of substances in the body is achieved by a matching of inputs and outputs. Total body balance of a substance may be negative, positive, or stable.

SECTION A KEY TERMS

homeostasis	steady state
homeostatic control system	operating point
	negative feedback
positive feedback	cyclic endoperoxide
error signal	prostaglandin
feedforward	thromboxane
reflex	leukotriene
learned reflex	phospholipase A_2
acquired reflex	cyclooxygenase
reflex arc	lipoxygenase
stimulus	adaptation
receptor (in reflex)	acclimatization
afferent pathway	critical period
integrating center	developmental acclimatization
effector	
efferent pathway	circadian rhythm
hormone	entrainment
endocrine gland	free-running rhythm
local homeostatic response	phase-shift
	pacemaker
target cell	apoptosis
neurotransmitter	pool
paracrine agent	negative balance
autocrine agent	positive balance
eicosanoid	stable balance
arachidonic acid	

SECTION A REVIEW QUESTIONS

1. Describe five important generalizations about homeostatic control systems.

2. Contrast negative-feedback systems and positive-feedback systems.

3. Contrast feedforward and negative feedback.

4. How do error signals develop, and why are they essential for maintaining homeostasis?

5. List the components of a reflex arc.

6. What is the basic difference between a local homeostatic response and a reflex?

7. List the general categories of intercellular messengers.

8. Describe two types of intercellular communication that do not depend on extracellular chemical messengers.

9. Draw a figure illustrating the various pathways for eicosanoid synthesis.

10. Describe the conditions under which acclimatization occurs. In what period of life might an acclimatization be irreversible? Are acclimatizations passed on to a person's offspring?

11. Under what conditions do circadian rhythms become free-running?

12. How do phase shifts occur?

13. What are the important environmental cues for entrainment of body rhythms?

14. What are the physiological manifestations of aging?

15. Draw a figure illustrating the balance concept in homeostasis.

16. What are the three possible states of total-body balance of any chemical?

SECTION B
MECHANISMS BY WHICH CHEMICAL MESSENGERS CONTROL CELLS

RECEPTORS

Most homeostatic systems require cell-to-cell communication via chemical messengers. The first step in the action of any intercellular chemical messenger is the binding of the messenger to specific target-cell proteins known as **receptors.** In the general language of Chapter 4, the chemical messenger is a "ligand" and the receptor is a "binding site." The term "receptor" can be the source of confusion because the same word is used to denote the "detectors" in a reflex arc, as described earlier in this chapter. The reader must keep in mind the fact that "receptor" has two totally distinct meanings, but the context in which the term is used usually makes it quite clear which is meant.

What is the nature of the receptors with which intercellular chemical messengers combine? They are proteins (or glycoproteins) located either in the cell's plasma membrane or inside the cell, mainly in the nucleus. The plasma membrane is the much more common location, applying to the very large number of messengers that are lipid-insoluble and so do not traverse the lipid-rich plasma membrane. In contrast, the much smaller number of lipid-soluble messengers readily pass through membranes to bind to their receptors inside the cell.

Plasma-membrane receptors are transmembrane proteins, that is, they span the entire membrane thickness. A typical plasma-membrane receptor is illustrated in Figure 7-11. Like other transmembrane proteins, a plasma-membrane receptor has segments within the membrane, one or more segments extending out from the membrane into the extracel-

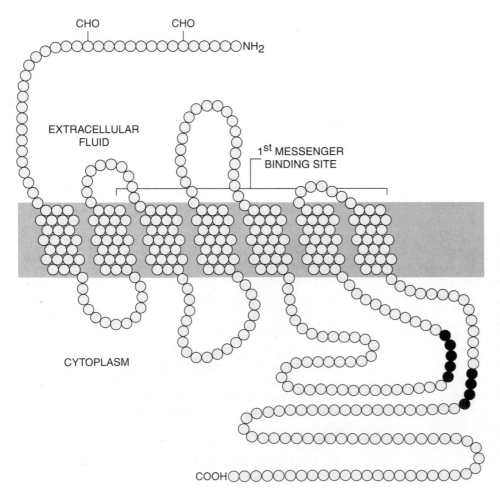

FIGURE 7-11

Structure of a human receptor, the beta$_2$-adrenergic receptor, which binds catecholamines. The seven clusters of amino acids in the plasma membrane represent hydrophobic portions of the protein's alpha helix. Note that the binding site for the first messenger includes several of the segments that extend into the extracellular fluid. The amino acids denoted by black circles represent sites at which the receptor can be phosphorylated, and thereby regulated, by cytoplasmic substances. (*Adapted from Dohlman et al.*)

TABLE 7-3 A GLOSSARY OF TERMS CONCERNING RECEPTORS

Receptor	A specific protein in either the plasma membrane or interior of a target cell with which a chemical messenger combines.
Specificity	The ability of a receptor to bind only one type or a limited number of structurally related types of chemical messengers.
Saturation	The degree to which receptors are occupied by a messenger. If all are occupied, the receptors are fully saturated; if half are occupied, the saturation is 50 percent, and so on.
Affinity	The strength with which a chemical messenger binds to its receptor.
Competition	The ability of different molecules very similar in structure to combine with the same receptor.
Antagonist	A molecule that competes for a receptor with a chemical messenger normally present in the body. The antagonist binds to the receptor but does not trigger the cell's response.
Agonist	A chemical messenger that binds to a receptor and triggers the cell's response; often refers to a drug that mimics a normal messenger's action.
Down-regulation	A decrease in the total number of target-cell receptors for a given messenger in response to chronic high extracellular concentration of the messenger.
Up-regulation	An increase in the total number of target-cell receptors for a given messenger in response to a chronic low extracellular concentration of the messenger.
Supersensitivity	The increased responsiveness of a target cell to a given messenger, resulting from up-regulation.

lular fluid, and other segments extending into the cytosol. It is to the extracellular portions that the messenger binds. Like other transmembrane proteins, a receptor is often composed of two or more nonidentical subunits bound together. Plasma-membrane receptors may be distributed over the entire surface of the cell or be confined to a particular region; the latter cases permit localization of the receptor's actions to that region of the cell.

It is the combination of chemical messenger and receptor that initiates the events leading to the cell's response. The existence of receptors explains a very important characteristic of intercellular communication— **specificity** (see Table 7-3 for a glossary of terms concerning receptors). Although a chemical messenger (hormone, neurotransmitter, paracrine/autocrine agent, or plasma-membrane-bound messenger) may come into contact with many different cells, it influences only certain cells and not others. The explanation is that cells differ in the types of receptors they contain. Accordingly,

only certain cell types, frequently just one, possess the receptor required for combination with a given chemical messenger (Figure 7-12). (In many cases the receptors for a group of messengers are closely related structurally; thus, for example, endocrinologists refer to "superfamilies" of hormone receptors.)

Where different types of cells possess the same receptors for a particular messenger, the responses of the various cell types to that messenger may differ from each other. For example, the neurotransmitter norepinephrine causes the smooth muscle of blood vessels to contract, but via the same type of receptor, norepinephrine causes cells in the pancreas to secrete less insulin. In essence, then, the receptor functions as a molecular "switch" that elicits the cell's response when "switched on" by the messenger binding to it. Just as identical types of switches can be used to turn on a light or a radio, a single type of receptor can be used to produce quite different responses in different cell types.

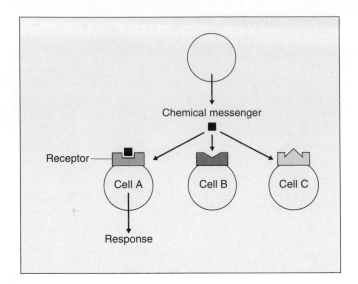

FIGURE 7-12
Specificity of receptors for chemical messengers.

Similar reasoning explains perhaps a more surprising phenomenon: A single cell may contain several different receptor types *for a single messenger*. Combination of the messenger with one of these receptor types may produce a cellular response quite different from, indeed sometimes opposite to, that produced when the messenger combines with the other receptors. For example, as we shall see in Chapter 14, there are several distinct types of receptors for the hormone epinephrine in the smooth muscle of certain blood vessels, and this hormone can cause either contraction or relaxation of the muscle depending on the relative degrees of binding to two different types. The degree to which the molecules of a particular messenger bind to different receptor types in a single cell is determined by the affinity of the different receptor types for the messenger.

It should not be inferred from these descriptions that a cell has receptors for only one messenger. In fact, a single cell usually contains many different receptors for different chemical messengers.

Other characteristics of messenger-receptor interactions are **saturation** and **competition.** These phenomena were described in Chapter 4 for ligands binding to binding sites on proteins and are fully applicable here. In most systems, a cell's response to a messenger increases as the extracellular concentration of messenger increases, because the number of receptors occupied by messenger molecules increases. There is an upper limit to this responsiveness, however, because only a finite number of receptors are available, and they become saturated at some point.

Competition is the ability of different messenger molecules that are very similar in structure to compete with each other for a receptor. Competition occurs physiologi-

cally with closely related messengers, and it also underlies the action of many drugs. If researchers or physicians wish to interfere with the action of a particular messenger, they can administer competing molecules, if available, that bind to the receptors for that messenger without activating them. This prevents the messenger from binding and does not trigger the cell's response. Such drugs are known as **antagonists** with regard to the usual chemical messenger. For example, so called "beta-blockers," used in the treatment of high blood pressure and other diseases, are drugs that antagonize the ability of epinephrine and norepinephrine to bind to one of their receptors—the beta-adrenergic receptor (Chapter 8). On the other hand, some drugs that bind to a particular receptor type do trigger the cell's response exactly as if the true chemical messenger had combined with the receptor; such drugs are known as **agonists** and are used to mimic the messenger's action. For example, the decongestant drug ephedrine mimics the action of epinephrine.

Regulation of Receptors

Receptors are themselves subject to physiological regulation. The number of receptors a cell has and the affinity of the receptors for their specific messenger can be increased or decreased, at least in certain systems. An important example of such regulation is the phenomenon of **down-regulation.** When a high extracellular concentration of a messenger is maintained for some time, the total number of the target-cell's receptors for that messenger may decrease, that is, down-regulate. Down-regulation has the effect of reducing the target cells' responsiveness to frequent or intense stimulation by a messenger and thus represents a local negative-feedback mechanism. For example, a prolonged high concentration of the hormone insulin, which stimulates glucose uptake by its target cells, causes down-regulation of its receptors, and this acts to dampen the ability of insulin to stimulate glucose uptake.

Change in the opposite direction (**up-regulation**) also occurs. Cells exposed for a prolonged period to very low concentrations of a messenger may come to have many more receptors for that messenger, thereby developing increased sensitivity to it. For example, days after the nerves to a muscle are cut, thereby eliminating the neurotransmitter released by those nerves, the muscle will contract in response to amounts of experimentally injected neurotransmitter much smaller than those to which an innervated muscle can respond.

Up-regulation and down-regulation are made possible because there is a continuous degradation and synthesis of receptors. The main cause of down-regulation of plasma-membrane receptors is as follows: The binding of a messenger to its receptor stimulates the internalization of the complex, that is, the messenger-receptor complex is taken into the cell by endocytosis (an example of so-called receptor-mediated endocytosis); this increases the rate of receptor degradation inside the cell. Thus, at high hormone concentrations, the number of plasma-membrane receptors of that type gradually decreases.

The opposite events also occur: The cell may contain stores of receptors bound in intracellular vesicles, and these are available for insertion into the membrane via exocytosis (Chapter 6).

Down-regulation and up-regulation are physiological responses, but there also are many disease processes in which the number of receptors or their affinity for messenger becomes abnormal. The result is unusually large or small responses to any given level of messenger. For example, the disease called **myasthenia gravis** is due to destruction of the skeletal muscle receptors for acetylcholine, the neurotransmitter that normally causes contraction of the muscle in response to nerve stimulation; the result is muscle weakness or paralysis.

SIGNAL TRANSDUCTION PATHWAYS

What are the mechanisms by which the binding of a chemical messenger (hormone, neurotransmitter, or paracrine/autocrine agent) to a receptor causes the cell to respond to the messenger.

The combination of messenger with receptor causes a change in the conformation of the receptor. This event, known as **receptor activation,** is always the initial step leading to the cell's ultimate responses to the messenger, which can take the form of changes in: (1) the permeability, transport properties, or electrical state of the cell's plasma membrane; (2) the cell's metabolism; (3) the cell's secretory activity; (4) the cell's rate of proliferation and differentiation; and (5) the cell's contractile activity.

Despite the seeming variety of these ultimate responses, there is a common denominator: They are all due directly to alterations of particular cell proteins. Let us take a few examples of messenger-induced responses, all of which are described fully in subsequent chapters. Generation of electric signals in nerve cells reflects the altered conformation of membrane proteins constituting ion channels through which ions can diffuse between extracellular fluid and intracellular fluid. Changes in the rate of glucose secretion by the liver reflect the altered activity and concentration of enzymes in the metabolic pathways for glucose synthesis. Muscle contraction results from the altered conformation of contractile proteins.

To repeat, receptor activation by a messenger is only the first step leading to the cell's ultimate response (contraction, secretion, and so on). The sequences of events, however, between receptor activation and the responses may be very complicated and are termed **signal trans-**

duction pathways. The "signal" is the receptor activation, and "transduction" denotes the process by which a stimulus is transformed into a response. The important question is: *How does receptor activation influence the cell's internal proteins, which are usually critical for the response but may be located far from the receptor?*

Signal transduction pathways differ at the very outset for lipid-soluble and lipid-insoluble messengers since, as described earlier, the receptors for these two broad chemical classes of messenger are in different locations—the former inside the cell and the latter in the plasma membrane of the cell. The rest of this chapter elucidates the general principles of the signal transduction pathways initiated by receptors.

Pathways Initiated by Intracellular Receptors

Most lipid-soluble messengers are hormones (to be described in Chapter 10)—steroid hormones, the thyroid hormones, and the steroid derivative 1,25-dihydroxyvitamin D_3. Structurally these hormones are all closely related, and their receptors constitute the steroid-hormone receptor "superfamily." The receptors are intracellular and are inactive when no messenger is bound to them; for certain lipid-soluble messengers the inactive receptors are in the cytosol, for others in the nucleus. In all cases, receptor activation leads to altered rates of gene transcription, the sequence of events being as follows.

The messenger diffuses across the cell's plasma membrane and enters the cytosol (Figure 7-13). Where the receptor is in the cytosol the messenger combines with it there, and the hormone/receptor complex then moves into the nucleus. With intranuclear receptors, the hormone diffuses by itself into the nucleus and binds to the receptor there. In all cases, the receptor, activated by the binding of hormone to it, then functions in the nucleus as a **transcription factor,** defined in Chapter 5 as any regulatory protein that directly influences gene transcription. The receptor binds to a specific sequence near a gene in DNA, termed a response element, and increases the rate of that gene's transcription into mRNA. The mRNA molecules formed enter the cytosol and direct the synthesis, on ribosomes, of the protein encoded by the gene. The result is an increase in the cellular concentration of the protein or its rate of secretion, and this accounts for the cell's ultimate response to the messenger. For example, if the protein encoded by the gene is an enzyme, the cell's response is an increase in the rate of the reaction catalyzed by that enzyme.

Two other points should be mentioned. First, more than one gene may be subject to control by a single receptor type, and second, in some cases the transcription of the gene(s) is *decreased* by the activated receptor rather than increased.

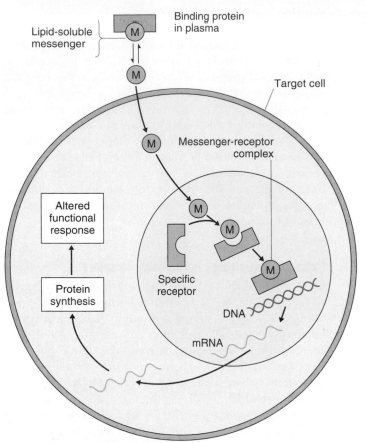

FIGURE 7-13

Mechanism of action of lipid-soluble messengers. This figure shows the receptor for these messengers as being in the nucleus. The receptor may also be in the cytosol, in which case the binding occurs there, and the messenger/receptor complex moves into the nucleus.

Pathways Initiated by Plasma-Membrane Receptors

On the basis of the signal transduction pathways they initiate, plasma-membrane receptors can be classified into the types listed in Table 7-4 and illustrated in Figure 7-14.

Three notes on general terminology are essential for this discussion. First, the intercellular chemical messengers (hormones, neurotransmitters, and paracrine/autocrine agents), which reach the cell from the extracellular fluid and bind to their specific receptors, are often referred to as **first messengers.** This usage applies to lipid-soluble messengers also. Next, substances that enter the cytoplasm or are enzymatically generated there as a result of plasma-membrane receptor activation and, in turn, act within the cell to trigger a response are termed **second messengers.** In essence, second messengers serve as chemical relays from the plasma membrane to the biochemical machinery inside the cell.

The third essential general term is **protein kinase.** As described in Chapter 4, protein kinase is the name for *any* enzyme that phosphorylates other proteins by transferring to them a phosphate group from ATP. Introduction of the phosphate group changes the conformation and/or activity of the protein, often itself an enzyme. There are many distinct protein kinases, each type being able to phosphorylate only certain proteins. The important point is that a variety of protein kinases are involved in signal transduction pathways. These pathways may involve long and complex series of reactions in which a particular inactive protein kinase is activated by phosphorylation and then catalyses the phosphorylation of another inactive protein kinase, and so on. At the ends of these sequences, the ultimate phosphorylation of key proteins (transporters, metabolic enzymes, ion channels, contractile proteins, and so on) underlies the cell's response to the original first messenger.

It should at least be mentioned that, as described in Chapter 4, protein phosphatases do the reverse of protein kinases; these enzymes also participate in signal transduction pathways, but their roles are much less well understood than those of the protein kinases and will not be described here.

Receptors Functioning as Ion Channels. The first type of plasma-membrane receptor listed in Table 7-4 (Figure 7-14A) is the simplest pathway since no molecules other than the receptor are involved. The protein that acts as the receptor itself constitutes an ion channel, and activation of the receptor by a first messenger causes the channel to open. The opening results in an increase in the net diffusion across the plasma membrane of the ion or ions specific to the channel. As we shall see in Chapter 8,

TABLE 7-4 CLASSIFICATION OF RECEPTORS BASED ON THEIR LOCATIONS AND THE SIGNAL TRANSDUCTION PATHWAYS THEY USE

1. INTRACELLULAR RECEPTORS (for lipid-soluble messengers)

   Function in the nucleus as transcription factors to alter the rate of transcription of particular genes.

2. PLASMA-MEMBRANE RECEPTORS (for lipid-insoluble messengers)

   a. Receptors that themselves function as ion channels.

   b. Receptors that themselves function as enzymes or are closely associated with cytoplasmic enzymes.

   c. Receptors that activate G proteins, which in turn act upon effector proteins—either ion channels or enzymes—in the plasma membrane.

FIGURE 7-14

Mechanisms of action of lipid-insoluble messengers (noted as "first messengers" in this and subsequent figures). (A) Signal transduction mechanism in which the receptor complex itself contains an ion channel; (B) Signal transduction mechanism in which the receptor itself functions as an enzyme, usually a protein kinase; (C) Signal transduction mechanism involving G proteins.

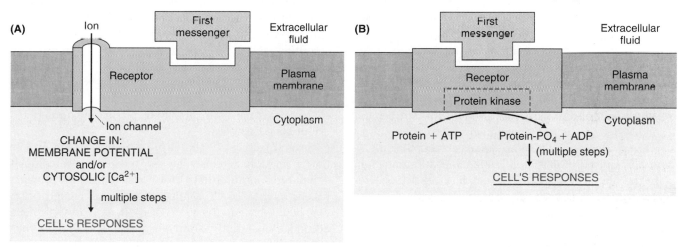

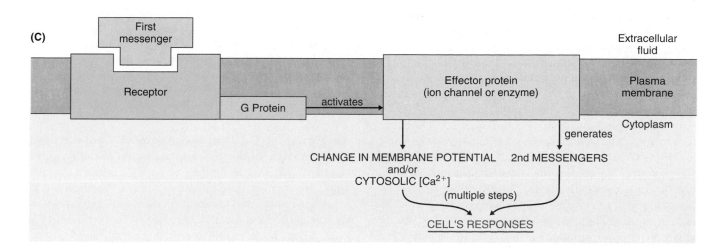

such a change in ion diffusion is usually associated with a change in the membrane potential, and this electric signal may be the essential event in the cell's response to the messenger. In addition, as described later in this chapter, when the channel is a calcium channel, its opening results in an increase, by diffusion, in the cytosolic calcium concentration, another essential event in the signal transduction pathway for many receptors.

Receptors Functioning as Enzymes. The second category of plasma-membrane receptor listed in Table 7-4 (Figure 7-14B), like the first, has the receptor playing a dual role, but in this case the receptor itself functions as an enzyme rather than an ion channel. The sequence of events is as follows: A first messenger binds to the receptor, and this changes the conformation of the receptor so that its enzyme portion, located on the cytoplasmic side of the plasma membrane, is activated. The activated enzyme then catalyses a cytosolic chemical reaction, the products of which carry on the signal transduction pathway.

Most of the receptors that have enzymatic activity function as protein kinases. When activated, such a receptor phosphorylates specific cytosolic and plasma-membrane proteins, including itself. Phosphorylation causes all these proteins to alter their activity, and this brings about the response of the cell. Most of the receptors that function as protein kinases phosphorylate specifically the tyrosine portions of proteins and so are termed tyrosine kinases. (Remember that "tyrosine kinase," like "protein kinase," is not the name of any particular protein but is a generic term that defines the function of many different enzymes.) The messengers that use tyrosine-kinase receptors are almost all involved in cell proliferation and differentiation.

In the above description, we emphasized that the receptor functions as both receptor and enzyme. In some cases, however, the enzymatic activity resides not in the receptor but in a separate cytoplasmic protein that is closely associated with the receptor. In such cases the activated receptor interacts with the adjacent inactive enzyme and activates it. The result is the same as if the receptor itself possessed the enzymatic activity.

Receptors Interacting with G Proteins. The third category of plasma-membrane receptors in Table 7-4 (Figure 7-14C) is by far the largest, including hundreds of distinct receptors. Such a receptor is normally bound to one or more proteins located on the inner (cytosolic) surface of the plasma membrane and belonging to the family known as **G proteins.** (The term "G protein" includes all the proteins, regardless of location, that share a particular chemical characteristic, to be described below, and the plasma-membrane G proteins encompass only a subset of all G proteins.) The binding of messenger to the receptor changes the conformation of the receptor. This change causes a subunit of the adjacent G protein to dissociate from the complex and diffuse along the inner surface of the plasma membrane to link up with still other plasma-membrane proteins, called **effector proteins.** These membrane effector proteins are either ion channels or enzymes, and they mediate the next steps in the sequences of events leading to the cell's response.

In essence, then, the G protein serves as a switch to "couple" a receptor to an ion channel or an enzyme in the plasma membrane. The G protein may cause the channel to open or close, or it may activate or inhibit the membrane enzyme. The enzymes acted upon by G proteins exert multiple actions, but there is a common denominator to these actions: The enzymes cause the generation, inside the cell, of second messengers, which serve as a relay from the plasma membrane—where the receptor, G protein, enzyme sequence has occurred—to the biochemical machinery inside the cell.

Because of its wide applicability, a word about the biochemical mechanism by which G proteins are activated ("switched on") by receptors is appropriate. In the inactive state G proteins are bound to guanosine diphosphate (GDP). The activated receptor causes the G protein to give up the GDP and bind instead guanosine triphosphate (GTP); it is this binding of GTP that activates the G protein, allowing it to interact with an effector protein. The activation is very brief because activated G protein also functions as an enzyme that splits off a phosphate from the GTP, thereby returning to its GDP-bound inactive state. This property of binding GDP and GTP is what characterizes the entire family of plasma-membrane and cytoplasmic G proteins.

Several features of the G-protein system help explain how a single first messenger can initiate a very complex cellular response. First, there are at least 20 distinct plasma-membrane G proteins, and a single receptor type may be associated with more than one type of G protein. Second, each of these G proteins may couple to more than one type of the many plasma-membrane effector proteins. Thus, a messenger-activated receptor, via its G-protein couplings, can call into action a variety of effector proteins, which in turn induce a variety of cellular events.

To illustrate some of the major points concerning G proteins, plasma-membrane effector proteins, second messengers, and protein kinases, the single most important pathway utilizing such components is described in the next section.

An Example: G$_S$ Protein, Adenylyl Cyclase, and Cyclic AMP. In this pathway, activation of the receptor (Figure 7-15) by the binding of first messenger (for example, the hormone epinephrine) allows the receptor to activate its associated G protein, in this example known as G$_S$ (the subscript s denotes "stimulatory"). This causes G$_S$ to activate its effector protein, the membrane enzyme called **adenylyl cyclase.** The activated adenylyl cyclase,

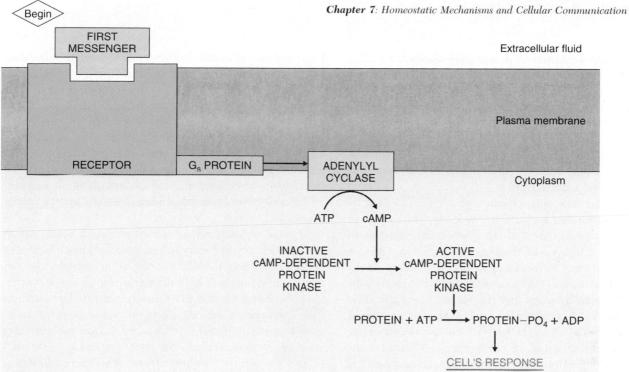

FIGURE 7-15

Cyclic AMP second-messenger system. Not shown in the figure is the existence of another regulatory protein, G_i, with which certain receptors can react to cause inhibition of adenylyl cyclase.

whose catalytic site is located on the cytosolic surface of the plasma membrane, then catalyzes the conversion of some cytosolic ATP molecules to cyclic 3′,5′-adenosine monophosphate, called simply **cyclic AMP (cAMP)** (Figure 7-16). Cyclic AMP then acts as a second messenger. It diffuses throughout the cell to trigger the sequences of events leading to the cell's ultimate response to the first messenger. The action of cAMP is eventually terminated by its breakdown to noncyclic AMP, a reaction catalyzed by the enzyme **phosphodiesterase** (Figure 7-16). This enzyme is also subject to physiological control (and influence by drugs such as caffeine) so that the cellular concentration of cAMP can be changed either by altering the rate of its messenger-mediated generation or the rate of its phosphodiesterase-mediated breakdown.

What does cAMP actually do inside the cell? It binds to and activates an enzyme known as **cAMP-dependent protein kinase** (also termed protein kinase A). As emphasized above, protein kinases phosphorylate other proteins—often enzymes—by transferring a phosphate group to them. The change in the activity of the proteins owing

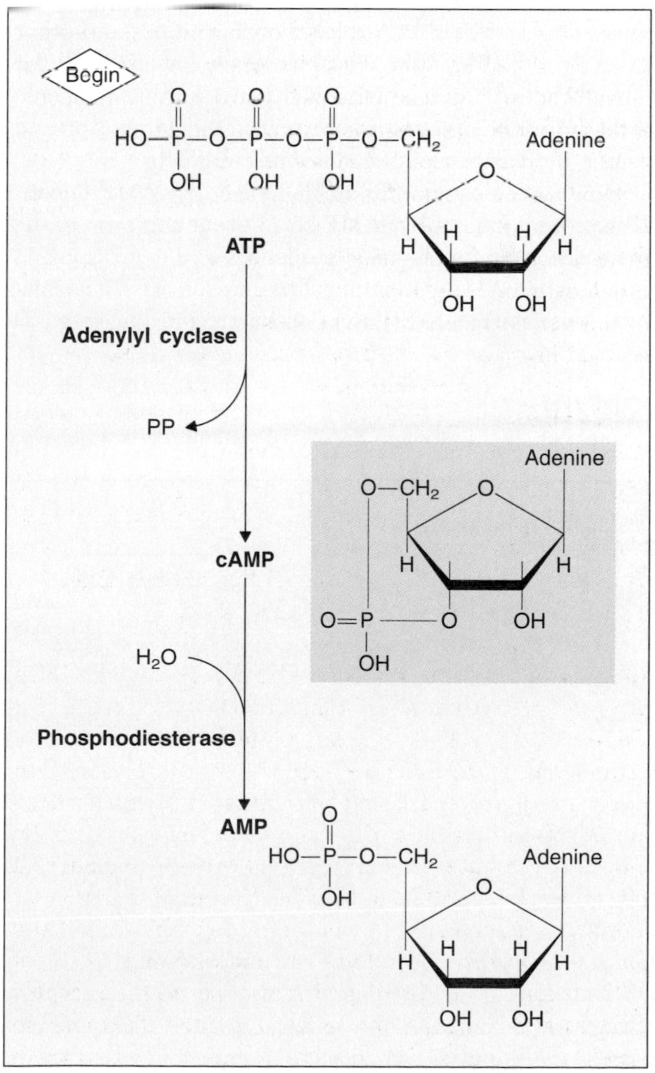

FIGURE 7-16

Structure of ATP, cAMP, and AMP, the last resulting from enzymatic alteration of cAMP.

to their phosphorylation by cAMP-dependent protein kinase brings about the response of the cell (secretion, contraction, and so on). The cAMP-dependent protein kinase is completely distinct from the membrane-receptor protein kinases mentioned earlier and from all the other protein kinases used in other signal transduction pathways. Again we emphasize that each of the various protein kinases that participate in signal transduction pathways has its own specific substrates.

In essence, then, the activation of adenylyl cyclase by G_S protein initiates a chain, or "cascade," of events in which proteins are converted in sequence from inactive to active forms. Figure 7-17 illustrates the benefit of such a cascade. While it is active, a single enzyme molecule is capable of transforming into product not one but many substrate molecules, let us say 100. Therefore, one active molecule of adenylyl cyclase may catalyze the generation of 100 cAMP molecules. At each of the two subsequent enzyme-activation steps in our example, another hundredfold amplification occurs. Therefore, the end result is that a single molecule of the first messenger could, in this example, cause the generation of 1 million product molecules. This fact helps to explain how hormones and other messengers can be effective at extremely low extracellular concentrations. To take an actual example, one molecule of the hormone epinephrine can cause the generation and release by the liver of 10^8 molecules of glucose.

How can activation of a single molecule, cAMP-dependent protein kinase, by cAMP be an event common to the great variety of biochemical sequences and cell responses initiated by cAMP-generating first messengers? The major answer is that cAMP-dependent protein kinase has a

FIGURE 7-17

Example of amplification in the cAMP system.

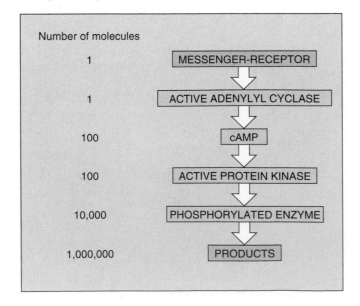

large number of distinct substrates—it can phosphorylate a large number of different proteins (Figure 7-18). In addition, some of this variety may be accounted for by cell-specific variants of cAMP-dependent protein kinase. These features allow activated cAMP-dependent protein kinase to have multiple actions within a single cell and different actions in different cells. For example, epinephrine acts via the cAMP pathway on fat cells to cause both glycogen breakdown (mediated by one phosphorylated enzyme) and triacylglycerol breakdown (mediated by another phosphorylated enzyme).

It must be emphasized that whereas phosphorylation mediated by cAMP-dependent protein kinase activates certain enzymes, it inhibits others. For example, the enzyme catalyzing the rate-limiting step in glycogen synthesis is inhibited by phosphorylation and this fact explains how epinephrine inhibits glycogen synthesis at the same time that it stimulates glycogen breakdown by activating the enzyme that catalyzes the latter response.

In summary, the biochemistry of cAMP-dependent protein kinase and its substrates (the proteins it phosphorylates) explains how a single molecule, cAMP, can produce so many different effects. Of course, the ability of a cell to respond to a cAMP-generating first messenger depends upon the presence of specific receptors for that messenger in the plasma membrane.

Not mentioned so far is the fact that receptors for some first messengers, upon activation by their messengers, cause adenylyl cyclase to be *inhibited*, resulting in less, rather than more, generation of cAMP. This occurs because these receptors are associated with a different G protein, known as G_i (the subscript i denotes "inhibitory"), and activation of G_i causes inhibition of adenylyl cyclase. The result is to decrease the concentration of cAMP in the cell and, thereby, to decrease the phosphorylation of key proteins inside the cell.

Control of Ion Channels by G Proteins.   A comparison of Figures 7-14C and 7-18 emphasizes one more important feature of G protein function—its ability to both directly and indirectly gate ion channels. As shown in Figure 7-14C and described earlier, an ion channel can be the effector protein for a G protein. This situation is known as *direct G-protein gating* of plasma-membrane ion channels because the G protein interacts directly with the channel (the term "gating" denotes control of the opening or closing of a channel). All the events occur in the plasma membrane and are independent of second messengers. Now look at Figure 7-18, and you will see that cAMP-dependent protein kinase can phosphorylate a plasma-membrane ion channel, thereby causing it to open. Since, as we have seen, the sequence of events leading to activation of cAMP-dependent protein kinase proceeds through a G protein, it should be clear that the opening of this

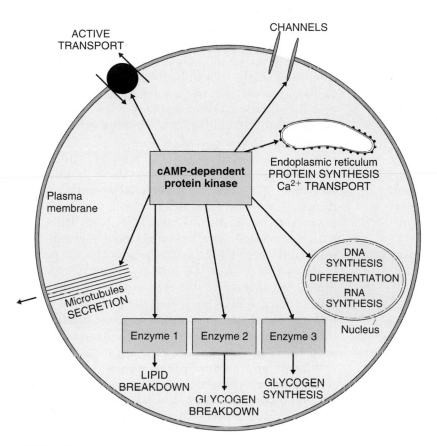

FIGURE 7-18

The variety of cellular responses induced by cAMP is due to the fact that activated cAMP-dependent protein kinase can phosphory-late many different proteins, activating them or inhibiting them. In this figure, the protein kinase is shown phosphorylating a micro-tubular protein, a Na,K-ATPase, an ion channel, a nuclear protein, a protein in the endoplasmic reticulum, and three enzymes. ◼️

channel is indirectly dependent on that G protein. To generalize, the *indirect gating* of ion channels by G proteins utilizes a second-messenger pathway for the opening (or closing) of the channel. Not just cAMP-dependent protein kinase but protein kinases involved in other signal transduction pathways can participate in reactions leading to such indirect gating. Table 7-5 summarizes the three ways we have described by which receptor activation by a first messenger leads to opening or closing of ion channels.

Calcium as a Second Messenger. The calcium ion (Ca^{2+}) functions as a second messenger in a great variety of cellular responses to stimuli, both chemical (first messenger) and electrical. The physiology of calcium as a second messenger requires an analysis of two broad questions: (1) How do stimuli cause the cytosolic calcium concentration to increase? (2) How does the increased calcium concentration elicit the cells' responses? Note that, for simplicity, our two questions are phrased in terms of an *increase* in cytosolic concentration. There are, in fact, first messengers that elicit a *decrease* in cytosolic calcium

concentration and therefore a decrease in calcium's second messenger effects. Now for the answer to the first question.

The regulation of cytosolic calcium concentration is described in Chapter 6. In brief, by means of active transport systems in the plasma membrane and cell organelles, Ca^{2+} is maintained at an extremely low concentration in the cytosol. Accordingly, there is always a large electro-chemical gradient favoring diffusion of calcium into the cytosol via calcium channels in both the plasma membrane and endoplasmic reticulum. A stimulus to the cell

---

TABLE 7-5 SUMMARY OF MECHANISMS BY WHICH RECEPTOR ACTIVATION INFLUENCES CHANNELS

1. The ion channel is part of the receptor.
2. A G protein directly gates the channel.
3. A G protein gates the channel indirectly via a second messenger.

TABLE 7-6 CALCIUM AS A SECOND MESSENGER

Most common mechanisms by which stimulation of a cell leads to an increase in cytosolic Ca^{2+} concentration:

1. Receptor activation
   a. Plasma-membrane calcium channels open in response to a first messenger; the receptor itself may contain the channel or the receptor may activate a G protein that directly or indirectly opens the channel.
   b. Calcium is released from the endoplasmic reticulum; this is mediated by second messengers.
   c. Active calcium transport out of the cell is inhibited by a second messenger.
2. Opening of voltage-sensitive calcium channels

Major mechanisms by which an increase in cytosolic Ca^{2+} concentration induces the cell's responses:

1. Calcium binds to calmodulin. On binding calcium, the calmodulin changes shape, which allows it to activate or inhibit a large variety of enzymes and other proteins. Many of these enzymes are protein kinases.
2. Calcium combines with calcium-binding intermediary proteins other than calmodulin. These proteins then act in a manner analogous to calmodulin.
3. Calcium combines with and alters response proteins directly, without the intermediation of any specific calcium-binding protein.

can alter this steady state by influencing the active transport systems and/or the ion channels, resulting in a change in cytosolic calcium concentration.

The most common ways that receptor activation by a first messenger increases the cytosolic Ca^{2+} concentration are summarized in the top part of Table 7-6. In essence, more calcium can enter the cell from the extracellular fluid via plasma-membrane calcium channels; more calcium can be released from intracellular organelles, mainly the endoplasmic reticulum, as a result of the action of second messengers; or less calcium can be transported out of the cell as a result of the action of second messengers. The mechanisms summarized in Table 7-6 for increasing cytosolic calcium concentration are not mutually exclusive; on the contrary, they are often triggered simultaneously by the same first messenger.

This section has thus far dealt exclusively with receptor-initiated sequences of events. This is a good place, however, to emphasize that there are calcium channels in the plasma membrane that are opened directly by an *electric* stimulus to the membrane (Chapter 6). Calcium can act as a second messenger, therefore, in response not only to chemical stimuli acting via receptors, but to electric stimuli acting via voltage-gated calcium channels as well.

Now we turn to the question of how the increased cytosolic calcium concentration elicits the cells' responses (bottom of Table 7-6). The common denominator of calcium's actions is its ability to bind tightly and specifically to various cytosolic calcium-binding proteins. These proteins are unusual in that they can bind calcium selectively at the very low concentrations of this ion present in the cytosol. One of the most important of these is a protein found in virtually all cells and known as **calmodulin** (Figure 7-19). On binding with calcium, calmodulin changes shape, and this allows calcium-calmodulin to activate or inhibit a large variety of enzymes and other proteins, many of which are protein kinases. Activation or inhibition of **calmodulin-dependent protein kinases** leads, via phosphorylation, to activation or inhibition of proteins involved in the cell's ultimate responses to the first messenger—contraction, secretion, and so on.

Calmodulin is not, however, the only intracellular protein influenced by calcium. Several other proteins bind calcium and then influence other proteins in turn. In still other cases, the calcium binds to and activates intracellular proteins immediately involved in the cell's responses.

Receptors and Gene Transcription

As described earlier in this chapter, the receptors for lipid-soluble messengers, once activated by hormone binding, act in the nucleus as transcription factors to increase or decrease the rate of gene transcription. We now emphasize that there are many other transcription factors inside cells and that the signal transduction pathways initiated by plasma-membrane receptors often result in the activation, by phosphorylation, of these transcription factors. Thus, many first messengers that bind to the plasma membrane can also alter gene transcription via second messengers.

Some of the genes influenced by transcription factors activated in response to first messengers are known collectively as **primary response genes,** or **PRGs** (also termed immediate-early genes). In many cases, especially those involving first messengers that influence the

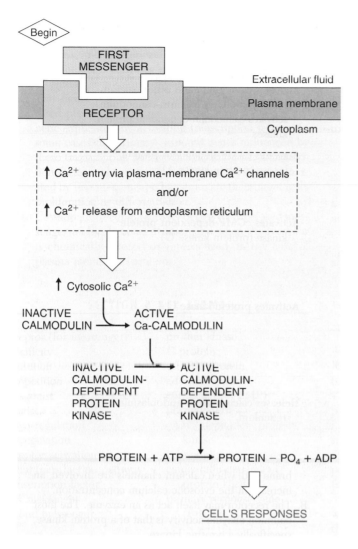

FIGURE 7-19

Calcium, calmodulin, and the calmodulin-dependent protein kinase system. The mechanisms for increasing cytosolic calcium concentration are summarized in Table 7-6.

proliferation or differentiation of their target cells, the story does not stop with a PRG and the protein it encodes. In these cases, the protein encoded by the PRG is itself a transcription factor for other genes. Thus, an initial transcription factor activated in the signal transduction pathway causes the synthesis of a different transcription factor, which in turn causes the synthesis of additional proteins, ones particularly important for the long-term biochemical events required for cellular pro-

liferation and differentiation. A great deal of research is being done on the transcription factors encoded by PRGs because of their relevance for the abnormal growth and differentiation typical of cancer.

Cessation of Activity in Signal Transduction Pathways

A word is needed about how signal transduction pathways are shut off. As expected, the key event is usually the cessation of receptor activation. Because organic second messengers are rapidly inactivated (for example, cAMP by phosphodiesterase) or broken down intracellularly, and because calcium is continuously being pumped out of the cell or back into the endoplasmic reticulum, increases in the cytosolic concentrations of all these components are transient events and continue only as long as the receptor is being activated by first messenger. Thus, a major way that receptor activation ceases is by a decrease in the concentration of messenger molecules in the region of the receptor. The mechanisms for this are discussed throughout Part 3. In addition, receptors are inactivated in two other ways: (1) The receptor becomes chemically altered (usually by phosphorylation), which lowers its affinity for messenger, and so the messenger is released; and (2) removal of plasma-membrane receptors occurs when the combination of first messenger and receptor is taken into the cell by endocytosis. All the processes described in this paragraph are physiologically controlled.

This concludes our description of the basic principles of signal transduction pathways. It is essential to recognize that the pathways do not exist in isolation but may be active simultaneously in a single cell, exhibiting complex interactions. This is possible because a single first messenger may trigger more than one pathway and, much more importantly, because a cell may be influenced simultaneously by many different first messengers—often dozens. Moreover, a great deal of cross-talk can occur at one or more levels among the various signal transduction pathways. For example, active molecules generated in the cAMP pathway can alter the ability of receptors that, themselves, function as protein kinases to activate transcription factors.

The biochemistry and physiology of plasma-membrane signal transduction pathways are among the most rapidly expanding fields in biology. Because of their widespread importance we did present specific information on two of the second messengers—cAMP and calcium; solely for reference purposes, Table 7-7 summarizes the biochemistry of several of the other prominent second messengers.

SECTION A
NEURAL TISSUE

The basic unit of the nervous system is the individual nerve cell, or **neuron.** Nerve cells operate by generating electric signals that pass from one part of the cell to another part of the same cell and by releasing chemical messengers—**neurotransmitters**—to communicate with other cells. Neurons serve as **integrators** because their output reflects the balance of inputs they receive from the thousands or even hundreds of thousands of other neurons that impinge upon them.

STRUCTURE AND MAINTENANCE OF NEURONS

Neurons occur in a variety of sizes and shapes; nevertheless, as shown in Figure 8-2, most of them contain four parts: (1) a cell body, (2) dendrites, (3) an axon, and (4) axon terminals.

As in other types of cells, a neuron's **cell body** contains the nucleus and ribosomes and thus has the genetic infor-

mation and machinery necessary for protein synthesis. The **dendrites** form a series of highly branched outgrowths from the cell body. They and the cell body receive most of the inputs from other neurons, the dendrites being vastly more important in this role than the cell body. The branching dendrites (some neurons may have as many as 400,000!) increase the cell's receptive surface area and thereby increase its capacity to receive signals from a myriad of other neurons.

The **axon,** sometimes also called a **nerve fiber,** is a single long process that extends from the cell body to its target cells. The portion of the axon closest to the cell body plus the part of the cell body where the axon is joined are known as the **initial segment,** or "trigger zone." The initial segment is where, in most neurons, the electric signals are generated that then propagate away from the cell body along the axon or, sometimes, back along the dendrites. The main axon may have branches, called **collaterals,** along its course; near their ends both the main axon and its collaterals undergo further branching (Figure 8-2).

FIGURE 8-2

(A) Diagrammatic representation of a neuron. The proportions shown here are misleading because the axon may be 5,000 to 10,000 times longer than the cell body is wide. The neuron shown here is a multipolar neuron, but there are several other types, one of which has no axon. (B) A neuron as observed through a microscope. The axon terminals cannot be seen at this magnification.

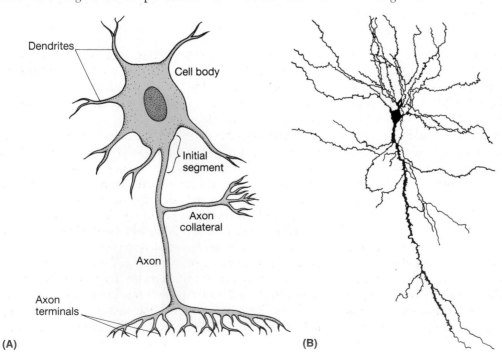

Dendrites

Cell body

Initial segment

Axon collateral

Axon

Axon terminals

(A)   (B)

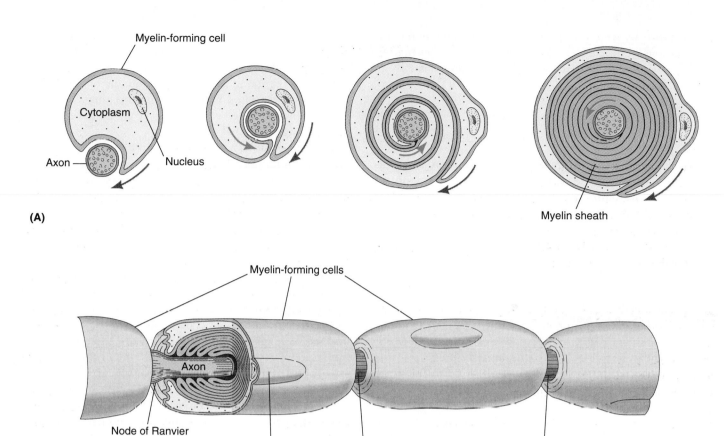

FIGURE 8-3

(A) Cross section of an axon in successive stages of myelinization. The myelin-forming cell may migrate around the axon, trailing successive layers of its plasma membrane (red arrows), or it may add to its tip, which lies against the axon, so that the tip is pushed around the axon, burrowing under the layers of myelin that are already formed (green arrows). The latter process must be used in the central nervous system where each myelin-forming cell may send branches to as many as 40 axons. (B) The myelin-forming cells are separated by a small space, the node of Ranvier.

The greater the degree of branching of the axon and axon collaterals, the greater the cell's sphere of influence.

Each branch ends in an **axon terminal,** which is responsible for releasing neurotransmitters from the axon. These chemical messengers diffuse across an extracellular gap to the cell opposite the terminal. In addition, some neurons release their chemical messengers from a series of bulging areas along the axon known as **varicosities.** Different parts of nerve cells serve different functions because of the segregated distribution of various membrane-bound channels and pumps as well as other molecules and organelles.

The axons of some neurons are covered by **myelin** (Figure 8-3), a fatty, membranous sheath formed by nearby supporting cells that wrap layers of their plasma membranes around the axon. In the central nervous system these myelin-forming cells are the **oligodendroglia** (glial cells will be described soon), and in the peripheral nervous system they are the **Schwann cells.** The spaces be-

tween adjacent myelin-forming cells where the axon's plasma membrane is exposed to extracellular fluid are the **nodes of Ranvier.** The myelin sheath speeds up conduction of the electric signals along the axon and conserves energy, as will be discussed later.

Various organelles and materials must be moved from the cell body, where they are made, to the axon and its terminals in order to maintain the structure and function of the cell axon. This movement is termed **axon transport.** Cytoskeletal filaments in the axon and cell body, which serve as the "rails" along which the transport occurs, are linked by proteins to the substances and organelles being moved. These same proteins act as the "motors" of axon transport and as ATPase enzymes, providing energy from split ATP to the "motors".

Axon transport of certain materials also occurs in the opposite direction, from the axon terminals to the cell body. By this route growth factors and other chemical signals picked up at the terminals can affect the neuron's morphology, biochemistry, and connectivity. This is also the route by which harmful substances, such as tetanus toxin and herpes and polio viruses, taken up by the peripheral nerve terminals enter the central nervous system.

FUNCTIONAL CLASSES OF NEURONS

Neurons can be divided into three functional classes: afferent neurons, efferent neurons, and interneurons. **Afferent neurons** convey information from the tissues and organs of the body *into* the central nervous system, **efferent neurons** transmit electric signals from the central nervous system *out* to effector cells (particularly muscle or gland cells or other neurons), and **interneurons** connect neurons *within* the central nervous system (Figure 8-4). As a rough estimate, for each afferent neuron entering the central nervous system, there are 10 efferent neurons and 200,000 interneurons. Thus, by far most of the neurons in the central nervous system are interneurons.

At their peripheral ends (the ends farthest from the central nervous system), afferent neurons have **receptors,** which respond to various physical or chemical changes in their environment by causing electric signals to be generated in the neuron. (Recall from Chapter 7 that the term "receptor" has two distinct meanings, the one defined here and the other referring to the specific proteins with which a chemical messenger combines to exert its effects on a target cell; both types of receptors will be referred to frequently in this chapter.) The receptor region may be a specialized portion of the plasma membrane or a separate cell closely associated with the neuron ending. Afferent neurons propagate electric signals from their receptors into the brain or spinal cord.

Afferent neurons are atypical in that they have no dendrites and only a single process, an axon. Shortly after leaving the cell body, the axon divides. One branch, the peripheral process, ends at the receptors; the other branch, the central process, enters the central nervous system to form junctions with other neurons. Note in Figure 8-4 that for afferent neurons both the cell body and the long peripheral process of the axon are *outside* the central nervous system and only a part of the central process enters the brain or spinal cord.

The cell bodies and dendrites of efferent neurons are within the central nervous system, but the axons extend out into the periphery. The axons of both the afferent and efferent neurons, except for the small part in the brain or spinal cord, form the **nerves** of the peripheral nervous system. (Note that a nerve fiber is a *single* axon, and a nerve is a *bundle* of axons bound together by connective tissue.)

Interneurons lie entirely within the central nervous system. They account for about 99 percent of all neurons and have a wide range of physiological properties, shapes, and

FIGURE 8-4

Three classes of neurons. Note that the interneurons form the connections between afferent and efferent neurons and lie entirely within the CNS. There is no real difference between the neurons of the central and peripheral nervous systems. The dendrites, which function as an extension of the neuron cell body, are not shown. The arrows indicate the direction of transmission of neural activity. The stylized neurons in this figure show the conventions that we will use throughout this book for the different parts of neurons.

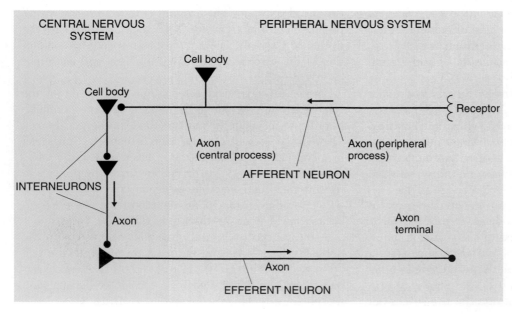

TABLE 8-1 THREE CLASSES OF NEURONS

I. Afferent neurons
 A. Transmit information into the central nervous system from receptors at their peripheral endings
 B. Are mostly (that is, the cell body and the long peripheral process of the axon) outside the central nervous system; only the short central process of the axon enters the central nervous system
 C. Have no dendrites

II. Efferent neurons
 A. Transmit information out of the central nervous system to effector cells, particularly muscles, glands, or other neurons
 B. Are mostly (that is, the cell body, dendrites, and a small segment of the axon) in the central nervous system; most of the axon is outside the central nervous system

III. Interneurons
 A. Function as integrators and signal changers
 B. Integrate groups of afferent and efferent neurons into reflex circuits
 C. Lie entirely within the central nervous system
 D. Account for 99 percent of all neurons

functions. The number of interneurons interposed between certain afferent and efferent neurons varies according to the complexity of the action. The knee-jerk reflex elicited by tapping below the kneecap has no interneurons—the afferent neurons end directly on the efferent neurons. In contrast, stimuli invoking memory or language may involve millions of interneurons.

Interneurons can serve as signal changers or gatekeepers, changing, for example, an excitatory input into an inhibitory output or into no output at all. The mechanisms used by interneurons to achieve these functions will be discussed at length throughout this chapter.

Characteristics of the three functional classes of neurons are summarized in Table 8-1.

The anatomically specialized junction between two neurons where one neuron alters the activity of another is called a **synapse.** At most synapses, the signal is transmitted from one neuron to another by neurotransmitters, a term that also includes the chemicals by which efferent neurons communicate with effector cells. The neurotransmitters released from one neuron alter the receiving neuron by binding with specific membrane receptors on the receiving neuron. (Once again, do not confuse this use of the term "receptor" with the receptors mentioned above that are at the peripheral ends of afferent neurons.)

Most synapses occur between the axon terminal of one neuron and the dendrite or cell body of a second neuron. In certain areas, however, synapses also occur between

two dendrites or between a dendrite and a cell body to modulate the input to a cell, or between an axon terminal and a second axon terminal to modulate its output. A neuron conducting signals toward a synapse is called a **presynaptic neuron,** whereas a neuron conducting signals away from a synapse is a **postsynaptic neuron.** Figure 8-5 shows how, in a multineuronal pathway, a single neuron can be postsynaptic to one cell and presynaptic to another. A postsynaptic neuron may have thousands of synaptic junctions on the surface of its dendrites and cell body, so that signals from many presynaptic neurons can affect it.

GLIAL CELLS

Neurons account for only about 10 percent of the cells in the central nervous system. The remainder are **glial cells** (also called neuroglia). The neurons branch more extensively than glia do, however, and therefore neurons occupy about 50 percent of the volume of the brain and spinal cord.

Glial cells physically and metabolically support neurons and, as noted earlier, some glia, the oligodendroglia, form

FIGURE 8-5

A neuron postsynaptic to one cell can be presynaptic to another.

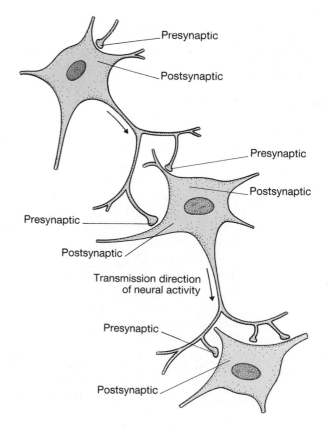

the myelin covering of axons. A second type of glial cell, the **astroglia,** helps regulate the composition of the extracellular fluid in the central nervous system by removing potassium ions and neurotransmitters around synapses. Astroglia sustain the neurons metabolically—for example, by providing glucose and removing ammonia. In development of the embryo, astroglia guide neurons as they migrate, and they stimulate the neurons' growth by secreting growth factors. In addition, astroglia have many neuron-like characteristics, for example, they have ion channels, receptors for certain neurotransmitters and the enzymes for processing them, and the capability of generating low-level electrical responses. Thus, in addition to all their other roles, it is speculated that astroglia may take part in information signaling in the brain. A third type of glia, the **microglia,** facilitate immune functions in the central nervous system.

NEURAL GROWTH AND REGENERATION

The elaborate networks of nerve cell processes that characterize the nervous system are remarkably similar in all human beings and depend upon the outgrowth of specific axons to specific targets.

Development of the nervous system in the embryo begins with a series of divisions of precursor cells called neuroblasts. After the last cell division, each daughter cell differentiates into a neuron, migrates to its final location, and begins to develop a spatial orientation, sending out processes that will become its axon and dendrites. A specialized enlargement, the growth cone, forms the tip of each extending process and is involved in finding the correct route and final target for the process.

As the cell process grows, it is guided along the surfaces of other cells, most commonly glial cells. Which particular route is followed depends largely on (1) the presence of recognition markers such as cell adhesion molecules (CAMs) on the membranes of embryonic cells, and (2) chemical messengers that affect growth, such as **nerve growth factor** or inhibitory proteins, in the extracellular region immediately surrounding the growth cone.

Once the target of the advancing growth cone is reached, synapses are formed. The synapses transmit activity, however, before their final maturation occurs, and this early activity, in part, determines their final use. During these intricate early stages of neural development, which occur during all trimesters of pregnancy and into infancy, alcohol and other drugs, radiation, malnutrition, and viruses can exert effects that cause permanent damage to the developing fetal nervous system.

A normal, although unexpected, aspect of development of the nervous system occurs after growth and projection of the axons. Many of the newly formed neurons and synapses *degenerate.* In fact, as many as 50 to 70 percent of neurons die in some regions of the developing nervous system! Exactly why this seemingly wasteful process occurs is unknown although neuroscientists speculate that in this way connectivity in the nervous system is refined, or "fine tuned."

Although the basic shape of existing neurons in the mature central nervous system does not change, the creation and removal of synaptic contacts begun during fetal development continue at a slower pace throughout life as part of normal growth, learning, and aging.

Division of neuron precursors is essentially complete before birth, and after early infancy virtually no new neurons are formed to replace those that die. Damaged neurons can repair themselves, however, and significant function may be regained, provided that the damage occurs *outside* the central nervous system and does not affect the neuron cell body. After repairable injury, the segment of the axon now separated from the cell body degenerates. The proximal part of the axon (the stump still attached to the cell body) then gives rise to a growth cone, which grows out to the effector organ so that in some cases function is restored.

In contrast, damage to axons *within* the central nervous system is followed by attempts at sprouting, but no significant regeneration of the axon occurs across the damaged site, and there are no well-documented reports of function return. Mature neurons of the central nervous system *are* capable of regrowing processes when isolated in tissue culture, but some property of their environment in vivo, such as inhibitory factors released from nearby glia, prevents this from happening in the body.

Researchers are attempting to create artificial tubes to provide an environment that will support axonal regeneration in the central nervous system. Attempts are also being made to restore function to damaged or diseased brains by the implantation of pieces of fetal brain or other tissues. For example, the adrenal medulla, which is part of the adrenal glands, synthesizes and secretes chemicals similar to some of the neurotransmitters found in the brain. When pieces of a patient's own adrenal medulla are inserted into damaged parts of his brain, the pieces continue to secrete these chemicals and replace the missing neurotransmitters.

We now turn to the mechanisms by which neurons and synapses function, beginning with the electrical properties that underlie all these events.

I. The nervous system is divided into two parts: the central nervous system (CNS) comprises the brain and spinal cord, and the peripheral nervous system consists of nerves extending from the CNS.

STRUCTURE AND MAINTENANCE OF NEURONS

I. The basic unit of the nervous system is the nerve cell, or neuron.

II. The cell body and dendrites receive information from other neurons.

III. The axon (nerve fiber), which may be covered with sections of myelin separated by nodes of Ranvier, transmits information to other neurons or effector cells.

FUNCTIONAL CLASSES OF NEURONS

I. Neurons are classified in three ways:
   A. Afferent neurons transmit information into the CNS from receptors at their peripheral endings.
   B. Efferent neurons transmit information out of the CNS to effector cells.
   C. Interneurons lie entirely within the CNS and form circuits with other interneurons or connect afferent and efferent neurons.

II. Information is transmitted across a synapse by neurotransmitters, which are released by a presynaptic neuron and combine with receptors on a postsynaptic neuron.

GLIAL CELLS

I. The CNS also contains glial cells, which help regulate the extracellular fluid composition, sustain the neurons metabolically, form myelin, serve as guides for developing neurons, and provide immune functions.

NEURAL GROWTH AND REGENERATION

I. Neurons develop from precursor cells, migrate to their final location, and send out processes to their target cells.

II. Cell division to form new neurons is virtually complete by birth.

III. After degeneration of a severed axon, damaged peripheral neurons may regrow the axon to their target organ. Damaged neurons of the CNS do not regenerate or restore function.

central nervous system (CNS)	Schwann cell
	node of Ranvier
peripheral nervous system	axon transport
neuron	afferent neuron
neurotransmitter	efferent neuron
integrator	interneuron
cell body	receptor
dendrite	nerve
axon	synapse
nerve fiber	presynaptic neuron
initial segment	postsynaptic neuron
collateral	glial cell
axon terminal	astroglia
varicosities	microglia
myelin	nerve growth factor
oligodendroglia	

1. Describe the direction of information flow through a neuron; through a network consisting of afferent neurons, efferent neurons, and interneurons.

2. Contrast the two uses of the word "receptor."

SECTION B
MEMBRANE POTENTIALS

BASIC PRINCIPLES OF ELECTRICITY

As discussed in Chapter 6, with the exception of water the major chemical substances in the extracellular fluid are sodium and chloride ions, whereas the intracellular fluid contains high concentrations of potassium ions and ionized nondiffusible molecules, particularly proteins with negatively-charged side chains and phosphate compounds. Electrical phenomena resulting from the distribution of these charged particles occur at the cell's plasma membrane and play a significant role in cell function.

Like charges repel each other; that is, positive charge repels positive charge, and negative charge repels negative charge. In contrast, an electric force draws oppositely charged substances together.

Separated electric charges of opposite sign have the potential of doing work if they are allowed to come together. This potential is called an **electric potential** or, because it is determined by the difference in charge between two points, a **potential difference,** which we shall often shorten to **potential.** The units of electric potential are volts, but since the total charge that can be separated in most biological systems is very small, the potential differ-

ences are small and are measured in millivolts (1 mV = 0.001 V).

The movement of electric charge is called a **current.** The electric force between charges tends to make them flow, producing a current. If the charges are of opposite sign, the current brings them toward each other; if the charges are alike, the current increases the separation between them. The amount of charge that moves—in other words, the current—depends on the potential difference between the charges and on the nature of the material through which they are moving. The hindrance to electric charge movement is known as **resistance.** The relationship between current I, voltage E (for electric potential), and resistance R is given by **Ohm's law:**

$$I = E/R$$

Materials that have a high electrical resistance are known as insulators, whereas materials that have a low resistance are conductors.

Water that contains dissolved ions is a relatively good conductor of electricity because the ions can carry the current. As we have seen, the intracellular and extracellular fluids contain numerous ions and can therefore carry cur-

FIGURE 8-6

(A) Apparatus for measuring membrane potentials. (B) The potential difference across a plasma membrane as measured by an intracellular microelectrode. The asterisk indicates the time the electrode entered the cell.

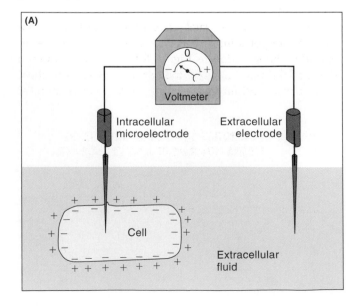

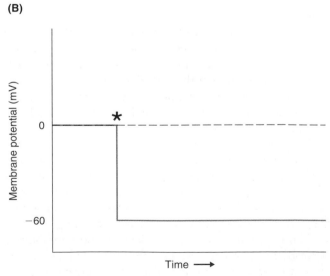

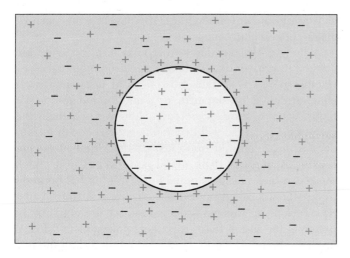

FIGURE 8-7

The excess of positive charges outside the cell and the excess of negative charges inside collect close to the plasma membrane. In reality, these excess charges are only an extremely small fraction of the total number of ions inside and outside the cell.

rent. Lipids, however, contain very few charged groups and cannot carry current. Therefore, the lipid layers of the plasma membrane are regions of high electrical resistance separating two water compartments—the intracellular fluid and the extracellular fluid—of low resistance.

THE RESTING MEMBRANE POTENTIAL

All cells under resting conditions have a potential difference across their plasma membranes oriented with the inside of the cell negatively charged with respect to the outside (Figure 8-6A). This potential is the **resting membrane potential.**

By convention, extracellular fluid is assigned a voltage of zero, and the polarity (positive or negative) of the membrane potential is stated in terms of the sign of the excess charge on the inside of the cell. For example, if the intracellular fluid has an excess of negative charge and the potential difference across the membrane has a magnitude of 70 mV, we say that the membrane potential is -70 mV.

The magnitude of the resting membrane potential varies from about -5 to -100 mV, depending upon the type of cell; in neurons, it is generally in the range of -40 to -75 mV (Figure 8-6B). The membrane potential of some cells can change rapidly in response to stimulation, an ability of key importance in their functioning.

The resting membrane potential exists because there is a small excess of negative ions inside the cell and an excess of positive ions outside. The excess negative charges

inside are electrically attracted to the excess positive charges outside the cell, and vice versa. Thus, the excess charges (ions) collect in a thin shell at the inner and outer surfaces of the plasma membrane (Figure 8-7), whereas the bulk of the intracellular and extracellular fluids are neutral. Unlike the diagrammatic representation in Figure 8-7, the number of positive and negative charges that have to be separated across a membrane to account for the potential is an infinitesimal fraction of the total number of charges in the two compartments.

The magnitude of the resting membrane potential is determined mainly by two factors (a third factor will be given later): (1) differences in specific ion concentrations in the intracellular and extracellular fluids, and (2) differences in membrane permeabilities to the different ions, which reflect the number of open channels for the different ions in the plasma membrane. The rest of this section analyzes how these two factors operate.

The concentrations of sodium, potassium, and chloride ions in the extracellular fluid and the intracellular fluid of a nerve cell are listed in Table 8-2. Although this table appears to contradict our earlier assertion that the bulk of the intra- and extracellular fluids are electrically neutral, there are many other ions, including Mg^{2+}, Ca^{2+}, H^+, HCO_3^-, HPO_4^{2-}, SO_4^{2-}, amino acids, and proteins, in both fluid compartments. Of the mobile ions, sodium, potassium, and chloride ions are present in the highest concentrations and therefore generally play the most important roles in generating the resting membrane potential. Note that the sodium and chloride concentrations are lower inside the cell than outside, and that the potassium concentration is greater inside the cell. As we described in Chapter 6, the concentration differences for sodium and potassium are due to the action of a plasma-membrane active-transport system that pumps sodium out of the cell and potassium into it. We will see later the reason for the chloride distribution.

To understand how such concentration differences for sodium and potassium create membrane potentials, let us consider the situation in Figure 8-8. The assumption in this model is that the membrane contains potassium channels but no sodium channels. *Initially,* compartment 1

TABLE 8-2 DISTRIBUTION OF MAJOR IONS ACROSS THE PLASMA MEMBRANE OF A TYPICAL NERVE CELL

Ion	Concentration, mmol/L	
	Extracellular	Intracellular
Na^+	150	15
Cl^-	110	10
K^+	5	150

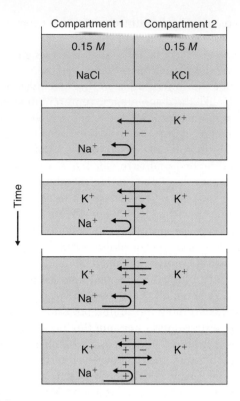

FIGURE 8-8

Generation of a diffusion potential across a membrane that contains only potassium channels. Arrows represent ion movements.

contains 0.15 *M* NaCl and compartment 2 contains 0.15 *M* KCl. There is no potential difference across the membrane because the two compartments contain equal numbers of positive and negative ions, that is, they are electrically neutral. The positive ions are different—sodium versus potassium, but the *total* numbers of positive ions in the two compartments are the same, and each positive ion is balanced by a chloride ion.

This initial state will not, however, last. Because of the potassium channels, potassium will diffuse down its concentration gradient from compartment 2 into compartment 1. After a few potassium ions have moved into compartment 1, that compartment will have an excess of positive charge, leaving behind an excess of negative charge in compartment 2. Thus, a potential difference has been created across the membrane.

Now we introduce a second factor that can cause net movement of ions across a membrane: an electrical potential. As compartment 1 becomes increasingly positive and compartment 2 increasingly negative, the membrane potential difference begins to influence the movement of the potassium ions. They are attracted by the negative

charge of compartment 2 and repulsed by the positive charge of compartment 1.

As long as the flux due to the potassium concentration gradient is greater than the flux due to the membrane potential, there will be net movement of potassium from compartment 2 to compartment 1, and the membrane potential will progressively increase. However, eventually the membrane potential will become negative enough to produce a flux equal but opposite the flux due to the concentration gradient. The membrane potential at which these two fluxes become equal in magnitude but opposite in direction is called the **equilibrium potential** for that type of ion—in this case, potassium. At the equilibrium potential for an ion there is no net movement of the ion because the opposing fluxes are equal, and the potential will undergo no further change.

The value of the equilibrium potential for any type of ion depends on the concentration gradient for that ion across the membrane. If the concentrations on the two sides were equal, the flux due to the concentration gradient would be zero and the equilibrium potential would also be zero. The larger the concentration gradient, the larger the equilibrium potential since a larger electrically driven movement of ions will be required to balance the larger movement due to the concentration difference.

If the membrane separating the two compartments is replaced with one that contains only sodium channels, a parallel situation will occur (Figure 8-9). A sodium equilibrium potential will eventually be established, but compartment 2 will be *positive* with respect to compartment 1, at which point net movement of sodium will cease.

FIGURE 8-9

Generation of a diffusion potential across a membrane that contains only sodium channels. Arrows represent ion movements.

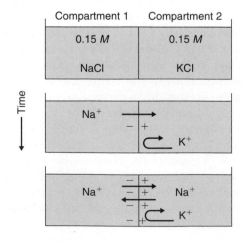

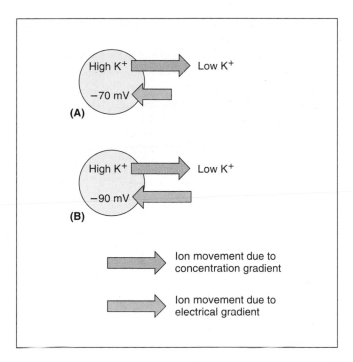

FIGURE 8-10

Forces acting on potassium when the membrane of a neuron is at (A) the resting potential (−70 mV, inside negative), and (B) the potassium equilibrium potential (−90 mV, inside negative).

Thus, the equilibrium potential for one ion species can be different in magnitude and direction from those for other ion species, depending on the concentration gradients for each ion. (Given the concentration gradient, the equilibrium potential for any ion can be calculated by means of the Nernst equation, Appendix D.)

Our examples were based on the membrane being permeable to only one ion at a time. When more than one ion species can diffuse across the membrane, the permeabilities and concentration gradients for all the ions must be considered when accounting for the membrane potential. For a given concentration gradient, the greater the membrane permeability to an ion species, the greater the contribution that ion species will make to the membrane potential. (Given the concentration gradients and membrane permeabilities for several ion species, the potential of a membrane permeable to these species can be calculated by the Goldman equation, Appendix D.)

It is not difficult to move from these hypothetical examples to a nerve cell at rest where (1) the potassium concentration is much greater inside than outside (Figure 8-10A) and the sodium concentration profile is just the opposite (Figure 8-11A); and (2) the plasma membrane contains about 50 to 75 times more open potassium channels than open sodium channels.

Given the actual potassium and sodium concentration differences, one can calculate that the potassium equilibrium potential will be approximately −90 mV (Figure 8-10B) and the sodium equilibrium potential about +60 mV (Figure 8-11B). However, since the membrane is permeable, to some extent, to both potassium and sodium, the resting membrane potential cannot be equal to either of these two equilibrium potentials. The resting potential will be much closer to the potassium equilibrium potential because the membrane is so much more permeable to potassium than to sodium.

In other words, a potential is generated across the plasma membrane largely because of the movement of potassium down its concentration gradient through open potassium channels, so that the inside of the cell becomes negative with respect to the outside. To repeat, the experimentally measured resting membrane potential is *not* equal to the potassium equilibrium potential, because a small number of sodium channels are open in the resting state, and some sodium ions continually move into the cell, canceling the effect of an equivalent number of potassium ions simultaneously moving out.

An actual resting membrane potential when recorded is about −70 mV, a typical value for neurons, and neither sodium nor potassium is at its equilibrium potential. Thus,

FIGURE 8-11

Forces acting on sodium when the membrane of a neuron is at (A) the resting potential (−70 mV, inside negative), and (B) the sodium equilibrium potential (+60 mV, inside positive).

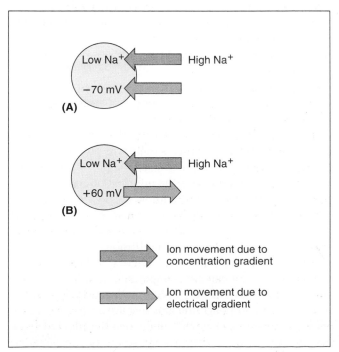

there is net movement through ion channels of sodium into the cell and potassium out. If such net ion movements occur, why does the concentration of intracellular sodium not progressively increase and that of intracellular potassium decrease? The reason is that active transport mechanisms in the plasma membrane utilize energy derived from cellular metabolism to pump the sodium back out of the cell and the potassium back in. Actually, the pumping of these ions is linked because they are both transported by the Na,K-ATPase pumps in the membrane (Chapter 6).

In a resting cell, the number of ions moved by the pump equals the number of ions that move in the opposite direction through membrane channels down their concentration and/or electrical gradients. Therefore the concentrations of sodium and potassium in the cell do not change. As long as the concentration gradients remain stable and the ion permeabilities of the plasma membrane do not change, the electric potential across the resting membrane will also remain constant.

Thus far, we have described the membrane potential as due purely and directly to the passive movement of ions down their electrical and concentration gradients, the concentration gradients having been established by membrane pumps. There is, however, another component to the membrane potential that reflects the *direct* separation of charge across the membrane by the transport of ions by the membrane Na,K-ATPase pump. This pump actually moves three sodium ions out of the cell for every two potassium ions that it brings in. This unequal transport of positive ions makes the inside of the cell more negative than it would be from ion diffusion alone. A pump that moves net charge across the membrane contributes directly to the membrane potential and is known as an **electrogenic pump.**

In most cells (but by no means all), however, the electrogenic contribution to the membrane potential is quite small. It must be reemphasized that even when the electrogenic contribution of the Na,K-ATPase pump is small, the pump always makes an essential *indirect* contribution to the membrane potential because it maintains the concentration gradients down which the ions diffuse to produce most of the charge separation that makes up the potential.

Figure 8-12 summarizes the information we have been presenting. This figure may mistakenly be seen to present a conflict: The *development* of the resting membrane potential depends predominately on the diffusion of potassium out of the cell, yet sodium diffusion, indicated by the blue Na$^+$ arrow in Figure 8-12, is greater than potassium diffusion. This imbalance compensates for the fact that the membrane pump moves three sodium ions out of the cell for every two potassium ions that are moved in. Figure 8-12 shows ion movements once steady state has been achieved.

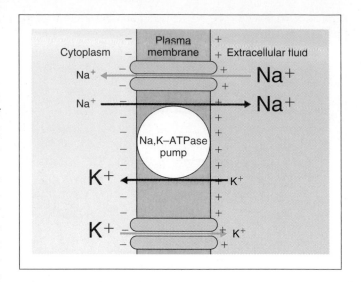

FIGURE 8-12

Movements of sodium and potassium ions across the plasma membrane of a resting neuron. The passive movements (blue arrows) are exactly balanced by the active transport (red arrows) of the ions in the opposite direction.

We have not yet dealt with chloride ions. The plasma membranes of many cells have chloride channels but do not contain chloride-ion pumps. Therefore, in these cells the membrane potential, set up by all the interactions just described, acts on chloride ions. The negative membrane potential moves chloride out of the cell, but the chloride concentration outside the cell is higher than that inside. This concentration gradient produces a diffusion of chloride back into the cell that exactly opposes the movement out because of the electric potential. Thus, chloride concentrations simply shift until the equilibrium potential for chloride is equal to the resting membrane potential. In this manner, chloride responds to the membrane potential, but makes no contribution to the magnitude of the potential.

In contrast, some cells have a non-electrogenic active transport system that moves chloride *out* of the cell. In these cells, the membrane potential is not at the chloride equilibrium potential, and net chloride diffusion *into* the cell contributes to the excess negative charge inside the cell; that is, net chloride diffusion makes the membrane potential more negative than it would otherwise be.

We noted earlier that most of the negative charge in neurons is accounted for not by chloride ions but by negatively charged organic molecules, such as proteins and phosphate compounds. Unlike chloride, however, these molecules do not readily cross the plasma membrane but remain inside the cell, where their charge contributes to the total negative charge within the cell.

GRADED POTENTIALS AND ACTION POTENTIALS

Transient changes in the membrane potential from its resting level produce electric signals. Such changes are the most important way that nerve cells process and transmit information. These signals occur in two forms: graded potentials and action potentials. Graded potentials are important in signaling over short distances, whereas action potentials are the long-distance signals of nerve and muscle membranes.

Adjectives such as "membrane," "resting," "action," and "graded" define the conditions under which the potential is measured or the way it develops (Table 8-3).

The terms "depolarize," "repolarize," and "hyperpolarize" are used to describe the direction of changes in the membrane potential relative to the resting potential (Figure 8-13). The membrane is said to be **depolarized** when its potential is less negative (closer to zero) than the resting level. **Overshoot** refers to a reversal of the membrane potential polarity, that is, when the inside of a cell becomes positive. When a membrane potential that has been depolarized returns toward the resting value, it is said to be **repolarizing.** The membrane is **hyperpolarized** when the potential is more negative than the resting level.

Graded Potentials

Graded potentials are changes in membrane potential that are confined to a relatively small region of the plasma membrane and die out within 1 to 2 mm of their site of

---

TABLE 8-3 A MINIGLOSSARY

Potential = potential difference	The voltage difference between two points.
Membrane potential = transmembrane potential	The voltage difference between the inside and outside of a cell.
Equilibrium potential	The voltage difference across a membrane that produces a flux of a given ion species that is equal but opposite the flux due to the concentration gradient affecting that same ion species.
Resting membrane potential = resting potential	The steady transmembrane potential of a cell that is not producing an electric signal.
Graded potential	A potential change of variable amplitude and duration that is conducted decrementally; it has no threshold or refractory period.
Action potential	A brief all-or-none depolarization of the membrane, reversing polarity in neurons; it has a threshold and refractory period and is conducted without decrement.
Postsynaptic potential	A graded potential change produced in the postsynaptic neuron in response to release of a neurotransmitter by a presynaptic terminal; it may be depolarizing (an excitatory postsynaptic potential or EPSP) or hyperpolarizing (an inhibitory postsynaptic potential or IPSP).
Receptor potential	A graded potential produced at the peripheral endings of afferent neurons (or in separate receptor cells) in response to a stimulus.
Pacemaker potential	A spontaneously occurring graded potential change that occurs in certain specialized cells.

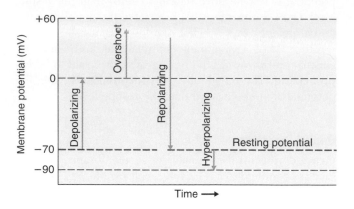

FIGURE 8-13

Depolarizing, repolarizing, hyperpolarizing, and overshoot changes in membrane potential.

origin. They are usually produced by some specific change in the cell's environment acting on a specialized region of the membrane, and they are called "graded potentials" simply because the magnitude of the potential change can vary (is graded). We shall encounter a number of graded potentials, which are given various names related to the location of the potential or to the function it performs: receptor potential, synaptic potential, end-plate potential, pacemaker potential.

Whenever a graded potential occurs, charge flows between the place of origin of the potential and adjacent regions of the plasma membrane, which are still at the resting potential. In Figure 8-14A, a small region of a membrane has been depolarized by a stimulus and therefore has a potential less negative than adjacent areas. Inside the cell (Figure 8-14B), positive charge (positive ions) will flow through the intracellular fluid away from

the depolarized region and toward the more negative, resting regions of the membrane. Simultaneously, outside the cell, positive charge will flow from the more positive region of the resting membrane toward the less positive region just created by the depolarization. The greater the potential change, the greater the currents. By convention, the direction in which positive ions move is designated the direction of the current, although negatively charged ions simultaneously move in the opposite direction.

Note that this local current moves positive charges toward the depolarization site along the *outside* of the membrane and away from the depolarization site along the *inside* of the membrane. Thus it produces a decrease in the amount of charge separation (depolarization) in the membrane sites adjacent to the originally depolarized region. The local current is carried by ions such as K^+, Na^+, Cl^-, and HCO_3^-.

Depending upon the initiation event, graded potentials can occur in either a depolarizing or hyperpolarizing direction (Figure 8-15A), and their magnitude is related to the magnitude of the initiating event (Figure 8-15B). Moreover, local current flows much like water flows through a leaky hose. Charge is lost across the membrane because the membrane is permeable to ions, just as water is lost from the leaky hose. The result is that the magnitude of the current decreases with the distance from the initial site of the potential change, just as water flow decreases the farther along the leaky hose you are from the faucet (Figure 8-16). In fact, plasma membranes are so leaky to ions that local currents die out almost completely within a few millimeters of their point of origin. There is another way of saying the same thing: Local current is **decremental;** that is, its amplitude decreases with increasing distance from the site of origin of

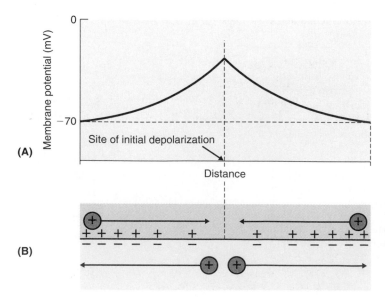

FIGURE 8-14

The membrane potential of a cell can be changed by using a stimulating current generator, and the potential can be recorded by a second pair of electrodes, one inside the cell and the other in the extracellular fluid, as in Figure 8-6. (A) Potential changes around a depolarization site. (B) Local current surrounding a depolarized region of a membrane produces a depolarization of adjacent regions.

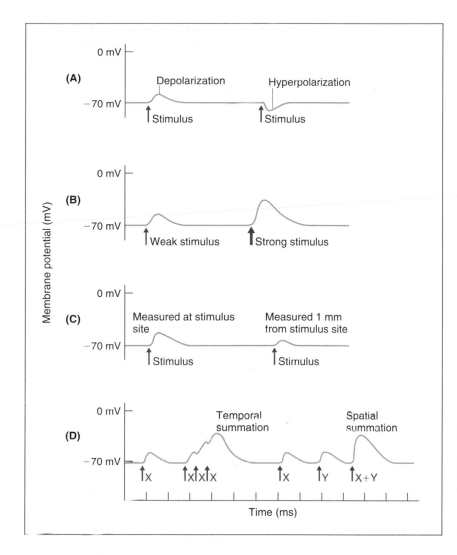

FIGURE 8-15

Graded potentials can be recorded under experimental conditions in which the stimulus or recording conditions are varied. Such experiments show that graded potentials (A) can be depolarizing or hyperpolarizing, (B) can vary in size, (C) are conducted decrementally, and (D) can be summed. Temporal and spatial summation are discussed later in the chapter.

the potential. The resulting change in membrane potential from resting level therefore also decreases with the distance from the potential's site of origin (Figures 8-14A and 8-15C).

Because the electric signal decreases with distance, graded potentials (and the local current they generate) can function as signals only over very short distances (a few millimeters). Nevertheless, graded potentials are the only means of communication used by some neurons and,

as we shall see, play very important roles in the initiation and integration of the long-distance signals by neurons and some other cells.

Action Potentials

Action potentials are very different from graded potentials. They are rapid, large alterations in the membrane potential during which time the membrane potential may

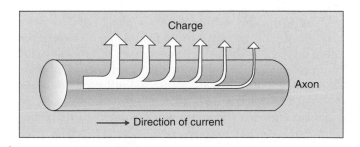

FIGURE 8-16

Leakage of charge across the plasma membrane reduces the local current at sites farther along the membrane.

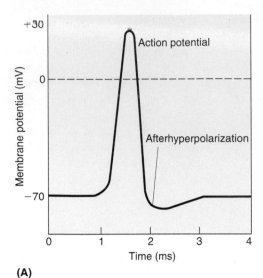

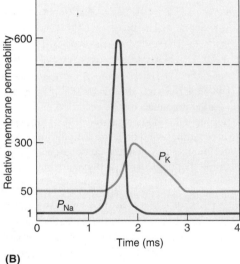

(B)

FIGURE 8-17

The changes during an action potential in (A) membrane potential and (B) membrane permeability (*P*) to sodium (red) and potassium (blue) ions.

change 100 mV, from −70 to +30 mV, and then repolarize to its resting membrane potential (Figure 8-17A). Nerve and muscle cells as well as some endocrine, immune, and reproductive cells have plasma membranes capable of producing action potentials. These membranes are called **excitable membranes,** and their ability to generate action potentials is known as **excitability.** The propagation of action potentials is the mechanism used by the nervous system to communicate over long distances.

How does an excitable membrane make rapid changes in its membrane potential? How is an action potential propagated along an excitable membrane? These questions are discussed in the following sections.

Ionic Basis of the Action Potential. Action potentials can be explained by the concepts already developed for describing the origins of resting membrane potentials. We have seen that the magnitude of the resting membrane potential depends upon the concentration gradients of and membrane permeabilities to different ions, particularly sodium and potassium. This is true for the action potential as well: The action potential results from a transient change in membrane ion permeability which allows selected ions to move down their concentration gradients. In the resting state, the open channels in the plasma membrane are predominantly those that are permeable to potassium (and chloride) ions. Very few sodium ion channels are open, and the resting potential is therefore close to the potassium equilibrium potential. During an action potential, however, the membrane permeabilities to sodium and potassium ions are markedly altered. (A review of voltage-gated ion channels, Chapter 6, may be helpful at this time.)

The depolarizing phase of the action potential is due to the opening of voltage-gated sodium channels, which increases the membrane permeability to sodium ions several hundredfold (red line in Figure 8-17B). This allows more sodium ions to move into the cell. During this period, therefore, more positive charge enters the cell in the form of sodium ions than leaves in the form of potassium ions, and the membrane depolarizes. It may even overshoot, becoming positive on the inside and negative on the outside of the membrane. In this phase the membrane potential approaches but does not quite reach the sodium equilibrium potential (+60 mV).

Action potentials in nerve cells last about 1 ms and typically show an overshoot. (They may last much longer in certain types of muscle cells.) What causes the membrane potential to return so rapidly to its resting level? The answer to this question is twofold: (1) The sodium channels that opened during the depolarization phase undergo inactivation, which causes them to close; and (2) a special set of potassium channels begins to open, albeit more slowly. The timing of these two events can be seen in Figure 8-17B.

Closure of the sodium channels alone would restore the membrane potential to its resting level since potassium flux out would then exceed sodium flux in. However, the process is speeded up by the simultaneous increase in potassium permeability. Potassium diffusion out of the cell is then much greater than the sodium diffusion in, rapidly returning the membrane potential to its resting level. In fact, after the sodium channels have closed, some of the additional potassium channels are still open, and in nerve cells there is generally a small hyperpolarizing overshoot of the membrane potential beyond the resting level (**afterhyperpolarization,** Figure 8-17A). Chloride permeability does not change during the action potential.

One might think that large movements of ions across the membrane would be required to produce such large changes in membrane potential. Actually, the number of sodium ions that enters a cell during an action potential and the number of potassium ions that diffuse out to return the membrane potential to its resting level are minute compared to the total number of ions in the cell. These ion movements are so small that they produce only infinitesimal changes in the intracellular ion concentrations, and that is why we said earlier that the action potential does *not* cause changes in ionic concentrations. Yet if this tiny number of additional ions crossing the membrane with each action potential were not eventually moved back across the membrane, the concentration gradients of sodium and potassium would gradually disappear and action potentials could no longer be generated. As might be expected, cellular accumulation of sodium and loss of potassium are prevented by the continuous action of the membrane Na,K-ATPase pumps.

What is achieved by letting sodium move into the neuron and then pumping it back out? Sodium movement down its electrochemical gradient into the cell creates the electric signal necessary for communication between parts of the cell, and pumping sodium out restores the concentration gradient so that, in response to a new stimulus, sodium will again enter the cell and create another signal.

Mechanism of Ion-channel Changes. In the above section, we described the various phases of the action potential as due to the opening and/or closing of voltage-gated ion channels. What causes these changes? The very first part of the depolarization, as we shall see later, is due to local current. Once depolarization starts, the depolarization itself causes voltage-gated sodium channels to open. In light of our discussion of the ionic basis of membrane potentials, it is very easy to confuse the cause-and-effect relationships of this last statement. Earlier we pointed out that an increase in sodium permeability *causes* membrane depolarization; now we are saying that depolarization *causes* an increase in sodium permeability. Combining these two distinct causal relationships yields the positive-feedback cycle (Figure 8-18) responsible for the depolarizing phase of the action potential: Depolarization opens voltage-gated sodium channels so that the membrane permeability to sodium increases. Because of increased

sodium permeability, sodium diffuses into the cell; this addition of positive charge to the cell further depolarizes the membrane, which in turn opens still more voltage-gated sodium channels, which produces a still greater increase in sodium permeability, etc. Many cells that have graded potentials cannot form action potentials because they have no voltage-gated sodium channels.

The potassium channels that open during an action potential are also voltage-gated. In fact, their opening is triggered by the same depolarization that opens the sodium channels, but the potassium channel opening is slightly delayed.

What about the inactivation of the voltage-gated sodium channels that opened during the rising phase of the action potential? This is the result of a voltage-induced change in the conformation of the proteins that constitute the channel, which closes the channel after its brief opening.

The generation of action potentials is prevented by **local anesthetics** such as procaine (Novocaine) and lidocaine (Xylocaine) because these drugs bind to the sodium channels and block them, preventing their opening in response to depolarization. Without action potentials, afferent signals generated in the periphery—in response to injury, for example—cannot reach the brain and give rise to the sensation of pain.

Some animals produce toxins that work by interfering with nerve conduction in the same way that local anesthetics do. For example, the puffer fish produces an extremely potent toxin, tetrodotoxin, that binds to voltage-gated sodium channels and prevents the sodium component of the action potential. The outward-potassium flux is not affected because this toxin's action is specific for sodium channels. This marked specificity makes tetrodotoxin an extremely useful tool for people studying membrane channels.

FIGURE 8-18

Positive-feedback relation between membrane depolarization and increased sodium permeability, which leads to the rapid depolarizing phase of the action potential.

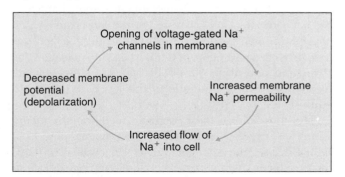

Although we have discussed only sodium and potassium channels, in the axon terminals of neurons and in various nonneural cells *calcium* channels open in response to membrane depolarization. In some of the nonneural cells, calcium diffusion into the cell through these voltage-gated channels generates action potentials, which are generally prolonged. The inward calcium diffusion also raises calcium concentration within the cell; as described in Chapter 7, this constitutes an essential part of the signal transduction pathway that couples membrane excitability to events within these cells.

Threshold.   Not all membrane depolarizations in excitable cells trigger the positive-feedback relationship that leads to an action potential. Action potentials occur only when the membrane is depolarized to about −50 mV, a depolarization sufficient to open enough voltage-gated sodium channels so that the positive-feedback cycle is able to take off.

The event that causes the membrane disturbance provides a current that adds positive charge to the inside of the cell, causing the initial depolarization from the resting membrane potential. As the depolarization begins, potassium efflux *increases* above its resting rate because the inside negativity, which tends to keep potassium in the cell, is less strong. The *initial* movement of sodium into the cell decreases because of this same lessened negativity; however, also in response to the depolarization, voltage-gated sodium channels open, which enhances sodium influx. All in all, at this stage potassium exit exceeds sodium entry. But, as the initial disturbing event continues to add current (positive charge) to the inside of the cell, the depolarization increases and more and more voltage-gated sodium channels open, allowing the influx of sodium ions to increase. Once the point is reached that this sodium influx exceeds potassium efflux, the positive-feedback cycle takes off and an action potential occurs. From this moment on, the membrane events are independent of the initial disturbing event and are driven entirely by the membrane properties.

In other words, during a stimulus action potentials occur only when the *net* movement of positive charge through ion channels is inward. The membrane potential at which the *net* movement of ions through the ion channels first changes from outward to inward and the positive feedback that produces an action potential takes off is called the **threshold potential.** Stimuli that are just strong enough to depolarize the membrane to this level are **threshold stimuli** (Figure 8-19).

The threshold of most excitable membranes is about 15 mV less negative inside than the resting membrane potential. Thus, if the resting potential of a neuron is −70 mV, the threshold potential may be −55 mV. At depolarizations

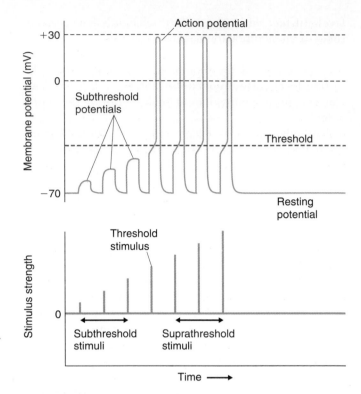

FIGURE 8-19
Changes in the membrane potential with increasing strength of depolarizing stimulus. When the membrane potential reaches threshold, action potentials are generated. Increasing the stimulus strength above threshold level does not cause larger action potentials. (The afterhyperpolarization has been omitted from this figure for clarity.)

less than threshold, outward potassium movement, increased because of the decreased membrane potential, exceeds sodium entry, and the positive-feedback cycle cannot get started despite the increase in sodium entry. In such cases, the membrane will return to its resting level as soon as the stimulus is removed, and no action potential is generated. These weak depolarizations are **subthreshold potentials,** and the stimuli that cause them are **subthreshold stimuli.**

Stimuli of *more than* threshold magnitude **(suprathreshold stimuli)** also elicit action potentials, but as can be seen in Figure 8-19, the action potentials resulting from such stimuli have exactly the same amplitude as those caused by threshold stimuli. This is because once threshold is reached, membrane events are no longer dependent upon stimulus strength. Rather, the depolarization generates an action potential because the positive-feedback cycle is operating. Action potentials either occur maximally or do not occur at all. Another way of saying this is that action potentials are **all-or-none.**

The firing of a gun is a mechanical analogy that shows the principle of all-or-none behavior. The magnitude of the explosion and the velocity at which the bullet leaves the gun do not depend on how hard the trigger is squeezed. Either the trigger is pulled hard enough to fire the gun, or it is not; the gun cannot be fired halfway.

Because of its all-or-none nature, a single action potential cannot convey information about the magnitude of the stimulus that initiated it. How then does one distinguish between a loud noise and a whisper, a light touch and a pinch? This information, as we shall see, depends in part upon the number and pattern of action potentials transmitted per unit of time and not upon their magnitude.

Refractory Periods. During the action potential, a second stimulus, no matter how strong, will not produce a second action potential, and the membrane is said to be in its **absolute refractory period.** This occurs because the sodium channels enter a closed, inactive state at the peak of the action potential. The membrane must repolarize before the sodium channel proteins return to the state in which they can be opened by depolarization.

Following the absolute refractory period, there is an interval during which a second action potential can be produced, but only if the stimulus strength is considerably greater than threshold. This is the **relative refractory period,** which can last 10 to 15 ms or longer in neurons and coincides roughly with the period of afterhyperpolarization. During the relative refractory period there is lingering inactivation of the voltage-gated sodium channels, and an increased number of potassium channels are open. If a depolarization exceeds the increased threshold or outlasts the relative refractory period, an additional action potential will be fired.

The refractory periods limit the number of action potentials that can be produced by an excitable membrane in a given period of time. Most nerve cells respond at frequencies of up to 100 action potentials per second, and some may produce much higher frequencies for brief periods.

Action-potential Propagation. Once generated, one particular action potential does not itself travel along the membrane. Rather, the local current produced by one action potential serves as the stimulus that depolarizes the adjacent membrane to its threshold potential. The sodium positive-feedback cycle takes over, and a new action potential occurs there.

The new action potential then produces local currents of its own, which depolarize the region adjacent to it, producing yet another action potential at the next site, and so on to cause **action-potential propagation** along the length of the membrane. Because each action potential depends on the sodium-feedback cycle of the membrane where the action potential is occurring, the action potential arriving at the end of the membrane is virtually identical in form to the initial one. Thus, action potentials are not conducted decrementally as are graded potentials.

Because the membrane areas that have just undergone an action potential are refractory and cannot immediately undergo another, the only direction of action potential propagation is away from a region of membrane that has recently been active (Figure 8-20).

If the membrane through which the action potential must travel is not refractory, excitable membranes are able to conduct action potentials in either direction, the direction of propagation being determined by the stimulus location. For example, the action potentials in skeletal muscle cells are initiated near the middle of these cylindrical

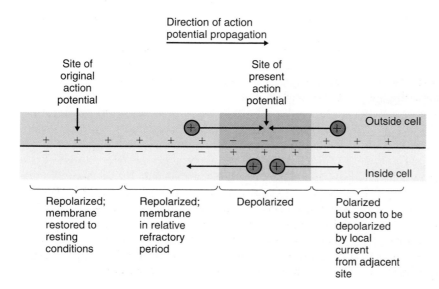

Direction of action potential propagation

Site of original action potential

Site of present action potential

Outside cell

Inside cell

Repolarized; membrane restored to resting conditions

Repolarized; membrane in relative refractory period

Depolarized

Polarized but soon to be depolarized by local current from adjacent site

FIGURE 8-20

Propagation of an action potential along a plasma membrane.

cells and propagate toward the two ends. In most nerve cells, however, action potentials are initiated physiologically at one end of the cell and propagate toward the other end, for reasons to be described in the next section. The propagation ceases when the action potential reaches the end of an axon.

The velocity with which an action potential propagates along a membrane depends upon fiber diameter and whether or not the fiber is myelinated. The larger the fiber diameter, the faster the action potential propagates. This is because a large fiber offers less resistance to local current; more ions will flow in a given time and the depolarization will be greater, bringing adjacent regions of the membrane to threshold faster.

Myelin is an insulator that makes it more difficult for charge to flow between intracellular and extracellular fluid compartments. Moreover, the concentration of sodium channels in the myelinated region of axons is low. Therefore, action potentials occur only at the nodes of Ranvier where the myelin coating is interrupted and the concentration of sodium channels is high. Thus, action potentials literally jump from one node to the next as they propagate along a myelinated fiber, and for this reason such propagation is called **saltatory conduction** (Latin *saltare*, to leap).

Propagation via saltatory conduction is faster than propagation in nonmyelinated fibers of the same axon diameter because less charge leaks out through the myelin-covered sections of the membrane (Figure 8-21). More charge arrives at the node adjacent to the active node,

and an action potential is generated there sooner than if the myelin were not present. Moreover, because ions cross the membrane only at the nodes of Ranvier, the membrane pumps need restore fewer ions. Myelinated axons are therefore metabolically more cost-effective than unmyelinated ones.

Conduction velocities range from about 0.5 m/s (1 mi/h) for small-diameter, unmyelinated fibers to about 100 m/s (225 mi/h) for large-diameter, myelinated fibers. At 0.5 m/s, an action potential would travel the distance from the head to the toe of an average-sized person in about 4 s; at a velocity of 100 m/s, it takes about 0.02 s.

Initiation of Action Potentials.   In afferent neurons, the initial depolarization to threshold is achieved by a graded potential generated in the receptors at the peripheral ends of the neurons. These are the ends farthest from the central nervous system, and where the nervous system functionally encounters the outside world. In all other neurons, the depolarization to threshold is due either to a graded potential generated by synaptic input to the neuron or to spontaneous changes in the neuron's membrane potential, known as a **pacemaker potential.** How synaptic potentials are produced is the subject of the next section. The production of receptor potentials is discussed in Chapter 9.

Spontaneous generation of pacemaker potentials occurs in the absence of any identifiable external stimulus and is an inherent property of certain neurons (and other

FIGURE 8-21

Current direction during an action potential in (A) a myelinated and (B) an unmyelinated axon.

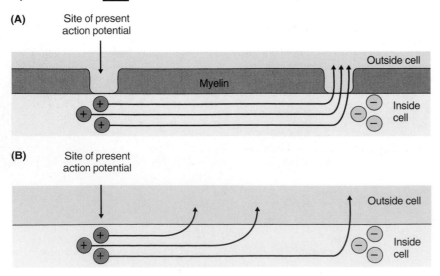

excitable cells, including certain smooth muscle and cardiac muscle cells). In these cells, the activity of different types of ion channels in the plasma membrane causes a graded depolarization of the membrane—the pacemaker potential. If threshold is reached, an action potential occurs; the membrane then repolarizes and again begins to depolarize (Figure 8-22). There is no stable, resting membrane potential in such cells because of the continuous change in membrane permeability. The rate at which the membrane depolarizes to threshold determines the action-potential frequency. Pacemaker potentials are implicated in many rhythmical behaviors, such as breathing, the heartbeat, and movements within the walls of the stomach and intestines.

The differences between graded potentials and action potentials are listed in Table 8-4.

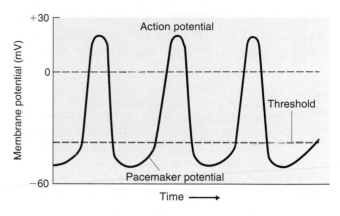

FIGURE 8-22

Action potentials resulting from pacemaker potentials.

TABLE 8-4   DIFFERENCES BETWEEN GRADED POTENTIALS AND ACTION POTENTIALS

Graded potentials	Action potentials
Graded response; amplitude varies with conditions of the initiating event	All-or-none response; once membrane is depolarized to threshold, amplitude is independent of initiating event
Graded response; can be summed	All-or-none response; cannot be summed
Has no threshold	Has a threshold that is usually about 15 mV depolarized relative to the resting potential
Has no refractory period	Has a refractory period
Is conducted decrementally; that is, amplitude decreases with distance	Is conducted without decrement; the amplitude is constant under constant conditions
Duration varies with initiating conditions	Duration constant for a given cell type under constant conditions
Can be a depolarization or a hyperpolarization	Is a depolarization (with overshoot in neurons)
Initiated by environmental stimulus (receptor), by neurotransmitter (synapse), or spontaneously	Initiated by membrane depolarization
Mechanism depends on receptor-operated channels or other chemical or physical changes	Mechanism depends on voltage-gated channels

THE RESTING MEMBRANE POTENTIAL

I. Membrane potentials are generated by the diffusion of ions and are determined by (a) the ionic concentration differences across the membrane, and (b) the membrane's relative permeabilities to different ions.
  A. Plasma membrane Na,K-ATPase pumps maintain intracellular sodium concentration low and potassium high.
  B. In almost all resting cells, the plasma membrane is much more permeable to potassium than to sodium, so the membrane potential is close to the potassium equilibrium potential—that is, the inside is negative relative to the outside.
  C. The Na,K-ATPase pumps also contribute directly a small component of the potential because they are electrogenic.

GRADED POTENTIALS AND ACTION POTENTIALS

I. Neurons signal information by graded potentials and action potentials (APs).
II. Graded potentials are local potentials whose magnitude can vary and that die out within 1 or 2 mm of their site of origin.
III. An AP is a rapid change in the membrane potential during which the potential rapidly depolarizes and repolarizes. In neurons the potential reverses and the membrane becomes positive inside. APs provide long-distance transmission of information through the nervous system.
  A. APs occur in excitable membranes because these membranes contain voltage-gated sodium channels, which open as the membrane depolarizes, causing a positive feedback toward the sodium equilibrium potential.
  B. The AP is ended as the sodium channels close and additional potassium channels open, which restores the resting conditions.
  C. Depolarization of excitable membranes triggers APs only when the membrane potential exceeds a threshold potential.
  D. Regardless of the size of the stimulus, if the membrane reaches threshold, the APs generated are all the same size.
  E. A membrane is refractory for a brief time even though stimuli that were previously effective are applied.
  F. APs are propagated without any change in size from one site to another along a membrane.
  G. In myelinated nerve fibers, APs manifest saltatory conduction.
  H. APs can be initiated by receptors at the ends of afferent neurons, at synapses, or in some cells, by pacemaker potentials.

electric potential
potential difference
potential
current
resistance
Ohm's law
resting membrane
  potential
equilibrium potential
electrogenic pump
depolarized
overshoot
repolarizing
hyperpolarized
graded potential
decremental

action potential
excitable membrane
excitability
afterhyperpolarization
threshold potential
threshold stimulus
subthreshold potential
subthreshold stimulus
suprathreshold stimulus
all-or-none
absolute refractory period
relative refractory period
action-potential
  propagation
saltatory conduction
pacemaker potential

1. Describe how negative and positive charges interact.
2. Contrast the abilities of intracellular and extracellular fluids and membrane lipids to conduct electric current.
3. Draw a simple cell; indicate where the concentrations of Na^+, $K+$, and Cl^- are high and low and the electric potential difference across the membrane when the cell is at rest.
4. Explain the conditions that give rise to the resting membrane potential. What effect does membrane permeability have on this potential? What is the role of Na,K-ATPase membrane pumps in the membrane potential? Is this role direct or indirect?
5. Which two factors involving ion diffusion determine the magnitude of the resting membrane potential?
6. Explain why the resting membrane potential is not equal to the potassium equilibrium potential.
7. Draw a graded potential and an action potential on a graph of membrane potential versus time. Indicate zero membrane potential, resting membrane potential, and threshold potential; indicate when the membrane is depolarized, repolarizing, and hyperpolarized.
8. List the differences between graded potentials and action potentials.
9. Describe the ionic basis of an action potential; include the role of voltage-gated channels and the positive-feedback cycle.
10. Explain threshold and the relative and absolute refractory periods in terms of the ionic basis of the action potential.
11. Describe the propagation of an action potential. Contrast this event in myelinated and unmyelinated axons.
12. List three ways in which action potentials can be initiated in neurons.

SECTION C
SYNAPSES

As defined earlier, a synapse is an anatomically specialized junction between two neurons, at which the electrical activity in one neuron, the presynaptic neuron, influences the electrical activity in the second, postsynaptic neuron. Anatomically, synapses include parts of the presynaptic and postsynaptic neurons and the extracellular space between these two cells. According to the latest estimate, there are approximately 10^{14} (100 quadrillion!) synapses in the CNS.

When active, synapses can increase or decrease the likelihood that the postsynaptic neuron will fire action potentials by producing a brief, graded potential there. The membrane potential of a postsynaptic neuron is brought closer to threshold at an **excitatory synapse,** and it is either driven farther from threshold or stabilized at its present level at an **inhibitory synapse.**

Thousands of synapses from many different presynaptic cells can affect a single postsynaptic cell (**convergence**), and a single presynaptic cell can send branches to affect many other postsynaptic cells (**divergence,** Figure 8-23).

The level of excitability of a postsynaptic cell at any moment, that is, how close its membrane potential is to threshold, depends on the number of synapses active at any one time and the number that are excitatory or inhibitory. If the membrane of the postsynaptic neuron reaches threshold, it will generate action potentials that are propagated along the axon to the terminal branches, which diverge to influence the excitability of other cells.

FIGURE 8-23

Convergence of neural input from many neurons onto a single neuron, and divergence of output from a single neuron onto many others. Presynaptic neurons are shown in orange, and postsynaptic neurons in brown. Arrows indicate the direction of transmission of neural activity.

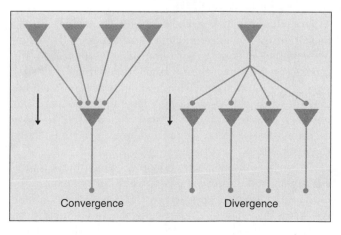

FUNCTIONAL ANATOMY OF SYNAPSES

There are two types of synapses: **electric** and **chemical.** At electric synapses, the plasma membranes of the pre- and postsynaptic cells are joined by gap junctions (Chapter 3). These allow the local currents resulting from action potentials to flow directly across the junction through the connecting channels in either direction from one neuron to the neuron on the other side of the junction, depolarizing the membrane to threshold and thus initiating an action potential in the second cell. Electric synapses are rare in the mammalian nervous system, and we shall henceforth discuss only the much more common chemical synapse.

Figure 8-24 shows the structure of a single typical chemical synapse. Note that in actuality the size and shape of the pre- and postsynaptic elements can vary greatly (Figure 8-25). The axon of the presynaptic neuron ends in a slight swelling, the axon terminal, and the postsynaptic membrane under the axon terminal appears denser. A 10- to 20-nm extracellular space, the **synaptic cleft,** separates

FIGURE 8-24

Diagram of a synapse. The "dust" in the synaptic cleft represents a fibrillar substance that lies within the cleft, not neurotransmitters.

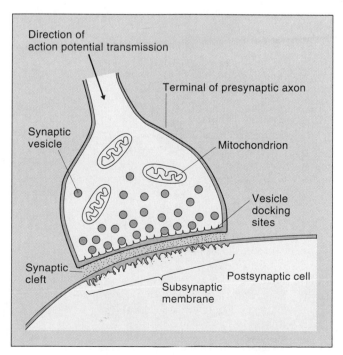

the pre- and postsynaptic neurons and prevents *direct* propagation of the current from the presynaptic neuron to the postsynaptic cell. Instead, signals are transmitted across the synaptic cleft by means of a chemical messenger—a neurotransmitter—released from the presynaptic axon terminal. Sometimes more than one neurotransmitter may be simultaneously released from an axon, in which case the additional neurotransmitter is called a **cotransmitter.**

The neurotransmitter in terminals is stored in membrane-bound **synaptic vesicles,** some of which are docked at specialized regions of the synaptic membrane. When an action potential in the presynaptic neuron reaches the end of the axon and depolarizes the axon terminal, voltage-gated calcium channels in the membrane open, and calcium diffuses from the extracellular fluid into the axon terminal near the docked vesicles. The calcium ions release a mechanism that allows some of the docked vesicles to fuse with the presynaptic plasma membrane and liberate their contents into the synaptic cleft by the process of exocytosis.

FIGURE 8-25

Variety of synaptic associations made upon a fiber of an ascending pathway in the spinal cord. Dendrites are shaded green, and axon terminals purple. (A) Simple axodendritic connection, which may include (B) one or more presynaptic synapses or (C) an arrangement known as a synaptic triad. (D) Synaptic arrangement ending on a spine extending from the postsynaptic cell. Such spines are common in the CNS and are thought to play a role in learning. The arrows indicate the direction of synaptic transmission. (*Redrawn from Maxwell and Rethelyi.*)

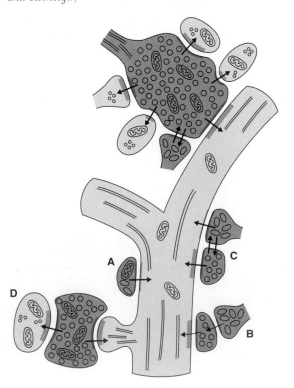

Once released from the presynaptic axon terminal, neurotransmitter and cotransmitter, if there is one, diffuse across the cleft. A fraction of these molecules bind to receptors on the plasma membrane of the postsynaptic cell (the fate of the others will be described later). The activated receptor may itself contain an ion channel, or it may act indirectly, via a G protein, on separate ion channels. In either case, the result of the binding of neurotransmitter to receptor is the opening or closing of specific ion channels in the postsynaptic plasma membrane. These channels belong, therefore, to the classes of chemically-gated channels whose function is controlled by receptors, as discussed in Chapter 7, and are distinct from voltage-gated channels.

Although Figure 8-25 shows a few exceptions, in general the neurotransmitter is stored on the presynaptic side of the synaptic cleft, whereas receptors are on the postsynaptic side. Therefore, most chemical synapses operate in only one direction. One-way conduction across synapses causes action potentials to be transmitted along a given multineuronal pathway in one direction.

Because of the sequence of events involved, there is a very brief synaptic delay—as short as $\frac{1}{5}$ ms—between the arrival of an action potential at a presynaptic terminal and the membrane-potential changes in the postsynaptic cell.

Neurotransmitter binding to the receptor is a transient event, and the ion channels in the postsynaptic membrane return to their resting state when the neurotransmitter is no longer bound. Unbound neurotransmitters are removed from the synaptic cleft when they (1) are enzymatically transformed into ineffective substances; (2) diffuse away from the receptor site; or (3) are actively transported back into the axon terminal or, in some cases, into nearby glial cells.

The two kinds of chemical synapses—excitatory and inhibitory—are differentiated by the effects of the neurotransmitter on the postsynaptic cell. Whether the effect is inhibitory or excitatory depends on the type of signal transduction mechanism brought into operation when the neurotransmitter binds to a receptor and the type of channel the receptor influences.

Excitatory Chemical Synapses

At an excitatory synapse, the postsynaptic response to the neurotransmitter is a depolarization, bringing the membrane potential closer to threshold. The usual effect of the activated receptor on the postsynaptic membrane at such synapses is to open postsynaptic-membrane ion channels that are permeable to sodium, potassium, and other small, positively charged ions. These ions then are free to move according to the electrical and chemical gradients across the membrane.

There is both an electrical and a concentration gradient driving sodium into the cell, while for potassium, the electrical gradient is opposed by the concentration gradient. Opening channels nonspecifically to all small positively charged ions, therefore, results in the simultaneous

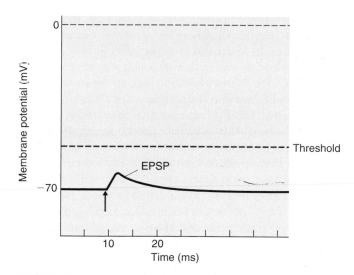

FIGURE 8-26

Excitatory postsynaptic potential (EPSP). Stimulation of the presynaptic neuron is marked by the arrow. ◁▭▷

movement of a relatively small number of potassium ions out of the cell and a larger number of sodium ions into the cell. Thus, the *net* movement of positive ions is into the postsynaptic cell, and this slightly depolarizes it. This potential change is called an **excitatory postsynaptic potential** (**EPSP,** Figure 8-26).

The EPSP is a graded potential that spreads decrementally away from the synapse by local current. Its only function is to bring the membrane potential of the postsynaptic neuron closer to threshold.

Inhibitory Chemical Synapses

At inhibitory synapses, the potential change in the postsynaptic neuron is a hyperpolarizing graded potential called an **inhibitory postsynaptic potential** (**IPSP,** Figure 8-27). Alternatively, there may be no IPSP but rather *stabilization* of the membrane potential at its existing value. In either case, activation of an inhibitory synapse lessens the likelihood that the postsynaptic cell will depolarize to threshold and generate an action potential.

At an inhibitory synapse the activated receptors on the postsynaptic membrane open chloride or, sometimes, potassium channels; sodium channels are not affected. In cells that actively transport chloride ions out of the cell, the chloride equilibrium potential (-80 mV) is more negative than the resting potential. Therefore, as chloride channels open, more chloride enters the cell, producing a hyperpolarization; that is, an IPSP. In cells that do not actively transport chloride, the equilibrium potential for chloride is equal to the resting membrane potential. A rise in chloride ion permeability therefore does not cause an IPSP but does increase chloride's influence on the membrane potential. This in turn makes it more difficult for other ion types to change the potential and results in a sta-

bilization of the membrane at the resting level without producing a hyperpolarization.

Increased potassium permeability, when it occurs in the postsynaptic cell, also produces an IPSP. Earlier it was noted that if a cell membrane were permeable only to potassium ions, the resting membrane potential would equal the potassium equilibrium potential, that is, the resting membrane potential would be -90 mV instead of -70 mV. Thus, with an increased potassium permeability more potassium ions leave the cell and cause a hyperpolarization.

ACTIVATION OF THE POSTSYNAPTIC CELL

A feature that makes postsynaptic integration possible is that in most neurons one excitatory synaptic event by itself is not enough to cause threshold to be reached in the postsynaptic neuron. For example, a single EPSP in some neurons is estimated to be only 0.5 mV, whereas changes of about 15 mV are necessary to depolarize the neuron's membrane to threshold. This being the case, an action potential can be initiated only by the combined effects of many excitatory synapses.

Of the thousands of synapses on any one neuron, probably hundreds are active simultaneously or close enough in time so that the effects can add together. The membrane potential of the postsynaptic neuron at any moment is, therefore, the resultant of all the synaptic activity affecting it at that time. There is a depolarization of the membrane toward threshold when excitatory synaptic input predominates, and either a hyperpolarization or stabilization when inhibitory input predominates (Figure 8-28).

Let us perform a simple experiment to see how EPSPs and IPSPs interact (Figure 8-29). Assume there are three

FIGURE 8-27

Inhibitory postsynaptic potential (IPSP). Stimulation of the presynaptic neuron is marked by the arrow.

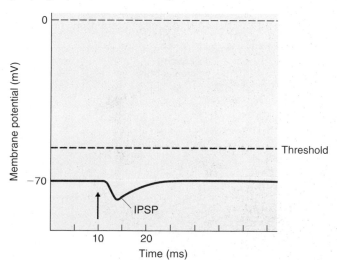

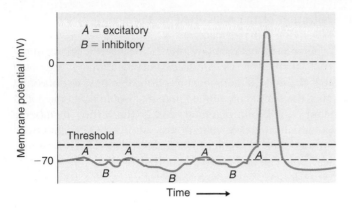

FIGURE 8-28

Intracellular recording from a postsynaptic cell during episodes when (A) excitatory synaptic activity predominates and the cell is facilitated, and (B) inhibitory synaptic activity dominates. ▄▅▆

synaptic inputs to the postsynaptic cell: The synapses from axons A and B are excitatory, and the synapse from axon C is inhibitory. There are laboratory stimulators on axons A, B, and C so that each can be activated individually. An electrode is placed in the cell body of the postsynaptic neuron and connected to record the membrane potential. In Part 1 of the experiment, we shall test the interaction of two EPSPs by stimulating axon A and then, after a short time, stimulating it again. Part 1 of Figure 8-29 shows that no interaction occurs between the two EPSPs. The reason is that the change in membrane potential associated with an EPSP is fairly short-lived. Within a few milliseconds (by the time we stimulate axon A for the second time), the postsynaptic cell has returned to its resting condition.

In Part 2, we stimulate axon A for the second time *before* the first EPSP has died away; the second synaptic potential adds to the previous one and creates a greater depolarization than from one input alone. This is called **temporal summation** since the input signals arrive at the same cell at different *times*. The potentials summate because there are a greater number of open ion channels and, therefore, a greater flow of positive ions into the cell. In Part 3, axon B is stimulated alone to determine its response, and then axons A and B are stimulated simultaneously. The two EPSPs that result also summate in the postsynaptic neuron; this is called **spatial summation** since the two inputs occurred at different *locations* on the same cell. The interaction of multiple EPSPs through ongoing spatial and temporal summation can increase the inward flow of positive ions and bring the postsynaptic membrane to threshold so that action potentials are initiated (Part 4).

So far we have tested only the patterns of interaction of excitatory synapses. What happens if an excitatory and an inhibitory synapse are activated so that their effects occur in the postsynaptic cell simultaneously? Since EPSPs and IPSPs are due to oppositely directed local currents, they tend to cancel each other, and there is little or no change in membrane potential (Figure 8-29, Part 5). Inhibitory potentials can also show spatial and temporal summation.

To repeat, the postsynaptic membrane is depolarized at an activated excitatory synapse and either hyperpolarized or stabilized at an activated inhibitory synapse. Via the local current mechanisms described earlier, the plasma membrane of the entire postsynaptic cell body and the initial segment reflect these changes. The membrane becomes slightly depolarized during activation of an excitatory synapse and slightly hyperpolarized or stabilized during activation of an inhibitory synapse (Figure 8-30).

In the above examples, we referred to the threshold of the postsynaptic neuron as though it were the same for all parts of the cell. However, different parts of the neuron have different thresholds. In many cells the initial segment has a lower threshold, that is, much closer to the resting potential than the threshold of the cell body and dendrites. In these cells the initial segment reaches thresh-

FIGURE 8-29

Interaction of EPSPs and IPSPs at the postsynaptic neuron. Arrows indicate time of stimulation. ▄▅▆

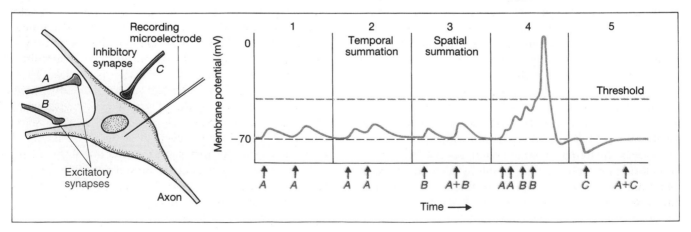

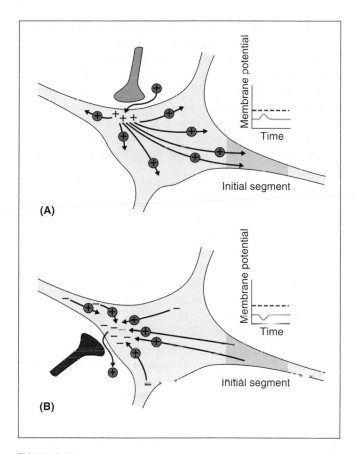

FIGURE 8-30

Comparison of excitatory and inhibitory synapses, showing current direction through the postsynaptic cell following synaptic activation. (A) Current through the postsynaptic cell is away from the excitatory synapse, depolarizing the cell. (B) Current through the postsynaptic cell is toward an inhibitory synapse, hyperpolarizing the cell.

old first whenever enough EPSPs summate, and the resulting action potential is initiated and then propagated from this point down the axon and, sometimes, back over the cell body and dendrites.

The fact that the initial segment usually has the lowest threshold explains why the location of individual synapses on the postsynaptic cell is important. A synapse located near the initial segment will produce a greater voltage change there than will a synapse on the outermost branch of a dendrite because it will expose the initial segment to a larger local current. In fact, some dendrites use propagated action potentials over portions of their length to convey information about the synaptic events occurring at their endings to the initial segment of the cell.

Postsynaptic potentials last much longer than action potentials do. In the event that cumulative EPSPs cause the initial segment to still be depolarized to threshold after an action potential has been fired and the refractory period is over, a second action potential will occur. In fact,

as long as the membrane is depolarized to threshold, action potentials will continue to arise. Neuronal responses at synapses almost always occur in bursts of action potentials rather than as single isolated events.

This discussion has no doubt left the impression that one neuron can influence another only at a synapse. This view requires qualification. Under certain circumstances, local currents from one neuron may directly affect the membrane potential of nearby neurons. This phenomenon is particularly important in areas of the central nervous system that contain a high density of unmyelinated neuronal processes.

SYNAPTIC EFFECTIVENESS

Individual synaptic events—whether excitatory or inhibitory—have been presented as though their effects are constant and reproducible. Actually, the variability in postsynaptic potentials following any particular presynaptic input is enormous. The effectiveness of a given synapse can be influenced by both presynaptic and postsynaptic mechanisms.

First, a presynaptic terminal does not release a constant amount of neurotransmitter every time it is activated. One reason for this variation involves calcium concentration. Calcium that has entered the presynaptic terminal during previous action potentials is pumped out of the cell or (temporarily) into intracellular organelles. If calcium removal does not keep up with entry, as can occur during high-frequency stimulation, calcium concentration in the terminal, and hence the amount of neurotransmitter released upon subsequent stimulation, will be greater than normal. The greater the amount of neurotransmitter released, the greater the number of ion channels opened (or closed) in the postsynaptic membrane, and the larger the amplitude of the EPSP or IPSP in the postsynaptic cell.

The neurotransmitter output of some presynaptic terminals is altered by input to membrane receptors like those of postsynaptic cells. These presynaptic receptors are often associated with a second synaptic ending known as an axon-axon synapse, or **presynaptic synapse,** in which an axon terminal of one neuron ends on an axon terminal of another. For example, in Figure 8-31 the neurotransmitter released by A combines with receptors on B, resulting in a change in the amount of neurotransmitter released from B in response to action potentials. Thus, neuron A has no *direct* effect on neuron C, but it has an important influence on the ability of B to influence C. Neuron A is said to be exerting a *presynaptic* effect on the synapse between B and C. Depending upon the nature of the neurotransmitter released from A and the type of receptors activated by that neurotransmitter on B, the presynaptic effect may decrease the amount of neurotransmitter released from B (**presynaptic inhibition**) or increase it (**presynaptic facilitation**).

Presynaptic synapses such as A in Figure 8-31 can alter the calcium concentration in axon terminal B or even affect neurotransmitter synthesis there. If the calcium concentration increases, the amount of neurotransmitter released from B increases; decreased calcium reduces the amount of transmitter released. Presynaptic synapses are important because they control one specific input to the postsynaptic neuron C.

Some receptors on the presynaptic terminal are not associated with axon-axon synapses. Rather they are activated by neurotransmitters released by other nearby neurons or glia, distant tissues, or the axon terminal itself. In the last case, the receptors are called **autoreceptors** and provide an important feedback mechanism by which the neuron can regulate its own neurotransmitter output.

Postsynaptic mechanisms for varying synaptic effectiveness also exist. For example, as described in Chapter 7, experiments with drugs that stimulate or block specific receptors have shown that there are many types and subtypes of receptors for each kind of neurotransmitter. All the different receptors and their various subtypes correspond to differences in the receptor proteins. The different receptor types operate by different signal transduction mechanisms and have different—sometimes even opposite—effects on the postsynaptic mechanisms they influence. Moreover, a given signal transduction mechanism may be regulated by multiple neurotransmitters, and the various second-messenger systems affecting a channel may interact with each other.

Recall, too, from Chapter 7 that the number of receptors is not constant, varying with up- and down-regulation, for example. Also, the ability of a given receptor to respond to its neurotransmitter can change. Thus, in many systems a receptor responds once and then temporarily fails to respond despite the continued presence of the receptor's neurotransmitter, a phenomenon known as receptor desensitization.

Imagine the complexity when a cotransmitter (or several cotransmitters) is released with the neurotransmitter to act upon postsynaptic receptors and maybe upon presynaptic receptors as well! Clearly, the possible variations in transmission at even a single synapse are enormous.

Modification of Synaptic Transmission by Drugs and Disease

The great majority of drugs that act on the nervous system do so by altering synaptic mechanisms and thus synaptic effectiveness. All the synaptic mechanisms labeled in Figure 8-32 are vulnerable.

The long-term effects of drugs are sometimes difficult to predict because the imbalances produced by the initial drug action are soon counteracted by feedback mechanisms that normally regulate the processes. For example, if a drug interferes with the action of a neurotransmitter by inhibiting the rate-limiting enzyme in its synthetic pathway, the neurons may respond by increasing the rate of precursor transport into the axon terminals to maximize the use of any enzyme that is available.

Recall from Chapter 7 that drugs that bind to a receptor and produce a response similar to the normal activation of that receptor are called **agonists,** and drugs that bind to the receptor but are unable to activate it are **antagonists.** By occupying the receptors, antagonists prevent binding of the normal neurotransmitter when it is released at the synapse. Specific agonists and antagonists can affect receptors on both pre- and postsynaptic membranes.

Diseases can also affect synaptic mechanisms. For example, the toxin produced by the bacillus *Clostridium tetani* (**tetanus toxin**) is a protease that destroys certain proteins in the synaptic-vesicle docking mechanism of neurons that provide inhibitory synaptic input to the neurons supplying skeletal muscles. The toxin of the *Clostridium botulinum* bacilli and the venom of the black widow spider also affect neurotransmitter release from synaptic vesicles by interfering with docking proteins, but they act on axon terminals of neurons different from those affected by tetanus toxin.

Table 8-5 summarizes the factors that determine synaptic effectiveness.

NEUROTRANSMITTERS AND NEUROMODULATORS

We have emphasized the role of neurotransmitters in eliciting EPSPs and IPSPs. In neurons, some chemical messengers elicit complex responses that cannot be simply de-

FIGURE 8-31

A presynaptic (axon-axon) synapse between axon terminal A and axon terminal B. C is the final postsynaptic cell body.

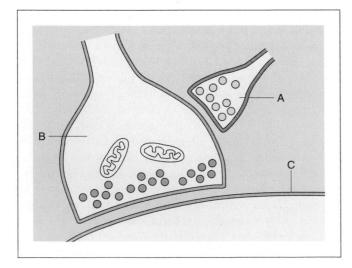

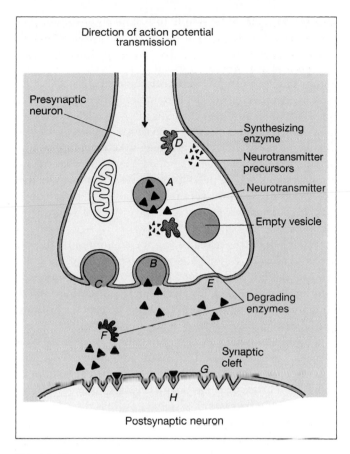

FIGURE 8-32

Drug actions at synapses. (A) Increases leakage of neurotransmitter from vesicle to cytoplasm, exposing it to enzyme breakdown; (B) increases transmitter release; (C) blocks transmitter release; (D) inhibits transmitter synthesis; (E) blocks transmitter reuptake; (F) blocks enzymes that metabolize transmitter; (G) binds to receptor to block (antagonist) or mimic (agonist) transmitter action; and (H) inhibits or facilitates second-messenger activity. (*Redrawn from DRUGS AND THE BRAIN by Solomon H. Snyder. Copyright © 1986 by Scientific American Books, Inc. Reprinted by permission of W. H. Freeman and Company.*)

scribed as EPSPs or IPSPs. The word "modulation" is used for these complex responses, and the messengers that cause them are called **neuromodulators.** The distinctions between neuromodulators and neurotransmitters are, however, far from clear. In fact, certain neuromodulators are often synthesized by the presynaptic cell and co-released with the neurotransmitter. To add to the complexity, certain hormones, paracrine agents, and messengers used by the immune system may serve as neuromodulators.

Neuromodulators often modify the *postsynaptic* cell's response to specific neurotransmitters, amplifying or dampening the effectiveness of ongoing synaptic activity. Alternatively, they may change the *presynaptic* cell's syn-

thesis, release, reuptake, or metabolism of a transmitter. In other words, they alter the effectiveness of the synapse.

In general, the receptors for neurotransmitters influence ion channels that directly affect excitation or inhibition of the postsynaptic cell. These mechanisms operate within milliseconds. Receptors for neuromodulators, on the other hand, more often bring about changes in metabolic processes in neurons, often via G proteins coupled to second-messenger systems. These changes, which can occur over minutes, hours, or even days, include alterations in enzyme activity or, by way of influences on DNA transcription, in protein synthesis. Thus, neurotransmitters are involved in rapid communication, whereas neuromodulators tend to be associated with slower events such as learning, development, motivational states, or even some sensory or motor activities.

Table 8-6 lists some of the major substances generally accepted as neurotransmitters or neuromodulators. A huge amount of information has accumulated concerning the synthesis, metabolism, and mechanisms of action of these messengers—material well beyond the scope of this book. The following sections will therefore present only some basic generalizations about certain of the neurotransmitters presently deemed most important. For simplicity's sake, in this discussion we use the term "neurotransmitter" in a general sense, realizing that sometimes the messenger may more appropriately be described as neuromodulator. A note on terminology should also be included

TABLE 8-5   FACTORS THAT DETERMINE SYNAPTIC EFFECTIVENESS

I. Presynaptic factors
  A. Availability of neurotransmitter
    1. Availability of precursor molecules
    2. Amount (or activity) of the rate-limiting enzyme in the pathway for neurotransmitter synthesis
  B. Axon terminal membrane potential
  C. Axon terminal residual calcium
  D. Activation of membrane receptors on presynaptic terminal
    1. Presynaptic (axon-axon) synapses
    2. Autoreceptors
    3. Other receptors
  E. Certain drugs and diseases
II. Postsynaptic factors
  A. Immediate past history of electrical state of postsynaptic membrane, that is, facilitation or inhibition from temporal or spatial summation
  B. Effects of other neurotransmitters or neuromodulators acting on postsynaptic neuron
  C. Certain drugs and diseases

here: Neurons are often referred to as -ergic, where the missing prefix is the type of neurotransmitter released by the neuron. For example, "dopaminergic" applies to neurons that release the neurotransmitter dopamine.

Acetylcholine

Acetylcholine (ACh) is synthesized from choline and acetyl coenzyme A in the cytoplasm of synaptic terminals and stored in synaptic vesicles. After it is released and activates receptors on the postsynaptic membrane, the concentration of ACh at the postsynaptic membrane is reduced (thereby stopping receptor activation) by the enzyme **acetylcholinesterase.** This enzyme is located on the pre- and postsynaptic membranes and rapidly destroys ACh, releasing choline. The choline is then transported back into the axon terminals where it is reused in the synthesis of new ACh. The ACh concentration at the recep-

---

TABLE 8-6   CLASSES OF SOME OF THE CHEMICALS KNOWN OR PRESUMED TO BE NEUROTRANSMITTERS OR NEUROMODULATORS

1. Acetylcholine (ACh)

---

2. Biogenic amines
   Catecholamines
       Dopamine (DA)
       Norepinephrine (NE)
       Epinephrine (Epi)
   Serotonin (5-hydroxytryptamine, 5-HT)
   Histamine

---

3. Amino acids
   Excitatory amino acids
       Glutamate
       Aspartate
   Inhibitory amino acids
       Gamma-aminobutyric acid (GABA)
       Glycine

---

4. Neuropeptides (only a few important examples are listed although some 85 have been identified)
       Endogenous opioids (beta-endorphin, enkephalins, dynorphins)
       Substance P
       Somatostatin
       Vasoactive intestinal peptide (VIP)
       Cholecystokinin
       Neurotensin
       Neuropeptide Y (NPY)

---

5. Miscellaneous
   Gases
       Nitric oxide
       Carbon monoxide
   Purines
       Adenosine
       Adenosine triphosphate (ATP)

---

tors is also reduced by simple diffusion away from the site and eventual breakdown of the molecule by an enzyme in the blood.

Some ACh receptors respond to the drug nicotine and have come to be known as **nicotinic receptors.** The nicotinic receptor is an excellent example of a receptor that itself contains an ion channel, in this case for positively charged ions. In essence the receptor is an allosteric protein, and the binding of acetylcholine to it changes the receptor's conformation so as to open the channel. Other cholinergic receptors are stimulated by the mushroom poison muscarine; they are called **muscarinic receptors.** These receptors couple with G proteins, which then alter the activity of a number of different enzymes and ion channels.

Acetylcholine is a major neurotransmitter in the peripheral nervous system, and it is also present in the brain. The cell bodies of the brain's cholinergic neurons are concentrated in relatively few areas, but their axons are widely distributed. Fibers that release ACh are called **cholinergic** fibers.

One cholinergic system in the brain (Figure 8-33) plays a major role in attention, learning, and memory by reinforcing the ability to detect and respond to meaningful stimuli. Neurons associated with this system degenerate in people with *Alzheimer's disease,* a brain disease that is usually age-related and is the most common cause of declining intellectual function in late life, affecting 10 to 15 percent of people over age 65 and 50 percent of people over age 85. Because of the degeneration of cholinergic neurons, this disease is associated with a decreased amount of ACh in certain areas of the brain and a loss of the postsynaptic neurons that would have responded to it. These defects and those in other neurotransmitter systems that are affected in this disease are related to the declining language and perceptual abilities, confusion, and memory loss that characterize Alzheimer's victims. The exact causes of this degeneration are unknown.

Biogenic Amines

The **biogenic amines** are neurotransmitters that are synthesized from amino acids and contain an amino group (R-NH_2). The most common biogenic amines in the nervous system are dopamine, norepinephrine, serotonin, and histamine. Epinephrine, another biogenic amine, is not a common neurotransmitter in the nervous system but is the major *hormone* secreted by the adrenal medulla.

Dopamine, norepinephrine (NE), and **epinephrine** all contain a catechol ring (a six-carbon ring with two adjacent hydroxyl groups) and an amine group; thus they are called **catecholamines.** The catecholamines are formed from the amino acid tyrosine and share the synthetic pathway shown in Figure 8-34, which begins with the uptake of tyrosine by the axon terminals. Depending on the enzymes present in the terminals, any one of the three catecholamines may be ultimately released. Release of cate-

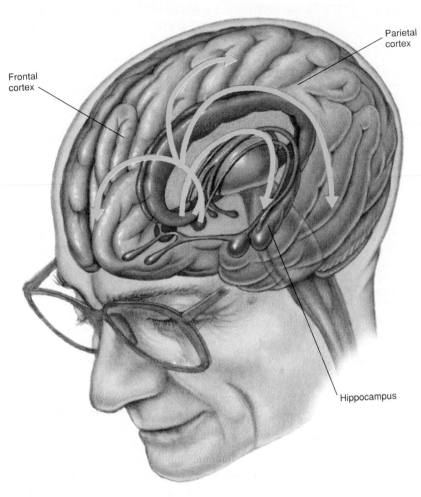

Frontal cortex

Parietal cortex

Hippocampus

FIGURE 8-33

The major cholinergic pathways implicated in Alzheimer's disease. As shown in Figure 8-42, the violet structure is the limbic system. (*Adapted from BRAIN, MIND, AND BEHAVIOR by Floyd E. Bloom and Arlyne Lazerson. Copyright © 1985, 1988 by Educational Broadcasting Corporation. Reprinted by permission of W. H. Freeman and Company.*)

cholamines from the presynaptic terminals is strongly modulated by receptors on the presynaptic terminals.

After activation of the receptors on the postsynaptic cell, the catecholamine concentration in the synaptic cleft declines because the catecholamine is actively transported back into the axon terminal. The catecholamine neurotransmitters are also broken down by enzymes in both the extracellular fluid and the axon terminal.·

Within the central nervous system, the cell bodies of the catecholamine-releasing neurons lie in the part of the brain called the brainstem, and although relatively few in number, their axons branch greatly and go to virtually all parts of the brain and spinal cord. The catecholamines exert a much greater influence in the central nervous system than the number of neurons alone would suggest, possibly because of their neuromodulator-like effects on post-synaptic neurons. These neurotransmitters play essential roles in states of consciousness, mood, motivation, and directed attention, all functions that will be covered in Chapter 13.

Norepinephrine and Epinephrine. Norepinephrine plays an important role in both the central nervous system and in the peripheral nervous system, where it serves as a hormone as well as a neurotransmitter.

During the early experiments on norepinephrine and epinephrine, norepinephrine was mistakenly taken to be epinephrine, and epinephrine was called by its British name "adrenaline." Consequently, nerve fibers that release epinephrine or norepinephrine came to be called **adrenergic** fibers. Norepinephrine releasing fibers are also called **noradrenergic.**

There are two major classes of receptors for norepinephrine and epinephrine: **alpha-adrenergic receptors** and **beta-adrenergic receptors** (these are also called alpha-

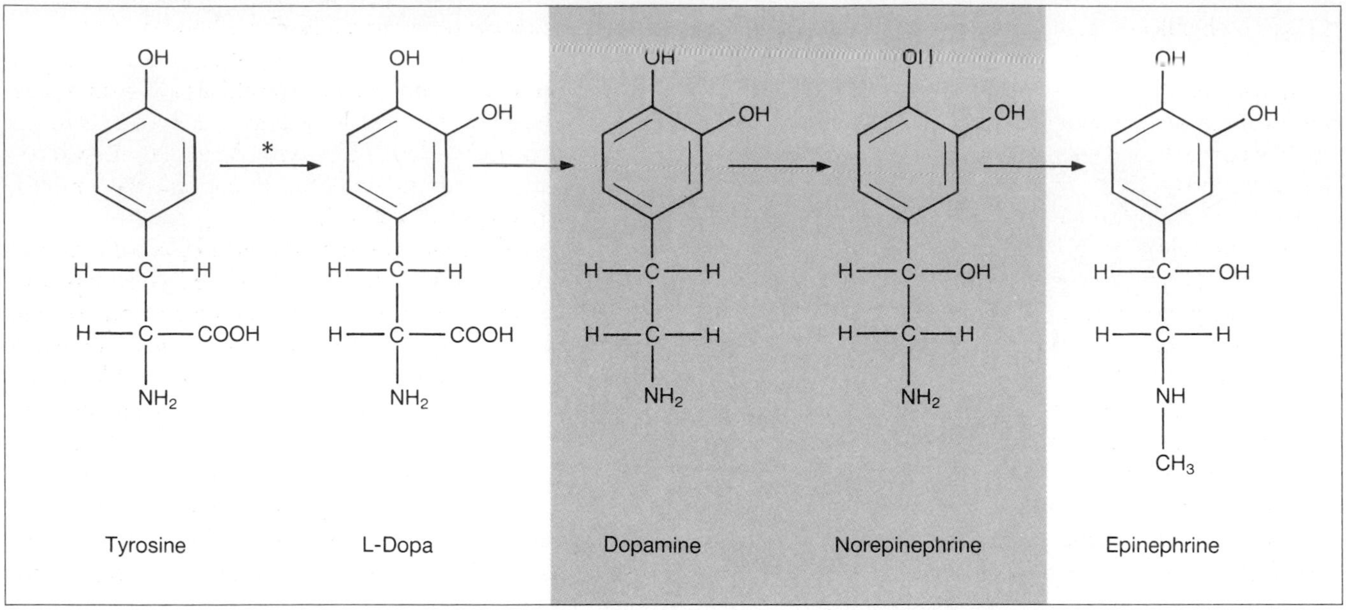

FIGURE 8-34

Catecholamine biosynthetic pathway. The asterisk indicates the site of action of tyrosine hydroxylase, the rate-limiting enzyme; dopamine and norepinephrine are the more common CNS catecholamine neurotransmitters.

adrenoceptors and beta-adrenoceptors). The major way of distinguishing between the two classes of receptors is that they are influenced by different drugs. Both alpha- and beta-adrenergic receptors can be subdivided still further (alpha$_1$ and alpha$_2$, for example), again according to the drugs that influence them and their second-messenger systems.

Serotonin. While not a catecholamine, **serotonin** (5-hydroxytryptamine, or 5-HT) is an important biogenic amine. It is produced from tryptophan, an essential amino acid. Its effects generally have a slow onset, indicating that it works as a neuromodulator. It is contained mainly in brainstem neurons that innervate virtually every structure in the brain and spinal cord. At least 16 different receptor types have been identified for this biogenic amine!

In general, serotonin has an excitatory effect on pathways that are involved in the control of muscles, and an inhibitory effect on pathways that mediate sensations. The activity of serotonergic neurons is lowest or absent during sleep and highest during states of alert wakefulness. Serotonergic pathways also function in the regulation of food intake and certain other homeostatic systems.

Serotonin is also present in many nonneural cells, for example, blood platelets, mast cells, and specialized cells lining the digestive tract. In fact, the brain contains only 1 to 2 percent of the body's serotonin.

Amino Acid Neurotransmitters

In addition to the neurotransmitters that are synthesized from amino acids, several amino acids themselves function as neurotransmitters. Although the amino acid neurotransmitters chemically fit the category of biogenic amines, neurophysiologists traditionally put them into a category of their own. The amino acid neurotransmitters are by far the most prevalent neurotransmitters in the central nervous system, and they affect virtually all neurons there.

Two so-called **excitatory amino acids, glutamate** and **aspartate,** serve as neurotransmitters at the vast majority of excitatory synapses in the central nervous system. The excitatory amino acids function in learning, memory, and neural development. They are also implicated in epilepsy, Alzheimer's and Parkinson's diseases, and the neural damage that follows strokes, brain trauma, and other conditions of low oxygen availability. One of the family of glutamate receptors is the site of action of a number of mind-altering drugs, such as phencyclidine ("angel dust").

GABA (gamma-aminobutyric acid) and the amino acid **glycine** are the major inhibitory neurotransmitters in the central nervous system. (GABA is not one of the twenty amino acids used to build proteins, but because it is a modified form of glutamate, it is classified with the amino acid neurotransmitters.) Drugs such as Valium that reduce anxiety enhance the action of GABA.

Neuropeptides

The **neuropeptides** are composed of two or more amino acids linked together by peptide bonds. Some 85 neuropeptides have been found, but their physiological roles are often unknown. It seems that evolution has selected the same chemical messengers for use in widely differing circumstances, and many of the neuropeptides had been previously identified in nonneural tissue where they function as hormones or paracrine agents. They generally retain the name they were given when first discovered in the nonneural tissue.

The neuropeptides are formed differently from the other neurotransmitters, which are synthesized in the axon terminals via a very few enzyme-mediated steps. The neuropeptides, in contrast, are derived from large precursor molecules called prohormones or preprohormones (the term "hormone" is used even when referring to peptide neurotransmitters), which in themselves have little, if any, inherent biological activity. The synthesis of these precursors is directed by mRNA and occurs on ribosomes, which exist only in the cell body and large dendrites of the neuron, often a considerable distance from axon terminals or varicosities where the peptides are released.

In the cell body, the prohormone is packaged into vesicles, which are then moved by axon transport into the terminals or varicosities where the vesicle contents are cleaved by specific peptidases. Many of the prohormones and preprohormones are actually "polyproteins" in that they contain multiple copies of peptides. These copies may be of the same or different peptides. Neurons that release one or more of the peptide neurotransmitters are collectively called **peptidergic.** In many cases neuropeptides are cosecreted with another type of neurotransmitter.

Certain neuropeptides, termed **endogenous opioids**—**beta-endorphin,** the **dynorphins,** and the **en-kephalins**—have attracted much interest because receptors for these neurotransmitters are the sites of action of the opiate drugs such as morphine and codeine. No physiological function of any of the opioid peptides has been demonstrated conclusively, but they are powerful *analgesics* (that is, they relieve pain without loss of consciousness) and undoubtedly play a role in regulating pain. The opioids, particularly beta-endorphin, have been implicated in the runner's "second wind," when the athlete feels a boost of energy and a decrease in pain and effort, and in the general feeling of well-being experienced after a bout of strenuous exercise, the so-called "jogger's high." There is also evidence that the opioids play a role in eating and drinking behavior, in regulation of the cardiovascular system, and in mood and emotion.

Substance P, another of the neuropeptides, is a neurotransmitter for afferent neurons that relay sensory information into the central nervous system.

Miscellaneous

Surprisingly, some gases serve as neurotransmitters, albeit unusual ones. Gases are not released from presynaptic vesicles, nor do they activate postsynaptic plasma-membrane receptors. They simply diffuse from their sites of origin in one cell into the cytoplasm of nearby cells.

Nitric oxide serves as a messenger between some neurons and between neurons and effector cells. It is produced in one cell from the amino acid arginine (in a reaction catalyzed by nitric oxide synthase), and it binds to and activates guanylyl cyclase in the recipient cell, thereby increasing the concentration of the second-messenger cyclic GMP in that cell (Chapter 7).

Nitric oxide is thought to play a role in a bewildering array of neural events—learning (Chapter 13), development, drug tolerance, and sensory and motor modulation, to name a few. Paradoxically, it is also implicated in neural damage that results, for example, from the stoppage of blood flow to the brain or from a head injury. In later chapters we shall see that nitric oxide is produced by a variety of nonneural cells and plays an important paracrine role in the cardiovascular and immune systems, among others.

Carbon monoxide, another gas, is suspected of serving as a neurotransmitter, with a role in learning. It too stimulates guanylyl cyclase.

The purines **adenosine** and **ATP** also serve as important neurotransmitters. Like glutamate, ATP is a very fast acting excitatory transmitter.

NEUROEFFECTOR COMMUNICATION

Thus far we have described the effects of neurotransmitters released at synapses. Many neurons of the peripheral nervous system end, however, not at synapses on other neurons but at neuroeffector junctions on muscle and gland cells. The neurotransmitters released by these efferent neurons' terminals or varicosities provide the link by which electrical activity of the nervous system is able to regulate effector cell activity.

The events that occur at neuroeffector junctions are similar to those at a synapse. The neurotransmitter is released from the efferent neuron upon the arrival of an action potential at the neuron's axon terminals or varicosities. The neurotransmitter then diffuses to the surface of the effector cell, where it binds to receptors on that cell's plasma membrane. The receptors may be directly under the axon terminal or varicosity, or they may be some distance away so that the diffusion path followed by the neurotransmitter is tortuous and long. The receptor of the effector cell may be associated with ion channels that alter the membrane potential of the cell, or it may be associated, via a G protein, with an enzyme that results in the formation of a second messenger in the effector cell. The response (altered muscle contraction or glandular secretion) of the effector cell to these changes will be described in later chapters. As we shall see below, the major neurotransmitters released at neuroeffector junctions are acetylcholine and norepinephrine.

I. An excitatory synapse brings the membrane of the postsynaptic cell closer to threshold. An inhibitory synapse hyperpolarizes the postsynaptic cell or stabilizes it at its resting level.

II. Whether a postsynaptic cell fires action potentials depends on the number of synapses that are active and whether they are excitatory or inhibitory.

FUNCTIONAL STATES OF SYNAPSES

I. The signal from a pre- to a postsynaptic neuron is a neurotransmitter stored in synaptic vesicles in the presynaptic axon terminal. Depolarization of the axon terminal, which raises the calcium concentration within the terminal, causes the release of neurotransmitter into the synaptic cleft.

II. The neurotransmitter diffuses across the synaptic cleft and binds to receptors on the postsynaptic cell; the activated receptors usually open ion channels.

   A. At an excitatory synapse the electrical response in the postsynaptic cell is called an excitatory postsynaptic potential (EPSP). At an inhibitory synapse, it is an inhibitory postsynaptic potential (IPSP).

   B. Usually at an excitatory synapse, channels in the postsynaptic cell that are permeable to sodium, potassium, and other small positive ions are opened; at inhibitory synapses, channels to chloride and/or potassium are opened.

   C. The postsynaptic cell's membrane potential is the result of temporal and spatial summation of the EPSPs and IPSPs at the many active excitatory and inhibitory synapses on the cell.

SYNAPTIC EFFECTIVENESS

I. Synaptic effects are influenced by pre- and postsynaptic events, drugs, and diseases (Table 8-5).

NEUROTRANSMITTERS AND NEUROMODULATORS

I. In general, neurotransmitters cause EPSPs and IPSPs, and neuromodulators cause, via second messengers, more complex metabolic effects in the postsynaptic cell.

II. The actions of neurotransmitters are usually faster than those of neuromodulators.

III. A substance can act as a neurotransmitter at one type of receptor and as a neuromodulator at another.

IV. The major known or suspected neurotransmitters and neuromodulators are listed in Table 8-6.

NEUROEFFECTOR COMMUNICATION

I. The junction between a neuron and an effector cell is called a neuroeffector junction.

II. The events at a neuroeffector junction (release of neurotransmitter into an extracellular space, diffusion of neurotransmitter to the effector cell, and binding with a receptor on the effector cell) are similar to those at a synapse.

excitatory synapse	dopamine
inhibitory synapse	norepinephrine (NE)
convergence	epinephrine
divergence	catecholamine
electric synapse	adrenergic
chemical synapse	noradrenergic
synaptic cleft	alpha-adrenergic receptor
cotransmitter	beta-adrenergic receptor
synaptic vesicle	serotonin
excitatory postsynaptic	excitatory amino acid
potential (EPSP)	glutamate
inhibitory postsynaptic	aspartate
potential (IPSP)	GABA (gamma-
temporal summation	aminobutyric acid)
spatial summation	glycine
presynaptic synapse	neuropeptide
presynaptic inhibition	peptidergic
presynaptic facilitation	endogenous opioid
autoreceptor	beta-endorphin
neuromodulator	dynorphin
acetylcholine (ACh)	enkephalin
acetylcholinesterase	substance P
nicotinic receptor	nitric oxide
muscarinic receptor	carbon monoxide
cholinergic	adenosine
biogenic amine	ATP

1. Contrast the postsynaptic mechanisms of excitatory and inhibitory synapses.

2. Explain how synapses allow neurons to act as integrators; include the concepts of facilitation, temporal and spatial summation, and convergence in your explanation.

3. List at least eight ways in which the effectiveness of synapses may be altered.

4. Discuss differences between neurotransmitters and neuromodulators.

5. Discuss the relationship between dopamine, norepinephrine, and epinephrine.

SECTION D
STRUCTURE OF THE NERVOUS SYSTEM

We shall now survey the anatomy and broad functions of the major structures of the nervous system; future chapters will describe these functions in more detail. First, we must deal with some potentially confusing terminology. Recall that a long extension from a *single* neuron is called an axon or a nerve fiber and that the term nerve refers to a group of *many* nerve fibers that are traveling together to the same general location in the peripheral nervous system. There are no nerves in the *central* nervous system. Rather, a group of nerve fibers traveling together in the central nervous system is called a **pathway,** a **tract,** or, when it links the right and left halves of the central nervous system, a **commissure.**

Information can pass through the central nervous system along two types of pathways: (1) **long neural pathways,** in which neurons with long axons carry information directly between the brain and spinal cord or between large regions of the brain, and (2) **multineuronal** or **multisynaptic pathways** (Figure 8-35). As their name suggests, the multineuronal pathways are made up of many neurons and many synaptic connections. Since synapses are the sites where new information can be integrated into neural messages, there are many opportunities for neural processing along the multineuronal pathways. The long pathways, on the other hand, consist of chains of only a few sequentially connected neurons. Because the long pathways contain few synapses, there is little opportunity for alteration in the information they transmit.

The cell bodies of neurons having similar functions are often clustered together. Groups of neuron cell bodies in the peripheral nervous system are called **ganglia** (singular *ganglion*), and in the central nervous system they are called **nuclei** (singular *nucleus*).

CENTRAL NERVOUS SYSTEM: SPINAL CORD

The spinal cord lies within the bony vertebral column (Figure 8-36). It is a slender cylinder of soft tissue about as big around as the little finger. The central butterfly-shaped area (in cross section) of **gray matter** is composed of interneurons, the cell bodies and dendrites of efferent neurons, the entering fibers of afferent neurons, and glial cells. It is called gray matter because the nerve fibers present lack myelin, and so appear gray.

The gray matter is surrounded by **white matter,** which consists of groups of myelinated axons of interneurons.

These groups of axons, called fiber tracts or pathways, run longitudinally through the cord, some descending to relay information from the brain to the spinal cord, others ascending to transmit information to the brain. Pathways also transmit information between different levels of the brain and spinal cord.

Groups of afferent fibers that enter the spinal cord from the peripheral nerves enter on the dorsal side of the cord (the side nearest the back of the body) via the **dorsal roots** (Figure 8-36). Small bumps on the dorsal roots, the **dorsal root ganglia,** contain the cell bodies of the afferent neurons. The axons of efferent neurons leave the

FIGURE 8-35

Relationship of long neural pathways and multineuronal (multisynaptic) pathways to the reticular formation.

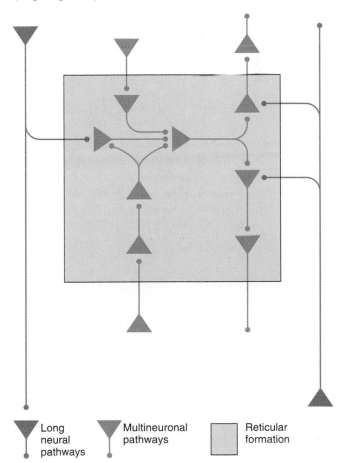

| ▼ Long neural pathways | ▼ Multineuronal pathways | ▢ Reticular formation |

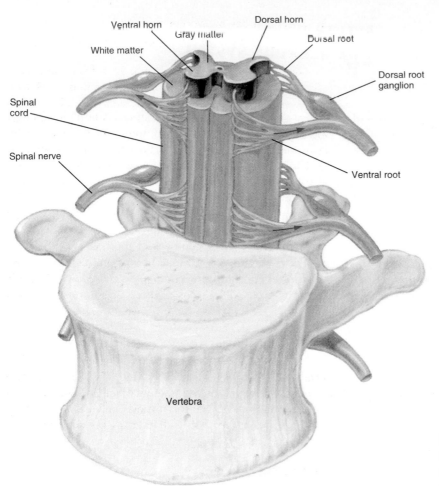

Ventral horn
Dorsal horn
Gray matter
White matter
Dorsal root
Spinal cord
Dorsal root ganglion
Spinal nerve
Ventral root
Vertebra

FIGURE 8-36

Section of the spinal cord, ventral view. The arrows indicate the direction of transmission of neural activity.

spinal cord on the ventral side (nearest the front surface of the body) via the **ventral roots.** A short distance from the cord, the dorsal and ventral roots from the same level combine to form a **spinal nerve,** one on each side of the spinal cord. The 31 pairs of spinal nerves are organized into four levels: cervical, thoracic, lumbar, and sacral (Figure 8-37).

CENTRAL NERVOUS SYSTEM: BRAIN

Anatomically, the brain is composed of four subdivisions: **cerebrum, diencephalon, brainstem,** and **cerebellum** (Figure 8-38). The cerebrum and diencephalon together constitute the **forebrain.** The brainstem consists of the **midbrain, pons,** and **medulla oblongata.**

The brain also contains four interconnected cavities, the **cerebral ventricles,** which are filled with circulating cerebrospinal fluid (Figure 8-39).

Brainstem

The brainstem is literally the stalk of the brain. Through it pass all the nerve fibers that relay signals between the spinal cord, forebrain, and cerebellum. Running through the core of the brainstem and consisting of loosely arranged neuron cell bodies intermingled with bundles of axons is the **reticular formation,** which is the one part of the brain absolutely essential for life. It receives and integrates input from all regions of the central nervous system and processes a great deal of neural information. It is involved in motor functions, cardiovascular and respiratory control, and the mechanisms that regulate sleep and wakefulness and focus attention. Most of the biogenic amine neurotransmitters are released from cell clusters in the brainstem.

Some reticular formation neurons send axons for considerable distances up or down the brainstem and beyond, to most regions of the brain and spinal cord. This pattern explains the very large scope of influence that the reticu-

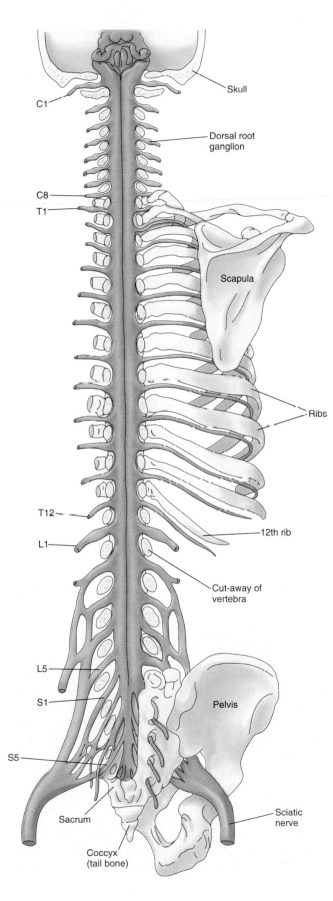

C1

Skull

Dorsal root
ganglion

C8

T1

Scapula

Ribs

T12

12th rib

L1

Cut-away of
vertebra

L5

S1

Pelvis

S5

Sacrum

Sciatic
nerve

Coccyx
(tail bone)

lar formation has over other parts of the central nervous system and explains the widespread effects of the biogenic amines.

The pathways that convey information from the reticular formation to the upper portions of the brain affect such things as wakefulness and the direction of attention to specific events by selectively facilitating neurons in some areas of the brain while inhibiting others. The fibers that descend to the spinal cord influence activity in both efferent and afferent neurons. There is considerable interaction between the reticular pathways that go up to the forebrain, down to the spinal cord, and to the cerebellum. For example, all three components function in controlling muscle activity.

Some reticular-formation neurons are clustered together, forming brainstem nuclei and integrating centers. These include the cardiovascular, respiratory, swallowing, and vomiting centers, all of which are discussed in subsequent chapters. The reticular formation also has nuclei important in eye-movement control and the reflex orientation of the body in space.

In addition, the brainstem contains nuclei involved in processing information for 10 of the 12 pairs of **cranial nerves.** These are the peripheral nerves that connect with the brain and innervate the muscles, glands, and sensory receptors of the head, as well as many organs in the thoracic and abdominal cavities (Table 8-7).

Cerebellum

The cerebellum consists of an outer layer of cells, the cerebellar cortex, and several deeper cell clusters. Although the cerebellum does not initiate voluntary movements, it is an important center for coordinating and learning movements and for controlling posture and balance. In order to carry out these functions, the cerebellum receives information from the muscles and joints, skin, eyes and ears, viscera, and the parts of the brain involved in control of movement.

FIGURE 8-37

Dorsal view of the spinal cord. Parts of the skull and vertebrae have been cut away. In general, the 8 cervical nerves (C) control the muscles and glands and receive sensory input from the neck, shoulder, arm, and hand. The 12 thoracic nerves (T) are associated with the chest and abdominal walls. The 5 lumbar nerves (L) are associated with the hip and leg, and the 5 sacral nerves (S) are associated with the genitals and lower digestive tract. (*Redrawn from FUNDAMENTAL NEUROANATOMY by Walle J. H. Nauta and Michael Fiertag. Copyright © 1986 by W. H. Freeman and Company. Reprinted by permission.*)

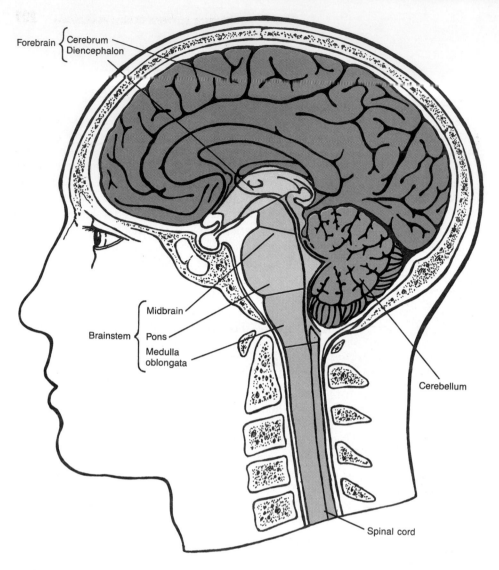

Forebrain { Cerebrum
            Diencephalon

Midbrain

Brainstem { Pons

Medulla
oblongata

Cerebellum

Spinal cord

FIGURE 8-38

The spinal cord and six divisions of the brain. 🦅

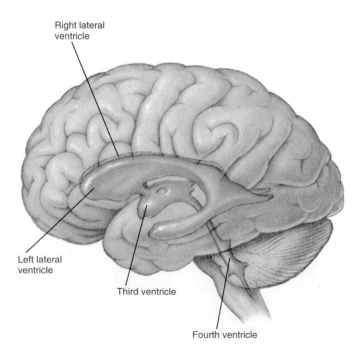

Right lateral
ventricle

Left lateral
ventricle

Third ventricle

Fourth ventricle

FIGURE 8-39

The four interconnected ventricles of the brain. 🦅

TABLE 8-7 THE CRANIAL NERVES

Name	Fibers	Comments
I. Olfactory	Afferent	Carries input from receptors in olfactory (smell) neuroepithelium. Not true nerve.
II. Optic	Afferent	Carries input from receptors in eye. Not true nerve.
III. Oculomotor	Efferent	Innervates skeletal muscles that move eyeball up, down, and medially and raise upper eyelid; innervates smooth muscles that constrict pupil and alter lens shape for near and far vision.
	Afferent	Transmits information from receptors in muscles.
IV. Trochlear	Efferent	Innervates skeletal muscles that move eyeball downward and laterally.
	Afferent	Transmits information from receptors in muscle.
V. Trigeminal	Efferent	Innervates skeletal chewing muscles.
	Afferent	Transmits information from receptors in skin; skeletal muscles of face, nose, and mouth; and teeth sockets.
VI. Abducens	Efferent	Innervates skeletal muscles that move eyeball laterally.
	Afferent	Transmits information from receptors in muscle.
VII. Facial	Efferent	Innervates skeletal muscles of facial expression and swallowing; innervates nose, palate, and lacrimal and salivary glands.
	Afferent	Transmits information from taste buds in front of tongue and mouth.
VIII. Vestibulocochlear	Afferent	Transmits information from receptors in ear.
IX. Glossopharyngeal	Efferent	Innervates skeletal muscles involved in swallowing and parotid salivary gland.
	Afferent	Transmits information from taste buds at back of tongue and receptors in auditory-tube skin.
X. Vagus	Efferent	Innervates skeletal muscles of pharynx and larynx and smooth muscle and glands of thorax and abdomen.
	Afferent	Transmits information from receptors in thorax and abdomen.
XI. Accessory	Efferent	Innervates neck skeletal muscles.
XII. Hypoglossal	Efferent	Innervates skeletal muscles of tongue.

Forebrain

The larger component of the forebrain, the cerebrum, consists of the right and left **cerebral hemispheres** as well as certain other structures on the underside of the brain. The central core of the forebrain is formed by the diencephalon.

The cerebral hemispheres (Figure 8-40) consist of the **cerebral cortex,** an outer shell of gray matter; underlying nuclei, which are also gray matter; and myelinated fiber tracts, which form white matter. The fiber tracts consist of the many nerve fibers that bring information into the cerebrum, carry information out, and connect different areas within a hemisphere. The cortex layers of the two cerebral hemispheres, although largely separated by a longitudinal division, are connected by a massive bundle of nerve fibers known as the **corpus callosum** (Figure 8-40).

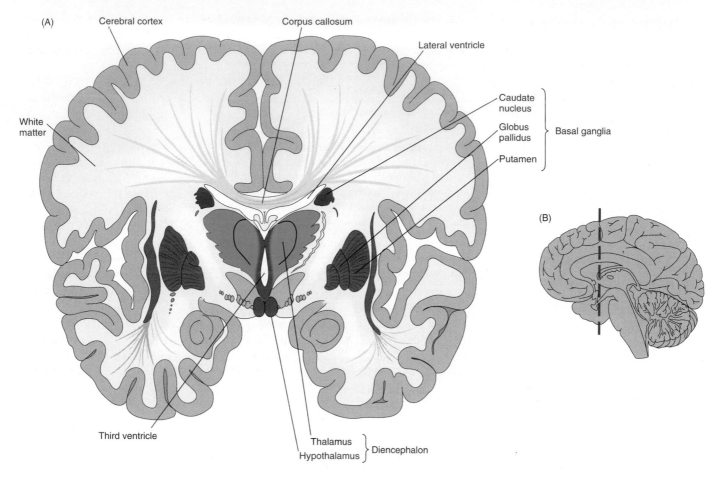

FIGURE 8-40

(A) Coronal (side-to-side) section of the brain, with the thalamus shaded in blue-green and the hypothalamus shaded in violet.
(B) The dashed line indicates the location of the cross section in A.

The cortex of each hemisphere is divided into four **lobes:** the **frontal, parietal, occipital,** and **temporal** (Figure 8-41). Although it averages only 3 mm in thickness, the cortex is highly folded, which results in an area for cortical neurons that is four times larger than it would be if unfolded, yet does not appreciably increase the volume of the brain. The cells of the cerebral cortex are organized in six layers. The cortical neurons are of two basic types: pyramidal cells (named for the shape of their cell bodies) and nonpyramidal cells. The pyramidal cells form the major output cells of the cortex, sending their axons to other parts of the cortex and to other parts of the central nervous system.

The cerebral cortex is the most complex integrating area of the nervous system. It is where basic afferent in-

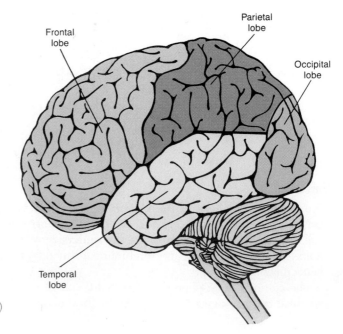

FIGURE 8-41

A lateral view of the brain. The outer layer of the forebrain (the cortex) is divided into four lobes, as shown. 𝍅

formation is collected and processed into meaningful perceptual images, and where the ultimate refinement of control over the systems that govern the movement of the skeletal muscles occurs. Nerve fibers enter the cortex predominately from a region known as the thalamus, other regions of the cortex, and the reticular formation of the brainstem. Some of the input fibers convey information about specific events in the environment, whereas others are more concerned with controlling levels of cortical excitability, determining states of arousal, and directing attention to specific stimuli.

The **subcortical nuclei** are heterogeneous groups of gray matter that lie deep within the cerebral hemispheres. Predominant among them are the **basal ganglia,** which play an important role in the control of movement and posture and in more complex aspects of behavior. (Note that the name basal ganglia is an exception to the generalization that ganglia lie outside the central nervous system.)

The diencephalon, which is divided in two by the slit-like third ventricle, is the second component of the forebrain. It contains two major parts: the thalamus and the hypothalamus (Figure 8-40). The **thalamus** is a collection of several large nuclei that serve as synaptic relay stations and important integrating centers for most inputs to the cortex. It also plays a key role in nonspecific arousal and focused attention.

The **hypothalamus** (Figure 8-40) lies below the thalamus. Although it is a tiny region that accounts for less than 1 percent of the brain's weight, it contains different cell groups and pathways that form the master command center for neural and endocrine coordination. Indeed, the hypothalamus is the single most important control area for homeostatic regulation of the internal environment and behaviors having to do with preservation of the individual—for example, eating and drinking—and preservation of the species—reproduction. The hypothalamus lies directly above the pituitary gland, an important endocrine structure, to which it is attached by a stalk (Chapter 10).

A collection of forebrain structures that consists of both gray and white matter forms the **limbic system.** This is an interconnected group of brain structures that includes portions of frontal-lobe cortex, temporal lobe, thalamus, and hypothalamus, as well as the circuitous fiber pathways that connect them (Figure 8-42). Besides being connected with each other, the parts of the limbic

FIGURE 8-42

Structures of the limbic system are shown shaded in violet in this partially transparent view of the brain. (*Redrawn from BRAIN, MIND, AND BEHAVIOR by Floyd E. Bloom and Arlyne Lazerson. Copyright ©1985, 1988 by Educational Broadcasting Corporation. Reprinted by permission of W. H. Freeman and Company.*)

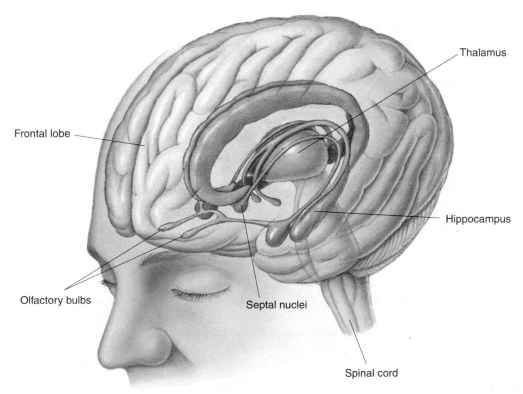

TABLE 8-8 SUMMARY OF FUNCTIONS OF THE MAJOR PARTS OF THE BRAIN

I. Brainstem
  A. Contains fibers passing between spinal cord, forebrain, and cerebellum
  B. Contains the reticular formation and its integrating centers for cardiovascular and respiratory activity (Chapters 14 and 15)
  C. Contains nuclei for most of the cranial nerves

II. Cerebellum
  A. Coordinates movements, including those for posture and balance (Chapter 12)
  B. Participates in some forms of learning (Chapter 13)

III. Cerebral hemispheres
  A. Contain the cerebral cortex, which participates in perception (Chapter 9), the generation of skilled movements (Chapter 12), reasoning, learning, and memory (Chapter 13)
  B. Contain subcortical nuclei, which participate in coordination of skeletal-muscle activity (Chapter 12)
  C. Contain interconnecting fiber pathways

IV. Thalamus
  A. Is a synaptic relay station for sensory pathways on their way to the cerebral cortex (Chapter 9)
  B. Participates in control of skeletal-muscle coordination (Chapter 12)
  C. Plays a key role in awareness

V. Hypothalamus
  A. Regulates anterior pituitary gland (Chapter 10)
  B. Regulates water balance (Chapter 16)
  C. Participates in regulation of autonomic nervous system (Chapters 8 and 18)
  D. Regulates eating and drinking behavior (Chapter 18)
  E. Regulates reproductive system (Chapters 10 and 19)
  F. Reinforces certain behaviors (Chapter 13)
  G. Generates and regulates circadian rhythms (Chapters 7, 9, 10, and 18)
  H. Regulates body temperature (Chapter 18)
  I. Participates in generation of emotional behavior (Chapter 13)

VI. Limbic system
  A. Participates in generation of emotions and emotional behavior (Chapter 13)
  B. Plays essential role in most kinds of learning (Chapter 13)

system are connected with many other parts of the central nervous system. Structures within the limbic system are associated with learning, emotional experience and behavior, and a wide variety of visceral and endocrine functions. In fact, much of the output of the limbic system is coordinated by the hypothalamus into behavioral and endocrine responses.

The functions of the major parts of the brain are listed in Table 8-8.

PERIPHERAL NERVOUS SYSTEM

Nerve fibers in the peripheral nervous system transmit signals between the central nervous system and receptors and effectors in all other parts of the body. As noted earlier, the nerve fibers are grouped into bundles called nerves. The peripheral nervous system consists of 43 pairs of nerves: 12 pairs of cranial nerves and 31 pairs that connect with the spinal cord as the spinal nerves. The cranial nerves and a summary of the information they transmit were listed in Table 8-7.

Each nerve fiber is surrounded by a Schwann cell. Some of the fibers are wrapped in layers of Schwann-cell membrane, and these tightly wrapped membranes form a myelin sheath (see Figure 8-3). Other fibers are unmyelinated.

A nerve contains nerve fibers that are the axons of efferent neurons or afferent neurons or both. Accordingly, fibers in a nerve may be classified as belonging to the **efferent** or the **afferent division** of the peripheral nervous system (Table 8-9). All the spinal nerves contain both afferent and efferent fibers, whereas some of the cranial nerves (the optic nerves from the eyes, for example) contain only afferent fibers.

As noted earlier, afferent neurons convey information from receptors at their peripheral endings to the central nervous system. The long part of their axon is outside the central nervous system and is part of the peripheral nervous system. Afferent neurons are sometimes called primary afferents or first-order neurons because they are the first cells entering the central nervous system in the synaptically linked chains of neurons that handle incoming information.

Recall that efferent neurons carry signals from the central nervous system out to muscles or glands. The efferent division of the peripheral nervous system is more complicated than the afferent, being subdivided into a **somatic nervous system** and an **autonomic nervous system.** These terms are somewhat misleading because they suggest additional nervous systems distinct from the central and peripheral systems. Keep in mind that the terms refer simply to the efferent division of the peripheral nervous system.

The simplest distinction between the somatic and autonomic systems is that the neurons of the somatic division innervate skeletal muscle, whereas the autonomic neurons innervate smooth and cardiac muscle, glands, and neurons in the gastrointestinal tract. Other differences are listed in Table 8-10.

TABLE 8-9   DIVISIONS OF THE PERIPHERAL NERVOUS SYSTEM

I. Afferent division
II. Efferent division
  A. Somatic nervous system
  B. Autonomic nervous system
    1. Sympathetic division
    2. Parasympathetic division
    3. Enteric division

TABLE 8-10   DIFFERENCES BETWEEN SOMATIC AND AUTONOMIC NERVOUS SYSTEMS

Somatic
1. Consists of a single neuron between central nervous system and skeletal-muscle cells
2. Innervates skeletal muscle
3. Can lead only to muscle excitation

Autonomic
1. Has two-neuron chain (connected by synapse) between central nervous system and effector organ
2. Innervates smooth and cardiac muscle, glands, and GI neurons
3. Can be either excitatory or inhibitory

The somatic portion of the efferent division of the peripheral nervous system is made up of all the nerve fibers going from the central nervous system to skeletal-muscle cells. The cell bodies of these neurons are located in groups in the brainstem or spinal cord. Their large diameter, myelinated axons leave the central nervous system and pass without any synapses to skeletal-muscle cells. The neurotransmitter released by these neurons is acetylcholine. Because activity in the somatic neurons leads to contraction of the innervated skeletal-muscle cells, these neurons are called **motor neurons.** Excitation of motor neurons leads only to the *contraction* of skeletal-muscle cells; there are no somatic neurons that inhibit skeletal muscles.

Autonomic Nervous System

The efferent innervation of all tissues other than skeletal muscle is by way of the autonomic nervous system. In the case of the gastrointestinal tract, the autonomic nervous system innervates the neurons that are part of the **enteric nervous system,** a nerve network in the wall of the intestinal tract that regulates the glands and smooth muscles found there. The enteric nervous system is a subdivision of the autonomic nervous system, but does not have all of the characteristics discussed below. It will be described in Chapter 17.

In the autonomic nervous system, parallel chains, each with two neurons, connect the central nervous system and the effector cells (Figure 8-43). (This is in contrast to the single neuron of the somatic system.) The first neuron has its cell body in the central nervous system. The synapse between the two neurons is outside the central nervous system, in a cell cluster called an **autonomic ganglion.** The nerve fibers passing between the central nervous system and the ganglia are called **preganglionic fibers;** those passing between the ganglia and the effector cells are **postganglionic fibers.**

Anatomical and physiological differences within the autonomic nervous system are the basis for its further subdivision into **sympathetic** and **parasympathetic** components (see Table 8-9). The nerve fibers of the sympathetic and parasympathetic components leave the central nervous system at different levels—the sympathetic fibers from the thoracic (chest) and lumbar regions of the spinal cord and the parasympathetic fibers from the brain and the sacral portion of the spinal cord (lower back, Figure 8-44). Therefore, the sympathetic division is also called

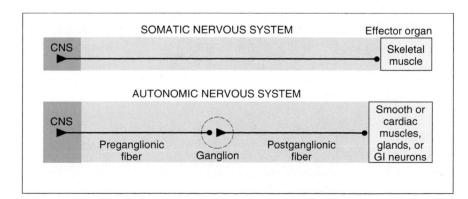

FIGURE 8-43
Efferent division of the peripheral nervous system. Overall plan of the somatic and autonomic nervous systems.

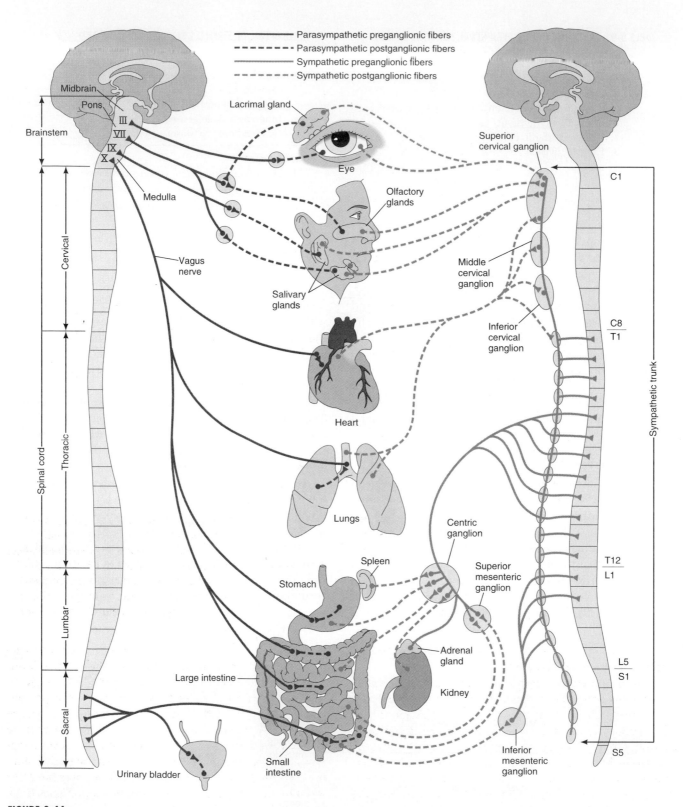

FIGURE 8-44

The parasympathetic (left) and sympathetic (right) divisions of the autonomic nervous system. The celiac, superior mesenteric, and inferior mesenteric ganglia are collateral ganglia. Only the sympathetic trunk on the right side of the spinal cord is shown, although there is also a trunk on the left. Not shown are the fibers passing to the liver, blood vessels, and skin glands from the sympathetic trunk.

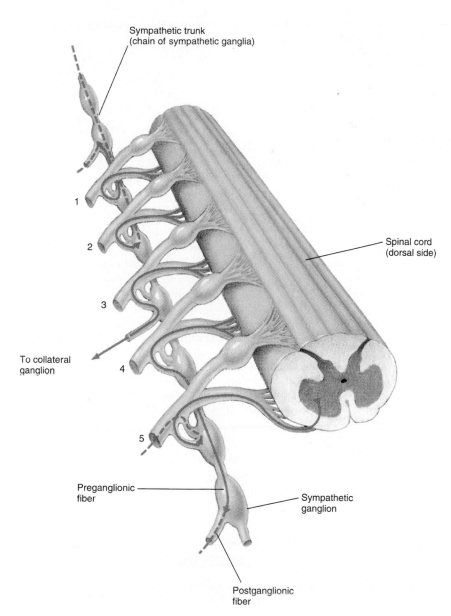

Sympathetic trunk
(chain of sympathetic ganglia)

1

2

3

To collateral
ganglion

4

5

Spinal cord
(dorsal side)

Preganglionic
fiber

Sympathetic
ganglion

Postganglionic
fiber

FIGURE 8-45

Relationship between a sympathetic trunk and spinal cord. (1 through 5) Various courses that preganglionic sympathetic fibers (solid lines) may take through the sympathetic trunk. Dashed lines represent postganglionic fibers. A mirror image of this exists on the opposite side of the spinal cord.

the thoracolumbar division, and the parasympathetic is called the craniosacral division.

The two divisions also differ in the location of ganglia. Most of the sympathetic ganglia lie close to the spinal cord and form the two chains of ganglia—one on each side of the cord—known as the **sympathetic trunks** (Figure 8-44). Other sympathetic ganglia, called collateral ganglia—the celiac, superior mesenteric, and inferior mesenteric ganglia—are in the abdominal cavity, closer to the innervated organ (Figure 8-44). In contrast, the parasympathetic ganglia lie within the organs innervated by the postganglionic neurons or close to them.

The anatomy of the sympathetic nervous system can be confusing. Preganglionic sympathetic *fibers* leave the spinal cord only between the first thoracic and third lum-

bar segments, whereas sympathetic *trunks* extend parallel to the entire length of the cord, from the cervical levels high in the neck down to the sacral levels. The ganglia in the extra lengths of sympathetic trunks receive preganglionic fibers from the thoracolumbar regions because some of the preganglionic fibers, once in the sympathetic trunks, turn to travel upward or downward for several segments before forming synapses with postganglionic neurons (Figure 8-45, numbers 1 and 4). Other possible paths taken by the sympathetic fibers are shown in Figure 8-45, numbers 2, 3, and 5.

The anatomical arrangements in the sympathetic nervous system to some extent tie the entire system together so it can act as a single unit, although small segments of the system can still be regulated independently. Furthermore,

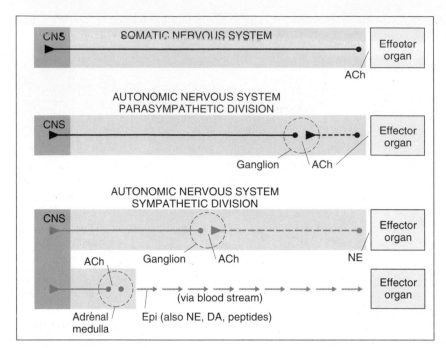

FIGURE 8-46

Transmitters used in the various components of the peripheral nervous system. In a few cases (to be described later) sympathetic neurons release a transmitter other than norepinephrine. ACh, Acetylcholine; NE, norepinephrine; Epi, epinephrine; DA, dopamine.

the ganglia can act either as relay stations with practically no change in the information they transmit to the effector organ or as important integrating centers capable of generating individualized responses. The parasympathetic system, in contrast, is made up of relatively independent components. Thus, overall autonomic responses, made up of many small parts, are quite variable and finely tailored to the specific demands of any given situation.

In both sympathetic and parasympathetic divisions, the major neurotransmitter released between pre- and postganglionic fibers in autonomic ganglia is acetylcholine (Figure 8-46). In the parasympathetic division, the major neurotransmitter between the postganglionic fiber and the effector cell is also acetylcholine. In the sympathetic division, the major transmitter between the postganglionic fiber and the effector cell is usually norepinephrine. We stress "major" and "usually" because acetylcholine is also released by some sympathetic postganglionic endings. Moreover, one or more cotransmitters are usually stored and released with the autonomic transmitters; these include ATP, dopamine, and several of the neuropeptides. These substances, however, play a relatively small role.

In addition to the classical plan of the autonomic postganglionic fibers just described, there is a widespread network of nerves recognized as nonadrenergic and noncholinergic that use nitric oxide and other neurotransmitters to mediate some forms of blood vessel dilation and to regulate various gastrointestinal, respiratory, urinary, and reproductive functions.

Many of the drugs that stimulate or inhibit various components of the autonomic nervous system affect receptors for acetylcholine and norepinephrine. Recall that there are several types of receptors for each neurotransmitter (Table 8-11). The great majority of acetylcholine receptors on autonomic postganglionic neurons are nicotinic receptors. In contrast, the acetylcholine receptors on smooth-muscle, cardiac-muscle, and gland cells are muscarinic receptors. To complete the story of the peripheral cholinergic receptors, it should be emphasized that the cholinergic re-

---

TABLE 8-11 CLASSES OF RECEPTORS FOR ACETYLCHOLINE, NOREPINEPHRINE, AND EPINEPHRINE

I. Receptors for acetylcholine
  A. Nicotinic receptors
    On postganglionic neurons in the autonomic ganglia
    At neuromuscular junctions of skeletal muscle
    On some central nervous system neurons
  B. Muscarinic receptors
    On smooth muscle
    On cardiac muscle
    On gland cells
    On some central nervous system neurons
    On some neurons of autonomic ganglia (although the great majority of receptors at this site are nicotinic)

II. Receptors for norepinephrine and epinephrine
    On smooth muscle
    On cardiac muscle
    On gland cells
    On some central nervous system neurons

ceptors on the neuromuscular junctions of skeletal-muscle fibers, innervated by the *somatic* motor neurons, not autonomic neurons, are nicotinic receptors.

One set of postganglionic neurons in the sympathetic division never develops axons; instead, upon activation by preganglionic axons, the cells of this "ganglion" release their transmitters into the bloodstream (Figure 8-46). This "ganglion," called the **adrenal medulla,** therefore functions as an endocrine gland whose secretion is controlled by the sympathetic preganglionic nerve fibers. It releases a mixture of about 80 percent epinephrine and 20 percent norepinephrine into the blood (plus small amounts of other substances, including dopamine, ATP, and neuropeptides). These catecholamines, properly called hormones rather than neurotransmitters in this circumstance, are transported via the blood to effector cells having receptors sensitive to them. The receptors may be the same adrenergic receptors that are located near the release sites of sympathetic postganglionic neurons and that are normally activated by the norepinephrine released from these neurons, or the receptors may be located at places that are not near the neurons and therefore activated only by the circulating epinephrine or norepinephrine.

Table 8-12 is a reference list of the effects of autonomic nervous system activity, which will be described in subsequent chapters. Note that the heart and many glands and smooth muscles are innervated by both sympathetic and parasympathetic fibers, that is, they receive **dual innervation.** Whatever effect one division has on the effector cells, the other division usually has the opposite effect. Moreover, the two divisions are usually activated reciprocally; that is, as the activity of one division is increased, the activity of the other is decreased. Dual innervation by nerve fibers that cause opposite responses provides a very fine degree of control over the effector organ.

A useful generalization is that the sympathetic system increases its response under conditions of physical or psychological stress. Indeed, a full-blown sympathetic response is called the **fight-or-flight response,** describing the situation of an animal forced to challenge an attacker or run from it. All resources are mobilized: heart rate and blood pressure increases, blood flow to the skeletal muscles, heart, and brain increase; the liver releases glucose; and the pupils dilate. Simultaneously, activity of the gastrointestinal tract and blood flow to the skin are decreased by inhibitory sympathetic effects.

The two divisions of the autonomic nervous system rarely operate alone, and autonomic responses generally represent the regulated interplay of both divisions. Autonomic responses usually occur without conscious control or awareness, as though they were indeed autonomous (in fact, the autonomic nervous system has been called the "involuntary" nervous system). However, it is wrong to assume that this is always the case, for it has been shown that discrete visceral or glandular responses can be learned and thus, to this extent, voluntarily controlled.

BLOOD SUPPLY, BLOOD-BRAIN BARRIER PHENOMENA, AND CEREBROSPINAL FLUID

As mentioned earlier, the brain lies within the skull, and the spinal cord within the vertebral column. Between the soft neural tissues and the bones that house them are three types of membranous coverings called **meninges:** the dura mater next to the bone, the arachnoid in the middle, and the pia mater next to the nervous tissue. A space, the subarachnoid space, between the arachnoid and pia is filled with **cerebrospinal fluid (CSF).** The meninges and their specialized parts protect and support the central nervous system, and they produce, circulate, and absorb the cerebrospinal fluid.

The cerebrospinal fluid circulates through the interconnected ventricular system to the brainstem, where it passes through small openings out to a space between the meninges on the surface of the brain and spinal cord (Figure 8–47). Aided by circulatory, respiratory, and postural pressure changes, the fluid ultimately flows to the top of the outer surface of the brain, where most of it enters the bloodstream through one-way valves in large veins. Thus, the central nervous system literally floats in a cushion of cerebrospinal fluid. Since the brain and spinal cord are soft, delicate tissues with a consistency similar to Jello, they are protected by the shock-absorbing fluid from sudden and jarring movements. If the flow is obstructed at any point, cerebrospinal fluid accumulates, causing *hydrocephalus* ("water on the brain"). In severe untreated cases, the resulting elevation of pressure in the ventricles leads to compression of the brain's blood vessels, which may in turn lead to neuronal damage and mental retardation.

Under normal conditions, glucose is the only substrate metabolized by the brain to supply its energy requirements, and most of the energy from the oxidative breakdown of glucose is transferred to ATP. The brain's glycogen stores being negligible, it is completely dependent upon a continuous blood supply of glucose. In fact, the most common form of brain damage is caused by a stoppage of the blood supply to a region of the brain. When neurons in the region are without a blood supply and deprived of nutrients and oxygen for even a few minutes, they cease to function and die. This neuronal death results in a *stroke.*

Although the adult brain makes up only 2 percent of the body weight, it receives 12 to 15 percent of the total blood supply, which supports its high oxygen utilization. If the blood flow to a region of the brain is reduced to 10 to

TABLE 8-12   SOME EFFECTS OF AUTONOMIC NERVOUS SYSTEM ACTIVITY

Effector organ	Sympathetic effect		Parasympathetic effect[†]
	Receptor type[*]		
Eyes			
Iris muscles	Alpha	Contracts radial muscle (widens pupil)	Contracts sphincter muscle (makes pupil smaller)
Ciliary muscle	Beta	Relaxes (flattens lens for far vision)	Contracts (allows lens to become more convex for near vision)
Heart			
SA node	Beta	Increases heart rate	Decreases heart rate
Atria	Beta	Increases contractility	Decreases contractility
AV node	Beta	Increases conduction velocity	Decreases conduction velocity
Ventricles	Beta	Increases contractility	Decreases contractility slightly
Arterioles			
Coronary	Alpha	Constricts	Dilates
	Beta	Dilates	
Skin	Alpha	Constricts	—[†]
Skeletal muscle	Alpha	Constricts	—
	Beta	Dilates	
Abdominal viscera	Alpha	Constricts	—
	Beta	Dilates	
Salivary glands	Alpha	Constricts	Dilates
Veins	Alpha	Constricts	
	Beta	Dilates	
Lungs			
Bronchial muscle	Beta	Relaxes	Contracts
Bronchial glands	Alpha	Inhibits secretion	Stimulates secretion
	Beta	Stimulates secretion	
Salivary glands	Alpha	Stimulates K^+ and H_2O secretion	Stimulates K^+ and H_2O secretion
	Beta	Stimulates enzyme secretion	
Stomach			
Motility, tone	Alpha Beta	Decreases	Increases
Sphincters	Alpha	Contracts	Relaxes
Secretion		Inhibits (?)	Stimulates

TABLE 8-12 *(continued)*

Effector organ	Sympathetic effect Receptor type[*]		Parasympathetic effect[†]
Intestine			
Motility	Alpha	Decreases	Increases
	Beta	Decreases	
Sphincters	Alpha	Contracts (usually)	Relaxes (usually)
Secretion	Alpha	Inhibits	Stimulates
Gallbladder	Beta	Relaxes	Contracts
Liver	Alpha and beta	Glycogenolysis and gluconeogenesis	—
Pancreas			
Exocrine glands	Alpha	Inhibits secretion	Stimulates secretion
Endocrine glands	Alpha	Inhibits secretion	Stimulates secretion
	Beta	Stimulates secretion	
Fat cells	Alpha and beta	Increases fat breakdown	
Kidneys	Beta	Increases renin secretion	—
Urinary bladder			
Bladder wall	Beta	Relaxes	Contracts
Sphincter	Alpha	Contracts	Relaxes
Uterus	Alpha	Contracts in pregnancy	Variable
	Beta	Relaxes	
Reproductive tract (male)	Alpha	Ejaculation	Erection
Skin			
Muscles causing hairs to stand erect	Alpha	Contracts	—
Sweat glands	Alpha	Localized secretion	Generalized secretion
Lacrimal glands	Alpha	Secretion	Secretion

[*]Note that in many effector organs, there are both alpha-adrenergic and beta-adrenergic receptors. Activation of these receptors may produce either the same or opposing effects. For simplicity, except for the arterioles and a few other cases, only the dominant sympathetic effect is given when the two receptors oppose each other.

[†]These effects are all mediated by muscarinic receptors.

[‡]A dash means these cells are not innervated by this branch of the autonomic nervous system or that these nerves do not play a significant physiological role.

Table adapted from "Goodman and Gilman's The Pharmacological Basis of Therapeutics," Joel G. Hardman, Lee E. Limbird, Perry B. Molinoff, Raymond W. Ruddon, and Alfred Goodman Gilman, eds., 9th edn., McGraw-Hill, New York, 1996.

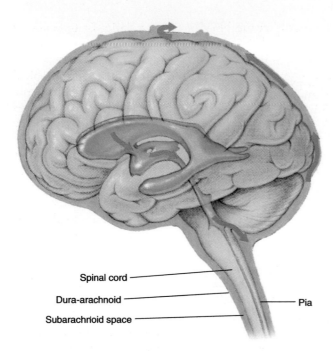

Spinal cord

Dura-arachnoid

Subarachnoid space

Pia

FIGURE 8-47

The ventricular system of the brain and the distribution of the cerebrospinal fluid, shown in blue. Cerebrospinal fluid is formed in the ventricles, passes to the subarachnoid space outside the brain and spinal cord, and then moves through small valvelike structures into the large veins of the head. ✗

25 percent of its normal level, energy stores are depleted, energy-dependent membrane ion pumps fail, membrane ion gradients decrease (that is, the membranes depolarize), and extracellular potassium concentrations increase.

The exchange of substances between blood and extracellular fluid in the central nervous system is different from the more-or-less unrestricted diffusion of nonprotein substances from blood to extracellular fluid in the other organs of the body. A complex group of **blood-brain barrier** mechanisms closely control both the kinds of substances that enter the extracellular fluid of the brain and the rate at which they enter. These mechanisms also minimize the ability of many harmful substances to reach the neurons. Therefore, the extracellular fluid of the brain and spinal cord is a product of, but chemically different from, the blood.

The blood-brain barrier, which comprises the cells that line the smallest blood vessels in the brain, has both anatomical structures, such as tight junctions, and physiological transport systems which handle different classes of substances in different ways (Figure 8-48). For example, substances that dissolve readily in the lipid components of the plasma membranes enter the brain quickly. The properties of the blood-brain barrier also explain why lipids cannot serve as a significant energy source for the brain when glucose supplies are low. Lipids are transported in

the plasma bound to albumin, a type of plasma protein, and the resulting lipoprotein aggregate cannot cross the blood-brain barrier.

The blood-brain barrier accounts for some drug actions, too, as can be seen from the following scenario: Morphine differs chemically from heroin only in that morphine has two hydroxyl groups where heroin has two acetyl groups ($-COCH_3$). This small difference renders morphine highly lipid insoluble and heroin highly lipid soluble. Thus, heroin crosses the blood-brain barrier more readily than morphine. As soon as heroin enters the brain, however, enzymes remove the acetyl groups from heroin and change it to morphine. The morphine, highly insoluble in lipid, is then effectively trapped in the brain where it continues to exert its effect. Other drugs that have rapid effects in the central nervous system because of their high lipid solubility are the barbiturates, nicotine, caffeine, and alcohol.

FIGURE 8-48

A diagram of the blood-brain barrier, which is formed by the endothelial cells that line the brain capillaries.

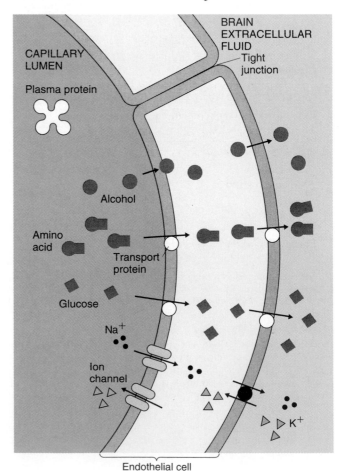

CAPILLARY LUMEN

BRAIN EXTRACELLULAR FLUID

Tight junction

Plasma protein

Alcohol

Amino acid

Transport protein

Glucose

Na$^+$

Ion channel

K$^+$

Endothelial cell

Many substances that do not dissolve readily in lipids, such as glucose and other important substrates of brain metabolism, nonetheless enter the brain quite rapidly by combining with membrane transport proteins in the cells that line the smallest brain blood vessels. Similar transport systems also move substances *out* of the brain and into the blood, preventing the buildup of molecules that could interfere with brain function.

In addition to its blood supply, the central nervous system is perfused by the cerebrospinal fluid. The cerebrospinal fluid is secreted into the ventricles by epithelial cells that cover the **choroid plexuses,** which form part of the lining of the four ventricles. A barrier is present here,

too, between the blood in the capillaries of the choroid plexuses and the cerebrospinal fluid; thus, cerebrospinal fluid is a selective secretion. For example, potassium and calcium concentrations are slightly lower in cerebrospinal fluid than in plasma, whereas the sodium and chloride concentrations are slightly higher. The choroid plexuses also trap toxic heavy metals such as lead, thus affording a degree of protection to the brain from these substances.

The cerebrospinal fluid and the extracellular fluid of the brain are, over time, in diffusion equilibrium. Thus, the extracellular environment of the brain and spinal cord neurons is regulated by restrictive, selective barrier mechanisms in the capillaries of the brain and choroid plexuses.

SECTION D SUMMARY

I.   Inside the skull and vertebral column, the brain and spinal cord are enclosed in and protected by the meninges.

CENTRAL NERVOUS SYSTEM: SPINAL CORD

I.   The spinal cord is divided into two areas: central gray matter, which contains nerve cell bodies and dendrites; and white matter, which surrounds the gray matter and contains myelinated axons organized into ascending or descending tracts.

II. The axons of the afferent and efferent neurons form the spinal nerves.

CENTRAL NERVOUS SYSTEM: BRAIN

I.   The brain is divided into six regions: cerebrum, diencephalon, midbrain, pons, medulla oblongata, and cerebellum.

II. The midbrain, pons, and medulla oblongata form the brainstem, which contains the reticular formation.

III. The cerebellum plays a role in posture, movement, and some kinds of memory.

IV. The cerebrum, made up of right and left cerebral hemispheres, and the diencephalon together form the forebrain. The cerebral cortex forms the outer shell of the cerebrum and is divided into parietal, frontal, occipital, and temporal lobes.

V.   The diencephalon contains the thalamus and hypothalamus.

VI. The limbic system is a set of deep forebrain structures associated with learning and emotions.

PERIPHERAL NERVOUS SYSTEM

I.   The peripheral nervous system consists of 43 paired nerves—12 pairs of cranial nerves and 31 pairs of spinal nerves. Most nerves contain axons of both afferent and efferent neurons.

II. The efferent division of the peripheral nervous system is divided into somatic and autonomic parts. The somatic fibers innervate skeletal muscle cells and release the neurotransmitter acetylcholine.

III. The autonomic nervous system innervates cardiac and smooth muscle, glands, and gastrointestinal-tract neurons. Each autonomic pathway consists of a preganglionic neuron with its cell body in the CNS and a postganglionic neuron with its cell body in an autonomic ganglion outside the CNS.
A.   The autonomic nervous system is divided into sympathetic and parasympathetic components. The preganglionic neurons in both sympathetic and parasympathetic divisions release acetylcholine; the postganglionic parasympathetic neurons release mainly acetylcholine, and the postganglionic sympathetics release mainly norepinephrine.
B.   The receptors that respond to acetylcholine are classified as nicotinic and muscarinic, and those that respond to norepinephrine or epinephrine as alpha- and beta-adrenergic types.
C.   The adrenal medulla is a hormone-secreting part of the sympathetic nervous system and secretes mainly epinephrine.
D.   Many effector organs innervated by the autonomic nervous system receive dual innervation.

BLOOD SUPPLY, BLOOD-BRAIN BARRIER PHENOMENA, AND CEREBROSPINAL FLUID

I.   Brain tissue depends on a continuous supply of glucose and oxygen for metabolism.

II. The brain ventricles and the space within the meninges are filled with cerebrospinal fluid, which is formed in the ventricles.

III. The chemical composition of the extracellular fluid of the CNS is closely regulated by the blood-brain barrier.

SECTION D KEY TERMS

pathway	subcortical nuclei
tract	basal ganglia
commissure	thalamus
long neural pathway	hypothalamus
multineuronal pathway	limbic system
multisynaptic pathway	efferent division of the
ganglia	peripheral nervous
nuclei	system
gray matter	afferent division of the
white matter	peripheral nervous
dorsal root	system
dorsal root ganglia	somatic nervous system
ventral root	autonomic nervous system
spinal nerve	motor neuron
cerebrum	enteric nervous system
diencephalon	autonomic ganglion
brainstem	preganglionic fiber
cerebellum	postganglionic fiber
forebrain	sympathetic division of the
midbrain	autonomic nervous
pons	system
medulla oblongata	parasympathetic division of
cerebral ventricle	the autonomic nervous
reticular formation	system
cranial nerve	sympathetic trunk
cerebral hemisphere	adrenal medulla
cerebral cortex	dual innervation
corpus callosum	fight-or-flight response
frontal lobe	meninges
parietal lobe	cerebrospinal fluid (CSF)
occipital lobe	blood-brain barrier
temporal lobe	choroid plexuses

SECTION D REVIEW QUESTIONS

1. Draw an organizational chart showing the central nervous system, peripheral nervous system, brain, spinal cord, spinal nerves, cranial nerves, forebrain, brainstem, cerebrum, diencephalon, midbrain, pons, medulla oblongata, and cerebellum.

2. Draw a cross section of the spinal cord showing the gray and white matter, dorsal and ventral roots, dorsal root ganglion, and spinal nerve. Indicate the general location of pathways.

3. List two functions of the thalamus.

4. List the functions of the hypothalamus.

5. Draw a peripheral nervous system chart indicating the relationships among afferent and efferent divisions, somatic and autonomic nervous systems, and sympathetic and parasympathetic divisions.

6. Contrast the somatic and autonomic nervous systems; mention at least three characteristics of each.

7. Name the neurotransmitter released at each synapse or neuroeffector junction in the somatic and autonomic systems.

8. Contrast the sympathetic and parasympathetic divisions; mention at least four characteristics of each.

9. Explain how the adrenal medulla can affect receptors on various effector organs despite the fact that its cells have no axons.

10. The chemical composition of the CNS extracellular fluid is different from that of blood. Explain how this difference is achieved.

CHAPTER 8 CLINICAL TERMS

local anesthetic	Alzheimer's disease
agonist	analgesic
antagonist	hydrocephalus
tetanus toxin	stroke

CHAPTER 8 THOUGHT QUESTIONS

(*Answers are given in Appendix A*)

1. Neurons are treated with a drug that instantly and permanently stops the Na,K-ATPase pumps. Assume for this question that the pumps are not electrogenic. What happens to the resting membrane potential immediately and over time?

2. Extracellular potassium concentration in a person is increased with no change in intracellular potassium concentration. What happens to the resting potential and the action potential?

3. A person has received a severe blow to the head but appears to be all right. Over the next weeks, however, he develops loss of appetite, thirst, and sexual capacity, but no loss in sensory or motor function. What part of the brain do you think may have been damaged?

4. A person is taking a drug that causes, among other things, dryness of the mouth and speeding of the heart rate but no impairment of the ability to use the skeletal muscles. What type of receptor does this drug probably block? (Table 8-11 will help you answer this.)

5. Some cells are treated with a drug that blocks chloride channels, and the membrane potential of these cells becomes slightly depolarized (less negative). From these facts, predict whether the plasma membrane of these cells actively transports chloride and in what direction.

6. If the enzyme acetylcholinesterase were blocked with a drug, what malfunctions would occur?

7. The compound tetraethylammonium (TEA) blocks the voltage-gated changes in potassium permeability that occur during an action potential. After administration of TEA, what changes would you expect in the action potential? In the afterhyperpolarization?

CHAPTER 9

The Sensory Systems

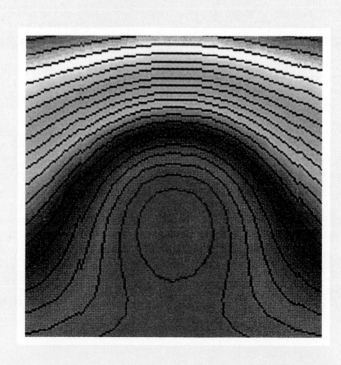

Awareness of our internal and external world is brought about by the neural mechanisms that process afferent information. The initial step of this processing is the transformation of stimulus energy first into graded potentials—the receptor potentials—and then into action potentials in nerve fibers. The pattern of action potentials in particular nerve fibers is a code that provides information about the world even though, as is frequently the case with symbols, the action potentials differ vastly from what they represent.

A **sensory system** is a part of the nervous system that consists of sensory receptors that receive stimuli from the external or internal environment, the neural pathways that conduct information from the receptors to the brain, and those parts of the brain that deal primarily with processing the information.

Information processed by a sensory system may or may not lead to conscious awareness of the stimulus. Regardless of whether the information reaches consciousness, it is called **sensory information.** If the information does reach consciousness, it can also be called a **sensation.** The understanding of the sensation's meaning is called **perception.** For example, feeling pain is a sensation, but my awareness that my tooth hurts is a perception. Perceptions occur by means of the neural processing of sensory information. At present we have little understanding of the final stages in the processing by which patterns of action potentials become sensations or perceptions.

Intuitively, it might seem that sensory systems operate like familiar electrical equipment, but this is true only up to a point. As an example, let us compare telephone transmission with our auditory (hearing) sensory system. The telephone changes sound waves into electric impulses, which are then transmitted along wires to the receiver. Thus far the analogy holds. (Of course, the mechanisms by which electric currents and action potentials are transmitted are quite different, but this does not affect our argument.) The telephone receiver then changes the coded electric impulses *back into sound waves.* Here is the crucial difference, for our brain does not physically translate the code into sound. Rather, the coded information itself or some correlate of it is what we perceive as sound.

RECEPTORS

Neural activity is initiated at the border between the nervous system and the outside world by **sensory receptors.** Since some receptors respond to changes in the internal environment, the "outside world" with regard to the re-

ceptors can also mean, for example, distension of a blood vessel in our body.

Information about the external world and about the body's internal environment exists in different energy forms—pressure, temperature, light, sound waves, and so on. Receptors at the peripheral ends of afferent neurons change these energy forms into graded potentials that can initiate action potentials, which travel into the central nervous system. The receptors are either specialized endings of afferent neurons (Figure 9-1A) or separate cells that affect the ends of afferent neurons (Figure 9-1B).

Regardless of the original form of the energy, information from receptors linking the nervous system with the outside world must be translated into the language of graded potentials or action potentials. The energy form that impinges upon and activates a receptor is known as a **stimulus.** The process by which a stimulus—a photon of light, say, or the mechanical stretch of a tissue—is transformed into an electrical response at a receptor is known as **stimulus transduction.**

There are many types of receptors, each of which is specific; that is, each responds much more readily to one form of energy than to others. The type of energy to which a receptor responds in normal functioning is known as its **adequate stimulus.**

Specificity exists at still another level. Within the general energy type that serves as a receptor's adequate stimulus, a particular receptor responds best, that is, at lowest threshold, to only a very narrow band of stimulus energy.

FIGURE 9-1

Sensory receptors. The sensitive membrane that responds to a stimulus is either (A) an ending of the afferent neuron or (B) on a separate cell adjacent to the afferent neuron (highly schematized).

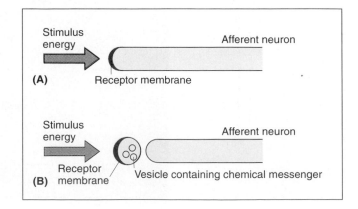

For example, individual receptors in the eye respond best to photic energy of one range of light wavelengths.

Most receptors are exquisitely sensitive to their specific energy forms. For example, some olfactory receptors respond to as few as three or four odor molecules in the inspired air, and visual receptors can respond to a single photon, the smallest quantity of light.

Virtually all receptors can be activated by several forms of energy if the intensity is sufficiently high. For example, the receptors of the eye normally respond to light, but they can be activated by an intense mechanical stimulus, like a poke in the eye. Note, however, that one still experiences the sensation of light in response to a poke in the eye. Regardless of how the receptor is stimulated, any given receptor gives rise to only one sensation.

The Receptor Potential

The transduction process in all receptors involves the opening or closing of ion channels that receive information about the outside world. These ion channels occur in a specialized receptor membrane and not on ordinary plasma membranes. The altered ion channels allow a change in the ion fluxes across the receptor membrane, which in turn produces a change in the membrane potential there. This change in potential is a graded potential called a **receptor potential** (sometimes called a generator potential). The different mechanisms by which ion channels are affected in the various types of sensory receptors are described throughout this chapter.

The specialized receptor membrane where the initial ion-channel changes occur, unlike ordinary axonal plasma membrane, does not generate action potentials. Instead, local current flows from the receptor membrane a short distance along the axon to a region where the membrane can generate action potentials. In myelinated afferent neurons, this region is usually at the first node of Ranvier of the myelin sheath (Figure 9-2).

In the case where the receptor membrane is on a separate cell, the receptor potential there causes the release of a neurotransmitter that diffuses across the extracellular cleft between the receptor cell and the afferent neuron and binds to specific sites on the afferent neuron. Thus, this junction is like a synapse. The combination of neurotransmitter with its binding sites on the afferent neuron generates a graded potential in the neuron's end analogous to an excitatory postsynaptic potential.

As is true of all graded potentials, the magnitude of a receptor potential (or a graded potential in the axon adjacent to the receptor cell) decreases with distance from its origin. However, if the amount of depolarization at the first node in the afferent neuron is large enough to bring the membrane there to threshold, action potentials are initiated, which then propagate along the nerve fiber. The only function of the graded potential is to trigger action potentials. (See Figure 8-15 to review the properties of graded potentials.)

As long as the afferent neuron remains depolarized to threshold, action potentials continue to fire and propagate along the afferent neuron. Moreover, for complex reasons, an increase in the graded potential magnitude causes an increase in the action-potential frequency in the afferent neuron (up to the limit imposed by the neuron's refractory

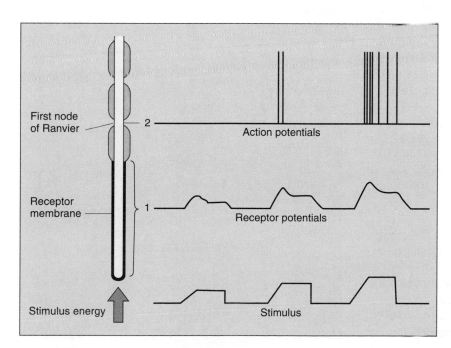

FIGURE 9-2

An afferent neuron with a receptor ending. The receptor potential arises at the nerve ending 1, and the action potential arises at the first node of the myelin sheath 2.

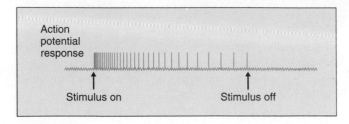

FIGURE 9-3

Action potentials in a single afferent nerve fiber showing adaptation to a stimulus of constant strength.

period). Although the graded-potential magnitude determines action-potential *frequency,* it does not determine action-potential *magnitude.* Since the action potential is all-or-none, its magnitude is independent of the strength of the initiating stimulus.

Factors that control the magnitude of the receptor potential include stimulus strength, rate of change of stimulus application, temporal summation of successive receptor potentials (Figure 8-15), and a process called **adaptation.** This last process is a decrease in the frequency of action potentials in an afferent neuron despite maintenance of the stimulus at constant strength (Figure 9-3). We shall see the significance of adaptation later when we discuss the coding of stimulus duration.

NEURAL PATHWAYS IN SENSORY SYSTEMS

The afferent neurons form the first link in a sensory pathway. A **sensory pathway** is made up of a group of neuron chains, each chain consisting of three or more neurons connected end to end by synapses. The chains in a given pathway run parallel to each other in the central nervous system and carry information to the part of the brain responsible for conscious recognition of the information, generally accepted to be the cerebral cortex. Sensory pathways are called **ascending pathways** because they go "up" to the brain.

Sensory Units

A single afferent neuron with all its receptor endings makes up a **sensory unit.** In a few cases, the afferent neuron has a single receptor, but generally the peripheral end of an afferent neuron divides into many fine branches, each terminating at a receptor.

The portion of the body that, when stimulated, leads to activity in a particular afferent neuron is called the **receptive field** for that neuron (Figure 9-4). Receptive

fields of neighboring afferent neurons overlap so that stimulation of a single point activates several sensory units; thus, activation at a single sensory unit almost never occurs. As we shall see, the degree of overlap varies in different parts of the body.

Ascending Pathways

The central processes of the afferent neurons enter the brain or spinal cord and synapse upon interneurons there. The central processes diverge to terminate on several, or many, interneurons (Figure 9-5A) and converge so that the processes of many afferent neurons terminate upon a single interneuron (Figure 9-5B). The interneurons upon which the afferent neurons synapse are termed second-order neurons, and these in turn synapse with third-order neurons, and so on, until the information (coded action potentials) reaches the cerebral cortex.

Some of the pathways convey information about only a single type of sensory information. Thus, one pathway may be influenced only by information from mechanoreceptors, whereas another may be influenced only by information from thermoreceptors. The ascending pathways in the spinal cord and brain that carry information about single types of stimuli are known as the **specific ascending pathways.** The specific pathways pass to the brainstem and thalamus, and the final neurons in the pathways go from there to different areas of the cerebral cortex (Figure 9-6). (The olfactory pathways are an exception because they go to parts of the limbic system before going to the thalamus.) By and large, the specific pathways cross to the side of the central nervous system that is opposite to the location of their receptors so that information from receptors on the right side of the body is transmitted to the left cerebral hemisphere and vice versa.

FIGURE 9-4

Sensory unit and receptive field.

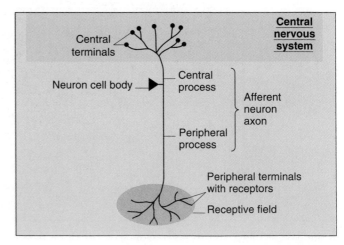

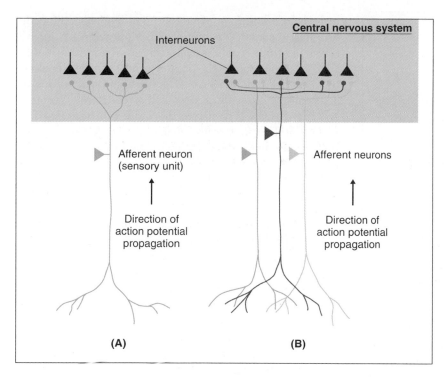

FIGURE 9-5

(A) Divergence of afferent neuron terminals. (B) Convergence of input from several afferent neurons onto single interneurons.

The specific ascending pathways that transmit information from **somatic receptors,** that is, the receptors in the framework or outer walls of the body, including skin, skeletal muscle, tendons, and joints, go to the **somatosensory cortex,** a strip of cortex that lies in the parietal lobe of the brain just behind the junction of the parietal and frontal lobes (Figure 9-6). The specific pathways from the eyes go to a different primary cortical receiving area, the **visual cortex,** which is in the occipital lobe, and the specific pathways from the ears go to the **auditory cortex,** which is in the temporal lobe (Figure 9-6). Specific pathways from the taste buds pass to a cortical area adjacent to the face region of the somatosensory cortex. The pathways serving olfaction have no representation in the cerebral cortex.

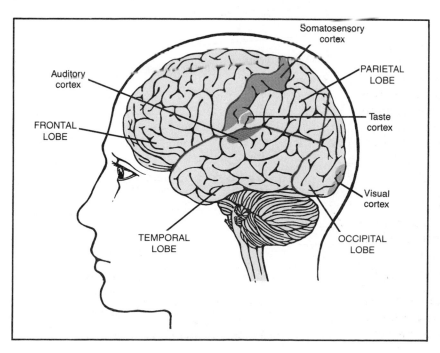

FIGURE 9-6

Primary sensory areas of the cerebral cortex.

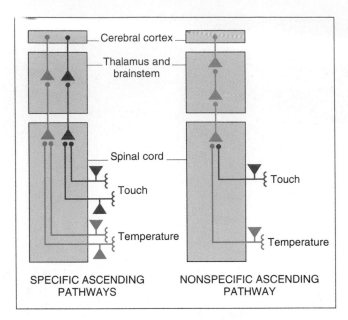

FIGURE 9-7
Diagrammatic representation of two specific sensory pathways and a nonspecific sensory pathway.

Finally, the processing of afferent information does not end in the primary cortical receiving areas but continues from these areas to association areas of the cerebral cortex.

In contrast to the specific ascending pathways, neurons in the **nonspecific ascending pathways** are activated by sensory units of several different types (Figure 9-7) and therefore signal general information. In other words, they indicate that *something* is happening, without specifying just what or where. A given neuron in a nonspecific pathway may respond, for example, to input from several afferent neurons, each activated by a different stimulus, such as maintained skin pressure, heating, and cooling. Such pathway neurons are called **polymodal neurons.** The nonspecific pathways, as well as collaterals from the specific pathways, feed into the brainstem reticular formation and regions of the thalamus and cerebral cortex that are not highly discriminative, but are important in the control of alertness and arousal.

ASSOCIATION CORTEX AND PERCEPTUAL PROCESSING

The **cortical association areas** (Figure 9-8) are brain areas that lie outside the primary cortical sensory or motor areas but are adjacent to them. The association areas are not considered part of the sensory pathways but rather play a role in the progressively more complex analysis of incoming information.

Information from the primary sensory cortical areas is elaborated after it is relayed to a cortical association area. The region of association cortex closest to the primary sensory cortical area processes the information in fairly simple ways and serves basic sensory-related functions. Regions farther from the primary sensory areas process the information in more complicated ways, including, for example, input from areas of the brain serving arousal, at-

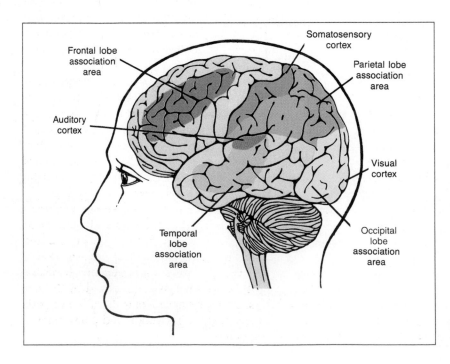

FIGURE 9-8
Areas of association cortex.

tention, memory, language, and emotions. Some of the neurons in these latter regions also receive input concerning two or more other types of sensory stimuli. Thus, a neuron receiving input from both the visual cortex and the "neck" region of the somatosensory cortex might be concerned with integrating visual information with sensory information about head position so that, for example, a tree is understood to be vertical even though the viewer's head is tipped sideways.

Fibers from neurons of the parietal and temporal lobes go to association areas in the frontal lobes that are part of the limbic system. Through these connections sensory information can be invested with emotional and motivational significance.

Further perceptual processing involves not only arousal, attention, learning, memory, language, and emotions, but also comparing the information presented via one type of sensation with that of another. For example, we may hear a growling dog, but our perception of the event and our emotional response vary markedly, depending upon whether our visual system detects the sound source to be an angry animal or a loudspeaker.

Factors That Distort Perception

We put great trust in our sensory-perceptual processes despite the inevitable modifications we know to exist. Some of the following factors are known to distort our perceptions of the real world:

1. Afferent information is distorted by receptor mechanisms, for example by adaptation, and by processing of the information along afferent pathways.
2. Factors such as emotions, personality, experience, and social background can influence perceptions so that two people can witness the same events and yet perceive them differently.
3. Not all information entering the central nervous system gives rise to conscious sensation. Actually, this is a very good thing because many unwanted signals are generated by the extreme sensitivity of our receptors. For example, under ideal conditions the rods of the eye can detect the flame of a candle 17 mi away. The hair cells of the ear can detect vibrations of an amplitude much lower than those caused by blood flow through the ears' blood vessels and can even detect molecules in random motion bumping against the ear drum. It is possible to detect one action potential generated by a certain type of mechanoreceptor. Although these receptors are capable of giving rise to sensations, much of their information is cancelled out by receptor or central mechanisms, which will be discussed later. Information in other receptors' afferent pathways is not canceled out—it simply does not feed

into parts of the brain that give rise to a conscious sensation. For example, stretch receptors in the walls of some of the largest blood vessels monitor blood pressure as part of reflex regulation of this pressure, but people have no conscious awareness of their blood pressure.

4. We lack suitable receptors for many energy forms. For example, we cannot directly detect ionizing radiation and radio or television waves.
5. Damaged neural networks may give faulty perceptions as in the bizarre phenomenon known as **phantom limb,** in which a limb that has been lost by accident or amputation is experienced as though it were still in place. The missing limb can be the "site" of tingling, touch, pressure, warmth, itch, wetness, pain, and even fatigue, and it is felt as though it were still a part of "self." It seems that the sensory neural networks in the central nervous system that exist genetically in everyone and are normally triggered by receptor activation are, instead, in the case of phantom limb, activated independently of peripheral input. The activated neural networks continue to generate the usual sensations, which are perceived as arising from the missing receptors. Moreover, somatosensory cortex undergoes marked reorganization after the loss of input from a part of the body so that a person whose arm has been amputated may perceive a touch on the cheek as though it were a touch on the phantom arm; because of the reorganization, the arm area of somatosensory cortex receives input normally directed to the face somatosensory area.
6. Some drugs alter perceptions. In fact, the most dramatic examples of a clear difference between the real world and our perceptual world can be found in illusions and drug- and disease-induced hallucinations, where whole worlds can be created.

In summary, for perception to occur, the three processes involved—transducing stimulus energy to action potentials by the receptor, transmitting data through the nervous system, and interpreting data—cannot be separated. Sensory information is processed at each synapse along the afferent pathways and at many levels of the central nervous system, with the more complex stages receiving input only after it has been processed by the more elementary systems. This hierarchical processing of afferent information along individual pathways is an important organizational principle of sensory systems. As we shall see, a second important principle is that information is processed by *parallel* pathways, each of which handles a limited aspect of the neural signals generated by the sensory transducers. Every synapse along the afferent pathways adds an element of organization and contributes to the sensory experience.

We turn now to how the particular characteristics of a stimulus are coded by the various receptors and sensory pathways.

PRIMARY SENSORY CODING

The sensory systems code four aspects of a stimulus: stimulus type, intensity, location, and duration.

Stimulus Type

Another term for stimulus type (heat, cold, sound, or pressure, for example) is stimulus **modality.** Modalities can be divided into submodalities: Cold and warm are submodalities of temperature, whereas salt, sweet, bitter, and sour are submodalities of taste. The type of receptor activated by a stimulus plays the primary role in coding the stimulus modality.

As mentioned earlier, a given receptor type is particularly sensitive to one stimulus modality because of the signal transduction mechanisms and ion channels incorporated in the receptor's plasma membrane. For example, receptors for vision contain pigment molecules whose shape is transformed by light; these receptors also have intracellular mechanisms by which changes in the pigment molecules alter the activity of membrane ion channels and generate a neural signal. Receptors in the skin have neither light-sensitive molecules nor plasma-membrane ion channels that can be affected by them; thus, receptors in the eyes respond to light and those in the skin do not.

All the receptors of a single afferent neuron are preferentially sensitive to the same type of stimulus. For example, they are all sensitive to cold or all to pressure.

Adjacent sensory units, however, may be sensitive to different types of stimuli. Since the receptive fields for different modalities overlap, a single stimulus, such as an ice cube on the skin, can give rise simultaneously to the sensations of touch and temperature.

Stimulus Intensity

How is a strong stimulus distinguished from a weak one when the information about both stimuli is relayed by action potentials that are all the same size? The *frequency* of action potentials in a single receptor is one way, since as described earlier, increased stimulus strength means a larger receptor potential and a higher frequency of action-potential firing.

In addition to an increased firing rate from individual receptors, receptors on other branches of the same afferent neuron also begin to respond. The action potentials generated by these receptors propagate along the branches to the main afferent nerve fiber and add to the train of action potentials there. Figure 9-9 is a record of an experiment in which increased stimulus intensity to the receptors of a single sensory unit is reflected in increased action-potential frequency in its afferent nerve fiber.

In addition to increasing the firing frequency in a single afferent neuron, stronger stimuli usually affect a larger area and activate similar receptors on the endings of *other* afferent neurons. For example, when one touches a surface lightly with a finger, the area of skin in contact with the surface is small, and only receptors in that skin area are stimulated. Pressing down firmly increases the area of skin stimulated. This "calling in" of receptors on additional afferent neurons is known as **recruitment** or, sometimes, as spatial summation (although this is different from synaptic spatial summation).

FIGURE 9-9

Action potentials from an afferent fiber leading from a pressure receptor as the receptor is subjected to pressures of different magnitudes.

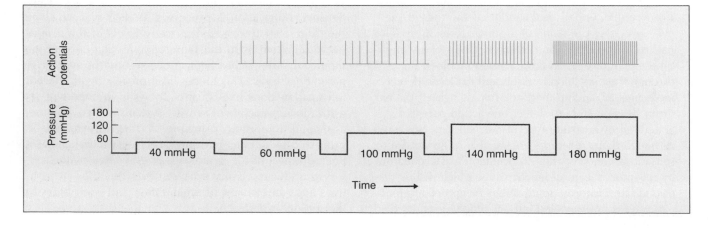

Stimulus Location

A third type of information to be signaled is the location of the stimulus, in other words, where the stimulus is being applied. (It should be noted that in vision, hearing, and smell, stimulus location is interpreted as arising from the site from which the stimulus *originated* rather than the place on our body where the stimulus was actually *applied*. For example, we interpret the sight and sound of a barking dog as occurring in that furry thing on the other side of the fence rather than in a specific region of our eyes and ears. More will be said of this later; we deal here with the senses in which the stimulus is located to a site on the body.)

The main factor coding stimulus location is the site of the stimulated receptor. The precision, or **acuity,** with which one stimulus can be located and differentiated from an adjacent one depends upon the amount of convergence of neuronal input in the specific ascending pathways. The greater the convergence, the less the acuity. Other factors affecting acuity are the size of the receptive field covered by a single sensory unit and the amount of overlap of nearby receptive fields. For example, it is easy to discriminate between two adjacent stimuli (two-point discrimination) applied to the skin on a finger, where the sensory units are small and the overlap considerable. It is harder to do so on the back, where the sensory units are large and widely spaced. Locating sensations from internal organs is less precise than from the skin because there are fewer afferent neurons in the internal organs and each has a larger receptive field.

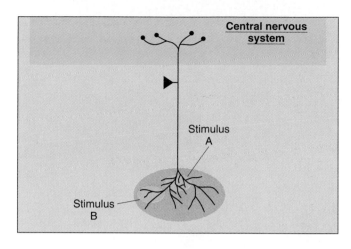

FIGURE 9-11

Two stimulus points, A and B, in the receptive field of a single afferent neuron. The density of nerve endings in area A is greater than in B, and therefore the frequency of action potentials in response to a stimulus in area A will be greater than the response to a similar stimulus at B.

It is fairly simple to see why a stimulus to a neuron that has a small receptive field can be located more precisely than a stimulus to a neuron with a large receptive field (Figure 9-10). The fact is, however, that even in the former case one cannot distinguish exactly where within the receptive field a stimulus has been applied; one can only tell that the afferent neuron has been activated. In this case, receptive-field overlap aids stimulus localization even though, intuitively, overlap would seem to "muddy" the image. Let us examine in the next two paragraphs how overlap improves stimulus localization.

An afferent neuron responds most vigorously to stimuli applied at the center of its receptive field because the receptor density, that is, the number of receptors in a given area, is greatest there. The response decreases as the stimulus is moved toward the receptive-field periphery. Thus, a stimulus activates more receptors and generates more action potentials if it occurs at the center of the receptive field (point A in Figure 9-11). The firing frequency of the afferent neuron is also related to stimulus strength, however, and a high frequency of impulses in the single afferent nerve fiber of Figure 9-11 could mean either that a moderately intense stimulus was applied to the center at A or that a strong stimulus was applied to the periphery at B. Thus, neither the intensity nor the location of the stimulus can be detected precisely with a single afferent neuron.

Since the receptor endings of different afferent neurons overlap, however, a stimulus will trigger activity in more than one sensory unit. In Figure 9-12, neurons A

FIGURE 9-10

The information from neuron A indicates the stimulus location more precisely than does that from neuron B because A's receptive field is smaller.

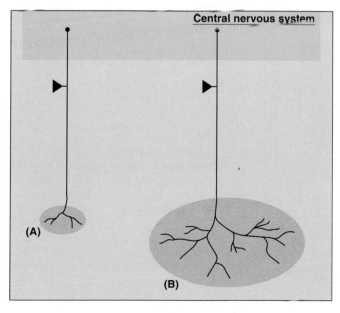

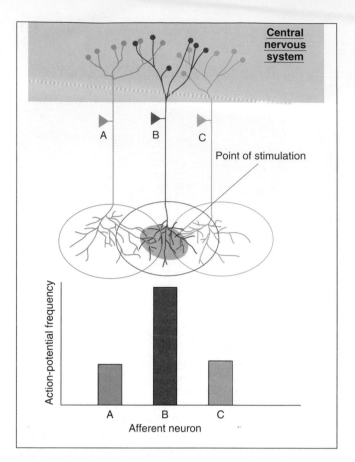

FIGURE 9-12

A stimulus point falls within the overlapping receptive fields of three afferent neurons. Note the difference in receptor response, that is, the action potential frequency in the three neurons, due to the difference in receptor distribution under the stimulus (low receptor density in A and C, high in B).

and C, stimulated near the edge of their receptive fields where the receptor density is low, fire at a lower frequency than neuron B, stimulated at the center of its receptive field. In the group of sensory units in Figure 9-12, a high action-potential frequency in neuron B occurring simultaneously with lower frequencies in A and C permits a more accurate localization of the stimulus near the center of neuron B's receptive field. Once this location is known, the firing frequency of neuron B can be used to indicate stimulus intensity.

Lateral Inhibition. Far more important in localization of the stimulus site than the different sensitivities of receptors throughout the receptive field is the phenomenon of **lateral inhibition.** In lateral inhibition, information from receptors at the edge of a stimulus is strongly inhibited compared to information from the stimulus's center. Thus, lateral inhibition increases the contrast between relevant and irrelevant information, thereby increasing the effectiveness of selected pathways and focusing sensory-processing mechanisms on "important" messages. Figure 9-13 shows one neuronal arrangement that accomplishes lateral inhibition. Lateral inhibition can occur at different levels of the sensory pathways but typically happens at an early stage.

Lateral inhibition can be demonstrated in the following way. While pressing the tip of a pencil against your finger with your eyes closed, you can localize the pencil point precisely, even though the region around the pencil tip is also indented and mechanoreceptors within this region are activated (Figure 9-14). Exact localization occurs because the information from the peripheral regions is removed by lateral inhibition.

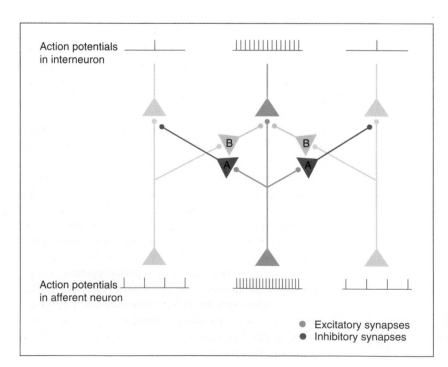

FIGURE 9-13

Afferent pathways showing lateral inhibition. The central fiber at the beginning of the pathway (bottom of figure) is firing at the highest frequency and inhibits, via inhibitory neurons A, the lateral neurons more strongly than the lateral pathways inhibit it, via inhibitory neurons B. Thus, because of the lateral inhibition, the lateral pathways are inhibited more strongly than the central pathway and the contrast between the activity in the adjacent pathways is enhanced.

Lateral inhibition is utilized to the greatest degree in the pathways providing the most accurate localization. For example, movement of skin hairs, which we can locate quite well, activates pathways that have significant lateral inhibition, but temperature and pain, which we can locate only poorly, activate pathways that use lateral inhibition to a lesser degree.

Stimulus Duration

Receptors differ in the way they respond to a constantly maintained stimulus, that is, in the way they undergo adaptation.

The response—the action-potential frequency—at the beginning of the stimulus indicates the stimulus strength, but after this initial response, the frequency differs widely in different types of receptors. Some receptors respond very rapidly at the stimulus onset, but, after their initial burst of activity, fire only very slowly or stop firing all together during the remainder of the stimulus. These are the **rapidly adapting receptors;** they are important in signaling rapid change, for example, vibrating or moving stimuli. Some receptors adapt so rapidly that they fire only

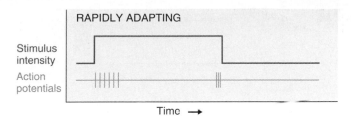

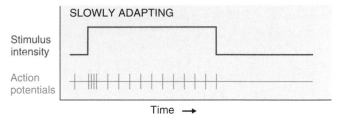

FIGURE 9-15

Rapidly and slowly adapting receptors. The top line indicates application of the stimulus, and the bottom line, the action potential firing of the afferent nerve fiber from the receptor.

FIGURE 9-14

(A) A pencil tip pressed against the skin depresses surrounding tissue. Receptors are activated under the pencil tip and in the adjacent tissue. (B) Because of lateral inhibition, the central area of excitation is surrounded by an area where the afferent information is inhibited. (C) The sensation is localized to a more restricted region than that in which mechanoreceptors were actually stimulated.

a single action potential at the onset of a stimulus—an on response—while others respond at the beginning of the stimulus and again at its removal—so-called on-off responses. The rapid fading of the sensation of our clothes pressing on our skin is due to rapidly adapting receptors.

Slowly adapting receptors maintain their response at or near the initial level of firing regardless of the stimulus duration (Figure 9-15). These receptors signal slow changes or prolonged events, such as occur in the joint and muscle receptors that participate in the maintenance of upright posture when we are standing or sitting for long periods of time.

Central Control of Afferent Information

All sensory information is subject to extensive control at the various synapses along the ascending pathways before it reaches higher levels of the central nervous system. Much of the incoming information is reduced or even abolished by inhibition from collaterals from other neurons in ascending pathways (lateral inhibition, discussed earlier) or by pathways descending from higher centers in the brain. The reticular formation and cerebral cortex, in particular, control much afferent information via descending pathways. The inhibitory controls may be exerted directly by synapses on the axon terminals of the primary afferent neurons (an example of presynaptic inhibition) or indirectly via interneurons that affect other neurons in the sensory pathways (Figure 9-16).

In some cases, for example in the pain pathways, the afferent input is continuously inhibited to some degree. This provides the flexibility of either removing the inhibition (disinhibition) so as to allow a greater degree of signal transmission or of increasing the inhibition so as to block the signal more completely.

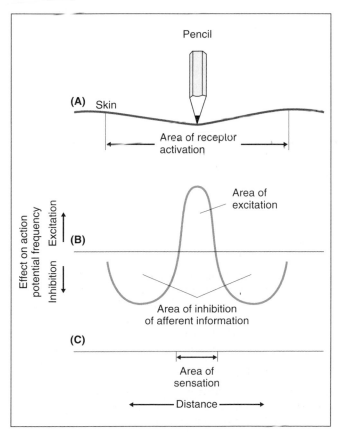

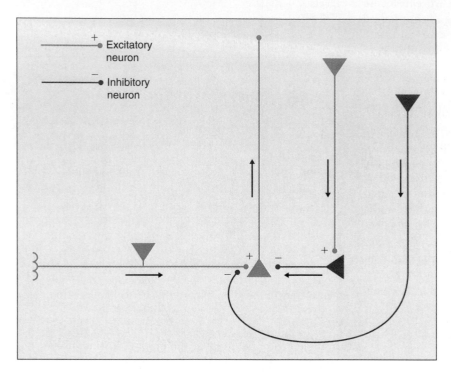

FIGURE 9-16

Descending pathways may control sensory information by directly inhibiting the central terminals of the afferent neuron (an example of presynaptic inhibition) or via an interneuron that affects the ascending pathway by inhibitory synapses. Arrows indicate the direction of action-potential transmission.

This completes our general introduction to sensory system pathways and coding. We now present the individual systems, covering the somatic sensations and vision in greater detail than the others because they serve as good models for the common principles of sensory system organization. These principles are (1) specific receptor types are sensitive to certain modalities, even to certain submodalities; (2) a specific sensory pathway codes for a particular modality; (3) information is organized such that initial cortical processing of the various modalities occurs in different parts of the brain; (4) the ascending pathways are crossed so that sensory information is generally processed by the side of the brain opposite the side of the body that was stimulated; and (5) in addition to other synaptic relay points, all ascending pathways, except for those involved in smell, synapse in the thalamus.

SECTION A SUMMARY

I. Sensory processing begins with the transformation of stimulus energy into graded potentials and then into action potentials in nerve fibers.

II. Information carried in a sensory system may or may not lead to a conscious awareness of the stimulus.

RECEPTORS

I. Receptors translate information from the external world and internal environment into graded potentials, which then generate action potentials.
   A. Receptors may be either specialized endings of afferent neurons or separate cells at the end of the neurons.
   B. Receptors respond best to one form of stimulus energy, but they may respond to other energy forms if the stimulus intensity is abnormally high.

C. Regardless of how a specific receptor is stimulated, activation of that receptor always leads to perception of one sensation. Not all receptor activations lead, however, to conscious sensations.

II. The transduction process in all sensory receptors involves the opening or closing of ion channels in the receptor. Ions then flow across the membrane, causing a receptor potential.
   A. Receptor-potential magnitude and action-potential frequency increase as stimulus strength increases.
   B. Receptor-potential magnitude varies with stimulus strength, rate of change of stimulus application, temporal summation of successive receptor potentials, and adaptation.

NEURAL PATHWAYS IN SENSORY SYSTEMS

I. A single afferent neuron with all its receptor endings is a sensory unit.

A. Afferent neurons, which usually have more than one receptor of the same type, are the first neurons in sensory pathways.

B. The area of the body that, when stimulated, causes activity in a sensory unit or other neuron in the ascending pathway of that unit is called the receptive field for that neuron.

II. Neurons in the specific ascending pathways convey information to specific primary receiving areas of the cerebral cortex about only a single type of stimulus.

III. Nonspecific ascending pathways convey information from more than one type of sensory unit to the brainstem reticular formation and regions of the thalamus that are not part of the specific ascending pathways.

ASSOCIATION CORTEX AND PERCEPTUAL PROCESSING

I. Information from the primary sensory cortical areas is elaborated after it is relayed to a cortical association area.

A. The region of association cortex closest to the primary sensory cortical area processes the information in fairly simple ways and serves basic sensory-related functions.

B. Regions of association cortex farther from the primary sensory areas process the sensory information in more complicated ways.

C. Processing in the association cortex includes input from areas of the brain serving other sensory modalities, arousal, attention, memory, language, and emotions.

PRIMARY SENSORY CODING

I. The type of stimulus perceived is determined primarily by the type of receptor activated. All receptors of a given sensory unit respond to the same stimulus modality.

II. Stimulus intensity is coded by the rate of firing of individual sensory units and by the number of sensory units activated.

III. Perception of the stimulus location depends on the size of the receptive field covered by a single sensory unit and on the overlap of nearby receptive fields. Lateral inhibition is a means by which ascending pathways emphasize wanted information and increase sensory acuity.

IV. Stimulus duration is coded by slowly adapting receptors.

V. Information coming into the nervous system is subject to control by both ascending and descending pathways.

SECTION A KEY TERMS

sensory system	somatic receptor
sensory information	somatosensory cortex
sensation	visual cortex
perception	auditory cortex
sensory receptor	nonspecific ascending
stimulus	pathway
stimulus transduction	polymodal neuron
adequate stimulus	cortical association area
receptor potential	modality
adaptation	recruitment
sensory pathway	acuity
ascending pathway	lateral inhibition
sensory unit	rapidly adapting receptor
receptive field	slowly adapting receptor
specific ascending pathway	

SECTION A REVIEW QUESTIONS

1. Distinguish between a sensation and a perception.

2. Describe the general process of transduction in a receptor that is a cell separate from the afferent neuron. Include in your description the terms specificity, stimulus, receptor potential, neurotransmitter, graded potential, action potential.

3. List several ways in which the magnitude of a receptor potential can be varied.

4. Describe the relationship between sensory information processing in the primary cortical sensory areas and in the cortical association areas.

5. List several ways in which sensory information can be distorted.

6. How does the nervous system distinguish between stimuli of different types?

7. How is information about stimulus intensity coded by the nervous system?

8. Make a diagram showing how a specific ascending pathway relays information from peripheral receptors to the cerebral cortex.

SECTION B
SPECIFIC SENSORY SYSTEMS

SOMATIC SENSATION

Sensation from the skin, body wall, muscles, bones, tendons, and joints is termed **somatic sensation** and is initiated by a variety of somatic receptors (Figure 9-17). Some respond to mechanical stimulation of the skin, hairs, and underlying tissues, whereas others respond to temperature or chemical changes. Activation of somatic receptors gives rise to the sensations of touch, pressure, warmth, cold, pain, and awareness of the position of the body parts and their movement. The receptors for visceral sensations, which arise in certain organs of the thoracic and abdominal cavities, are the same as the receptors that give rise to somatic sensations. Some organs, such as the liver, have no sensory receptors at all.

Each sensation is associated with a specific receptor type. In other words, there are distinct receptors for heat, cold, touch, pressure, limb position or movement, and pain. After entering the central nervous system, the afferent nerve fibers from the somatic receptors synapse on neurons that form the specific ascending pathways going primarily to the somatosensory cortex via the brainstem and thalamus. They also synapse on interneurons that give rise to the nonspecific pathways. For reference, the location of some important ascending pathways is shown in a cross section of the spinal cord (Figure 9-18A), and two are diagramed as examples in Figure 9-18B and C.

Note that the pathways cross from the side where the afferent neurons enter the central nervous system to the opposite side either in the spinal cord (Figure 9-18B) or brainstem (Figure 9-18C). Thus, the sensory pathways from somatic receptors on the left side of the body go to the somatosensory cortex of the right cerebral hemisphere, and vice versa.

FIGURE 9-17

Skin receptors. Some nerve fibers have free endings not related to any apparent receptor structures. Thicker, myelinated axons, on the other hand, end in receptors that are complex mechanoreceptors sensitive to displacement of the skin, for example, the Merkel, Pacinian, Meissner, and Ruffini corpuscles. The receptors are not drawn to scale; for example, Pacinian corpuscles are actually four to five times larger than Meissner's corpuscles.

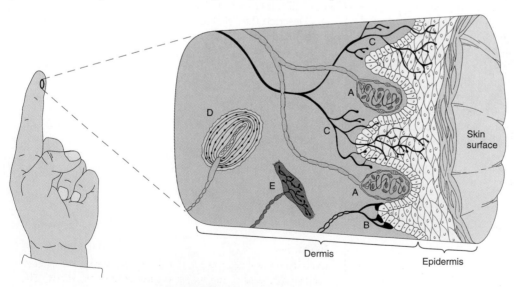

Dermis   Epidermis

A – Tactile (Meissner's) corpuscle (light touch)
B – Tactile (Merkle's) corpuscles (touch)
C – Free terminal (pain)
D – Lamellated (Pacinian) corpuscle (deep pressure)
E – Ruffini corpuscle (warmth)

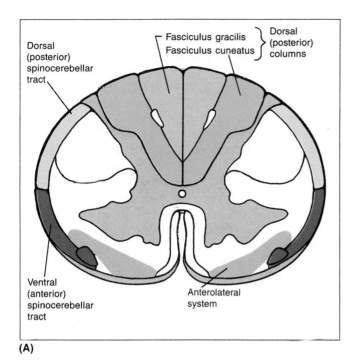

(A)

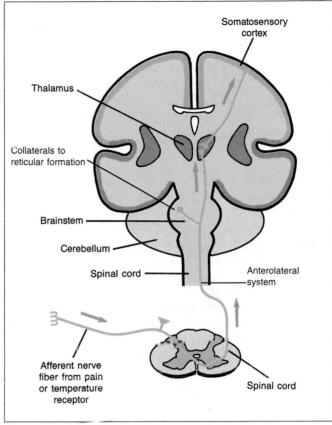

(B)

(C)

FIGURE 9-18

(A) A reference cross section of the spinal cord showing the relative locations of the major ascending fiber tracts. (B) The anterolateral system. (C) The dorsal columns. Information carried over collaterals to the reticular formation in (B) and (C) contribute to alertness and arousal mechanisms. (*Parts B and C adapted from Gardner.*)

In the somatosensory cortex, the endings of the axons of the specific somatic pathways are grouped according to the location of the receptors giving rise to the pathways (Figure 9-19). The parts of the body that are most densely innervated—fingers, thumb, and lips—are represented by the largest areas of the somatosensory cortex. There are qualifications, however, to this seemingly precise picture: The sizes of the areas can be modified with changing sensory experience, and there is considerable overlap of the body-part representations.

Touch-Pressure

Stimulation of the variety of mechanoreceptors in the skin (Figure 9-17) leads to a wide range of touch-pressure experiences—hair bending, deep pressure, vibrations, and superficial touch, for example. These mechanoreceptors are highly specialized nerve endings encapsulated in elaborate cellular structures. The details of the mechanoreceptors vary, but generally the nerve endings are linked to collagen-fiber networks within the capsule. The capsule receives the mechanical energy of the stimulus, and the capsule-nerve ending links focus the energy and transmit it to the nerve endings, which change the mechanical energy to receptor potentials.

The receptors for the senses of touch and pressure depend on mechanosensitive ion channels, which open (or close) more frequently when pressure increases the tension in the plasma membrane of the receptor.

The skin mechanoreceptors adapt at different rates, about half adapting rapidly (that is, they fire only when the stimulus is changing), and the others adapting slowly. Activation of rapidly adapting receptors gives rise to the sensations of touch, movement, and vibration, whereas slowly adapting receptors give rise to the sensation of pressure.

In both categories, some receptors have small, well-defined receptive fields and are able to provide precise information about the contours of objects indenting the skin. As might be expected, these receptors are concentrated at the fingertips. In contrast, other receptors have large receptive fields with obscure boundaries, sometimes covering a whole finger or a large part of the palm. These receptors are not involved in detailed spatial discrimination but signal information about vibration, skin stretch, and joint movement.

Sense of Posture and Movement

The senses of posture and movement are complex. The major receptors responsible for these senses are the

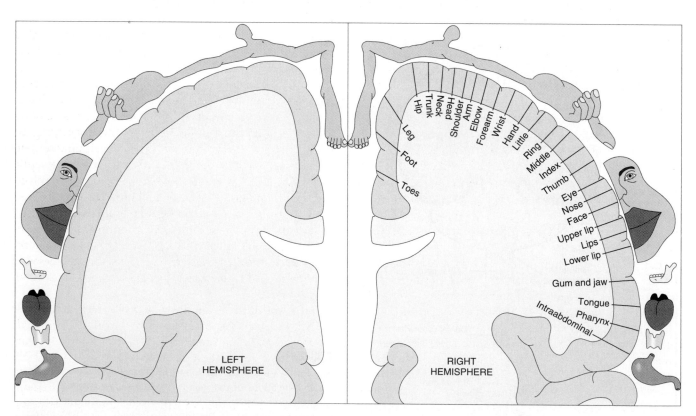

FIGURE 9-19

Location of pathway terminations for different parts of the body in the somatosensory cortex, although there is actually much overlap between the cortical regions. The left half of the body is represented on the right hemisphere of the brain, and the right half of the body is represented on the left hemisphere.

muscle-spindle stretch receptors, which occur in skeletal muscles and respond both to the absolute magnitude of muscle stretch and to the rate at which the stretch occurs (to be described in Chapter 12). The senses of posture and movement are also supported by vision and the vestibular organs (the "sense organs of balance," described below). Mechanoreceptors in the joints, tendons, ligaments, and skin also play a role. The term **kinesthesia** is often used for the sense of movement at a joint.

Temperature

There are two types of thermoreceptors in the skin, each of which responds to a limited range of temperature. Warmth receptors respond to temperatures between 30 and 43°C with an increased discharge rate upon warming, whereas receptors for cold are stimulated by temperatures between 35 and 20°C and increase their discharge rate upon cooling. It is not known how heat or cold alter the endings of the thermosensitive afferent neurons to generate receptor potentials.

Pain

A stimulus that causes (or is on the verge of causing) tissue damage usually elicits a sensation of pain. Pain differs from the other somatosensory modalities, however, in that the first stimuli set off a train of changes in the pain pathways that alter the way other pathway components respond to subsequent stimuli. Thus, the pain experienced in response to stimuli occurring even a short time after the original stimulus (and the reactions to that pain) can be very different from the pain experienced initially.

Moreover, probably more than any other type of sensation, pain can be altered by past experiences, suggestion, emotions (particularly anxiety), and the simultaneous activation of other sensory modalities. Thus, the level of pain is not solely a physical property of the stimulus.

The receptors known as **nociceptors**, whose stimulation gives rise to pain, are on the peripheral ends of small unmyelinated or lightly myelinated afferent neurons. Nociceptors respond to intense mechanical or thermal stimulation or to many chemicals, including neuropeptide transmitters, bradykinin, histamine, cytokines, prostaglandins, and growth factors. These substances act by combining with chemical binding sites on the nociceptors.

If there is actually tissue damage and inflammation at the painful site, a great deal of interaction occurs between the damaged tissue, cells of the immune system that move into the area of injury (described in Chapter 20), and nearby afferent pain neurons. All three of these—the tissue, immune cells, and afferent neurons—release chemicals such as those named above and are, in turn, affected by them.

Some of the substances alter the ion permeabilities of neurons in the pain pathways so that their threshold is decreased and their response to later painful stimuli enhanced. This increase in sensitivity to painful stimuli, known as **hyperalgesia,** can last hours after the original stimulus is over. On the other hand, opioid messengers synthesized and released by nearby tissues decrease receptor sensitivity. Note that this complex scenario occurs in the periphery before action potentials in the afferent neuron even enter the central nervous system!

The primary afferents having nociceptor endings synapse on interneurons after entering the central nervous system. The neuropeptide, substance P, is one of the transmitters released at this site. Pain information is mediated by several ascending pathways that together convey information that leads to both the localization of pain and its sensory and emotional components.

The activation of interneurons by incoming nociceptive afferents may lead to the phenomenon of **referred pain,** in which the sensation of pain is experienced at a site other than the injured or diseased part. For example, during a heart attack, pain is often experienced in the left arm. This type of referred pain occurs because both visceral and somatic afferents often converge on the same interneurons in the pain pathways. Excitation of the somatic afferent fibers is the more usual source of afferent discharge, so we "refer" the location of receptor activation to the somatic source even though, in the case of visceral pain, the perception is incorrect.

Electrical stimulation of specific areas of the central nervous system can produce a profound reduction in pain, a phenomenon called **stimulation-produced analgesia,** by inhibiting pain pathways. Analgesia occurs from such stimulation because descending pathways that originate in these brain areas selectively inhibit the transmission of information originating in nociceptors. The descending axons end at lower brainstem and spinal levels on interneurons in the pain pathways as well as on the synaptic terminals of the afferent nociceptor neurons themselves. Some of the neurons in these inhibitory pathways release or are sensitive to certain endogenous opioids (Chapter 8). Thus, infusion of morphine, which stimulates opioid receptors, into the spinal cord at the level of entry of the active nociceptor fibers can provide relief in many cases of intractable pain. This is separate from morphine's effect on the brain.

Transcutaneous electric nerve stimulation, or **TENS,** in which the painful site itself or the nerves leading from it are stimulated by electrodes placed on the surface of the skin, is often useful in lessening pain. TENS works because the stimulation of nonpain, low-threshold afferent fibers (for example, the fibers from touch receptors) leads to inhibition of neurons in the pain pathways. We often apply our own type of TENS therapy when we rub or press hard on a painful area.

Under certain circumstances the ancient Chinese therapy, **acupuncture,** prevents or alleviates pain. During

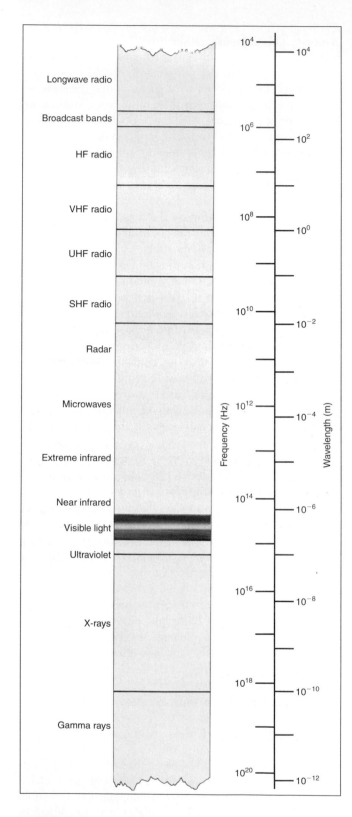

FIGURE 9-20
Electromagnetic spectrum.

acupuncture analgesia, needles are introduced into specific parts of the body to stimulate afferent fibers, which causes analgesia in the same way as TENS. The endogenous opioid neurotransmitters may be involved in acupuncture analgesia.

Stimulation-produced analgesia, TENS, and acupuncture work by exploiting the body's built-in mechanisms that control pain.

VISION

The eyes are composed of an optical portion, which focuses the visual image on the receptor cells, and a neural component, which transforms the visual image into a pattern of neural discharges.

Light

The receptors of the eye are sensitive only to that tiny portion of the vast spectrum of electromagnetic radiation that we call visible light (Figure 9-20). Radiant energy is described in terms of wavelengths and frequencies. The **wavelength** is the distance between two successive wave peaks of the electromagnetic radiation (Figure 9-21). Wavelengths vary from several kilometers at the long-wave radio end of the spectrum to minute fractions of a millimeter at the gamma-ray end. The **frequency,** or the number of cycles per second, of the radiation wave varies inversely with wavelength. Those wavelengths capable of stimulating the receptors of the eye—the **visible spectrum**—are between 400 and 700 nm. Light of different wavelengths within this band is perceived as having different colors.

The Optics of Vision

A light wave can be represented by a line drawn in the direction in which the wave is traveling. Light waves are propagated in all directions from every point of a visible object. Before an accurate image of a point on the object

FIGURE 9-21

Properties of a wave. The frequency of this wave is 2 Hz (cycles/s).

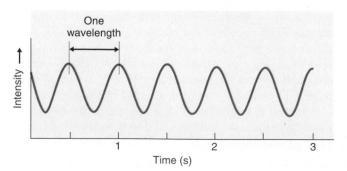

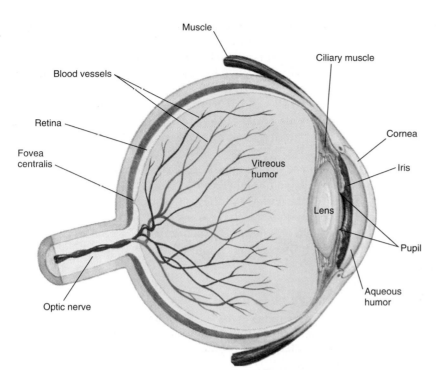

FIGURE 9-22

The human eye. The blood vessels depicted run along the back of the eye between the retina and the vitreous humor, not through the vitreous humor.

is achieved, these divergent light waves must pass through an optical system that focuses them back into a point. In the eye, the image of the object being viewed is focused upon the **retina,** a thin layer of neural tissue lining the back of the eyeball (Figure 9-22). The retina contains the light-sensitive receptor cells, the rods and cones, as well as several types of neurons.

The **lens** and **cornea** of the eye are the optical systems that focus impinging light rays into an image upon the retina. At a boundary between two substances of different densities, such as the cornea and the air, light rays are bent so that they travel in a new direction. The cornea plays a larger quantitative role than the lens in focusing light rays because the rays are bent more in passing from air into the cornea than they are when passing into and out of the lens or any other transparent structure of the eye.

The surface of the cornea is curved so that light rays coming from a single point source hit the cornea at different angles and are bent different amounts, directing the light rays back to a point after emerging from the lens.

FIGURE 9-23

Refraction (bending) of light by the lens system of the eye. The focusing of light rays from a single point, for example, a, forms a single point, *a′*, on the retina. The focusing of light rays from more than one point forms an image that is inverted and, not shown here, reversed right to left. In reality, the image is not focused where the optic nerve leaves the eye, which is a "blind spot." Rather, the image is focused on the fovea centralis, the area of the retina with greatest clarity, which is near the exit of the nerve (Figure 9-22). In this figure and in Figure 9-25, for simplicity we show light refraction only at the surface of the cornea where the greatest effect occurs. Refraction also occurs in the lens and at other sites in the eye.

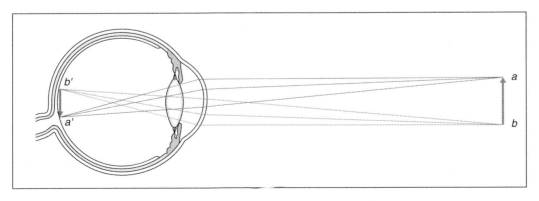

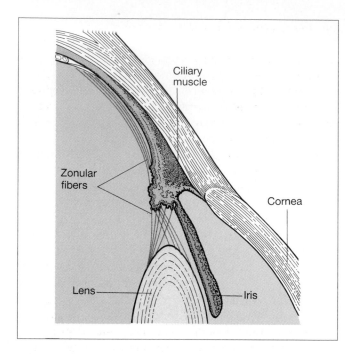

FIGURE 9-24
Ciliary muscle, zonular fibers, and lens of the eye.

The image on the retina is upside down relative to the original light source (Figure 9-23), and it is also reversed right to left.

Light rays from objects close to the eye strike the cornea at greater angles and must be bent more in order to reconverge on the retina. Although, as noted above, the cornea performs the greater part quantitatively of focusing the visual image on the retina, all *adjustments* for distance are made by changes in lens shape. Such changes are part of the process known as **accommodation.**

The shape of the lens is controlled by the **ciliary muscle** and the tension it applies to the **zonular fibers,** which attach this smooth muscle to the lens (Figure 9-24). To focus on distant objects, the zonular fibers pull the lens into a flattened, oval shape. When their pull is removed for near vision, the natural elasticity of the lens causes it to become more spherical. This more spherical shape provides additional bending of the light rays, which is important when near objects are focused on the retina. The ciliary muscle, which is stimulated by parasympathetic nerves, is circular, like a sphincter, so that it draws nearer to the lens as it contracts and therefore removes tension on the zonular fibers, resulting in accommodation for viewing near objects (Figure 9-25). Accommodation also includes other mechanisms that move the lens slightly toward the back of the eye, turn the eyes inward toward the nose (convergence), and constrict the pupil. The sequence of events for accommodation is reversed when distant objects are viewed.

The cells that make up most of the lens lose all their internal membranous organelles early in life and are thus transparent, but they lack the ability to replicate. The only lens cells that retain the capacity to divide are on the surface of the lens, and as new cells are formed, older cells come to lie deeper within the lens. With increasing age, the central part of the lens becomes increasingly dense and stiff and acquires a coloration that progresses from yellow to black.

Since the lens must be elastic to assume a more spherical shape during accommodation for near vision, the increasing stiffness of the lens that occurs with aging makes accommodation for near vision increasingly difficult. This condition, known as ***presbyopia,*** is a normal part of the aging process and is the reason that people around 45

FIGURE 9-25
Accommodation of the lens for near vision. Again, for simplicity, refraction is shown only at the surface of the cornea even though it occurs at several sites. The change in refraction during accommodation is a function of the lens, not the cornea.

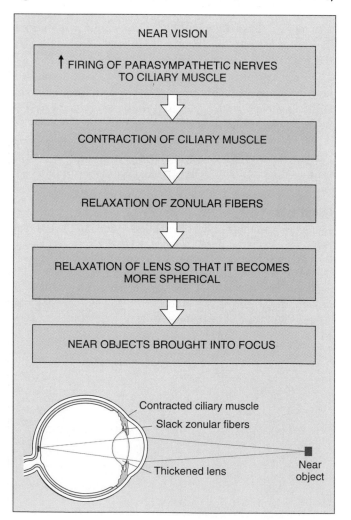

years of age may have to begin wearing reading glasses or bifocals for close work.

The changes in lens color that occur with aging are responsible for **cataract,** which is an opacity of the lens and one of the most common eye disorders. Early changes in lens color do not interfere with vision, but vision is impaired as the process slowly continues. The opaque lens can be removed surgically. With the aid of an implanted artificial lens or compensating eyeglasses, effective vision can be restored, although the ability to accommodate is lost.

Cornea and lens shape and eyeball length determine the point where light rays reconverge. Defects in vision occur if the eyeball is too long in relation to the lens size. In this case, the images of near objects fall on the retina

FIGURE 9-26

In the nearsighted eye, light rays from a distant source are focused in front of the retina. A concave (cupping inward) lens placed before the eye bends the light rays out sufficiently to move the focused image back onto the retina. In contrast, in the farsighted eye, near objects are focused behind the retina, and they are blurred. A convex (bulging outward) lens converges light rays before they enter the eye so that images of near objects can be focused on the retina.

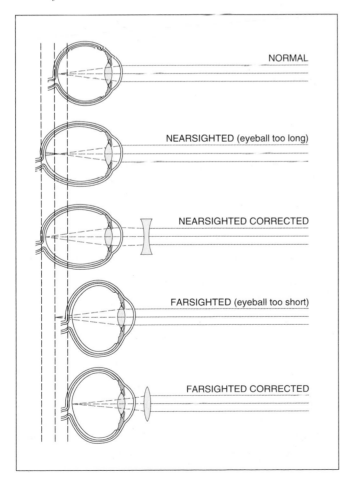

but the images of far objects focus at a point in front of the retina. This is a **nearsighted,** or **myopic,** eye, which is unable to see distant objects clearly. If the eye is too short for the lens, images of distant objects are focused on the retina but those of near objects are focused behind it. This eye is **farsighted,** or **hyperopic,** and near vision is poor. The use of corrective lenses for near- and farsighted vision is shown in Figure 9-26.

Defects in vision also occur where the lens or cornea does not have a smoothly spherical surface, a condition known as **astigmatism.** These surface imperfections can usually be compensated for by eyeglasses.

The lens separates two fluid-filled chambers in the eye, the anterior chamber, which contains aqueous humor, and the posterior chamber, which contains the more viscous vitreous humor (Figure 9-22). These two fluids are colorless and permit the transmission of light from the front of the eye to the retina. The aqueous humor is formed by special vascular tissue that overlies the ciliary muscle. In some instances the aqueous humor is formed faster than it is removed, which results in increased pressure within the eye, a condition known as **glaucoma.** This can be associated with damage to nerve fibers of the eye and loss of vision.

The amount of light entering the eye is controlled by muscles in the ringlike, pigmented tissue known as the **iris** (Figure 9-22), the color being of no importance as long as the tissue is sufficiently opaque to prevent the passage of light. The hole in the center of the iris through which light enters the eye is the **pupil.** The iris is composed of smooth muscles, which are innervated by autonomic nerves. Stimulation of sympathetic nerves to the iris enlarges the pupil by causing the radially arranged muscle fibers to contract. Stimulation of parasympathetic fibers to the iris makes the pupil smaller by causing the sphincter muscle fibers, which circle around the pupil, to contract.

These neurally induced changes occur in response to light-sensitive reflexes. Bright light causes a decrease in the diameter of the pupil, which reduces the amount of light entering the eye and directs the light to the central part of the lens for more accurate vision. Conversely, the sphincter fibers relax in dim light, when maximal illumination is needed. Changes also occur as a result of emotion or pain.

Photoreceptor Cells

The photoreceptor cells in the retina are called **rods** and **cones** because of the shapes of their light-sensitive tips (Figure 9-27). Note in Figures 9-27 and 9-28 that the light-sensitive portion of the photoreceptor cells faces *away* from the incoming light. The rods are extremely sensitive and respond to very low levels of illumination, whereas the cones are considerably less sensitive and respond only when the light is brighter than, for example, twilight.

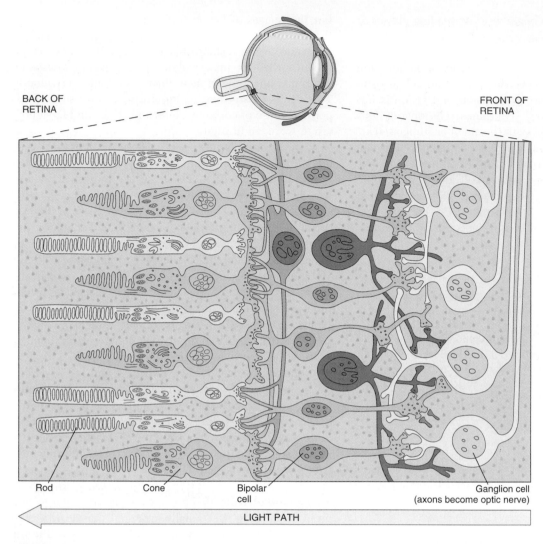

BACK OF RETINA

FRONT OF RETINA

Rod   Cone   Bipolar cell   Ganglion cell (axons become optic nerve)

LIGHT PATH

FIGURE 9-27

Organization of the retina. Light enters the eye through the cornea, passes through the aqueous humor, pupil, vitreous humor, and the front surface of the retina. Since the tips of the rods and cones are on the side of retina opposite the light entry, the light must pass through all the cell layers before reaching the photoreceptors and stimulating them. The photopigments, which are sensitive to light, are embedded in membranes that are layered repeatedly across the outer ends of the rods and cones, that is, the ends farthest from the light source. In the rods these membranes form discrete disks, but in the cones they are continuous with the plasma membrane and account for the comb-like appearance of these cells. The pigmented layer of the eye (the choroid), which lies behind the retina, absorbs light and prevents its reflection back to the rods and cones, which would cause the visual image to be blurred. Two other cell types, here depicted in lavender and red, provide lateral interaction in the retina. (*Redrawn from Dowling and Boycott.*)

The photoreceptors contain molecules called **photopigments,** which absorb light. There are four different photopigments in the retina, one (**rhodopsin**) in the rods and one in each of the three cone types. Each photopigment contains an opsin and a chromophore. **Opsin** is a collective term for a group of integral membrane proteins, one of which surrounds and binds a **chromophore** molecule (Figure 9-28). The chromophore, which is the actual light-sensitive part of the photopigment, is the same in each of the four photopigments and is **retinal,** a derivative of vitamin A. The opsin differs in each of the four photopigments. Since each type of opsin binds to the chromophore in a different way and filters light differently, each of the four photopigments absorbs light most effectively at a different part of the visible spectrum. For example, one photopigment absorbs wavelengths in the range of red light best, whereas another absorbs green light best.

Within the photoreceptor cells the photopigments lie in specialized membranes that are arranged in highly ordered stacks, or disks, parallel to the retina (Figures 9-27 and 9-28). The repeated layers of membranes in each photoreceptor may contain over a billion molecules of photopigment, providing an effective trap for light.

Light activates a chromophore (retinal), causing it to change shape. This change triggers a sequence of biochemical events that lead to *hyperpolarization* of the photoreceptor cell's plasma membrane and, thereby, *decreased* release of the neurotransmitter glutamate from the cell. Note that in the case of photoreceptors the response of the cell to a stimulus (light) is a hyperpolarizing receptor potential and a decrease in neurotransmitter release. The decrease in neurotransmitter then causes the bipolar cells that synapse with the photoreceptor cell to undergo a change in membrane potential (see Essay, page 260).

After its activation by light, the retinal changes back to its resting shape by several mechanisms that do not depend on light but are enzyme mediated. Thus, in the dark, retinal has its resting shape, the photoreceptor cell is partially depolarized, and *more* glutamate is being released.

High sensitivity of the visual receptors is not always an advantage. For example, when one steps from a dark place into bright daylight, the light is blinding at first. Moreover, when the receptors are saturated by high light levels, they lose their ability to respond to further small increases or decreases in light intensity. Over a period of several seconds in bright light, however, the eyes adjust; this phenomenon, known as **light adaptation,** is partly due to the fact that light exposure initiates a sequence of events involving intracellular calcium that decrease the sensitivity of the phototransduction process to further light stimulation.

The opposite adjustment—**dark adaptation**—also occurs. When one steps back from a place of bright sunlight into a darkened room, a temporary "blindness" takes place. In the low levels of illumination of the darkened room, vision can only be supplied by the rods, which have greater sensitivity than the cones. During the exposure to bright light, however, the rods' rhodopsin has been completely activated. It cannot respond again until it is restored to its resting state, a process requiring some tens of minutes. Dark adaptation occurs, in part, as enzymes regenerate the initial form of rhodopsin that can respond to light.

FIGURE 9-28

The arrangement of opsin and retinal (the chromophore) in the membrane of the photoreceptor disks of a rod. The opsin actually crosses the membrane seven times, not three as shown here. Note that the retinal is perpendicular to the incoming light rays so that its ability to "catch" light is maximal.

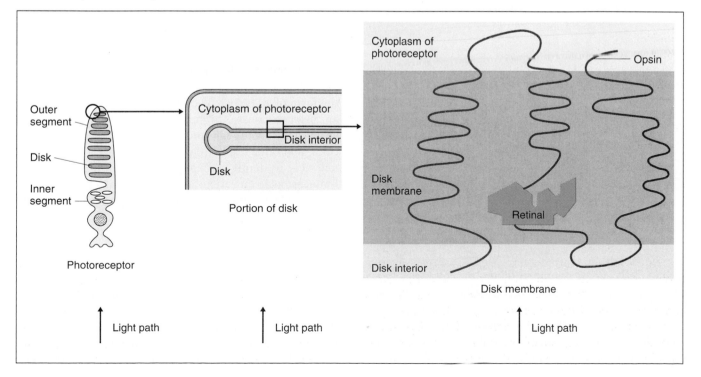

Neural Pathways of Vision

The neural pathways of vision begin with the rods and cones. These photoreceptors communicate by way of electrical synapses with each other and with second-order neurons, the only one of which we shall mention is the **bipolar cell** (Figures 9-27 and 29). The bipolar cells synapse (still within the retina) both upon neurons that pass information horizontally from one part of the retina to another and upon the **ganglion cells.** Via these synapses, the ganglion cells are caused to respond differentially to the various characteristics of visual images, such as color, intensity, form, and movement. Note that a great deal of information processing takes place at this early stage of the sensory pathway.

The distinct characteristics of the visual image are transmitted through the visual system along multiple, parallel pathways by two types of ganglion cells, each type concerned with different aspects of the visual stimulus. Parallel processing of information continues all the way to the cerebral cortex, to the highest stages of visual neural networks. (Note that in this discussion of the visual pathway, the term "respond" and "response" denote not a direct response to a light stimulus—only the rods and cones show such responses—but rather to the synaptic input reaching the relevant pathway neuron as a consequence of the original stimulus to the rods and cones.)

FIGURE 9-29

Diagrammatic representation of the visual pathway.

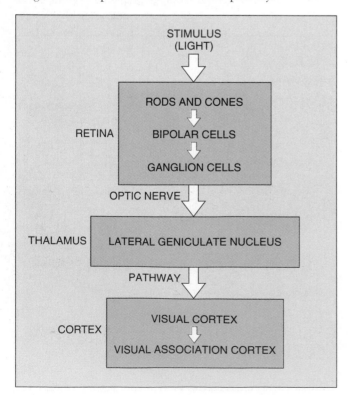

Ganglion cells respond to activation by producing action potentials, whereas the rods and cones and almost all other retinal neurons produce only graded potentials. The axons of the ganglion cells form the output from the retina—the optic nerve, or cranial nerve II. The two optic nerves meet at the base of the brain to form the optic chiasm, where some of the fibers cross to the opposite side of the brain, providing both cerebral hemispheres with input from each eye.

Optic nerve fibers project to several structures in the brain, the largest number passing to the **lateral geniculate nucleus** in the thalamus (Figure 9-29), where the information from the different ganglion cell types is kept distinct. In addition to the input from the retina, many neurons of the lateral geniculate nucleus also receive input from the brainstem reticular formation and input relayed *back* from the visual cortex. These nonretinal inputs can control the transmission of information from the retina to the visual cortex and may be involved in the ability to shift attention between vision and the other sensory modalities or between different objects in the visual field.

The lateral geniculate nucleus sends action potentials on to the visual cortex, the primary visual area of cerebral cortex (Figure 9-6). The visual cortex has several subdivisions, each representing the complete **visual field,** that is, that part of the visual world that stimulates the retina at any given time. The relay of information from the retina to the various areas of the visual cortex follows a precise point-to-point projection pattern so that adjacent retinal regions project to adjacent regions in each subdivision of the visual cortex.

While there is much interaction between the different areas of the visual cortex, the separate functions of cells, which began in the retina, are maintained. Thus, most neurons in one subdivision of the visual cortex are responsive only to stimuli oriented in a particular direction in the visual field, a property important in elaboration of the form of an object. The neurons in another subdivision are most responsive to movement of an object across the visual field. Other neuronal groups may respond best to color, and others only to input from both eyes, an important clue to depth perception. Again, the functional differences between various areas of the visual cortex are determined by the input these areas receive from the distinct classes of ganglion cells.

Some of the information is kept separate even after it leaves the visual cortex. For example, some pathways transmit information from the visual cortex to the temporal lobe, where the identification of objects in the visual field is processed—analyzing the "what" of the visual stimulus. Other pathways travel to the parietal lobe, where information about the location of objects in the visual field is processed—analyzing the "where" of the visual stimulus. Output from both the temporal and parietal lobes then

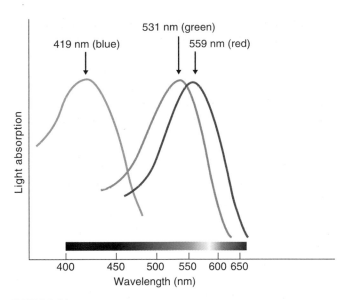

FIGURE 9-30

The sensitivities of the photopigments in the three types of cones in the normal human retina. Action potential frequency in the optic nerve is directly related to absorption of light by a photopigment.

passes to the frontal cortex, which uses the visual information, for example, in directing actions. Thus, different aspects of visual information are carried in parallel pathways and are processed simultaneously in a number of independent ways in different parts of the cerebral cortex before they are reintegrated to produce the conscious sensation of sight.

Note that the cells of the visual pathways are organized to handle information about line, contrast, movement, and color. They do not, however, form a picture in the brain. Rather, they form a spatial and temporal pattern of electrical activity.

We mentioned that a substantial number of fibers of the visual pathway project to regions of the brain other than the visual cortex. For example, visual information is transmitted to the **suprachiasmatic nucleus,** which lies just above the optic chiasm and functions as a "biological clock," as described in Chapter 7. Information about diurnal cycles of light intensity is used to synchronize this neuronal clock. Other visual information is passed to the brainstem and cerebellum, where it is used in the coordination of eye and head movements, fixation of gaze, and change in pupil size.

Color Vision

The colors we perceive are related to the wavelengths of light that are reflected, absorbed, or transmitted by the pigments in the objects of our visual world. For example, an object appears red because shorter wavelengths, which would be perceived as blue, are absorbed by the object, while the longer wavelengths, perceived as red, are reflected from the object to excite the photopigment of the retina most sensitive to red. Light perceived as white is a mixture of all wavelengths, and black is the absence of all light.

Color vision begins with activation of the photopigments in the cone receptor cells. Human retinas have three kinds of cones, which contain red-, green-, or blue-sensitive photopigments. As their names imply, these pigments absorb and hence respond optimally to light of different wavelengths. Because the red pigment is actually more sensitive to the wavelengths that correspond to yellow, this pigment is sometimes called the yellow photopigment.

Although each type of cone is excited most effectively by light of one particular wavelength, it responds to other wavelengths as well. Thus, for any given wavelength, the three cone types are excited to different degrees (Figure 9-30). For example, in response to light of 530-nm wavelength, the green cones respond maximally, the red cones less, and the blue cones not at all. Our sensation of color depends upon the relative outputs of these three types of cone cells and their comparison by higher-order cells in the visual system.

The pathways for color vision follow those described in Figure 9-29. Ganglion cells of one type respond to a broad band of wavelengths. In other words, they receive input from all three types of cones, and they signal not specific color but general brightness. Ganglion cells of a second type code specific colors. These latter cells are also called **opponent color cells** because they have an excitatory input from one type of cone receptor and an inhibitory input from another. For example, the cell in Figure 9-31

FIGURE 9-31

Response of a single opponent color ganglion cell to blue, red, and white lights. (*Redrawn from Hubel and Wiesel.*)

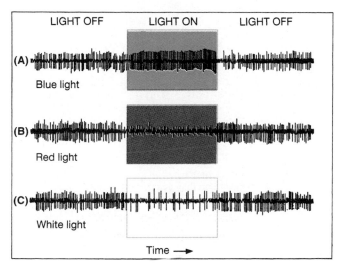

increases its rate of firing when the retina is stimulated by a blue light but decreases it when a red light replaces the blue. The cell gives a weak response when stimulated with a white light because the light contains both blue and red wavelengths. Other more complicated patterns also exist. Opponent cells are also found in the lateral geniculate nucleus and visual cortex.

At high light intensities, as in daylight vision, most people—92 percent of the male population and over 99 percent of the female population—have normal color vision. People with the most common kind of *color blindness* lack either the red or green cone pigments and, as a result, have trouble perceiving red versus green.

Eye Movement

The cones are most concentrated in a specialized area of the retina known as the **fovea centralis** (Figure 9-22), and images focused there are seen with the greatest acuity. In order to focus the most important point in the visual image (the fixation point) on the fovea and keep it there, the eyeball must be able to move. Six skeletal muscles attached to the outside of each eyeball (Figure 9-32) control its movement. These muscles perform two basic movements, fast and slow.

The fast movements, called **saccades,** are small, jerking movements that rapidly bring the eye from one fixation point to another to allow search of the visual field. In addition, saccades move the visual image over the receptors, thereby preventing adaptation. Saccades also occur during certain periods of sleep when the eyes are closed and may be associated with "watching" the visual imagery of dreams.

Slow eye movements are involved both in tracking visual objects as they move through the visual field and during compensation for movements of the head. The control centers for these compensating movements obtain their information about head movement from the vestibular system, which will be described shortly. Control systems for the other slow movements of the eyes require the continuous feedback of visual information about the moving object.

FIGURE 9-32

A superior view of the muscles that move the eye to direct the gaze and give convergence.

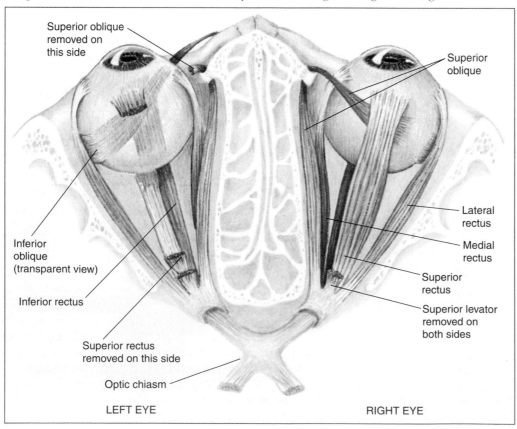

Superior oblique removed on this side

Superior oblique

Inferior oblique (transparent view)

Inferior rectus

Superior rectus removed on this side

Optic chiasm

Lateral rectus

Medial rectus

Superior rectus

Superior levator removed on both sides

LEFT EYE

RIGHT EYE

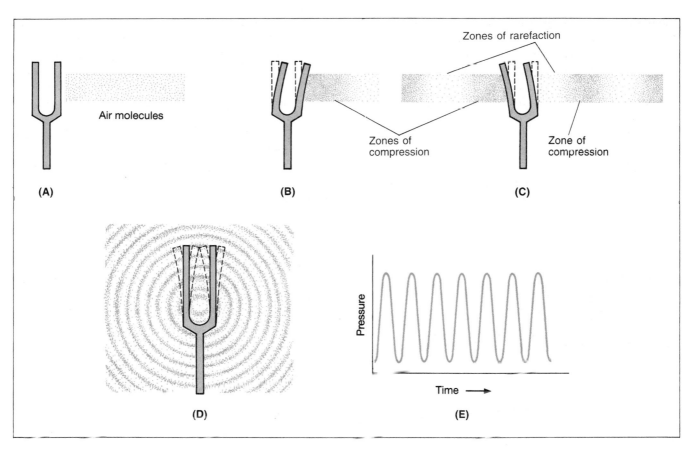

FIGURE 9-33

Formation of sound waves from a vibrating tuning fork.

HEARING

The sense of hearing is based on the physics of sound and the physiology of the external, middle, and inner ear, the nerves to the brain, and the brain parts involved in processing acoustic information

Sound

Sound energy is transmitted through a gaseous, liquid, or solid medium by setting up a vibration of the medium's molecules; air is the most common medium for sound transmission. When there are no molecules, as in a vacuum, there can be no sound. Anything capable of creating a disturbance of molecules can serve as a sound source, but a common one is a vibrating object such as a tuning fork (Figure 9-33). The disturbance of molecules—air molecules in our example—that makes up a sound wave consists of zones of compression, in which the molecules are close together and the pressure is high, alternating with zones of rarefaction, where the molecules are farther apart and the pressure is lower (Figure 9-33A through D).

A sound wave measured over time (Figure 9-33E) consists of rapidly alternating pressures that vary continuously from a high during compression of molecules, to a low during rarefaction, and back again. The difference between the pressure of molecules in zones of compression and rarefaction determines the wave's amplitude, which is related to the loudness of the sound; the greater the amplitude, the louder the sound. The frequency of vibration of the sound source determines the pitch we hear; the faster the vibration, the higher the pitch. The sounds heard most keenly by human ears are those from sources vibrating at frequencies between 1000 and 4000 Hz (hertz, or cycles per second), but the entire range of frequencies audible to human beings extends from 20 to 20,000 Hz. Sound waves with sequences of pitches are generally perceived as musical, the complexity of the individual waves giving the sound its characteristic quality, or timbre.

We can distinguish about 400,000 different sounds. For example, we can distinguish the note A played on a piano from the same note on a violin. We can also selectively *not* hear sounds, tuning out the babble of a party to concentrate on a single voice.

Sound Transmission in the Ear

The first step in hearing is the entrance of sound waves into the **external auditory canal** (Figure 9-34). The shapes of the outer ear (the pinna, or auricle) and the external auditory canal help to amplify and direct the sound. The sound waves reverberate from the sides and end of the external auditory canal, filling it with the continuous vibrations of pressure waves.

The **tympanic membrane** (eardrum) is stretched across the end of the external auditory canal, and air molecules push against the membrane, causing it to vibrate at the same frequency as the sound wave. Under higher pressure during a zone of compression, the tympanic membrane bows inward. The distance the membrane moves, although always very small, is a function of the force with which the air molecules hit it and is related to the sound pressure and therefore its loudness. During the subsequent zone of rarefaction, the membrane returns to its original position. The exquisitely sensitive tympanic membrane responds to all the varying pressures of the sound waves, vibrating slowly in response to low-frequency sounds and rapidly in response to high-frequency ones.

The tympanic membrane separates the external auditory canal from the **middle-ear cavity,** an air-filled cavity in the temporal bone of the skull. The pressures in the external auditory canal and middle-ear cavity are normally equal to atmospheric pressure. The middle-ear cavity is exposed to atmospheric pressure through the **auditory (eustachian) tube,** which connects the middle ear to the pharynx. The slitlike ending of this tube in the pharynx is normally closed, but muscle movements open the tube during yawning, swallowing, or sneezing, and the pressure in the middle ear equilibrates with atmospheric pressure. A difference in pressure can be produced with sudden changes in altitude (as in an ascending or descending elevator or airplane), when the pressure outside the ear and in the ear canal changes while the pressure in the middle ear remains constant because of the closed auditory tube. This pressure difference can stretch the tympanic membrane and cause pain.

FIGURE 9-34

The human ear. In this and the following drawing, violet indicates the outer ear, green the middle ear, and blue the inner ear. The malleus, incus, and stapes are bones even though they are colored green in this figure to indicate that they are components of the middle ear compartment. Actually, the auditory tube is closed except during movements of the pharynx, such as swallowing or yawning.

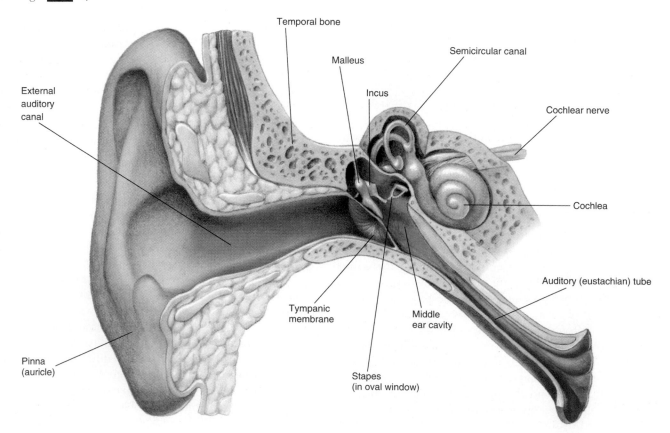

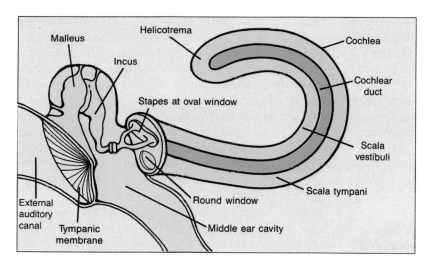

FIGURE 9-35

Relationship between the middle ear bones and the cochlea. Movement of the stapes against the membrane covering the oval window sets up pressure waves in the fluid-filled scala vestibuli. These waves cause vibration of the cochlear duct and the basilar membrane. Some of the pressure is transmitted around the helicotrema directly into the scala tympani. (*Redrawn from Kandel and Schwartz.*)

The second step in hearing is the transmission of sound energy from the tympanic membrane through the middle-ear cavity to the **inner ear.** The inner ear, called the **cochlea,** is a *fluid-filled,* spiral-shaped passage in the temporal bone. The temporal bone also houses other passages, including the semicircular canals, which contain the sensory organs for balance and movement. These passages are connected to the cochlea but will be discussed later.

Because liquid is more difficult to move than air, the sound pressure transmitted to the inner ear must be amplified. This is achieved by a movable chain of three small bones, the **malleus, incus,** and **stapes** (Figure 9-35); these bones act as a piston and couple the motions of the tympanic membrane to the **oval window,** a membrane-covered opening separating the middle and inner ear (Figure 9-36).

The *total* force of a sound wave applied to the tympanic membrane is transferred to the oval window, but because the oval window is much smaller than the tympanic membrane, the *force per unit area* (that is, the pressure) is increased 15 to 20 times. Additional advantage is gained through the lever action of the middle-ear bones. The amount of energy transmitted to the inner ear can be lessened by the contraction of two small skeletal muscles in the middle ear that alter the tension of the tympanic membrane and the position of the stapes in the oval window. These muscles help to protect the delicate receptor apparatus of the inner ear from continuous intense sound stimuli and improve hearing over certain frequency ranges.

The entire system described thus far has been concerned with the transmission of sound energy into the cochlea, where the receptor cells are located. The cochlea is almost completely divided lengthwise by a fluid-filled membranous tube, the **cochlear duct,** which follows the cochlear spiral (Figure 9-35). On either side of the cochlear duct are fluid-filled compartments: the **scala vestibuli,** which is on the side of the cochlear duct that ends at the oval window; and the **scala tympani,** which is below the cochlear duct and ends in a second membrane-covered opening to the middle ear, the round window. The scala vestibuli and scala tympani meet at the end of the cochlear duct at the helicotrema (Figure 9-35).

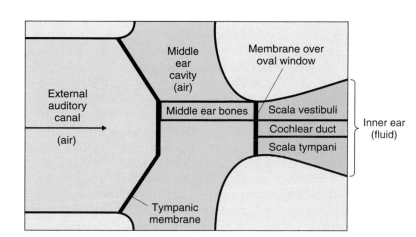

FIGURE 9-36

Diagrammatic representation showing that the middle ear bones act as a piston against the fluid of the inner ear. (*Redrawn from von Bekesy.*)

Sound waves in the ear canal cause in-and-out movement of the tympanic membrane, which moves the chain of middle-ear bones against the membrane covering the oval window, causing it to bow into the scala vestibuli and back out (Figure 9-37), creating waves of pressure there. The wall of the scala vestibuli is largely bone, and there are only two paths by which the pressure waves can be dissipated. One path is to the helicotrema, where the waves pass around the end of the cochlear duct into the scala tympani and back to the round-window membrane, which is then bowed out into the middle-ear cavity. However, most of the pressure is transmitted from the scala vestibuli across the cochlear duct.

One side of the cochlear duct is formed by the **basilar membrane** (Figure 9-38), upon which sits the **organ of Corti** (also called the spiral organ), which contains the ear's sensitive receptor cells. Pressure waves crossing the cochlear duct cause the basilar membrane to vibrate.

The region of maximal displacement of the vibrating basilar membrane varies with the frequency of the sound source. The properties of the membrane nearest the middle ear are such that this region vibrates most easily, that is, undergoes the greatest movement, in response to high-frequency (high-pitched) tones. As the frequency of the sound is lowered, vibration waves travel out along the membrane for greater distances. Progressively more distant regions of the basilar membrane vibrate maximally in response to progressively lower tones.

Hair Cells of the Organ of Corti

The receptor cells of the organ of Corti, the **hair cells,** are mechanoreceptors that have hairlike **stereocilia** protruding from one end (Figure 9-38C). The hair cells transform the pressure waves in the cochlea into receptor potentials. Movements of the basilar membrane stimulate the hair cells because they are attached to the membrane.

The stereocilia are in contact with the overhanging **tectorial membrane** (Figure 9-38), which projects inward from the side of the cochlea. As the basilar membrane is displaced by pressure waves, the hair cells move in relation to the tectorial membrane, and, consequently, the stereocilia are bent. Whenever the stereocilia bend, ion channels in the plasma membrane of the hair cell open, and the resulting ion movements depolarize the membrane and create a receptor potential.

Efferent fibers from the brainstem regulate the activity of certain of the hair cells and dampen their response to protect them. Despite this protective action, the hair cells are easily damaged or even completely destroyed by exposure to high-intensity noises such as amplified rock music concerts, engines of jet planes, and revved-up motorcycles. Lesser noise levels also cause damage if exposure is chronic.

Hair cell depolarization leads to release of the neurotransmitter glutamate (the same neurotransmitter released by photoreceptor cells), which activates receptors (binding sites) on the terminals of the ten or so afferent neurons that contact it. This causes the generation of action potentials in the neurons, the axons of which form the cochlear nerve, a component of cranial nerve VIII. The greater the energy (loudness) of the sound wave, the greater the frequency of action potentials generated in the afferent nerve fibers. Because of its position on the basilar membrane, each hair cell and, therefore, the nerve fibers that synapse upon it respond to a limited range of sound frequency and intensity, but they respond best to a single frequency.

Neural Pathways in Hearing

Cochlear nerve fibers enter the brainstem and synapse with interneurons there, fibers from both ears often converging on the same neuron. Many of these neurons are sensitive to the different arrival times and intensities of the input from the two ears. The different arrival times of low-frequency sounds and the difference in intensities of high-frequency sounds are used to determine the direction of the sound source. If, for example, a sound is louder in the right ear or arrives sooner at the right ear than at the left, we assume that the sound source is on the right.

From the brainstem, the information is transmitted via a multineuron pathway to the thalamus and on to auditory cortex (Figure 9-6). The neurons responding to different pitches are arranged along the auditory cortex in an orderly manner in much the same way that signals from different regions of the body are represented at different sites in the somatosensory cortex. Different areas of the auditory system are further specialized, some neurons re-

FIGURE 9-37

Transmission of sound vibrations through the middle and inner ear. (*Redrawn from Davis and Silverman.*)

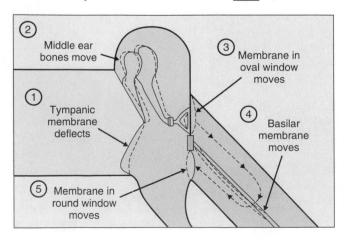

② Middle ear bones move

③ Membrane in oval window moves

① Tympanic membrane deflects

④ Basilar membrane moves

⑤ Membrane in round window moves

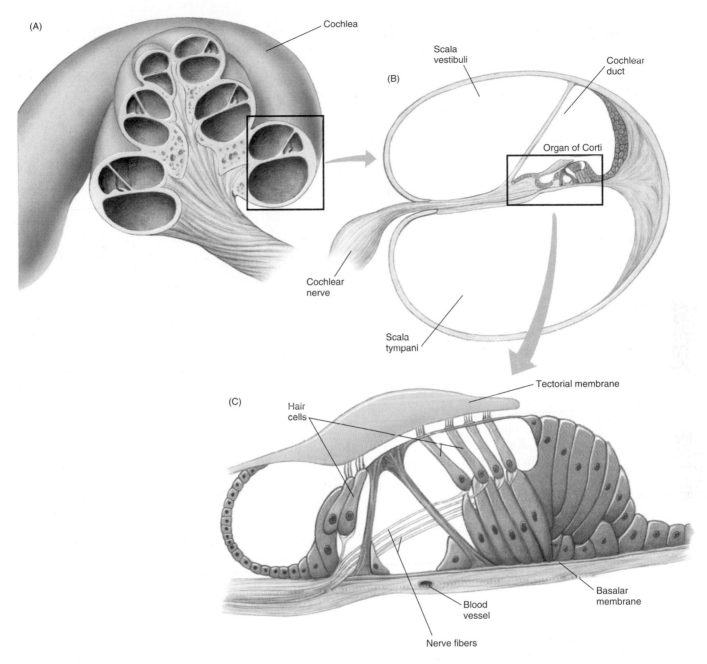

(A)

Cochlea

(B)

Scala vestibuli

Cochlear duct

Organ of Corti

Cochlear nerve

Scala tympani

Tectorial membrane

(C)

Hair cells

Basalar membrane

Blood vessel

Nerve fibers

FIGURE 9-38

Cross section of the membranes and compartments of the inner ear with detailed view of the hair cells and other structures on the basilar membrane as shown with increasing magnifications in views (A), (B), and (C). (*Redrawn from Rasmussen.*)

sponding best to complex sounds such as those used in verbal communication whereas others signal the location, movement, duration, or loudness of a sound.

Electronic devices can help compensate for damage to the intricate middle ear, cochlea, or neural structures. **Hearing aids** amplify incoming sounds, which then pass via the ear canal to the same cochlear mechanisms used by normal sound. When substantial damage has occurred, however, and hearing aids cannot correct the deafness, electronic devices known as **cochlear implants** may restore functional hearing. Cochlear implants stimulate the auditory nerves (that is, the cochlear branch of cranial nerve VIII, the vestibulocochlear nerve, Table 8-7) with tiny electric currents so that sound signals are transmitted directly to the auditory pathways, bypassing the cochlea.

VESTIBULAR SYSTEM

Changes in the motion and position of the head are detected by hair cells in the **vestibular apparatus** of the inner ear (Figure 9-39), a series of fluid-filled membranous tubes that connect with each other and with the cochlear duct. The vestibular apparatus consists of three **semicircular ducts** and two saclike swellings, the **utricle** and **saccule,** all of which lie in tunnels in the temporal bone on each side of the head. The bony canals of the inner ear in which the vestibular apparatus and cochlea are housed have such a complicated shape that they are sometimes called the **labyrinth.**

The Semicircular Canals

The bony **semicircular canals** house the membranous semicircular ducts, which detect angular acceleration during *rotation* of the head along three perpendicular axes. (The term "semicircular canal" is often also used to denote the semicircular ducts.) The three axes of the semicircular canals are those activated while nodding the head up and down as in signifying "yes," shaking the head from side to side as in signifying "no," and tipping the head so the ear touches the shoulder (Figure 9-40).

Receptor cells of the semicircular canals, like those of the organ of Corti, contain hair-like stereocilia. These stereocilia are closely ensheathed by a gelatinous mass,

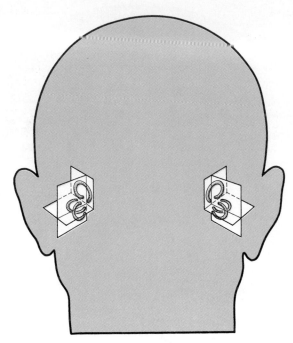

FIGURE 9-40

Relation of the two sets of semicircular canals.

the **cupula,** which extends across the lumen of each semicircular duct at the **ampulla,** a slight bulge in the wall of each duct (Figure 9-41). Whenever the head is moved, the bony semicircular canal, its enclosed duct, and the attached bodies of the hair cells all turn with it. The fluid filling the duct, however, is not attached to the skull, and because of inertia, tends to retain its original position, that is, to be "left behind." Thus, the moving ampulla is pushed against the stationary fluid, which causes bending of the stereocilia and alteration in the rate of release of a chemical transmitter from the hair cells. This transmitter activates the nerve terminals synapsing with the hair cells.

The speed and magnitude of rotational head movements determine the way in which the stereocilia are bent and the hair cells stimulated. Neurotransmitter is released from the hair cells at rest, and the release changes from this resting rate according to the direction in which the hairs are bent. Each hair cell receptor has one direction of maximum sensitivity, and when its stereocilia are bent in this direction, the receptor cell depolarizes. When the stereocilia are bent in the opposite direction, the cell hyperpolarizes (Figure 9-42). The frequency of action potentials in the afferent nerve fibers that synapse with the hair cells is related to both the amount of force bending the stereocilia on the receptor cells and to the direction in which this force is applied.

At a constant velocity, the duct fluid begins to move at the same rate as the rest of the head, and the stereocilia slowly

FIGURE 9-39

A tunnel in the temporal bone contains a fluid-filled membranous duct system, part of which can be seen in this cutaway view. The semicircular duct, utricle, and saccule make up the vestibular apparatus. The vestibular duct is connected to the cochlear duct. The purple region on the ducts indicates the positions of hair (receptor) cells. (*Redrawn from Hudspeth.*)

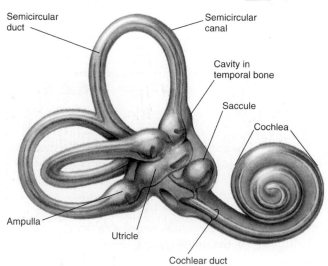

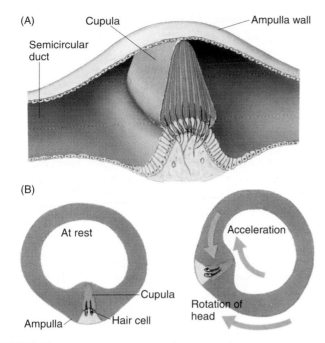

FIGURE 9-41

(A) Organization of a cupula and ampulla. (B) Relation of the cupula to the ampulla when the head is at rest and when it is accelerating. ⨍

return to their resting position. For this reason, the hair cells are stimulated only during *changes* in the rate of rotation, that is, during acceleration or deceleration, of the head.

The Utricle and Saccule

The utricle and saccule provide information about *linear* acceleration and changes in head position relative to the forces of gravity. Here, too, the receptor cells are mechanoreceptors sensitive to the displacement of projecting hairs. The patch of hair cells in the utricle is nearly horizontal in a standing person, and that in the saccule is vertical.

The stereocilia are covered by a gelatinous substance in which tiny stones, or otoliths, are embedded; the otoliths, which are calcium carbonate crystals, make the gelatinous substance heavier than the surrounding fluid. In response to any linear acceleration, the gelatinous-otolithic material changes its position and is pulled against the hair cells so that the stereocilia are bent and the receptor cells stimulated.

Vestibular Information and Dysfunction

Information about hair-cell stimulation is relayed from the vestibular apparatus to the brainstem via the vestibular branch of cranial nerve VIII (the same cranial nerve that carries acoustic information). It is transmitted via a multineuronal pathway to the parietal lobe to an area adjacent to the somatosensory cortex. Vestibular information is in-

tegrated with information from the joints, tendons, and skin, leading to the sense of posture and movement.

Vestibular information is used in three ways. One is to control the eye muscles so that, in spite of changes in head position, the eyes can remain fixed on the same point. *Nystagmus* is a large, jerky, back-and-forth movement of the eyes that can occur in response to vestibular input in normal people but can also be a pathological sign.

The second use of vestibular information is in reflex mechanisms for maintaining upright posture. The vestibular apparatus play a role in the support of the head during movement, orientation of the head in space, and reflexes accompanying locomotion. Very few postural reflexes, however, depend exclusively on input from the vestibular system despite the fact that the vestibular organs are sometimes called the sense organs of balance.

The third use of vestibular information is in providing conscious awareness of the position and acceleration of the body.

Unexpected inputs from the vestibular system and other sensory systems can induce *vertigo,* defined as an illusion of movement, which is often accompanied by feelings of nausea and lightheadedness. This occurs when there is a mismatch in information from the various sensory systems as, for example, when one is in a high place looking down to the ground: The visual input indicates that you are floating in space while the vestibular system signals that you are not moving at all. *Motion sickness* also involves the vestibular system, occurring when unfamiliar patterns of

FIGURE 9-42

Relationship between position of hairs and activity in afferent neurons. (A) Resting activity. (B) Movement of hairs in one direction increases the action potential frequency in the afferent nerve activated by the hair cell. (C) Movement in the opposite direction decreases the rate relative to the resting state. (*Redrawn from Wersall, Gleisner, and Lundquist.*) ▭ ⨍

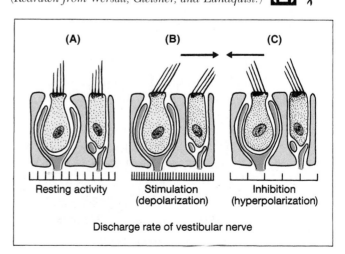

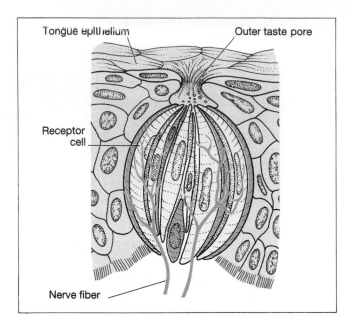

FIGURE 9-43

Structure and innervation of a taste bud. Unlike the taste bud here, each taste bud contains 100 or so receptor cells. The three different types of cells are actually receptor cells in different stages of development. They give rise to new receptors when older, damaged ones die.

linear and rotational acceleration are experienced and adaptation to them has not occurred.

Ménière's disease involves the vestibular system and is associated with episodes of abrupt and often severe dizziness, ringing in the ears, and bouts of hearing loss. It is due to an increased fluid pressure in the membranous duct system of the inner ear. The dizziness occurs because the inputs from the two ears are not balanced, either because only one ear is affected or because the two are affected to different degrees.

CHEMICAL SENSES

Receptors sensitive to specific chemicals are **chemoreceptors.** Some of these respond to chemical changes in the internal environment, two examples being the oxygen and hydrogen-ion receptors in certain large blood vessels. Others respond to external chemical changes, and in this category are the receptors for taste and smell, which affect a person's appetite, saliva flow, gastric secretions, and avoidance of harmful substances.

Taste

The specialized sense organs for taste are the 10,000 or so **taste buds** that are found primarily on the tongue. The receptor cells are arranged in the taste buds like the seg-

ments of an orange (Figure 9-43). A long narrow process on the upper surface of each receptor cell extends into a small pore at the surface of the taste bud, where the process is bathed by the fluids of the mouth.

Taste sensations (modalities) are traditionally divided into four basic groups: sweet, sour, salty, and bitter, the receptor cells of each group having a distinct transductional system. For example, salt taste begins with sodium entry into the cell through ion channels, which depolarizes the plasma membrane; depolarization causes neurotransmitter release from the receptor cell at synapses with afferent nerve fibers.

Although more than one afferent fiber synapses with each receptor cell, the taste system is organized into independent coded pathways into the central nervous system. Single receptor cells respond in varying degrees to substances that fall into more than one taste category, however, and awareness of the specific taste of a substance depends also upon the pattern of firing in a group of neurons.

The pathways for taste in the central nervous system project to the parietal cortex, near the "mouth" region of the somatosensory cortex (see Figure 9-6).

Sensations of pain ("hot"), texture, and temperature contribute to taste. The odor of the substance clearly helps to identify it, too, as is attested by the common experience that food lacks taste when one has a head cold.

Smell

The olfactory receptor cells, the first cells in the pathways that give rise to the sense of smell, lie in a small patch of membrane, the **olfactory epithelium,** in the upper part of the nasal cavity (Figure 9-44A). These cells are specialized afferent neurons that have an enlarged extension, analogous to a dendrite. Several long hairlike processes extend out from this extension along the surface of the olfactory epithelium (Figure 9-44B) where they are bathed in mucus. The hairlike processes contain the receptor proteins (binding sites) for olfactory stimuli. The axons of these neurons form the olfactory nerve, which is cranial nerve I.

For an odorous substance, that is, an **odorant,** to be detected, molecules of the substance must first diffuse into the air and pass into the nose to the region of the olfactory epithelium. Once there, they dissolve in the mucus that covers the epithelium and then bind to specific receptor proteins on the cilia. Proteins in the mucus may interact with the odorant molecules, transport them to the receptors, and facilitate their binding to the receptors.

Although there are many thousands of olfactory neurons, each contains one, or at most a few, of the 1,000 or so different receptor types, each of which responds only to a specific chemically related group of odorant molecules. Each odorant has characteristic chemical groups that distinguish it from other odorants, and each of these

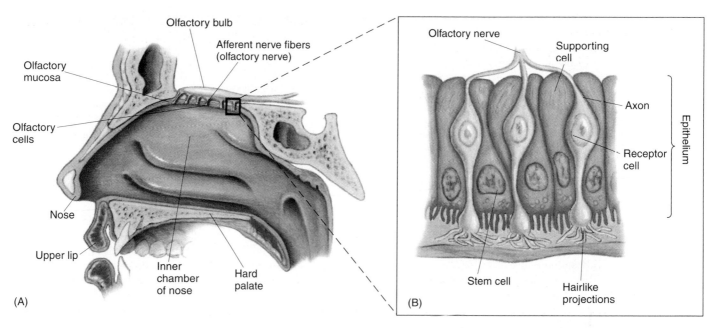

FIGURE 9-44

(A) Location and (B) enlargement of a portion of the mucosa showing the structure of the olfactory receptors. In addition to the receptors, the olfactory epithelium contains stem cells, which give rise to new receptors (an exception to our statement in Chapter 8 that no new neurons are formed after birth), and supporting cells.

groups activates a different receptor type. Thus, the identity of a particular odorant is determined by the activation of a precise combination of receptors, each of which is contained in a distinct group of olfactory neurons.

The axons of the olfactory neurons synapse in the brain structures known as olfactory bulbs, which lie on the undersurface of the frontal lobes. Axons from olfactory neurons sharing a common receptor specificity synapse together on certain olfactory-bulb neurons, thereby maintaining the specificity of the original stimulus.

Information is passed from the olfactory bulbs to olfactory cortex, which is in the limbic system (see Figure 8-42), a part of the brain intimately associated with emotional, food-getting, and sexual behavior. As in other sensory systems, much refinement and enhancement of the incoming information occurs in neurons along the afferent pathways, different odors eliciting different patterns of electrical activity in the brain, allowing humans to discriminate between some 10,000 different odorants even though they have only 1,000 different olfactory receptor types.

Olfactory discrimination varies with attentiveness, state of the olfactory mucosa—the sense of smell decreases when the mucosa is congested, as in a head cold; hunger—sensitivity is greater in hungry subjects; gender—women in general have keener olfactory sensitivities than men; smoking—decreased sensitivity has been repeatedly associated with smoking; and age—the ability to identify odors decreases with age, and a large percentage of elderly persons cannot detect odors at all.

SECTION B SUMMARY

SOMATIC SENSATION

I. Sensory function of the skin and underlying tissues is served by a variety of receptors sensitive to one (or a few) stimulus types.

II. Information about somatic sensation enters both specific and nonspecific ascending pathways. The specific pathways cross to the opposite side of the brain.

III. The somatic sensations include touch-pressure, the senses of posture and movement, temperature, and pain.

A. Rapidly adapting mechanoreceptors of the skin give rise to sensations such as vibration, touch, and movement, whereas slowly adapting ones give rise to the sensation of pressure.

B. Skin receptors having small receptive fields are involved in fine spatial discrimination, whereas receptors having larger receptive fields signal less spatially precise touch-pressure sensations.

C. The major receptor type responsible for the senses of posture and kinesthesia is the muscle-spindle stretch receptor.

D. Cold receptors are sensitive to decreasing temperature; warmth receptors signal information about increasing temperature.

E. Tissue damage stimulates specific receptors that give rise to the sensation of pain, which may also induce emotional and reflex responses.

F. Stimulation-produced analgesia, transcutaneous nerve stimulation (TENS), and acupuncture control pain by blocking transmission in the pain pathways.

VISION

I. Light is defined by its wavelength or frequency.

II. The light that falls on the retina must be focused by the cornea and lens.

A. Lens shape is changed to permit viewing near or distant objects (accommodation) so that they are focused on the retina.

B. Stiffening of the lens with aging interferes with accommodation. Cataracts decrease the amount of light reaching the retina.

C. An eyeball too long or too short relative to the focusing power of the lens causes nearsighted or farsighted vision, respectively.

III. The photopigments of the rods and cones are made up of protein component (opsin) and a chromophore (retinal).

A. The rods and each of the three cone types have different opsins, which make each of the four receptor types sensitive to a different wavelength of light.

B. When light falls upon the chromophore, the photic energy causes the chromophore to change shape, which triggers a cascade of events leading to hyperpolarization of the photoreceptors and decreased neurotransmitter release from them. When exposed to darkness, the rods and cones are depolarized and release more neurotransmitter.

IV. The rods and cones synapse on bipolar cells, which synapse on ganglion cells.

A. Ganglion-cell axons form the optic nerves, which lead into the brain.

B. The optic-nerve fibers from half of each retina cross to the opposite side of the brain in the optic chiasm. The fibers from the optic nerves terminate in the lateral geniculate nuclei of the thalamus, which send fibers to visual cortex.

C. Visual information is also relayed to areas of the brain dealing with biological rhythms.

V. Coding in the visual system occurs along parallel pathways, in which different aspects of visual information, such as color, form, movement, and depth, are kept separate from each other.

VI. The colors we perceive are related to the wavelength of light. Different wavelengths excite one of the three cone photopigments most strongly.

A. Certain ganglion cells are excited by input from one type of cone cell and inhibited by input from a different cone type.

B. Our sensation of color depends on the output of the various opponent-color cells and the processing of this output by brain areas involved in color vision.

VII. Six skeletal muscles control eye movement to scan the visual field for objects of interest, keep the fixation point focused on the fovea centralis despite movements of the object or the head, prevent adaptation of the photoreceptors, and move the eyes during accommodation.

HEARING

I. Sound energy is transmitted by movements of pressure waves.

A. Sound wave frequency determines pitch.

B. Sound wave amplitude determines loudness.

II. The sequence of sound transmission is as follows:

A. Sound waves enter the external auditory canal and press against the tympanic membrane, causing it to vibrate.

B. The vibrating membrane causes movement of the three small middle-ear bones; the stapes vibrates against the oval-window membrane.

C. Movements of the oval-window membrane set up pressure waves in the fluid-filled scala vestibuli, which cause vibrations in the cochlear duct wall, setting up pressure waves in the fluid there.

D. These pressure waves cause vibrations in the basilar membrane, which is located on one side of the cochlear duct.

E. As this membrane vibrates, the hair cells of the organ of Corti move in relation to the tectorial membrane.

F. Movement of the hair cells' stereocilia stimulates them to release neurotransmitter, which activates receptors on the peripheral ends of the afferent nerve fibers.

III. Each part of the basilar membrane vibrates maximally in response to one particular sound frequency.

VESTIBULAR SYSTEM

I. A vestibular apparatus lies in the temporal bone on each side of the head and consists of three semicircular ducts, a utricle, and a saccule.

II. The semicircular ducts detect angular acceleration during rotation of the head, which causes bending of the stereo-cilia on their hair cells.

III. Otoliths in the gelatinous substance of the utricle and saccule move in response to changes in linear acceleration and the position of the head relative to gravity and stimulate the stereocilia on the hair cells.

CHEMICAL SENSES

I. The receptors for taste lie in taste buds throughout the mouth, principally on the tongue. Different types of taste receptors operate by different mechanisms.

II. Olfactory receptors, which are part of the afferent olfactory neurons, lie in the upper nasal cavity.

A. Odorant molecules, once dissolved in the mucus that bathes the olfactory receptors, bind to specific receptors (protein binding sites). Each receptor cell has one of the 1,000 different receptor types.

B. Olfactory pathways go to the limbic system.

SECTION B KEY TERMS

somatic sensation	external auditory canal
kinesthesia	tympanic membrane
nociceptor	middle-ear cavity
wavelength	auditory tube
frequency	inner ear
visible spectrum	cochlea
retina	malleus
lens	incus
cornea	stapes
accommodation	oval window
ciliary muscle	cochlear duct
zonular fiber	scala vestibuli
iris	scala tympani
pupil	basilar membrane
rod	organ of Corti
cone	hair cell
photopigment	stereocilia
rhodopsin	tectorial membrane
opsin	vestibular apparatus
chromophore	semicircular duct
retinal	utricle
light adaptation	saccule
dark adaptation	labyrinth
bipolar cell	semicircular canal
ganglion cell	cupula
lateral geniculate nucleus	ampulla
visual field	chemoreceptor
suprachiasmatic nucleus	taste bud
opponent color cell	olfactory epithelium
fovea centralis	odorant
saccade	

SECTION B REVIEW QUESTIONS

1. Describe the similarities between pain and the other somatic sensations. Describe the differences.

2. List the structures through which light must pass before it reaches the photopigment in the rods and cones.

3. Describe the events that take place during accommodation for far vision.

4. What changes take place in neurotransmitter release from the rods or cones when they are exposed to light?

5. Beginning with the ganglion cells of the retina, describe the visual pathway.

6. List the sequence of events that occur between entry of a sound wave into the external auditory canal and the firing of action potentials in the cochlear nerve.

7. Describe the anatomical relationship between the cochlea and the cochlear duct.

8. What is the relationship between head movement and cupula movement in a semicircular canal?

9. What causes the release of neurotransmitter from the utricle and saccule receptor cells?

10. In what ways are the sensory systems for taste and olfaction similar? In what ways are they different?

CHAPTER 9 CLINICAL TERMS

phantom limb	myopic
hyperalgesia	farsighted
referred pain	hyperopic
stimulation-produced	astigmatism
analgesia	glaucoma
transcutaneous electric	color blindness
nerve stimulation	hearing aid
(TENS)	cochlear implant
acupuncture	nystagmus
presbyopia	vertigo
cataract	motion sickness
nearsighted	Ménière's disease

CHAPTER 9 THOUGHT QUESTIONS

(Answers are given in Appendix A.)

1. Describe several mechanisms by which pain could theoretically be controlled medically or surgically.

2. At what two sites would central nervous system injuries interfere with the perception of heat applied to the right side of the body? At what single site would a central nervous system injury interfere with the perception of heat applied to either side of the body?

3. What would vision be like after a drug has destroyed all the cones in the retina?

4. Damage to what parts of the cerebral cortex could explain the following behaviors? (a) A person walks into a chair placed in her path. (b) The person does not walk into the chair but does not know what the chair can be used for.

The endocrine system is one of the body's two major communication systems, the nervous system being the other. The **endocrine system** consists of all those glands, termed **endocrine glands,** that secrete hormones. **Hormones,** as noted in Chapter 7, are chemical messengers that are carried *by the blood* from endocrine glands to the cells upon which they act. The cells influenced by a particular hormone are the **target cells** for that hormone.

Table 10-1 summarizes, for reference and orientation, the endocrine glands, the hormones they secrete, and the major functions the hormones control. The endocrine system differs from most of the other organ systems of the body in that the various glands are not anatomically continuous; however, they do form a system in the functional sense. The reader may be puzzled to see listed as endocrine glands some organs—the heart, for instance—that

TABLE 10-1 SUMMARY OF THE HORMONES

Site produced (endocrine gland)	Hormone	Major function* is control of:
Adipose tissue cells	Leptin	Food intake; metabolic rate
Adrenal:		
Adrenal cortex	Cortisol	Organic metabolism; response to stress; immune system
	Androgens	Sex drive in women
	Aldosterone	Sodium, potassium, and acid excretion by kidneys
Adrenal medulla	Epinephrine Norepinephrine	Organic metabolism; cardiovascular function; response to stress
Gastrointestinal tract	Gastrin Secretin Cholecystokinin Glucose-dependent insulinotropic peptide (GIP)† Somatostatin	Gastrointestinal tract; liver; pancreas; gallbladder
Gonads:		
Ovaries: female	Estrogen Progesterone	Reproductive system; breasts; growth and development
	Inhibin	FSH secretion
	Relaxin	? Relaxation of cervix and pubic ligaments
Testes: male	Testosterone	Reproductive system; growth and development; sex drive
	Inhibin	FSH secretion
	Müllerian-inhibiting hormone	Regression of Müllerian ducts
Heart	Atrial natriuretic factor (ANF, atriopeptin, auriculin)	Sodium excretion by kidneys; blood pressure
Hypothalamus	Hypophysiotropic hormones	Secretion of hormones by the anterior pituitary
	Corticotropic releasing hormone (CRH)	Secretion of ACTH (stimulation)
	Thyrotropin releasing hormone (TRH)	Secretion of TSH (stimulation)
	Growth hormone releasing hormone (GHRH)	Secretion of GH (stimulation)
	Somatostatin (SS)	Secretion of GH (inhibition)
	Gonadotropin releasing hormone (GnRH)	Secretion of LH and FSH (stimulation)
	Dopamine (DA, also called prolactin-inhibiting hormone, PIH)	Secretion of prolactin (inhibition)
	Posterior pituitary hormones	See posterior pituitary
Kidneys	Renin ($\rightarrow$ angiotensin II)§ Erythropoietin 1,25-dihydroxyvitamin D_3	Aldosterone secretion; blood pressure Erythrocyte production Plasma calcium

TABLE 10-1 *(continued)*

Site produced (endocrine gland)	Hormone	Major function° is control of:
Leukocytes, macrophages, endothelial cells, and fibroblasts	Cytokines¶ (these include the interleukins, colony-stimulating factors, interferons, tumor necrosis factors)	Immune defenses
Liver (and other cells)	Insulin-like growth factors (IGF-1 and II)	Growth
Pancreas	Insulin Glucagon Somatostatin Pancreatic polypeptide	Organic metabolism; plasma glucose
Parathyroids	Parathyroid hormone (PTH, PH, parathormone)	Plasma calcium and phosphate
Pineal	Melatonin	? Sexual maturity; body rhythms
Pituitary glands: **Anterior pituitary**	Growth hormone (GH, somatotropin)	Growth, mainly via secretion of IGF-I; protein, carbohydrate, and lipid metabolism
	Thyroid-stimulating hormone (TSH, thyrotropin)	Thyroid gland
	Adrenocorticotropic hormone (ACTH, corticotropin)	Adrenal cortex
	Prolactin	Breast growth and milk synthesis; may be permissive for certain reproductive functions in the male

	Gonadotropic hormones:
	  Follicle-stimulating hormone (FSH)
	  Luteinizing hormone (LH)
		Gonads (gamete production and sex hormone secretion)

	β-lipotropin and β-endorphin	Unknown
Posterior pituitary‡	Oxytocin	Milk let-down; uterine motility
	Vasopressin (antidiuretic hormone, ADH)	Water excretion by the kidneys; blood pressure
Placenta	Chorionic gonadotropin (CG)	Secretion by corpus luteum
	Estrogens	See Gonads: ovaries
	Progesterone	See Gonads: ovaries
	Placental lactogen	Breast development; organic metabolism
Thymus	Thymopoietin	T-lymphocyte function
Thyroid	Thyroxine (T₄) Triiodothyronine (T₃)	Metabolic rate; growth; brain development and function
	Calcitonin	? Plasma calcium
Multiple cell types	Growth factors¶ (e.g., epidermal growth factor)	Growth and proliferation of specific cell types

°This table does not list all functions of the hormones.

†The names and abbreviations in parentheses are synonyms.

‡The posterior pituitary stores and secretes these hormones; they are made in the hypothalamus.

§Renin is an enzyme that initiates reactions in blood that generate angiotensin II.

¶Some classifications include the cytokines under the category of growth factors.

clearly have other functions. The explanation is that, in addition to the cells that carry out the organ's other functions, the organ also contains cells that secrete hormones. This illustrates the fact that organs are made up of different types of cells.

Note also in Table 10-1 that the hypothalamus, a part of the brain, is considered part of the endocrine system too. This is because the chemical messengers released by certain neuron terminals in both the hypothalamus and its extension, the posterior pituitary, do not function as neurotransmitters affecting adjacent cells but rather enter the blood, which carries them to their sites of action.

Table 10-1 demonstrates that there are a large number of endocrine glands and hormones. One way of describing the physiology of the individual hormones is to present all relevant material, gland by gland, in a single chapter. In keeping with our emphasis on hormones as messengers in homeostatic control mechanisms, however, we have chosen to describe the physiology of specific hormones and the glands that secrete them in subsequent chapters, in the context of the control systems in which they participate. For example, the pancreatic hormones are described in Chapter 18, which is on organic metabolism, parathyroid hormone in Chapter 16 in the context of calcium metabolism, and so on.

The aims of the present chapter are therefore limited to presenting (1) the general principles of endocrinology, that is, a structural and functional analysis of hormones in general that transcends individual glands; and (2) an analysis of the hypothalamus-pituitary hormonal system. The control systems for the hormones of this particular system are so interconnected that they are best described as a unit to lay the foundation for subsequent descriptions in other chapters.

Before turning to these analyses, however, several additional general points should be made concerning Table 10-1. One phenomenon evident from this table is that a single gland may secrete multiple hormones. The usual pattern in such cases is that a single cell type secretes only one hormone, so that multiple hormone secretion reflects the presence of different types of endocrine cells in the same gland. In a few cases, however, a single cell may secrete more than one hormone (for example, the secretion of follicle-stimulating hormone and luteinizing hormone by the anterior pituitary).

Another point of interest illustrated by Table 10-1 is that a particular hormone may be produced by more than one type of endocrine gland. For example, somatostatin is secreted by endocrine cells in both the gastrointestinal tract and the pancreas and is also one of the hormones secreted by the hypothalamus.

HORMONE STRUCTURES AND SYNTHESIS

Hormones fall into three chemical classes: (1) amines, (2) peptides and proteins, and (3) steroids.

Amine Hormones

The **amine hormones** are all derivatives of the amino acid tyrosine. They include the thyroid hormones, epinephrine and norepinephrine (produced by the adrenal medulla), and dopamine (produced by the hypothalamus).

Thyroid Hormones.   The **thyroid gland** is located in the lower part of the neck wrapped around the front of the trachea (windpipe). It is composed of many spherical structures called follicles, each consisting of a single layer of epithelial cells surrounding an extracellular central space filled with a glycoprotein colloid called **thyroglobulin.** The follicles secrete the two iodine-containing amine hormones—**thyroxine (T_4)** and **triiodothyronine (T_3)** (Figure 10-1), collectively known as the **thyroid hormones (TH).** Parafollicular cells, which are located between follicles, secrete a third hormone—a peptide called calcitonin; this hormone does not contain iodine and is not included in the term "thyroid hormones."

Iodine is an essential element that functions as a component of T_4 and T_3. Most of the iodine ingested in food is absorbed into the blood from the gastrointestinal tract by active transport; in the process it is converted to the ionized form, iodide. Iodide is actively transported from the blood into the thyroid follicular cells. Once in the cells, the iodide is converted back to iodine, which is then coupled to the side chains of tyrosine molecules that had previously been incorporated into thyroglobulin precursor. The result is the formation of thyroglobulin, which is stored in the central space of the follicles.

During hormone secretion thyroglobulin is moved into the follicular cells by endocytosis, T_4 and T_3 are enzymatically split from it, and the freed hormones cross the cells' plasma membranes to enter the blood. T_4 is secreted in much larger amounts than is T_3. However, a variety of tissues, particularly the liver and kidneys, convert most of

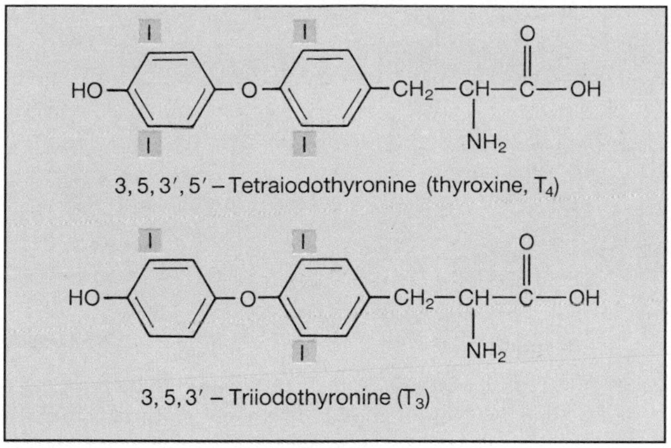

FIGURE 10-1
Chemical structures of the thyroid hormones, thyroxine and triiodothyronine. The two molecules differ by only one iodine atom, a difference noted in the abbreviations T_3 and T_4.

this T_4 into T_3 by enzymatic removal of one iodine atom; this constitutes the major source of plasma T_3. This is an important point because T_3 is a much more active hormone than is T_4. Indeed it is likely that T_4 has little or no action unless it is converted into T_3.

Virtually every tissue in the body is affected by the thyroid hormones. These effects, which are described in Chapter 18, include regulation of metabolic rate, growth, and brain development and function.

Adrenal Medullary Hormones and Dopamine.
There are two adrenal glands, one on the top of each kidney. Each **adrenal gland** constitutes two distinct endocrine glands, an inner **adrenal medulla,** which secretes amine hormones, and a surrounding **adrenal cortex,** which secretes steroid hormones. As described in Chapter 8, the adrenal medulla is really a modified sympathetic ganglion whose cell bodies do not have axons but instead release their secretions into the blood, thereby fulfilling a criterion for an endocrine gland.

The adrenal medulla secretes mainly two amine hormones, **epinephrine (E)** and **norepinephrine (NE).** Recall from Chapter 8 that these molecules constitute, with dopamine, the chemical family of **catecholamines.** The structures and pathways for synthesis of the catecholamines were described in Chapter 8, when they were dealt with as neurotransmitters. In humans, the adrenal medulla secretes approximately four times more epinephrine than norepinephrine. Epinephrine and norepinephrine exert actions similar to those of the sympathetic nerves. These effects are described in various chapters and summarized in Chapter 20 in the section on stress.

The adrenal medulla also secretes small amounts of dopamine and several substances other than cate-cholamines, but whether any of these adrenal secretions other than epinephrine and norepinephrine actually serve hormonal functions is unknown. In contrast, as described below, the dopamine secreted by certain cells in the hypothalamus definitely functions as a hormone.

Peptide Hormones
The great majority of hormones are either peptides or proteins. They range in size from small peptides having only three amino acids to small proteins (some of which are glycoproteins). For convenience, we shall follow a common practice of endocrinologists and refer to all these hormones as **peptide hormones.**

In many cases, they are initially synthesized on the ribosomes of the endocrine cells as larger proteins known as preprohormones, which are then cleaved to **prohormones** by proteolytic enzymes in the granular endoplasmic reticulum (Figure 10-2). The prohormone is then packaged into secretory vesicles by the Golgi apparatus. In this process, the prohormone is cleaved to yield the active hormone and other peptide chains found in the prohormone. Therefore, when the cell is stimulated to release the contents of the secretory vesicles by exocytosis, the other peptides are cosecreted with the hormone. In certain cases they, too, may exert hormonal effects. In other words, instead of just one peptide hormone, the cell may be secreting multiple peptide hormones that differ in their effects on target cells.

One more point about peptide hormones: As mentioned in Chapters 7 and 8, many peptides serve as both neurotransmitters (or neuromodulators) and as hormones. For example, most of the hormones secreted by the endocrine glands in the gastrointestinal tract (for example, cholecystokinin) are also produced by neurons in the brain where they function as neurotransmitters.

Steroid Hormones
The third family of hormones is the steroids, the lipids whose ring-like structure was described in Chapter 2. **Steroid hormones** are produced by the adrenal cortex and the **gonads** (testes and ovaries), as well as by the placenta during pregnancy. Examples are shown in Figure 10-3. In addition, the hormone 1,25-dihydroxyvitamin D_3, the active form of vitamin D, is a steroid derivative.

Cholesterol is the precursor of all steroid hormones. The many biochemical steps in steroid synthesis beyond cholesterol involve small changes in the molecules and are mediated by specific enzymes. The steroids produced by a particular cell depend therefore on the types and concentrations of enzymes present. Because steroids are highly lipid-soluble, once they are synthesized they simply diffuse across the plasma membrane of the steroid-producing cell and enter the interstitial fluid and then the blood.

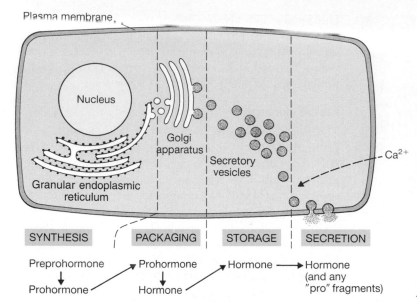

SYNTHESIS | PACKAGING | STORAGE | SECRETION

Preprohormone → Prohormone → Hormone → Hormone
Prohormone   Hormone   (and any "pro" fragments)

FIGURE 10-2

Typical synthesis and secretion of peptide hormones. Not all peptide hormones go through all these stages. In some cells the calcium that causes exocytosis is released from the endoplasmic reticulum by action of a second messenger rather than entering from the extracellular fluid. (*Adapted from Hedge et al.*)

The next sections describe the pathways for steroid synthesis by the adrenal cortex and gonads. Those for the placenta are somewhat unusual and are discussed in Chapter 19.

Hormones of the Adrenal Cortex. Steroid synthesis by the adrenal cortex is illustrated in Figure 10-4. The five hormones normally secreted in physiologically significant amounts by the adrenal cortex are aldosterone, cortisol, corticosterone, dehydroepiandrosterone, and androstenedione. **Aldosterone** is known as a **mineralocorticoid** because its effects are on salt (mineral) balance, mainly on the kidneys' handling of sodium, potassium, and hydrogen ions. **Cortisol** and corticosterone are called **glucocorticoids** because they have important effects on the metabolism of glucose and other organic nutrients. Cortisol is by far the more important of the two glucocorticoids in humans, and so we shall deal only with it in future discussions. In addition to its effects on organic metabolism (described in Chapter 18), cortisol exerts many other effects, including facilitation of the body's responses to stress and regulation of the immune system (Chapter 20).

Dehydroepiandrosterone and androstenedione belong to the class of hormones known as **androgens,** which also includes the male sex hormone, testosterone, produced by the testes. All androgens have actions similar to those of testosterone. Because the adrenal androgens are much less potent than testosterone, they are of little physiological significance in the male; they do, however, play roles in the female, as described in Chapter 19.

The adrenal cortex is not a homogeneous gland but is composed of three distinct layers (Figure 10-5). The outer layer—the zona glomerulosa—possesses very high concentrations of the enzymes required to convert corticosterone to aldosterone but lacks the enzymes required for the formation of cortisol and androgens. Accordingly, this layer synthesizes and secretes aldosterone but not the other major adrenal cortical hormones. In contrast, the zona fasciculata and zona reticularis have just the opposite enzyme profile. They, therefore, secrete no aldosterone but much cortisol and androgen.

In certain disease states, the adrenal cortex may secrete decreased or increased amounts of various steroids. For example, an absence of the enzymes for the formation of cortisol by the adrenal cortex can result in the shunting of the cortisol precursors into the androgen pathway. In a woman, the result of the large increase in androgen

FIGURE 10-3

Structures of representative steroid hormones.

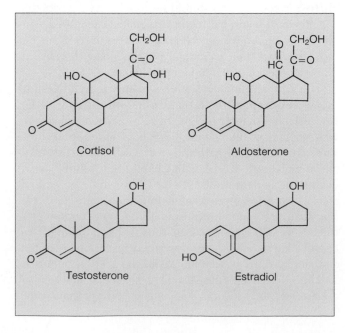

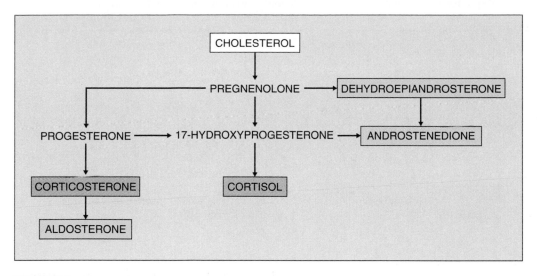

FIGURE 10-4

Simplified flow sheet for synthesis of steroid hormones by the adrenal cortex; many intermediate steps have been left out. The various steps are mediated by specific enzymes. The five hormones in colored boxes are the major steroid hormones secreted. Dehydroepiandrosterone and androstenedione are androgens, that is, testosterone-like hormones. Cortisol and corticosterone are glucocorticoids, and aldosterone is a mineralocorticoid.

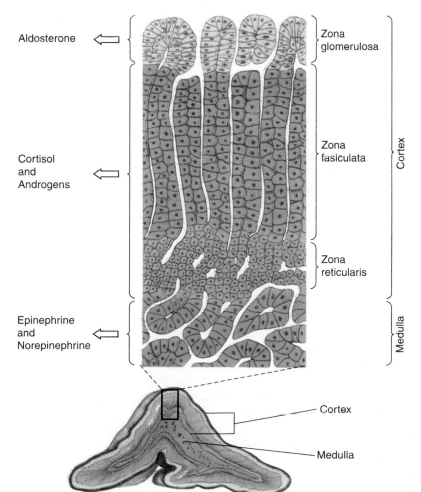

FIGURE 10-5

Section through an adrenal gland showing both the medulla and cortex, as well as the hormones they secrete.

TABLE 10-2 CATEGORIES OF HORMONES

Types	Major form in plasma	Location of receptors	Signal transduction mechanisms	Rate of excretion/metabolism
Peptides and catecholamines	Free	Plasma membrane	Receptors alter: Receptor channels Channels gated by G proteins Cellular concentration of cAMP Cellular concentration of cGMP Cellular concentrations of Ca^{2+}, IP_3, DAG	Fast (minutes to a few hours)
Steroids and thyroid hormones	Protein-bound	Cell interior	Receptors directly alter gene transcription	Slow (hours to days)

secretion would be **masculinization,** in addition to the effects of cortisol deficiency (see Chapter 20).

Hormones of the Gonads. Compared to the adrenal cortex, the gonads have very different concentrations of key enzymes in their steroid pathways. Endocrine cells in both testes and ovaries lack the enzymes needed to produce aldosterone and cortisol. They possess high concentrations of enzymes in the androgen pathways leading to androstenedione, as in the adrenal cortex. In addition, the endocrine cells in the testes contain a high concentration of the enzyme that converts androstenedione to **testosterone,** which is therefore the major androgen secreted by the testes. The ovarian endocrine cells that synthesize the major female sex hormone, **estradiol,** have a high concentration of the enzyme required to go one step further, that is, to transform androgens to estradiol. Accordingly, estradiol, rather than androgens, is secreted by the ovaries.

Very small amounts of testosterone do leak out of ovarian endocrine cells, however, and very small amounts of estradiol are produced from testosterone in the testes. Moreover, following their secretion into the blood by the gonads and the adrenal cortex, steroid hormones may undergo further interconversion in either the blood or other organs. For example, testosterone is converted to estradiol in some of its target cells. Thus, the major male and female sex hormones—testosterone and estradiol, respectively—are not unique to males and females, although, of course, the relative concentrations of the hormones are quite different in the two sexes.

Finally, certain ovarian endocrine cells secrete another major steroid hormone, **progesterone.**

HORMONE TRANSPORT IN THE BLOOD

Peptide and catecholamine hormones are water-soluble. Therefore, with the exception of a few peptides, these hormones are transported simply dissolved in plasma (Table 10-2). In contrast, the steroid hormones and the thyroid hormones circulate in the blood largely bound to plasma proteins.

Even though the steroid and thyroid hormones exist in plasma mainly bound to large proteins, small concentrations of these hormones do exist dissolved in the plasma. The dissolved, or free, hormone is in equilibrium with the bound hormone:

Free hormone + binding protein $\rightleftharpoons$
hormone-protein complex

The total hormone concentration in plasma is the sum of the free and bound hormone. It is important to realize, however, that only the *free* hormone can diffuse across capillary walls and encounter its target cells. Accordingly, the concentration of the free hormone is what is physiologically important rather than the concentration of the total hormone, most of which is bound. As we shall see, the degree of protein binding also influences the rate of metabolism and the excretion of the hormone.

HORMONE METABOLISM AND EXCRETION

A hormone's concentration in the plasma depends not only upon its rate of secretion by the endocrine gland but also upon its rate of removal from the blood, either by excretion or by metabolic transformation. The liver and the kidneys are the major organs that excrete or metabolize hormones.

The liver and kidneys, however, are not the only routes for eliminating hormones. Sometimes the hormone is metabolized by the cells upon which it acts. Very importantly, in the case of peptide hormones, endocytosis of hormone-receptor complexes on plasma membranes enables cells to remove the hormones rapidly from their surfaces and catabolize them intracellularly. The receptors are then often recycled to the plasma membrane.

In general, catecholamine and peptide hormones are excreted comparatively easily or attacked by enzymes in the blood and tissues. These hormones therefore tend to remain in the bloodstream for only brief periods—minutes to a few hours. In contrast, because protein-bound hormones are less vulnerable to excretion or metabolism by enzymes, removal of the circulating steroid and thyroid hormones generally takes many hours or even several days.

In some cases, metabolism of the hormone after its secretion *activates* the hormone rather than inactivates it. In other words, the secreted hormone may be relatively or completely unable to act upon a target cell until metabolism elsewhere in the body transforms it into a substance that can act. We have already seen one example of hormone activation—the conversion of circulating T_4 to the far more active T_3 by various organs. Another example is provided by testosterone, which is converted either to estradiol or dihydrotestosterone in certain of its target cells. These molecules, rather than testosterone itself, then bind to receptors inside the target cell and elicit the cell's response.

There is another kind of "activation" that applies to a few hormones. Instead of the hormone itself being activated after secretion, it acts enzymatically on a completely different plasma protein to split off a peptide that functions as the active hormone. The best known example of this is the renin-angiotensin system, described in Chapters 14 and 16.

Figure 10-6 summarizes the fates of hormones after their secretion.

MECHANISMS OF HORMONE ACTION

Hormone Receptors

Because they travel in the blood, hormones are able to reach virtually all tissues. Yet the body's response to a hormone is not all-inclusive, but is highly specific, involving only the target cells for that hormone. The ability to respond depends upon the presence on (or in) the target cells of specific receptors for those hormones.

As emphasized in Chapter 7, the response of a target cell to a chemical messenger is the final event in a sequence that begins when the messenger binds to specific cell receptors. As described in that chapter, the receptors for peptide hormones and catecholamines are proteins located in the plasma membranes of the target cells. In contrast, the receptors for steroid hormones and the thyroid hormones are proteins located mainly *inside* the target cells. Because these hormones are lipid-soluble, they readily cross the plasma membrane and combine with their specific intracellular receptors.

Hormones can influence the ability of target cells to respond by regulating hormone receptors. Basic concepts of receptor modulation (up-regulation and down-regulation) were described in Chapter 7. In the context of hormones, **up-regulation** is an increase in the number of a hormone's receptors on a cell, often resulting from prolonged exposure to a low concentration of the hormone. **Down-regulation** is a decrease in receptor number, often from exposure to high concentrations of the hormone.

FIGURE 10-6

Possible fates and actions of a hormone following its secretion by an endocrine cell. Not all paths apply to all hormones.

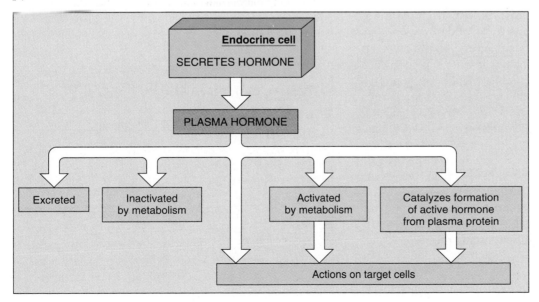

Hormones can down-regulate or up-regulate not only their own receptors but the receptors for other hormones as well. If one hormone induces a loss of a second hormone's receptors, the result will be a reduction of the second hormone's effectiveness; in such cases, the one hormone is said to antagonize the action of the other. On the other hand, a hormone may induce an increase in the number of receptors for a second hormone. In this case the effectiveness of the second hormone is increased.

This latter phenomenon, in some cases, underlies the important hormone-hormone interaction known as permissiveness. In general terms, **permissiveness** means that hormone A must be present for the full strength of hormone B's effect. A low concentration of hormone A is usually all that is needed for this permissive effect, which is due to A's positive effect on B's receptors. For example (Figure 10-7), epinephrine causes a large release of fatty acids from adipose tissue, but only in the presence of permissive amounts of thyroid hormone; the reason is that thyroid hormone facilitates the synthesis of receptors for epinephrine in adipose tissue and so the tissue becomes much more sensitive to epinephrine. It should be noted, however, that receptor alteration does not explain all cases of permissiveness; often the explanation is not known.

Events Elicited by Hormone-Receptor Binding

The events initiated by the binding of a hormone to its receptor, that is, the mechanisms by which the hormone elicits a cellular response, are the signal transduction pathways that apply to all chemical messengers, as described in Chapter 7 (Figures 7-13 and 7-14). In other words, there is nothing unique about the mechanisms initiated by hormones as compared to those utilized by neurotransmitters and paracrine agents, and so they are only briefly reviewed at this point (Table 10-2).

Effects of Peptide Hormones and Catecholamines.

As stated above, the receptors for peptide hormones and the catecholamine hormones are located on the outer surface of the target cell's plasma membrane. When activated by hormone binding, the receptors trigger one or more of the signal transduction pathways described for plasma-membrane receptors in Chapter 7 (Figure 7-14). That is, the activated receptors directly influence: (1) ion channels that are part of the receptors; (2) enzyme activity that is part of the receptor; or (3) G proteins coupled in the plasma membrane to effector proteins—ion channels and enzymes. The opening or closing of ion channels causes changes in the electrical potential across the membrane and, when a calcium channel is involved, a change in the cytosolic concentration of this important ionic second messenger. The changes in enzyme activity produce—most commonly by phosphorylation catalysed by protein kinase enzymes—changes in the conformation and hence the activity of various cellular proteins. In some cases the signal transduction pathways also lead to stimulation (or inhibition) of the synthesis of new proteins by the cell.

Effects of Steroid and Thyroid Hormones.

Structurally, the steroid hormones, the thyroid hormones, and the steroid derivative 1,25-dihydroxyvitamin D_3 are all closely related, and their receptors, which are intracellular, constitute the steroid-hormone receptor superfamily. As described in Chapter 7 (Figure 7-13), the binding of hormone to one of these receptors leads to activation (or, in some cases, inhibition) of particular genes, causing a change in the rate of synthesis of the proteins coded for by those genes. The ultimate result of changes in the concentrations of these proteins is an enhancement or inhibition of particular processes carried out by the cell, or a change in the rate of protein secretion by the cell.

Surprisingly, in addition to having intracellular receptors, some target cells probably also have *plasma-membrane receptors* for certain of the steroid hormones. In such cases the signal-transduction pathways initiated by the plasma-membrane receptors elicit rapid cell responses while the intracellular receptors mediate delayed responses, which require new protein synthesis.

Pharmacological Effects of Hormones

Administration of very large quantities of a hormone for medical purposes may have effects that are never seen in a normal person. They are called ***pharmacological effects,*** and they can also occur in diseases when excessive amounts of hormones are secreted. Pharmacological effects are of great importance in medicine, since hormones

FIGURE 10-7

Ability of thyroid hormone to "permit" epinephrine-induced release of fatty acids from adipose-tissue cells. The thyroid hormones exert this effect by causing an increased number of epinephrine receptors on the cell.

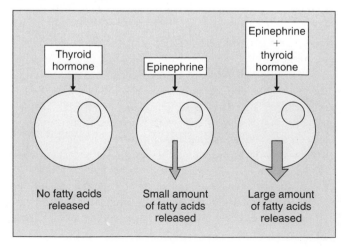

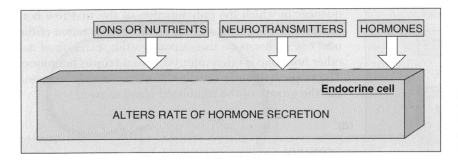

FIGURE 10-8

Inputs that act directly on endocrine-gland cells to stimulate or inhibit hormone secretion.

are often used in large doses as therapeutic agents. Perhaps the most common example is that of the adrenal cortical hormone cortisol, which is administered in large amounts to suppress allergic and inflammatory reactions.

INPUTS THAT CONTROL HORMONE SECRETION

Most hormones are released in short bursts, with little or no release occurring between bursts. Accordingly, the plasma concentrations of hormones may fluctuate rapidly over brief time periods. Hormones also manifest 24-h cyclical variations in their secretory rates, the circadian patterns being different for different hormones. Some are clearly linked to sleep; for example, growth hormone's secretion is markedly increased during the early period of sleep and is quite low or absent during the rest of the day and night. The mechanisms underlying these cycles are ultimately traceable to cyclical variations in the activity of neural pathways involved in the hormone's release.

Hormone secretion is controlled mainly by three types of inputs to endocrine cells (Figure 10-8): (1) changes in the plasma concentrations of mineral ions or organic nutrients; (2) neurotransmitters released from neurons impinging on the endocrine cell; and (3) another hormone (or, in some cases, a paracrine agent) acting on the endocrine cell. There is actually a fourth type of input—chemical and physical factors in the lumen of the gastrointestinal tract—but it applies only to the hormones secreted by the gastrointestinal tract and will be described in Chapter 17.

Before we look more closely at each category, it must be stressed that, in many cases, hormone secretion is influenced by more than one input. For example, insulin secretion is controlled by the extracellular concentrations of glucose and other nutrients, by both sympathetic and parasympathetic neurons to the insulin-secreting endocrine cells, and by several hormones acting on these cells. Thus, endocrine cells, like neurons, may be subject to multiple, simultaneous, often opposing inputs, and the resulting output—the rate of hormone secretion—reflects the integration of all these inputs.

One more point should be made to avoid misunderstanding. The term "secretion" applied to a hormone denotes its synthesis and release from the cell. Some inputs to endocrine cells specifically stimulate or inhibit only synthesis, for example, by altering the expression of the gene for that hormone, with changes in release occurring as a secondary result. In contrast, other inputs directly influence only the actual release of the hormone from the cell, and some inputs influence both synthesis and release. For simplicity in this chapter and the rest of the book, we will generally not distinguish between these possibilities when we refer to stimulation or inhibition of hormone "secretion."

Control by Plasma Concentrations of Mineral Ions or Organic Nutrients

There are at least five hormones whose secretion is directly controlled, at least in part, by the plasma concentrations of specific mineral ions or organic nutrients. In each case, a major function of the hormone is to regulate, in a negative-feedback manner, the plasma concentration of the ion or nutrient controlling its secretion. For example, insulin secretion is stimulated by an elevated plasma glucose concentration, and the additional insulin then causes, by several actions, the plasma glucose concentration to decrease (Figure 10-9).

Control by Neurons

The adrenal medulla behaves like a sympathetic ganglion and thus is stimulated by sympathetic preganglionic fibers. In addition to its control of the adrenal medulla, the autonomic nervous system has influences on other endocrine glands (Figure 10-10B). Both parasympathetic and sympathetic inputs to these other glands may occur, some inhibitory and some stimulatory. Examples are the secretion of insulin and the gastrointestinal hormones.

Thus far, our discussion of neural control of hormone release has been limited to the role of the *autonomic* nervous system (Figure 10-10B). However, one large group of hormones—those secreted by the hypothalamus and its extension, the posterior pituitary—are under the direct control not of autonomic neurons but of neurons in the brain itself (Figure 10-10A). This category will be described in detail later in this chapter.

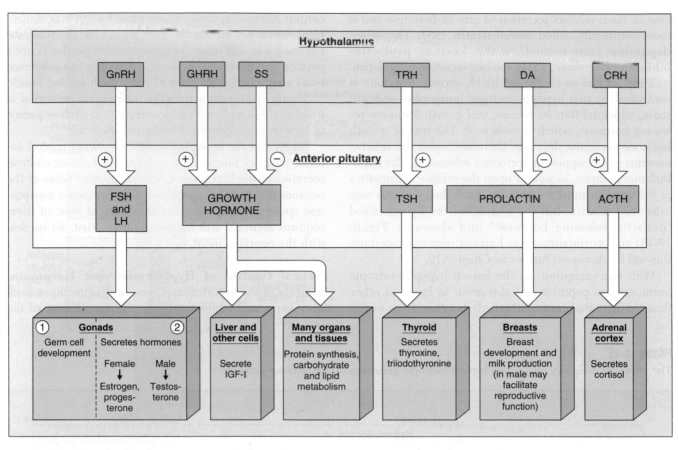

FIGURE 10-16

A combination of Figures 10-13 and 10-15 summarizes the hypothalamic-anterior-pituitary system.

central nervous system, and specific neural pathways influence secretion of the individual hypophysiotropic hormones. A large number of neurotransmitters, including the catecholamines and acetylcholine, are found at the synapses on the hormone-secreting hypothalamic neurons, and this explains why the secretion of the hypophysiotropic hormones can be altered by drugs that influence these neurotransmitters.

Figure 10-17 illustrates one example of the role of neural input to the hypothalamus. Corticotropin releasing hormone (CRH) from the hypothalamus stimulates the anterior pituitary to secrete ACTH, which in turn stimulates the adrenal cortex to secrete cortisol. A wide variety of stresses, both physical and emotional, act via neural pathways to the hypothalamus to increase CRH secretion and, hence, ACTH and cortisol secretion, markedly above basal values. Thus, stress is the common denominator of reflexes leading to increased cortisol secretion. Cortisol then functions to facilitate an individual's response to stress. Even in an unstressed person, however, the secretion of cortisol varies in a highly stereotyped manner during a 24-h period because neural rhythms within the cen-

tral nervous system also impinge upon the hypothalamic neurons that secrete CRH.

Hormonal Feedback Control of the Hypothalamus and Anterior Pituitary. A prominent feature of each of the hormonal sequences initiated by a hypophysiotropic hormone is negative feedback exerted upon the hypothalamo-pituitary system by one or more of the hormones in its sequence. For example, in the CRH-ACTH-cortisol sequence (Figure 10-17), the final hormone, cortisol, acts upon the hypothalamus to reduce secretion of CRH by causing a decrease in the frequency of action potentials in the neurons secreting CRH. In addition, cortisol acts directly on the anterior pituitary to reduce the response of the ACTH-secreting cells to CRH. Thus, by a double-barreled action, cortisol exerts a negative-feedback control over its own secretion.

Such a system is effective in dampening hormonal responses, that is, in limiting the extremes of hormone secretory rates. For example, when a painful stimulus elicits increased secretion, in turn, of CRH, ACTH, and cortisol, the resulting elevation in plasma cortisol concentration feeds back to inhibit the hypothalamus and anterior pituitary. Therefore, cortisol secretion does not rise as much as it would without these negative feedbacks.

Another adaptive function of these negative-feedback mechanisms is that they maintain the plasma concentration of the final hormone in a sequence relatively constant whenever a disease-induced primary change occurs in the secretion or metabolism of that hormone. An example of this is shown in Figure 10-18 for cortisol.

FIGURE 10-17

CRH-ACTH-cortisol sequence. The "stress" input to the hypothalamus is via neural pathways. Cortisol exerts a negative-feedback control over the system by acting on (1) the hypothalamus to inhibit CRH secretion and (2) the anterior pituitary to reduce responsiveness to CRH.

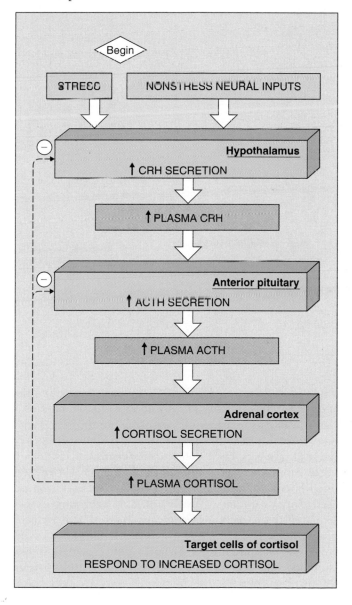

Another example is provided by the TRH-TSH-TH system. The thyroid hormones (TH) exert a feedback inhibition on the hypothalamo-pituitary system (mainly by decreasing the response of TSH-secreting cells to the stimulatory effects of TRH). Since iodine is essential for the synthesis of TH, individuals with iodine deficiencies tend to have a deficient production of TH. The resulting decrease in plasma TH concentration relieves some of the feedback inhibition TH exerts on the pituitary. Therefore, more TSH is secreted in response to the TRH coming from the hypothalamus, and the increased plasma TSH stimulates the thyroid gland to enlarge and to utilize more efficiently whatever iodine is available (the enlarged gland is known as *iodine-deficient goiter*). In this manner, plasma TH concentration can be kept quite close to normal.

The situations described above for cortisol and the thyroid hormones, in which the hormone secreted by the third endocrine gland in a sequence exerts a negative-feedback effect over the anterior pituitary and/or hypothalamus, is known as a **long-loop negative feedback** (Figure 10-19). This type of feedback exists for each of the five three-hormone sequences initiated by a hypophysiotropic hormone.

Long-loop feedback cannot exist for prolactin since this is the one anterior pituitary hormone that does not have major control over another endocrine gland—that is, it does not participate in a three-hormone sequence. Nonetheless, there is negative feedback in the prolactin system, for this hormone itself acts upon the hypothalamus to *stimulate* the secretion of dopamine, which then, you will recall, *inhibits* the secretion of prolactin. The influence of an anterior pituitary hormone on the hypothalamus is known as a **short-loop negative feedback** (Figure 10-19). Like prolactin, several other anterior pituitary hormones, including growth hormone, also exert such feedback on the hypothalamus.

The Role of "Non-Sequence" Hormones on the Hypothalamus and Anterior Pituitary. It must be emphasized that there are many stimulatory and inhibitory hormonal influences on the hypothalamus and/or anterior pituitary other than those that fit the feedback patterns just described. In other words, a hormone that is not itself in a particular hormonal sequence may nevertheless exert important influences on the secretion of the hypophysiotropic or anterior pituitary hormones in that sequence. For example, estrogen markedly enhances the secretion of prolactin by the anterior pituitary, even though estrogen secretion is not controlled by prolactin. Thus, one should not view the sequences we have been describing as isolated units.

A Summary Example: Control of Growth Hormone Secretion. The three-hormone sequences beginning in

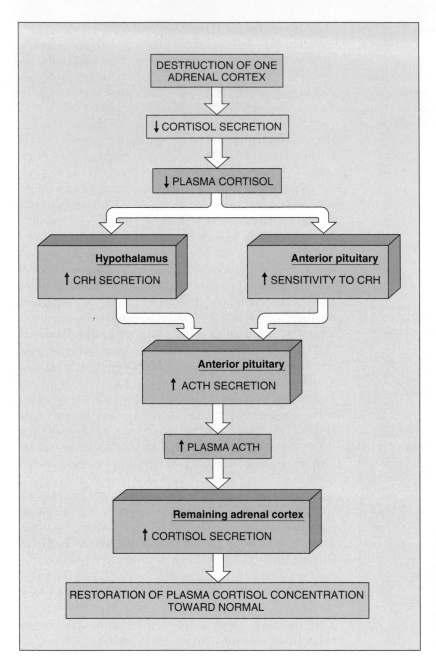

FIGURE 10-18

How negative feedback in a hormonal sequence helps maintain the plasma concentration of the final hormone when disease-induced changes in hormone secretion occur. The same analysis would apply if the original reduction in plasma cortisol were due to excessive metabolism of cortisol rather than deficient secretion.

the hypothalamus can be extremely complex, incorporating multiple sites of feedback, both long-loop and short-loop, as well as other hormones not in the sequence. Purely for the sake of illustrating this complexity, we describe here the control of growth hormone secretion (Figure 10-20), building upon the information already presented in this chapter.

Recall from Figure 10-15 that the secretion of GH by the anterior pituitary is controlled mainly by two hormones from the hypothalamus: (1) somatostatin, which inhibits GH secretion; and (2) GHRH, which stimulates it. With such a dual control system, the rate of GH secretion at any moment reflects the relative amounts of simultane-

ous stimulation by GHRH and inhibition by somatostatin. For example, very little growth hormone is secreted during the day in nonstressed persons who are eating normally, but whether this is due to a very low secretion of GHRH or a very high secretion of somatostatin is not yet clear. Similarly, a large number of physiological states (exercise, stress, fasting, a low plasma glucose concentration, and sleep) stimulate growth hormone secretion by decreasing the secretion of somatostatin and/or increasing that of GHRH.

As shown in Figures 10-13 and 10-16, growth hormone stimulates the secretion of another hormone, IGF-I, from the liver and many other target cells of GH. This makes possible a variety of feedbacks by which an increase in plasma GH results in inhibition of GH secretion (Figure 10-20): (1) a short-loop negative feedback exerted by GH on the hypothalamus, and (2) a long-loop negative feedback exerted by IGF-I on the hypothalamus. Both of these feedbacks operate by inhibiting secretion of GHRH and/or stimulating secretion of somatostatin, the result in either case being less stimulation of GH secretion. (3) IGF-I also acts directly on the pituitary to inhibit the stimulatory effect of GHRH on GH secretion; again, the result is decreased GH secretion. Finally, the secretion of growth hormone is influenced by several other hormones not in the sequence, for example, by the sex hormones. Some of these hormones may affect the sequence indirectly by altering the release of somatostatin and/or GHRH from the hypothalamus, whereas others may act directly upon the anterior pituitary.

CANDIDATE HORMONES

Many substances, termed **candidate hormones,** are suspected of being hormones in humans but are not considered classical hormones for one of two reasons: Either (1) their functions have not been conclusively documented; or (2) they have well-documented functions as paracrine and autocrine agents, but it is not certain that they ever reach additional target cells via the blood, an essential criterion for classification as a hormone. This second category includes certain of the eicosanoids (Chapter 7) and a large number of what are called growth factors (Chapter 18) that are secreted by multiple cell types and stimulate specific cells to undergo cell division and differentiation.

A substance that fits the first category described above is the amino acid derivative **melatonin,** which is synthesized from serotonin. This candidate hormone is produced by the **pineal gland,** an outgrowth from the roof of the diencephalon of the brain (shown but not labeled in Figure 8-37). The exact functions of melatonin in humans are uncertain, but this hormone probably plays an important role in the setting of the body's circadian rhythms (Chapter 7). Its secretion is stimulated by sympathetic postganglionic neurons that constitute the last link in a neuronal chain primarily triggered by receptors in the eyes; darkness stimulates melatonin secretion, and light inhibits it. Melatonin secretion, therefore, undergoes a marked 24-h cycle, being high at night and low during the day. Melatonin's ability to reduce the symptoms of *jet lag* when administered in small amounts at the proper time, its relationship to *seasonal affective disorder* ("winter depression"), its potential use as a "natural" sleeping pill, its ability to scavenge damaging free radicals, and its possible role in the control of the reproductive system are all being studied.

TYPES OF ENDOCRINE DISORDERS

Most endocrine disorders fall into one of four categories: (1) too little hormone (*hyposecretion*); (2) too much hormone (*hypersecretion*); (3) reduced response of the target cells (*hyporesponsiveness*); and (4) increased response of the target cells (*hyperresponsiveness*). In the first two categories, the phrases "too little hormone" and "too much hormone" here mean too little or too much for any given physiological situation. For example, as we shall

FIGURE 10-19

Short-loop and long-loop feedbacks. Long-loop feedback is exerted on the hypothalamus and/or anterior pituitary by the third hormone in the sequence. Short-loop feedback is exerted by the anterior pituitary hormone on the hypothalamus.

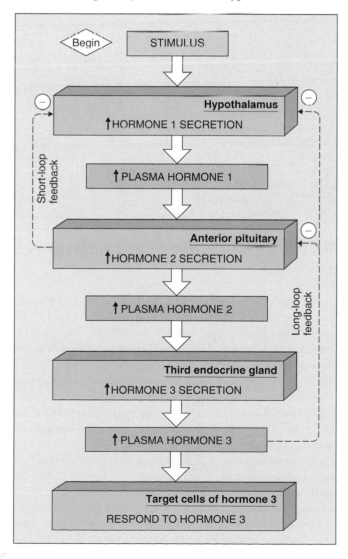

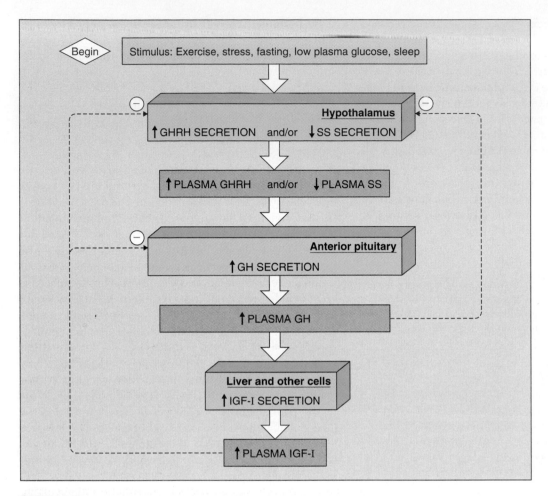

FIGURE 10-20

Hormonal pathways controlling the secretion of growth hormone (GH) and insulin-like growth factor I (IGF-I). At the hypothalamus, the minus sign (−) denotes that the input inhibits the secretion of growth hormone releasing hormone (GHRH) and/or stimulates the release of somatostatin (SS). Not shown in the figure is that several hormones not in the sequence (for example, the sex hormones) influence growth hormone secretion via effects on the hypothalamus and/or anterior pituitary.

see, insulin secretion decreases during fasting, and this decrease is an adaptive physiological response, not too little insulin. In contrast, insulin secretion should increase after eating, and if its increase is less than normal, this is too little hormone secretion.

Hyposecretion

An endocrine gland may be secreting too little hormone because the gland is not able to function normally. This is termed *primary hyposecretion.* Examples of primary hyposecretion include (1) genetic absence of a steroid-forming enzyme leading to decreased cortisol secretion, and (2) dietary deficiency of iodine leading to decreased secretion of thyroid hormones. There are many other causes—infections, toxic chemicals, and so on—all having the common denominator of damaging the endocrine gland.

In contrast to primary hyposecretion, a gland may be secreting too little hormone not because the gland is abnormal but because there is not enough of its tropic hormone. This is termed *secondary hyposecretion.* For example, there may be nothing wrong with the thyroid gland, but it may be secreting too little thyroid hormone because the secretion of TSH by the anterior pituitary is abnormally low. Thus, the hyposecretion by the thyroid gland in this case is secondary to inadequate secretion by the anterior pituitary.

This example raises the next question applicable to any of the other anterior pituitary hormones as well: Is the hyposecretion of TSH primary, that is, due to a defect in the anterior pituitary, or is it secondary to a hypothalamic defect causing too little secretion of TRH? If the latter were true, then we would have the following sequence: primary hyposecretion of TRH leading to secondary hyposecretion of TSH leading to secondary (sometimes called, in such cases, tertiary) hyposecretion of TH.

In diagnosing the presence of hyposecretion, a basic measurement to be made is the concentration of the hormone in either plasma or, for some hormones, urine. The finding of a low concentration will not distinguish between primary and secondary hyposecretion, however. To do this, the concentration of the relevant tropic hormone must also be measured. Thus, in our example, if the hyposecretion of TH is secondary to hyposecretion of TSH, then the plasma concentrations of both will be decreased. If the hyposecretion of TH is primary, then TH concentration will be decreased and TSH concentration will be *increased* because of less negative-feedback inhibition by TH over TSH secretion.

Another diagnostic approach is to attempt to stimulate the gland in question by administering either its tropic hormone or some other substance known to elicit increased secretion. The increase in hormone secretion elicited by the stimulus will be normal if the original hyposecretion is secondary, but less than normal if primary hyposecretion is the problem.

The most common means of treating hormone hyposecretion is to administer the hormone that is missing or present in too small amounts. In cases of secondary hyposecretion, there is a choice since at least two hormones are involved. In our example, the TH deficiency resulting from primary hyposecretion of TSH could theoretically be eliminated by administering either TH or TSH.

Hypersecretion

A hormone can also undergo either ***primary hypersecretion*** (the gland is secreting too much of the hormone on its own) or ***secondary hypersecretion*** (there is excessive stimulation of the gland by its tropic hormone). One of the most common causes of primary hypersecretion is the presence of a hormone-secreting endocrine-cell tumor.

The diagnosis of primary versus secondary hypersecretion of a particular hormone is analogous to that of hyposecretion. The concentrations of the hormone and, if relevant, its tropic hormone are measured in plasma or urine. If both concentrations are elevated, then the hormone in question is being secondarily hypersecreted. If the hypersecretion is primary, there will be a decreased concentration of the tropic hormone because of negative feedback by the high concentration of the hormone being hypersecreted. Again as with hyposecretion, one can get hypersecretion of a hypophysiotropic hormone, leading to secondary hypersecretion of an anterior pituitary hormone, leading to tertiary hypersecretion of the peripheral endocrine gland.

When an endocrine tumor is the cause of hypersecretion, it can often be removed surgically or destroyed with radiation. In many cases hypersecretion can also be blocked by drugs that inhibit the hormone's synthesis. Alternatively, the situation can be treated with drugs that do not alter the hormone's secretion but instead block the hormone's actions on its target cells.

Hyporesponsiveness and Hyperresponsiveness

In some cases, the endocrine system may be dysfunctioning even though there is nothing wrong with hormone secretion. The problem is that the target cells do not respond normally to the hormone. This condition is termed hyporesponsiveness. An important example of a disease resulting from hyporesponsiveness is the major form of ***diabetes mellitus*** ("sugar diabetes"), in which the target cells of the hormone insulin are hyporesponsive to this hormone.

One cause of hyporesponsiveness is either a lack or deficiency of receptors for the hormone. For example, certain men have a genetic defect manifested by the ***absence of receptors for dihydrotestosterone,*** the form of testosterone active in many target cells. In such men, these cells are unable to bind dihydrotestosterone, and the result is lack of development of certain male characteristics, just as though the hormone were not being produced.

In a second type of hyporesponsiveness the receptors for a hormone may be normal, but some event occurring after the hormone binds to receptors may be defective. For example, the activated receptor might be unable to stimulate formation of cyclic AMP or open a plasma-membrane channel.

A third cause of hyporesponsiveness applies to hormones that require metabolic activation by some other tissue after secretion. There may be a lack or deficiency of the enzymes that catalyze the activation. For example, some men secrete testosterone normally and have normal receptors for dihydrotestosterone but are missing the enzyme that converts testosterone to dihydrotestosterone.

In situations characterized by hyporesponsiveness to a hormone, the plasma concentration of the hormone in question is normal or elevated, but the response to administered hormone is diminished.

Finally, hyperresponsiveness to a hormone can also occur and cause problems. For example, the thyroid hormones cause an up-regulation of certain receptors for epinephrine, and so a primary or secondary hyperthyroidism causes, in turn, a hyperresponsiveness to epinephrine.

21. List the major hypophysiotropic hormones and the hormone whose release each controls.

22. What kinds of inputs control secretion of the hypophysiotropic hormones?

23. Diagram the CRH-ACTH-cortisol system.

24. What is the difference between long-loop and short-loop negative feedback in the hypothalamo-anterior pituitary system?

25. How would you distinguish between primary and secondary hyposecretion of a hormone? Between hyposecretion and hyporesponsiveness?

CLINICAL TERMS

masculinization of a female
pharmacological effect
iodine-deficient goiter
jet lag
seasonal affective disorder
hyposecretion
hypersecretion
hyporesponsiveness

hyperresponsiveness
primary hyposecretion
secondary hyposecretion
primary hypersecretion
secondary hypersecretion
diabetes mellitus
absence of receptors for
    dihydrotestosterone

THOUGHT QUESTIONS

(Answers are given in Appendix A)

1. In an experimental animal the sympathetic preganglionic fibers to the adrenal medulla are cut. What happens to the plasma concentration of epinephrine at rest and during stress?

2. During pregnancy there is an increase in the production (by the liver) and, hence, the plasma concentration of the major plasma binding protein for the thyroid hormones (TH). This causes a sequence of events involving feedback that results in an increase in the plasma concentration of TH, but no evidence of hyperthyroidism. Describe the sequence of events.

3. A child shows the following symptoms: deficient growth; failure to show sexual development; decreased ability to respond to stress. What is the most likely cause of all these symptoms?

4. If all the neural connections between the hypothalamus and pituitary were severed, the secretion of which pituitary hormones would be affected? Which pituitary hormones would not be affected?

5. An antibody to a peptide combines with the peptide and renders it nonfunctional. If an animal were given an antibody to somatostatin, the secretion of which anterior pituitary hormone would change and in what direction?

6. A drug that blocks the action of norepinephrine is injected directly into the hypothalamus of an experimental animal, and the secretion rates of several anterior pituitary hormones are observed to change. How is this possible, since norepinephrine is not a hypophysiotropic hormone?

7. A person is receiving very large doses of a cortisol-like drug to treat her arthritis. What happens to her secretion of cortisol?

8. A person with symptoms of hypothyroidism (for example, sluggishness and intolerance to cold) is found to have abnormally low plasma concentrations of T_4, T_3, and TSH. After an injection of TRH the plasma concentrations of all three hormones increase. Where is the site of the defect leading to the hypothyroidism?

CHAPTER 11

Muscle

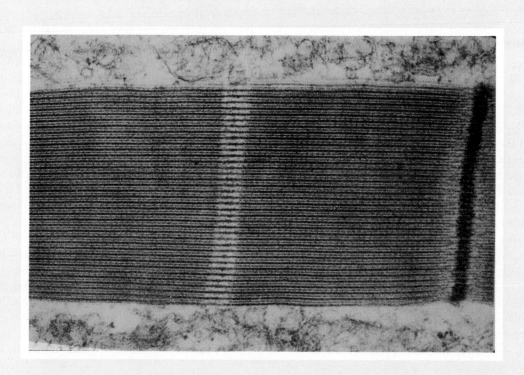

287

The ability to use chemical energy to produce force and movement is present to a limited extent in most cells. It is, however, in muscle cells that this process has become the dominant cell function. The primary function of these specialized cells is to generate the forces and movements used by multicellular organisms in the regulation of their internal environments and to produce the movements of the entire organism in the external environment. In humans, the ability to communicate, whether by speech, writing, or artistic expression, also depends on muscle contractions. Indeed, it is only by controlling the activity of muscles that the human mind ultimately expresses itself.

Three types of muscle tissue can be identified on the basis of structure, contractile properties, and control mechanisms: (1) **skeletal muscle,** (2) **smooth muscle,** and (3) **cardiac muscle.** Most skeletal muscle, as the name implies, is attached to bone, and its contraction is responsible for supporting and moving the skeleton. The contraction of skeletal muscle is initiated by impulses in the neurons to the muscle and is usually under voluntary control.

Sheets of smooth muscle surround various hollow organs and tubes, including the stomach, intestines, urinary bladder, uterus, blood vessels, and airways in the lungs. Contraction of the smooth muscle surrounding hollow organs may propel the luminal contents through the organ, or it may regulate internal flow by changing the tube diameter. In addition, small bundles of smooth muscle cells are attached to the hairs of the skin and iris of the eye. Smooth-muscle contraction is controlled by the autonomic nervous system, hormones, autocrine/paracrine agents, and other local chemical signals. Some smooth muscles contract spontaneously, however, even in the absence of such signals. In contrast to skeletal muscle, smooth muscle is not normally under direct voluntary control.

Cardiac muscle is the muscle of the heart. Its contraction propels blood through the circulatory system. Like smooth muscle, it is regulated by the autonomic nervous system, hormones, and autocrine/paracrine agents, and certain portions of it can undergo spontaneous contractions.

Although there are significant differences in these three types of muscle, the force-generating mechanism is similar in all of them. Skeletal muscle will be described first, followed by a discussion of smooth muscle. Cardiac muscle, which combines some of the properties of both skeletal and smooth muscle, will be described in Chapter 14 in association with its role in the circulatory system.

SECTION A
SKELETAL MUSCLE

STRUCTURE

A single skeletal-muscle cell is known as a **muscle fiber.** Each muscle fiber is formed during development by the fusion of a number of undifferentiated, mononucleated cells, known as **myoblasts,** into a single cylindrical, multinucleated cell. Skeletal muscle differentiation is completed around the time of birth, and these differentiated fibers continue to increase in size during growth from infancy to adult stature, but no new fibers are formed from myoblasts. If skeletal-muscle fibers are destroyed after birth as a result of injury, they cannot be replaced by the division of other existing muscle fibers. New fibers can be formed, however, from undifferentiated cells known as **satellite cells,** which are located adjacent to the muscle fibers and undergo differentiation similar to that followed by embryonic myoblasts. This capacity for forming new skeletal-muscle fibers is considerable but will not restore a severely damaged muscle to full strength. Much of the compensation for a loss of muscle tissue occurs through an increase in the size of the remaining muscle fibers.

Skeletal muscle fibers have diameters between 10 and 100 μm and lengths that may extend up to 20 cm. The term **muscle** refers to a number of muscle fibers bound together by connective tissue (Figure 11-1). Muscles are usually linked to bones by bundles of collagen fibers known as **tendons,** which are located at each end of the muscle. The relationship between a single muscle fiber and a muscle is analogous to that between a single neuron and a nerve, which is composed of the axons of many neurons.

In some muscles, the individual fibers extend the entire length of the muscle, but in most, the fibers are shorter, often oriented at an angle to the longitudinal axis of the muscle. The transmission of force from muscle to bone is like a number of people pulling on a rope, each person corresponding to a single muscle fiber and the rope corresponding to the connective tissue and tendons.

Some tendons are very long, with the site of tendon attachment to bone far removed from the end of the muscle. For example, some of the muscles that move the fingers are in the forearm, as one can observe by wiggling one's fingers and feeling the movement of the muscles in the lower arm. These muscles are connected to the fingers by long tendons.

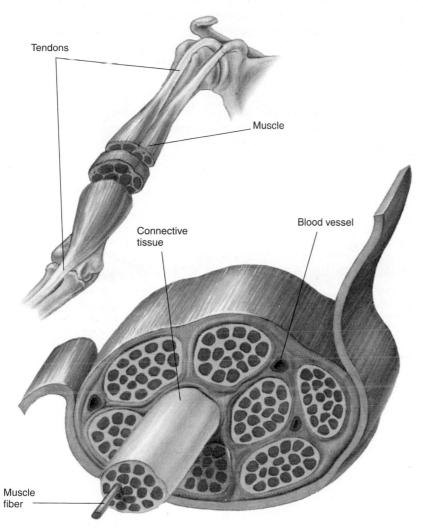

FIGURE 11-1

Organization of cylindrical skeletal-muscle fibers in a muscle that is attached to bones by tendons.

The most striking feature seen when observing a muscle fiber through a light microscope (Figure 11-2) is a series of light and dark bands perpendicular to the long axis of the fiber. The fibers of skeletal and cardiac muscle (Figure 11-3) have this characteristic banding, so both types are known as **striated muscle.** Smooth-muscle cells do not show a banding pattern. The striated pattern in skeletal and cardiac fibers results from the arrangement in the cytoplasm of numerous thick and thin filaments which are organized into approximately cylindrical bundles (1 to 2 μm in diameter) known as **myofibrils** (Figure 11-4). Most of the cytoplasm of a fiber is filled with myofibrils, each of which extends from one end of the fiber to the other and is linked to the tendons at the ends of the fiber.

The thick and thin filaments in each myofibril (Figures 11-4 and 11-5) are arranged in a repeating pattern along the length of the myofibril. One unit of this repeating pattern is known as a **sarcomere** (*sarco,* muscle; *mere,* small). The **thick filaments** are composed almost entirely of the contractile protein **myosin.** The **thin filaments** (which are about half the diameter of the thick filaments) contain the contractile protein **actin,** as well as to two other proteins—troponin and tropomyosin—that play important roles in regulating contraction, as we shall see.

The thick filaments are located in the middle of each sarcomere, where their orderly parallel arrangement produces a wide, dark band known as the **A band** (Figure 11-4). Each sarcomere contains two sets of thin filaments, one at each end. One end of each thin filament is anchored to a network of interconnecting proteins known as the **Z line,** whereas the other end overlaps a portion of the thick filaments. Two successive Z lines define the

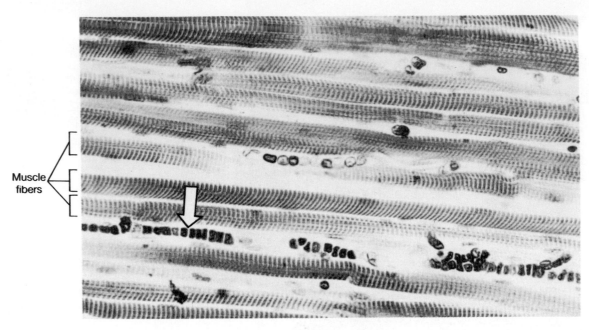

FIGURE 11-2

Skeletal-muscle fibers viewed through a light microscope. Arrow indicates a blood vessel containing red blood cells. (*From Edward K. Keith and Michael H. Ross, "Atlas of Descriptive Histology," Harper & Row, New York, 1968.*) ⚡

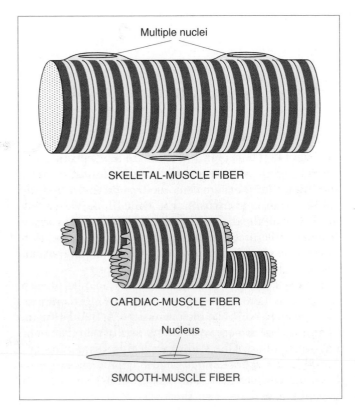

FIGURE 11-3

The three types of muscle fibers. Note the differences in their sizes. (The single nucleus in the center of the cardiac-muscle fiber is not shown.) ⚡

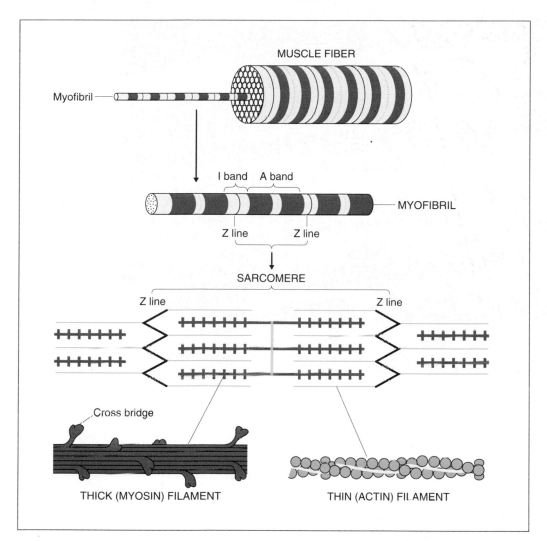

FIGURE 11-4

Arrangement of filaments in a skeletal-muscle fiber that produces the striated banding pattern.

limits of one sarcomere. Thus, thin filaments from two adjacent sarcomeres are anchored to the two sides of each Z line.

The light band, known as the **I band** (Figure 11-4), lies between the ends of the A bands of two adjacent sarcomeres and contains those portions of the thin filaments that do not overlap the thick filaments. It is bisected by the Z line.

Two additional bands are present in the A-band region of each sarcomere (Figure 11-5). The **H zone** is a relatively light band in the center of the A band. It corresponds to the space between the ends of the two sets of thin filaments in each sarcomere; hence, only thick filaments, specifically their central parts, are found in the H zone. The narrow, dark band in the center of the H zone is known as the **M line** and corresponds to proteins that

link together the central region of the thick filaments. In addition, filaments composed of the protein **titin** extend from the Z line to the M line and are linked to both the M-line proteins and the thick filaments. The M-line linkage between thick filaments and the titin filaments acts to maintain the regular array of thick filaments centered in the middle of each sarcomere. Thus, neither the thick nor the thin filaments are free-floating.

A cross section through the A bands of six adjacent myofibrils shows the regular, almost crystalline, arrangement of overlapping thick and thin filaments (Figure 11-6). Each thick filament is surrounded by a hexagonal array of six thin filaments, and each thin filament is surrounded by a triangular arrangement of three thick filaments. Altogether there are twice as many thin as thick filaments in the region of filament overlap.

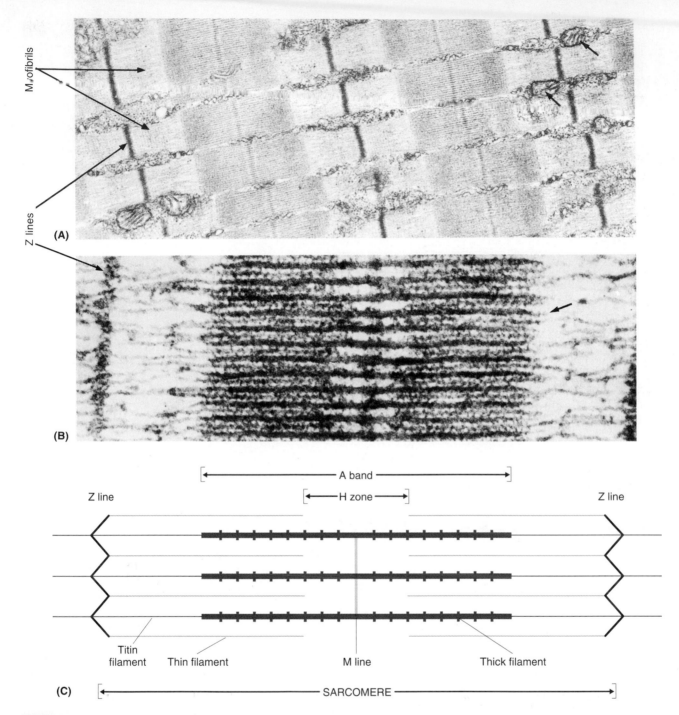

FIGURE 11-5

(A) Numerous myofibrils in a single skeletal-muscle fiber (arrows in upper right corner indicate mitochondria located between the myofibrils). (B) High magnification of a single sarcomere within a single myofibril (arrow at the right of A band indicates end of a thick filament). (C) Arrangement of the thick and thin filaments in the single sarcomere shown in B.

The space between overlapping thick and thin filaments is bridged by projections known as **cross bridges.** These are portions of myosin molecules that extend from the surface of the thick filaments toward the thin filaments (Figures 11-4 and 11-7). During muscle contraction, the cross bridges make contact with the thin filaments and exert force on them.

MOLECULAR MECHANISMS OF CONTRACTION

The term **contraction,** as used in muscle physiology, does not necessarily mean "shortening"; rather it refers only to the turning on of the force-generating sites—the cross bridges—in a muscle fiber. In order for shortening to

(A)

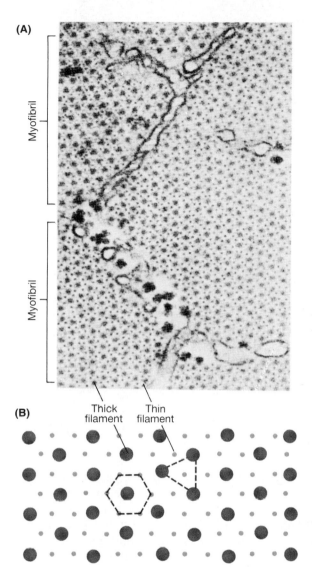

Myofibril

Myofibril

(B)

Thick filament   Thin filament

FIGURE 11-6

(A) Electron micrograph of a cross section through several myofibrils in a single skeletal-muscle fiber. [*From H. E. Huxley, J. Mol. Biol., 37:507–520 (1968).*] (B) Hexagonal arrangements of the thick and thin filaments in the overlap region in a single myofibril. Six thin filaments surround each thick filament and three thick filaments surround each thin filament. ⚘

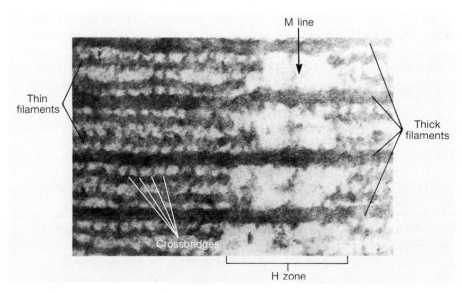

M line

Thin filaments

Thick filaments

Crossbridges

H zone

FIGURE 11-7

High-magnification electron micrograph in the filament-overlap region near the middle of a sarcomere. Cross bridges between the thick and thin filaments can be seen at regular intervals along the filaments. [*From H. E. Huxley and J. Hanson, in G. H. Bourne (ed.), "The Structure and Function of Muscle," Vol. I, Academic Press, New York, 1960.*] ⚘

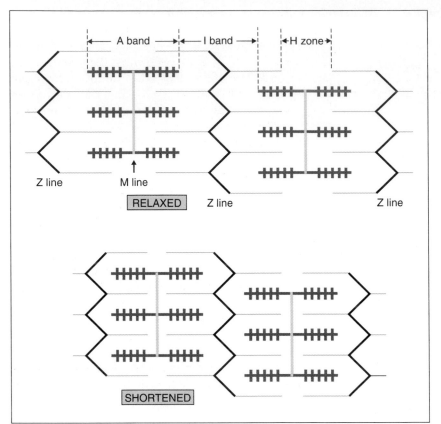

FIGURE 11-8

The sliding of thick filaments past overlapping thin filaments produces shortening with no change in thick or thin filament length.

occur, the forces exerted on the thin filaments by the cross bridges must be greater than the force opposing shortening. Following contraction, the mechanisms that initiate force generation are turned off, and tension declines, allowing **relaxation** of the muscle fiber.

Sliding-Filament Mechanism

When force generation produces shortening of a skeletal-muscle fiber, the overlapping thick and thin filaments in each sarcomere move past each other, propelled by movements of the cross bridges. During this shortening of the sarcomeres, there is no change in the lengths of either the thick or thin filaments (Figure 11-8). This is known as the **sliding-filament mechanism** of muscle contraction.

During shortening, each cross bridge attached to a thin filament moves in an arc much like an oar on a boat. This swiveling motion of many cross bridges forces the thin filaments at either end of the A band toward the center of the sarcomere, thereby shortening the sarcomere (Figure 11-9). One stroke of a cross bridge produces only a very small movement of a thin filament relative to a thick filament. As long as a muscle fiber remains "turned on," however, the cross bridges repeat their swiveling motion many times, resulting in large displacements of the filaments.

Let us look more closely at these events. A muscle fiber's ability to generate force and movement depends on the interactions of the two so-called contractile proteins—

FIGURE 11-9

Cross bridges in the thick filaments bind to actin in the thin filaments and undergo a conformational change that propels the thin filaments toward the center of a sarcomere. (Only 2 of the approximately 200 cross bridges in each thick filament are shown.)

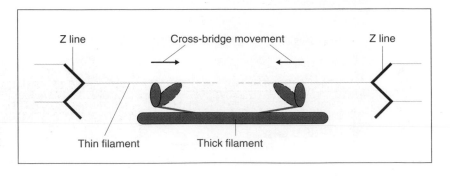

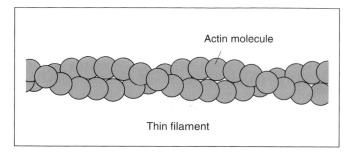

FIGURE 11-10

Two intertwined helical chains of actin molecules form the primary structure of the thin filaments.

myosin in the thick filaments and actin in the thin filaments—and energy provided by ATP.

An actin molecule is a globular protein composed of a single polypeptide chain that polymerizes with other actins to form the two intertwined helical chains (Figure 11-10) that make up the core of a thin filament. Each myosin molecule, on the other hand, is composed of two large polypeptide heavy chains and four smaller light chains. These polypeptides combine to form a molecule that consists of two globular heads (containing heavy and light chains) and a long tail formed by the two intertwined heavy chains (Figure 11-11). The tails of each myosin molecule lie along the axis of the thick filament, and the two globular heads extend out to the sides, forming the cross bridges.

The myosin molecules in the two ends of each thick filament are oriented in opposite directions, such that all their tail ends are directed toward the center of the filament (Figure 11-11). Because of this arrangement, the power strokes of the cross bridges move the attached thin filaments at the two ends of the sarcomere toward the center during shortening (Figure 11-9).

The sequence of events that occurs between the time a cross bridge binds to a thin filament, moves, and then is set to repeat the process is known as a **cross-bridge cy-**

cle. Each cycle consists of four steps: (1) attachment of the cross bridge to a thin filament, (2) movement of the cross bridge, producing tension in the thin filament, (3) detachment of the cross bridge from the thin filament, and (4) energizing of the cross bridge so that it can again attach to a thin filament and repeat the cycle. Each cross bridge undergoes its own cycle of movements independently of the other cross bridges, and at any one instant during contraction only about 50 percent of the cross bridges overlapping a thin filament are attached to the thin filaments and producing tension.

The chemical and physical events occurring during the four steps of a cross-bridge cycle are illustrated in Figure 11-12. At the conclusion (step 4) of the preceding cycle, the ATP bound to myosin is split, releasing chemical energy. This energy is transferred to myosin (M), producing an energized form of myosin (M°) to which the products of ATP hydrolysis, ADP and inorganic phosphate (P_i), are still bound.

Step 4 $\qquad M \cdot ATP \longrightarrow M° \cdot ADP \cdot P_i$

ATP hydrolysis

A new cross-bridge cycle begins with the binding of an energized myosin cross bridge to actin (A) in a thin filament (step 1):

Step 1 $\quad A + M° \cdot ADP \cdot P_1 \longrightarrow A \cdot M° \cdot ADP \cdot P_i$

Actin
binding

The binding of this energized myosin to actin triggers the release of the energy stored in this configuration of the myosin head, resulting in the change in protein conformation that produces the movement of the bound cross bridge (step 2) and the release of ADP and P_i:

Step 2 $\quad A \cdot M° \cdot ADP \cdot P_i \longrightarrow A \cdot M + ADP + P_i$

Cross-bridge
movement

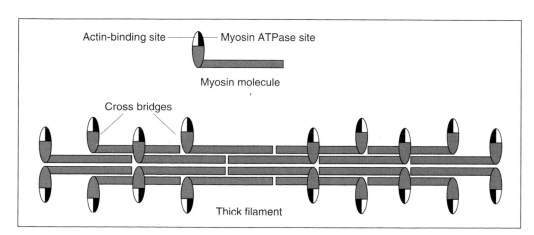

FIGURE 11-11

Orientation of myosin molecules in one thick filament. The globular heads of myosin form cross bridges and the myosin tails are located in the core of the filament. (The two globular heads per myosin molecule are not shown on the diagram.)

tension at each length is increased up to a maximum at l_o. Further lengthening leads to a *drop* in tension. At lengths of 175 percent l_o or beyond, the fiber develops no tension when stimulated.

The lengths of most fibers when all the skeletal muscles in the body are relaxed are near l_o and thus near the optimal lengths for force generation. The length of a relaxed fiber can be altered by the load on the muscle or the contraction of other muscles that stretch the relaxed fibers, but the extent to which the relaxed length can be changed is limited by the muscle's attachments to bones. It rarely exceeds a 30 percent change from l_o and is often much less. Over this range of lengths, the ability to develop tension never falls below about half of the tension that can be developed at l_o (Figure 11-25).

The relationship between fiber length and the fiber's capacity to develop tension can be partially explained in terms of the sliding-filament mechanism. Stretching a relaxed muscle fiber pulls the thin filaments past the thick filaments, changing the amount of overlap between them. Stretching a fiber to 1.75 l_o pulls the filaments apart to the point where there is no overlap. At this point there can be no cross-bridge binding to actin and no development of tension. Between 1.75 l_o and l_o, there is more and more filament overlap, and the tension developed upon stimulation increases in proportion to the increased number of cross bridges in the overlap region. Filament overlap is greatest at l_o, allowing the maximal number of cross bridges to bind to the thin filaments, thereby producing maximal tension.

The tension decline at lengths less than l_o is the result of several factors. For example, (1) the overlapping sets of thin filaments from opposite ends of the sarcomere may interfere with the cross bridges' ability to bind and exert force, and (2) for unknown reasons, the affinity of troponin for calcium decreases at short fiber lengths.

SKELETAL-MUSCLE ENERGY METABOLISM

As we have seen, ATP performs three functions directly related to muscle-fiber contraction and relaxation (Table 11-1). In no other cell type does the rate of ATP breakdown increase so much from one moment to the next as in a skeletal muscle fiber (20 to several hundredfold depending on the type of muscle fiber) when it goes from rest to a state of contractile activity. The small supply of preformed ATP that exists at the start of contractile activity would only support a few twitches. If a fiber is to sustain contractile activity, molecules of ATP must be supplied by metabolism as rapidly as they are broken down during the contractile process.

There are three ways a muscle fiber can form ATP during contractile activity (Figure 11-26): (1) phosphorylation of ADP by **creatine phosphate,** (2) oxidative phosphorylation of ADP in the mitochondria, and (3) substrate-level phosphorylation of ADP by the glycolytic pathway in the cytosol.

Phosphorylation of ADP by creatine phosphate (CP) provides a very rapid means of forming ATP at the onset of contractile activity. When the chemical bond between creatine (C) and phosphate is broken, the amount of energy released is about the same as that released when the terminal phosphate bond in ATP is broken. This energy, along with the phosphate group, can be transferred to ADP to form ATP in a reversible reaction catalyzed by creatine kinase:

$$CP + ADP \overset{\text{Creatine}}{\underset{\text{kinase}}{\rightleftharpoons}} C + ATP$$

In a resting muscle fiber, the concentration of ATP is much greater than that of ADP, leading by mass action to the formation of creatine phosphate. During periods of rest, muscle fibers build up a concentration of creatine phosphate approximately five times that of ATP. At the beginning of contraction, when the concentration of ATP begins to fall and that of ADP to rise owing to the increased rate of ATP breakdown, mass action favors the formation of ATP from creatine phosphate. This transfer of energy from creatine phosphate to ATP is so rapid that the concentration of ATP in a muscle fiber changes very little at the start of contraction, whereas the concentration of creatine phosphate falls rapidly.

Although the formation of ATP from creatine phosphate is very rapid, requiring only a single enzymatic reaction, the amount of ATP that can be formed by this process is limited by the initial concentration of creatine phosphate in the cell. If contractile activity is to be continued for more than a few seconds, the muscle must be able to form ATP from the other two sources listed above. The use of creatine phosphate at the start of contractile activity provides the few seconds necessary for the slower, multienzyme pathways of oxidative phosphorylation and glycolysis to increase their rates of ATP formation to levels that match the rates of ATP breakdown.

At moderate levels of muscular activity, most of the ATP used for muscle contraction is formed by oxidative phosphorylation, and during the first 5 to 10 min of such exercise, muscle glycogen is the major fuel contributing to oxidative phosphorylation. For the next 30 min or so, blood-borne fuels become dominant, blood glucose and fatty acids contributing approximately equally; beyond this period, fatty acids become progressively more important and glucose utilization decreases.

If the intensity of exercise exceeds about 70 percent of the maximal rate of ATP breakdown, however, glycolysis

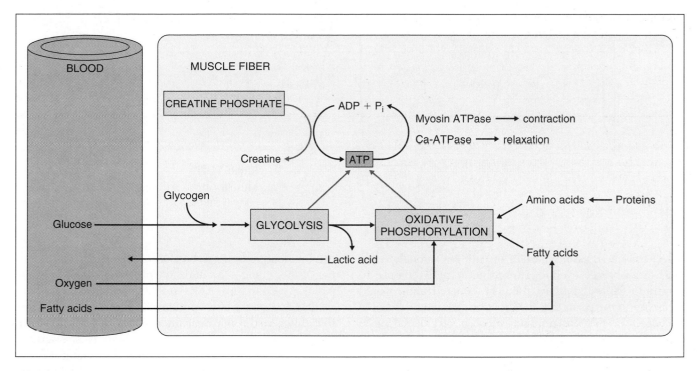

FIGURE 11-26

The three sources of ATP production during muscle contraction: (1) creatine phosphate, (2) oxidative phosphorylation, and (3) glycolysis.

contributes an increasingly significant fraction of the total ATP generated by the muscle. The glycolytic pathway, although producing only small quantities of ATP from each molecule of glucose metabolized, can produce large quantities of ATP when enough enzymes and substrate are available, and it can do so in the absence of oxygen. The glucose for glycolysis can be obtained from two sources: the blood or the stores of glycogen within the contracting muscle fibers. As the intensity of muscle activity increases, more and more of the ATP is formed by anaerobic glycolysis, with a corresponding increase in the production of lactic acid (which dissociates to yield lactate ions and hydrogen ions).

At the end of muscle activity, creatine phosphate and glycogen levels in the muscle have decreased. To return a muscle fiber to its original state, these energy-storing compounds must be replaced. Both processes require energy, and so a muscle continues to consume increased amounts of oxygen for some time after it has ceased to contract, as evidenced by the fact that one continues to breathe deeply and rapidly for a period of time immediately following intense exercise. This elevated consumption of oxygen following exercise repays what has been called the **oxygen debt,** that is the increased production of ATP by oxidative phosphorylation following exercise that is used to restore the energy reserves. The longer and more intense the exercise, the longer it takes to repay the oxygen debt.

Muscle Fatigue

When a skeletal-muscle fiber is repeatedly stimulated, the tension developed by the fiber eventually decreases even though the stimulation continues (Figure 11-27). This decline in muscle tension as a result of previous contractile activity is known as **muscle fatigue.** Additional characteristics of fatigued muscle are a decreased shortening velocity and a slower rate of relaxation. The onset of fatigue and its rate of development depend on the type of skeletal-muscle fiber that is active and on the intensity and duration of contractile activity.

If a muscle is allowed to rest after the onset of fatigue, it can recover its ability to contract upon restimulation (Figure 11-27). The rate of recovery depends upon the duration and intensity of the previous activity. Some muscle fibers fatigue rapidly if continuously stimulated at a high frequency but also recover rapidly after a brief rest. This is the type of fatigue that accompanies high-intensity, short-duration, exercise, such as weight lifting. Fatigue that develops more slowly with low-intensity, long-duration exercise, such as long-distance running, during which there are cyclical periods of contraction and relaxation, requires much longer periods of rest, often up to 24 h, before the muscle achieves complete recovery.

It might seem logical that depletion of energy in the form of ATP would account for fatigue, but the ATP concentration in fatigued muscle is found to be only slightly

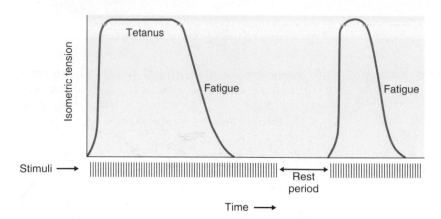

FIGURE 11-27

Muscle fatigue during a maintained isometric tetanus and recovery following a period of rest.

lower than in a resting muscle, and not low enough to impair cross-bridge cycling. If contractile activity were to continue without fatigue, the ATP concentration would decrease to the point that the cross bridges would become linked in a rigor configuration, which is very damaging to muscle fibers. Thus, muscle fatigue may have evolved as a mechanism for preventing the onset of rigor.

Depending on the intensity and duration of the activity, different factors contribute to the fatigue of skeletal muscle. These include failure of excitation-contraction coupling which is primarily responsible for fatigue during high intensity, short duration exercise. During longer durations of moderate activity the increase in muscle acidity due to lactic acid, the rise in phosphate that accompanies the hydrolysis of ATP and creatine phosphate, and the depletion of muscle glycogen contribute to fatigue. Recovery from short-duration, high-intensity activity is relatively rapid, on the order of minutes, while recovery from the fatigue resulting for long-duration activity requires hours of rest. The specific molecular mechanisms responsible for fatigue are still unclear, especially those responsible for the long recovery period from prolonged activity.

Another type of fatigue quite different from muscle fatigue is due to failure of the appropriate regions of the cerebral cortex to send excitatory signals to the motor neurons. This is called **central command fatigue,** and it may cause an individual to stop exercising even though the muscles are not fatigued. An athlete's performance depends not only on the physical state of the appropriate muscles but also upon the "will to win," that is, the ability to initiate central commands to muscles during a period of increasingly distressful sensations.

TYPES OF SKELETAL-MUSCLE FIBERS

All skeletal-muscle fibers do not have the same mechanical and metabolic characteristics. Different types of fibers can be identified on the basis of (1) their maximal velocities of shortening—fast and slow fibers—and (2) the major pathway used to form ATP—oxidative and glycolytic fibers.

Fast and slow fibers contain different myosin enzymes that differ in the maximal rates at which they split ATP, which in turn determine the maximal rate of cross-bridge cycling and hence the fibers' maximal shortening velocity. Fibers containing myosin with high ATPase activity are classified as **fast fibers,** and those containing myosin with lower ATPase activity are **slow fibers.**

The second means of classifying skeletal-muscle fibers is according to the type of enzymatic machinery available for synthesizing ATP. Some fibers contain numerous mitochondria and thus have a high capacity for oxidative phosphorylation. These fibers are classified as **oxidative fibers.** Most of the ATP produced by such fibers is dependent upon blood flow to deliver oxygen and fuel molecules to the muscle, and these fibers are surrounded by numerous small blood vessels. They also contain large amounts of an oxygen-binding protein known as **myoglobin,** which increases the rate of oxygen diffusion within the fiber and provides a small store of oxygen. The large amounts of myoglobin present in oxidative fibers give the fibers a dark-red color, and thus oxidative fibers are often referred to as **red muscle fibers.**

In contrast, **glycolytic fibers** have few mitochondria but possess a high concentration of glycolytic enzymes and a large store of glycogen. Corresponding to their limited use of oxygen, these fibers are surrounded by relatively few blood vessels and contain little myoglobin. The lack of myoglobin is responsible for the pale color of glycolytic fibers and their designation as **white muscle fibers.**

On the basis of these two general characteristics, three types of skeletal-muscle fibers can be distinguished:

1. **Slow-oxidative fibers** (type I) combine low myosin-ATPase activity with high oxidative capacity.
2. **Fast-oxidative fibers** (type IIa) combine high myosin-ATPase activity with high oxidative capacity.

3. **Fast-glycolytic fibers** (type IIb) combine high myosin-ATPase activity with high glycolytic capacity.

In addition to these biochemical differences, there are also size differences, glycolytic fibers generally having much larger diameters than oxidative fibers (Figure 11-28). This fact has significance for tension development. The number of thick and thin filaments per unit of cross-sectional area is about the same in all types of skeletal-muscle fibers. Therefore, the larger the diameter of a muscle fiber, the greater the number of thick and thin filaments acting in parallel to produce force, and the greater the maximum tension it can develop (greater strength). Accordingly, the average glycolytic fiber, with its larger diameter, develops more tension when it contracts than does an average oxidative fiber.

These three types of fibers also differ in their capacity to resist fatigue. Fast-glycolytic fibers fatigue rapidly, whereas slow-oxidative fibers are very resistant to fatigue, which allows them to maintain contractile activity for long periods with little loss of tension. Fast-oxidative fibers have an intermediate capacity to resist fatigue (Figure 11-29).

The characteristics of the three types of skeletal-muscle fibers are summarized in Table 11-3.

WHOLE-MUSCLE CONTRACTION

As described earlier, whole muscles are made up of many muscle fibers organized into motor units. All the muscle fibers in a single motor unit are of the same fiber type. Thus, one can apply the fiber type designation to the motor unit and refer to slow-oxidative motor units, fast-glycolytic motor units, and fast-oxidative motor units.

Most muscles are composed of all three motor unit types interspersed with each other (Figure 11-30). No muscle has only a single fiber type. Depending on the proportions of the three fiber types present, muscles can differ considerably in their maximal contraction speed, strength, and fatigability. For example, the muscles of the back and legs, which must be able to maintain their activity for long periods of time without fatigue while supporting an upright posture, contain large numbers of slow-oxidative and fast-oxidative fibers. In contrast, the muscles in the arms may be called upon to produce large amounts of tension over a short time period, as when lifting a heavy object, and these muscles have a greater proportion of fast-glycolytic fibers.

FIGURE 11-28

Cross sections of skeletal muscle. (A) The capillaries surrounding the muscle fibers have been stained. Note the large number of capillaries surrounding the small-diameter oxidative fibers. (B) The mitochondria have been stained indicating the large numbers of mitochondria in the small-diameter oxidative fibers. (*Courtesy of John A. Faulkner.*)

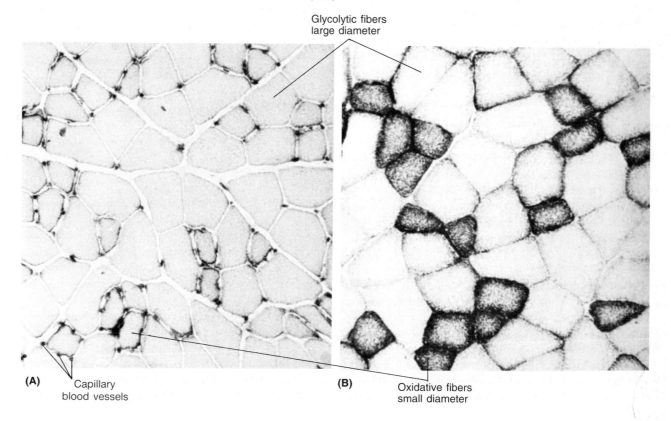

Glycolytic fibers
large diameter

(A) Capillary
blood vessels

(B) Oxidative fibers
small diameter

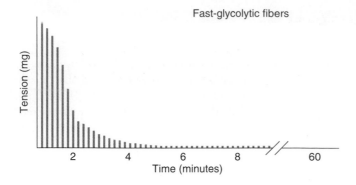

Fast-glycolytic fibers

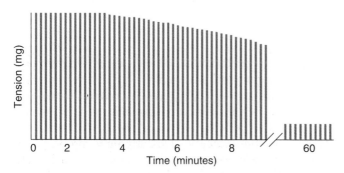

Fast-oxidative fibers

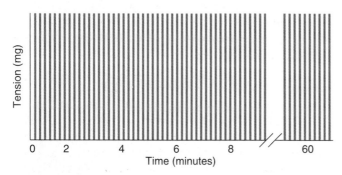

Slow-oxidative fibers

FIGURE 11-29

The rate of fatigue development in the three fiber types. Each vertical line is the contractile response to a brief tetanic stimulus and relaxation. The contractile responses occurring between about 9 minutes and 60 minutes are not shown on the figure.

We will now use the characteristics of single fibers to describe whole-muscle contraction and its control.

Control of Muscle Tension

The total tension a muscle can develop depends upon two factors: (1) the amount of tension developed by each fiber, and (2) the number of fibers contracting at any time. By controlling these two factors, the nervous system controls whole-muscle tension, as well as shortening velocity. The conditions that determine the amount of tension developed in a single fiber have been discussed previously.

The number of fibers contracting at any time depends

on: (1) the number of fibers in each motor unit (motor unit size), and (2) the number of active motor units (Table 11-4).

The size of a motor unit varies considerably from one muscle to another. The muscles in the hand and eye, which produce very delicate movements, contain small motor units. For example, one motor neuron innervates only about 13 fibers in an eye muscle. In contrast, in the more coarsely controlled muscles of the back and legs, each motor unit is large, containing hundreds and in some cases several thousand fibers. When a muscle is composed of small motor units, the total tension produced by the muscle can be increased in small steps by activating additional motor units. If the motor units are large, large increases in tension will occur as each additional motor unit is activated. Thus, finer control of muscle tension is possible in muscles with small motor units.

FIGURE 11-30

(A) Diagram of a cross section through a muscle composed of three types of motor units. (B) Tetanic muscle tension resulting from the successive recruitment of the three types of motor units. Note that motor unit 3, composed of fast-glycolytic fibers, produces the greatest rise in tension because it is composed of the largest-diameter fibers and contains the largest number of fibers per motor unit.

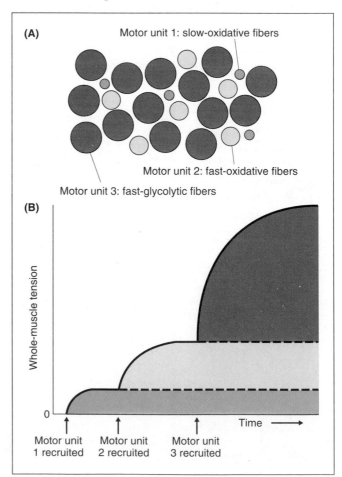

TABLE 11-3 CHARACTERISTICS OF THE THREE TYPES OF SKELETAL MUSCLE FIBERS

	Slow-oxidative fibers	**Fast-oxidative fibers**	**Fast-glycolytic fibers**
Primary source of ATP production	Oxidative phosphorylation	Oxidative phosphorylation	Glycolysis
Mitochondria	Many	Many	Few
Capillaries	Many	Many	Few
Myoglobin content	High (red muscle)	High (red muscle)	Low (white muscle)
Glycolytic enzyme activity	Low	Intermediate	High
Glycogen content	Low	Intermediate	High
Rate of fatigue	Slow	Intermediate	Fast
Myosin-ATPase activity	Low	High	High
Contraction velocity	Slow	Fast	Fast
Fiber diameter	Small	Intermediate	Large
Motor unit size	Small	Intermediate	Large
Size of motor neuron innervating fiber	Small	Intermediate	Large

The force produced by a single fiber, as we have seen earlier, depends in part on the fiber diameter—the greater the diameter, the greater the force. We have also noted that fast-glycolytic fibers have the largest diameters. Thus, a motor unit composed of 100 fast-glycolytic fibers produces more force than a motor unit composed of 100 slow-oxidative fibers. In addition, fast-glycolytic motor units tend to have more muscle fibers. For both of these reasons, activating a fast-glycolytic motor unit will produce more force than activating a slow-oxidative motor unit.

The process of increasing the number of motor units that are active in a muscle at any given time is called **recruitment.** It is achieved by increasing the excitatory synaptic input to the motor neurons. The greater the number of active motor neurons, the more motor units recruited, and the greater the muscle tension.

Motor neuron size plays an important role in the recruitment of motor units (the size of a motor neuron refers to the diameter of the nerve cell body, which is usually correlated with the diameter of its axon, and does not refer to the size of the motor unit the neuron controls).

TABLE 11-4 FACTORS DETERMINING MUSCLE TENSION

I. Tension developed by each individual fiber
  A. Action-potential frequency (frequency-tension relation)
  B. Fiber length (length-tension relation)
  C. Fiber diameter
  D. Fatigue
II. Number of active fibers
  A. Number of fibers per motor unit
  B. Number of active motor units

Given the same number of sodium ions entering a cell during synaptic activity in a large and in a small motor neuron, the small neuron will undergo a greater depolarization because these ions will be distributed over a smaller membrane surface area. Accordingly, given the same level of synaptic input, the smallest neurons will be recruited first. The larger neurons will be recruited only as the level of synaptic input increases. Since the smallest motor neurons innervate the slow-oxidative motor units (Table 11-3), these motor units are recruited first, followed by fast-oxidative motor units, and finally, during very strong contractions, by fast-glycolytic motor units (Figure 11-30).

Thus, during moderate-strength contractions, such as are used in most endurance types of exercise, relatively few fast-glycolytic motor units are recruited, and most of the activity occurs in oxidative fibers, which are more resistant to fatigue. The large fast-glycolytic motor units, which fatigue rapidly, begin to be recruited when the intensity of contraction exceeds about 40 percent of the maximal tension that can be produced by the muscle.

In conclusion, the neural control of whole-muscle tension involves both the frequency of action potentials in individual motor units (to vary the tension generated by the fibers in that unit) and the recruitment of motor units (to vary the number of active fibers). Most motor neuron activity occurs in bursts of action potentials, which produce tetanic contractions of individual motor units rather than single twitches. Recall that the tension of a single fiber increases only three- to fivefold when going from a twitch to a maximal tetanic contraction. Therefore, varying the frequency of action potentials in the neurons supplying them provides a way to make only three- to fivefold adjustments in the tension of the recruited motor units. The force a whole muscle exerts can be varied over a much wider range than this, from very delicate movements to extremely powerful contractions, by the recruitment of motor units. Thus recruitment provides the primary means of varying tension in a whole muscle. Recruitment is controlled by the central commands from the motor centers in the brain to the various motor neurons.

Control of Shortening Velocity

As we saw earlier, the velocity at which a *single* muscle fiber shortens is determined by (1) the load on the fiber and (2) whether the fiber is a fast fiber or a slow fiber. Translated to a *whole* muscle, these characteristics become (1) the load on the whole muscle and (2) the types of motor units in the muscle. For the whole muscle, however, recruitment becomes a third very important factor so that the shortening velocity can be varied from very fast to very slow even though the load on the muscle remains constant. Consider, for the sake of illustration, a muscle composed of only two motor units of the same size and fiber type. One motor unit by itself will lift a 4-g load more slowly than a 2-g load because the shortening velocity decreases with increasing load. When both units are active and a 4-g load is lifted, each motor unit bears only half the load, and its fibers will shorten as if it were lifting only a 2-g load. In other words, the muscle will lift the 4-g load at a higher velocity when both motor units are active. Thus both force and velocity of shortening are controlled primarily by the recruitment of motor units.

Muscle Adaptation to Exercise

The regularity with which a muscle is used, as well as the duration and intensity of its activity, affects the properties of the muscle. If the neurons to a skeletal muscle are severed or otherwise destroyed, the denervated muscle fibers will become progressively smaller in diameter, and the amount of contractile proteins they contain will decrease. This condition is known as ***denervation atrophy.*** A muscle can also atrophy with its nerve supply intact if the muscle is not used for a long period of time, as when a broken arm or leg is immobilized in a cast. This condition is known as ***disuse atrophy.***

In contrast to the decrease in muscle mass that results from a lack of neural stimulation, increased amounts of contractile activity—in other words, exercise—can produce an increase in the size (**hypertrophy**) of muscle fibers as well as changes in their capacity for ATP production. Since the number of fibers in a muscle remains essentially constant throughout adult life, the changes in muscle size with atrophy and hypertrophy do not result from changes in the *number* of muscle fibers but in the metabolic capacity and size of each fiber.

Exercise that is of relatively low intensity but of long duration (popularly called "aerobic exercise"), such as running and swimming, produces increases in the number of mitochondria in the fast-oxidative and slow-oxidative fibers, which are recruited in this type of activity. In addition, there is an increase in the number of capillaries around these fibers. All these changes lead to an increase in the capacity for endurance activity with a minimum of fatigue. (Surprisingly, fiber diameter decreases slightly, and thus there is a small decrease in the maximal strength of muscles as a result of endurance exercise.) As we shall see in later chapters, endurance exercise produces changes not only in the skeletal muscles but also in the respiratory and circulatory systems, changes that improve the delivery of oxygen and fuel molecules to the muscle.

In contrast, short-duration, high-intensity exercise (popularly called "strength training"), such as weight lifting, affects primarily the fast-glycolytic fibers, which are briefly recruited during strong contractions. These fibers undergo an increase in fiber diameter due to the increased synthesis of actin and myosin filaments, which form more myofibrils. In addition, the glycolytic activity is increased by increasing the synthesis of glycolytic enzymes. The result of such high-intensity exercise is an increase in the strength of the muscle and the bulging muscles of a conditioned weight lifter. Such muscles, although very powerful, have little capacity for endurance and fatigue rapidly.

Exercise produces little change in the types of myosin enzymes formed by the fibers and thus little change in the proportions of fast and slow fibers in a muscle. As described above, however, exercise does change the rates at which metabolic enzymes are synthesized, leading to changes in the proportion of oxidative and glycolytic fibers within a muscle. With endurance training there is a decrease in the number of fast-glycolytic fibers and an increase in the number of fast-oxidative fibers as the oxidative capacity of the fibers is increased. The reverse occurs with strength training as fast-oxidative fibers are converted to fast-glycolytic fibers.

The signals responsible for all these changes in muscle with different types of activity are unknown. They are related to the frequency and intensity of the contractile activity in the muscle fibers and thus to the pattern of action potentials produced in the muscle over an extended period of time. Similar adaptive changes in cell size, number, or capacity for functional activity are seen in many other organs in the body in response to increased demands.

Because different types of exercise produce quite different changes in the strength and endurance capacity of a muscle, an individual performing regular exercises to improve muscle performance must choose a type of exercise that is compatible with the type of activity he or she ultimately wishes to perform. Thus, lifting weights will not improve the endurance of a long-distance runner, and jogging will not produce the increased strength desired by a weight lifter. Most exercises, however, produce some effects on both strength and endurance.

These changes in muscle in response to repeated periods of exercise occur slowly over a period of weeks. If regular exercise is stopped, the changes in the muscle that occurred as a result of the exercise will slowly revert to their state before exercise began.

The maximum force generated by a muscle decreases by 30 to 40 percent between the ages of 30 and 80. This decrease in tension-generating capacity is due primarily to a decrease in average fiber diameter. Some of the change is simply the result of diminishing physical activity with age and can be prevented by exercise programs. The ability of a muscle to adapt to exercise, however, decreases with age: The same intensity and duration of exercise in an older individual will not produce the same amount of change as in a younger person. This decreased ability to adapt to increased activity is seen in most organs as one ages (Chapter 7).

This effect of aging, however, is only partial, and there is no question that even in the elderly, exercise can produce significant adaptation. Aerobic training has received major attention because of its effect on the cardiovascular system (Chapter 14). More recently, however, strength training of a modest degree has also been strongly recommended because it can partially prevent the loss of muscle tissue that occurs with aging. [Moreover, it helps maintain stronger bones (Chapter 18).]

Extensive exercise by an individual whose muscles have not been used in performing a particular type of exercise leads to muscle soreness the next day. This soreness is the result of a mild inflammation in the muscle, which occurs whenever tissues are damaged (Chapter 20). The most severe inflammation occurs following a period of lengthening contractions, suggesting that the lengthening of a muscle fiber by an external force produces greater muscle damage than do either isotonic or isometric contractions. Thus, exercising by gradually lowering weights will produce greater muscle soreness than an equivalent amount of weight lifting.

The effects of anabolic steroids on skeletal-muscle growth and strength are described in Chapter 18.

Lever Action of Muscles and Bones

A contracting muscle exerts a force on bones through its connecting tendons. When the force is great enough, the bone moves as the muscle shortens. A contracting muscle exerts only a pulling force, so that as the muscle shortens, the bones to which it is attached are pulled toward each other. **Flexion** refers to the *bending* of a limb at a joint, whereas **extension** is the *straightening* of a limb (Figure 11-31). These opposing motions require at least two muscles, one to cause flexion and the other extension. Groups of muscles that produce oppositely directed movements at a joint are known as **antagonists.** For example, from Figure 11-31 it can be seen that contraction of the biceps causes flexion of the arm at the elbow, whereas contraction of the antagonistic muscle, the triceps, causes the arm to extend. Both muscles exert only a pulling force upon the forearm when they contract.

Sets of antagonistic muscles are required not only for flexion-extension, but also for side-to-side movements or rotation of a limb. The contraction of some muscles leads to two types of limb movement, depending on the contractile state of other muscles acting on the same limb. For example, contraction of the gastrocnemius muscle in the leg causes a flexion of the leg at the knee, as in walking (Figure 11-32). However, contraction of the gastrocnemius muscle with the simultaneous contraction of the quadriceps femoris (which causes extension of the lower leg) prevents the knee joint from bending, leaving only the ankle joint capable of moving. The foot is extended, and the body rises on tiptoe.

The muscles, bones, and joints in the body are arranged in lever systems. The basic principle of a lever is illustrated by the flexion of the arm by the biceps muscle (Figure 11-33), which exerts an upward pulling force on the forearm about 5 cm away from the elbow joint. In this example, a 10-kg weight held in the hand exerts a downward force of 10 kg about 35 cm from the elbow. A law of physics tells us that the forearm is in mechanical equilibrium (no net forces acting on the system) when the product of the downward force (10 kg) and its distance from the elbow (35 cm) is equal to the product of the isometric tension exerted by the muscle (X), and its distance from the elbow (5 cm); that is, $10 \times 35 = 5 \times X$. Thus $X = 70$ kg. The important point is that this system is working at a mechanical disadvantage since the force exerted by the muscle (70 kg) is considerably greater than the load (10 kg) it is supporting.

However, the mechanical disadvantage under which most muscle lever systems operate is offset by increased maneuverability. In Figure 11-34, when the biceps shortens 1 cm, the hand moves through a distance of 7 cm. Since the muscle shortens 1 cm in the same amount of time that the hand moves 7 cm, the velocity at which the hand moves is seven times greater than the rate of muscle shortening. The lever system amplifies the velocity of muscle shortening so that short, relatively slow movements of the muscle produce faster movements of the hand. Thus, a pitcher can throw a baseball at 90 to 100 mi/h even though his muscles shorten at only a small fraction of this velocity.

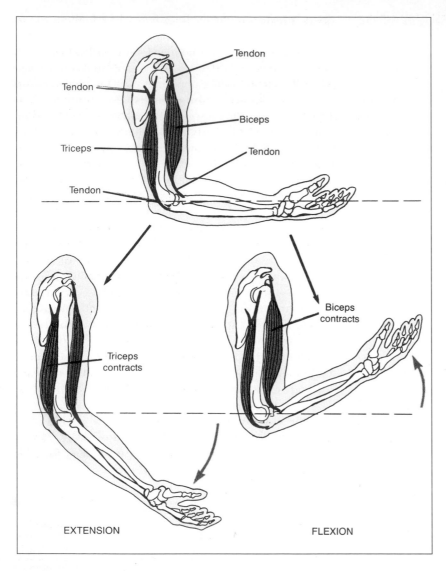

FIGURE 11-31

Antagonistic muscles for flexion and extension of the forearm. ⟨⟩

Skeletal-Muscle Disease

A number of diseases can affect the contraction of skeletal muscle. Many of them are due to defects in the parts of the nervous system that control contraction of the muscle fibers, however, rather than to defects in the muscle fibers themselves. For example, *poliomyelitis* is a viral disease that destroys motor neurons, leading to the paralysis of skeletal muscle, and may result in death due to respiratory failure.

Muscle Cramps. Involuntary tetanic contraction of skeletal muscles produces *muscle cramps.* During cramping, action potentials fire at rates as high as 300/s, a much greater rate than occurs during maximal voluntary contraction. The specific cause of this high activity is unknown but is probably related to electrolyte imbalances in the extracellular fluid surrounding both the muscle and

nerve fibers and changes in extracellular osmolarity, especially hypoosmolarity.

Hypocalcemic Tetany. Similar in symptoms to muscular cramping is *hypocalcemic tetany,* the involuntary tetanic contraction of skeletal muscles that occurs when the extracellular calcium concentration falls to about 40 percent of its normal value. This may seem surprising since we have seen that calcium is required for excitation-contraction coupling. However, recall that this calcium is sarcoplasmic reticulum calcium, not extracellular calcium. The effect of changes in extracellular calcium is exerted not on the sarcoplasmic reticulum calcium, but directly on the plasma membrane. Low extracellular calcium (hypocalcemia) increases the opening of sodium channels in excitable membranes, leading to membrane depolarization and the spontaneous firing of action potentials. It is this

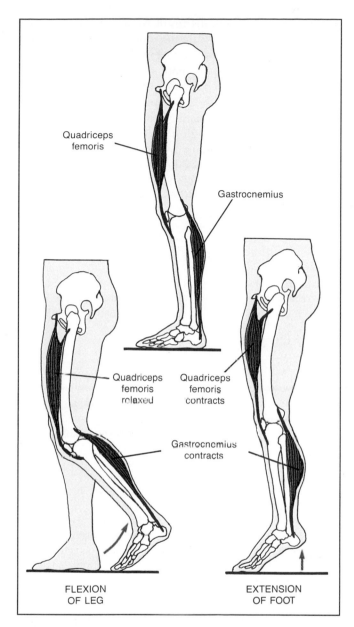

FIGURE 11-32

Contraction of the gastrocnemius muscle in the calf can lead either to flexion of the leg, if the quadriceps femoris muscle is relaxed, or to extension of the foot, if the quadriceps is contracting, preventing bending of the knee joint. 𝒳

that causes the increased muscle contractions. The mechanisms controlling the extracellular concentration of calcium ions are discussed in Chapter 16.

Muscular Dystrophy. This disease is one of the most frequently encountered genetic diseases, affecting one in every 4000 boys (but much less commonly girls) born in America. ***Muscular dystrophy*** is associated with the progressive degeneration of skeletal- and cardiac-muscle

fibers, weakening the muscles and leading ultimately to death from respiratory or cardiac failure. While exercise strengthens normal skeletal muscle, it weakens dystrophic muscle. The symptoms become evident at about 2 to 6 years of age, and most affected individuals do not survive much beyond the age of 20. In addition, about one-third of dystrophic patients show signs of mental retardation.

The recessive gene responsible for a major form of muscular dystrophy has been identified on the X chromosome. (As described in Chapter 19, girls have two X chromosomes and boys only one. Accordingly, a girl with one abnormal X chromosome and one normal one will not develop the disease. This is why the disease is so much more common in boys.) This gene codes for a protein known as dystrophin, which is absent or present in a nonfunctional form in patients with the disease. Dystrophin is located on the inner surface of the plasma membrane in normal muscle. It resembles other known cytoskeletal proteins and may be involved in maintaining the structural integrity of the plasma membrane or of elements within the membrane, such as ion channels, in fibers subjected to repeated structural deformation during contraction. Preliminary attempts are being made to treat the disease by inserting the normal gene into dystrophic muscle cells.

Myasthenia Gravis. ***Myasthenia gravis*** is characterized by muscle fatigue and weakness that progressively worsens as the muscle is used. It affects about 12,000

FIGURE 11-33

Mechanical equilibrium of forces acting on the forearm while supporting a 10-kg load. 𝒳

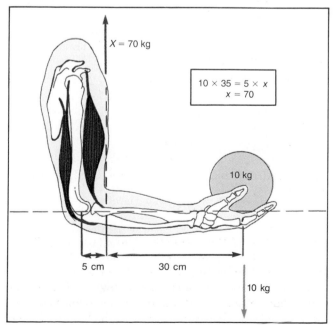

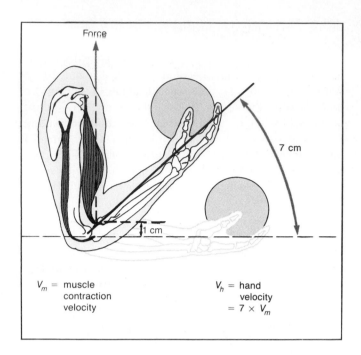

V_m = muscle contraction velocity

V_h = hand velocity = 7 × V_m

Americans. The symptoms result from a decrease in the number of ACh receptors on the motor end plate. The release of ACh from the nerve terminals is normal, but the magnitude of the end-plate potential is markedly reduced because of the decreased number of receptors. After a few motor nerve impulses, the magnitude of the EPP falls below the threshold for initiating a muscle action potential. As described in Chapter 20, the destruction of the ACh receptors is brought about by the body's own defense mechanisms gone awry, specifically because of the formation of antibodies to the ACh-receptor proteins.

FIGURE 11-34

Velocity of the biceps muscle is amplified by the lever system of the arm, producing a greater velocity of the hand. The range of movement is also amplified (1 cm of shortening by the muscle produces 7 cm of movement by the hand).

SECTION A SUMMARY

I. There are three types of muscle—skeletal, smooth, and cardiac. Skeletal muscle is attached to bones and moves and supports the skeleton. Smooth muscle surrounds hollow cavities and tubes. Cardiac muscle is the muscle of the heart.

STRUCTURE OF SKELETAL MUSCLE

I. Skeletal muscles, composed of cylindrical muscle fibers (cells), are linked to bones by tendons at each end of the muscle.

II. Skeletal-muscle fibers have a repeating, striated pattern of light and dark bands due to the arrangement of the thick and thin filaments within the myofibrils.

III. Actin-containing thin filaments are anchored to the Z lines at each end of a sarcomere, while their free ends partially overlap the myosin-containing thick filaments in the A band at the center of the sarcomere.

MOLECULAR MECHANISMS OF CONTRACTION

I. When a skeletal-muscle fiber actively shortens, the thin filaments are propelled toward the center of their sarcomere by movements of the myosin cross bridges that bind to actin.
  A. The two globular heads of each cross bridge contains a binding site for actin and an enzymatic site that splits ATP.
  B. The four steps occurring during each cross-bridge cycle are summarized in Figure 11-12. The cross bridges undergo repeated cycles during a contraction, each producing only a small increment of movement.
  C. The three functions of ATP in muscle contraction are summarized in Table 11-1.

II. In a resting muscle, attachment of cross bridges to actin is prevented by tropomyosin molecules that are in contact with the actin subunits of the thin filaments.

III. Contraction is initiated by an increase in cytosolic calcium concentration. The calcium ions bind to troponin, producing a change in its shape that is transmitted via tropomyosin uncovering the binding sites on actin, allowing the cross bridges to bind to the thin filaments.
  A. The rise in cytosolic calcium concentration is triggered by an action potential in the plasma membrane. The action potential is propagated into the interior of the fiber along the transverse tubules to the region of the sarcoplasmic reticulum, where it produces a release of calcium ions from the reticulum.
  B. Relaxation of a contracting muscle fiber occurs as a result of the active transport of cytosolic calcium ions back into the sarcoplasmic reticulum.

IV. Branches of a motor neuron axon form neuromuscular junctions with many muscle fibers in a muscle. Each muscle fiber is innervated by a branch from only one motor neuron. A motor unit consists of a single motor neuron and the muscle fibers it innervates.
  A. Acetylcholine released by an action potential in a motor neuron binds to receptors on the motor end plate of the muscle membrane, opening ion channels that allow the passage of sodium and potassium ions, which depolarize the end-plate membrane.

B. A single action potential in a motor neuron is sufficient to produce an action potential in a skeletal-muscle fiber.

V. Table 11-2 summarizes the events leading to the contraction of a skeletal-muscle fiber.

MECHANICS OF SINGLE-FIBER CONTRACTION

I. Contraction refers to the turning on of the cross-bridge cycle. Whether there is an accompanying change in muscle length depends upon the external forces acting on the muscle.

II. Three types of contractions can occur following activation of a muscle fiber: (1) an isometric contraction in which the muscle generates tension but does not change length; (2) an isotonic contraction in which the muscle shortens, moving a load; and (3) a lengthening contraction in which the external load on the muscle is greater than the muscle tension, causing the muscle to lengthen during the period of contractile activity.

III. Increasing the frequency of action potentials in a muscle fiber increases the mechanical response (tension or shortening), up to the level of maximal tetanic tension.

IV. Maximum isometric tetanic tension is produced when there is a maximal overlap of thick and thin filaments, that is, at the optimal length l_o. Stretching a fiber beyond its optimal length decreases the filament overlap and decreases the tension produced, whereas decreasing the fiber length below l_o also decreases the tension generated.

V. The velocity of muscle fiber shortening decreases with increases in load. Maximum velocity occurs at zero load.

SKELETAL-MUSCLE ENERGY METABOLISM

I. Muscle fibers form ATP by the transfer of phosphate from creatine phosphate to ADP, by oxidative phosphorylation of ADP in mitochondria, and by substrate-level phosphorylation of ADP in the glycolytic pathway.

II. At the beginning of exercise, muscle glycogen is the major fuel consumed. As the exercise proceeds, glucose and fatty acids from the blood provide most of the fuel, fatty acids becoming progressively more important during prolonged exercise. When the intensity of exercise exceeds about 70 percent of maximum, glycolysis begins to contribute an increasing fraction of the total ATP generated.

III. Muscle fatigue is caused by a variety of factors, including internal changes in acidity, phosphate concentration, glycogen depletion, and excitation-contraction coupling failure, not by a lack of ATP.

TYPES OF SKELETAL-MUSCLE FIBERS

I. Three types of skeletal-muscle fibers can be distinguished by their maximal shortening velocities and the predominate pathway used to form ATP: slow-oxidative, fast-oxidative, and fast-glycolytic fibers.

A. Differences in maximal shortening velocities are the result of different myosin enzymes with high or low ATPase activities, giving rise to fast and slow fibers.

B. Fast-glycolytic fibers have a larger average diameter than oxidative fibers and therefore produce greater tension, but they also fatigue more rapidly.

II. All the muscle fibers in a single motor unit belong to the same fiber type, and most muscles contain all three types.

III. Table 11-3 summarizes the characteristics of the three types of skeletal-muscle fibers.

WHOLE-MUSCLE CONTRACTION

I. The tension produced by whole-muscle contraction depends on the amount of tension developed by each fiber and the number of active fibers in the muscle (Table 11-4).

II. Muscles that produce delicate movements have a small number of fibers per motor unit, whereas large postural muscles have much larger motor units.

III. Fast-glycolytic motor units not only have large-diameter fibers but also tend to have large numbers of fibers per motor unit.

IV. Increases in muscle tension are controlled primarily by increasing the number of active motor units in a muscle, a process known as recruitment. Slow-oxidative motor units are recruited first during weak contractions, then fast-oxidative motor units, and finally fast-glycolytic motor units during very strong contractions.

V. Increasing motor-unit recruitment increases the velocity at which a muscle will move a given load.

VI. The strength and susceptibility to fatigue of a muscle can be altered by exercise.

A. Long-duration, low-intensity exercise increases a fiber's capacity for oxidative ATP production by increasing the number of mitochondria and blood vessels in the muscle, resulting in increased endurance.

B. Short-duration, high-intensity exercise increases fiber diameter as a result of increased synthesis of actin and myosin, resulting in increased strength.

VII. Movement around a joint requires two antagonistic groups of muscles: one flexes the limb at the joint, and the other extends the limb.

VIII. The lever system of muscles and bones requires muscle tensions far greater than the load in order to sustain a load in an isometric contraction. The advantage of the lever system is that it produces a shortening velocity at the end of the lever arm that is greater than the muscle-shortening velocity.

SECTION A KEY TERMS

skeletal muscle	myoblast
smooth muscle	satellite cell
cardiac muscle	muscle
muscle fiber	tendon

striated muscle
myofibril
sarcomere
thick filament
myosin
thin filament
actin
A band
Z line
I band
H zone
M line
titin
cross bridge
contraction
relaxation
sliding-filament mechanism
cross-bridge cycle
rigor mortis
troponin
tropomyosin
excitation-contraction
   coupling
sarcoplasmic reticulum
lateral sac
transverse tubule (T tubule)
motor neuron
motor unit
motor end plate
neuromuscular junction
acetylcholine (ACh)
end-plate potential (EPP)

acetylcholinesterase
tension
load
isometric contraction
isotonic contraction
lengthening contraction
twitch
latent period
contraction time
summation
tetanus
optimal length (l_o)
creatine phosphate
oxygen debt
muscle fatigue
central command fatigue
fast fiber
slow fiber
oxidative fiber
myoglobin
red muscle fiber
glycolytic fiber
white muscle fiber
slow-oxidative fiber
fast-oxidative fiber
fast-glycolytic fiber
recruitment
hypertrophy
flexion
extension
antagonist

SECTION A REVIEW QUESTIONS

1. List the three types of muscle cells and their locations.
2. Diagram the arrangement of thick and thin filaments in a striated-muscle sarcomere, and label the major bands that give rise to the striated pattern.
3. Describe the organization of myosin and actin molecules in the thick and thin filaments.
4. Describe the four steps of one cross-bridge cycle.
5. Describe the physical state of a muscle fiber in rigor mortis and the conditions that produce this state.
6. What three events in skeletal-muscle contraction and relaxation are dependent on ATP?
7. What prevents cross bridges from attaching to sites on the thin filaments in a resting skeletal muscle?

8. Describe the role and source of calcium ions in initiating contraction in skeletal muscle.
9. Describe the location, structure, and function of the sarcoplasmic reticulum in skeletal-muscle fibers.
10. Describe the structure and function of the transverse tubules.
11. Describe the events that result in the relaxation of skeletal-muscle fibers.
12. Define a motor unit and describe its structure.
13. Describe the sequence of events by which an action potential in a motor neuron produces an action potential in the plasma membrane of a skeletal-muscle fiber.
14. What is an end-plate potential and what ions produce it?
15. Compare and contrast the transmission of electrical activity at a neuromuscular junction with that at a synapse.
16. Describe isometric, isotonic, and lengthening contractions.
17. What factors determine the duration of an isotonic twitch in skeletal muscle? An isometric twitch?
18. What effect does increasing the frequency of action potentials in a skeletal-muscle fiber have upon the force of contraction? Explain the mechanism responsible for this effect.
19. Describe the length-tension relationship in striated-muscle fibers.
20. Describe the effect of increasing the load on a skeletal-muscle fiber on the velocity of shortening.
21. What is the function of creatine phosphate in skeletal-muscle contraction?
22. What fuel molecules are metabolized to produce ATP during skeletal-muscle activity?
23. List the factors responsible for skeletal-muscle fatigue.
24. What component of skeletal-muscle fibers accounts for the differences in the fibers' maximal shortening velocities?
25. Summarize the characteristics of the three types of skeletal-muscle fibers.
26. Upon what two factors does the amount of tension developed by a whole skeletal muscle depend?
27. Describe the process of motor-unit recruitment in controlling (a) whole-muscle tension, and (b) velocity of whole-muscle shortening.
28. During increases in the force of skeletal-muscle contraction, what is the order of recruitment of the different types of motor units?
29. What happens to skeletal-muscle fibers when the motor neuron to the muscle is destroyed?
30. Describe the changes that occur in skeletal muscles following a period of (a) long-duration, low-intensity exercise training; and (b) short-duration, high-intensity exercise training.
31. How are skeletal muscles arranged around joints so that a limb can push or pull?
32. What are the advantages and disadvantages of the muscle-bone-joint lever system?

SECTION B
SMOOTH MUSCLE

Having described the properties and control of skeletal muscle, we now examine the second of the three types of muscle found in the body—smooth muscle. Two characteristics are common to all smooth muscles: they lack the cross-striated banding pattern found in skeletal and cardiac fibers (hence the name "smooth" muscle), and the nerves to them are derived from the autonomic division of the nervous system rather than the somatic division. Thus, smooth muscle is not normally under direct voluntary control.

Smooth muscle, like skeletal muscle, uses cross-bridge movements between actin and myosin filaments to generate force, and calcium ions to control cross-bridge activity. However, the organization of the contractile filaments and the process of excitation-contraction coupling are quite different in these two types of muscle. Furthermore, there is considerable diversity among various smooth muscles with respect to the mechanism of excitation-contraction coupling.

STRUCTURE

Each smooth-muscle fiber is a spindle-shaped cell with a diameter ranging from 2 to 10 μm, as compared to a range of 10 to 100 μm for skeletal-muscle fibers (Figure 11-3). Unlike skeletal-muscle fibers, the precursor cells of smooth-muscle fibers do not fuse during embryological development. While skeletal-muscle fibers are multinucleate cells that are unable to divide once they have differentiated during embryonic development, smooth-muscle fibers have a single nucleus and have the capacity to divide throughout the life of an individual.

Two types of filaments are present in the cytoplasm of smooth-muscle fibers (Figure 11-35): thick myosin-containing filaments and thin actin-containing filaments. The latter are anchored either to the plasma membrane or to cytoplasmic structures known as **dense bodies,** which are functionally similar to the Z lines in skeletal-muscle fibers. Note in Figure 11-35 that the filaments are oriented slightly diagonally to the long axis of the cell. When the fiber shortens, the regions of the plasma membrane between the points where actin is attached to the membrane balloon out. The thick and thin filaments are not organized into myofibrils, as in striated muscles, and there is no regular alignment of these filaments into sarcomeres, which accounts for the absence of a banding pattern (Figure 11-36).

The concentration of myosin in smooth muscle is only about one-third of that in striated muscle, whereas the actin content can be twice as great. In spite of these differences, the maximal tension per unit of cross-sectional area developed by smooth muscles is similar to that developed by skeletal muscle.

The isometric tension produced by smooth-muscle fibers varies with muscle length in a manner qualitatively similar to that observed in skeletal muscle. There is an optimal length at which tension development is maximal, and less tension is generated at lengths shorter or longer than this optimal length. The range of muscle lengths over which smooth muscle is able to develop tension is greater, however, than it is in skeletal muscle. This property is highly adaptive since most smooth muscle surrounds hollow organs that undergo changes in volume with accompanying

FIGURE 11-35

Thick and thin filaments in smooth muscle are arranged in slightly diagonal chains that are anchored to the plasma membrane or to dense bodies within the cytoplasm. When activated, the thick and thin filaments slide past each other causing the smooth-muscle fiber to shorten and thicken.

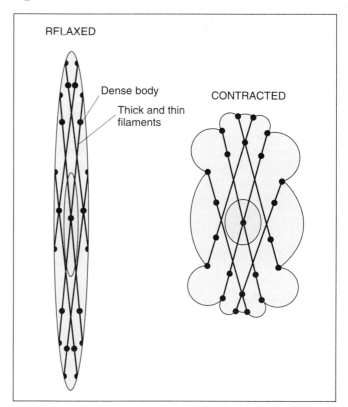

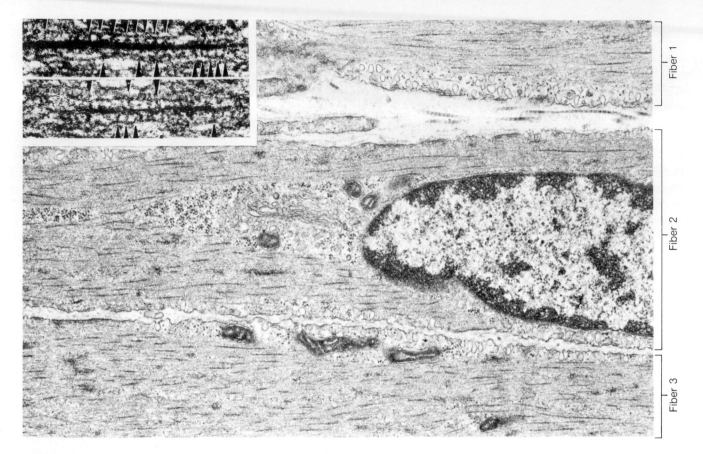

Fiber 1

Fiber 2

Fiber 3

FIGURE 11-36
Electron micrograph of portions of three smooth-muscle fibers. Higher magnification of thick filaments (insert) with arrows indicating cross bridges connecting to adjacent thin filaments. [*From A. P. Somlyo, C. E. Devine, Avril V. Somlyo, and R. V. Rice, Phil. Trans. R. Soc. Lond. B, 265:223–229, (1973).*]

changes in the lengths of the smooth-muscle fibers in their walls. Even with relatively large increases in volume, as during the accumulation of large amounts of urine in the bladder, the smooth-muscle fibers in the wall retain some ability to develop tension, whereas such distortion might stretch skeletal-muscle fibers beyond the point of thick- and thin-filament overlap.

The presence of actin and myosin, overlapping thick and thin filaments with cross bridges, and the length-tension relationship all provide evidence that smooth-muscle contraction occurs by a sliding-filament mechanism similar to that in skeletal and cardiac muscle.

CONTRACTION AND ITS CONTROL

Changes in cytosolic calcium concentration control the contractile activity in smooth-muscle fibers, as in striated muscle. However, there are significant differences between the two types of muscle in the way in which calcium exerts its effects on cross-bridge activity and in the mechanisms by which stimulation leads to alterations in calcium concentration.

Cross-bridge Activation

The thin filaments in smooth muscle do not have the calcium-binding protein troponin that mediates calcium-triggered cross-bridge activity in both skeletal and cardiac muscle. Instead, cross-bridge cycling in smooth muscle is controlled by a calcium-regulated enzyme that phosphorylates myosin. Only the phosphorylated form of smooth-muscle myosin is able to bind to actin and undergo cross-bridge cycling.

The following sequence of events occurs after a rise in cytosolic calcium in a smooth-muscle fiber (Figure 11-37): (1) Calcium binds to calmodulin, a calcium-binding protein that is present in most cells (Chapter 7) and whose structure is related to that of troponin. (2) The calcium-calmodulin complex binds to a protein kinase, **myosin light-chain kinase,** thereby activating the enzyme. (3) The active protein kinase then uses ATP to phosphorylate myosin light chains in the globular head of myosin. Hence, cross-bridge activity in smooth muscle is turned on by calcium-mediated changes in the thick filaments, whereas in striated muscle, calcium mediates changes in the thin filaments.

The smooth-muscle myosin enzyme has a very low maximal rate of ATPase activity, on the order of 10 to 100 times less than that of skeletal-muscle myosin. Since the rate of ATP splitting determines the rate of cross-bridge cycling and thus shortening velocity, smooth-muscle shortening is much slower than that of skeletal muscle. The overall low rate of ATP utilization may contribute to the fact that smooth muscle does not undergo fatigue during prolonged periods of activity.

In addition, if the cytosolic calcium concentration remains elevated, the rate of ATP splitting by the cross bridges declines even though isometric tension is maintained. When a phosphorylated cross bridge is dephosphorylated while still attached to actin, it can maintain tension in a rigor-like state without movement. Dissociation of these cross bridges from actin by the binding of ATP occurs at a much slower rate than dissociation of phosphorylated bridges. The net result is the ability to maintain tension for long periods of time with a very low rate of ATP consumption.

To relax a contracted smooth muscle, myosin must be dephosphorylated because in its dissociated state dephosphorylated myosin is unable to bind to actin. This dephosphorylation is mediated by the enzyme myosin light-chain phosphatase, which is continuously active in smooth muscle during periods of rest and contraction. When cytosolic calcium rises, the rate of myosin phosphorylation

by the activated kinase exceeds the rate of dephosphorylation by the phosphatase, and the amount of phosphorylated myosin in the cell increases, producing a rise in tension. When the cytosolic calcium concentration decreases, the rate of dephosphorylation exceeds the rate of phosphorylation, and the amount of phosphorylated myosin decreases, producing relaxation.

Sources of Cytosolic Calcium

Two sources of calcium contribute to the rise in cytosolic calcium that initiates smooth-muscle contraction: (1) the sarcoplasmic reticulum and (2) extracellular calcium entering the cell through plasma-membrane calcium channels. The amount of calcium contributed by these two sources differs among various smooth muscles, some being more dependent on extracellular calcium than the internal stores of calcium in the sarcoplasmic reticulum, and vice versa.

Let us look first at the sarcoplasmic reticulum. The total quantity of this organelle in smooth muscle is smaller than in skeletal muscle, and it is not arranged in any specific pattern in relation to the thick and thin filaments. Moreover, there are no T tubules connected to the plasma membrane in smooth muscle. The small fiber diameter and the slow rate of contraction do not require such a

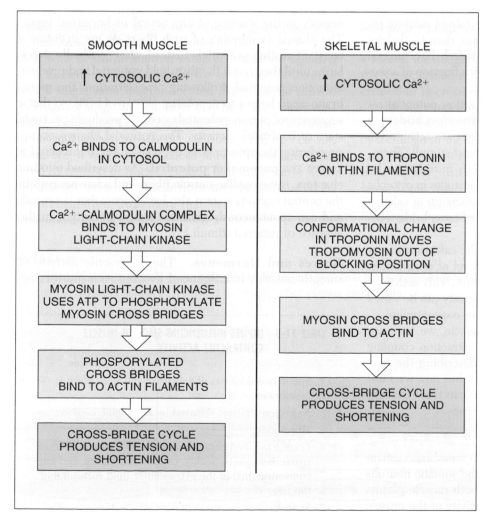

FIGURE 11-37

Pathways leading from increased cytoplasmic calcium to cross-bridge cycling in smooth- and skeletal-muscle fibers.

the specialization is not so extreme that certain areas perform only certain functions; (2) neurons in different regions are activated more or less in parallel and there is no clear evidence for any one area initiating motor activity; (3) neurons in the different motor areas show a great deal of flexibility in their responses; and (4) movements are ultimately performed as a result of activity in large ensembles of neurons. Nevertheless, just as researchers have found it useful to retain the notion of a motor control hierarchy despite its flaws, you the reader should also find the hierarchical model conceptually helpful.

Voluntary and Involuntary Actions

Given such a highly interconnected and complicated neuroanatomical basis for the motor system, it is difficult to use the phrase **voluntary movement** with any real precision. We shall use it, however, to refer to those actions that have the following characteristics: (1) The movement is accompanied by a conscious awareness of what we are doing and why we are doing it rather than the feeling that it "just happened," and (2) our attention is directed toward the action or its purpose.

The term "involuntary," on the other hand, describes actions that do not have these characteristics. "Unconscious," "automatic," and "reflex" are often taken to be synonyms for "involuntary," although in the motor system the term "reflex" has a more precise meaning (Chapter 7).

Despite our attempts to distinguish between voluntary and involuntary actions, almost all motor behavior involves both components, and the distinction between the two cannot be made easily. Even such a highly conscious act as threading a needle involves the unconscious postural support of the hand and forearm and inhibition of the antagonistic muscles—those muscles whose activity would oppose the intended action, in this case, the muscles that straighten the fingers.

Thus, most motor behavior is neither purely voluntary nor purely involuntary but falls somewhere between these two extremes. Moreover, actions shift along this continuum according to the frequency with which they are performed. When a person first learns to drive a car with a standard transmission, for example, shifting gears is a fairly complicated process and requires a great deal of conscious attention. With practice, the same actions become automatic, and a complicated pattern of muscle movements is shifted from the highly conscious end of the spectrum toward the involuntary end by the process of learning. On the other hand, reflex behaviors, which are all the way at the involuntary end of the spectrum, can with special effort be modified or even prevented.

We now turn to an analysis of the individual components of the motor control system, beginning with local control mechanisms because their activity serves as a base upon which the pathways descending from the brain exert their influence.

LOCAL CONTROL OF MOTOR NEURONS

The local control systems are the relay points for instructions from centers higher in the motor control hierarchy to the motor neurons. In addition, the local control systems play a major role in adjusting motor unit activity to unexpected obstacles to movement and to harmful factors in the surrounding environment.

In describing local control, we must add one more element to our motor control system. Afferent fibers in the muscles, tendons, joints, and skin of the body part to be moved enter the central nervous system and transmit information not only to higher levels of the hierarchy, as noted earlier, but to the local level as well.

Interneurons

Most of the synaptic input to motor neurons from the descending pathways and afferent neurons does not go *directly* to motor neurons but rather to interneurons that synapse with the motor neurons. These interneurons are of several types. Some are confined to the general region of the motor neuron upon which they synapse and thus are called local interneurons. Others have processes that extend up or down short distances in the spinal cord and brainstem, or even throughout much of the length of the central nervous system. The interneurons with longer processes are important in movements that involve the coordinated interaction of, for example, a shoulder and arm or an arm and a leg.

The local interneurons are important elements of the lowest level of the motor control hierarchy, integrating inputs not only from higher centers and peripheral receptors but from other interneurons as well (Figure 12-3). They are crucial in determining which muscles are activated when by distributing control to specific motor neurons. Moreover, interneurons can act as "switches" that enable a movement to be turned on or off under the command of higher motor centers. For example, if we pick up a hot object, a local reflex arc will be activated, normally resulting in our dropping the object. But if it were something we had spent a great deal of effort to prepare, descending commands could influence the local activity, and we would continue to hold onto the object until we could put it down safely.

Local Afferent Input

As noted earlier, afferent fibers usually impinge upon the local interneurons (in one case, they synapse directly on motor neurons). The afferent fibers bring information

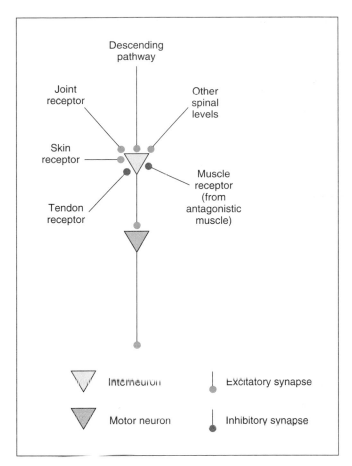

FIGURE 12-3
Convergence of axons onto a local interneuron.

from receptors in three areas: (1) the very muscles controlled by the motor neurons, (2) other nearby muscles, and (3) the tendons, joints, and skin surrounding the muscles.

These receptors monitor the length and tension of the muscles, movement of the joints, and the effect of movements on the overlying skin. As we shall see next, their input both provides negative-feedback control over the muscles and contributes to the conscious awareness of limb and body position.

Length-Monitoring Systems and the Stretch Reflex. Absolute muscle length and changes in muscle length are monitored by stretch receptors embedded within the muscle. These receptors consist of endings of afferent nerve fibers that are wrapped around modified muscle fibers, several of which are enclosed in a connective-tissue capsule. The entire structure is called a **muscle spindle** (Figure 12-4). The modified muscle fibers within the spindle are known as **intrafusal fibers,** whereas the skeletal-muscle fibers that form the bulk of the mus-

cle and generate its force and movement are the **extra-fusal fibers.**

Within a given spindle, there are two kinds of stretch receptors: One responds best to how much the muscle has been stretched, the other to both the magnitude of the stretch and the speed with which it occurs. Although the

FIGURE 12-4

A muscle spindle and Golgi tendon organ. Note that the muscle spindle is parallel to the extrafusal muscle fibers. The Golgi tendon organ will be discussed later in the chapter. (*Adapted from Elias, Pauly, and Burns.*)

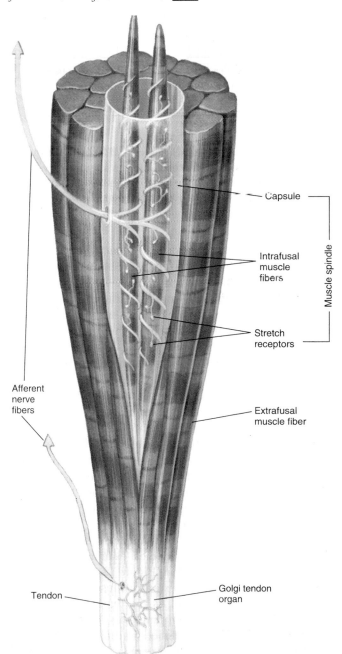

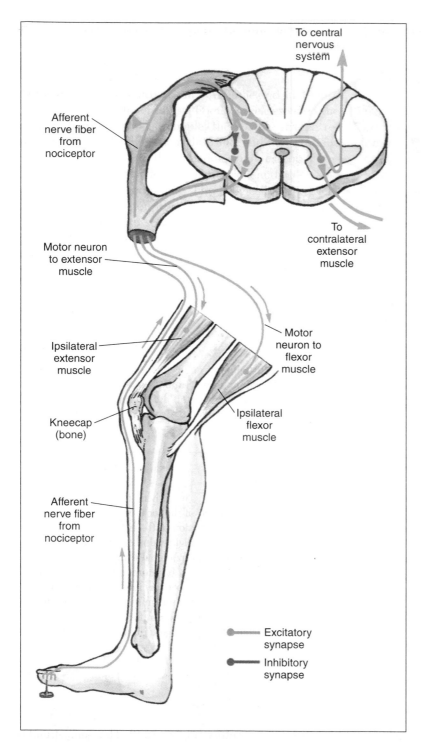

Afferent nerve fiber from nociceptor

To central nervous system

Motor neuron to extensor muscle

To contralateral extensor muscle

Motor neuron to flexor muscle

Ipsilateral extensor muscle

Kneecap (bone)

Ipsilateral flexor muscle

Afferent nerve fiber from nociceptor

Excitatory synapse

Inhibitory synapse

FIGURE 12-9

In response to pain, the ipsilateral flexor's motor neuron is stimulated (withdrawal reflex). In the case illustrated, the opposite limb is extended (crossed-extensor reflex) to support the body's weight. Arrows indicate direction of action potential propagation.

THE BRAIN MOTOR CENTERS AND THE DESCENDING PATHWAYS THEY CONTROL

As stated earlier, the motor neurons and interneurons at the local levels of motor control are influenced by descending pathways, and these pathways are themselves controlled by various motor centers in the brain (Figure 12-1).

Cerebral Cortex

The cerebral cortex plays a critical role in both the planning and ongoing control of voluntary movements, functioning in both the highest and middle levels of the motor control hierarchy. The term **sensorimotor cortex** is used to include all those parts of the cerebral cortex that act together in the control of muscle movement. A large number of nerve fibers that give rise to some of the descending pathways come from two areas of sensorimotor cortex on the posterior part of the frontal lobe: the **primary motor cortex** (sometimes called simply the **motor cortex**) and the **premotor area** (Figure 12-10). Other areas of sensorimotor cortex are the **supplementary motor cortex,** which lies mostly on the surface of the frontal lobe where the cortex folds down between the two hemispheres (Figure 12-10B), the **somatosensory cortex,** and parts of the **parietal-lobe association cortex** (Figure 12-10A and B).

Although these areas are distinct, they are heavily interconnected, and individual muscles or movements are represented at multiple sites.

Despite the fact that we will emphasize overlapping functions of the different areas of sensorimotor cortex, certain areas do serve specialized functions. For example, the primary motor cortex is involved in the guidance of complex actions requiring the coordination of several muscles. A single neuron here may help control movement of several joints—the wrist, elbow, and shoulder, for example, which are all active in reaching for an object; however, different regions of the primary motor cortex are concerned with movements of smaller areas of the body (Figure 12-11).

Neurons in other parts of sensorimotor cortex are involved in functions such as a change from one task to another, a movement made in response to visual control or a spoken command, two-handed coordination, and postural support for a wide variety of detailed movements. Neurons in still other areas show sustained preparatory activity that begins long before the actual movement or are active when one is performing a memorized sequence of movements.

We have described the various areas of sensorimotor cortex as giving rise, either directly or indirectly, to pathways descending to the motor neurons. As we have stated, however, these neurons are part of the middle hierarchical level and are not the prime *initiators* of movement. They are simply tremendously important relay and integrative stations. We presently don't know just where or how movements are initiated except in the case of reflexes.

Association areas of the cerebral cortex also play a role in motor control. For example, neurons of parietal association cortex are important in the visual control of hand action for reaching and grasping. These neurons play an important role in matching motor signals con-

FIGURE 12-10

(A) The major motor areas of cerebral cortex. (B) Midline view of the brain showing the supplementary motor cortex, which lies in the part of the cerebral cortex that is folded down between the two cerebral hemispheres. Other cortical motor areas also extend onto this area. The premotor, supplementary motor, primary motor, somatosensory, and parietal-lobe association cortexes together make up the sensorimotor cortex.

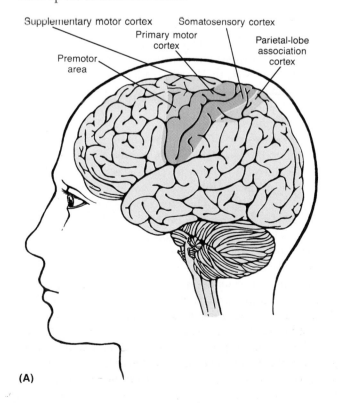

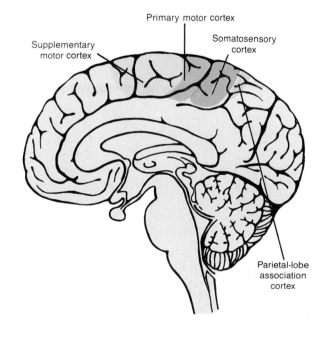

(A)

(B)

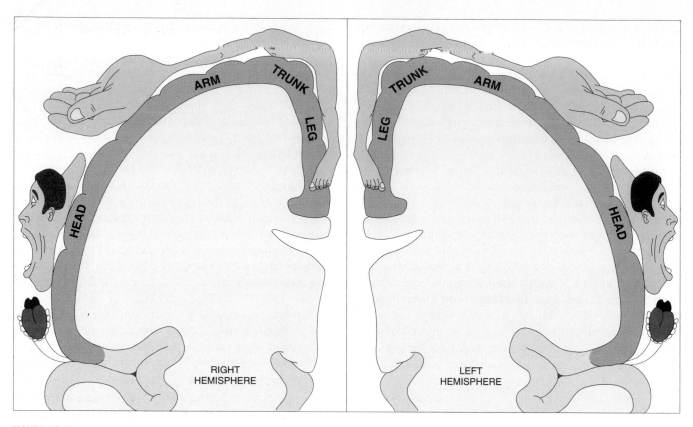

FIGURE 12-11

Representation of major body areas in primary motor cortex. Within the broad areas, however, no one area exclusively controls the movement of a single body region, and there is much overlap of cortical representation.

cerning the pattern of hand action with signals from the visual system concerning the three-dimensional features of the objects to be manipulated.

During activation of the cortical areas involved in motor control, subcortical mechanisms also become active, and it is to these areas of the motor control system that we now turn.

Subcortical and Brainstem Nuclei

A dozen or so highly interconnected structures lie within the cerebrum beneath the cerebral cortex and in the brainstem, and they interact with the cortex to control movements. Their influence is transmitted to the motor neurons both by pathways that go to the cerebral cortex and by descending pathways that arise directly from some of the brainstem nuclei.

It is not known to what extent, if any, these structures *initiate* movements, but they definitely play a prominent role in planning and monitoring them, establishing the programs that determine the specific sequence of movements needed to accomplish a desired action. They are also important in learning skilled movements.

Prominent among the subcortical nuclei are the paired **basal ganglia** (Figure 12-2B). (Despite their name, these

neuronal clusters are technically nuclei because they are within the central nervous system.) They form a link in some of the parallel circuits through which activity in the motor system is transmitted from a specific region of sensorimotor cortex to the basal ganglia, from there to the thalamus, and then back to the cortical area from which the circuit started. Some of these circuits facilitate movements and others suppress them.

In *Parkinson's disease,* in which the amount of dopamine in the basal ganglia is diminished, the interplay of the facilitory and inhibitory circuits is unbalanced, and activation of the motor cortex (via the basal ganglia–thalamus limb of the circuit mentioned above) is reduced. Clinically, Parkinson's disease is characterized by a reduced amount of movement (*akinesia*), slow movements (*bradykinesia*), muscular rigidity, and a 3- to 5-Hz tremor at rest. Other motor and nonmotor abnormalities may also be present. For example, a common set of symptoms include a change in facial expression resulting in a masklike, unemotional appearance; a shuffling gait with loss of arm swing; and a stooped and unstable posture.

Although the symptoms of Parkinson's disease reflect inadequate functioning of the basal ganglia, the initial defect arises in neurons of the **substantia nigra** ("black substance"), a subcortical nucleus that gets its name from the dark pigment in its cells. These neurons, which degenerate in Parkinson's disease, project to the basal ganglia where they normally liberate dopamine from their axon terminals. As substantia nigra neurons degenerate, the amount of dopamine they deliver to the basal ganglia is reduced and the basal ganglia function decreases.

The most powerful drugs currently available for Parkinson's disease are those that mimic the action of dopamine or increase its availability. One such drug is L-dopa, a precursor of dopamine. L-dopa enters the bloodstream, crosses the blood-brain barrier, and is converted to dopamine (dopamine itself is not used as medication because it cannot cross the blood-brain barrier). The newly formed dopamine activates the receptors in the basal ganglia and relieves the symptoms of the disease. Another drug inhibits the brain enzyme that breaks down dopamine so that more of the neurotransmitter reaches the neurons in the basal ganglia. Other therapies include the electrical destruction ("lesioning") of overactive areas of the basal ganglia or stimulation of the underactive ones. Still highly controversial is an experimental treatment that involves the transplantation into the basal ganglia of neurons from either human fetuses or an animal such as a pig, or cells that have been genetically engineered. Another tissue used for implant into the basal ganglia is a piece of the patient's own adrenal medulla, which secretes dopamine. Regardless of their source, the implanted cells then synthesize dopamine and growth factors or form tyrosine hydroxylase, the enzyme that catalyzes dopamine production (see Figure 8-34).

Cerebellum

The cerebellum is behind the brainstem, as can be seen in Figure 12-2A. Although it does not initiate behavior, it exerts a very important influence on posture and movement. It does this by means of input to brainstem nuclei and (by way of the thalamus) to regions of the sensorimotor cortex that give rise to pathways that descend to the motor neurons. The cerebellum receives information both from the sensorimotor cortex (relayed via brainstem nuclei) and from the vestibular system, eyes, ears, skin, muscles, joints, and tendons, that is, from the major receptors affected by movement.

The cerebellum's role in motor functioning includes providing timing signals to the cerebral cortex and spinal cord for precise execution of the different phases of a motor program, in particular the timing of the agonist/antagonist components of a movement. It also helps coordinate movements and stores the memories of them so they can be achieved more easily the next time they are tried.

The cerebellum participates in planning movements—integrating information about the nature of an intended movement with information about the space outside the person into which the movement will be made. The cerebellum then provides this as a "feedforward" signal to the brain areas responsible for refining the motor program.

Moreover, during the course of the movement the cerebellum compares information about what the muscles *should* be doing with information about what they actually *are* doing. If there is a discrepancy between the intended movement and the actual one, the cerebellum sends an error signal to the motor cortex and subcortical centers to correct the ongoing program.

The role of the cerebellum in programming movements can best be appreciated when seeing the absence of this function in individuals with **cerebellar disease,** who cannot perform limb or eye movements smoothly but move with a tremor—a so-called **intention tremor** that increases as the course of the movement nears its final destination and becomes more precise. People with cerebellar disease also cannot start or stop movements quickly or easily, and cannot combine the movements of several joints into a single smooth, coordinated motion.

Unstable posture and awkward gait are two other symptoms characteristic of cerebellar dysfunction. For example, persons with cerebellar damage walk with the feet well apart, and they have such difficulty maintaining balance that their gait appears drunken. A final symptom involves difficulty in learning new motor skills, and individuals with cerebellar dysfunction find it hard to modify movements in response to new situations. Unlike damage to areas of sensorimotor cortex, cerebellar damage does not cause paralysis.

Descending Pathways

The influence exerted by the various brain regions on posture and movement is via descending pathways to the motor neurons and the interneurons that affect these neurons. The pathways are of two types: the **corticospinal pathways,** which, as their name implies, originate in the cerebral cortex; and a second group called the **brainstem pathways,** which originate in the brainstem.

Fibers from both types of descending pathways end at synapses on alpha and gamma motor neurons or on interneurons that affect the alpha motor neurons either directly or via still other interneurons. Sometimes, as mentioned earlier, these are the same interneurons that function in reflex arcs, thereby ensuring that the descending signals are fully integrated with local information before the activity of the motor neurons is altered. The ultimate effect of the descending pathways on the alpha motor neurons may be excitatory or inhibitory.

Importantly, some of the descending fibers affect *afferent* systems. They do this via (1) presynaptic synapses on the terminals of afferent neurons as these fibers enter the central nervous system, or (2) synapses on interneurons in the ascending pathways. The overall effect of this descending input to afferent systems is to limit their influence on either the local or brain motor control areas, thereby altering the importance of a particular bit of afferent information or sharpening its focus. Because of this descending (motor) control over ascending (sensory) information, there is clearly no real functional separation of the motor and sensory systems.

Corticospinal Pathway.   The nerve fibers of the corticospinal pathways, as mentioned before, have their cell bodies in sensorimotor cortex and terminate in the spinal cord. The corticospinal pathways are also called the **pyramidal tracts,** or **pyramidal system** (perhaps because of their shape as they pass along the surface of the medulla oblongata or because they were formerly thought to arise solely from the pyramidal cells of the motor cortex). In the medulla oblongata near the junction of the spinal cord and brainstem, most of the corticospinal fibers cross the spinal cord to descend on the opposite side (Figure 12-12). Thus,

FIGURE 12-12

The corticospinal and brainstem pathways. Most of the corticospinal fibers cross in the brainstem to descend in the opposite side of the spinal cord, but the brainstem pathways are mostly uncrossed. Arrows indicate direction of action potential propagation. (*Adapted from Gardner.*)

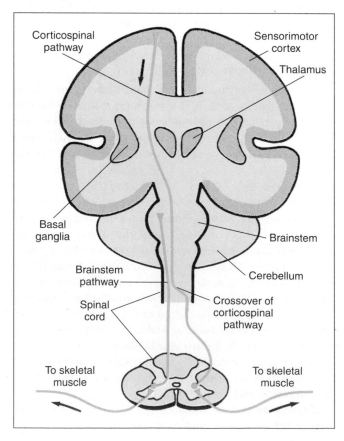

the skeletal muscles on the left side of the body are controlled largely by neurons in the right half of the brain, and vice versa.

As the corticospinal fibers descend through the brain from the cerebral cortex, they are accompanied by fibers of the **corticobulbar pathway** (bulbar means "pertaining to the brainstem"), a pathway that begins in the sensorimotor cortex and ends in the brainstem. The corticobulbar fibers control, directly or indirectly via interneurons, the motor neurons that innervate muscles of the eye, face, tongue, and throat. These fibers are the main source of control for voluntary movement of the muscles of the head and neck, whereas the corticospinal fibers serve this function for the muscles of the rest of the body. For convenience, we shall henceforth include the corticobulbar pathway in the general term "corticospinal pathways."

Axons of single corticospinal neurons diverge to influence motor neurons of several muscles; single neurons, in fact, often influence synergistic sets of muscles. Moreover, neurons from wide areas of the sensorimotor cortex converge onto single motor neurons so that most muscles are controlled by multiple brain areas.

The corticospinal pathways have their greatest influence on rapid, fine movements of the distal extremities, such as those made when an object is manipulated by the fingers. After damage to the corticospinal pathways, all movements are slower and weaker, individual finger movements are absent, and it is difficult to release a grip.

Brainstem Pathways.   Axons from neurons in the brainstem also form pathways that descend into the spinal cord to influence motor neurons. These pathways are sometimes referred to as the **extrapyramidal system,** or indirect pathways, to distinguish them from the corticospinal (pyramidal) pathways. However, no general term is widely accepted for these pathways, and for convenience we shall refer to them collectively as the brainstem pathways.

Axons of some of the brainstem pathways cross from their side of origin in the brainstem to affect muscles on the opposite side of the body, but most remain uncrossed. In the spinal cord the fibers of the brainstem pathways descend as distinct clusters, named according to their sites of origin. For example, the vestibulospinal pathway descends to the spinal cord from the vestibular nuclei in the brainstem, whereas the reticulospinal pathway descends from neurons in the brainstem reticular formation.

The brainstem pathways are especially important in the control of upright posture, balance, and walking. These pathways are anatomically suited to invoke the large groups of muscles characteristic of postural support in three ways: (1) They generally exert greater influence on motor neurons that control muscles in the neck, trunk, and upper part of the limbs than on motor neurons that control muscles in the fingers and toes; (2) as they descend through the spinal cord, they give rise to collateral branches that affect motor neurons at many levels of the cord; and (3) rather than ending on interneurons that are part of the local reflex arcs, as corticospinal neurons generally do, fibers of the brainstem descending pathways are apt to end on interneurons that send long branches that interconnect different levels of the spinal cord.

Concluding Comments on the Descending Pathways.

As stated above, the corticospinal neurons generally have their greatest influence over motor neurons that control muscles involved in fine, isolated movements, particularly those of the fingers and hands. The brainstem descending pathways, in contrast, are more involved with coordination of the large muscle groups used in the maintenance of upright posture, in locomotion, and in head and body movements when turning toward a specific stimulus.

There is, however, much interaction between the descending pathways. For example, some fibers of the corticospinal pathway end on interneurons that play important roles in posture, whereas fibers of the brainstem descending pathways sometimes end directly on the alpha motor neurons to control discrete muscle movements. Because of this redundancy, loss of function resulting from damage to one system may be compensated for by the remaining system, although the compensation is generally not complete.

The distinctions between the corticospinal and brainstem descending pathways are not clear-cut. All movements, whether automatic or voluntary, require the continuous coordinated interaction of both types of pathways.

In conclusion, the complete control network of any movement—complex or simple—is presently understood only imperfectly. Some interesting additional features of motor control that are becoming clear, however, are listed in Table 12-2.

The Role of Serotonin

While various transmitters are involved in the motor system—we have mentioned dopamine in conjunction with the basal ganglia—serotonin plays a particularly important role. In fact, one current hypothesis states that the primary function of the widely spread serotonergic system is to facilitate motor output. Neurons that release serotonin increase their firing rate above baseline levels during, and just preceding, increased motor activity. Serotonergic neurons also coordinate the autonomic and neuroendocrine activities, such as the redistribution of blood flow to the exercising muscles, that support a given motor activity. In addition, they inhibit information processing in the sensory pathways. This serotonergic facilitation of the motor system is briefly removed during times of highly focused attention, experienced for example, as freezing in one's tracks when hearing a loud crash, and it is decreased with sleep, experienced as a nodding head and slumped posture.

MUSCLE TONE

The resistance of skeletal muscle to stretch as an examiner moves the relaxed limb or neck of a subject is known as *muscle tone.* Under such circumstances in a normal person, the resistance to passive movement is slight and uniform, regardless of the speed of the movement.

Muscle tone is due both to the viscoelastic properties of the muscles and joints and to whatever degree of alpha motor neuron activity exists. When a person is deeply relaxed, the alpha motor neuron activity probably makes no contribution to the resistance to stretch. As the person becomes increasingly alert, however, some activation of the alpha motor neurons occurs and muscle tone increases. Active motor neurons play an ever-increasing role until the myofibrillar cross-bridge activity they stimulate accounts for much of the resistance to stretch.

Abnormal Muscle Tone

Abnormally high muscle tone, called *hypertonia,* occurs in individuals with certain disease processes and is seen particularly clearly when a joint is moved passively at high speeds. The increased resistance is due to a greater-than-normal level of alpha motor neuron activity that keeps a muscle contracted despite the individual's attempt to relax it. Hypertonia is usually found when there are disorders of the descending pathways that result in decreased inhibitory influence exerted by them on the motor neurons.

Clinically, the descending pathways—primarily the corticospinal pathways—and neurons of the motor cortex are often referred to as the "upper motor neurons" (a serious misnomer because they are not really motor neurons at all). Abnormalities due to their dysfunction are classed, therefore, as *upper motor neuron disorders.* Thus, hypertonia indicates an upper motor neuron disorder. In this clinical classification, the alpha motor neurons are termed lower motor neurons.

Spasticity is a form of hypertonia in which the muscles do not develop increased tone until they are stretched a bit, and after a brief increase in tone, the contraction subsides for a short time. The period of "give" occurring after a time of resistance is called the *clasp-knife phenomenon.* Spasticity is accompanied by increased responses of motor reflexes such as the knee jerk, and by decreased coordination and strength of voluntary actions. *Rigidity* is a form of hypertonia in which the increased muscle contraction is continual and the resistance to passive stretch is constant. Two other forms of hypertonia that can occur suddenly in individual or multiple muscles are *spasms,* which are brief contractions, and *cramps,* which are prolonged and painful.

Hypotonia is a condition of abnormally low muscle tone, accompanied by weakness, atrophy (a decrease in

TABLE 12-2 SOME CHARACTERISTICS OF MOTOR CONTROL SYSTEMS

1. Even though it seems obvious to think of the motor system as controlling the contraction of *muscles*, much of the motor system deals more with *actions*. Thus, to take a book from a shelf, more of the motor control system is involved with calculating the direction the arm must move and how far than with selecting which motor neurons to activate and when.

2. Before descending pathways can activate specific motor neurons, a great deal of neural activity takes place in the brain. Whether the intention of an action arises from within the brain (I think I'll study now so I won't have to do it later) or as an outside stimulus (the telephone rings and I answer it), the intention has to be translated into the direction and distance the body must be moved. This translation process is completed gradually by many components of the brain's motor system acting simultaneously, each component adding a special contribution.

3. The motor control system must select from a variety of ways of achieving the purpose, since most motor acts can be executed in many ways with different sets of muscles. A simple example of this is a rat trained to press a lever whenever a light flashes. Depending upon its original position in the cage, its movements can be quite varied. If the rat is to the left of the lever, it moves to the right; if it is to the right, it moves to the left. If its paw is on the floor, it raises the paw; if its paw is above the lever, it lowers the paw.

4. The brain operates pretty much as a whole in producing movement; that is, many brain structures are involved.

5. The motor cortex is unable to generate coordinated voluntary movements without input from the subcortical motor centers.

6. Not all voluntary movements are channeled through the motor cortex. For example, tongue movements during speech are controlled by special speech areas (Chapter 13).

7. A given set of muscles can be controlled by several neural systems, and the central nervous system can select from them to execute a movement.

8. Movements that involve learning, movements to be made in response to an expected signal, highly skilled movements that have become almost automatic—all these are controlled by different components of the motor control systems.

9. The limbic system (emotion and motivation) plays an important role in motor control (we are all aware that it is usually difficult to make well-controlled, detailed movements when in a highly emotional state).

10. The motor cortex is more important in controlling skilled, accurate movements than in producing automatic or rhythmical ones.

11. Movements performed to achieve a specific task are variable and seemingly inconsequential. For example, it usually does not matter whether you comb your hair with the right or left hand. It is the end result that matters, but only the intervening acts or movements can be programmed by the nervous system.

12. In certain circumstances a spinal reflex, for example, the withdrawal reflex, can override a command from a higher hierarchical level, for example, "extend leg to bear body's weight."

13. Movements are generally accompanied by changes in posture and alterations in sensory input. Anticipatory adjustments are made to prevent loss of balance and to compensate for the changed sensory information.

---

muscle bulk), and decreased or absent reflex responses. Dexterity and coordination are generally preserved unless profound weakness is present. While hypotonia may develop after cerebellar disease, it more frequently accompanies disorders of the alpha motor neurons ("lower motor neurons"), neuromuscular junctions, or muscles. The term *flaccid,* which means "weak" or "soft," is often used to describe hypotonic muscles.

MAINTENANCE OF UPRIGHT POSTURE AND BALANCE

The skeleton supporting the body is a system of long bones and a many-jointed spine that cannot stand erect against the forces of gravity without the support given by coordinated muscle activity. The muscles that maintain upright posture, that is, support the body's weight against gravity, are controlled by the brain and by reflex mechanisms that are "wired into" the neural networks of the brainstem and spinal cord. Many of the reflex pathways previously introduced, for example, the stretch and crossed-extensor reflexes, are used in posture control.

Added to the problem of maintaining upright posture is that of maintaining balance. A human being is a very tall structure balanced on a relatively small base, and its center of gravity is quite high, being situated just above the pelvis. For stability, the center of gravity must be kept within the base of support provided by the feet (Figure 12-13). Once the center of gravity has moved beyond this base, the body will fall unless one foot is shifted to broaden the base of support. Yet people can operate under conditions of unstable equilibrium because their balance is protected by complex interacting **postural reflexes,** all the components of which we have met previously.

The afferent pathways of the postural reflexes come from three sources: the eyes, the vestibular apparatus, and the somatic receptors. The efferent pathways are the alpha motor neurons to the skeletal muscles, and the integrating centers are neuron networks in the brainstem and spinal cord.

In addition to these integrating centers, there are centers in the brain that form an internal representation of the body's geometry, its support conditions, and its orientation with respect to verticality. This internal representation serves two purposes: It serves as a reference frame for the perception of the body's position and orientation in space and for planning actions, and it contributes to stability via the motor controls involved in maintenance of upright posture.

There are many familiar examples of using reflexes to maintain upright posture, one being the crossed-extensor reflex. As one leg is flexed and lifted off the ground, the other is extended more strongly to support the added weight of the body, and the positions of various parts of the body are shifted to move the center of gravity over the single, weight-bearing leg. This shift in the center of gravity, demonstrated in Figure 12-14, is an important component in the stepping mechanism of locomotion.

Twisting or bending the hips or ankles and stepping are other ways of restoring balance, the strategy used depending on the intensity of the disturbance and any constraints on movement. For example, when leaning forward more and more, "ankle" mechanisms are activated first, but they gradually give way to "hip" mechanisms, which in turn give way to stepping. Postural adjustments also occur in anticipation of voluntary movements in a feedforward way and minimize the disturbances that would accompany the movement.

It is clear that afferent input from several sources is necessary for effective postural adjustments, yet interfering with any one of these inputs alone does not cause a person to topple over. Blind people maintain their balance quite well with only a slight loss of precision, and people whose vestibular mechanisms have been destroyed have very little disability in everyday life as long as their visual system and somatic receptors are functioning.

The conclusion to be drawn from such examples is that the postural control mechanisms are not only effective and flexible, they are also highly adaptable.

WALKING

Walking requires the coordination of literally hundreds of muscles in different parts of the body, each activated to a precise degree at a precise time. The cyclical, alternating movements of walking are controlled by networks of interneurons that generate patterns of motor-neuronal activation in the spinal cord at the level of the motor neurons themselves. The networks coordinate the output of the various motor-neuron pools controlling muscles of the arms, shoulders, trunk, hips, and legs. In addition to activating those alpha motor neurons necessary for locomotion, the neuronal networks regulate afferent input so that it does not hinder the motor neurons' activity.

The networks rely on both plasma-membrane pacemaker properties and patterned synaptic activity to establish their rhythms. At the same time, however, the networks are remarkably flexible; a single network can generate many different patterns of neural activity, depending upon the synaptic activity induced in it by its inputs. This flexibility seems to be due in part to the neuromodulatory effects of amine and neuropeptide transmitters at synapses in the network where they alter the effectiveness of synapses and electrical properties of the neurons' plasma membranes.

On a larger physical scale, walking is initiated by allowing the body to fall forward to an unstable position and then moving one leg forward to regain equilibrium. The

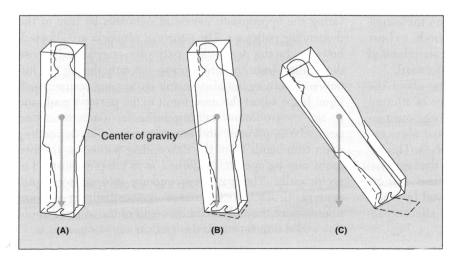

(A) Center of gravity **(B)** **(C)**

FIGURE 12-13

The center of gravity is the point in an object at which, if a string were attached to the object at this point and pulled up, all the downward force due to gravity would be exactly balanced. (A) The center of gravity must remain within the upward vertical projections of the object's base (the tall box outlined in the drawing) if stability is to be maintained. (B) Stable conditions: Box tilts a bit, but the center of gravity remains within the base area and so the box will return to its upright position. (C) Unstable conditions: The box tilts so far that its center of gravity is not above any part of the object's base—the dashed rectangle on the floor—and the object will fall.

The actual laying down of the memory in a more-or-less permanent form is called **memory consolidation.** Just where the memory trace occurs, what form it takes, and how memories are consolidated are the paramount questions in the study of learning.

New scientific facts about memory are being generated at a tremendous pace, and the difficulty comes when one tries to fit this information into an overall, workable scheme. First, memory can be viewed in two broad categories: **Implicit memory,** in which the person has no previous awareness of the memory, cannot describe the learned information except through behavior, and cannot necessarily remember how, when, or where the learning occurred; and **declarative memories,** which are involved in remembering facts and events. An example of an implicit memory would be entering a house and feeling you had been there before, even knowing the locations of the various rooms, but having no awareness of a previous visit. An example of a declarative memory would be knowing a friend's phone number.

To get a better grasp of the various facets of memory, one currently proposed plan further categorizes memory into five basic types: procedural, perceptual representation, semantic, working, and episodic. As we shall see, procedural memory involves knowing how to do something, whereas the other four memory types involve a thought or a conscious awareness that may or may not lead to an action. An example of these latter memory types would be knowing the name of someone you meet at the library regardless of whether you speak the name or not.

Procedural Memory. Procedural memory can be thought of as the memory of how to do things. In other words, it is the memory for skilled behaviors independent of any understanding, as for example, riding a bicycle. Individuals can suffer severe deficits in other types of memory but have intact procedural memory. One case study describes a pianist who learned a new piece to accompany a singer at a concert but had no recollection the following morning of having performed the composition. He could remember *how* to play the music but could not remember having done so.

Regions of the sensorimotor cortex, basal ganglia, and cerebellum are the primary areas of the brain involved in procedural learning.

Perceptual Representation. Perceptual representation, a form of implicit memory, is simply the memory of perceiving an object or event, a memory that is not tied to knowing the name of the perception or even understanding it. Because of this type of memory, the next time the object or event, or one similar to it, is encountered, it is identified more quickly and easily—that is, its identification relies on less stimulus information.

This type of memory was profited from in the now illegal practice of subliminal advertising, in which, for example, the name and picture of a product were presented on television too quickly for the viewer to be aware of *perceiving* the event. The next time the viewer was at the supermarket, however, he could be more attracted to that particular brand.

Semantic Memory. Semantic memory, a form of declarative memory, refers basically to "general knowledge of the world." This type of memory provides the knowledge of something when it is no longer actually being perceived. Semantic memory provides the basic material required for thinking.

The hippocampus and deep, inner part of the temporal lobe, both parts of the limbic system, are required for the formation of semantic memories and other types of declarative memory.

Working Memory. **Working memory,** also known as primary or short-term memory, registers and retains incoming information for a short time after its input. In other words, it is the memory that we use when we keep information consciously "in mind." Working memory makes possible a temporary impression of one's present environment in a readily accessible form.

Even working memory seems not to be a single entity but rather to have several components: one part deals with storing incoming visual information; a second deals both with memories for sounds and with attaching names to objects or events being dealt with in the memory; and the third provides a link between the first two components and other aspects of consciousness such as focused attention. The third component becomes increasingly active with difficult memory tasks and problem solving. Concerning the second component, it is interesting that successful working memory requires only the motor "programming" aspect of speech and not the actual behavior of physically speaking.

The component of working memory that focuses attention is essential for many memory-based skills. The longer the span of attention in working memory, the better the chess player, the greater the ability to reason, and the better a student is at understanding complicated sentences and drawing inferences from texts. In fact, there is a strong correlation between working memory and standard measures of intelligence. Conversely, the specific memory deficit in victims of Alzheimer's disease, a condition marked by serious memory losses, may be in this attention-focusing component of working memory.

Although all forms of working memory seem to involve the **prefrontal cortex** of the frontal lobes (Figure 13-9), discrete regions seem to be specialized for dealing with specific kinds of information. Thus, one area of the prefrontal cortex encodes information about the location of objects, whereas a different area encodes information about the objects' color, size, and shape.

Other cortical areas are involved in working memory as well. For example, neurons of those cortical association areas that deal with the high-level integration of sensory input are active during working memory, especially during simpler memory tasks. Thus, neurons of the temporal lobe association areas are active when remembering the "what is it" aspect of a visual stimulus (Chapter 9), whereas neurons of the parietal association areas are active when remembering the location, that is, the "where is it," of that same stimulus. The more complex the task, as for example when trying to remember several aspects of a stimulus or when holding on to the memory for a longer time, the more prefrontal cortex is involved.

Prefrontal cortex is a major target of not only the dopamine (mesolimbic) system but also an acetylcholine system that originates in the brainstem as a component of the reticular activating system. Afferents from the dopamine system often synapse on the same prefrontal neurons that receive excitatory sensory inputs, and drugs that are dopamine antagonists interfere with working memory.

Episodic Memory.   Episodic memory is engaged by people when they remember their own experiences as they happened in a specific place and time of their personal history. These memories include a unique, subjective aspect that distinguishes them from perceptual representations. Episodic memories, like semantic memories, are declarative memories and involve the hippocampus and medial temporal lobe.

Additional Facts Concerning Learning and Memory.   Certain types of learning depend not only on factors such as attention, motivation, and various neurotransmitters but also on certain hormones. For example, the hormones epinephrine, ACTH, and vasopressin affect the retention of learned experiences. These hormones are normally released in stressful or even mildly stimulating experiences, suggesting that the hormonal consequences of our experiences affect our memories of them.

Two of the opioid peptides, enkephalin and endorphin, interfere with learning and memory, particularly when the lesson involves a painful stimulus. They may inhibit learning simply because they decrease the emotional (fear, anxiety) component of the painful experience associated with the learning situation, thereby decreasing the motivation necessary for learning to occur.

Memories can be consolidated very rapidly, sometimes after just one trial, and they can be retained over extended periods. Information can be retrieved from memory stores after long periods of disuse, and the common notion that memory, like muscle, atrophies with lack of use is not always true. Also, unlike working memory, some types of memory apparently have an unlimited capacity because people's memories never seem to be so full that they cannot learn something new.

Although we have mentioned specific areas of the brain that are active in learning, we want to stress at this time the following point. Memory traces are laid down in specific neural systems throughout the brain, and different types of memory tasks utilize different systems. For even a simple memory task, such as trying to recall a certain word from a previously seen word list, different specific parts of the brain are activated in sequence. It is as though there are several small "processors" linked together in a memory system for that type of memory task.

The Neural Basis of Learning and Memory

Not all types of memory have the same neural mechanisms. Conditions such as coma, deep anesthesia, electroconvulsive shock, and insufficient blood supply to the brain, all of which interfere with the electrical activity of the brain, interfere with working memory. Thus, it is assumed that working memory exists in the form of ongoing graded or action potentials. Working memory is interrupted when a person becomes unconscious from a blow on the head, and memories are abolished for all that happened for a variable period of time before the blow, so-called **retrograde amnesia.** (**Amnesia** is defined as the loss of memory.) Working memory is also susceptible to external interference, such as an attempt to learn conflicting information. On the other hand, other forms of memory can survive deep anesthesia, trauma, or electroconvulsive shock, all of which disrupt the normal patterns of neural conduction in the brain.

As stated earlier, memory consolidation involves cellular or molecular changes in one or more neurons, and these changes constitute the memory trace. The changes are initiated by electrical activity in the neurons, but the question remains: What happens next?

One model for memory consolidation is **long-term potentiation (LTP),** in which certain synapses undergo a long-lasting increase in their effectiveness when they are heavily used. This increased efficacy, which results in increased firing in the *postsynaptic* cell, can last minutes, hours, or even weeks. Although LTP occurs at many excitatory synapses in the central nervous system, it takes place most prominently in the hippocampus which, as we have just seen, is essential for some types of learning. Long-term potentiation results from increased activation of glutamate receptors on the postsynaptic cell, which opens Ca^{2+} channels in the receptor, and an increase of intracellular calcium. The calcium enhances the enzymatic formation in the postsynaptic cell of a substance—probably nitric oxide—that diffuses back across the synapse to

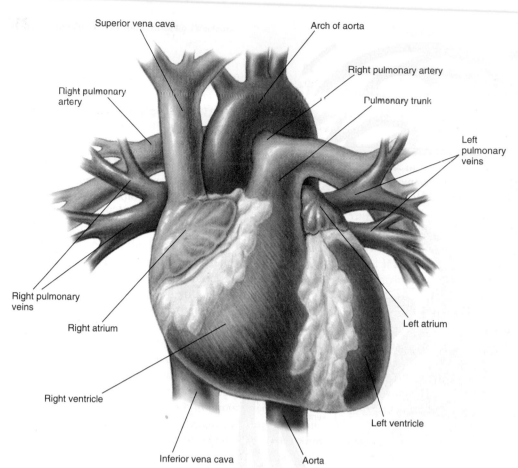

Superior vena cava
Arch of aorta
Right pulmonary artery
Right pulmonary artery
Pulmonary trunk
Left pulmonary veins
Right pulmonary veins
Right atrium
Left atrium
Right ventricle
Left ventricle
Inferior vena cava
Aorta

FIGURE 14-8
Blood leaves each of the ventricles via a single artery, the pulmonary trunk from the right ventricle and the aorta from the left ventricle. [Because the aorta and pulmonary trunk cross each other before emerging from the heart (Figure 14-12), one gets the mistaken notion that both arise from the right ventricle.] Blood enters the right atrium via two large veins, the superior vena cava and inferior vena cava; it enters the left atrium via four pulmonary veins. ✗

	Rest ml/min
Brain	650 (13%)
Heart	215 (4%)
Muscle	1030 (20%)
Skin	430 (9%)
Kidney	950 (20%)
Abdominal organs	1200 (24%)
Other	525 (10%)
Total	5000

FIGURE 14-9

Distribution of systemic blood flow to the various organs and tissues of the body at rest. (*Adapted from Chapman and Mitchell.*)

It must be emphasized that it is not the absolute pressure at any point in the cardiovascular system that determines flow rate but the *difference* in pressure between the relevant points (Figure 14-10).

Knowing only the pressure difference between two points will not tell you the flow rate, however. For this, you also need to know the **resistance (R)** to flow, that is, how difficult it is for blood to flow between two points *at any given pressure difference*. Resistance is the measure of the friction that impedes flow. The basic equation relating these variables is:

$$F = \Delta P/R \qquad (14\text{-}1)$$

In words, flow rate is directly proportional to the pressure difference between two points and inversely proportional to the resistance. This equation applies not only to the cardiovascular system but to any system in which liquid or air moves by bulk flow, for example, in the urinary and respiratory systems, respectively.

Resistance cannot be measured directly, but it can be calculated from the directly measured F and ΔP. For example, in Figure 14-10 the resistances in both tubes can be calculated to be 90 mmHg ÷ 10 ml/min = 9 mmHg/ml per minute.

This example illustrates how resistance can be *calculated*, but what is it that actually *determines* the resis-

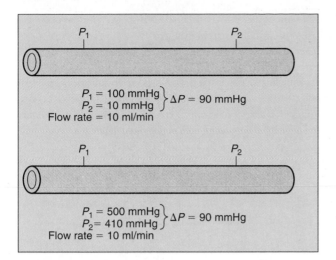

FIGURE 14-10

Flow between two points within a tube is proportional to the pressure difference between the points. The flows in these two identical tubes are the same, because the pressure differences are the same.

tance? (The distinction between how a thing is calculated or measured and its determinants may seem confusing, but consider the following: By standing on a scale you *measure* your weight, but your weight is not *determined* by the scale but rather by how much you eat and exercise, and so on.) One determinant of resistance is the fluid property known as **viscosity,** which is a function of the friction between adjacent layers of a flowing fluid; the greater the friction, the greater the viscosity. The other determinants of resistance are the length and radius of the structure through which the fluid is flowing, since these characteristics determine the amount of friction between the fluid and the structure's wall. The following equation defines the contributions of these three determinants:

$$R = (\eta L/r^4)\,(8/\pi) \qquad (14\text{-}2)$$

where η = fluid viscosity

L = length of the structure

r = inside radius of the structure

$8/\pi$ = a constant

In other words, resistance is directly proportional to both the fluid viscosity and the structure's length, and inversely proportional to the fourth power of the structure's radius, that is, the radius multiplied by itself four times.

Blood viscosity is not fixed but increases as hematocrit increases, and changes in hematocrit, therefore, can have significant effects on the resistance to flow in certain situations. Under most physiological conditions, however, the hematocrit and, hence, viscosity of blood is relatively constant and does not play a role in the *control* of resistance.

Similarly, since the lengths of the blood vessels remain constant in the body, length is also not a factor in the control of resistance along these vessels. In contrast, as we shall see, the radii of the blood vessels do not remain constant, and so vessel radius—the $1/r^4$ term in our equation—is the most important determinant of changes in resistance along the blood vessels. Just how important changes in radius can be is illustrated in Figure 14-11: Decreasing the radius of a tube *twofold* increases its resistance *sixteenfold*. If ΔP is held constant in this example, flow through the tube decreases sixteenfold since $F = \Delta P/R$.

Because resistance in the cardiovascular system is so often discussed in the context of blood vessels—the "tubes"—it is easy to forget that the equation relating

FIGURE 14-11

Effect of tube radius (r) on resistance (R) and flow.

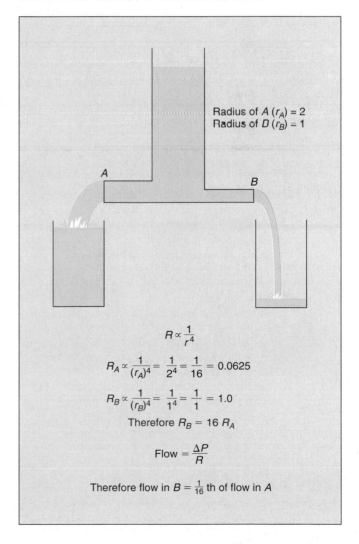

Radius of A (r_A) = 2
Radius of B (r_B) = 1

$$R \propto \frac{1}{r^4}$$

$$R_A \propto \frac{1}{(r_A)^4} = \frac{1}{2^4} = \frac{1}{16} = 0.0625$$

$$R_B \propto \frac{1}{(r_B)^4} = \frac{1}{1^4} = \frac{1}{1} = 1.0$$

Therefore $R_B = 16\,R_A$

$$\text{Flow} = \frac{\Delta P}{R}$$

Therefore flow in $B = \frac{1}{16}$ th of flow in A

pressure, flow, and resistance applies not only to flow through blood vessels but to the flows into and out of the various chambers of the heart. As we shall see, these flows occur through valves, and the resistance offered by a valvular opening determines the flow through the valve at any given pressure difference across it.

This completes our introductory survey of the cardiovascular system. We now turn to a description of its components and their control. In so doing, we might very easily lose sight of the forest for the trees if we do not persistently ask of each section: How does this component of the circulation contribute to adequate blood flow through the capillaries of the various organs or to an adequate exchange of materials between blood and cells? Refer to the summary in Table 14-4 as you read the description of each component to keep focused on this question.

TABLE 14-4 THE CARDIOVASCULAR SYSTEM

Component	Function
Heart	
Atria	Chambers through which blood flows from veins to ventricles. Atrial contraction adds to ventricular filling but is not essential for it.
Ventricles	Chambers whose contractions produce the pressures that drive blood through the pulmonary and systemic vascular systems and back to the heart.
Vascular system	
Arteries	Low-resistance tubes conducting blood to the various organs with little loss in pressure. They also act as pressure reservoirs for maintaining blood flow during ventricular relaxation.
Arterioles	Major sites of resistance to flow; responsible for the pattern of blood-flow distribution to the various organs; participate in the regulation of arterial blood pressure.
Capillaries	Sites of nutrient, metabolic end product, and fluid exchange between blood and tissues.
Venules	Sites of nutrient, metabolic end product, and fluid exchange between blood and tissues.
Veins	Low-resistance conduits for blood flow back to the heart. Their capacity for blood is adjusted to facilitate this flow.

SECTION B SUMMARY

I. The cardiovascular system consists of two circuits: the pulmonary circulation, from the right ventricle to the lungs and then to the left atrium, and the systemic circulation, from the left ventricle to all peripheral organs and tissues and then to the right atrium.

II. Arteries carry blood away from the heart, and veins carry blood toward the heart.
   A. In the systemic circuit, the large artery leaving the left heart is the aorta, and the large veins emptying into the right heart are the superior vena cava and inferior vena cava. The analogous vessels in the pulmonary circulation are the pulmonary trunk and the four pulmonary veins.
   B. The microcirculation consists of the vessels between arteries and veins: the arterioles, capillaries, and venules.

III. Flow between two points in the cardiovascular system is directly proportional to the pressure difference between the points and inversely proportional to the resistance: $F = \Delta P/R$.

IV. Resistance is directly proportional to the viscosity of a fluid and to the length of the tube. It is inversely proportional to the fourth power of the tube's radius, which is the major variable controlling changes in resistance.

SECTION B KEY TERMS

bulk flow	venule
atrium	microcirculation
ventricle	inferior vena cava
pulmonary circulation	superior vena cava
systemic circulation	pulmonary trunk
artery	pulmonary arteries
vein	pulmonary veins
aorta	hydrostatic pressure
arteriole	resistance (R)
capillary	viscosity

SECTION B REVIEW QUESTIONS

1. State the formula relating flow, pressure difference, and resistance.

2. What are the three determinants of resistance?

SECTION C
THE HEART

ANATOMY

The heart is a muscular organ enclosed in a fibrous sac, the **pericardium,** and located in the chest (thorax). The narrow space between the pericardium and the heart is filled with a watery fluid that serves as a lubricant as the heart moves within the sac.

The walls of the heart are composed primarily of cardiac muscle cells and are termed the **myocardium.** The inner surface of the walls, that is, the surface in contact with the blood within the cardiac chambers, is lined by a thin layer of cells known as **endothelial cells,** or **endothelium.** (As we shall see, endothelial cells line not only the heart chambers, but the entire vascular system.)

As noted earlier, the human heart is divided into right and left halves, each consisting of an atrium and a ventricle. Located between the atrium and ventricle in each half of the heart are the **atrioventricular (AV) valves,** which permit blood to flow from atrium to ventricle but not from ventricle to atrium (Figure 14-12). The right AV valve is called the **tricuspid valve,** and the left is called the **mitral valve.**

The opening and closing of the AV valves is a passive process resulting from pressure differences across the valves. When the blood pressure in an atrium is greater than that in the ventricle separated from it by a valve, the valve is pushed open and flow proceeds from atrium to ventricle. In contrast, when a contracting ventricle achieves an internal pressure greater than that in its connected atrium, the AV valve between them is forced closed. Therefore, blood does not normally move back into the atria but is forced into the pulmonary trunk from the right ventricle and into the aorta from the left ventricle.

To prevent the AV valves from being pushed up into the atrium, the valves are fastened to muscular projections (**papillary muscles**) of the ventricular walls by fibrous strands (chordae tendinae). The papillary muscles do *not* open or close the valves. They act only to limit the valves' movements and prevent them from being everted.

The opening of the right ventricle into the pulmonary trunk and of the left ventricle into the aorta also contain valves, the **pulmonary** and **aortic valves** (Figure 14-12) (these valves are also collectively referred to as the semilunar valves). These valves permit blood to flow into the arteries during ventricular contraction but prevent blood

FIGURE 14-12

Diagrammatic section of the heart. The arrows indicate the direction of blood flow.

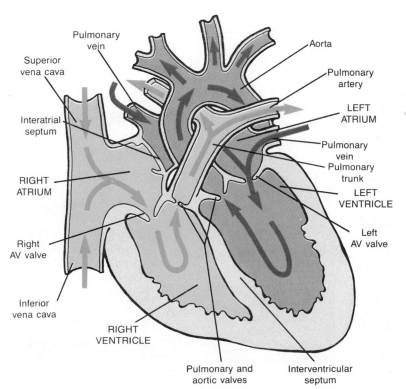

Pulmonary vein
Superior vena cava
Interatrial septum
RIGHT ATRIUM
Right AV valve
Inferior vena cava
RIGHT VENTRICLE
Pulmonary and aortic valves
Interventricular septum
Aorta
Pulmonary artery
LEFT ATRIUM
Pulmonary vein
Pulmonary trunk
LEFT VENTRICLE
Left AV valve

from moving in the opposite direction during ventricular relaxation (Figure 14-10). Like the AV valves, they act in a purely passive manner: Their being open or closed depends upon the pressure differences across them.

Another important point concerning the heart valves is that, when open, they offer very little resistance to flow. Accordingly, very small pressure differences across them suffice to produce large flows. In disease states, however, a valve may become narrowed so that even when open it offers a high resistance to flow; in such a state, the contracting cardiac chamber must produce an unusually high pressure to cause flow across the valve.

There are no valves at the entrances of the superior and inferior venae cavae (plural of vena cava) into the right atrium, and of the pulmonary veins into the left atrium. However, atrial contraction pumps very little blood back into the veins because atrial contraction compresses the veins at their sites of entry into the atria, greatly increasing the resistance to backflow. (Actually, a little blood is ejected back into the veins, and this accounts for the venous pulse that can often be seen in the neck veins when the atria are contracting.)

Figure 14-14 summarizes the path of blood flow through the entire cardiovascular system.

Cardiac Muscle

The cardiac-muscle cells of the myocardium are arranged in layers that are tightly bound together and completely encircle the blood-filled chambers. When the walls of a chamber contract, they come together like a squeezing fist and exert pressure on the blood they enclose.

Cardiac muscle combines properties of both skeletal and smooth muscle (Chapter 11). The cells are striated (Figure 14-15) as the result of an arrangement of thick myosin and thin actin filaments similar to that of skeletal muscle. Cardiac-muscle cells are considerably shorter than skeletal-muscle fibers, however, and have several branching processes. Adjacent cells are joined end to end at structures called intercalated disks, within which are desmosomes that hold the cells together and to which the myofibrils are attached. Adjacent to the intercalated disks are gap junctions, similar to those in many smooth muscles.

Approximately 1 percent of the cardiac-muscle cells do not function in contraction, but have specialized features

FIGURE 14-13

Photographs of the pulmonary valve viewed from the top, that is, from the pulmonary trunk looking down into the right ventricle. On the left the valve is partly open and blood would be flowing through it from the right ventricle into the pulmonary trunk, that is, toward the viewer. When completely open the cusps of the valve are fully flattened against the inside of the vessel by the flowing blood. On the right, the valve is almost completely closed; this occurs when the cusps are forced together by the downward pressure of the blood, that is, by the pressure of the blood in the pulmonary trunk being greater than the pressure in the right ventricle. *[From R. Carola, J.P. Harley, and C.R. Noback, "Human Anatomy and Physiology," McGraw-Hill, New York, 1990 (photos by Dr. Wallace McAlpine).]*

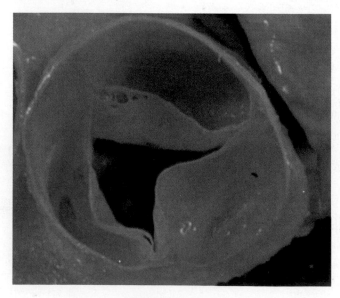

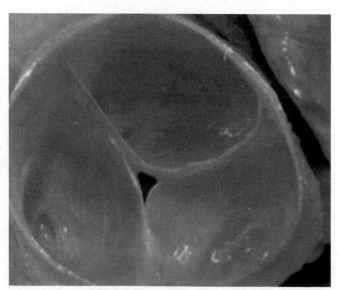

Valve partly open   Valve almost completely closed

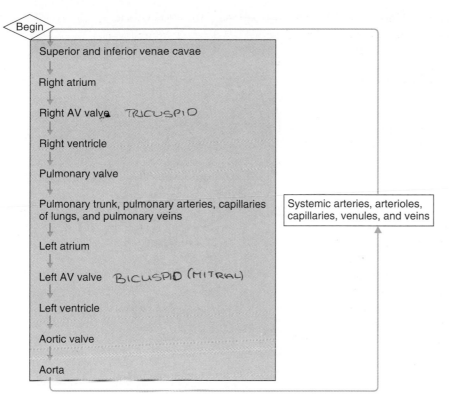

FIGURE 14-14

Path of blood flow through the entire cardiovascular system. All the structures within the colored box are located in the chest.

Begin

Superior and inferior venae cavae

Right atrium

Right AV valve   *TRICUSPID*

Right ventricle

Pulmonary valve

Pulmonary trunk, pulmonary arteries, capillaries of lungs, and pulmonary veins

Left atrium

Left AV valve   *BICUSPID (MITRAL)*

Left ventricle

Aortic valve

Aorta

Systemic arteries, arterioles, capillaries, venules, and veins

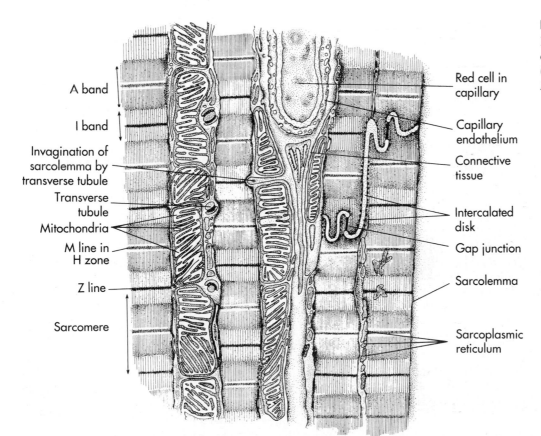

FIGURE 14-15

Diagram of an electron micrograph of cardiac muscle. (*From R. M. Berne and M. N. Levy.*)

A band

I band

Invagination of sarcolemma by transverse tubule

Transverse tubule

Mitochondria

M line in H zone

Z line

Sarcomere

Red cell in capillary

Capillary endothelium

Connective tissue

Intercalated disk

Gap junction

Sarcolemma

Sarcoplasmic reticulum

that are essential for normal heart excitation. These cells constitute a network known as the **conducting system** of the heart and are in contact with the other cardiac-muscle cells via gap junctions. The conducting system initiates the heartbeat and helps spread the impulse rapidly throughout the heart.

One final point about the cardiac-muscle cells is that certain cells in the atria secrete the family of peptide hormones collectively called atrial natriuretic factor, described in Chapter 16.

Innervation. The heart receives a rich supply of sympathetic and parasympathetic nerve fibers, the latter contained in the vagus nerves. The sympathetic postganglionic fibers release primarily norepinephrine, and the parasympathetics release primarily acetylcholine. The receptors for norepinephrine on cardiac muscle are mainly beta-adrenergic. The hormone epinephrine, from the adrenal medulla, combines with the same receptors as norepinephrine and exerts the same actions on the heart. The receptors for acetylcholine are of the muscarinic type.

Blood Supply. The blood being pumped through the heart chambers does not exchange nutrients and metabolic end products with the myocardial cells. They, like the cells of all other organs, receive their blood supply via arteries that branch from the aorta. The arteries supplying the myocardium are the **coronary arteries,** and the blood flowing through them is termed the **coronary blood flow.** The coronary arteries exit from the very first part of the aorta and lead to a branching network of small arteries, arterioles, capillaries, venules, and veins similar to those in all other organs. Most of the coronary veins drain into a single large vein, the coronary sinus, which empties into the right atrium.

HEARTBEAT COORDINATION

The heart is, in essence, a dual pump in that the atria contract first, followed almost immediately by the ventricles. Contraction of cardiac muscle, like that of skeletal muscle and many smooth muscles (Chapter 11), is triggered by depolarization of the plasma membrane. As described earlier, myocardial cells are connected to each other by gap junctions that allow action potentials to spread from one cell to another. Thus, the initial excitation of one cardiac cell eventually results in the excitation of all cardiac cells. (Because of this, the heart is said to constitute a functional syncytium.) This initial depolarization normally arises in a small group of conducting-system cells, the **sinoatrial (SA) node,** located in the right atrium near the entrance of the superior vena cava (Figure 14-16). The action potential then spreads from the SA node throughout the

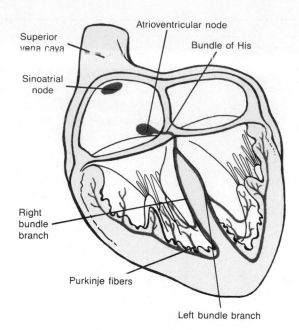

FIGURE 14-16
Conducting system of the heart.

atria and then into and throughout the ventricles. This pattern raises two questions: (1) What causes the SA node to "fire," and (2) precisely what is the path of spread of excitation? We'll deal first with the second question and then return to the first question in the next section.

Sequence of Excitation

To reiterate, the SA node is the normal pacemaker for the entire heart. Its depolarization normally generates the current that leads to depolarization of all other cardiac muscle cells, and so its discharge rate determines the **heart rate,** the number of times the heart contracts per minute.

The action potential initiated in the SA node spreads throughout the myocardium, passing from cell to cell by way of gap junctions. The spread throughout the right atrium and from the right atrium to the left atrium does not depend on fibers of the conducting system. The spread is rapid enough that the two atria are depolarized and contract at essentially the same time.

The spread of the action potential to the ventricles is more complicated and involves the rest of the conducting system (Figures 14-16 and 14-17). The link between atrial depolarization and ventricular depolarization is a portion of the conducting system called the **atrioventricular (AV) node,** which is located at the base of the right atrium. The action potential spreading through the right atrium causes depolarization of the AV node. This node manifests a particularly important characteristic: For several reasons related to the electrical properties of the AV-node cells, the

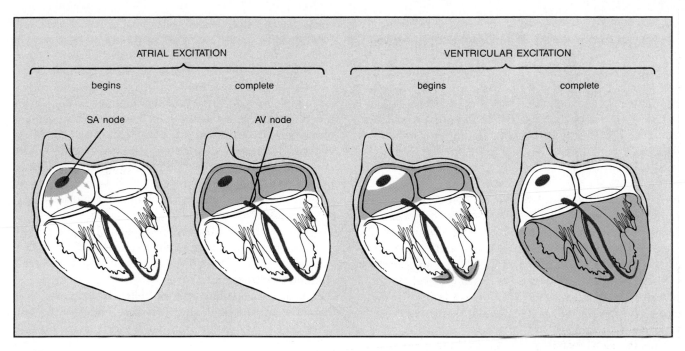

FIGURE 14-17

Sequence of cardiac excitation. The blue color denotes areas that are depolarized. Impulse spread from right atrium to left atrium is via the atrial muscle cells where the atria contact each other in their shared wall. (*Adapted from Rushmer.*)

propagation of action potentials through the AV node is relatively slow (requiring approximately 0.1 s). This delay allows atrial contraction to be completed before ventricular excitation occurs.

After leaving the AV node, the impulse enters the wall between the two ventricles (the interventricular septum) via the conducting-system fibers termed the **bundle of His** (or atrioventricular bundle) after its discoverer (pronounced Hiss). It should be emphasized that the AV node and the bundle of His constitute the only electrical link between the atria and the ventricles. There are no others because a layer of nonconducting connective tissue, pierced by the bundle of His, completely separates each atrium from its ventricle.

Within the interventricular septum the bundle of His divides into **right** and **left bundle branches,** which eventually leave the septum to enter the walls of both ventricles. These fibers in turn make contact with **Purkinje fibers,** large conducting cells that rapidly distribute the impulse throughout much of the ventricles. Finally, the Purkinje fibers make contact with non-conducting-system ventricular myocardial cells, via which the impulse spreads through the rest of the ventricles.

The rapid conduction along the Purkinje fibers and the diffuse distribution of these fibers cause depolarization of all right and left ventricular cells more or less simultaneously and ensure a single coordinated contraction. Actually, depolarization and contraction begin slightly earlier in the bottom (apex) of the ventricles and spread upward. The result is a more efficient contraction, like squeezing a tube of toothpaste from the bottom up.

Cardiac Action Potentials

A typical ventricular myocardial cell action potential is illustrated in Figure 14-18A. The membrane permeability changes that underlie it are shown in Figure 14-18B. As in skeletal-muscle cells and neurons, the resting membrane is much more permeable to potassium than to sodium. Therefore, the resting membrane potential is much closer to the potassium equilibrium potential (−90 mV) than to the sodium equilibrium potential (+60 mV). Similarly, the depolarizing phase of the action potential is due mainly to a positive-feedback increase in sodium permeability caused by the opening of voltage-gated sodium channels; that is, the channels are opened by depolarization. At almost the same time, the permeability to potassium decreases as potassium channels close, and this also contributes to the membrane depolarization.

Again as in skeletal-muscle cells and neurons, the increased sodium permeability is very transient, since the sodium channels quickly close again. Unlike the case in these other excitable tissues, however, in cardiac muscle the return of sodium permeability toward its resting value is *not* accompanied by membrane repolarization. The membrane remains depolarized at a plateau of about 0 mV (Figure 14-18A). The reasons for this continued depolarization are: (1) potassium permeability stays below the resting

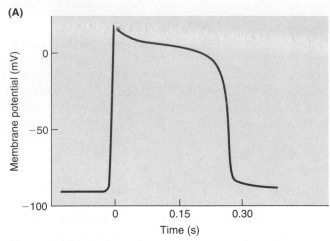

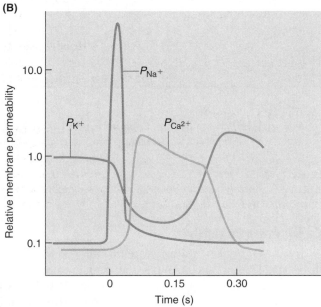

FIGURE 14-18

(A) Membrane potential recording from a ventricular muscle cell. (B) Simultaneously measured permeabilities P to potassium, sodium, and calcium during the action potential of (A).

value, that is, the potassium channels remain closed, and (2) there is a marked increase in the membrane permeability to *calcium*. The second reason is the more important of the two, and the explanation for it is as follows.

In myocardial cells, the original membrane depolarization causes voltage-gated calcium channels in the plasma membrane to open, which results in a flow of calcium ions down their electrochemical gradient into the cell. These channels are referred to as **slow channels** because there is a delay in their opening (Figure 14-18B). The flow of positive calcium ions into the cell just balances the flow of positive potassium charge out of the cell and keeps the membrane depolarized at the plateau value.

Ultimately, repolarization does occur when the permeabilities of calcium and potassium return to their original state, that is, when the slow channels close and the potassium channels reopen.

The action potentials of atrial cells, except those of the SA node, are similar in shape to those just described for ventricular cells, but the duration of their plateau phase is shorter.

In contrast, there are extremely important differences between action potentials of the vast majority of the atrial and ventricular myocardial cells, as just described, and those in the conducting system. Figure 14-19B illustrates the action potentials of a myocardial cell from the SA node. Note that the resting potential of the SA-node cell is not steady but instead manifests a slow depolarization. This gradual depolarization is known as a **pacemaker potential;** it brings the membrane potential to threshold, at which point an action potential occurs. Following the peak of the action potential, the membrane repolarizes, and the gradual depolarization begins again.

Thus, the pacemaker potential provides the SA-node with **automaticity,** the capacity for spontaneous, rhythmical self-excitation. The slope of the pacemaker potential, that is, how quickly the membrane potential changes per unit time, determines how quickly threshold is reached and the next action potential elicited. The inherent rate of

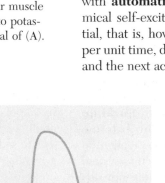

FIGURE 14-19

Comparison of action potentials in a ventricular muscle cell (A, from Figure 14-18) and (B) a sinoatrial (SA)-node cell. The most important difference is the presence of the pacemaker potential in the SA node.

the SA node—the rate exhibited in the total absence of any neural or hormonal input to the node—is approximately 100 depolarizations per minute.

What is responsible for the pacemaker potential? The major cause of this gradual depolarization is movement of sodium ions into the cells through a special set of voltage-gated plasma-membrane channels that are opened by the *repolarizing* phase of the *preceding* action potential. (Recall that the more common voltage-gated sodium channels of nerve, skeletal muscle, and non-conducting-system cardiac muscle are opened by depolarization, not repolarization.)

Several other portions of the conducting system are capable of generating pacemaker potentials, but the inherent rate of these other areas is slower than that of the SA node, and so they normally are "captured" by the SA node and do not manifest their own rhythm. However, they can do so under certain circumstances and are then termed **ectopic pacemakers.** For example, recall that excitation travels from the SA node to both ventricles only through the AV node; therefore, drug- or disease-induced malfunction of the AV node may reduce or completely eliminate the transmission of action potentials from the atria to the ventricles. If this occurs, autorhythmic cells in the bundle of His, no longer driven by the SA node, begin to initiate excitation at their own inherent rate and become the pacemaker for the ventricles. Their rate is quite slow, generally 25 to 40 beats/min, and it is completely out of synchrony with the atrial contractions, which continue at the normal, higher rate of the SA node. Under such conditions, the atria are ineffective as pumps since they are often contracting against closed AV valves. Fortunately, atrial pumping, as we shall see, is relatively unimportant for cardiac function except during strenuous exercise. The current treatment for all severe **AV conduction disorders,** as well as for many other abnormal rhythms is permanent surgical implantation of an electrical device, a pacemaker, that stimulates the ventricular cells at a normal rate.

The Electrocardiogram

The **electrocardiogram** (**ECG** or EKG—the k is from the German "kardio" for heart) is primarily a tool for evaluating the *electrical* events within the heart. The action potentials of cardiac muscle cells can be viewed as batteries that cause charge to move throughout the body fluids. These moving charges—currents, in other words—represent the sum of the action potentials occurring simultaneously in many individual myocardial cells and can be detected by recording electrodes at the surface of the skin. Figure 14-20 illustrates a typical normal ECG recorded as the potential difference between the right and left wrists. The first deflection, the **P wave,** corresponds to current flows during atrial depolarization. The second deflection, the **QRS complex,** occurring approximately 0.15 s later, is

the result of ventricular depolarization. It is a complex deflection because the paths taken by the wave of depolarization through the thick ventricular walls differ from instant to instant, and the currents generated in the body fluids change direction accordingly. Regardless of its form, for example, the Q and/or S portions may be absent, the deflection is still called a QRS complex. The final deflection, the **T wave,** is the result of ventricular repolarization. Atrial repolarization is usually not evident on the ECG because it occurs at the same time as the QRS complex.

A typical clinical ECG makes use of multiple combinations of recording locations on the limbs and chest so as to obtain as much information as possible concerning different areas of the heart. The shapes and sizes of the P wave, QRS complex, and T wave vary with the electrode locations.

To reiterate, the ECG is not a direct record of the changes in membrane potential across individual cardiac muscle cells but is rather a measure of the currents generated

FIGURE 14-20

(Top) Typical electrocardiogram recorded from electrodes connecting the arms. P, atrial depolarization; QRS, ventricular depolarization; T, ventricular repolarization. (Bottom) Ventricular action potential recorded from a single ventricular muscle cell. Note the correspondence of the QRS complex with depolarization and the correspondence of the T wave with repolarization.

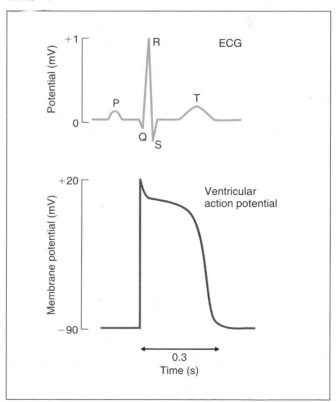

in the extracellular fluid by the changes occurring simultaneously in many cells. To emphasize this point, Figure 14-20 shows, in addition to an ECG, the simultaneously occurring changes in membrane potential in a single ventricular cell.

Because many myocardial defects alter normal impulse propagation, and thereby the shapes and timing of the waves, the ECG is a powerful tool for diagnosing certain types of heart disease. Figure 14-21 gives one example. It must be emphasized, however, that the ECG provides information concerning only the electrical activity of the heart. Thus, if something is wrong with the heart's mechanical activity, but this defect does not give rise to altered electrical activity, then the ECG will not be of diagnostic value.

Excitation-Contraction Coupling

As described in Chapter 11, the mechanism that couples excitation—an action potential in the plasma membrane of the muscle cell—and contraction is an increase in the cell's cytosolic calcium concentration. As is true for skeletal muscle, the increase in cytosolic calcium concentration in cardiac muscle is due mainly to release of calcium from the sarcoplasmic reticulum; this calcium combines with the regulator protein troponin, and cross-bridge formation between actin and myosin is initiated.

FIGURE 14-21

Electrocardiograms from a healthy person and from two persons suffering from atrioventricular block. (A) A normal ECG. (B) Partial block. Damage to the AV node permits only one-half of the atrial impulses to be transmitted to the ventricles. Note that every second P wave is not followed by a QRS and T. (C) Complete block. There is absolutely no synchrony between atrial and ventricular electrical activities, and the ventricles are being driven by a pacemaker in the bundle of His. ◀▭▶

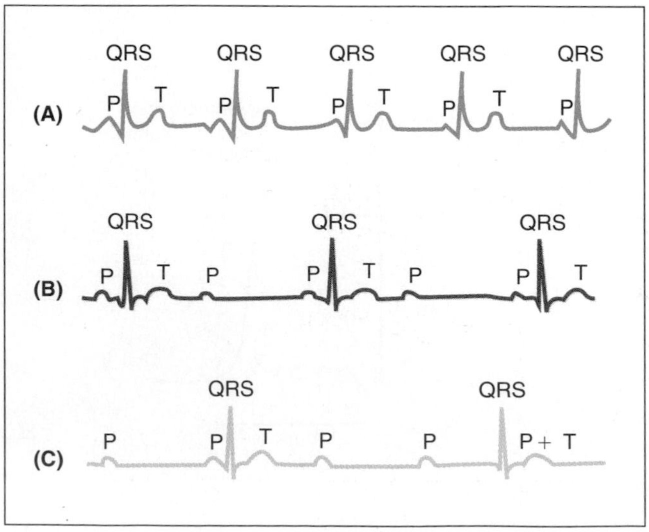

But there is a difference between skeletal and cardiac muscle in the sequence of events by which the action potential leads to increased release of calcium from the sarcoplasmic reticulum. In both muscle types, the plasma-membrane action potential spreads into the interior of muscle cells via the T tubules (the lumen of each tubule is continuous with the extracellular fluid). In skeletal muscle, as we saw in Chapter 11, the action potential in the T tubules then causes the opening of voltage-sensitive calcium channels *in the sarcoplasmic reticulum* adjacent to the T tubules. In contrast, in cardiac muscle (Figure 14-22): (1) The action potential in the T tubule opens voltage-sensitive calcium channels *in the T tubule membrane;* calcium diffuses from the extracellular fluid through these channels into the cells, causing a small increase in cytosolic calcium concentration in the region of the T tubules and adjacent sarcoplasmic reticulum. (2) This small increase in calcium concentration then causes the opening of *calcium-sensitive channels* in the sarcoplasmic reticulum, resulting in the release of a large amount of calcium from this organelle. (3) It is mainly this calcium that causes the contraction. Thus, even though most of the calcium causing contraction comes from the sarcoplasmic reticulum, the process—unlike that in skeletal muscle—is dependent on the movement of *extracellular* calcium into the muscle, the calcium then acting as the signal for release of the sarcoplasmic reticulum calcium.

Contraction ends when the cytosolic calcium concentration is restored to its original extremely low value by active transport of calcium back into the sarcoplasmic reticulum. Also, an amount of calcium equal to the small amount that had entered the cell from the extracellular fluid during excitation is transported out of the cell, so that the total cellular calcium content remains constant.

As we shall see, how much cytosolic calcium concentration increases during excitation is a major determinant of the strength of cardiac muscle contraction. In this regard, cardiac muscle differs importantly from skeletal muscle, in which the increase in cytosolic calcium occurring during membrane excitation is always adequate to produce maximal "turning-on" of cross bridges by calcium binding to all troponin sites. In cardiac muscle, the amount of calcium released from the sarcoplasmic reticulum is not usually sufficient to saturate all troponin sites. Therefore, the number of active cross bridges and thus the strength of contraction can be increased still further if more calcium is released from the sarcoplasmic reticulum.

Refractory Period of the Heart

Ventricular muscle, unlike skeletal muscle, is incapable of any significant degree of summation of contractions, and this is a very good thing. Imagine that cardiac muscle is able to undergo a prolonged tetanic contraction. During this period no ventricular filling could occur since filling

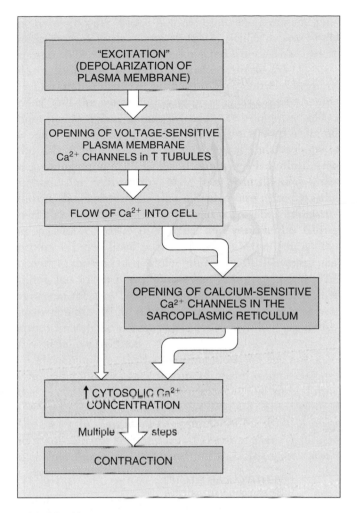

FIGURE 14-22

Excitation-contraction coupling in cardiac muscle. Calcium released from the sarcoplasmic reticulum is the major source of the increased cytosolic calcium. The signal for this release is extracellular calcium diffusing into the cell across voltage-sensitive calcium channels in the T tubules.

can occur only when the ventricular muscle is relaxed, and the heart would therefore cease to function as a pump.

The inability of the heart to generate tetanic contractions is the result of the long **refractory period** of cardiac muscle, defined as the period during and following an action potential when an excitable membrane cannot be re-excited. As described in Chapter 11, the absolute refractory periods of skeletal muscle are much shorter (1 to 2 ms) than the duration of contraction (20 to 100 ms), and a second contraction can therefore be elicited before the first is over (summation of contractions). In contrast, because of the long plateau in the cardiac muscle action potential, the absolute refractory period of cardiac muscle lasts almost as long as the contraction (250 ms), and the muscle cannot be re-excited in time to produce summation (Figure 14-23).

MECHANICAL EVENTS OF THE CARDIAC CYCLE

The orderly process of depolarization described in the previous sections triggers a recurring **cardiac cycle** of atrial and ventricular contractions and relaxations (Figure 14-24). For orientation, we shall first merely name the parts of this cycle and their key events. Then we shall go through the cycle again, this time describing the pressure and volume changes that cause the events.

The cycle is divided into two major phases, both named for events in the ventricles: the period of ventricular contraction and blood ejection, **systole,** followed by the period of ventricular relaxation and blood filling, **diastole.** At an average heart rate of 72 beats/min, each cardiac cycle lasts approximately 0.8 s, with 0.3 s in systole and 0.5 s in diastole.

As illustrated in Figure 14-24, both systole and diastole can be subdivided into two discrete periods. During the first part of systole, the ventricles are contracting but all valves in the heart are closed and so no blood can be ejected. This period is termed **isovolumetric ventricular contraction** because the ventricular volume is constant. The ventricular walls are developing tension and squeezing on the blood they enclose, raising the ventricular blood pressure, but because the volume of blood in the ventricles is constant and because blood, like water, is essentially incompressible, the ventricular muscle fibers cannot shorten. Thus, isovolumetric ventricular contraction is analogous to an isometric skeletal-muscle contraction: the muscle develops tension, but does not shorten.

Once the rising pressure in the ventricles exceeds that in the aorta and pulmonary trunk, the aortic and pulmonary valves open and the **ventricular ejection** period

FIGURE 14-23

Relationship between membrane potential changes and contraction in a ventricular muscle cell. The refractory period lasts almost as long as the contraction.

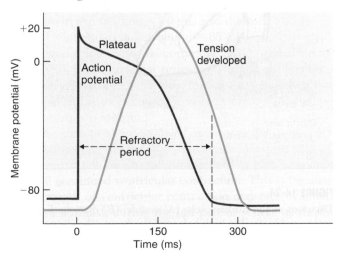

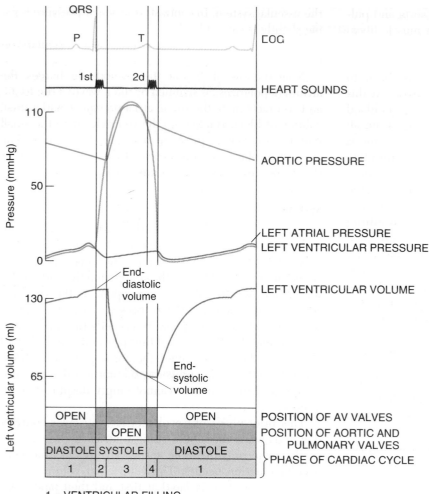

FIGURE 14-25
Summary of events in the left atrium, left ventricle, and aorta during the cardiac cycle.

1 = VENTRICULAR FILLING
2 = ISOVOLUMETRIC VENTRICULAR CONTRACTION
3 = VENTRICULAR EJECTION
4 = ISOVOLUMETRIC VENTRICULAR RELAXATION

less than the rate at which blood is leaving the aorta. Accordingly, the volume and therefore the pressure in the aorta begin to decrease.

Early Diastole

Diastole begins as ventricular contraction and ejection stop and the ventricular muscle begins to relax (recall that the T waves of the ECG corresponds to the end of the plateau phase of ventricular action potentials, and to the onset of ventricular repolarization). Immediately, the ventricular pressure falls significantly below aortic pressure, and the aortic valve closes. However, at this time, ventricular pressure still exceeds atrial pressure, so that the AV valve also remains closed. This early diastolic phase of isovolumetric ventricular relaxation ends as the rapidly decreasing ventricular pressure falls below atrial pressure, the AV valve opens, and rapid ventricular filling begins.

The ventricle's previous contraction compressed the elastic elements of this chamber in such a way that the ventricle actually tends to recoil outward once systole is over. This expansion, in turn, lowers ventricular pressure more rapidly than would otherwise occur and may even create a negative pressure in the ventricle, which enhances filling. Thus, some energy is stored within the myocardium during contraction, and its release during the subsequent relaxation aids filling.

The fact that ventricular filling is almost complete during early diastole is of the greatest importance. It ensures that filling is not seriously impaired during periods when the heart is beating very rapidly and the duration of diastole and therefore total filling time are reduced. However, when rates of approximately 200 beats/min or more are reached, filling time does become inadequate, and the volume of blood pumped during each beat is decreased. The significance of this will be described in the section on exercise.

Early ventricular filling also explains why the conduction defects that eliminate the atria as efficient pumps do not seriously impair ventricular filling, at least in otherwise normal individuals at rest. One example of this is **atrial fibrillation,** a state in which the atria contract in a continuous and completely disordered manner and so fail to serve as effective pumps. Thus, the atrium may be conveniently viewed as merely a continuation of the large veins.

Pulmonary Circulation Pressures

The pressure changes in the right ventricle and pulmonary arteries (Figure 14-26) are qualitatively similar to those just described for the left ventricle and aorta. There are striking quantitative differences, however; typical pulmonary artery systolic and diastolic pressures are 24 and 8 mmHg, respectively, compared to systemic arterial pressures of 120 and 70 mmHg. Thus, the pulmonary circulation is a low-pressure system, for reasons to be described in a later section. This difference is clearly reflected in the ventricular architecture, the right ventricular wall being much thinner than the left. Despite its lower pressure during contraction, however, the right ventricle ejects the same amount of blood as the left over a given period of time.

Heart Sounds

Two sounds, termed **heart sounds,** stemming from cardiac contraction are normally heard through a stethoscope

FIGURE 14-26

Pressures in the right ventricle and pulmonary artery during the cardiac cycle. This figure is done on the same scale as Figure 14-25 to facilitate comparison. The pressure profiles are qualitatively identical in the pulmonary and systemic circuits, but the pressures are much lower in the former.

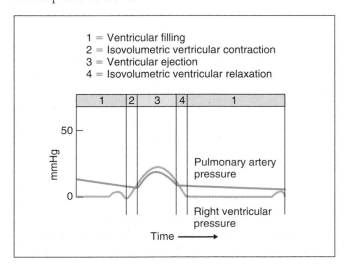

placed on the chest wall. The first sound, a soft low-pitched *lub,* is associated with closure of the AV valves at the onset of systole and isovolumetric ventricular contraction (Figure 14-24); the second sound, a louder *dup,* is associated with closure of the pulmonary and aortic valves at the onset of diastole and isovolumetric ventricular relaxation (Figure 14-24). These sounds, which result from vibrations caused by the closing valves, are perfectly normal, but other sounds, known as **heart murmurs,** are frequently a sign of heart disease.

Murmurs can be produced by blood flowing rapidly in the usual direction through an abnormally narrowed valve (**stenosis**), by blood flowing backward through a damaged, leaky valve (**insufficiency**), or by blood flowing between the two atria or two ventricles via a small hole in the wall separating them.

The exact timing and location of the murmur provide the physician with a powerful diagnostic clue. For example, a murmur heard throughout systole suggests a stenotic pulmonary or aortic valve, an insufficient AV valve, or a hole in the interventricular septum. In contrast, a murmur heard during diastole suggests a stenotic AV valve or an insufficient pulmonary or aortic valve.

THE CARDIAC OUTPUT

The volume of blood pumped by *each* ventricle per minute is called the **cardiac output (CO),** usually expressed in liters per minute. It is also the volume of blood flowing through *either* the systemic *or* the pulmonary circuit per minute.

The cardiac output is determined by multiplying the heart rate (HR)—the number of beats per minute—and the stroke volume (SV)—the blood volume ejected by each ventricle with each beat:

$$CO = HR \times SV$$

Thus, if each ventricle has a rate of 72 beats/min and ejects 70 ml of blood with each beat, the cardiac output is:

$$CO = 72 \text{ beats/min} \times 0.07 \text{ L/beat} = 5.0 \text{ L/min}$$

These values are approximately normal for a resting adult. Since, by coincidence, total blood volume is also approximately 5 L, this means that essentially all the blood is pumped around the circuit once each minute. During periods of strenuous exercise in well-trained athletes, the cardiac output may reach 35 L/min, that is, the entire blood volume is pumped around the circuit seven times a minute. Even sedentary, untrained individuals can reach cardiac outputs of 20–25 L/min during exercise.

The following description of the factors that alter the two determinants of cardiac output — heart rate and stroke volume—applies in all respects to both the right and left heart since stroke volume and heart rate are the same for both under steady-state conditions. It must also be emphasized that heart rate and stroke volume do not always change in the same direction. For example, as we shall see, stroke volume decreases following blood loss while heart rate increases. These changes produce opposing effects on cardiac output.

Control of Heart Rate

Rhythmical beating of the heart at a rate of approximately 100 beats/min will occur in the complete absence of any nervous or hormonal influences on the SA node. This is, as we have seen, the inherent autonomous discharge rate of the SA node. The heart rate may be much lower or higher than this, however, since the SA node is normally under the constant influence of nerves and hormones.

As mentioned earlier, a large number of parasympathetic and sympathetic postganglionic fibers end on the SA node. Activity in the parasympathetic (vagus) nerves causes the heart rate to decrease, whereas activity in the sympathetic nerves increases the heart rate. In the resting state, there is considerably more parasympathetic activity to the heart than sympathetic, and so the normal resting heart rate of about 70 beats/min is well below the inherent rate of 100 beats/min.

Figure 14-27 illustrates how sympathetic and parasympathetic activity influences SA-node function. Sympathetic stimulation increases the slope of the pacemaker potential, causing the SA-node cells to reach threshold more rapidly and the heart rate to increase. Stimulation of the parasympathetics has the opposite effect—the slope of the pacemaker potential decreases, threshold is reached more slowly, and heart rate decreases. Parasympathetic stimulation also hyperpolarizes the plasma membrane of the SA-node cells so that the pacemaker potential starts from a lower value.

How do the neurotransmitters released by the autonomic neurons change the slope of the pacemaker potential? They influence the ion channels through which sodium ions move into the cell to cause the diastolic depolarization; norepinephrine, the sympathetic neurotransmitter, enhances this current by opening more channels whereas acetylcholine, the parasympathetic neurotransmitter, does the opposite.

Factors other than the cardiac nerves can also alter heart rate. Epinephrine, the main hormone liberated from the adrenal medulla, speeds the heart by acting on the same beta-adrenergic receptors in the SA node as norepinephrine released from neurons. The heart rate is also

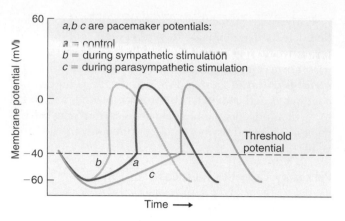

FIGURE 14-27

Effects of sympathetic and parasympathetic nerve stimulation on the slope of the pacemaker potential of an SA-nodal cell. Note that parasympathetic stimulation not only reduces the slope of the pacemaker potential but also causes the membrane potential to be more negative before the pacemaker potential begins. (*Adapted from Hoffman and Cranefield.*)

sensitive to changes in body temperature, plasma electrolyte concentrations, hormones other than epinephrine, and a metabolite—adenosine—produced by myocardial cells. These factors are normally of lesser importance, however, than the cardiac nerves. Figure 14-28 summarizes the major determinants of heart rate.

As stated in the section on innervation, sympathetic and parasympathetic neurons innervate other parts of the conducting system as well as the SA node. Thus, sympathetic stimulation not only speeds the SA node but increases conduction through the AV node. Parasympathetic stimulation, in contrast, decreases the rate of spread of excitation along almost all of the conducting system of the heart.

Control of Stroke Volume

The second variable that determines cardiac output is stroke volume, the volume of blood ejected by each ventricle during each contraction. As stated earlier, the ventricles do not completely empty themselves of blood during contraction. Therefore, a more forceful contraction can produce an increase in stroke volume by causing greater emptying. Changes in the force of contraction can be produced by a variety of factors, but two are dominant under most physiological conditions: (1) changes in the end-diastolic volume, that is, the volume of blood in the ventricles just before contraction; and (2) changes in the magnitude of sympathetic nervous system input to the ventricles.

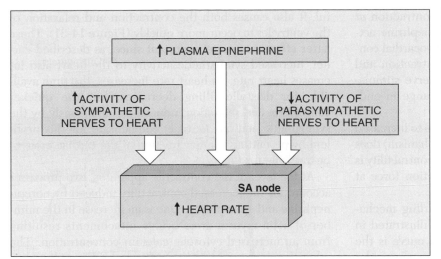

FIGURE 14-28

Major factors that influence heart rate. All effects are exerted upon the SA node. The figure shows how heart rate is increased; reversal of all the arrows in the boxes would illustrate how heart rate is decreased.

Relationship Between Ventricular End-diastolic Volume and Stroke Volume: The Frank-Starling Mechanism.

The mechanical properties of cardiac muscle are the basis for an inherent mechanism for altering stroke volume: The ventricle contracts more forcefully during systole when it has been filled to a greater degree during diastole. In other words, all other factors being equal, the stroke volume increases as the end-diastolic volume increases, as illustrated in Figure 14-29, termed a **ventricular function curve.** This relationship between stroke volume and end-diastolic volume is known as the **Frank-Starling mechanism** (also called Starling's law of the heart) in recognition of the two physiologists who identified it.

What accounts for the Frank-Starling mechanism? Basically it is simply a length-tension relationship, as described for skeletal muscle in Chapter 11, in that end-diastolic volume is a major determinant of how stretched the ventricular sarcomeres are just before contraction. Thus, the greater the end-diastolic volume, the greater the stretch, and the more forceful the contraction. A comparison of Figure 14-29 with Figure 11-25 reveals several important differences between the length-tension relationships in skeletal and cardiac muscle. First, as normal cardiac muscle is stretched, contractile force continuously rises to reach a maximum, but further stretch causes no decline in contractile force, as occurs in skeletal muscle. Second, the normal point for cardiac muscle in a resting individual is not at its optimal length for contraction, as it is for most resting skeletal muscles, but is on the rising phase of the curve; for this reason, additional stretching of the cardiac-muscle fibers by greater filling causes increased force of contraction.

The significance of the Frank-Starling mechanism is as follows: At any given heart rate, an increase in the **venous return**—the flow of blood from the veins into the heart—automatically forces an increase in cardiac output by increasing end-diastolic volume and hence stroke volume. One important function of this relationship is maintaining the equality of right and left cardiac outputs. Should the right heart, for example, suddenly begin to pump more blood than the left, the increased blood flow to the left ventricle would automatically produce an equivalent increase in left ventricular output. This ensures that blood will not accumulate in the lungs.

The Sympathetic Nerves.

Sympathetic nerves are distributed not only to the conducting system, as described in the section on heart rate, but to the entire myocardium. The effect of the sympathetic mediator norepinephrine acting on beta-adrenergic receptors is to increase ventricular

FIGURE 14-29

A ventricular function curve, which expresses the relationship between ventricular end-diastolic volume and stroke volume (the Frank-Starling mechanism). The data were obtained by progressively increasing ventricular filling pressure. The horizontal axis could have been labeled "sarcomere length," and the vertical "contractile force." In other words, this is a length-tension curve, analogous to that for skeletal muscle (Chapter 11, Figure 11-25).

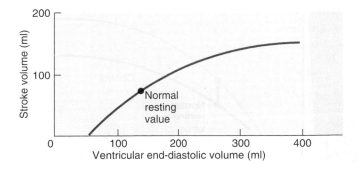

contractil

any given

ing on thes

tractility.

stroke volu

tion or ep

diastolic v

Note th

end-diasto

not reflect

specificall

any given

The re

nism and

Figure 1

same as t

function

sympathe

anism sti

volume i

other wo

complete

One

fraction

to end-

Express

average

contrac

Not

the my

FIGURE

Effects

to the h

volume

tricular

200

Stroke volume (ml)

10

MECHANICAL EVENTS OF THE CARDIAC CYCLE

I. The cardiac cycle is divided into systole (ventricular contraction) and diastole (ventricular relaxation).

   A. At the onset of systole, ventricular pressure rapidly exceeds atrial pressure, and the AV valves close. The aortic and pulmonary valves are not yet open, however, and so no ejection occurs during this isovolumetric ventricular contraction.

   B. When ventricular pressures exceed aortic and pulmonary trunk pressures, the aortic and pulmonary valves open, and ventricular ejection of blood occurs.

   C. When the ventricles relax at the beginning of diastole, the ventricular pressures fall significantly below those in the aorta and pulmonary trunk, and the aortic and pulmonary valves close. Because the AV valves are also still closed, no change in ventricular volume occurs during this isovolumetric ventricular relaxation.

   D. When ventricular pressures fall below the pressures in the right and the left atria, the AV valves open, and the ventricular filling phase of diastole begins.

   E. Filling occurs very rapidly at first so that atrial contraction, which occurs at the very end of diastole, usually adds only a small amount of additional blood to the ventricles.

II. The amount of blood in the ventricles just before systole is the end-diastolic volume. The volume remaining after ejection is the end-systolic volume, and the volume ejected is the stroke volume.

III. Pressure changes in the systemic and pulmonary circulations have similar patterns, but the pulmonary pressures are much lower.

IV. The first heart sound is due to the closing of the AV valves, and the second to the closing of the aortic and pulmonary valves.

THE CARDIAC OUTPUT

I. The cardiac output is the volume of blood pumped by each ventricle and equals the product of heart rate and stroke volume.

   A. Heart rate is increased by stimulation of the sympathetic nerves to the heart and by epinephrine; it is decreased by stimulation of the parasympathetic nerves to the heart.

   B. Stroke volume is increased by an increase in end-diastolic volume (the Frank-Starling mechanism) and by an increase in contractility due to sympathetic-nerve stimulation or to epinephrine.

SECTION C KEY TERMS

pericardium
myocardium
endothelial cell
endothelium
atrioventricular (AV) valve
tricuspid valve
mitral valve
papillary muscles
pulmonary valve
aortic valve

conducting system
coronary artery
coronary blood flow
sinoatrial (SA) node
heart rate
atrioventricular (AV) node
bundle of His
right and left bundle
   branches
Purkinje fibers

slow channel
pacemaker potential
automaticity
electrocardiogram (ECG)
P wave
QRS complex
T wave
refractory period (of cardiac
   muscle)
cardiac cycle
systole
diastole
isovolumetric ventricular
   contraction

ventricular ejection
stroke volume (SV)
isovolumetric ventricular
   relaxation
ventricular filling
end-diastolic volume
end-systolic volume
heart sounds
cardiac output (CO)
ventricular function curve
Frank-Starling mechanism
venous return
contractility
ejection fraction (EF)

SECTION C REVIEW QUESTIONS

1. List the structures through which blood passes from the systemic veins to the systemic arteries.

2. Contrast and compare the structure of cardiac muscle with skeletal and smooth muscle.

3. Describe the autonomic innervation of the heart, including the types of receptors involved.

4. Draw a ventricular action potential. Describe the changes in membrane permeability that underlie the potential changes.

5. Contrast action potentials in ventricular cells with SA-node action potentials. What is the pacemaker potential due to, and what is its inherent rate? By what mechanism does the SA node function as the pacemaker for the entire heart?

6. Describe the spread of excitation from the SA node through the rest of the heart.

7. Draw and label a normal ECG. Relate the P, QRS, and T waves to the atrial and ventricular action potentials.

8. Describe the sequence of events leading to excitation-contraction coupling in cardiac muscle.

9. What prevents the heart from undergoing summation of contractions?

10. Draw a diagram of the pressure changes in the left atrium, left ventricle, and aorta throughout the cardiac cycle. Show when the valves open and close, when the heart sounds occur, and the pattern of ventricular ejection.

11. Contrast the pressures in the right ventricle and pulmonary trunk with those in the left ventricle and aorta.

12. What causes heart murmurs in diastole? In systole?

13. Write the formula relating cardiac output, heart rate, and stroke volume; give normal values for a resting adult.

14. Describe the effects of the sympathetic and parasympathetic nerves on heart rate. Which is dominant at rest?

15. What are the two major factors influencing force of contraction?

16. Draw a ventricular function curve illustrating the Frank-Starling mechanism.

17. Describe the effects of the sympathetic nerves on cardiac muscle during contraction and relaxation.

18. Draw a family of curves relating end-diastolic volume and stroke volume during different levels of sympathetic stimulation.

19. Summarize the effects of the autonomic nerves on the heart.

20. Draw a flow diagram summarizing the factors determining cardiac output.

SECTION D
THE VASCULAR SYSTEM

The functional and structural characteristics of the blood vessels change with successive branching. Yet the entire cardiovascular system, from the heart to the smallest capillary, has one structural component in common, a smooth, single-celled layer of endothelial cells, or endothelium, which lines the inner (blood-contacting) surface of the vessels. Capillaries consist only of endothelium, whereas all other vessels have, in addition, layers of connective tissue and smooth muscle. Endothelial cells have a large number of active functions. These are summarized for reference in Table 14-6 and are described in relevant sections of this chapter or subsequent chapters.

We have previously described the pressures in the aorta and pulmonary arteries during the cardiac cycle. Figure 14-35 illustrates the pressure changes that occur along the rest of the systemic and pulmonary vascular systems. Text sections below dealing with the individual vascular segments will describe the reasons for these changes in pressure. For the moment, note only that by the time the blood has completed its journey back to the atrium in each circuit, virtually all the pressure originally generated by the ventricular contraction has been dissipated. The reason pressure at any point in the vascular system is less than that at an earlier point is that the blood vessels offer resistance to the flow from one point to the next.

ARTERIES

The aorta and other systemic arteries have thick walls containing large quantities of elastic tissue. Although they also have smooth muscle, arteries can be viewed most conveniently as elastic tubes. Because the arteries have large radii, they serve as low-resistance tubes conducting blood to the various organs. Their second major function, related to their elasticity, is to act as a "pressure reservoir" for maintaining blood flow through the tissues during diastole, as described below.

Arterial Blood Pressure

What are the factors determining the pressure within an elastic container, such as a balloon filled with water? The pressure inside the balloon depends on (1) the volume of water, and (2) how easily the balloon walls can be stretched. If the walls are very stretchable, large quantities of water can be added with only a small rise in pressure. Conversely, the addition of a small quantity of water causes a large pressure rise in a balloon that is difficult to stretch. The term used to denote how easily a structure can be stretched is **compliance:**

$$\text{Compliance} = \Delta \text{ volume}/\Delta \text{ pressure}$$

TABLE 14-6   FUNCTIONS OF ENDOTHELIAL CELLS

1. Serve as a physical lining of heart and blood vessels to which blood cells do not normally adhere.

2. Serve as a permeability barrier for the exchange of nutrients, metabolic end-products, and fluid between plasma and interstitial fluid; regulate transport of macromolecules and other substances.

3. Secrete paracrine agents that act on adjacent vascular smooth-muscle cells; these include vasodilators—prostacyclin and nitric oxide (endothelium-derived relaxing factor, EDRF)—and vasoconstrictors—notably endothelin-1.

4. Mediate angiogenesis (new capillary growth).

5. Play a central role in vascular remodeling by detecting signals and releasing paracrine agents that act on adjacent cells in the blood vessel wall.

6. Contribute to the formation and maintenance of extracellular matrix (Chapter 1).

7. Produce growth factors in response to damage.

8. Secrete substances that regulate platelet clumping, clotting, and anticlotting.

9. Synthesize active hormones from inactive precursors (Chapter 16).

10. Extract or degrade hormones and other mediators (Chapter 15).

11. Secrete cytokines during immune responses (Chapter 20).

12. Influence vascular smooth-muscle proliferation in the disease atherosclerosis.

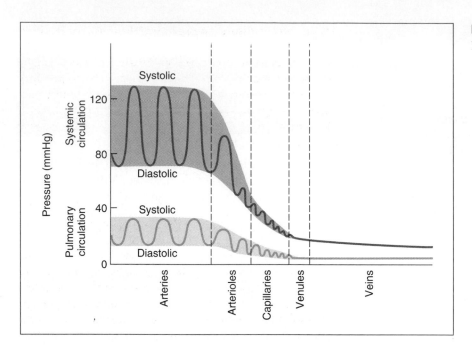

FIGURE 14-35
Pressures in the vascular system. ⫪

The *higher* the compliance of a structure, the *more easily* it can be stretched.

These principles can be applied to an analysis of arterial blood pressure. The contraction of the ventricles ejects blood into the pulmonary and systemic arteries during systole. If a precisely equal quantity of blood were to flow simultaneously out of the arteries, the total volume of blood in the arteries would remain constant and arterial pressure would not change. Such is not the case, however. As shown in Figure 14-36, a volume of blood equal to only about one-third the stroke volume leaves the arteries during systole. The rest of the stroke volume remains in the arteries during systole, distending them and raising the arterial pressure. When ventricular contraction ends, the stretched arterial walls recoil passively, like a stretched rubberband being released, and blood continues to be driven into the arterioles during diastole. As blood leaves the arteries, the arterial volume and therefore the arterial pressure slowly fall, but the next ventricular contraction occurs while there is still adequate blood in the arteries to stretch them partially. Therefore, the arterial pressure does not fall to zero.

The aortic pressure pattern shown in Figure 14-37A is typical of the pressure changes that occur in all the large systemic arteries. The maximum pressure reached during peak ventricular ejection is called **systolic pressure (SP).** The minimum pressure occurs just before ventricular ejection begins and is called **diastolic pressure (DP).** Arterial pressure is generally recorded as systolic/diastolic, that is, 125/75 mmHg in our example (see Figure 14-37B for average values at different ages in the population of the United States).

The difference between systolic pressure and diastolic pressure (125 − 75 = 50 mmHg in the example) is called the **pulse pressure.** It can be felt as a pulsation or throb in the arteries of the wrist or neck with each heartbeat. During diastole, nothing is felt over the artery, but the rapid rise in pressure at the next systole pushes out the

FIGURE 14-36
Movement of blood into and out of the arteries during the cardiac cycle. The lengths of the arrows denote relative quantities flowing into and out of the arteries and remaining in the arteries.

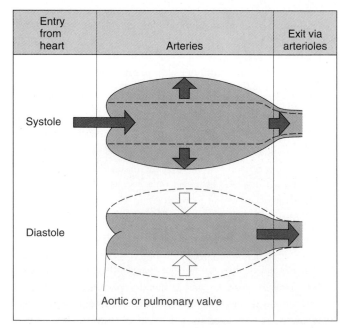

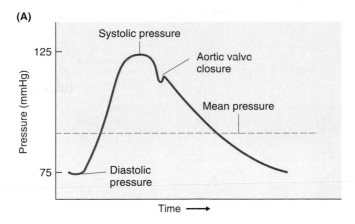

(A)

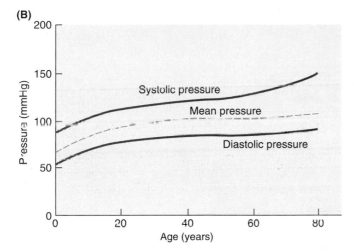

(B)

FIGURE 14-37

(A) Typical arterial pressure fluctuations during the cardiac cycle. (B) Changes in arterial pressure with age in the United States population. (*Adapted from Guyton.*)

artery wall, and it is this expansion of the vessel that produces the detectable throb.

The most important factors determining the magnitude of the pulse pressure, that is, how much greater systolic pressure is than diastolic, are (1) stroke volume, (2) speed of ejection of the stroke volume, and (3) arterial compliance. Specifically, the pulse pressure produced by a ventricular ejection is greater if the volume of blood ejected is increased, if the speed at which it is ejected is increased, or if the arteries are less compliant. This last phenomenon occurs in atherosclerosis, the "hardening" of the arteries that progresses with age and accounts for the increasing pulse pressure seen so often in older people.

It is evident from Figure 14-37A that arterial pressure is continuously changing throughout the cardiac cycle.

The *average* pressure (**mean arterial pressure, MAP**) in the cycle is not merely the value halfway between systolic pressure and diastolic pressure because diastole usually lasts longer than systole. The true mean arterial pressure can be obtained by complex methods, but for most purposes it is approximately equal to the diastolic pressure plus one-third of the pulse pressure (SP − DP), largely because diastole lasts about twice as long as systole:

$$MAP = DP + 1/3 \ (SP - DP)$$

Thus, in our example: MAP = 75 + 1/3 (50) = 92 mmHg.

The MAP is the most important of the pressures described because it is the pressure driving blood into the tissues averaged over the entire cardiac cycle. We can say mean "arterial" pressure without specifying to which artery we are referring because the aorta and other large arteries have such large diameters that they offer only negligible resistance to flow, and the mean pressures are therefore similar everywhere in the large arteries.

One additional important point should be made: We have stated that arterial compliance is an important determinant of *pulse* pressure, but for complex reasons, compliance does *not* influence the *mean* arterial pressure. Thus, for example, a person with a low arterial compliance (due to atherosclerosis) but an otherwise normal cardiovascular system will have a large pulse pressure but a normal mean arterial pressure.

Measurement of Systemic Arterial Pressure

Both systolic and diastolic blood pressure are readily measured in human beings with the use of a sphygmomanometer. An inflatable cuff is wrapped around the upper arm, and a stethoscope is placed in a spot on the arm just below the cuff and beneath which the major artery to the lower arm runs.

The cuff is then inflated with air to a pressure greater than systolic blood pressure (Figure 14-38). The high pressure in the cuff is transmitted through the tissue of the arm and completely compresses the artery under the cuff, thereby preventing blood flow through the artery. The air in the cuff is then slowly released, causing the pressure in the cuff and on the artery to drop. When cuff pressure has fallen to a value just below the systolic pressure, the artery opens slightly and allows blood flow for a brief time at the peak of systole. During this interval, the blood flow through the partially compressed artery occurs at a very high velocity because of the small opening and the large pressure difference across the opening. The high-velocity blood flow is turbulent and, therefore, produces vibrations that can be heard through the stethoscope. Thus, the pressure, measured on the gauge attached to the cuff, at which sounds are first heard as the cuff pressure is lowered is identified as the systolic blood pressure.

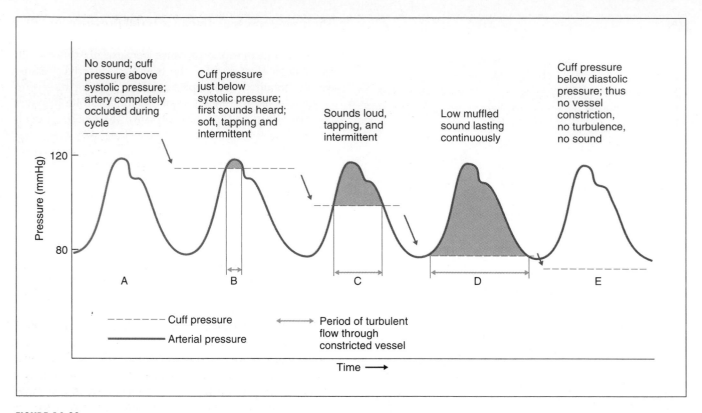

FIGURE 14-38

Sounds heard through a stethoscope while the cuff pressure of a sphygmomanometer is gradually lowered. Sounds are first heard at systolic pressure, and they disappear at diastolic pressure.

As the pressure in the cuff is lowered further, the duration of blood flow through the artery in each cycle becomes longer. When the cuff pressure reaches the diastolic blood pressure, all sound stops because flow is now continuous and nonturbulent through the open artery. Thus, diastolic pressure is identified as the cuff pressure at which sounds disappear.

It should be clear from this description that the sounds heard during measurement of blood pressure are *not* the same as the *heart* sounds described earlier, which are due to closing of cardiac valves.

ARTERIOLES

The arterioles play two major roles: (1) The arterioles in individual organs are responsible for determining the relative blood flows to those organs at any given mean arterial pressure, and (2) the arterioles, as a whole, are a major factor in determining mean arterial pressure itself. The first function will be described in this section, and the second in the section on control of arterial pressure.

Figure 14-39 illustrates the major principles of blood-flow distribution in terms of a simple model, a fluid-filled tank with a series of compressible outflow tubes. What determines the rate of flow through each exit tube? As stated in Section B of this chapter,

$$F = \Delta P/R$$

Since the driving pressure (the height of the fluid column in the tank) is identical for each tube, differences in flow are completely determined by differences in the resistance to flow offered by each tube. The lengths of the tubes are approximately the same, and the viscosity of the fluid is constant; therefore, differences in resistance offered by the tubes are due solely to differences in their radii. Obviously, the widest tubes have the greatest flows. If we equip each outflow tube with an adjustable cuff, we can obtain various combinations of flows.

This analysis can now be applied to the cardiovascular system. The tank is analogous to the arteries, which serve as a pressure reservoir, the major arteries themselves being so large that they contribute little resistance to flow. Therefore, all the large arteries of the body can be considered a single pressure reservoir.

The arteries branch within each organ into progressively smaller arteries, which then branch into arterioles. The smallest arteries are narrow enough to offer signifi-

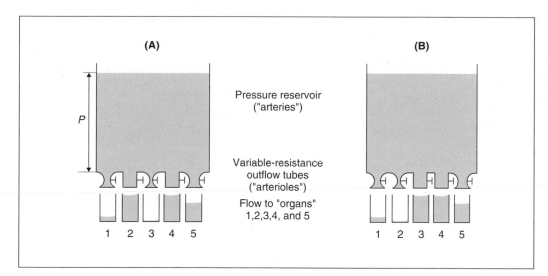

FIGURE 14-39
Physical model of the relationship between arterial pressure, arteriolar radius in different organs, and blood-flow distribution. The shift in blood flow through tubes 2 and 3 in going from (A) to (B) was achieved by constricting tube 2 and dilating tube 3.

In figure: (A) (B)

Pressure reservoir ("arteries")

P

Variable-resistance outflow tubes ("arterioles")

Flow to "organs" 1,2,3,4, and 5

1 2 3 4 5

1 2 3 4 5

cant resistance to flow, but the still narrower arterioles are the major sites of resistance in the vascular tree and are therefore analogous to the outflow tubes in the model. This explains the large decrease in mean pressure—from about 90 mmHg to 35 mmHg—as blood flows through the arterioles (Figure 14-35). Pulse pressure also diminishes to the point that flow beyond the arterioles, that is, through capillaries, venules, and veins, is no longer pulsatile.

Like the model's outflow tubes (Figure 14-39), the arteriolar radii in individual organs are subject to independent adjustment. The blood flow through any organ is given by the following equation:

$$F_{organ} = (MAP - \text{venous pressure})/R_{organ}$$

Since venous pressure is normally approximately zero we may write:

$$F_{organ} = MAP/R_{organ}$$

Since the MAP, the driving force for flow through each organ, is identical throughout the body, differences in flows between organs depend entirely on the relative resistances offered by the arterioles of each organ. Arterioles contain smooth muscle, which can either relax and cause the vessel radius to increase (**vasodilation**) or contract and decrease the vessel radius (**vasoconstriction**). Thus the pattern of blood-flow distribution depends upon the degree of arteriolar smooth-muscle contraction within each organ and tissue. Look back at Figure 14-9, which illustrates the distribution of blood flows at rest; these are due to differing resistances in the various locations. Such distribution can be changed markedly, as during exercise, for example, by changing the resistances.

How can resistance be changed? Arteriolar smooth muscle possesses a large degree of spontaneous activity, that is, contraction independent of any neural, hormonal, or paracrine input. This spontaneous contractile activity is called **myogenic tone** (also termed intrinsic or basal tone). It sets a baseline level of contraction that can be increased or decreased by external signals, such as neurotransmitters. These signals act by inducing changes in the muscle cell's cytosolic calcium concentration (see Chapter 11 for a description of excitation-contraction coupling in smooth muscle). An increase in contractile force above the vessel's myogenic tone causes vasoconstriction, whereas a decrease in contractile force causes vasodilation. The mechanisms controlling vasoconstriction and vasodilation in arterioles fall into two general categories: (1) local controls, and (2) extrinsic (or reflex) controls.

Local Controls

The term **local controls** denotes mechanisms independent of nerves or hormones by which organs and tissues alter their own arteriolar resistances, thereby self-regulating their blood flows. It does include changes caused by autocrine/paracrine agents. This self-regulation includes the phenomena of active hyperemia, flow autoregulation, reactive hyperemia, and local response to injury.

Active Hyperemia. Most organs and tissues manifest an increased blood flow (**hyperemia**) when their metabolic activity is increased (Figure 14-40); this is termed **active hyperemia.** For example, the blood flow to exercising skeletal muscle increases in direct proportion to the increased activity of the muscle. Active hyperemia is the direct result of arteriolar dilation in the more active organ or tissue.

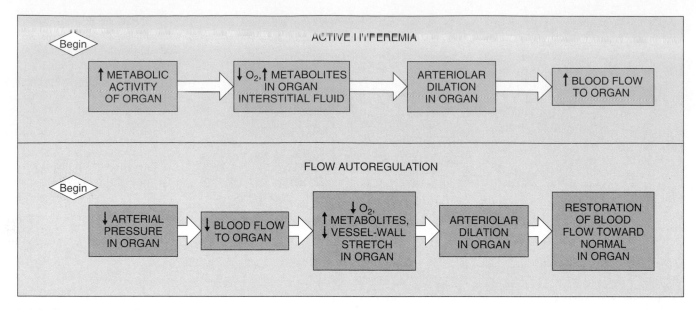

FIGURE 14-40

Local control of organ blood flow. Decreases in metabolic activity or increases in blood pressure would produce changes opposite those shown here.

The factors acting upon arteriolar smooth muscle in active hyperemia to cause it to relax are local chemical changes in the extracellular fluid surrounding the arterioles. These result from the increased metabolic activity in the cells near the arterioles. The relative contributions of the various factors implicated vary, depending upon the organs involved and on the duration of the increased activity. Therefore, we shall name but not quantify some of these local chemical changes that occur in the extracellular fluid: decreased oxygen concentration; increased concentrations of carbon dioxide, hydrogen ion, and metabolites such as adenosine; increased concentration of potassium as a result of enhanced potassium movement out of muscle cells during the more frequent action potentials; increased osmolarity resulting from the increased breakdown of high-molecular-weight substances; increased concentrations of eicosanoids (Chapter 7); and, in some glands, increased concentration of a peptide known as **bradykinin.** This last substance is generated locally from a circulating protein called **kininogen** by the action of an enzyme, **kallikrein,** secreted by the active gland cells.

Local changes in all these chemical factors have been shown to cause arteriolar dilation under controlled experimental conditions, and they all probably contribute to the active-hyperemia response in one or more organs. It is likely, moreover, that additional important local factors remain to be discovered. It must be emphasized that all these chemical changes in the extracellular fluid act locally upon the arteriolar smooth muscle, causing it to relax. No nerves or hormones are involved.

It should not be too surprising that active hyperemia is most highly developed in skeletal muscle, cardiac muscle, and glands, tissues that show the widest range of normal metabolic activities in the body. It is highly efficient, therefore, that their supply of blood be primarily determined locally.

Flow Autoregulation. During active hyperemia, increased metabolic activity of the tissue or organ is the initial event leading to local vasodilation. However, locally mediated changes in arteriolar resistance can also occur when a tissue or organ suffers a change in its blood supply resulting from a change in blood pressure (Figure 14-40). The change in resistance is in the direction of maintaining blood flow nearly constant in the face of the pressure change and is therefore termed **flow autoregulation.** For example, when arterial pressure in an organ is reduced, say, because of a partial occlusion in the artery supplying the organ, local controls cause arteriolar vasodilation, which tends to maintain flow relatively constant.

What is the mechanism of flow autoregulation? One mechanism is the same metabolic factors described for active hyperemia. When an arterial pressure reduction lowers blood flow to an organ, the supply of oxygen to the organ is diminished and the local extracellular oxygen concentration decreases. Simultaneously, the extracellular concentrations of carbon dioxide, hydrogen ion, and metabolites all increase because they are not removed by the blood as fast as they are produced. Also, eicosanoid synthesis is increased by still unclear stimuli. Thus, the local metabolic changes occurring during decreased blood supply at constant metabolic activity are similar to those that occur during increased metabolic activity. This is because both situations reflect an initial imbalance between blood supply and level of cellular metabolic activity. Note then that the vasodilations of active hyperemia and of flow autoregulation in response to low arterial pressure do not differ in their major mechanisms, which involve local metabolic factors, but in the event—altered metabolism or altered blood pressure—that brings these mechanisms into play.

Flow autoregulation is not limited to circumstances in which arterial pressure decreases. The opposite events occur when, for various reasons, arterial pressure increases: The initial increase in flow due to the increase in pressure removes the local vasodilator chemical factors faster than they are produced and also increases the local concentration of oxygen. This causes the arterioles to constrict, thereby maintaining local flow relatively constant in the face of the increased pressure.

Although our description has emphasized the role of local *chemical* factors in flow autoregulation, it should be noted that another mechanism also participates in this phenomenon in certain tissues and organs. Some arteriolar smooth muscle responds directly to increased stretch, caused by increased arterial pressure, by contracting to a greater extent. Conversely, decreased stretch, due to decreased arterial pressure, causes this vascular smooth muscle to decrease its tone. These direct responses of arteriolar smooth muscle to stretch are termed **myogenic responses.** They are due to stretch-dependent changes in calcium movement into the smooth-muscle cells.

Reactive Hyperemia. When an organ or tissue has had its blood supply completely occluded, a profound transient increase in its blood flow occurs as soon as the occlusion is released. This phenomenon, known as **reactive hyperemia,** is essentially an extreme form of flow autoregulation. During the period of no blood flow, the arterioles in the affected organ or tissue dilate, owing to the local factors described above. Blood flow, therefore, is very great through these wide-open arterioles as soon as the occlusion to arterial inflow is removed.

Response to Injury. Tissue injury causes a variety of substances to be released locally from cells or generated from plasma precursors. These substances make arteriolar smooth muscle relax and cause vasodilation in an injured area. This phenomenon, a part of the general process known as inflammation, will be described in detail in Chapter 20.

Extrinsic Controls

Sympathetic Nerves. Most arterioles receive a rich supply of sympathetic postganglionic nerve fibers. These neurons release mainly norepinephrine, which combines with alpha-adrenergic receptors on the vascular smooth muscle to cause vasoconstriction.

In contrast, recall that the receptors for norepinephrine on heart muscle, including the conducting system, are mainly beta-adrenergic. This permits the use of beta-adrenergic antagonists to block the actions of norepinephrine on the heart but not the arterioles, and vice versa for alpha-adrenergic antagonists.

Control of the sympathetic nerves can also be used to produce *vasodilation*. Since the sympathetic nerves are seldom completely quiescent but discharge at some finite rate that varies from organ to organ, they always are causing some degree of tonic constriction in addition to the vessels' myogenic tone. Dilation can be achieved by *decreasing* the rate of sympathetic activity below this basal level.

The skin offers an excellent example of the role of the sympathetic nerves. At room temperature, skin arterioles are already under the influence of a moderate rate of sympathetic discharge. An appropriate stimulus—fear or loss of blood, for example—causes reflex enhancement of this sympathetic discharge, and the arterioles constrict further. In contrast, an increased body temperature reflexly inhibits the sympathetic nerves to the skin, the arterioles dilate, and the skin flushes. (The control of skin blood flow by the sympathetic nerves in temperature-regulating reflexes is actually more complex than this, as described in Chapter 18.)

In contrast, to active hyperemia and flow autoregulation, the primary functions of sympathetic nerves to blood vessels are concerned not with the coordination of local metabolic needs and blood flow but with reflexes that serve whole body "needs." The most common reflex employing these nerves, as we shall see, is that which regulates arterial blood pressure by influencing arteriolar resistance throughout the body. Other reflexes redistribute blood flow to achieve a specific function (for example, to increase heat loss from the skin).

Parasympathetic Nerves. With several exceptions, to be described in later chapters, there is little or no important parasympathetic innervation of arterioles. In other words, the great majority of blood vessels receive sympathetic but not parasympathetic input.

Noncholinergic, Nonadrenergic Autonomic Neurons. As described in Chapter 8, there is a population of autonomic postganglionic neurons that are labeled noncholinergic, nonadrenergic neurons because they release neither acetylcholine nor norepinephrine. Instead they release **nitric oxide,** which is a vasodilator, and, possibly, other noncholinergic vasodilator substances. These neurons are particularly prominent in the enteric nervous system, which plays a significant role in the control of the gastrointestinal system's blood vessels (Chapter 17) These neurons also innervate arterioles in certain other locations, for example, in the penis, where they mediate erection (Chapter 19).

Hormones.   Epinephrine, like norepinephrine released from sympathetic nerves, can bind to alpha-adrenergic receptors on arteriolar smooth muscle and cause vasoconstriction. The story is more complex, however, because many arteriolar smooth-muscle cells possess beta-adrenergic receptors as well as alpha-adrenergic receptors, and the binding of epinephrine to these beta-adrenergic receptors causes the muscle cells to relax rather than contract (Figure 14-41).

In most vascular beds, the existence of beta-adrenergic receptors on vascular smooth muscle is of little if any importance since they are greatly outnumbered by the alpha-adrenergic receptors. The arterioles in skeletal muscle are an important exception, however. Because they have large numbers of beta-adrenergic receptors, circulating epinephrine, usually causes dilation in this vascular bed.

Another hormone important for arteriolar control is **angiotensin II,** which directly constricts most arterioles and also increases the activity of the sympathetic nervous system. This peptide is part of the renin-angiotensin system (Chapter 16).

Yet another important hormone that, when present at high plasma concentrations, causes arteriolar constriction is **vasopressin,** which is released into the blood by the posterior pituitary gland (Chapter 10). The functions of vasopressin will be described more fully in Chapter 16.

Finally, the hormone secreted by the cardiac atria—**atrial natriuretic factor**—is a potent vasodilator. Whether this hormone, whose actions on the kidneys are described in Chapter 16, plays a widespread role in control of arterioles is unsettled.

Endothelial Cells and Vascular Smooth Muscle

It should be clear from the previous sections that a large number of substances can induce the contraction or relaxation of vascular smooth muscle. Many of these substances do so by acting directly on the arteriolar smooth muscle, but others act indirectly via the endothelial cells adjacent to the smooth muscle. Endothelial cells, in response to these latter substances as well as certain mechanical stimuli, secrete several paracrine agents that diffuse to the adjacent vascular smooth muscle and induce either relaxation or contraction, resulting in vasodilation or vasoconstriction, respectively.

One very important paracrine vasodilator released by endothelial cells is nitric oxide. (Before the identity of the vasodilator paracrine agent released by the endothelium was determined to be nitric oxide it was called **endothelium-derived relaxing factor (EDRF),** and this name is still often used because there may be one or more substances other than nitric oxide that fit this general definition.) Nitric oxide is released continuously in significant amounts by endothelial cells in the arterioles and contributes to arteriolar vasodilation in the basal state. In addition, its secretion is rapidly and markedly increased in response to a large number of the chemical mediators involved in both reflex and local control of arterioles. For example, nitric oxide release is stimulated by bradykinin and histamine,

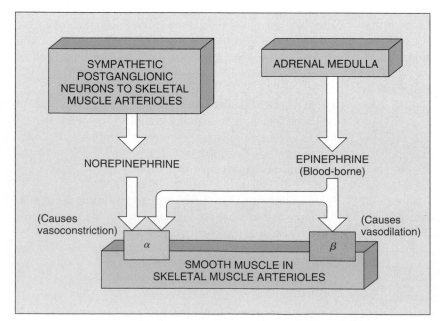

FIGURE 14-41

Effects of sympathetic nerves and plasma epinephrine on the arterioles in skeletal muscle. After its release from neuron terminals, norepinephrine diffuses to the arterioles, whereas epinephrine, a hormone, is blood-borne. Note that activation of alpha-adrenergic receptors and beta-adrenergic receptors produces opposing effects. For simplicity, norepinephrine is shown binding only to alpha-adrenergic receptors; it also can bind to beta-adrenergic receptors on the arterioles, but this occurs to a lesser extent.

substances produced locally during inflammation (Chapter 20).

This is the second time we have encountered nitric oxide in this chapter, the first being as a neurotransmitter released by certain postganglionic autonomic neurons supplying a few vascular beds. In addition, a third possible route for nitric oxide distribution to blood vessels involves its transport by hemoglobin (Chapter 15).

Another vasodilator released by endothelial cells is the eicosanoid **prostacyclin (PGI₂).** Unlike the case for nitric oxide, there is little basal secretion of PGI_2, but secretion can increase markedly in response to various inputs. The roles of PGI_2 in the vascular responses to blood clotting are described in Section G of this chapter.

One of the important vasoconstrictor paracrine agents released by endothelial cells in response to certain mechanical and chemical stimuli is **endothelin-1 (ET-1).** ET-1 is a member of the endothelin family of peptide paracrine agents secreted by a variety of cells in diverse tissues, including the brain, kidneys, and lungs. Not only does ET-1 serve as a paracrine agent but under certain circumstances it can also achieve high enough concentrations in the blood to serve as a hormone, causing widespread arteriolar vasoconstriction.

This discussion has so far focused only on arterioles. However, endothelial cells in *arteries* can also secrete various paracrine agents that influence the arteries' smooth muscle and, hence, their diameters and resistances to flow. The force exerted on the inner surface of the arterial wall, specifically on the endothelial cells, by the flowing blood is termed **shear stress;** it increases as the blood flow through the vessel increases. In response to this increased shear stress, arterial endothelium releases PGI_2, increased amounts of nitric oxide, and less endothelin-1. All these

changes cause the arterial vascular smooth muscle to relax and the artery to dilate. This **flow-induced arterial vasodilation** (which should be distinguished from *arteriolar* flow autoregulation) may be important in remodeling of arteries and in optimizing the blood supply to tissues under certain conditions. As an example of the latter, the increase in flow that occurs in a contracting skeletal muscle as a result of *arteriolar* vasodilation (active hyperemia) will increase shear stress in the arteries supplying those arterioles, and this will induce an *arterial* vasodilation that further enhances blood flow to the skeletal muscle. Note that in such cases the arterial changes are secondary to the arteriolar changes.

Arteriolar Control in Specific Organs

Figure 14-42 summarizes the factors that determine arteriolar radius. The importance of local and reflex controls varies from organ to organ, and Table 14-7 lists for reference the key features of arteriolar control in specific organs.

CAPILLARIES

As mentioned at the beginning of Section B, at any given moment, approximately 5 percent of the total circulating blood is flowing through the capillaries, and it is this 5 percent that is performing the ultimate function of the entire cardiovascular system—the exchange of nutrients and metabolic end products. Some exchange also occurs in the venules, which can be viewed as extensions of capillaries.

The capillaries permeate almost every tissue of the body. Since most cells are no more than 0.01 cm (only a

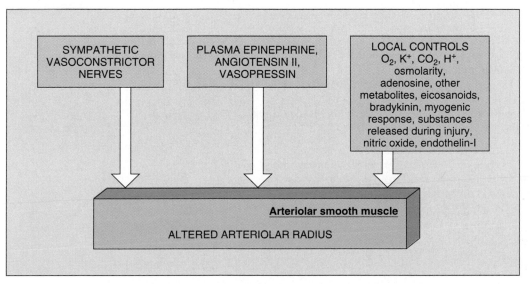

FIGURE 14-42

Major factors affecting arteriolar radius.

SYMPATHETIC VASOCONSTRICTOR NERVES

PLASMA EPINEPHRINE, ANGIOTENSIN II, VASOPRESSIN

LOCAL CONTROLS O_2, K^+, CO_2, H^+, osmolarity, adenosine, other metabolites, eicosanoids, bradykinin, myogenic response, substances released during injury, nitric oxide, endothelin-I

Arteriolar smooth muscle

ALTERED ARTERIOLAR RADIUS

TABLE 14-7 REFERENCE SUMMARY OF VASCULAR CONTROL IN SPECIFIC ORGANS

Heart

High intrinsic tone; oxygen extraction is very high at rest and flow must increase when oxygen consumption increases.

Controlled mainly by local metabolic factors, particularly adenosine, and flow autoregulation; direct sympathetic influences are minor and normally overridden by local factors.

Vessels are compressed during systole, and so coronary flow occurs mainly during diastole.

Skeletal Muscle

Controlled by local metabolic factors during exercise.

Sympathetic nerves cause vasoconstriction (mediated by alpha-adrenergic receptors) in reflex response to decreased arterial pressure.

Epinephrine causes vasodilation, via beta-adrenergic receptors, when present in low concentration and vasoconstriction, via alpha-adrenergic receptors, when present in high concentration.

GI Tract, Spleen, Pancreas, and Liver

Actually two capillary beds partially in series with each other; blood from the capillaries of the GI tract, spleen, and pancreas flows via the portal vein to the liver. In addition, the liver also receives a separate arterial blood supply.

Sympathetic nerves cause vasoconstriction, mediated by alpha-adrenergic receptors, in reflex response to decreased arterial pressure and during stress. The venous constriction causes displacement of a large volume of blood from the liver to the veins of the thorax.

Increased blood flow occurs following ingestion of a meal and is mediated by local metabolic factors, neurons, and hormones secreted by the GI tract.

Kidneys

Flow autoregulation is a major factor.

Sympathetic nerves cause vasoconstriction, mediated by alpha-adrenergic receptors, in reflex response to decreased arterial pressure and during stress. Angiotensin II is also a major vasoconstrictor. These reflexes help conserve sodium and water.

Brain

Excellent flow autoregulation.

Distribution of blood within the brain is controlled by local metabolic factors.

Vasodilation occurs in response to increased concentration of carbon dioxide in arterial blood.

Influenced relatively little by the autonomic nervous system.

Skin

Controlled mainly by sympathetic nerves, mediated by alpha-adrenergic receptors; reflex vasoconstriction occurs in response to decreased arterial pressure and cold, whereas vasodilation occurs in response to heat.

Substances released from sweat glands and noncholinergic, nonadrenergic neurons also cause vasodilation.

Venous plexus contains large volume of blood, which contributes to skin color.

Lungs

Very low resistance compared to systemic circulation.

Controlled mainly by gravitational forces and passive physical forces within the lung.

Constriction, mediated by local factors, occurs in response to low oxygen concentration—just opposite that which occurs in the systemic circulation.

few cell widths) from a capillary, diffusion distances are very small, and exchange is highly efficient. There are an estimated 25,000 miles of capillaries in an adult, each individual capillary being only about 1 mm long with an inner diameter of 5 μm, just wide enough for an erythrocyte to squeeze through. (For comparison, a human hair is about 100 μm in diameter.)

The essential role of capillaries in tissue function has stimulated many questions concerning how capillaries develop and grow **(angiogenesis).** For example, how do tumors stimulate growth of the new capillaries required for continued tumor growth, and what turns on angiogenesis during wound healing? It is known that the vascular endothelial cells play a central role in the building of a new capillary network by cell locomotion and cell division. They are stimulated to do so by a variety of **angiogenic factors** secreted locally by various tissue cells (fibroblasts, for example) and the endothelial cells themselves. Tumor cells also secrete angiogenic factors, and development of methods to interfere with the secretion or action of these factors is a promising research area in anticancer therapy.

Anatomy of the Capillary Network

Capillary structure varies considerably from organ to organ, but the typical capillary (Figure 14-43) is a thin-walled tube of endothelial cells one layer thick resting on a basement membrane, without any surrounding smooth muscle or elastic tissue. Capillaries in several organs (for example, the brain) have a second set of cells that adhere to the opposite side of the basement membrane and influence the ability of substances to penetrate the capillary wall.

The flat cells that constitute the endothelial tube are not attached tightly to each other but are separated by narrow water-filled spaces termed **intercellular clefts.** The endothelial cells generally contain large numbers of endocytotic and exocytotic vesicles, and sometimes these fuse to form continuous **fused-vesicle channels** across the cell (Figure 14-43).

Blood flow through capillaries depends very much on the state of the other vessels that constitute the microcirculation (Figure 14-44). Thus, vasodilation of the arterioles supplying the capillaries causes increased capillary flow, whereas arteriolar vasoconstriction reduces capillary flow. In addition, in some tissues and organs blood does not enter capillaries directly from arterioles but from vessels called **metarterioles,** which connect arterioles to venules. Metarterioles, like arterioles, contain scattered smooth muscle cells. The site at which a capillary exits from a metarteriole is surrounded by a ring of smooth muscle, the **precapillary sphincter,** which relaxes or contracts in response to local metabolic factors. When contracted, the precapillary sphincter closes the entry to the capillary completely. The more active the tissue, the more precapillary sphincters are open at any moment and the more capillaries in the network are receiving blood. Precapillary sphincters may also exist at the site of capillary exit from arterioles.

Velocity of Capillary Blood Flow

Figure 14-45 illustrates a simple mechanical model of a series of 1-cm-diameter balls being pushed down a single tube that branches into narrower tubes. Although each

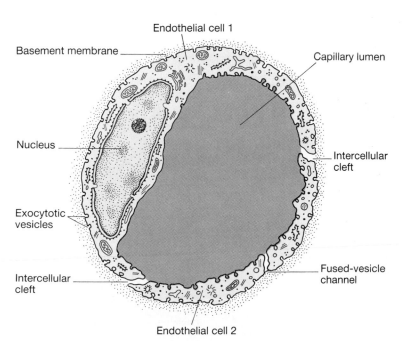

Basement membrane

Endothelial cell 1

Capillary lumen

Nucleus

Exocytotic vesicles

Intercellular cleft

Intercellular cleft

Fused-vesicle channel

Endothelial cell 2

FIGURE 14-43

Capillary cross section. There are two endothelial cells in the figure, but the nucleus of only one is seen because the other is out of the plane of section. The fused-vesicle channel is part of endothelial cell 2. (*Adapted from Lentz.*)

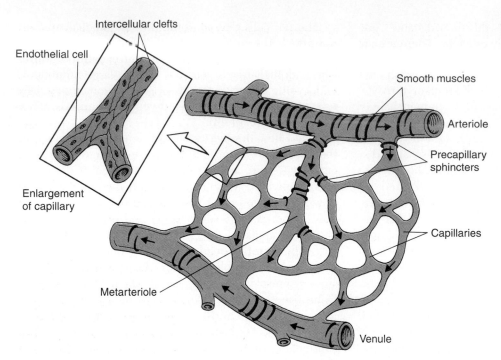

FIGURE 14-44
Diagram of microcirculation. Note the absence of smooth muscle in the capillaries. (*Adapted from Chaffee and Lytle.*)

tributary tube has a smaller cross section than the wide tube, the *sum* of the tributary cross sections is much greater than that of the wide tube. Let us assume that in the wide tube each ball moves 3 cm/min. If the balls are 1 cm in diameter and they move two abreast, six balls leave the wide tube per minute and enter the narrow tubes, and six balls leave the narrow tubes per minute. At what speed does each ball move in the small tubes? The answer is 1 cm/min.

This example illustrates the following important principle: When a continuous stream moves through consecutive sets of tubes, the velocity of flow decreases as the sum of the cross-sectional areas of the tubes increases. This is precisely the case in the cardiovascular system (Figure 14-46). The blood velocity is very great in the aorta, slows progressively in the arteries and arterioles, and then slows markedly as the blood passes through the huge cross-

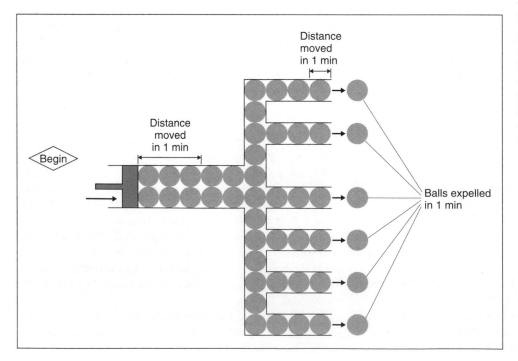

FIGURE 14-45
Relationship between total cross-sectional area and flow velocity. The total cross-sectional area of the small tubes is three times greater than that of the large tube. Accordingly, velocity of flow is one-third as great in the small tubes.

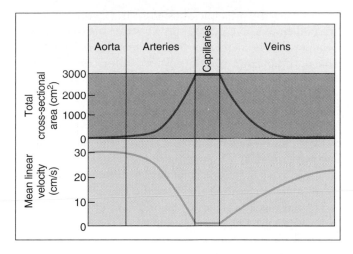

FIGURE 14-46

Relationship between total cross-sectional area and flow velocity in the systemic circulation. (*Adapted from Lytle.*)

sectional area of the capillaries. The velocity of flow then progressively increases in the venules and veins because the cross-sectional area decreases. To reemphasize, flow velocity is not dependent on proximity to the heart but rather on total cross-sectional area of the vessel type.

The huge cross-sectional area of the capillaries accounts for another important feature of capillaries: Because each capillary is very narrow it offers considerable resistance to flow, but the huge total number of capillaries provides such a large cross-sectional area that the *total* resistance of *all* the capillaries is only about 40 percent of that of the arterioles.

Diffusion Across the Capillary Wall: Exchanges of Nutrients and Metabolic End Products

There are three basic mechanisms by which substances move across the capillary walls in most organs and tissues to enter or leave the interstitial fluid: diffusion, vesicle transport, and bulk flow. Mediated transport constitutes a fourth mechanism in the capillaries of the brain. Diffusion and vesicle transport are described in this section, and bulk flow in the next.

In all capillaries, excluding those in the brain, diffusion constitutes the only important means by which net movement of nutrients, oxygen, and metabolic end products occurs across the capillary walls. As described in the next section, there is some movement of these substances by bulk flow, but it is of negligible importance.

The factors determining diffusion rates were described in Chapter 6. Lipid-soluble substances, including oxygen and carbon dioxide, easily diffuse through the plasma membranes of the capillary endothelial cells. In contrast, ions and polar molecules are poorly soluble in lipid and

must pass through small water-filled channels in the endothelial lining.

The presence of water-filled channels in the capillary walls causes the permeability of ions and small polar molecules to be quite high, although still much lower than that of lipid-soluble molecules. One location of these channels is the intercellular clefts, that is, the narrow water-filled spaces between adjacent cells. Another set of water-filled channels is provided by the fused-vesicle channels that penetrate the endothelial cells.

The water-filled channels allow only very small amounts of protein to diffuse through them. Very small amounts of protein may also cross the endothelial cells by endocytosis of plasma at the luminal border and exocytosis of the endocytotic vesicle at the interstitial side.

Variations in the size of the water-filled channels account for great differences in the "leakiness" of capillaries in different organs. At one extreme are the "tight" capillaries of the brain, which have no intercellular clefts, only tight junctions. Therefore, water-soluble substances, even those of low molecular weight, can gain access to or exit from brain interstitial space only by carrier-mediated transport through the blood-brain barrier (Chapter 8).

At the other end of the spectrum are liver capillaries, which have large intercellular clefts as well as "windows" in the plasma membranes of the endothelial cells so that even protein molecules can readily pass across them. This is important because two of the major functions of the liver are the synthesis of plasma proteins and the metabolism of protein-bound substances.

The leakiness of capillaries in most organs and tissues lies between these extremes of brain and liver capillaries.

What is the sequence of events involved in transfers of nutrients and metabolic end-products between capillary blood and cells? Nutrients diffuse first across the capillary wall into the interstitial fluid, from which they gain entry to cells. Conversely, metabolic end products from the tissues move across the cells' plasma membranes into interstitial fluid, from which they diffuse across the capillary endothelium into the plasma.

Transcapillary diffusion gradients for oxygen, nutrients, and metabolic end products occur as a result of cellular utilization or production of the substance. Let us take two examples: glucose and carbon dioxide in muscle. Glucose is continuously transported from interstitial fluid into the muscle cell by carrier-mediated transport mechanisms. The removal of glucose from interstitial fluid lowers the interstitial-fluid glucose concentration below the glucose concentration in capillary plasma and creates the gradient for diffusion of glucose from the capillary into the interstitial fluid.

Simultaneously, carbon dioxide, which is continuously produced by muscle cells, diffuses into the interstitial fluid. This causes the carbon dioxide concentration in interstitial fluid to be greater than that in capillary plasma, producing a gradient for carbon dioxide diffusion from the interstitial fluid into the capillary. Note that in both examples metabolism of the substance is the event that ultimately establishes the transcapillary diffusion gradients.

If a tissue is to increase its metabolic rate, it must obtain more nutrients from the blood and it must eliminate more metabolic end products. One mechanism for achieving this is active hyperemia. The second important mechanism is increased diffusion gradients between plasma and tissue: Increased cellular utilization of oxygen and nutrients lowers their tissue concentrations, whereas increased production of carbon dioxide and other end products raises their tissue concentrations. In both cases the substance's transcapillary concentration difference is increased, which increases the rate of diffusion.

Bulk Flow Across the Capillary Wall: Distribution of the Extracellular Fluid

At the same time that the diffusional exchange of nutrients, oxygen, and metabolic end products is occurring across the capillaries, another, completely distinct process is also taking place across the capillary—the bulk flow of protein-free plasma. The function of this process is *not* exchange of nutrients and metabolic end-products but rather distribution of the extracellular fluid. As described in Chapter 1, extracellular fluid comprises the blood plasma and the interstitial fluid. Normally, there is approximately three times more interstitial fluid than plasma, 10 L versus 3 L in a 70-kg person. This distribution is not fixed, however, and the interstitial fluid functions as a reservoir that can supply fluid to the plasma or receive fluid from it.

As described in the previous section, the capillary wall is highly permeable to water and to almost all plasma solutes, except plasma proteins. Therefore, in the presence of a hydrostatic pressure difference across it, the capillary wall behaves like a porous filter through which protein-free plasma (**ultrafiltrate**) moves by bulk flow from capillary plasma to interstitial fluid through the water-filled channels. The concentrations of all the plasma solutes except protein are virtually the same in the filtering fluid as in plasma.

The magnitude of the bulk flow is determined, in part, by the difference between the capillary blood pressure and the interstitial-fluid hydrostatic pressure. Normally, the former is much larger than the latter. Therefore, a considerable hydrostatic pressure difference exists to filter protein-free plasma out of the capillaries into the interstitial fluid, the protein remaining behind in the plasma.

Why then does all the plasma not filter out into the interstitial space? The explanation is that the hydrostatic pressure difference favoring filtration is offset by an osmotic force opposing filtration. To understand this we

must review the principle of osmosis. In Chapter 6, we described how a net movement of water occurs across a semipermeable membrane from a solution of high water concentration to a solution of low water concentration, that is, from a region with a low concentration of solute to which the membrane is impermeable (nonpermeating solute) to a region of high nonpermeating-solute concentration. Moreover, this osmotic flow of water "drags" along with it any dissolved solutes to which the membrane is highly permeable (permeating solute). Thus, a difference in water concentration secondary to different concentrations of *nonpermeating solute* on the two sides of a membrane can result in the movement of a solution containing both water and permeating solutes in a manner similar to the bulk flow produced by a hydrostatic-pressure difference. Units of pressure can be used in expressing this osmotic flow across a membrane.

This analysis can now be applied to osmotically induced flow across capillaries. The plasma within the capillary and the interstitial fluid outside it contain large quantities of low-molecular-weight permeating solutes (also termed **crystalloids**), for example, sodium, chloride, and glucose. Since the capillary lining is highly permeable to all these crystalloids, their concentrations in the two solutions are essentially identical (as we have seen, there actually are small concentration differences for substances that are consumed or produced by the cells, but these tend to cancel each other). Accordingly, the presence of the crystalloids causes no significant difference in water concentration. In contrast, the plasma proteins (also termed **colloids**), being essentially nonpermeating, have a very low concentration in the interstitial fluid. This difference in protein concentration between plasma and interstitial fluid means that the water concentration of the plasma is lower than that of interstitial fluid, inducing an osmotic flow of water from the interstitial compartment into the capillary. Since the crystalloids in the interstitial fluid move along with the water, osmotic flow of fluid, like flow driven by a hydrostatic-pressure difference, does not alter their concentrations in either plasma or interstitial fluid.

A key word in this last sentence is "concentrations." Because the interstitial fluid and plasma have virtually identical crystalloid concentrations to start with, a transfer of fluid across the capillary wall does not change the crystalloid concentrations in either location. However, the *amount* of water (the *volume*) and the *amount* of crystalloids in the two locations do change. Thus, an increased filtration of fluid from plasma to interstitial fluid increases the volume of the interstitial fluid and decreases the volume of the plasma, even though no changes in crystalloid concentrations occur.

In summary (Figure 14-47), two opposing forces act to move fluid across the capillary wall: (1) the difference between capillary blood hydrostatic pressure and interstitial-

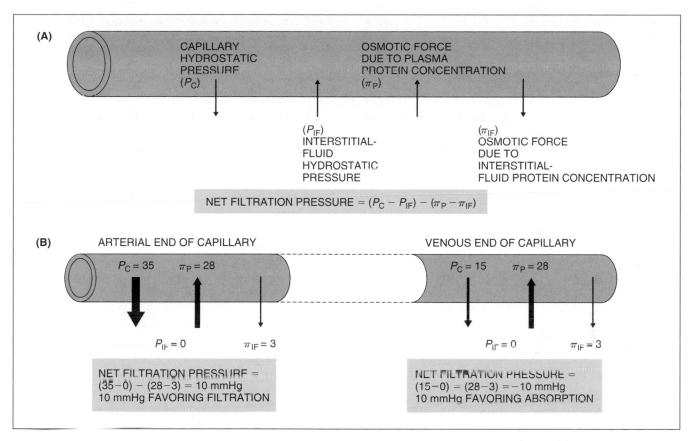

FIGURE 14-47

(A) The four factors determining fluid movement across capillaries. (B) Quantitation of forces causing filtration at the arterial end of the capillary and absorption at the venous end. Arrows in (B) denote magnitude of forces. No arrow is shown for interstitial-fluid hydrostatic pressure (P_{IF}) in (B) because it is zero.

fluid hydrostatic pressure favors filtration out of the capillary; and (2) the water-concentration difference between plasma and interstitial fluid, which results from differences in protein concentration, favors the flow of interstitial fluid into the capillary. Accordingly, the movement of fluid depends directly upon four variables: capillary hydrostatic pressure (favoring fluid movement out of the capillary), interstitial hydrostatic pressure (favoring fluid movement into the capillary), plasma protein concentration (favoring fluid movement into the capillary), and interstitial-fluid protein concentration (favoring fluid movement out of the capillary). These four factors are termed the **Starling forces** (because Starling, the same physiologist who helped elucidate the Frank-Starling mechanism of the heart, was the first to develop the ideas).

We may now consider this movement quantitatively in the systemic circulation (Figure 14-47B). Much of the arterial blood pressure has already been dissipated as the blood flows through the arterioles, so that pressure at the beginning of the capillary (the part closest to the arteriole) is about 35 mmHg. Since the capillary also offers resis-

tance to flow, the pressure continuously decreases to approximately 15 mmHg at the end of the capillary (the part farthest from the arteriole). The interstitial pressure is very low, and we shall assume it to be zero. The plasma protein concentration would produce an osmotic flow of water into the capillary equivalent to that produced by a hydrostatic pressure of 28 mmHg. The interstitial protein concentration would produce a flow of water out of the capillary equivalent to that produced by a hydrostatic pressure of 3 mmHg. Therefore, the difference in protein concentrations induces a flow of fluid into the capillary equivalent to that produced by a hydrostatic pressure difference of $28 - 3 = 25$ mmHg.

Thus, in the beginning of the capillary the hydrostatic pressure difference across the capillary wall (35 mmHg) is greater than the opposing osmotic force (25 mmHg), and a net movement of fluid out of the capillary (**filtration**) occurs. In the end of the capillary, however, the osmotic force (25 mmHg) is greater than the hydrostatic pressure difference (15 mmHg), and fluid moves into the capillary (**absorption**). The result is that the early and late capillary events tend to cancel each other out. For the aggregate of capillaries in the body, however, there is small net filtration of approximately 4 L/day (this number does not include the capillaries in the kidneys). The fate of this fluid will be described in the section on the lymphatic system.

A very important point is that capillary pressure in any vascular bed is subject to physiological regulation, mediated mainly by changes in the resistance of the arterioles in that bed. As shown in Figure 14-48, dilating the arterioles in a particular vascular bed raises capillary pressure in that bed because less pressure is lost overcoming resistance between the arteries and the capillaries. Because of the increased capillary pressure, filtration is increased, and more protein-free fluid is lost to the interstitial fluid. In contrast, marked arteriolar constriction produces decreased capillary pressure and hence favors net movement of interstitial fluid into the vascular compartment. Indeed, the arterioles supplying a capillary bed may be so dilated or so constricted that the capillaries manifest only filtration or only reabsorption, respectively, along their entire length.

It should be stated again that capillary filtration and absorption play no significant role in the exchange of nutrients and metabolic end products between capillary and tissues. The reason is that the total quantity of a substance, such as glucose or carbon dioxide, moving into or out of a capillary as a result of net bulk flow is extremely small in comparison with the quantities moving by net diffusion.

Finally, this analysis of capillary fluid dynamics has been in terms of the systemic circulation. Precisely the same Starling forces apply to the capillaries in the pulmonary circulation, but the values of the four variables differ. In particular, because the pulmonary circulation is a low-resistance, low-pressure circuit, the normal pulmonary capillary pressure—the major force favoring movement of fluid out of the pulmonary capillaries into the interstitium—is only 15 mmHg. Therefore, normally, net absorption of fluid occurs along the entire length of lung capillaries.

VEINS

Blood flows from capillaries into venules and then into veins. Some exchange of materials occurs between the interstitial fluid and the venules. Indeed, permeability to macromolecules is often greater for venules than for capillaries, particularly in damaged areas.

The veins outside the chest, the **peripheral veins,** contain valves that permit flow only toward the heart. Why are these valves necessary if the pressure gradient created by cardiac contraction pushes blood only toward the heart anyway? The answer will be given below.

The veins are the last set of tubes through which blood flows on its way back to the heart. In the systemic circulation the force driving this venous return is the pressure difference between the peripheral veins and the right atrium. The pressure in the first portion of the peripheral veins is generally quite low—only 5 to 10 mmHg—because most of the pressure imparted to the blood by the heart is dissipated by resistance as blood flows through the arterioles, capillaries, and venules. The right atrial pressure is normally approximately 0 mmHg. Therefore, the total driving pressure for flow from the peripheral veins to the right atrium is only 5 to 10 mmHg. This pressure difference is adequate because of the low resistance to flow offered by the veins, which have large diameters. Thus, a major function of the veins is to act as low-resistance conduits for blood flow from the tissues to the heart.

In addition to their function as low-resistance conduits, the veins perform a second important function: Their diameters are reflexly altered in response to changes in blood volume, thereby maintaining peripheral venous pressure and venous return to the heart. In a previous section, we emphasized that the rate of venous return to the heart is a major determinant of end-diastolic ventricular volume and thereby stroke volume. Thus, we now see that peripheral venous pressure is an important determinant of stroke volume.

Determinants of Venous Pressure

The factors determining pressure in any elastic tube are, as we know, the volume of fluid within it and the compliance of its wall. Accordingly, total blood volume is one important determinant of venous pressure since, at any given moment, most blood is in the veins. Also, the walls of

FIGURE 14-48

Effects of arteriolar vasodilation or vasoconstriction in an organ on capillary blood pressure in that organ.

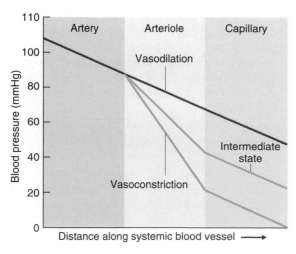

veins are thinner and much more compliant than those of arteries. Thus, veins can accommodate large volumes of blood with a relatively small increase in internal pressure. Approximately 60 percent of the total blood volume is present in the systemic veins at any given moment (Figure 14-49), but the venous pressure averages less than 10 mmHg. In contrast, the systemic arteries contain less than 15 percent of the blood, at a pressure of approximately 100 mmHg.

The walls of the veins contain smooth muscle innervated by sympathetic neurons. Stimulation of these neurons releases norepinephrine, which causes contraction of the venous smooth muscle, decreasing the diameter and compliance of the vessels and raising the pressure within them. Increased venous pressure then drives more blood out of the veins into the right heart. Although the sympathetic nerves are the most important input, venous smooth muscle, like arteriolar smooth muscle, is also influenced by hormonal and paracrine vasodilators and vasoconstrictors.

Two other mechanisms, in addition to contraction of venous smooth muscle, can increase venous pressure and facilitate venous return. These mechanisms are the **skeletal-muscle pump** and the **respiratory pump.** During skeletal muscle contraction, the veins running through the muscle are partially compressed, which reduces their diameter and forces more blood back to the heart. Now we can describe a major function of the peripheral-vein valves: When the skeletal-muscle pump raises venous pressure locally, the valves permit blood flow only toward the heart and prevent flow back toward the tissues (Figure 14-50).

The respiratory pump is somewhat more difficult to visualize. As will be described in Chap. 15, during inspiration of air, the diaphragm descends, pushes on the abdominal contents, and increases abdominal pressure. This pressure increase is transmitted passively to the intraabdominal veins. Simultaneously, the pressure in the thorax

FIGURE 14-50

The skeletal-muscle pump. During muscle contraction, venous diameter decreases and venous pressure rises. The resulting increase in blood flow can occur only toward the heart because the valves in the veins are forced closed by any backward flow.

FIGURE 14-49

Distribution of the total blood volume in different parts of the cardiovascular system. (*Adapted from Guyton.*)

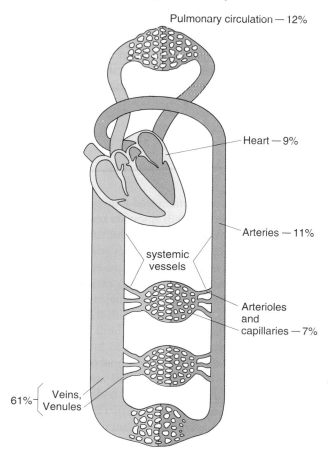

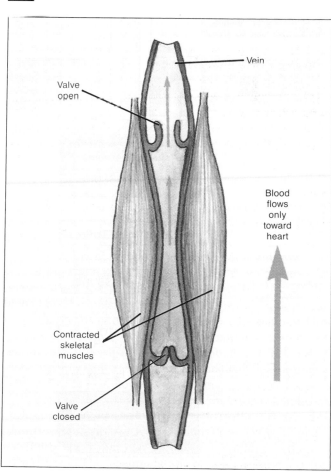

Massive liberation of endogenous substances that relax arteriolar smooth muscle may also cause hypotension by reducing total peripheral resistance. An important example is the hypotension that occurs during severe allergic responses (Chapter 20).

Shock

The term **shock** denotes any situation in which a decrease in blood flow to the organs and tissues damages them. One important cause of shock is severe hemorrhage, and arterial pressure is usually, but not always, low in shock. The cardiovascular system, especially the heart, suffers damage if shock is prolonged. As the heart deteriorates, cardiac output declines even more and shock becomes progressively worse and ultimately irreversible even though blood transfusions and other appropriate therapy may temporarily restore blood pressure.

Just as hemorrhage is not the only cause of hypotension and decreased cardiac output, so it is not the only cause of shock. Loss of fluid other than blood, excessive release of vasodilators as in allergy and infection, loss of sympathetic tone to the cardiovascular system, and extensive bodily damage can all lead to severe reductions of tissue blood flow and the positive-feedback cycles culminating in irreversible shock.

THE UPRIGHT POSTURE

A decrease in the *effective* circulating blood volume occurs in the circulatory system when going from a lying, horizontal position to a standing, vertical one. Why this is so requires an understanding of the action of gravity upon the long, continuous columns of blood in the vessels between the heart and the feet.

The pressures we have given in previous sections of this chapter are for an individual in the horizontal position, in which all blood vessels are at approximately the same level as the heart. In this position, the *weight* of the blood produces negligible pressure. In contrast, when a person is vertical, the intravascular pressure everywhere becomes equal to the pressure generated by cardiac contraction *plus* an additional pressure equal to the weight of a column of blood from the heart to the point of measurement. In an average adult, for example, the weight of a column of blood extending from the heart to the feet amounts to 80 mmHg. In a foot capillary, therefore, the pressure increases from 25 (the pressure resulting from cardiac contraction) to 105 mmHg, the extra 80 mmHg being due to the weight of the column of blood.

This increase in pressure due to gravity influences the effective circulating blood volume in several ways. First, the increased hydrostatic pressure that occurs in the legs (as well as the buttocks and pelvic area) when a person is quietly standing pushes outward on the highly distensible vein walls, causing marked distension. The result is pooling of blood in the veins; that is, much of the blood emerging from the capillaries simply goes into expanding the veins rather than returning to the heart. Simultaneously, the increase in capillary pressure caused by the gravitational force produces increased filtration of fluid out of the capillaries into the interstitial space. This accounts for the fact that our feet swell during prolonged standing. The combined effects of venous pooling and increased capillary filtration reduce the *effective* circulating blood volume very similarly to the effects caused by a mild hemorrhage. This explains why a person may sometimes feel faint upon standing up suddenly. This feeling is usually very transient, however, since the decrease in arterial pressure immediately causes reflex baroreceptor-mediated compensatory adjustments virtually identical to those shown in Figure 14-60 for hemorrhage.

The effects of gravity can be offset by contraction of the skeletal muscles in the legs. Even gentle contractions of the leg muscles without movement produce intermittent, complete emptying of the leg veins so that uninterrupted columns of venous blood from the heart to the feet no longer exist (Figure 14-64). The result is a decrease in both venous distension and pooling plus a marked reduction in capillary hydrostatic pressure and fluid filtration out of the capillaries. The importance of this phenomenon is illustrated by the fact that soldiers may faint while standing at attention for long periods of time because they have minimal contractions of the leg muscles. Here fainting may be considered adaptive in that the venous and capillary pressure changes induced by gravity are eliminated once the person is prone. The pooled venous blood is mobilized, and the filtered fluid is absorbed back into the capillaries. Thus, the wrong thing to do to a person who has fainted for whatever reason is to hold him or her upright.

EXERCISE

During exercise, cardiac output may increase from a resting value of 5 L/min to a maximal value of 35 L/min in trained athletes. The distribution of this cardiac output during strenuous exercise is illustrated in Figure 14-65. As expected, most of the increase in cardiac output goes to the exercising muscles, but there are also increases in flow to skin, required for dissipation of heat, and to the heart, required for the additional work performed by the heart in pumping the increased cardiac output. The increases in flow through these three vascular beds are the result of arteriolar vasodilation in them. In both skeletal and cardiac

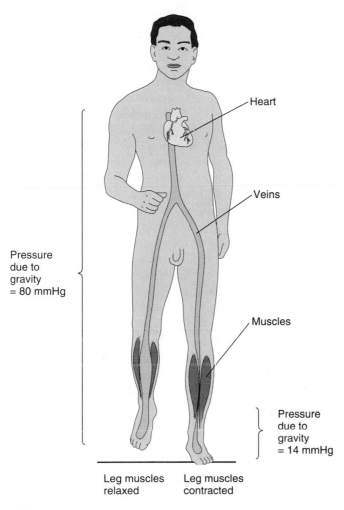

FIGURE 14-64

Role of contraction of the leg skeletal muscles in reducing capillary pressure and filtration in the upright position. The skeletal muscle contraction compresses the veins, causing intermittent emptying so that the columns of blood are interrupted.

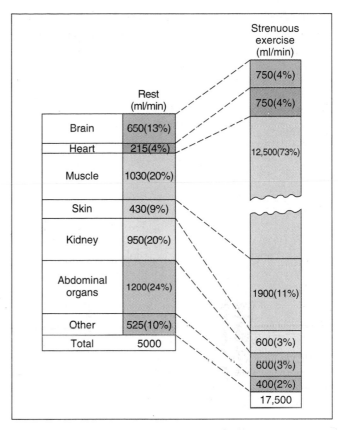

FIGURE 14-65

Distribution of the systemic cardiac output at rest and during strenuous exercise. The values at rest were previously presented in Figure 14-9. (*Adapted from Chapman and Mitchell.*)

muscle, the vasodilation is mediated by local metabolic factors, whereas the vasodilation in skin is achieved mainly by a *decrease* in the firing of the sympathetic neurons to the skin (additional mechanisms are described in Chapter 18). At the same time that arteriolar vasodilation is occurring in these three beds, arteriolar vasoconstriction—manifested as decreased blood flow in Figure 14-65—is occurring in the kidneys and gastrointestinal organs, secondary to *increased* activity of the sympathetic neurons supplying them.

Vasodilation of arterioles in skeletal muscle, cardiac muscle, and skin causes a decrease in total peripheral resistance to blood flow. This decrease is partially offset by vasoconstriction of arterioles in other organs. Such resistance "juggling," however, is quite incapable of compensating for the huge dilation of the muscle arterioles, and

the net result is a marked decrease in total peripheral resistance.

What happens to arterial blood pressure during exercise? As always, the mean arterial pressure is simply the arithmetic product of cardiac output and total peripheral resistance. During most forms of exercise (Figure 14-66 illustrates the case for mild exercise), the cardiac output tends to increase somewhat more than the total peripheral resistance decreases, so that mean arterial pressure usually increases a small amount. Pulse pressure, in contrast, markedly increases, mainly because of the increased speed at which the stroke volume is ejected.

The cardiac output increase during exercise is due to a large increase in heart rate and a small increase in stroke volume. The heart rate increase is caused by a combination of decreased parasympathetic activity to the SA node and increased sympathetic activity. The increased stroke volume is due mainly to an increased ventricular contractility, manifested by an increased ejection fraction and mediated by the sympathetic nerves to the ventricular myocardium. Note, however, in Figure 14-66 that there is a small increase (10 percent) in end-diastolic ventricular

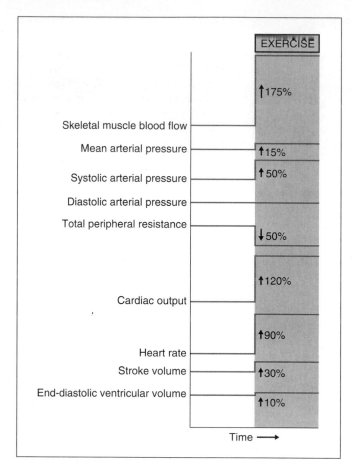

FIGURE 14-66

Summary of cardiovascular changes during mild upright exercise. The person was sitting quietly prior to the exercise.

volume; because of this increased filling, the Frank-Starling mechanism also contributes to the increased stroke volume, although not to the same degree as the increased contractility does. The increased contractility also accounts for the greater speed at which the stroke volume is ejected, as noted in the previous paragraph.

We have focused our attention on factors that act directly upon the heart to alter cardiac output during exercise, but it would be incorrect to leave the impression that these factors, by themselves, are sufficient to account for the elevated cardiac output. The fact is that cardiac output can be increased to high levels only if the peripheral processes favoring venous return to the heart are simultaneously activated to the same degree. Otherwise, the shortened filling time resulting from the high heart rate would lower end-diastolic volume and stroke volume by the Frank-Starling mechanism. Factors promoting venous return during exercise are: (1) increased activity of the skeletal-muscle pump, (2) increased depth and frequency of inspiration, (3) sympathetically mediated increase in venous

tone, and (4) greater ease of blood flow from arteries to veins through the dilated skeletal muscle arterioles.

What are the control mechanisms by which the cardiovascular changes in exercise are elicited? As described previously, vasodilation of arterioles in skeletal and cardiac muscle once exercise is underway represents active hyperemia secondary to local metabolic factors within the muscle. But what drives the enhanced sympathetic outflow to most other arterioles, the heart, and the veins, and the decreased parasympathetic outflow to the heart? The control of this autonomic outflow during exercise offers an excellent example of what we earlier referred to as a pre-programmed pattern, modified by continuous afferent input. One or more discrete control centers in the brain are activated during exercise by output from the cerebral cortex, and according to this "central command," descending pathways from these centers to the appropriate autonomic preganglionic neurons elicit the firing pattern typical of exercise. Indeed, these centers begin to "direct traffic" even before the exercise begins, since a person just about to begin exercising already manifests many of the changes in cardiac and vascular function; thus, this constitutes a feedforward system.

Once exercise is underway, local chemical changes in the muscle can develop, particularly during high levels of exercise, because of imperfect matching between flow and metabolic demands. These changes activate chemoreceptors in the muscle. Afferent input from these receptors goes to the medullary cardiovascular center and facilitates the output reaching the autonomic neurons from higher brain centers (Figure 14-67). The result is a further increase in heart rate, myocardial contractility, and vascular resistance in the nonactive organs. Such a system permits a fine degree of matching between cardiac pumping and total oxygen and nutrients required by the exercising muscles. Mechanoreceptors in the exercising muscles are also stimulated and provide input to the medullary cardiovascular center.

Finally, the arterial baroreceptors also play a role in the altered autonomic outflow. Knowing that the mean and pulsatile pressures rise during exercise, you might logically assume that the arterial baroreceptors will respond to these elevated pressures and signal for increased parasympathetic and decreased sympathetic outflow, a pattern designed to counter the rise in arterial pressure. In reality, however, exactly the opposite occurs; the arterial baroreceptors play an important role in *elevating* the arterial pressure over that existing at rest. The reason is that one neural component of the central command output goes to the arterial baroreceptors and "resets" them upward as exercise begins. This resetting causes the baroreceptors to respond as though arterial pressure had decreased, and their output (decreased action-potential frequency) signals for decreased parasympathetic and increased sympathetic outflow.

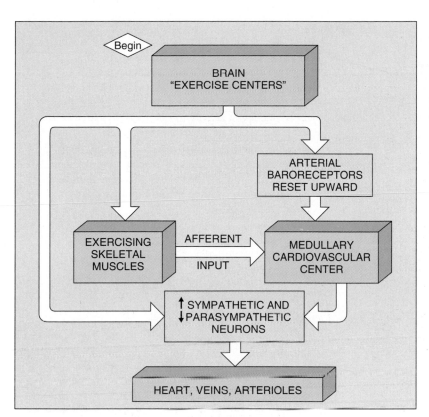

FIGURE 14-67
Control of the autonomic nervous system during exercise. The primary outflow to the sympathetic and parasympathetic neurons is via pathways from "exercise centers" in the brain. Afferent input from chemoreceptors in the exercising muscles and from reset arterial baroreceptors also influence the autonomic neurons by way of the medullary cardiovascular center.

Table 14-9 summarizes the changes that occur during moderate endurance exercise, that is, exercise (like jogging, swimming, or fast walking) that involves large muscle groups for an extended period of time.

In closing, a few words should be said about the other major category of exercises—those involving maintained *isometric* contractions, as in weight-lifting. Here, too, cardiac output and arterial blood pressure increase, and the arterioles in the exercising muscles undergo vasodilation due to local metabolic factors. However, there is a crucial difference: During isometric contractions, once the contracting muscles exceed 10 to 15 percent of their maximal force, the blood flow to the muscle is greatly reduced because the muscles are physically compressing the blood vessels that run through them. Thus, the cardiovascular changes are ineffective in causing increased blood flow to the muscles, and isometric contractions can be maintained only briefly before fatigue sets in.

Maximal Oxygen Consumption and Training

As the magnitude of any endurance exercise increases, oxygen consumption also increases in exact proportion until a point is reached when it fails to rise despite a further increment in work load. This is known as **maximal oxygen consumption ($\dot{V}_{O_2}$max).** After $\dot{V}_{O_2}$max has been reached, work can be increased and sustained only very briefly by anaerobic metabolism in the exercising muscles.

It is important to distinguish $\dot{V}_{O_2}$max from *endurance* (also termed work capacity). To take an example from cars: $\dot{V}_{O_2}$max is analogous to the maximal power the engine can develop, but maximal power says nothing about how far the car can go on a certain amount of fuel—its work capacity or endurance.

Theoretically, $\dot{V}_{O_2}$max could be limited by (1) the cardiac output, (2) the respiratory system's ability to deliver oxygen to the blood, or (3) the exercising muscles' ability to use oxygen. In fact, in normal people (except for a few very highly trained athletes), cardiac output is the factor that determines $\dot{V}_{O_2}$max. With increasing work load (Figure 14-68), heart rate increases progressively and markedly until it reaches a maximum. Stroke volume increases much less and tends to level off when 75 percent of $\dot{V}_{O_2}$max has been reached (it actually starts to go back down in elderly people). The major factors responsible for limiting the rise in stroke volume and, hence, cardiac output are (1) the very rapid heart rate, which decreases diastolic filling time, and (2) inability of the peripheral factors favoring venous return (skeletal-muscle pump, respiratory pump, venoconstriction, arteriolar vasodilation) to increase ventricular filling further during the very short time available.

A person's $\dot{V}_{O_2}$max is not fixed at any given value but can be altered by the habitual level of physical activity. For example, prolonged bed rest may decrease maximal exercise capacity by 15 to 25 percent, whereas intense

TABLE 14-9 CARDIOVASCULAR CHANGES IN MODERATE ENDURANCE EXERCISE

Variable	Change	Explanation
Cardiac output	Increases	Heart rate and stroke volume both increase, the former to a greater extent.
Heart rate	Increases	Sympathetic nerve activity to the SA node increases, and parasympathetic nerve activity decreases.
Stroke volume	Increases	Contractility increases due to increased sympathetic nerve activity to the ventricular myocardium; increased ventricular end-diastolic volume also contributes to increased stroke volume by the Frank-Starling mechanism.
Total peripheral resistance	Decreases	Resistance in heart and skeletal muscles decreases more than resistance in other vascular beds increases.
Mean arterial pressure	Increases	Cardiac output increases more than total peripheral resistance decreases.
Pulse pressure	Increases	Stroke volume and velocity of ejection of the stroke volume increase.
End-diastolic volume	Increases	Filling time is decreased by the high heart rate, but this is more than compensated for by the factors favoring venous return—venoconstriction, skeletal-muscle pump, and increased inspiratory movements.
Blood flow to heart and skeletal muscle	Increases	Active hyperemia occurs in both vascular beds, mediated by local metabolic factors.
Blood flow to skin	Increases	Sympathetic nerves to skin vessels are inhibited reflexly by the increase in body temperature.
Blood flow to viscera	Decreases	Sympathetic nerves to the blood vessels in the abdominal organs and the kidneys are stimulated.
Blood flow to brain	Unchanged	Autoregulation of brain arterioles maintains constant flow despite the increased mean arterial pressure.

long-term physical training may increase it by a similar amount. To be effective, the training must be of an endurance type and must include certain minimal levels of duration, frequency, and intensity. For example, jogging 20 to 30 min three times weekly at 5 to 8 mi/h definitely produces a significant training effect in most people.

At rest, compared to his or her values prior to training, the trained individual has an increased stroke volume and decreased heart rate with no change in cardiac output. At $\dot{V}_{O_2}$max, he or she has an increased cardiac output, compared to pretraining values; this is due entirely to an increased maximal stroke volume since maximal heart rate is not altered by training (Figure 14-68). The increase in stroke volume is due to a combination of (1) effects on the heart (the mechanism is unknown but may include a thicker myocardium and increased ventricular contractility), and (2) peripheral effects, including increased blood volume and increases in the number of blood vessels in skeletal muscle, which permit increased muscle blood flow and venous return.

Training also increases the concentrations of oxidative enzymes and mitochondria in the exercised muscles (Chapter 12). These changes increase the speed and efficiency of metabolic reactions in the muscles and permit large increases—200 to 300 percent—in exercise *endurance*, but they do not increase $\dot{V}_{O_2}$max because they were not limiting it in the untrained individuals.

HYPERTENSION

Hypertension is defined as a chronically increased systemic arterial pressure. The dividing line between normal pressure and hypertension is set at approximately 140/90 mmHg.

Theoretically, hypertension could result from an increase in cardiac output or in total peripheral resistance, or both. In reality, however, the major abnormality in most cases of well-established hypertension is increased total

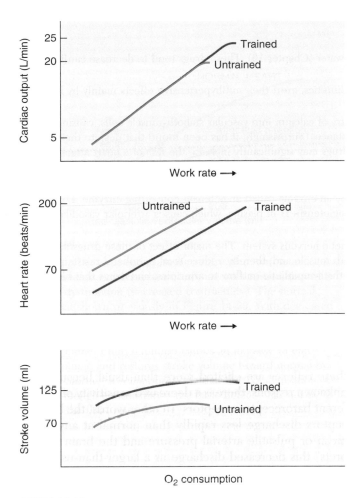

FIGURE 14-68

Changes in cardiac output, heart rate, and stroke volume with increasing work in untrained and trained persons.

peripheral resistance caused by abnormally reduced arteriolar radius.

What causes the arteriolar constriction? In only a small fraction of cases is the cause known. For example, diseases that damage a kidney or decrease its blood supply are often associated with **renal hypertension;** the cause of the hypertension is increased release of renin from the kidney, with subsequent increased generation of the potent vasoconstrictor angiotensin II. However, for more than 95 percent of the individuals with hypertension, the cause of the arteriolar constriction is unknown. Hypertension of unknown cause is called **primary hypertension** (formerly "essential hypertension").

Many hypotheses have been proposed to explain the increased arteriolar constriction of primary hypertension. At present, much evidence suggests that excessive sodium retention is a contributing factor in genetically predisposed ("salt-sensitive") persons. Although this relationship remains controversial, it is certainly true that many persons with hypertension show a drop in blood pressure after being on low-sodium diets or receiving drugs, termed **diuretics,** that cause increased sodium loss via the urine. Low dietary intake of calcium has also been implicated as a possible contributor to primary hypertension, but this, too, remains controversial. Obesity is a definite risk factor for primary hypertension, and weight reduction and exercise are frequently effective in causing some reduction of blood pressure in overweight sedentary persons with hypertension. Cigarette smoking, too, is a definite risk factor.

Hypertension causes a variety of problems. One of the organs most affected is the heart. Because the left ventricle in a hypertensive person must chronically pump against an increased arterial pressure, it develops an adaptive increase in muscle mass (**left ventricular hypertrophy**). In the early phases of the disease, this helps maintain the heart's function as a pump. With time, however, changes in the organization and properties of myocardial cells occur, and these result in diminished contractile function and heart failure (see below). The presence of hypertension also enhances the development of atherosclerosis and heart attacks (see below), kidney damage, and rupture of a cerebral blood vessel, which causes localized brain damage—a **stroke.**

The major categories of drugs used to treat hypertension are summarized in Table 14-10. These drugs all act in ways that reduce cardiac output and/or total peripheral resistance. You will note in subsequent sections of this chapter that these same drugs are also used in the treatment of heart failure and in both the prevention and treatment of heart attacks. One reason for this overlap is that these three diseases are causally interrelated; for example, as noted in this section, hypertension is a major risk factor for the development of heart failure and heart attacks. But in addition, drugs such as the **beta-adrenergic receptor blockers, calcium-channel blockers,** and **angiotensin-converting enzyme inhibitors** have multiple cardiovascular effects, which may play different roles in the treatment of the different diseases.

HEART FAILURE

Heart failure (also termed congestive heart failure) is a complex of signs and symptoms that occurs when the heart fails to pump an adequate cardiac output. This may happen for many reasons; two examples are pumping against a chronically elevated arterial pressure in hypertension, and structural damage due to decreased coronary blood flow. It has become standard practice to separate persons with heart failure into two categories: (1) those with diastolic dysfunction (problems with ventricular filling) and

intrinsic pathway
extrinsic pathway
tissue factor
vitamin K
tissue factor pathway
   inhibitor (TFPI)
thrombomodulin
protein C

antithrombin III
heparin
fibrinolytic system
plasminogen
plasmin
plasminogen activators
tissue plasminogen activator
   (t-PA)

transient ischemic attacks
   (TIAs)
embolus
embolism
hematoma
hemophilia

hypercoagulability
aspirin
oral anticoagulants
thrombolytic therapy
recombinant t-PA
streptokinase

SECTION G REVIEW QUESTIONS

1. Describe the sequence of events leading to platelet activation and aggregation, and the formation of a platelet plug. What helps keep this process localized?

2. Diagram the clotting pathway beginning with prothrombin.

3. What is the role of platelets in clotting?

4. List all the procoagulant effects of thrombin.

5. How is the clotting cascade initiated? How does the extrinsic pathway recruit the intrinsic pathway?

6. Describe the roles of the liver and vitamin K in clotting.

7. List three ways in which clotting is limited.

8. Diagram the fibrinolytic system.

9. How does fibrin help initiate the fibrinolytic system?

CHAPTER 14 CLINICAL TERMS

anemia
iron-deficiency anemia
pernicious anemia
hemorrhage
sickle-cell anemia
microcytosis
normocytosis
macrocytosis
polycythemia
ectopic pacemakers
AV conduction disorders
atrial fibrillation
heart murmurs
stenosis
insufficiency
echocardiography
cardiac angiography
elephantiasis
edema
hypotension
shock
hypertension
renal hypertension
primary hypertension
diuretics

beta-adrenergic receptor
   blockers
calcium-channel blockers
angiotensin-converting
   enzyme inhibitors
left ventricular hypertrophy
stroke
heart failure
diastolic dysfunction
systolic dysfunction
pulmonary edema
digitalis
coronary artery disease
myocardial infarction
heart attack
angina pectoris
ventricular fibrillation
cardiopulmonary
   resuscitation (CPR)
defibrillation
atherosclerosis
coronary thrombosis
nitroglycerin
coronary balloon angioplasty
coronary bypass

CHAPTER 14 THOUGHT QUESTIONS

(Answers are given in Appendix A)

1. A person is found to have a hematocrit of 35 percent. Can you conclude from this that there is a decreased volume of erythrocytes in the blood?

2. Which would cause a greater increase in resistance to flow, a doubling of blood viscosity or a halving of tube radius?

3. If all plasma-membrane calcium channels in contractile cardiac-muscle cells were blocked with a drug, what would happen to the muscle's action potentials and contraction?

4. A person with a heart rate of 40 has no P waves but normal QRS complexes on the ECG. What is the explanation?

5. A person has a left ventricular systolic pressure of 180 mmHg and an aortic systolic pressure of 110 mmHg. What is the explanation?

6. A person has a left atrial pressure of 20 mmHg and a left ventricular pressure of 5 mmHg during ventricular filling. What is the explanation?

7. A patient is taking a drug that blocks beta-adrenergic receptors. What changes in cardiac function will the drug cause?

8. What is the mean arterial pressure in a person whose systolic and diastolic pressures are, respectively, 160 and 100 mmHg?

9. A person is given a drug that doubles the blood flow to her kidneys but does not change the mean arterial pressure. What must the drug be doing?

10. A blood vessel removed from an experimental animal dilates when exposed to acetylcholine. After the endothelium is scraped from the lumen of the vessel, it no longer dilates in response to this mediator. Explain.

11. A person is accumulating edema throughout the body. Average capillary pressure is 25 mmHg, and lymphatic function is normal. What is the most likely cause of the edema?

12. A person's cardiac output is 7 L/min and mean arterial pressure is 140 mmHg. What is the person's total peripheral resistance?

13. The following data are obtained for an experimental animal before and after a drug. Before: Heart rate = 80 beats/min, and stroke volume = 80 ml/beat. After: Heart rate = 100 beats/min, and stroke volume = 64 ml/beat. Total peripheral resistance remains unchanged. What has the drug done to mean arterial pressure?

14. When the nerves from all the arterial baroreceptors are cut in an experimental animal, what happens to mean arterial pressure?

15. What happens to the hematocrit within several hours after a hemorrhage?

CHAPTER 15

Respiration

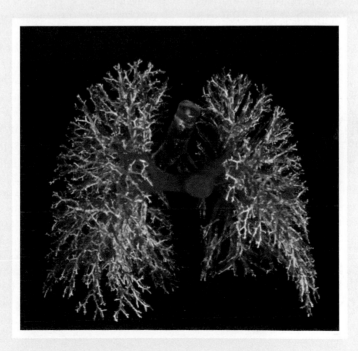

Respiration has two quite different meanings: (1) utilization of oxygen in the metabolism of organic molecules by cells (often termed internal or cellular respiration), as described in Chapter 4, and (2) the exchanges of oxygen and carbon dioxide between an organism and the external environment. The second meaning is the subject of this chapter.

Human cells obtain most of their energy from chemical reactions involving oxygen. In addition, cells must be able to eliminate carbon dioxide, the major end product of oxidative metabolism. A unicellular organism can exchange oxygen and carbon dioxide directly with the external environment, but this is obviously impossible for most cells of a complex organism like a human being. Therefore, the evolution of large animals required the development of specialized structures to exchange oxygen and carbon dioxide for the entire animal with the external environment. In humans (and other mammals), the **respiratory system** includes the lungs, the series of tubes leading to the lungs, and the chest structures responsible for moving air into and out of the lungs during breathing.

In addition to the provision of oxygen and elimination of carbon dioxide, the respiratory system serves other functions, as listed in Table 15-1.

ORGANIZATION OF THE RESPIRATORY SYSTEM

There are two lungs, the right and left, each divided into several lobes. **Pulmonary** is the adjective referring to "lungs." The lungs consist mainly of tiny air-containing sacs called **alveoli** (singular, **alveolus**), which number approximately 300 million. The alveoli are the sites of gas exchange with the blood. The **airways** are all the tubes through which air flows between the external environment and the alveoli.

Inspiration is the movement of air from the external environment through the airways into the alveoli during breathing. **Expiration** is movement in the opposite direction. An inspiration and an expiration constitute a respiratory cycle. During the entire respiratory cycle, the right ventricle of the heart pumps blood through the capillaries surrounding each alveolus. At rest, in a normal adult, approximately 4 L of fresh air enters and leaves the alveoli per minute, while 5 L of blood, the entire cardiac output, flows through the pulmonary capillaries. During heavy exercise, the air flow can increase twenty-fold, and the blood flow five- to six-fold.

The Airways and Blood Vessels

During inspiration air passes through either the nose (the most common site) or mouth into the **pharynx** (throat), a passage common to both air and food (Figure 15-1). The pharynx branches into two tubes, the esophagus through which food passes to the stomach, and the **larynx,** which is part of the airways. The larynx houses the **vocal cords,** two folds of elastic tissue stretched horizontally across its lumen. The flow of air past the vocal cords causes them to vibrate, producing sounds. The nose, mouth, pharynx, and larynx are termed the upper airways.

The larynx opens into a long tube, the **trachea,** which in turn branches into two **bronchi** (singular, **bronchus**), one of which enters each lung. Within the lungs, there are more than 20 generations of branchings, each resulting in narrower, shorter, and more numerous tubes, the names of which are summarized in Figure 15-2. The walls of the trachea and bronchi contain cartilage, which gives them their cylindrical shape and supports them. The first airway branches that no longer contain cartilage are termed **bronchioles.** Alveoli first begin to appear, attached to the walls of respiratory bronchioles. The number of alveoli increases in the alveolar ducts (Figure 15-2), and the airways then end in grapelike clusters consisting entirely of alveoli (Figure 15-3). The airways, like blood vessels, are surrounded by smooth muscle, the contraction or relaxation of which can alter airway radius.

The airways beyond the larynx can be divided into two zones: (1) The **conducting zone** extends from the top of the trachea to the beginning of the respiratory bronchi-

---

TABLE 15-1 FUNCTIONS OF THE RESPIRATORY SYSTEM

1. Provides oxygen
2. Eliminates carbon dioxide
3. Regulates the blood's hydrogen-ion concentration (pH)
4. Forms speech sounds (phonation)
5. Defends against microbes
6. Influences arterial concentrations of chemical messengers by removing some from pulmonary capillary blood and producing and adding others to this blood
7. Traps and dissolves blood clots

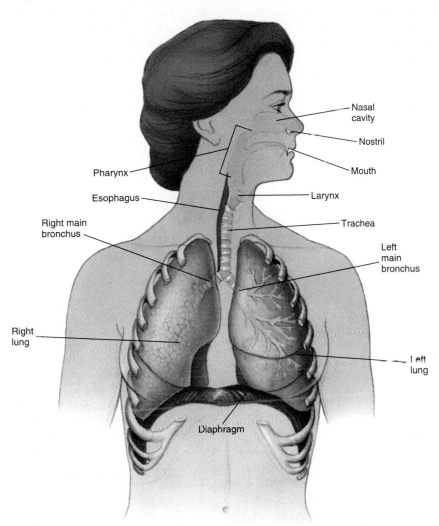

Pharynx

Esophagus

Right main
bronchus

Right
lung

Diaphragm

Nasal
cavity

Nostril

Mouth

Larynx

Trachea

Left
main
bronchus

Left
lung

FIGURE 15-1
Organization of the respiratory system. The ribs have been removed in front, and the left lung is shown in cross-section. The esophagus, cut off in the figure, is not part of the respiratory system.

oles; it contains no alveoli and there is no gas exchange with the blood (Table 15-2). (2) The **respiratory zone,** which extends from the respiratory bronchioles on down, contains alveoli and there is an exchange of gases with the blood.

The blood vessels supplying the lung generally accompany the airways and also undergo numerous branchings. The smallest of these vessels branch into networks of capillaries that richly supply the alveoli (Figure 15-3).

The epithelial surfaces of the airways, to the end of the respiratory bronchioles, contain cilia that constantly beat toward the pharynx. They also contain glands and individual epithelial cells that secrete mucus. Particulate matter, such as dust contained in the inspired air, sticks to the mucus, which is continually and slowly moved by the cilia to the pharynx and then swallowed. This mucus escalator is important in keeping the lungs clear of particulate matter and the many bacteria that enter the body on dust particles. Ciliary activity can be inhibited by many noxious agents. For example, smoking a single cigarette can immobilize the cilia for several hours.

The airway epithelium also secretes a watery fluid upon which the mucus can ride freely. The production of this fluid is impaired in the disease called **cystic fibrosis,** the most common lethal genetic disease of Caucasians, and the mucus layer becomes thick and dehydrated. The impaired secretion is due to a defect in the chloride channels involved in the secretory process.

A second protective mechanism against infection is provided by cells that are present in the airways and alveoli and are termed macrophages. These cells, described in detail in Chapter 20, engulf inhaled particles and bacteria,

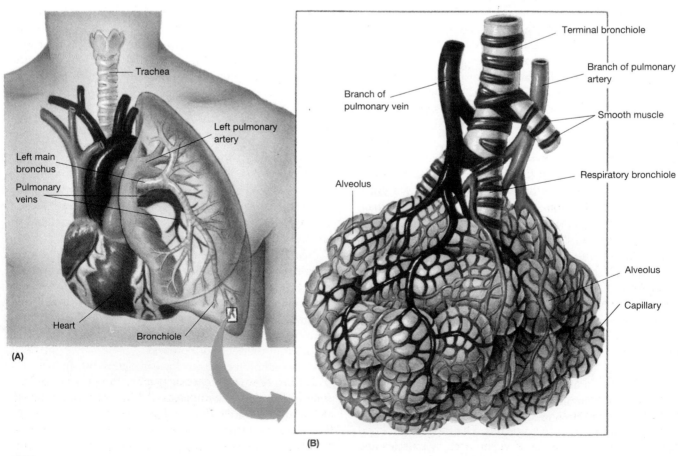

	Name of branches	Number of tubes in branch
Conducting zone	Trachea	1
	Bronchi	2
		4
		8
	Bronchioles	16
		32
	Terminal bronchioles	6×10^4
Respiratory zone	Respiratory bronchioles	5×10^5
	Alveolar ducts	
	Alveolar sacs	8×10^6

FIGURE 15-2
Airway branching. 𝔛

FIGURE 15-3
Relationships between blood vessels and airways. (A) The lung appears transparent so that the relationships can be seen. The airways beyond the bronchiole are too small to be seen. (B) An enlargement of a small section of Figure 15-3A to show the continuation of the airways and the clusters of alveoli at their ends. Virtually the entire lung, not just the surface, consists of such clusters. 𝔛

Trachea

Left pulmonary artery

Left main bronchus

Pulmonary veins

Heart

Bronchiole

(A)

Terminal bronchiole

Branch of pulmonary artery

Branch of pulmonary vein

Smooth muscle

Alveolus

Respiratory bronchiole

Alveolus

Capillary

(B)

TABLE 15-2 FUNCTIONS OF THE CONDUCTING ZONE OF THE AIRWAYS

1. Provides a low-resistance pathway for air flow; resistance is physiologically regulated by changes in contraction of airway smooth muscle and by physical forces acting upon the airways.
2. Defends against microbes and other toxic chemicals and foreign matter; cilia, mucus, and phagocytes perform this function.
3. Warms and moistens the air.
4. Phonates (vocal cords).

rendering them harmless. Macrophages, like cilia, are injured by cigarette smoke and air pollutants.

Site of Gas Exchange: The Alveoli

The alveoli are tiny hollow sacs whose open ends are continuous with the lumens of the airways (Figures 15-2 and 15-4A). Typically, the air in two adjacent alveoli is separated by a single alveolar wall (Figure 15-4A). Most of the air-facing surface(s) of the wall are lined by a continuous layer, one cell thick, of flat epithelial cells termed **type I alveolar cells.** Interspersed between these cells are thicker specialized cells termed **type II alveolar cells** (Figure 15-4B) that produce a detergent-like substance, surfactant, to be discussed below.

The alveolar walls contain capillaries and a very small interstitial space, which consists of interstitial fluid and a loose meshwork of connective tissue (Figure 15-4B). In many places, the interstitial space is absent altogether, and the basement membranes of the alveolar-surface epithelium and the capillary-wall endothelium fuse. Thus the blood within an alveolar-wall capillary is separated from the air within the alveolus by an extremely thin barrier (0.2 μm, compared with the 7-μm diameter of an average red blood cell). The total surface area of alveoli in contact with capillaries is roughly the size of a tennis court. This extensive area and the thinness of the barrier permit the rapid exchange of large quantities of oxygen and carbon dioxide by diffusion.

In some of the alveolar walls there are pores that permit the flow of air between alveoli. This route can be very important when the airway leading to an alveolus is occluded by disease, since some air can still enter the alveolus by way of the pores between it and adjacent alveoli.

Relation of the Lungs to the Thoracic (Chest) Wall

The lungs, like the heart, are situated in the **thorax,** the compartment of the body between the neck and abdomen. "Thorax" and "chest" are synonyms. The thorax is a closed compartment that is bounded at the neck by muscles and

connective tissue and completely separated from the abdomen by a large, dome-shaped sheet of skeletal muscle, the **diaphragm.** The wall of the thorax is formed by the spinal column, the ribs, the breastbone (sternum), and several groups of muscles that run between the ribs (collectively termed the **intercostal muscles**). The thoracic wall also contains large amounts of elastic connective tissue.

Each lung is surrounded by a completely closed sac, the **pleural sac,** consisting of a thin sheet of cells called **pleura.** The two pleural sacs, one on each side of the midline, are completely separate from each other. The relationship between a lung and its pleural sac can be visualized by imagining what happens when you push a fist into a balloon (Figure 15-5): The arm represents the major bronchus leading to the lung, the fist is the lung, and the balloon is the pleural sac. The fist becomes coated by one surface of the balloon. In addition, the balloon is pushed back upon itself so that its opposite surfaces lie close together. Unlike the hand and balloon, however, the pleural surface coating the lung is firmly attached to the lung by connective tissue. Similarly, the outer layer is attached to and lines the interior thoracic wall and diaphragm. The two layers of pleura are so close to each other that normally they are always in virtual contact, but they are *not* attached to each other. Rather, they are separated by an extremely thin layer of **intrapleural fluid,** the total volume of which is only a few milliliters. The intrapleural fluid totally surrounds the lungs and lubricates the pleural surfaces so that they can slide over each other during breathing. More important, as we shall see in the next section, changes in the hydrostatic pressure of the intrapleural fluid—the **intrapleural pressure (P_{ip}),** or intrathoracic pressure—cause the lungs and thoracic wall to move in and out together during normal breathing.

VENTILATION AND LUNG MECHANICS

An inventory of steps involved in respiration (Figure 15-6) is provided for orientation before beginning the detailed descriptions of each step, beginning with ventilation.

Ventilation is defined as the exchange of air between the atmosphere and alveoli. Like blood, air moves by *bulk flow,* from a region of high pressure to one of low pressure. We saw in Chapter 14 that bulk flow can be described by the equation

$$F = \Delta P/R \qquad (15\text{-}1)$$

That is, flow (F) is proportional to the pressure difference (ΔP) between two points and inversely proportional to the resistance (R). For air flow into or out of the lungs, the relevant pressures are the gas pressure in the alveoli—the **alveolar pressure (P_{alv})**—and the gas pressure at the

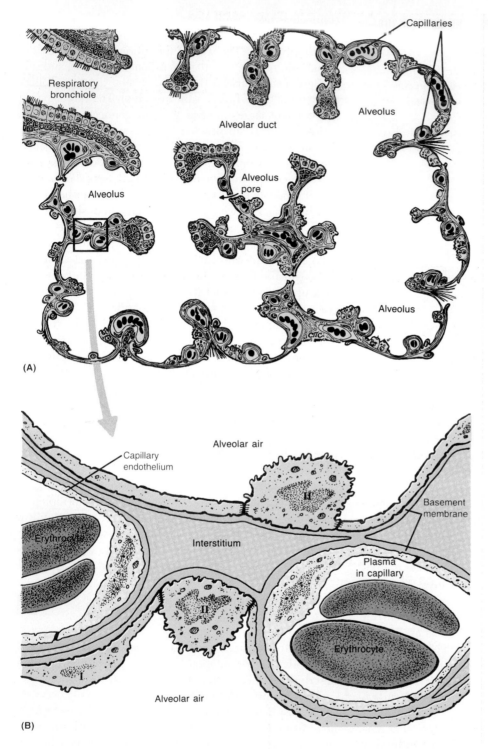

Respiratory bronchiole

Alveolar duct

Capillaries

Alveolus

Alveolus

Alveolus pore

Alveolus

(A)

Capillary endothelium

Alveolar air

Basement membrane

Erythrocyte

Interstitium

Plasma in capillary

Erythrocyte

Alveolar air

(B)

FIGURE 15-4

(A) Cross section through an area of the respiratory zone. There are 18 alveoli in this figure, only 4 of which are labelled. Two frequently share a common wall. (*From R. O. Greep and L. Weiss, "Histology," 3d ed., McGraw-Hill, New York, 1973.*)
(B) Schematic enlargement of a portion of an alveolar wall. I = nucleus of a type I cell; II = nucleus of a type II cell. (*Adapted from Gong and Drage.*)

nose and mouth, normally **atmospheric pressure (P_{atm})**, the pressure of the air surrounding the body:

$$F = (P_{atm} - P_{alv})/R \qquad (15\text{-}2)$$

A very important point must be made at this point: All pressures in the respiratory system, as in the cardiovascular system, are given *relative to atmospheric pressure*, which is 760 mmHg at sea level. For example, the alveolar pressure between breaths is said to be 0 mmHg, which means that it is the same as atmospheric pressure. During ventilation, air moves into and out of the lungs because the alveolar pressure is alternately made less than and greater than atmospheric pressure (Figure 15-7). These

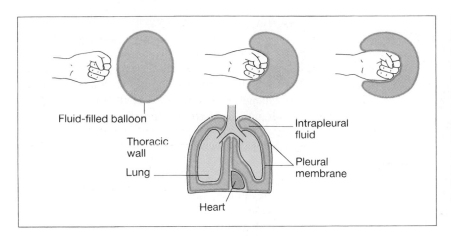

Relationship of lungs, pleura, and thoracic wall, shown as analogous to pushing a fist into a fluid-filled balloon. Note that there is no communication between the right and left intrapleural fluids. For purposes of illustration, in this figure and other similar ones in this chapter, the volume of intrapleural fluid is greatly exaggerated. It normally consists of an extremely thin layer of fluid between the pleura membrane lining the inner surface of the thoracic wall and that lining the outer surface of the lungs.

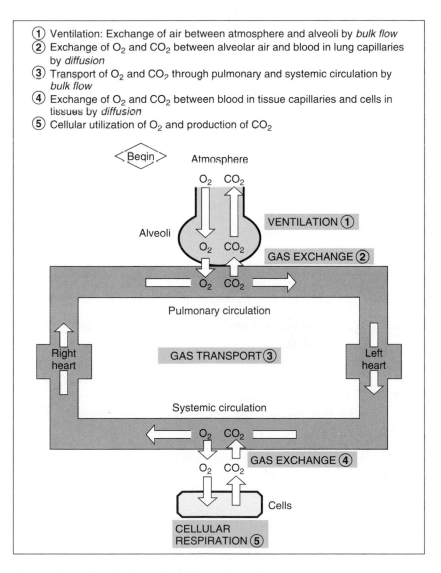

① Ventilation: Exchange of air between atmosphere and alveoli by *bulk flow*
② Exchange of O_2 and CO_2 between alveolar air and blood in lung capillaries by *diffusion*
③ Transport of O_2 and CO_2 through pulmonary and systemic circulation by *bulk flow*
④ Exchange of O_2 and CO_2 between blood in tissue capillaries and cells in tissues by *diffusion*
⑤ Cellular utilization of O_2 and production of CO_2

FIGURE 15-6
The steps of respiration.

alveolar pressure changes are caused, as we shall see, by changes in the dimensions of the lungs.

To understand how a change in lung dimensions causes a change in alveolar pressure you need to learn one more basic concept—**Boyle's law** (Figure 15-8). At constant temperature, the relationship between the pressure exerted by a fixed number of gas molecules and the volume of their container is as follows: An increase in the volume

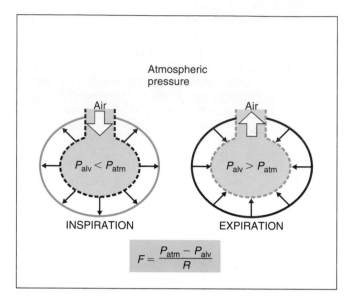

FIGURE 15-7

Relationships required for ventilation. When the alveolar pressure (P_{alv}) is less than atmospheric pressure (P_{atm}), air enters the lungs. Flow (F) is directly proportional to the pressure difference and inversely proportional to airway resistance (R). Black lines show lung's position at beginning of inspiration or expiration, and blue lines at end. ✗

FIGURE 15-8

Boyle's law: The pressure exerted by a constant number of gas molecules in a container is inversely proportional to the volume of the container, that is, P is proportional to $1/V$. ✗

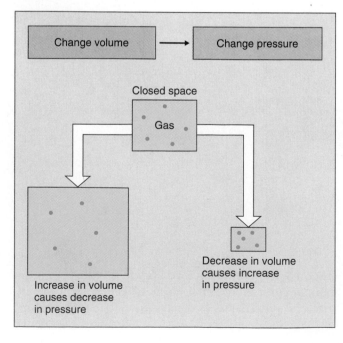

of the container decreases the pressure of the gas, whereas a decrease in the container volume increases the pressure.

It is essential to recognize the correct causal sequences in ventilation: During inspiration and expiration the volume of the "container"—the lungs—is made to change, and these changes then cause, by Boyle's law, the alveolar pressure changes that drive air flow into or out of the lungs. Our descriptions of ventilation must focus, therefore, on how the changes in lung dimensions are brought about.

There are no muscles attached to the lung surface to pull the lungs open or push them shut. Rather, the lungs are passive elastic structures—like balloons, and their volume, therefore, depends upon: (1) the *difference* in pressure—termed the **transpulmonary pressure**—between the inside and the outside of the lungs; and (2) how stretchable the lungs are. The rest of this section and the next three sections focus only on transpulmonary pressure; stretchability will be discussed later in the section on lung compliance.

The pressure inside the lungs is the *air pressure* inside the alveoli (P_{alv}), and the pressure outside the lungs is the pressure of the *intrapleural fluid* surrounding the lungs (P_{ip}). Thus,

$$\text{Transpulmonary pressure} = P_{alv} - P_{ip} \quad (15\text{-}3)$$

To repeat, the respiratory muscles are not attached to the lung surface. Rather, these muscles are part of the chest wall. When they contract or relax, they directly change the dimensions of the *thorax*, which in turn causes the transpulmonary pressure to change. The changes in transpulmonary pressure then cause the changes in lung volume.

The Stable Balance Between Breaths

Figure 15-9 illustrates the situation that normally exists at the end of an unforced expiration, that is, between breaths when the respiratory muscles are relaxed and no air is flowing. As noted above, the alveolar pressure (P_{alv}) is 0 mmHg, that is, it is the same as atmospheric pressure. The intrapleural pressure (P_{ip}) is approximately 4 mmHg less than atmospheric pressure, that is, -4 mmHg, using the standard convention of giving all pressures relative to atmospheric pressure. Therefore, the transpulmonary pressure ($P_{alv} - P_{ip}$) equals [0 mmHg$-(-4$ mmHg)] = 4 mmHg. As emphasized in the previous section, this transpulmonary pressure is the force acting to inflate the lungs; it is opposed by the elastic recoil of the partially expanded and, therefore, partially stretched lungs. In other words, inherent elastic recoil tending to collapse the lungs is exactly balanced by the transpulmonary pressure tending to expand them, and the volume of the lungs is stable at this point.

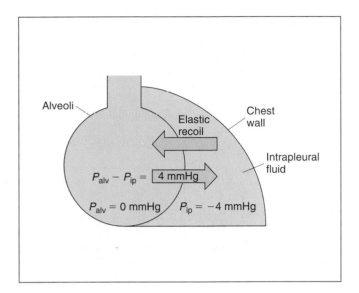

FIGURE 15-9

Alveolar (P_{alv}), intrapleural (P_{ip}), and transpulmonary ($P_{alv} - P_{ip}$) pressures at the end of an unforced expiration. The transpulmonary pressure exactly opposes the elastic recoil of the lung, and the lung volume (and chest wall) remain stable.

At the same time there is also a pressure difference of 4 mmHg pushing *inward* on the chest wall for the following reason. The atmospheric pressure against the *outside* of the chest wall (P_{atm}) is 0 mmHg, and the intrapleural pressure against the *inside* of the chest wall is −4 mmHg; the difference is 4 mmHg directed inward. This pressure difference across the chest wall just balances the tendency of the partially compressed elastic chest wall to move outward, and so the chest wall also is stable in the absence of any muscular contraction. [Notice that even though the value of the pressure acting across the chest wall—4 mmHg—between breaths is the same as the value of the transpulmonary pressure, the force acting across the chest wall is *not* the transpulmonary pressure ($P_{alv} - P_{ip}$) but is rather ($P_{atm} - P_{ip}$).]

Clearly, the subatmospheric intrapleural pressure is the essential factor keeping the lungs partially expanded and the chest wall partially compressed. The important question now is: What causes the intrapleural pressure to be subatmospheric? As the lungs (tending to move inward from their stretched position because of their elastic recoil) and the thoracic wall (tending to move outward from its compressed position because of its elastic recoil) move ever so slightly away from each other, there occurs an infinitesimal enlargement of the fluid-filled intrapleural space between them. But fluid cannot expand the way air can, and so even this tiny enlargement of the intrapleural space—so small that the pleural surfaces still remain in contact with each other—drops the intrapleural pressure below atmospheric pressure. In this way, the elastic recoil

of both the lungs and thoracic wall creates the subatmospheric intrapleural pressure that keeps them from moving apart more than a very tiny amount.

The importance of the transpulmonary pressure in achieving this stable balance can be seen when, during surgery or trauma, the chest wall is pierced without damaging the lung. Atmospheric air rushes through the wound into the intrapleural space (a phenomenon called **pneumothorax**), and the intrapleural pressure goes from −4 mmHg to 0 mmHg. The transpulmonary pressure acting to hold the lung open is thus eliminated, and the lung collapses. At the same time, the chest wall moves outward since its elastic recoil is also no longer opposed.

Inspiration

Figures 15-10 and 15-11 summarize the events during normal inspiration at rest. Inspiration is initiated by the neurally induced contraction of the diaphragm and the "inspiratory" intercostal muscles located between the ribs

FIGURE 15-10

Sequence of events during inspiration. Figure 15-11 illustrates these events quantitatively.

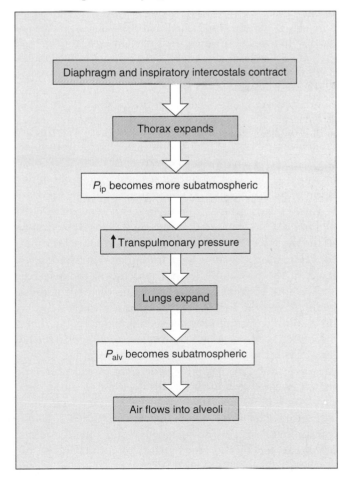

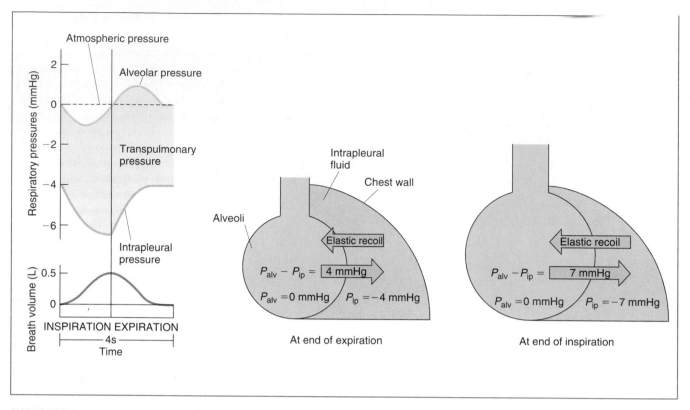

FIGURE 15-11

Summary of alveolar, intrapleural, and transpulmonary pressure changes and air flow during inspiration and expiration of 500 ml of air. The transpulmonary pressure is the gray area between the alveolar pressure and intrapleural pressure. Note that atmospheric pressure (760 mmHg at sea level) has a value of zero on the respiratory pressure scale. Note also on the right side of the figure that the transpulmonary pressure exactly opposes the elastic recoil of the lungs at the end of both inspiration and expiration.

(the adjective "inspiratory" here is a functional term, not an anatomical one; it denotes the several groups of intercostal muscles that contract during inspiration). The diaphragm is the most important inspiratory muscle during normal quiet breathing. When activation of the nerves to it causes it to contract, its dome moves downward into the abdomen, enlarging the thorax. Simultaneously, activation of the nerves to the inspiratory intercostal muscles causes them to contract, leading to an upward and outward movement of the ribs and a further increase in thoracic size.

The crucial point is that contraction of these muscles, by increasing the size of the thorax, upsets the stability set up by purely elastic forces between breaths. As the thorax enlarges, the thoracic wall moves ever so slightly further away from the lung surface, and the intrapleural fluid pressure therefore becomes even more subatmospheric than it was between breaths. This decrease in intrapleural pressure *increases* the transpulmonary pressure. Therefore, the force acting to expand the lungs—the transpulmonary pressure—is now greater than the elastic recoil exerted by the lungs at this moment, and so the lungs expand further. Note in Figure 15-11 that, by the

end of inspiration, equilibrium *across the lungs* is once again established since the more inflated lungs exert a greater elastic recoil, which equals the increased transpulmonary pressure.

Thus, when contraction of the inspiratory muscles actively increases the thoracic dimensions, the lungs are also forced to enlarge virtually to the same degree because of the change in intrapleural pressure and hence transpulmonary pressure. The enlargement of the lungs causes an increase in the sizes of the alveoli throughout the lungs. Therefore, by Boyle's law, the pressure within the alveoli drops to less than atmospheric. This produces the difference in pressure ($P_{alv} < P_{atm}$) that causes a bulk flow of air from the atmosphere through the airways into the alveoli. By the end of the inspiration, the pressure in the alveoli again equals atmospheric pressure.

Expiration

Figures 15-11 and 15-12 summarize the sequence of events during expiration. At the end of inspiration, the nerves to the diaphragm and inspiratory intercostal muscles cease firing, and so these muscles relax. The chest wall is no

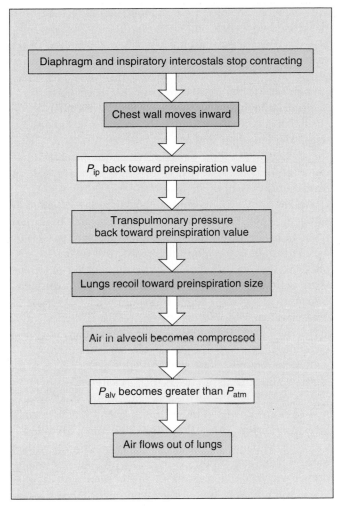

FIGURE 15-12

Sequence of events during expiration. Figure 15-11 illustrates these events quantitatively.

longer being actively being pulled outward and upward by the muscle contractions and so it starts to recoil to its original smaller dimensions existing between breaths. This immediately makes the intrapleural pressure less negative and hence *decreases* the transpulmonary pressure. Therefore, the transpulmonary pressure acting to expand the lungs is now smaller than the elastic recoil exerted by the more expanded lungs, and the lungs also passively recoil to their original dimensions. Thus, just as during inspiration, primary changes in the dimensions of the thorax during expiration force virtually identical changes in the dimensions of the lungs.

As the lungs become smaller, air in the alveoli becomes temporarily compressed so that, by Boyle's law, alveolar pressure exceeds atmospheric pressure. Therefore, air flows from the alveoli through the airways out into the atmosphere. Thus, expiration at rest is completely *passive,*

depending only upon the relaxation of the inspiratory muscles and recoil of the chest wall and stretched lungs.

Under certain conditions, during exercise, for example, expiration of larger volumes is achieved by contraction of a different set of intercostal muscles and the abdominal muscles, which *actively* decreases thoracic dimensions. The "expiratory" (again a functional term, not an anatomical one) intercostal muscles insert on the ribs in such a way that their contraction pulls the chest wall downward and inward. Contraction of the abdominal muscles increases intraabdominal pressure and forces the diaphragm up into the thorax.

Lung Compliance

To repeat, the degree of lung expansion at any instant is proportional to the transpulmonary pressure, that is, $P_{alv} - P_{ip}$. But just how much any given transpulmonary pressure expands the lung depends upon the stretchability, or compliance, of the lung. **Lung compliance (C_L)** is defined as the magnitude of the change in lung volume (ΔV_L) produced by a given change in the transpulmonary pressure:

$$C_L = \Delta V_L / \Delta(P_{alv} - P_{ip}) \qquad (15\text{-}4)$$

Thus, the greater the lung compliance, the easier it is to expand the lungs at any given transpulmonary pressure. A low lung compliance means that a greater-than-normal transpulmonary pressure must be developed across the lung to produce a given amount of lung expansion. In other words, when lung compliance is low, intrapleural pressure must be made more subatmospheric than usual during inspiration to achieve lung expansion. This requires more vigorous contractions of the diaphragm and inspiratory intercostal muscles. Thus, the less compliant the lung, the more energy is required for a given amount of expansion. Persons with low lung compliance due to disease therefore tend to breathe shallowly and must breathe rapidly to inspire an adequate volume of air.

Determinants of Lung Compliance. There are two major determinants of lung compliance. One is the stretchability of the lung tissues, particularly their elastic connective tissues. Thus a thickening of the lung tissues decreases lung compliance. However, an equally important determinant of lung compliance is not the elasticity of the lung tissues, but the surface tension at the air-water interfaces within the alveoli.

The surface of the alveolar cells is moist, and so they can be pictured as air-filled sacs lined with water. At an air-water interface, the attractive forces between the water molecules, known as **surface tension,** make the water lining like a stretched balloon that constantly tries to shrink and resists further stretching. Thus, expansion of the lung requires energy not only to stretch the connective tissue of the lung but also to overcome the surface tension of the water layer lining the alveoli.

Indeed, the surface tension of pure water is so great that were the alveoli lined with pure water, lung expansion would require exhausting muscular effort and the lungs would tend to collapse. It is extremely important, therefore, that the type II alveolar cells secrete a detergent-like substance known as pulmonary **surfactant,** which markedly reduces the cohesive forces between water molecules on the alveolar surface. Therefore, surfactant lowers the surface tension, which increases lung compliance and makes it easier to expand the lungs (Table 15-3).

Surfactant is a complex of both lipids and proteins, but its major component is a phospholipid that forms a monomolecular layer between the air and water at the alveolar surface. The amount of surfactant tends to decrease when breaths are small and constant. A deep breath, which people normally intersperse frequently in their breathing pattern, stretches the type II cells, which stimulates the secretion of surfactant. This is why patients who have had thoracic or abdominal surgery and are breathing shallowly because of the pain must be urged to take occasional deep breaths.

A striking example of what occurs when surfactant is deficient is the disease known as *respiratory-distress syndrome of the newborn.* This is the second leading cause of death in premature infants, in whom the surfactant-synthesizing cells may be too immature to function adequately. Because of low lung compliance, the infant is able to inspire only by the most strenuous efforts, which may ultimately cause complete exhaustion, inability to breathe, lung collapse, and death. Therapy in such cases is assisted breathing with a mechanical ventilator and the administration of natural or synthetic surfactant via the infant's trachea.

Airway Resistance

As previously stated, the volume of air that flows into or out of the alveoli per unit time is directly proportional to the pressure difference between the atmosphere and alveoli and inversely proportional to the resistance to flow offered by the airways. The factors that determine airway resistance are analogous to those determining vascular resistance in the circulatory system: tube length, tube radius, and interactions between moving molecules (gas molecules, in this case). As in the circulatory system the most important factor by far is the tube radius: Airway resistance is inversely proportional to the fourth power of the airway radii.

Total pulmonary resistance to air flow is normally so small that very small pressure differences suffice to produce large volumes of air flow. As we have seen (Figure 15-11), the average atmosphere-to-alveoli pressure difference during a normal breath at rest is less than 1 mmHg; yet, approximately 500 ml of air is moved by this tiny difference.

Airway radii and therefore resistance are affected by physical, neural, and chemical factors. One important physical factor is the transpulmonary pressure, which exerts a distending force on the airways, just as on the alveoli. This is a major factor keeping the smaller airways—those without cartilage to support them—from collapsing. Because, as we have seen, transpulmonary pressure increases during inspiration, airway radius becomes larger and airway resistance smaller as the lungs expand during inspiration. The opposite occurs during expiration.

A second physical factor holding the airways open is the elastic connective tissue fibers that attach to the airway exteriors and, because of their arrangement, continuously pull outward on the sides of the airways. This is termed **lateral traction.** Since these fibers become stretched as the lungs expand, they help pull the airways open even more during inspiration. Thus, both the transpulmonary pressure and lateral traction act in the same direction.

Such physical factors also explain why the airways become narrower and airway resistance increases during a forced expiration. Indeed, because of increased airway resistance, there is a limit as to how much one can increase the air flow rate during a forced expiration no matter how intense the effort.

In addition to these physical factors, a variety of neuroendocrine and paracrine factors can influence airway smooth muscle and thereby airway resistance. For example, the hormone epinephrine relaxes airway smooth muscle (via an effect on beta-adrenergic receptors) whereas the leukotrienes (members of the eicosanoid family) produced in the lungs during inflammation contract the muscle.

One might wonder why physiologists are concerned with all the many physical and chemical factors that *can* influence airway resistance when we earlier stated that airway resistance is *normally* so low that it is no impediment to air flow. The reason is that, under *abnormal* circumstances, changes in these factors may cause serious increases in airway resistance. Asthma and chronic obstructive pulmonary disease provide important examples.

Asthma.   *Asthma* is a disease characterized by intermittent attacks in which airway smooth muscle contracts

---

TABLE 15-3   SOME IMPORTANT FACTS ABOUT PULMONARY SURFACTANT

1. Pulmonary surfactant is a mixture of phospholipids and protein.
2. It is secreted by type II alveolar cells.
3. It lowers surface tension of the water layer at the alveolar surface, which increases lung compliance, that is, makes the lungs easier to expand.
4. Its concentration decreases when breaths are small and constant.

strongly, markedly increasing airway resistance. The basic defect in asthma is inflammation of the airways, the causes of which vary from person to person and include, among others, allergy and virus infections. The important point is that the underlying inflammation causes the airway smooth muscle to be hyperresponsive and to contract strongly when, depending upon the individual, he or she exercises or is exposed to cold air, cigarette smoke, environmental pollutants, viruses, allergens, normally released bronchoconstrictor chemical messengers, and a variety of other potential triggers.

The therapy for asthma is twofold: (1) to reduce the inflammation and hence the airway hyperresponsiveness with so-called anti-inflammatory drugs, particularly glucocorticoids (Chapter 20) taken by inhalation; and (2) to overcome the excessive airway smooth-muscle contraction with **bronchodilator drugs,** that is, drugs that relax the airways. The latter drugs work on the airways either by enhancing the actions of bronchodilator neuroendocrine or paracrine messengers or by blocking the actions of bronchoconstrictors. For example, one class of bronchodilator drug mimics the action of epinephrine on beta-adrenergic receptors.

Chronic Obstructive Pulmonary Disease. The term *chronic obstructive pulmonary disease* refers to emphysema or chronic bronchitis or a combination of the two. These diseases, which cause severe difficulties not only in ventilation but in oxygenation of the blood, are among the major causes of disability and death in the United States. In contrast to asthma, increased smooth-muscle contraction is *not* the cause of airway obstruction in these diseases.

Emphysema is characterized by destruction of the alveolar walls and consequently a marked enlargement of the alveolar air spaces and loss of pulmonary capillaries. These changes impair gas exchange, as will be described later. In addition, the small airways—those from the terminal bronchioles on down—are reduced in number and have atrophied walls. How all these changes occur is still poorly understood, but it is hypothesized that the lung undergoes self-destruction by proteolytic enzymes secreted by leukocytes in the lung in response to a variety of factors. Cigarette smoking is by far the most important cause of emphysema; it probably does so by stimulating the release of these proteolytic enzymes and by destroying other enzymes that normally protect the lung against the proteolytic enzymes.

The airway obstruction in emphysema is due to collapse of the airways. To understand this, recall that two physical factors passively holding the airways open are the transpulmonary pressure and the lateral traction of connective-tissue fibers attached to the airway exteriors. Both of these factors are diminished in emphysema because of the destruction of the lung elastic tissues, and so the airways collapse.

Chronic bronchitis is characterized by excessive mucus production in the bronchi and chronic inflammatory changes in the small airways. The cause of obstruction is an accumulation of mucus in the airways and thickening of the inflamed airways. The same agents—smoking, for example—that cause emphysema also cause chronic bronchitis, which is why the two diseases frequently coexist.

The Heimlich Maneuver. The *Heimlich maneuver* is used to aid a person choking on foreign matter caught in the upper airways. A sudden increase in abdominal pressure is produced as the rescuer's fists, placed against the victim's abdomen slightly above the navel and well below the tip of the sternum, are pressed into the abdomen with a quick upward thrust (Figure 15-13). The increased abdominal pressure forces the diaphragm upward into the thorax, reducing thoracic size and, by Boyle's Law, increasing alveolar pressure. The forceful expiration produced

FIGURE 15-13

The Heimlich maneuver. The rescuer's fists are placed against the victim's abdomen. A quick upward thrust of the fists causes elevation of the diaphragm and a forceful expiration. As this expired air is forced through the trachea and larynx, the foreign object in the airway is expelled.

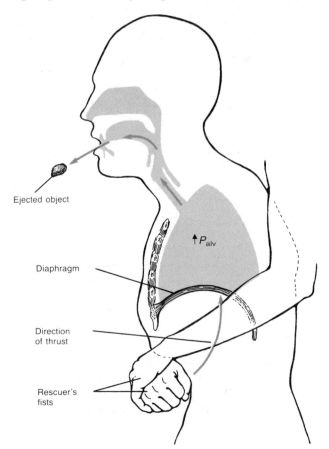

Ejected object

$\uparrow P_{alv}$

Diaphragm

Direction of thrust

Rescuer's fists

by the increased alveolar pressure often expels the object caught in the respiratory tract.

Lung Volumes and Capacities

Normally the volume of air entering the lungs during a single inspiration is approximately equal to the volume leaving on the subsequent expiration and is called the **tidal volume.** The tidal volume during normal quiet breathing is termed the resting tidal volume and is approximately 500 ml. As illustrated in Figure 15-14, the maximal amount of air that can be increased above this value during deepest inspiration is termed the **inspiratory reserve volume** (and is about 3000 ml, that is, six-fold greater that resting tidal volume).

After expiration of a resting tidal volume, the lungs still contain a very large volume of air. As described earlier this is the resting position of the lungs and chest wall, that is, the position that exists when there is no contraction of the respiratory muscles; it is termed the functional residual capacity (and averages about 2500 ml). Thus, the 500 ml of air inspired with each resting breath adds to and mixes with the much larger volume of air already in the lungs, and then 500 ml of the total is expired. Through maximal active contraction of the expiratory muscles, it is possible to expire much more of the air remaining after the resting tidal volume has been expired; this additional volume is termed the **expiratory reserve volume** (about 1500 ml). Even after a maximal active expiration, approximately 1000 ml of air still remains in the lungs and is termed the **residual volume.**

A useful clinical measurement is the **vital capacity,** the maximal volume of air that a person can expire after a maximal inspiration. Under these conditions, the person is expiring both the resting tidal volume and inspiratory reserve volume just inspired, plus the expiratory reserve volume (Figure 15-14). In other words, the vital capacity is the sum of these three volumes.

A variant on this method is the **_forced expiratory volume in 1 s, (FEV$_1$),_** in which the person takes a maximal inspiration and then exhales maximally *as fast as possible.* The important value is the fraction of the total "forced" vital capacity expired in 1 s. Normal individuals can expire approximately 80 percent.

These measurements are useful diagnostic tools. For example, people with **_obstructive lung diseases_** (increased airway resistance) typically have a FEV$_1$ which is less than 80 percent of the vital capacity because it is difficult for them to expire air rapidly through the narrowed airways. In contrast to obstructive lung diseases, **_restrictive lung diseases_** are characterized by normal airway resistance but impaired respiratory movements because of abnormalities in the lung tissue, the pleura, the chest wall, or the neuromuscular machinery. Restrictive lung diseases are characterized by a reduced vital capacity but a normal ratio of FEV$_1$ to vital capacity.

Alveolar Ventilation

The total ventilation per minute, termed the **minute ventilation,** is equal to the tidal volume multiplied by the respiratory rate:

Minute ventilation =
(ml/min)

$$\text{tidal volume} \times \text{respiratory rate} \quad (15\text{-}5)$$
$$\text{(ml/breath)} \quad \text{(breaths/min)}$$

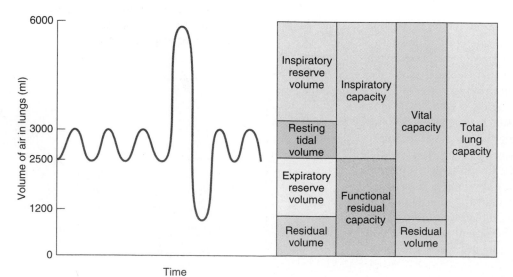

FIGURE 15-14

Lung volumes and capacities recorded on a spirometer, an apparatus for measuring inspired and expired volumes. When the subject inspires, the pen moves up; with expiration, it moves down. The capacities are the sums of two or more lung volumes. The lung volumes are the four distinct components of total lung capacity. Note that residual volume and total lung capacity cannot be measured with a spirometer.

For example, at rest, a normal person moves approximately 500 ml of air in and out of the lungs with each breath and takes 10 breaths each minute. The minute ventilation is therefore 500 ml/breath × 10 breaths/minute = 5000 ml of air per minute. However, because of dead space, not all this air is available for exchange with the blood.

Dead Space.   The conducting airways have a volume of about 150 ml. Exchanges of gases with the blood occur only in the alveoli and not in this 150 ml of the airways. Picture, then, what occurs during expiration of a tidal volume, which in this instance is 450 ml. The 450 ml of air is forced out of the alveoli and through the airways. Approximately 300 ml of this alveolar air is exhaled at the nose or mouth, but approximately 150 ml still remains in the airways at the end of expiration. During the next inspiration (Figure 15-15), 450 ml of air flows into the alveoli, but the first 150 ml entering the alveoli is not atmospheric air but the 150 ml left behind from the last breath. Thus, only 300 ml of new atmospheric air enters the alveoli during the inspiration. The end result is that 150 ml of the 450 ml of atmospheric air entering the respiratory system during each inspiration never reaches the alveoli but is merely moved in and out of the airways. Because these airways do not permit gas exchange with the blood, the space within them is termed the **anatomic dead space.**

Thus the volume of *fresh* air entering the alveoli during each inspiration equals the tidal volume minus the volume of air in the anatomic dead space. For the example:

> Tidal volume = 450 ml
>
> Anatomic dead space = 150 ml
>
> Fresh air entering alveoli in one inspiration =
> 450 ml − 150 ml = 300 ml

The total volume of fresh air entering the alveoli per minute is called the **alveolar ventilation:**

Alveolar ventilation =
   (ml/min)
   (tidal volume − dead space) × respiratory rate   (15-6)
   (ml/breath)   (ml/breath)   (breath/min)

When evaluating the efficacy of ventilation one should always focus on the alveolar ventilation, not the minute ventilation. This generalization is demonstrated readily by the data in Table 15-4. In this experiment, subject A breathes rapidly and shallowly, B normally, and C slowly and deeply. Each subject has exactly the same minute ventilation, that is, each is moving the same amount of air in and out of the lungs per minute. Yet, when we subtract the anatomic-dead-space ventilation from the minute ventilation, we find marked differences in alveolar ventilation. Subject A has no alveolar ventilation and would become unconscious in several minutes, whereas C has a considerably greater alveolar ventilation than B, who is breathing normally.

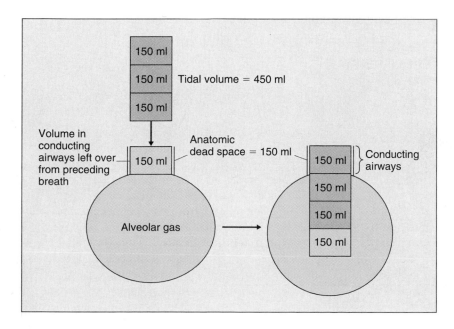

FIGURE 15-15

Effects of anatomic dead space on alveolar ventilation. Anatomic dead space is the volume of the conducting airways.

TABLE 15-4   EFFECT OF BREATHING PATTERNS ON ALVEOLAR VENTILATION

Subject	Tidal volume, ml/breath	×	Frequency, breaths/min	=	Minute ventilation, ml/min	Anatomic dead-space ventilation, ml/min	Alveolar ventilation, ml/min
A	150		40		6000	$150 \times 40 = 6000$	0
B	500		12		6000	$150 \times 12 = 1800$	4200
C	1000		6		6000	$150 \times \ 6 = \ 900$	5100

Another important generalization to be drawn from this example is that increased *depth* of breathing is far more effective in elevating alveolar ventilation than is an equivalent increase in breathing *rate*. Conversely, a decrease in depth can lead to a critical reduction in alveolar ventilation. This is because a fixed volume of *each* tidal volume goes to the dead space. If the tidal volume decreases, the fraction of the tidal volume going to the dead space increases until, as in subject A, it may represent the entire tidal volume. On the other hand, any increase in tidal volume goes entirely toward increasing alveolar ventilation. These concepts have important physiological implications. Most situations that produce an increased ventilation, such as exercise, reflexly call forth a relatively greater increase in breathing depth than rate.

The anatomic dead space is not the only type of dead space. Some fresh inspired air is not used for gas exchange with the blood even though it reaches the alveoli because some alveoli, for various reasons, have little or no blood supply. This volume of air is known as **alveolar dead space.** It is quite small in normal persons but may be very large in several kinds of lung disease. As we shall see, it is minimized by local mechanisms that match air and blood flows. The sum of the anatomic and alveolar dead spaces is known as the **physiologic dead space.**

EXCHANGE OF GASES IN ALVEOLI AND TISSUES

We have now completed our discussion of the lung mechanics that produce alveolar ventilation, but this is only the first step in the respiratory process. Oxygen must move across the alveolar membranes into the pulmonary capillaries, be transported by the blood to the tissues, leave the tissue capillaries and enter the extracellular fluid, and finally cross plasma membranes to gain entry into cells. Carbon dioxide must follow a similar path in reverse.

In the steady state, the volume of oxygen that leaves the tissue capillaries and is consumed by the body cells per unit time is exactly equal to the volume of oxygen added to the blood in the lungs during the same time period. Similarly, in

the steady state, the rate at which carbon dioxide is produced by the body cells and enters the systemic blood is identical to the rate at which carbon dioxide leaves the blood in the lungs and is expired. Note that these statements apply to the steady state; *transiently,* oxygen utilization in the tissues *can* differ from oxygen uptake in the lungs and carbon dioxide production can differ from elimination in the lungs, but within a short time these imbalances automatically produce changes in diffusion gradients in the lungs and tissues that reestablish steady state balances.

The amounts of oxygen consumed by cells and carbon dioxide produced, however, are not necessarily identical to each other. The balance depends primarily upon which nutrients are being used for energy. The ratio of CO_2 produced/O_2 consumed is known as the **respiratory quotient (RQ).** On a mixed diet the RQ is approximately 0.8, that is, 8 molecules of CO_2 are produced for every 10 molecules of O_2 consumed. (The RQ is 1 for carbohydrate, 0.7 for fat, and 0.8 for protein.)

For purposes of illustration, Figure 15-16 presents typical exchange values during 1 min for a person at rest, assuming a cellular oxygen consumption of 250 ml/min, a carbon dioxide production of 200 ml/min, an alveolar ventilation (supply of fresh air to the alveoli) of 4000 ml/min, and a cardiac output of 5000 ml/min. Since only 21 percent of the atmospheric air is oxygen, the total oxygen entering the alveoli per min is 21 percent of 4000 ml, or 840 ml/min. Of this inspired oxygen, 250 ml crosses the alveoli into the pulmonary capillaries and the rest is subsequently exhaled. Note that blood entering the lungs already contains a large quantity of oxygen, to which the new 250 ml is added. The blood then flows from the lungs to the left heart and is pumped by the left ventricle through the tissue capillaries, where 250 ml of oxygen leaves the blood to be taken up and utilized by cells. Thus, the quantities of oxygen added to the blood in the lungs and removed in the tissues are identical. The story reads in reverse for carbon dioxide: There is already a good deal of carbon dioxide in systemic arterial blood; to this is added an additional 200 ml, the amount produced by the cells, as blood flows through tissue capillaries; this 200 ml leaves the blood as blood flows through the lungs, and is expired.

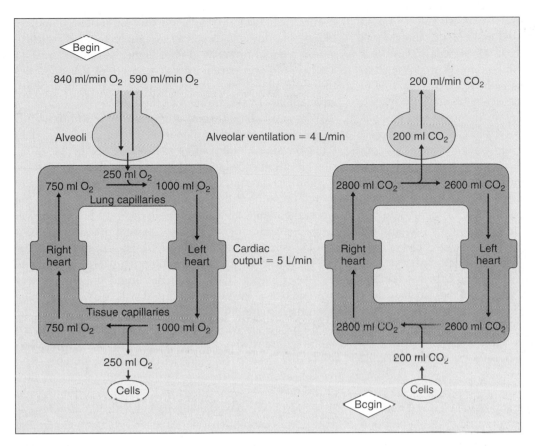

FIGURE 15-16

Summary of typical oxygen and carbon dioxide exchanges between atmosphere, lungs, blood, and tissues during 1 min in a resting individual. Note that the values given in this figure for oxygen and carbon dioxide in blood are *not* the values per liter of blood but rather the amounts transported per minute in the cardiac output (5 L in this example). The volume of oxygen in 1 L of arterial blood = 200 ml O_2/L of blood, that is, 1000 ml O_2/5 L of blood.

Blood pumped by the heart carries oxygen and carbon dioxide between the lungs and tissues by bulk flow, but diffusion is responsible for the net movement of these molecules between the alveoli and blood, and between the blood and the cells of the body. Understanding the mechanisms involved in these diffusional exchanges depends upon some basic chemical and physical properties of gases, which we will now discuss.

Partial Pressures of Gases

Gas molecules undergo continuous random motion. These rapidly moving molecules exert a pressure, the magnitude of which is increased by anything that increases the rate of movement. The pressure a gas exerts is proportional to (1) the temperature (because heat increases the speed at which molecules move) and (2) the concentration of the gas, that is, the number of molecules per unit volume.

As stated by **Dalton's law,** in a mixture of gases, the pressure exerted by each gas is independent of the pressure exerted by the others. This is because gas molecules are normally so far apart that they do not interfere with each other. Since each gas in a mixture behaves as though no other gases are present, the total pressure of the mixture is simply the sum of the individual pressures. These individual pressures, termed **partial pressures,** are denoted by a *P* in front of the symbol for the gas. For example, the partial pressure of oxygen is represented by P_{O_2}. Net diffusion of a gas will occur from a region where its partial pressure is high to a region where it is low.

Atmospheric air consists primarily of nitrogen (approximately 79 percent) and oxygen (approximately 21 percent), with very small quantities of water vapor, carbon dioxide, and inert gases. The sum of the partial pressures of all these gases is termed atmospheric pressure, or barometric pressure. It varies in different parts of the world as a result of differences in altitude (it also varies with local weather conditions), but at sea level it is 760 mmHg. Since the partial pressure of any gas in a mixture is the fractional concentration of that gas times the total pressure of all the gases, the P_{O_2} of atmospheric air is 0.21 × 760 mmHg = 160 mmHg at sea level.

Diffusion of Gases in Liquids. When a liquid is exposed to air containing a particular gas, molecules of the gas will enter the liquid and dissolve in it. **Henry's law** states that the amount of gas dissolved will be directly proportional to the partial pressure of the gas with which the liquid is in equilibrium. A corollary is that, at equilibrium, the partial pressures of the gas molecules in the liquid and gaseous phases must be identical. Suppose, for example, that a closed container contains both water and gaseous oxygen. Oxygen molecules from the gas phase constantly bombard the surface of the water, some entering the water and dissolving. Since the number of molecules striking the surface is directly proportional to the P_{O_2} of the gas phase, the number of molecules entering the water is also directly proportional to the P_{O_2}. As long as the P_{O_2} in the gas phase is higher than the P_{O_2} in the liquid, there will be a net diffusion of oxygen into the liquid. Diffusion equilibrium will be reached when the P_{O_2} in the liquid is equal to the P_{O_2} in the gas phase, and there will be no further net diffusion between the two phases.

Conversely, if a liquid containing a dissolved gas at high partial pressure is exposed to a lower partial pressure of that same gas in a gas phase, gas molecules will diffuse out of the liquid into the gas phase until the partial pressures in the two phases become equal.

The exchanges *between* gas and liquid phases described in the last two paragraphs are precisely the phenomena occurring between alveolar air and pulmonary capillary blood. In addition, dissolved gas molecules also diffuse *within* a liquid from a region of higher partial pressure to a region of lower partial pressure, an effect that underlies the exchange of gases between cells, extracellular fluid, and capillary blood throughout the body.

Why must the diffusion of gases into or within liquids be presented in terms of partial pressures rather than "concentrations," the values used to deal with the diffusion of all other solutes? The reason is that the *concentration* of a gas in a liquid is proportional not only to the partial pressure of the gas but also to the solubility of the gas in the liquid; the more soluble the gas, the greater will be its concentration at any given partial pressure. Thus, if a liquid is exposed to two different gases at the same partial pressures, at equilibrium the *partial pressures* of the two gases will be identical in the liquid but the *concentrations* of the gases in the liquid will differ, depending upon their solubilities in that liquid.

With these basic gas properties as the foundation, we can now discuss the diffusion of oxygen and carbon dioxide across alveolar, capillary, and plasma membranes. The partial pressures of these gases in air and in various sites of the body are given in Figure 15-17 for a resting person at sea level. We start our discussion with the *alveolar* gas pressures because their values set those of systemic arterial

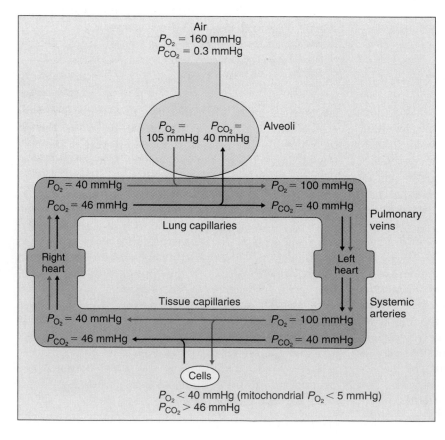

FIGURE 15-17

Partial pressures of carbon dioxide and oxygen in inspired air at sea level and various places in the body. The reason that the alveolar P_{O_2} and pulmonary vein P_{O_2} are not exactly the same is described later in the text. Note also that the P_{O_2} in the systemic arteries is shown as identical to that in the pulmonary veins; for reasons involving the anatomy of the blood flow to the lungs, the systemic arterial value is actually slightly less, but we have ignored this for the sake of clarity.

blood. This fact cannot be emphasized too strongly: The alveolar P_{O_2} and P_{CO_2} determine the systemic arterial P_{O_2} and P_{CO_2}.

Alveolar Gas Pressures

Normal alveolar gas pressures are $P_{O_2} = 105$ mmHg and $P_{CO_2} = 40$ mmHg. (We do not deal with nitrogen, even though it is the most abundant gas in the alveoli, because nitrogen is biologically inert under normal conditions and does not undergo any net exchange in the alveoli.) Compare these values with the gas pressures in the air being breathed: $P_{O_2} = 160$ mmHg and $P_{CO_2} = 0.3$ mmHg, a value so low that we will simply assume it to be zero. The alveolar P_{O_2} is lower than atmospheric P_{O_2} because some of the oxygen in the air entering the alveoli leaves them to enter the pulmonary capillaries. Alveolar P_{CO_2} is higher than atmospheric P_{CO_2} because carbon dioxide enters the alveoli from the pulmonary capillaries.

The factors that determine the precise value of alveolar P_{O_2} are (1) the P_{O_2} of atmospheric air, (2) the rate of alveolar ventilation, and (3) the rate of cellular oxygen consumption. Although there are equations for calculating the alveolar gas pressures from these variables, we will describe the interactions in a qualitative manner (Table 15-5). To start, we will assume that only one of the factors changes at a time.

First, a decrease in the P_{O_2} of the inspired air (at high altitude, for example) will decrease alveolar P_{O_2}. A decrease in alveolar ventilation will do the same thing (Figure 15-18) since less fresh air is entering the alveoli per unit time. Finally, an increase in the cells' utilization of oxygen will lower alveolar P_{O_2} because a larger fraction of the oxygen in the entering fresh air will leave the alveoli to enter the blood and be used by the tissues (recall that in the steady state the volume of oxygen entering the blood in the lungs per unit time is always equal to the volume utilized by the tissues). This discussion has been in terms of things that *lower* alveolar P_{O_2}; just reverse the direction of change of the three factors to see how to increase P_{O_2}.

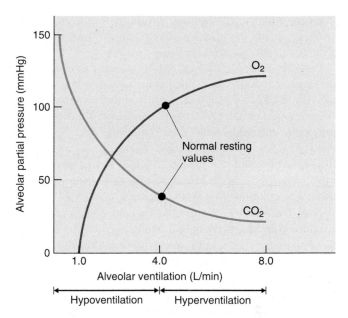

FIGURE 15-18

Effects of increasing or decreasing alveolar ventilation on alveolar partial pressures in a person having a constant metabolic rate (cellular oxygen consumption and carbon dioxide production). Note that alveolar P_{O_2} approaches zero when alveolar ventilation is about 1 L/min. At this point all the oxygen entering the alveoli crosses into the blood, leaving virtually no oxygen in the alveoli.

The story for P_{CO_2} is analogous, again assuming that only one factor changes at a time. There is normally essentially no carbon dioxide in inspired air and so we can ignore that factor. A decreased alveolar ventilation will increase the alveolar P_{CO_2} (Figure 15-18) because there is less inspired fresh air to dilute the carbon dioxide entering the alveoli from the blood. An increased production of carbon dioxide will also increase the alveolar P_{CO_2} because more carbon dioxide will be entering the alveoli from the blood per unit time (recall that in the steady state the volume of carbon dioxide entering the alveoli per unit time

TABLE 15-5 EFFECTS OF VARIOUS CONDITIONS ON ALVEOLAR GAS PRESSURES		
Condition	**Alveolar P_{O_2}**	**Alveolar P_{CO_2}**
Breathing air with low P_{O_2}	Decreases	No change°
↑Alveolar ventilation and unchanged metabolism	Increases	Decreases
↓Alveolar ventilation and unchanged metabolism	Decreases	Increases
↑Metabolism and unchanged alveolar ventilation	Decreases	Increases
Proportional increases in metabolism and alveolar ventilation	No change	No change

°Breathing air with low P_{O_2} has no direct effect on alveolar P_{CO_2}. However, as described later in the text, people in this situation will reflexly increase their ventilation and that will lower P_{CO_2}.

is always equal to the volume produced by the tissues). Just reverse the direction of changes in this paragraph to cause a *decrease* in alveolar P_{CO_2}.

For simplicity we allowed only one factor to change at a time, but if more than one factor changes, the effects will either add to or subtract from each other. For example, if oxygen consumption and alveolar ventilation both increase at the same time, their opposing effects on alveolar P_{O_2} will tend to cancel each other out.

This last example emphasizes that, at any particular atmospheric P_{O_2}, it is the *ratio* of oxygen consumption to alveolar ventilation (that is, O_2 consumption/alveolar ventilation) that determines alveolar P_{O_2}—the higher the ratio, the *lower* the alveolar P_{O_2}. Similarly, alveolar P_{CO_2} is determined by a ratio, in this case the *ratio* of carbon dioxide production to alveolar ventilation (that is, CO_2 production/alveolar ventilation); the higher the ratio, the *higher* the alveolar P_{CO_2}.

Two terms that denote the adequacy of ventilation, that is, the relationship between metabolism and alveolar ventilation, can now be defined. Physiologists state these definitions in terms of carbon dioxide rather than oxygen. **Hypoventilation** exists when there is an *increase* in the ratio of CO_2 production to alveolar ventilation. In other words, a person is said to be hypoventilating if his alveolar ventilation cannot keep pace with his carbon dioxide production. The result is that alveolar P_{CO_2} increases above the normal value of 40 mmHg. **Hyperventilation** exists when there is a *decrease* in the ratio of CO_2 production to alveolar ventilation, that is, when alveolar ventilation is actually too great for the amount of carbon dioxide being produced. The result is that alveolar P_{CO_2} decreases below the normal value of 40 mmHg.

Alveolar-Blood Gas Exchange

The blood that enters the pulmonary capillaries is, of course, systemic venous blood pumped to the lungs via the pulmonary arteries. Having come from the tissues, it has a relatively high P_{CO_2} (46 mmHg in a normal person at rest) and a relatively low P_{O_2} (40 mmHg) (Table 15-6). The differences in the partial pressures of oxygen and carbon dioxide on the two sides of the alveolar-capillary membrane result in the net diffusion of oxygen from alveoli to

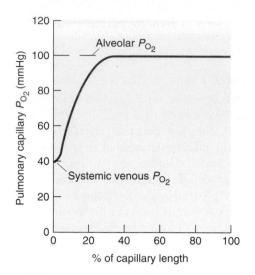

FIGURE 15-19

Equilibration of blood P_{O_2} with alveolar P_{O_2} along the length of the pulmonary capillaries.

blood and of carbon dioxide from blood to alveoli. As this diffusion occurs, the capillary blood P_{O_2} rises and its P_{CO_2} falls. The net diffusion of these gases ceases when the capillary partial pressures become equal to those in the alveoli.

In a normal person, the rates at which oxygen and carbon dioxide diffuse are so rapid and the blood flow through the capillaries so slow that complete equilibrium is reached well before the end of the capillaries (Figure 15-19). Only during the most strenuous exercise, when blood flows through the lung capillaries very rapidly, is there insufficient time for complete equilibration.

Thus, the blood that leaves the pulmonary capillaries to return to the heart and be pumped into the systemic arteries has essentially the same P_{O_2} and P_{CO_2} as alveolar air. (They are not exactly the same, for reasons given later.) Accordingly, the factors described in the previous section—atmospheric P_{O_2}, cellular oxygen consumption and carbon dioxide production, and alveolar ventilation—determine the alveolar gas pressures, which then determine the systemic arterial gas pressures.

TABLE 15-6	NORMAL GAS PRESSURES			
	Venous blood	**Arterial blood**	**Alveoli**	**Atmosphere**
P_{O_2}	40 mmHg	100 mmHg°	105 mmHg°	160 mmHg
P_{CO_2}	46 mmHg	40 mmHg	40 mmHg	0.3 mmHg

°The reason that the arterial P_{O_2} and alveolar P_{O_2} are not exactly the same is described later in the text.

Given that diffusion between alveoli and pulmonary capillaries normally achieves complete equilibration, the more capillaries participating in this process the more total oxygen and carbon dioxide can be exchanged. Many of the pulmonary capillaries are normally closed at rest. During exercise, these capillaries open and receive blood, thereby enhancing gas exchange. The mechanism by which this occurs is a simple physical one; the pulmonary circulation at rest is at such a low pressure that the pressure in many capillaries is inadequate to keep them open, but the increased cardiac output of exercise raises pulmonary vascular pressures, which opens these capillaries.

The diffusion of gases between alveoli and capillaries may be impaired in a number of ways, resulting in inadequate oxygen diffusion into the blood, particularly during exercise when the time for equilibration is reduced. For one thing, the surface area of the alveoli in contact with pulmonary capillaries may be decreased. In lung infections or pulmonary edema, for example, some of the alveoli may become filled with fluid. Diffusion may also be impaired if the alveolar walls become thickened, as, for example, in the disease (of unknown cause) called **diffuse interstitial fibrosis.** Very importantly, diffusion problems in the lung are restricted to oxygen and do not affect elimination of carbon dioxide, which is much more diffusible than oxygen.

Matching of Ventilation and Blood Flow in Alveoli

The major disease-induced cause of inadequate oxygen movement between alveoli and pulmonary capillary blood is the mismatching of the air supply and blood supply in individual alveoli.

The lungs are composed of approximately 300 million discrete alveoli, each capable of receiving carbon dioxide from, and supplying oxygen to, the pulmonary capillary blood. To be most efficient, the right proportion of alveolar air flow (ventilation) and capillary blood flow (perfusion) should be available to each alveolus. Any mismatching is termed **ventilation-perfusion inequality.**

The major effect of ventilation-perfusion inequality is to lower the P_{O_2} of systemic arterial blood. Indeed, largely because of gravitational effects on ventilation and perfusion there is enough ventilation-perfusion inequality in normal people to lower the arterial P_{O_2} about 5 mmHg. This is the major explanation of the fact, given earlier, that the P_{O_2} of blood in the pulmonary veins and systemic arteries is normally 5 mmHg less than that of average alveolar air.

In disease states, regional changes in lung compliance, airway resistance, and vascular resistance can cause marked ventilation-perfusion inequalities. The extremes of this phenomenon are easy to visualize: (1) There may be ventilated alveoli with no blood supply at all, or (2) there may be blood flowing through areas of lung that have no ventilation (this is termed a **shunt**). But the inequality need not be all-or-none to be quite significant.

A striking example of severe ventilation-perfusion inequality is provided by emphysema, in which systemic arterial P_{O_2} is usually much lower than alveolar P_{O_2}. First, because of the destruction of lung tissue in this disease, as well as the obstruction of airways, some areas of lung may receive large amounts of air while others receive little or none. Second, there is also a maldistribution of blood flow because capillaries are lost when the alveolar walls are destroyed.

Carbon dioxide elimination is also impaired by ventilation-perfusion inequality but, for complex reasons, not nearly to the same degree as oxygen uptake. Nevertheless, severe ventilation-perfusion inequalities in diseases such as emphysema can lead to some elevation of arterial P_{CO_2}.

There are several local homeostatic responses within the lungs to minimize the mismatching of ventilation and blood flow. One of the most important operates on the blood vessels to alter blood-flow distribution. If an alveolus is receiving too little air relative to its blood supply, the P_{O_2} in the alveolus and its surrounding area will be low. This decreased P_{O_2} causes vasoconstriction of the small pulmonary blood vessels. (Note that this local effect of oxygen on pulmonary blood vessels is precisely the opposite of that exerted on systemic arterioles.) The net adaptive effect is to supply less blood to poorly ventilated areas and more blood to well ventilated areas.

Gas Exchange in the Tissues

As the systemic arterial blood enters capillaries throughout the body, it is separated from the interstitial fluid by only the thin capillary wall, which is highly permeable to both oxygen and carbon dioxide. The interstitial fluid in turn is separated from intracellular fluid by the plasma membranes of the cells, which are also quite permeable to oxygen and carbon dioxide. Metabolic reactions occurring within cells are constantly consuming oxygen and producing carbon dioxide. Therefore, as shown in Figure 15-17, intracellular P_{O_2} is lower and P_{CO_2} higher than in blood. The lowest P_{O_2} of all—less than 5 mmHg—is in the mitochondria, the site of oxygen utilization. As a result, there is a net diffusion of oxygen from blood into cells (and, within the cells, into the mitochondria), and a net diffusion of carbon dioxide from cells into blood. In this manner, as blood flows through systemic capillaries, its P_{O_2} decreases and its P_{CO_2} increases. This accounts for the systemic venous blood values shown in Figure 15-17 and Table 15-6.

The mechanisms that enhance diffusion of oxygen and carbon dioxide between cells and blood when a tissue increases its metabolic activity were discussed in Chapter 14.

In summary, the supply of new oxygen to the alveoli and the consumption of oxygen in the cells create P_{O_2}

gradients that produce net diffusion of oxygen from alveoli to blood in the lungs and from blood to cells in the rest of the body. Conversely, the production of carbon dioxide by cells and its elimination from the alveoli via expiration create P_{CO_2} gradients that produce net diffusion of carbon dioxide from cells to blood in the rest of the body and from blood to alveoli in the lungs.

TRANSPORT OF OXYGEN IN BLOOD

Table 15-7 summarizes the oxygen content of systemic arterial blood (we shall henceforth refer to systemic arterial blood simply as arterial blood). Each liter normally contains the number of oxygen molecules equivalent to 200 ml of pure gaseous oxygen at atmospheric pressure. The oxygen is present in two forms: (1) dissolved in the plasma and erythrocyte water and (2) reversibly combined with hemoglobin molecules.

As predicted by Henry's law, the amount of oxygen dissolved in blood is directly proportional to the P_{O_2} of the blood. Because oxygen is relatively insoluble in water, only 3 ml can be dissolved in 1 L of blood at the normal arterial P_{O_2} of 100 mmHg. The other 197 ml of oxygen in a liter of arterial blood, more than 98 percent of the oxygen content in the liter, is transported in the erythrocytes reversibly combined with hemoglobin.

Each **hemoglobin** molecule is a protein made up of four subunits bound together. Each subunit consists of a molecular group known as **heme** and a polypeptide attached to the heme. [Hemoglobin is not the only heme-containing protein in the body; the others are the cytochromes (Chapter 4).] The four polypeptides of a hemoglobin molecule are collectively called **globin.** Each of the four heme groups in a hemoglobin molecule (Figure 15-20) contains one atom of **iron (Fe),** to which oxygen binds. Since each iron atom can bind one molecule of oxygen, a single hemoglobin molecule can bind four molecules of oxygen. However, for simplicity, the equation for the reaction between oxygen and hemoglobin is usually written in terms of a single polypeptide-heme chain of a hemoglobin molecule:

$$O_2 + Hb \rightleftharpoons HbO_2 \qquad (15\text{-}7)$$

Thus this chain can exist in one of two forms—**deoxyhemoglobin (Hb)** and **oxyhemoglobin (HbO$_2$).** In a blood sample containing many hemoglobin molecules, the fraction of all the hemoglobin in the form of oxyhemoglobin is expressed as the **percent hemoglobin saturation:**

$$\text{Percent saturation} = \frac{O_2 \text{ bound to Hb}}{\text{maximal capacity of Hb to bind } O_2} \times 100 \qquad (15\text{-}8)$$

For example, if the amount of oxygen bound to hemoglobin is 40 percent of the maximal capacity of hemoglobin to bind oxygen, the sample is said to be 40 percent saturated. The denominator in this equation is also termed the **oxygen-carrying capacity** of the blood.

What factors determine the percent hemoglobin saturation? By far the most important is the blood P_{O_2}. Before turning to this subject, however, it must be stressed that the total amount of oxygen carried by hemoglobin in blood depends not only on the percent saturation of hemoglobin but also on how much hemoglobin there is in each liter of blood. For example, if a person's blood contained only half as much hemoglobin per liter as normal, then at any given percent saturation the oxygen content of the blood would be only half as much.

FIGURE 15-20

Heme. Oxygen binds to the iron atom (Fe). Heme attaches to a polypeptide chain by a nitrogen atom to form one subunit of hemoglobin. Four of these subunits bind to each other to make a single hemoglobin molecule.

TABLE 15-7 OXYGEN CONTENT OF SYSTEMIC ARTERIAL BLOOD AT SEA LEVEL

1 liter (L) arterial blood contains
3 ml   O$_2$ physically dissolved (1.5%)
197 ml O$_2$ bound to hemoglobin (98.5%)
Total 200 ml O$_2$
Cardiac output = 5 L/min
O$_2$ carried to tissues/min = 5 L/min × 200 ml O$_2$/L
= 1000 ml O$_2$/min

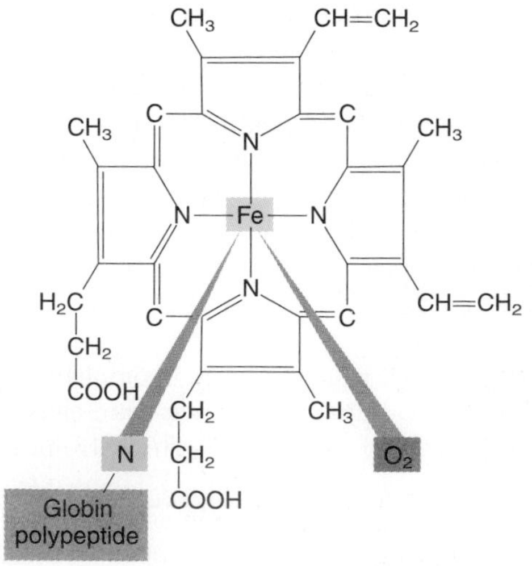

Effect of P_{O_2} on Hemoglobin Saturation

From inspection of Equation 15-7 and the law of mass action, one can see that raising the blood P_{O_2} should increase the combination of oxygen with hemoglobin. The experimentally determined quantitative relationship between these variables is shown in Figure 15-21, which is called an **oxygen-hemoglobin dissociation curve.** (The term "dissociate" means "to separate," in this case oxygen from hemoglobin; it could just as well have been called an oxygen-hemoglobin association curve.) The curve is S-shaped because, as stated earlier, each hemoglobin molecule contains four subunits; each subunit can combine with one molecule of oxygen, and the reactions of the four subunits occur sequentially, with each combination facilitating the next one. (This is an example of cooperativity, as described in Chapter 4.)

Note that the curve has a steep slope between 10 and 60 mmHg P_{O_2} and a relatively flat portion (or plateau) between 70 and 100 mmHg P_{O_2}. Thus, the extent to which oxygen combines with hemoglobin increases very rapidly as the P_{O_2} increases from 10 to 60 mmHg, so that at a P_{O_2} of 60 mmHg, 90 percent of the total hemoglobin is combined with oxygen. From this point on, a further increase in P_{O_2} produces only a small increase in oxygen binding.

The importance of this plateau at higher P_{O_2} values is as follows. Many situations, including high altitude and pulmonary disease, are characterized by a moderate reduction in alveolar and therefore arterial P_{O_2}. Even if the P_{O_2} fell from the normal value of 100 to 60 mmHg, the total quantity of oxygen carried by hemoglobin would decrease by only 10 percent since hemoglobin saturation is still close to 90 percent at a P_{O_2} of 60 mmHg. The plateau therefore provides an excellent safety factor in the supply of oxygen to the tissues.

The plateau also explains another fact: In a normal person at sea level, raising the alveolar (and therefore the arterial) P_{O_2} either by hyperventilating or by breathing 100 percent oxygen adds very little additional oxygen to the blood. A small additional amount dissolves, but because hemoglobin is already almost completely saturated with oxygen at the normal arterial P_{O_2} of 100 mmHg, it simply cannot pick up any more oxygen when the P_{O_2} is elevated beyond this point. But this applies only to normal people at sea level. If the person initially has a low arterial P_{O_2} because of lung disease or high altitude, then there would be a great deal of deoxyhemoglobin initially present in the arterial blood. Therefore, raising the alveolar and thereby the arterial P_{O_2} would result in significantly more oxygen transport.

We now retrace our steps and reconsider the movement of oxygen across the various membranes, this time including hemoglobin in our analysis. It is essential to recognize that the oxygen bound to hemoglobin does *not* contribute directly to the P_{O_2} of the blood. Only dissolved oxygen does so. Therefore, oxygen diffusion is governed only by the dissolved portion, a fact that permitted us to ignore hemoglobin in discussing transmembrane partial pressure gradients. However, the presence of hemoglobin plays a critical role in determining the *total amount* of oxygen that will diffuse, as illustrated by a simple example (Figure 15-22).

FIGURE 15-21

Oxygen-hemoglobin dissociation curve. This curve applies to blood at 37°C and a normal arterial hydrogen-ion concentration. At any given blood hemoglobin concentration, the vertical axis could also have plotted oxygen content, in milliliters of oxygen. At 100 percent saturation, the hemoglobin in 1L of normal blood carries 200 ml of oxygen. (*Adapted from Comroe.*)

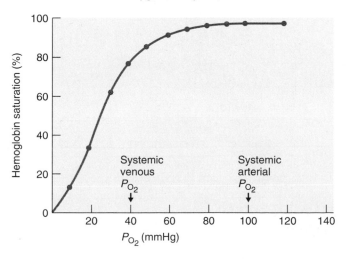

FIGURE 15-22

Effect of added hemoglobin on oxygen distribution between two compartments containing a fixed number of oxygen molecules and separated by a semipermeable membrane. At the new equilibrium, the P_{O_2} values are again equal to each other but lower than before the hemoglobin was added. However, the total oxygen, in other words, that dissolved plus that combined with hemoglobin, is now much higher on the right side of the membrane. (*Adapted from Comroe.*)

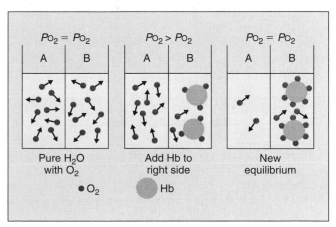

Two solutions separated by a semipermeable membrane contain equal quantities of oxygen, the gas pressures are equal, and no net diffusion occurs. Addition of hemoglobin to compartment B destroys this equilibrium because much of the oxygen combines with hemoglobin. Despite the fact that the total quantity of oxygen in compartment B is still the same, the number of dissolved oxygen molecules has decreased. Therefore, the P_{O_2} of compartment B is less than that of A, and so there is a net diffusion of oxygen from A to B. At the new equilibrium, the oxygen pressures are once again equal, but almost all the oxygen is in compartment B and is combined with hemoglobin.

Let us now apply this analysis to capillaries of the lungs and tissues (Figure 15-23). The plasma and erythrocytes entering the lungs have a P_{O_2} of 40 mmHg. As we can see from Figure 15-21, hemoglobin saturation at this P_{O_2} is 75 percent. The alveolar P_{O_2}—105 mmHg—is higher than the blood P_{O_2} and so oxygen diffuses from the alveoli into the plasma. This increases plasma P_{O_2} and induces diffusion of oxygen into the erythrocytes, elevating erythrocyte P_{O_2} and causing increased combination of oxygen and hemoglobin. The vast preponderance of the oxygen diffusing into the blood from the alveoli does not remain dissolved but combines with hemoglobin. Therefore, the blood P_{O_2} normally remains less than the alveolar P_{O_2} until hemoglobin is virtually 100 percent saturated. Thus the diffusion gradient favoring oxygen movement into the blood is maintained despite the very large transfer of oxygen.

In the tissue capillaries, the procedure is reversed: As the blood enters the capillaries, plasma P_{O_2} is greater than interstitial fluid P_{O_2} and net oxygen diffusion occurs out of the capillary from the plasma. Plasma P_{O_2} is now lower than erythrocyte P_{O_2}, and oxygen diffuses out of the erythrocytes into the plasma. The lowering of erythrocyte P_{O_2} causes the dissociation of some of the oxygen from hemoglobin, thereby liberating oxygen. Simultaneously, the oxygen that had diffused into the interstitial fluid is moving into cells along the concentration gradient generated by cellular utilization of oxygen. The net result is a transfer of large quantities of oxygen from hemoglobin into tissue cells purely by diffusion.

To repeat, in most tissues under resting conditions, hemoglobin is still 75 percent saturated as the blood leaves the tissue capillaries. This fact underlies an important automatic mechanism by which cells can obtain more oxygen whenever they increase their activity. An exercising muscle consumes more oxygen, thereby lowering its tissue P_{O_2}. This increases the blood-to-tissue P_{O_2} gradient and hence the diffusion of oxygen from blood to cell. In turn, the resulting reduction in erythrocyte P_{O_2} causes additional dissociation of hemoglobin and oxygen. In this manner an exercising muscle can extract almost all the oxygen from its blood supply, not just the usual 25 percent. Of course, an increased blood flow to the muscles also contributes greatly to the increased oxygen supply.

FIGURE 15-23

Oxygen movement in the lungs and tissues. Movement of inspired air into the alveoli is by bulk flow; all movement across membranes is by diffusion. ✗

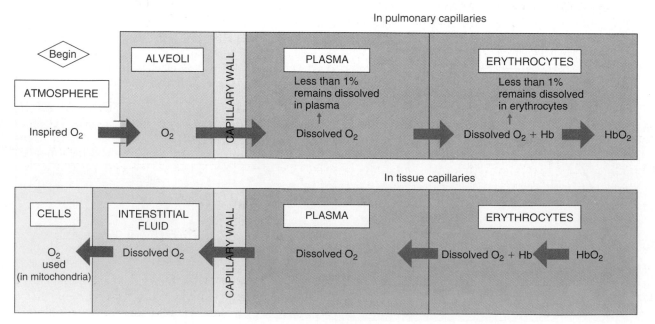

Effects of Blood P_{CO_2}, H^+ Concentration, Temperature, and DPG Concentration on Hemoglobin Saturation

At any given P_{O_2}, a variety of other factors influence the degree of hemoglobin saturation: blood P_{CO_2}, H^+ concentration, temperature, and the concentration of a substance—**2,3-diphosphoglycerate (DPG)**—produced by the erythrocytes. As illustrated in Figure 15-24, an increase in any of these factors causes the dissociation curve to shift to the right, which means that, at any given P_{O_2}, hemoglobin has less affinity for oxygen. In contrast, a decrease in any of these factors causes the dissociation curve to shift to the left, which means that, at any given P_{O_2}, hemoglobin has a greater affinity for oxygen.

The effects of increased P_{CO_2}, H^+ concentration, and temperature are continuously exerted on the blood in tissue capillaries, because each of these factors is higher in tissue-capillary blood than in arterial blood: The P_{CO_2} is increased because of the carbon dioxide entering the blood from the tissues. For reasons to be described later, the H^+ concentration is elevated because of the elevated P_{CO_2} and the release of metabolically produced acids such as lactic acid. The temperature is increased because of the heat produced by tissue metabolism. Therefore, hemoglobin exposed to this elevated blood P_{CO_2}, H^+ concentration, and temperature as it passes through the tissue capillaries has its affinity for oxygen decreased, and therefore hemoglobin gives up even more oxygen than it would have if the decreased tissue-capillary P_{O_2} had been the only operating factor.

The more active a tissue is, the greater its P_{CO_2}, H^+ concentration, and temperature. At any given P_{O_2}, this causes hemoglobin to release more oxygen during passage through the tissue's capillaries and provides the more active cells with additional oxygen. Here is another local mechanism that increases oxygen delivery to tissues that have increased metabolic activity.

What is the mechanism by which these factors influence hemoglobin's affinity for oxygen? Carbon dioxide and hydrogen ions do so by combining with the protein portion of hemoglobin and altering its molecular configuration. Thus, these effects are a form of allosteric modulation (Chapter 4). An elevated temperature also decreases hemoglobin's affinity for oxygen by altering its molecular configuration.

Erythrocytes contain large quantities of DPG, which is present in only trace amounts in other mammalian cells. DPG, which is produced by the erythrocytes during glycolysis, binds reversibly with hemoglobin, causing it to have a lower affinity for oxygen (Figure 15-24). The net result is that whenever DPG levels are increased, there is enhanced unloading of oxygen from hemoglobin as blood flows through the tissues. Such an increase in DPG concentration is triggered by a variety of conditions associated with inadequate oxygen supply to the tissues and helps to maintain oxygen delivery. Examples include anemia and exposure to high altitude.

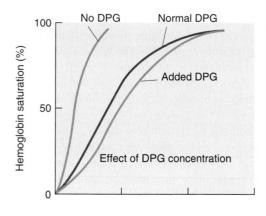

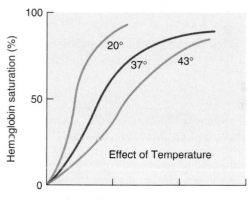

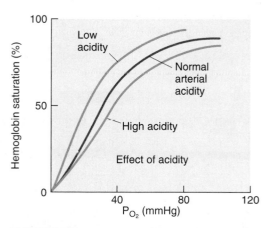

FIGURE 15-24

Effects of DPG concentration, temperature, and acidity on the relationship between P_{O_2} and hemoglobin saturation. The temperature of normal blood, of course, never diverges from 37° as much as shown in the figure, but the principle is still the same when the changes are within the physiological range. (*Adapted from Comroe.*)

TRANSPORT OF CARBON DIOXIDE IN BLOOD

In a resting person, metabolism generates about 200 ml of carbon dioxide per minute. When arterial blood flows through tissue capillaries, this volume of carbon dioxide diffuses from the tissues into the blood (Figure 15-25). Carbon dioxide is much more soluble in water than is oxygen, and so more dissolved carbon dioxide than dissolved oxygen is carried in blood. Even so, only a small amount of blood carbon dioxide is transported in this way; only 10 percent of the carbon dioxide entering the blood remains physically dissolved in the plasma and erythrocytes.

Another 30 percent of the carbon dioxide molecules entering the blood reacts reversibly with the amino groups of proteins, particularly hemoglobin, to form **carbamino compounds.** For simplicity, this reaction with hemoglobin is written as:

$$CO_2 + Hb \rightleftharpoons HbCO_2 \qquad (15\text{-}9)$$

This reaction is aided by the fact that deoxyhemoglobin, formed as blood flows through the tissue capillaries, re-

acts more readily with carbon dioxide than does oxyhemoglobin.

The remaining 60 percent of the carbon dioxide molecules entering the blood in the tissues is converted to bicarbonate:

$$\overset{\text{carbonic anhydrase}}{CO_2 + H_2O \rightleftharpoons \underset{\substack{\text{Carbonic} \\ \text{acid}}}{H_2CO_3} \rightleftharpoons \underset{\text{Bicarbonate}}{HCO_3^- + H^+}}$$

$$(15\text{-}10)$$

The first reaction in Equation 15-10 is rate-limiting and is catalyzed by the enzyme **carbonic anhydrase,** which is present in the erythrocytes but not in the plasma. Therefore, this reaction occurs mainly in the erythrocytes. In contrast, carbonic acid dissociates into a bicarbonate ion and a hydrogen very rapidly without any enzyme as-

FIGURE 15-25

Summary of CO_2 movement. All movement of CO_2 across membranes is by diffusion. Thickness of the arrows reflects relative proportions of the fates of the CO_2. Note that about two-thirds of the CO_2 entering the blood in the tissues ultimately is converted to HCO_3^-. This occurs almost entirely in the erythrocytes because the carbonic anhydrase is located there, but most of the HCO_3^- then moves out of the erythrocytes into the plasma in exchange for chloride ions (the "chloride shift"). See Figure 15-26 for the fate of the hydrogen ions generated in the erythrocytes. 🏃

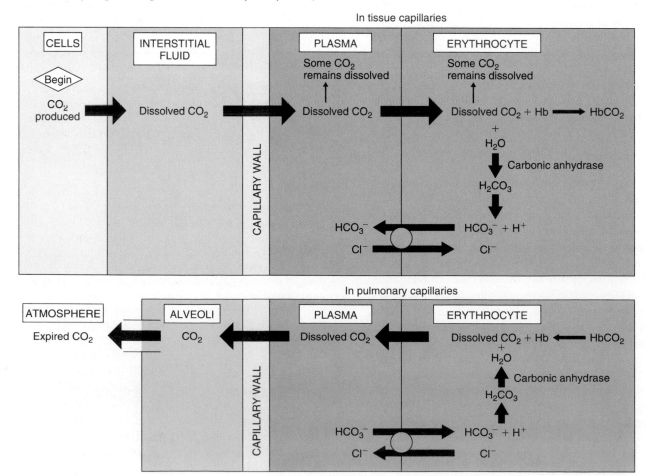

sistance. Once formed, most of the bicarbonate moves out of the erythrocytes into the plasma via a transporter that exchanges one bicarbonate for one chloride ion (this is called the "chloride shift").

The reactions shown in Equation 15-10 also explain why, as mentioned earlier, the H^+ concentration in tissue capillary blood and systemic venous blood is higher than that of the arterial blood and increases as metabolic activity increases. The fate of these hydrogen ions will be discussed in the next section.

Because carbon dioxide undergoes these various fates in blood, it is customary to add up the amounts of dissolved carbon dioxide, bicarbonate, and carbon dioxide in carbamino form and call this sum the **total blood carbon dioxide.** In *arterial* blood about 90 percent of the "total blood carbon dioxide" is bicarbonate, about 5–10 percent is carbamino, and about 5–10 percent is dissolved carbon dioxide. (These numbers for bicarbonate and carbamino would seem to conflict with our earlier statements that 60 percent of the carbon dioxide entering the blood is converted to bicarbonate and 30 percent to carbamino. There is no conflict, however. The *new* carbon dioxide added to blood in tissue capillaries does not distribute among bicarbonate and carbamino in the same proportions as those already present in *arterial blood.* The reason is that the concentration of deoxyhemoglobin is much higher in the tissue capillary blood than in the arterial blood and is available for binding carbon dioxide to form carbamino.)

Just the opposite events occur as systemic venous blood flows through the lung capillaries (Figure 15-25). Because the blood P_{CO_2} is higher than alveolar P_{CO_2}, a net diffusion of CO_2 from blood into alveoli occurs. This loss of carbon dioxide from the blood lowers the blood P_{CO_2} and drives reactions 15-10 and 15-9 to the left: HCO_3^- and H^+ combine to give H_2CO_3, which then dissociates to CO_2 and H_2O. Similarly, $HbCO_2$ generates Hb and free CO_2. Normally, as fast as CO_2 is generated from HCO_3^- and H^+ and from $HbCO_2$, it diffuses into the alveoli. In this manner, all the CO_2 delivered into the blood in the tissues now is delivered into the alveoli; it is eliminated from the alveoli and from the body during expiration.

TRANSPORT OF HYDROGEN IONS BETWEEN TISSUES AND LUNGS

To repeat, as blood flows through the tissues, a fraction of oxyhemoglobin loses its oxygen to become deoxyhemoglobin, while simultaneously a large quantity of carbon dioxide enters the blood and undergoes the reactions that generate bicarbonate and hydrogen ions. What happens to these hydrogen ions?

Deoxyhemoglobin has a much greater affinity for H^+ than does oxyhemoglobin, and so it binds (buffers) most of the hydrogen ions (Figure 15-26). Indeed, deoxyhemoglobin is often abbreviated HbH rather than Hb to denote its binding of H^+. In effect, the reaction is $HbO_2 + H^+ \rightleftharpoons HbH + O_2$. In this manner only a small number of the hydrogen ions generated in the blood remain free. This explains why the acidity of venous blood (pH = 7.36) is only slightly greater than that of arterial blood (pH = 7.40).

As the venous blood passes through the lungs, all these reactions are reversed. Deoxyhemoglobin becomes converted to oxyhemoglobin and, in the process, releases the hydrogen ions it had picked up in the tissues. The hydrogen ions react with bicarbonate to give carbonic acid, which dissociates to form carbon dioxide and water, and the carbon dioxide diffuses into the alveoli to be expired. Normally all the hydrogen ions that are generated in the tissue capillaries from the reaction of carbon dioxide and water recombine with bicarbonate to form carbon dioxide and water in the pulmonary capillaries. Therefore, none of these hydrogen ions appear in the *arterial* blood.

But what if the person is hypoventilating or has a lung disease that prevents normal elimination of carbon dioxide? Not only would arterial P_{CO_2} rise as a result but so would arterial H^+ concentration. Increased arterial H^+ concentration due to carbon dioxide retention is termed **respiratory acidosis.** Conversely, hyperventilation would lower the arterial values of both P_{CO_2} and H^+ concentration, producing **respiratory alkalosis.**

In the course of describing the transport of oxygen, carbon dioxide, and H^+ in blood, we have presented multiple factors that influence the binding of these substances by hemoglobin. They are all summarized in Table 15-8.

FIGURE 15-26

Binding of hydrogen ions by hemoglobin as blood flows through tissue capillaries. This reaction is facilitated because deoxyhemoglobin, formed as oxygen dissociates from hemoglobin, has a greater affinity for hydrogen ions than does oxyhemoglobin. For this reason, Hb and HbH are both abbreviations for deoxyhemoglobin.

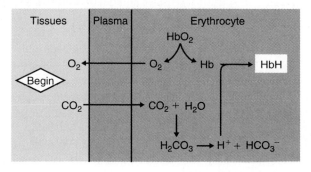

TABLE 15-8 EFFECTS OF VARIOUS FACTORS ON HEMOGLOBIN

The affinity of hemoglobin for oxygen is decreased by:
1. Increased hydrogen-ion concentration
2. Increased P_{CO_2}
3. Increased temperature
4. Increased DPG concentration

The affinity of hemoglobin for both hydrogen ions and carbon dioxide is decreased by increased P_{O_2}.

One more aspect of the remarkable hemoglobin molecule should at least be mentioned—its ability to transport nitric oxide. As described in Chapters 8 and 14, respectively, nitric oxide is an important neurotransmitter and is also released by endothelial cells. A present hypothesis is that as blood passes through the lungs, hemoglobin picks up and binds nitric oxide synthesized there, carries it to the peripheral tissues, and releases it along with oxygen. Simultaneously, hemoglobin picks up and catabolizes nitric oxide produced in the peripheral tissues. Theoretically this cycle could play an important role in determining the peripheral concentration of nitric oxide and, thereby, the overall effect of this agent on function. For example, by supplying net nitric oxide to the periphery the process could cause additional vasodilation by systemic blood vessels; this would have effects on both local blood flow and systemic arterial blood pressure.

CONTROL OF RESPIRATION

Neural Generation of Rhythmical Breathing

The diaphragm and intercostal muscles are skeletal muscles and therefore do not contract unless stimulated to do so by nerves. Thus, breathing depends entirely upon cyclical respiratory muscle excitation of the diaphragm and the intercostal muscles by their motor nerves. Destruction of these nerves or the areas from which they originate, as in the viral disease **poliomyelitis,** for example, results in paralysis of the respiratory muscles and death, unless some form of artificial respiration can be instituted.

Inspiration is initiated by a burst of action potentials in the nerves to the inspiratory muscles. Then the action potentials cease, the inspiratory muscles relax, and expiration occurs as the elastic lungs recoil. In situations when expiration is facilitated by contraction of expiratory muscles, the nerves to these muscles, which were quiescent during inspiration, begin firing during expiration.

By what mechanism are nerve impulses to the respiratory muscles alternately increased and decreased? Control of this neural activity resides primarily in neurons in the medulla oblongata, the same area of brain that contains the major cardiovascular control centers. (For the rest of this chapter we shall refer to the medulla oblongata simply as the medulla.) In several nuclei of the medulla, neurons called **medullary inspiratory neurons** discharge in synchrony with inspiration and cease discharging during expiration. They provide, through either direct or interneuronal connections, the rhythmic input to the motor neurons innervating the inspiratory muscles. The alternating cycles of firing and quiescence in the medullary inspiratory neurons are generated by a cooperative interaction between synaptic input from other medullary neurons and intrinsic pacemaker potentials of the inspiratory neurons themselves.

The medullary inspiratory neurons receive a rich synaptic input from neurons in various areas of the pons, the part of the brainstem just above the medulla. This input modulates the output of the medullary inspiratory neurons and may help terminate inspiration by inhibiting them. It is likely that an area of the lower pons called the apneustic center is the major source of this output, whereas an area of the upper pons called the pneumotaxic center modulates the activity of the apneustic center.

Another cutoff signal for inspiration comes from **pulmonary stretch receptors,** which lie in the airway smooth-muscle layer and are activated by a large lung inflation. Action potentials in the afferent nerve fibers from the stretch receptors travel to the brain and inhibit the medullary inspiratory neurons. (This is known as the Hering-Breur inflation reflex.) Thus, feedback from the lungs helps to terminate inspiration. However, this pulmonary stretch-receptor reflex plays a role in setting respiratory rhythm only under conditions of very large tidal volumes, as in rigorous exercise.

One last point about the medullary inspiratory neurons should be made: They are quite sensitive to depression by drugs such as barbiturates and morphine, and death from an overdose of these drugs is often due directly to a cessation of ventilation.

Control of Ventilation by P_{O_2}, P_{CO_2}, and H^+ Concentration

Respiratory rate and tidal volume are not fixed but can be increased or decreased over a wide range. For simplicity, we shall describe the control of total ventilation without discussing whether rate or depth makes the greater contribution to the change.

There are many inputs to the medullary inspiratory neurons, but the most important for the automatic control of ventilation *at rest* are from peripheral chemoreceptors and central chemoreceptors.

The **peripheral chemoreceptors,** located high in the neck at the bifurcation of the common carotid arteries and in the thorax on the arch of the aorta (Figure 15-27), are called the **carotid bodies** and **aortic bodies.** In both locations they are quite close to, but distinct from, the arterial

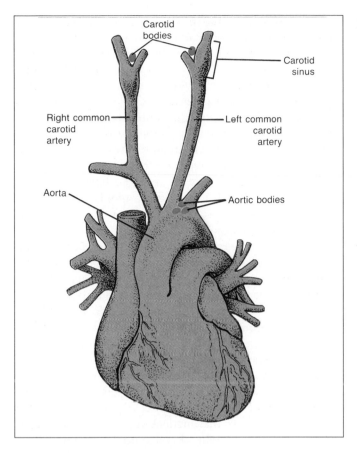

FIGURE 15-27

Location of the carotid and aortic bodies. Note that the carotid body is quite close to the carotid sinus, the major arterial baroreceptor. Both right and left common carotid bifurcations contain a carotid sinus and a carotid body.

baroreceptors described in Chapter 14 and are in intimate contact with the arterial blood. The peripheral chemoreceptors are composed of specialized receptor cells that communicate synaptically with neuron terminals. Afferent nerve fibers arising from these terminals pass to the brainstem, where they provide excitatory synaptic input to the medullary inspiratory neurons. The peripheral chemoreceptors are stimulated mainly by a decrease in the arterial P_{O_2} and an increase in the arterial H^+ concentration (Table 15-9).

The **central chemoreceptors** are located in the medulla and, like the peripheral chemoreceptors, provide excitatory synaptic input to the medullary inspiratory neurons. They are stimulated by an increase in the H^+ concentration of the brain's extracellular fluid. As we shall see, such changes result mainly from changes in blood P_{CO_2}.

Control by P_{O_2}. Figure 15-28 illustrates an experiment in which healthy subjects breathe low-P_{O_2} gas mixtures for several minutes. (The experiment is performed in a way that keeps arterial P_{CO_2} constant so that the pure effects of changing only P_{O_2} can be studied.) Little increase in

TABLE 15-9   MAJOR STIMULI FOR THE CENTRAL AND PERIPHERAL CHEMORECEPTORS

Peripheral chemoreceptors—that is, carotid bodies and aortic bodies—respond to changes in the *arterial blood*. They are stimulated by:

1. Decreased P_{O_2}
2. Increased hydrogen-ion concentration

Central chemoreceptors—that is, located in the medulla oblongata—respond to changes in the *brain extracellular fluid*. They are stimulated by increased P_{CO_2}, via associated changes in hydrogen-ion concentration. (See Equation 15-10.)

ventilation is observed until the oxygen content of the inspired air is reduced enough to lower arterial P_{O_2} to 60 mmHg. Beyond this point, any further reduction in arterial P_{O_2} causes a marked reflex increase in ventilation.

This reflex is mediated by the peripheral chemoreceptors (Figure 15-29). The low arterial P_{O_2} increases the rate at which the receptors discharge, resulting in an increased number of action potentials traveling up the afferent nerve fibers and stimulating the medullary inspiratory neurons. The resulting increase in ventilation provides more oxygen to the alveoli and minimizes the drop in alveolar and arterial P_{O_2} produced by the low P_{O_2} gas mixture.

It may seem surprising that we are so insensitive to smaller reductions of arterial P_{O_2}, but look again at the oxygen-hemoglobin dissociation curve (Figure 15-21). Total oxygen transport by the blood is not really reduced very much until the arterial P_{O_2} falls below about 60 mmHg. Therefore, increased ventilation would not result in very much more oxygen being added to the blood until that point is reached.

FIGURE 15-28

The effect on ventilation of breathing low-oxygen mixtures. The arterial P_{CO_2} was maintained at 40 mmHg throughout the experiment.

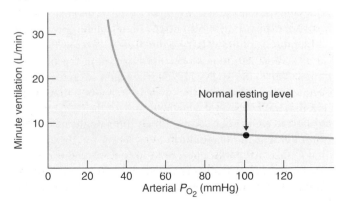

Control by P_{CO_2}. Figure 15-30 illustrates an experiment in which subjects breathe air to which variable quantities of carbon dioxide have been added. The presence of carbon dioxide in the inspired air causes an elevation of alveolar P_{CO_2} and thereby an elevation of arterial P_{CO_2}. Note that even a very small increase in arterial P_{CO_2} causes a marked reflex increase in ventilation. Experiments like this have documented that small increases in arterial P_{CO_2} are resisted by the reflex mechanisms controlling ventilation to a much greater degree than are equivalent decreases in arterial P_{O_2}.

Of course we don't usually breathe bags of gas containing carbon dioxide. What is the physiological role of this reflex? If a defect in the respiratory system (emphysema, for example) causes a retention of carbon dioxide in the body, the increase in arterial P_{CO_2} stimulates ventilation, which promotes the elimination of carbon dioxide. Conversely, if arterial P_{CO_2} decreases below normal levels for whatever reason, this removes some of the stimulus for ventilation, thereby reducing ventilation and allowing metabolically produced carbon dioxide to accumulate and return the P_{CO_2} to normal. In this manner, the arterial P_{CO_2} is stabilized at the normal value of 40 mmHg.

The ability of changes in arterial P_{CO_2} to control ventilation reflexly is largely due to associated changes in H^+ concentration (Equation 15-10). As summarized in Figure 15-31, the pathways that mediate these reflexes are initiated by both the peripheral and central chemoreceptors. The peripheral chemoreceptors are stimulated by the increased arterial H^+ concentration resulting from the increased P_{CO_2}. At the same time, because carbon dioxide diffuses rapidly across the membranes separating capillary

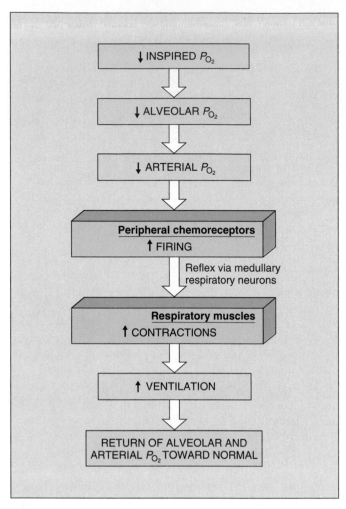

FIGURE 15-29

Sequence of events by which a low arterial P_{O_2} causes increased ventilation, which maintains alveolar (and, hence, arterial) P_{O_2} at a value higher than would exist if the ventilation had remained unchanged.

FIGURE 15-30

Effects on respiration of increasing arterial P_{CO_2} achieved by adding carbon dioxide to inspired air.

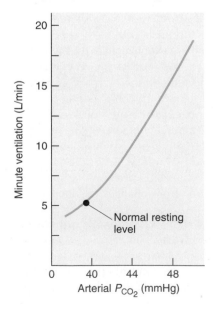

To reiterate, the peripheral chemoreceptors respond to decreases in arterial P_{O_2}, as occurs in lung disease or exposure to high altitude. However, the peripheral chemoreceptors are *not* stimulated in situations in which there are modest reductions in the oxygen *content* of the blood but no change in arterial P_{O_2}. An example of this is anemia, where there is a decrease in the amount of hemoglobin present in the blood (Chapter 14) but no decrease in arterial P_{O_2}, because the concentration of dissolved oxygen in the blood is normal. This same analysis holds true when oxygen content is reduced moderately by the presence of carbon monoxide, a gas that has an extremely high affinity for the oxygen binding sites in hemoglobin and reduces the amount of oxygen combined with hemoglobin by competing for these sites. Since carbon monoxide does not affect the amount of oxygen that can dissolve in blood, the arterial P_{O_2} is unaltered, and no increase in peripheral chemoreceptor output occurs.

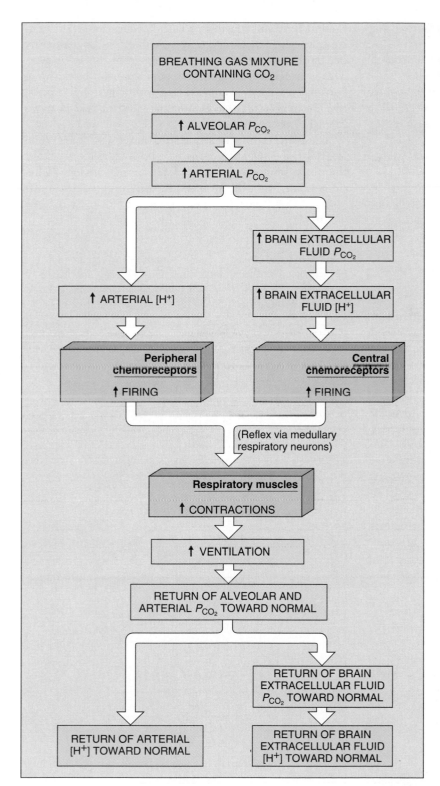

FIGURE 15-31
Pathways by which increased arterial P_{CO_2} stimulates ventilation. Note that the peripheral chemoreceptors are stimulated by an *increase* in H^+ concentration, whereas they are stimulated by a *decrease* in P_{O_2} (Figure 15-29).

blood and brain tissue, the increase in arterial P_{CO_2} causes a rapid increase in brain extracellular fluid P_{CO_2}. This increased P_{CO_2} increases *brain extracellular-fluid* H^+ concentration, which stimulates the central chemoreceptors. Inputs from both the peripheral and central chemoreceptors stimulate the medullary inspiratory neurons to increase ventilation. The end result is a return of arterial and brain extracellular fluid P_{CO_2} and H^+ concentration toward normal. Of the two sets of receptors involved in this reflex response to elevated P_{CO_2}, the central chemoreceptors are the more important, accounting for about 70 percent of the increased ventilation.

It should also be noted that the effects of increased P_{CO_2} and decreased P_{O_2} not only exist as independent inputs to the medulla but manifest synergistic interactions as well. Acute ventilatory response to combined low P_{O_2} and high P_{CO_2} is considerably greater than the sum of the individual responses.

Throughout this section, we have described the stimulatory effects of carbon dioxide on ventilation via reflex input to the medulla, but very high levels of carbon dioxide actually *inhibit* ventilation and may be lethal. This is because such concentrations of carbon dioxide act *directly* on the medulla to inhibit the respiratory neurons by an anesthesia-like effect. Other symptoms caused by very high blood P_{CO_2} include severe headaches, restlessness, and dulling or loss of consciousness.

Control by Changes in Arterial H$^+$ Concentration Not Due to Altered Carbon Dioxide. We have seen that retention or excessive elimination of carbon dioxide causes respiratory acidosis and respiratory alkalosis, respectively. There are, however, many normal and pathological situations in which a change in arterial H$^+$ concentration is due to some cause other than a primary change in P_{CO_2}. These are termed **metabolic acidosis** when H$^+$ concentration is increased and **metabolic alkalosis** when it is decreased. In such cases, the peripheral chemoreceptors play the major role in altering ventilation.

For example, addition of lactic acid to the blood, as in strenuous exercise, causes hyperventilation almost entirely by stimulation of the peripheral chemoreceptors (Figures 15-32 and 15-33). The central chemoreceptors are only

minimally stimulated in this case because brain H$^+$ concentration is increased to only a small extent, at least early on, by the hydrogen ions generated from the lactic acid. This is because hydrogen ions penetrate the blood-brain barrier very slowly. Contrast this with how easily carbon dioxide penetrates the blood-brain barrier and changes brain H$^+$ concentration.

The converse of the above situation is also true: When arterial H$^+$ concentration is lowered by any means other than by a reduction in P_{CO_2} (for example, by loss of hydrogen ions from the stomach in vomiting), ventilation is

FIGURE 15-33

Reflexly induced hyperventilation minimizes the change in arterial hydrogen-ion concentration when non-CO_2 acids are produced in excess in the body. Note that under such conditions, arterial P_{CO_2} is reflexly reduced below its normal value.

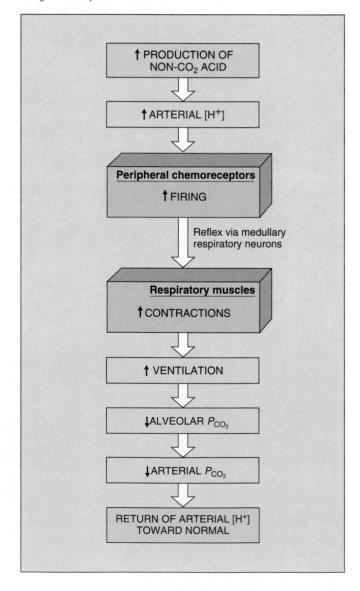

FIGURE 15-32

Changes in ventilation in response to an elevation of plasma hydrogen-ion concentration produced by the administration of lactic acid. (*Adapted from Lambertsen.*)

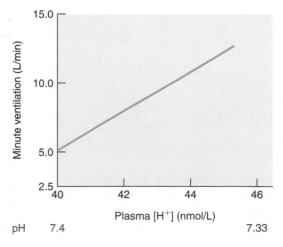

reflexly depressed because of decreased peripheral chemoreceptor output.

The adaptive value such reflexes have in regulating arterial H^+ concentration is shown in Figure 15-33. The hyperventilation induced by a metabolic acidosis lowers arterial P_{CO_2}, which lowers arterial H^+ concentration back toward normal. Similarly, hypoventilation induced by a metabolic alkalosis results in an elevated arterial P_{CO_2} and a restoration of H^+ concentration toward normal.

Notice that when a change in arterial H^+ concentration due to some acid unrelated to carbon dioxide influences ventilation via the peripheral chemoreceptors, P_{CO_2} is displaced from normal. This is a reflex that regulates arterial H^+ concentration at the expense of changes in arterial P_{CO_2}.

Figure 15-34 summarizes the control of ventilation by P_{O_2}, P_{CO_2}, and H^+ concentration.

FIGURE 15-34

Summary of the major chemical inputs that stimulate ventilation. This is a combination of Figs. 15-29, 15-31, and 15-33. When arterial P_{O_2} increases or when P_{CO_2} or hydrogen-ion concentration decreases, ventilation is reflexly decreased.

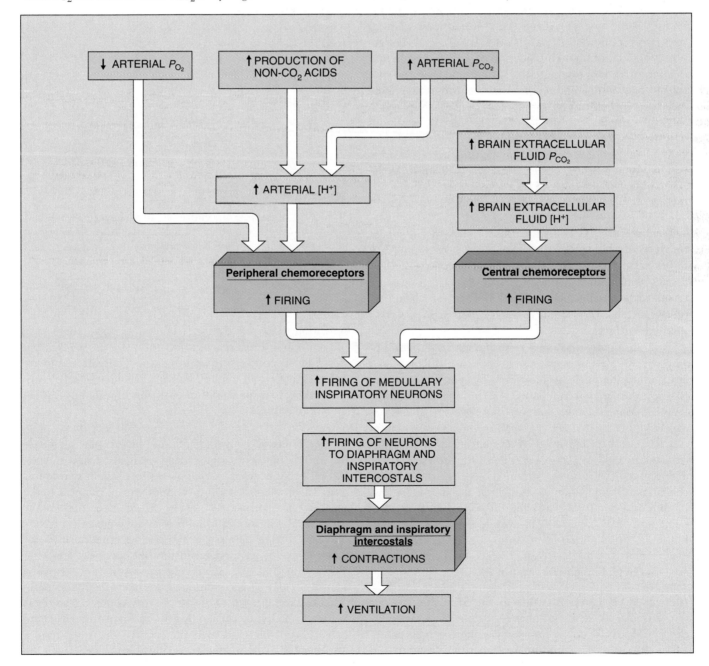

Control of Ventilation During Exercise

During exercise, the alveolar ventilation may increase as much as twentyfold. On the basis of our three variables—P_{O_2}, P_{CO_2}, and H^+ concentration—it might seem easy to explain the mechanism that induces this increased ventilation. Unhappily, such is not the case, and the major stimuli to ventilation during exercise, at least that of moderate degree, remain unclear.

Increased P_{CO_2} as the Stimulus?   It would seem logical that, as the exercising muscles produce more carbon dioxide, blood P_{CO_2} would increase. This is true, however, only for systemic *venous* blood but not for systemic *arterial* blood. Why doesn't arterial P_{CO_2} increase during exercise? Recall two facts from the section on alveolar gas pressures: (1) alveolar P_{CO_2} sets arterial P_{CO_2}, and (2) alveolar P_{CO_2} is determined by the *ratio* of carbon dioxide production to alveolar ventilation. During moderate exercise, the alveolar ventilation increases in exact proportion to the increased carbon dioxide production, and so alveolar and therefore arterial P_{CO_2} do not change. Indeed, for reasons described below, in very strenuous exercise the alveolar ventilation increases relatively more than carbon dioxide production. In other words, during severe exercise the person hyperventilates, and thus alveolar and systemic arterial P_{CO_2} actually decrease (Figure 15-35)!

Decreased P_{O_2} as the Stimulus?   The story is similar for oxygen. Although systemic *venous* P_{O_2} decreases during exercise, alveolar P_{O_2} and hence systemic *arterial* P_{O_2} usually remain unchanged (Figure 15-35). This is because cellular oxygen consumption and alveolar ventilation increase in exact proportion to each other, at least during moderate exercise.

Note in Figure 15-35 that even though the hyperventilation of very strenuous exercise lowers arterial P_{CO_2} it does not produce the increase in arterial P_{O_2} you would expect to occur during hyperventilation. In fact, not shown in the figure, *alveolar* P_{O_2} does increase during severe exercise, as expected, but for a variety of reasons, the increase in alveolar P_{O_2} is not accompanied by an increase in *arterial* P_{O_2}. (Indeed, in highly conditioned athletes the arterial P_{O_2} may decrease somewhat during strenuous exercise and may also contribute to stimulation of ventilation at that time.)

This is a good place to remind you of an important point made in Chapter 14: In normal individuals, ventilation is not the limiting factor in endurance exercise, cardiac output is. Ventilation can, as has just been seen, increase enough to maintain arterial P_{O_2}.

Increased H^+ Concentration as the Stimulus?   Since the arterial P_{CO_2} does not change during moderate exercise and decreases in severe exercise, there is no accumulation of excess H^+ resulting from carbon dioxide accu-

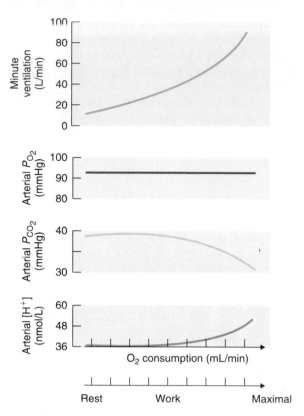

FIGURE 15-35

The effect of exercise on ventilation, arterial gas pressures, and hydrogen-ion concentration. All these variables remain constant during moderate exercise; any change occurs only during strenuous exercise when the person is actually hyperventilating. (*Adapted from Comroe.*)

mulation. However, during strenuous exercise, there *is* an increase in arterial H^+ concentration (Figure 15-35), but for quite a different reason, namely, generation and release of lactic acid into the blood. This change in H^+ concentration is responsible, in part, for stimulating the hyperventilation of severe exercise.

Other Factors.   A variety of other factors play some role in stimulating ventilation during exercise. These include (1) reflex input from mechanoreceptors in joints and muscles; (2) an increase in body temperature; (3) inputs to the respiratory neurons via branches from axons descending from the brain to motor neurons supplying the exercising muscles; (4) an increase in the plasma epinephrine concentration; (5) an increase in the plasma potassium concentration (because of movement out of the exercising muscles); and (6) a conditioned (learned) response mediated by neural input to the respiratory centers. The operation of this last factor can be seen in Figure 15-36: There is an abrupt increase—within seconds—in ventilation at the onset of exercise and an equally abrupt decrease at the

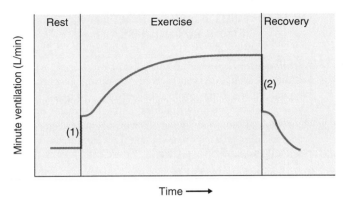

FIGURE 15-36

Ventilation changes during exercise. Note (1) the abrupt increase at the onset of exercise and (2) the equally abrupt but larger decrease at the end of exercise.

end; these changes occur too rapidly to be explained by alteration of chemical constituents of the blood or altered body temperature.

Figure 15-37 summarizes various factors that influence ventilation during exercise. The possibility that *oscillatory* changes in arterial P_{O_2}, P_{CO_2}, or H^+ concentration occur, despite unchanged average levels of these variables, and play a role has been proposed but remains unproven.

Other Ventilatory Responses

Protective Reflexes. A group of responses protect the respiratory system from irritant materials. Most familiar are the cough and the sneeze reflexes, which originate in receptors located between airway epithelial cells. The receptors for the sneeze reflex are in the nose or pharynx, and those for cough are in the trachea. When these receptors are stimulated, the medullary respiratory neurons reflexly cause a deep inspiration and a violent expiration. In this manner, particles can be literally exploded out of the respiratory tract.

The cough reflex is inhibited by alcohol, which may contribute to the susceptibility of alcoholics to choking and pneumonia.

Another example of a protective reflex is the immediate cessation of respiration that is frequently triggered when noxious agents are inhaled. Chronic smoking causes a loss of this reflex.

Voluntary Control of Breathing. Although we have discussed in detail the involuntary nature of most respiratory reflexes, it is quite obvious that considerable voluntary control of respiratory movements exists. Voluntary control is accomplished by descending pathways from the cerebral cortex to the motor neurons of the respiratory muscles. This voluntary control of respiration cannot be maintained when the involuntary stimuli, such as an elevated P_{CO_2} or H^+ concentration, become intense. An example is the inability to hold one's breath for very long.

The opposite of breath-holding—deliberate hyperventilation—lowers alveolar and arterial P_{CO_2} and increases P_{O_2}. Unfortunately, swimmers sometimes voluntarily hyperventilate immediately before underwater swimming to

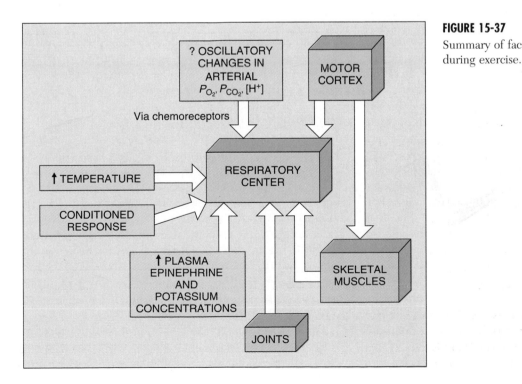

FIGURE 15-37

Summary of factors that stimulate ventilation during exercise.

be able to hold their breath longer. We say "unfortunately" because the low P_{CO_2} may still permit breath-holding at a time when the exertion is lowering the arterial P_{O_2} to levels that can cause unconsciousness and lead to drowning.

Besides the obvious forms of voluntary control, respiration must also be controlled during such complex actions as speaking and singing.

Reflexes from J Receptors.   In the lungs, either in the capillary walls or the interstitium, are a group of receptors called **J receptors.** They are normally dormant but are stimulated by an increase in lung interstitial pressure caused by the collection of fluid in the interstitium. Such an increase occurs during the vascular congestion caused by either occlusion of a pulmonary vessel (*pulmonary embolus*) or left ventricular failure (Chapter 14), as well as by strong exercise in healthy people. The main reflex effects are rapid breathing (tachypnea) and a dry cough. In addition, neural input from J receptors gives rise to sensations of pressure in the chest and *dyspnea*—the feeling that breathing is labored or difficult.

HYPOXIA

Hypoxia is defined as a deficiency of oxygen at the tissue level. There are many potential causes of hypoxia, but they can be classed in four general categories: (1) *hypoxic hypoxia* (also termed *hypoxemia*), in which the arterial P_{O_2} is reduced; (2) *anemic hypoxia,* in which the arterial P_{O_2} is normal but the total oxygen *content* of the blood is reduced because of inadequate numbers of erythrocytes, deficient or abnormal hemoglobin, or competition for the hemoglobin molecule by carbon monoxide; (3) *ischemic hypoxia* (also called hypoperfusion hypoxia), in which blood flow to the tissues is too low; and (4) *histotoxic hypoxia,* in which the quantity of oxygen reaching the tissues is normal but the cell is unable to utilize the oxygen because a toxic agent—cyanide, for example—has interfered with the cell's metabolic machinery.

The primary causes of hypoxic hypoxia in disease are listed in Table 15-10. Exposure to the reduced P_{O_2} of high altitude also causes hypoxic hypoxia but is, of course, not a "disease." The brief summaries in Table 15-10 provide a review of many of the key aspects of respiratory physiology and pathophysiology described in this chapter.

This table also emphasizes that some of the diseases that produce hypoxia also produce carbon dioxide retention and an increased arterial P_{CO_2} (*hypercapnea*). In such cases, treating only the oxygen deficit by administering oxygen may constitute inadequate therapy because it does nothing about the hypercapnea. (Indeed, such therapy may be dangerous. The primary respiratory drive in

TABLE 15-10   CAUSES OF A DECREASED ARTERIAL P_{O_2} (HYPOXIC HYPOXIA) IN DISEASE

1. *Hypoventilation* may be caused (a) by a defect anywhere along the respiratory control pathway, from the medulla through the respiratory muscles, (b) by severe thoracic cage abnormalities, and (c) by major obstruction of the upper airway. The hypoxemia of hypoventilation is always accompanied by an increased arterial P_{CO_2}.

2. *Diffusion impairment* results from thickening of the alveolar membranes or a decrease in their surface area. In turn, it causes failure of equilibration of blood P_{O_2} with alveolar P_{O_2}. Often it is apparent only during exercise. Arterial P_{CO_2} is either normal, since carbon dioxide diffuses more readily than oxygen, or reduced, if the hypoxemia reflexly stimulates ventilation.

3. A *shunt* is an abnormality of the cardiovascular system or lungs that causes blood to bypass ventilated alveoli in passing from the right heart to the left heart. The most common causes are cardiac defects that permit blood to flow directly from the right heart to the left heart. Arterial P_{CO_2} generally does not rise since the effect of the shunt on it is counterbalanced by the increased ventilation reflexly stimulated by the hypoxemia.

4. *Ventilation-perfusion inequality* is by far the most common cause of hypoxemia. It occurs in chronic obstructive lung diseases and many other lung diseases. Arterial P_{CO_2} may be normal or increased, depending upon how much ventilation is reflexly stimulated.

such patients is the hypoxia, since for several reasons the response to an increased P_{CO_2} may be lost in chronic situations; the administration of pure oxygen may cause such patients to stop breathing entirely.)

Acclimatization to High Altitude

Atmospheric pressure progressively decreases as altitude increases. Thus, at the top of Mt. Everest (approximately 29,000 ft, or 9000 m), the atmospheric pressure is 253 mmHg (recall that it is 760 mmHg at sea level). The air is still 21 percent oxygen, which means that the P_{O_2} is 53 mmHg (0.21×253 mmHg). Obviously, the alveolar and arterial P_{O_2} must decrease as one ascends unless pure oxygen is breathed. The highest villages permanently inhabited by people are in the Andes at 19,000 ft (5700 m). These villagers work quite normally, and the only major precaution they take is that the women come down to lower altitudes during late pregnancy.

The effects of oxygen lack vary from one individual to another, but most people who ascend rapidly to altitudes above 10,000 ft experience some degree of *mountain sickness.* This disorder consists of breathlessness, headache, nausea, vomiting, insomnia, fatigue, and impairment of mental processes. Much more serious is the appearance, in some individuals, of life-threatening pulmonary edema, the leakage of fluid from the pulmonary capillaries into the alveolar walls and eventually the airspaces, themselves. Brain edema can also occur.

Over the course of several days, the symptoms of mountain sickness disappear, although maximal physical capacity remains reduced. Acclimatization to high altitude is achieved by the compensatory mechanisms given in Table 15-11.

Finally, it should be noted that the responses to high altitude are essentially the same as the responses to hypoxia from any other cause. Thus, a person with severe hypoxia from lung disease may show many of the same changes—increased hematocrit, for example—as a high-altitude sojourner.

NONRESPIRATORY FUNCTIONS OF THE LUNGS

The lungs have a variety of functions in addition to their roles in gas exchange and regulation of H^+ concentration. Most notable are the influences they have on the arterial concentrations of a large number of biologically active substances. Many substances (neurotransmitters and paracrine agents, for example) released locally into interstitial fluid may diffuse into capillaries and thus make their way into the systemic venous system. The lungs partially or completely remove some of these substances from the blood and thereby prevent them from reaching other locations in the body via the arteries. The cells that perform this function are the endothelial cells lining the pulmonary capillaries.

In contrast, the lungs may also produce and add new substances to the blood. Some of these substances play local regulatory roles within the lungs, but if produced in large enough quantity, they may diffuse into the pulmonary capillaries and be carried to the rest of the body. For example, inflammatory responses (Chapter 20) in the lung may lead, via excessive release of potent chemicals such as histamine, to profound alterations of systemic blood pressure or flow. In at least one case, the lungs contribute a hormone, angiotensin II, to the blood (Chapter 16).

Finally, the lungs also act as a "sieve" that traps and dissolves small blood clots generated in the systemic circulation, thereby preventing them from reaching the systemic arterial blood where they could occlude blood vessels in other organs.

TABLE 15-11   ACCLIMATIZATION TO THE HYPOXIA OF HIGH ALTITUDE

1. The peripheral chemoreceptors stimulate ventilation.

2. Erythropoietin, secreted by the kidneys, stimulates erythrocyte synthesis, resulting in increased erythrocyte and hemoglobin concentration of blood.

3. DPG increases and shifts the hemoglobin dissociation curve to the right, facilitating oxygen unloading in the tissues. However, this DPG change is not always adaptive and may be maladaptive. For example, at very high altitudes, a right shift in the curve impairs oxygen *loading* in the lungs, an effect that outweighs any benefit from facilitation of *unloading* in the tissues.

4. Increases in capillary density, mitochondria, and muscle myoglobin occur, all of which increase oxygen transfer.

5. The peripheral chemoreceptors stimulate an increased loss of sodium and water in the urine. This reduces plasma volume, resulting in a concentration of the erythrocytes and hemoglobin in the blood.

SUMMARY

ORGANIZATION OF THE RESPIRATORY SYSTEM

I.   The respiratory system comprises the lungs, the airways leading to them, and the chest structures responsible for movement of air into and out of them.
   A. The conducting zone of the airways consists of the trachea, bronchi, and terminal bronchioles.
   B. The respiratory zone of the airways consists of the alveoli, which are the sites of gas exchange, and those airways to which alveoli are attached.
   C. The alveoli are lined by type I cells and some type II cells, which produce surfactant.
   D. The lungs and interior of the thorax are covered by pleura; between the two pleural layers is an extremely thin layer of intrapleural fluid.

II. The lungs are elastic structures whose volume depends upon the pressure difference across the lungs—the transpulmonary pressure—and how stretchable the lungs are.

SUMMARY OF STEPS INVOLVED IN RESPIRATION

I.   The steps involved in respiration are summarized in Figure 15-6. In the steady state, the net volumes of oxygen and carbon dioxide exchanged in the lungs per unit time are equal to the net volumes exchanged in the tissues. Typical volumes per minute are 250 ml for oxygen consumption and 200 ml for carbon dioxide production.

VENTILATION AND LUNG MECHANICS

I. Bulk flow of air between the atmosphere and alveoli is proportional to the difference between the atmospheric and alveolar pressures and inversely proportional to the airway resistance: $F = (P_{atm} - P_{alv})/R$.

II. Between breaths at the end of an unforced expiration $P_{atm} = P_{alv}$, no air is flowing, and the dimensions of the lungs and thoracic cage are stable as the result of opposing elastic forces. The lungs are stretched and are attempting to recoil, whereas the chest wall is compressed and attempting to move outward. This creates a subatmospheric intrapleural pressure and hence a transpulmonary pressure that opposes the forces of elastic recoil.

III. During inspiration, the contractions of the diaphragm and inspiratory intercostal muscles increase the volume of the thoracic cage.
 A. This makes intrapleural pressure more subatmospheric, increases transpulmonary pressure, and causes the lungs to expand to a greater degree than between breaths.
 B. This expansion initially makes alveolar pressure subatmospheric, which creates the pressure difference between atmosphere and alveoli to drive air flow into the lungs.

IV. During expiration, the inspiratory muscles cease contracting, allowing the elastic recoil of the chest wall and lungs to return them to their original between-breath size.
 A. This initially compresses the alveolar air, raising alveolar pressure above atmospheric pressure and driving air out of the lungs.
 B. In forced expirations, the contraction of expiratory intercostal muscles and abdominal muscles actively decreases thoracic dimensions.

V. Lung compliance is determined by the elastic connective tissues of the lungs and the surface tension of the fluid lining the alveoli. The latter is greatly reduced, and compliance increased, by surfactant, produced by the type II cells of the alveoli.

VI. Airway resistance determines how much air flows into the lungs at any given pressure difference between atmosphere and alveoli. The major determinant of airway resistance is the radii of the airways.

VII. The vital capacity is the sum of resting tidal volume, inspiratory reserve volume, and expiratory reserve volume. The volume expired during the first second of a forced vital capacity (FVC) measurement is the FEV_1 and normally averages 80 percent of FVC.

VIII. Minute ventilation is the product of tidal volume and respiratory rate. Alveolar ventilation = (tidal volume − anatomic dead space) × (respiratory rate).

EXCHANGE OF GASES IN ALVEOLI AND TISSUES

I. Exchange of gases in lungs and tissues is by diffusion, as a result of differences in partial pressures. Gases diffuse from a region of higher partial pressure to one of lower partial pressure.

II. Normal alveolar gas pressure for oxygen is 105 mmHg and for carbon dioxide 40 mmHg.
 A. At any given inspired P_{O_2}, the ratio of oxygen consumption to alveolar ventilation determines alveolar P_{O_2}—the higher the ratio, the lower the alveolar P_{O_2}.
 B. The higher the ratio of carbon dioxide production to alveolar ventilation, the higher the alveolar P_{CO_2}.

III. The average value at rest for systemic venous P_{O_2} is 40 mmHg and for P_{CO_2} is 46 mmHg.

IV. As systemic venous blood flows through the pulmonary capillaries, there is net diffusion of oxygen from alveoli to blood and of carbon dioxide from blood to alveoli. By the end of the pulmonary capillaries, the blood gas pressures have become equal to those in the alveoli.

V. Inadequate gas exchange between alveoli and pulmonary capillaries may occur when the alveolus-capillary surface area is decreased, when the alveolar walls thicken or when there are ventilation-perfusion inequalities.

VI. Significant ventilation-perfusion inequalities cause the systemic arterial P_{O_2} to be reduced. An important mechanism for opposing mismatching is that a low local P_{O_2} causes local vasoconstriction, thereby shunting blood away from poorly ventilated areas.

VII. In the tissues, net diffusion of oxygen occurs from blood to cells, and net diffusion of carbon dioxide from cells to blood.

TRANSPORT OF OXYGEN IN THE BLOOD

I. Each liter of systemic arterial blood normally contains 200 ml of oxygen, more than 98 percent bound to hemoglobin and the rest dissolved.

II. The major determinant of the degree to which hemoglobin is saturated with oxygen is blood P_{O_2}.
 A. Hemoglobin is almost 100 percent saturated at the normal systemic arterial P_{O_2} of 100 mmHg. The fact that saturation is already more than 90 percent at a P_{O_2} of 60 mmHg permits relatively normal uptake of oxygen by the blood even when alveolar P_{O_2} is moderately reduced.
 B. Hemoglobin is 75 percent saturated at the normal systemic venous P_{O_2} of 40 mmHg. Thus, only 25 percent of the oxygen has dissociated from hemoglobin and entered the tissues.

III. The affinity of hemoglobin for oxygen is decreased by an increase in P_{CO_2}, hydrogen-ion concentration, and temperature. All these conditions exist in the tissues and facilitate the dissociation of oxygen from hemoglobin.

IV. The affinity of hemoglobin for oxygen is also decreased by binding DPG, which is synthesized by the erythrocytes. DPG increases in situations associated with inadequate oxygen supply and helps maintain oxygen release in the tissues.

TRANSPORT OF CARBON DIOXIDE IN BLOOD

I. When carbon dioxide molecules diffuse from the tissues into the blood, 10 percent remains dissolved in plasma and erythrocytes, 30 percent combines in the erythrocytes with deoxyhemoglobin to form carbamino compounds, and 60 percent combines in the erythrocytes with water to form carbonic acid, which then dissociates to yield bicarbonate and hydrogen ions. Most of the bicarbonate then moves out of the erythrocytes into the plasma in exchange for chloride ions.

II. As venous blood flows through lung capillaries, P_{CO_2} decreases because of diffusion of carbon dioxide out of the blood into the alveoli, and the above reactions are reversed.

TRANSPORT OF HYDROGEN IONS BETWEEN TISSUES AND LUNGS

I. Most of the hydrogen ions generated in the erythrocytes from carbonic acid during blood passage through tissue capillaries bind to deoxyhemoglobin because deoxyhemoglobin, formed as oxygen unloads from oxyhemoglobin, has a high affinity for hydrogen ions.

II. As the blood flows through the lung capillaries, hydrogen ions bound to deoxyhemoglobin are released and combine with bicarbonate to yield carbon dioxide and water.

CONTROL OF RESPIRATION

I. Breathing depends upon cyclical inspiratory muscle excitation by the nerves to the diaphragm and intercostal muscles. This neural activity is triggered by the medullary inspiratory neurons.

II. The most important inputs to the medullary inspiratory neurons for the involuntary control of minute ventilation are from the peripheral chemoreceptors—the carotid and aortic bodies—and the central chemoreceptors.

III. Ventilation is reflexly stimulated, via the peripheral chemoreceptors, by a decrease in arterial P_{O_2}, but only when the decrease is large.

IV. Ventilation is reflexly stimulated, via both the peripheral and central chemoreceptors, when the arterial P_{CO_2} goes up even a slight amount. The stimulus for this reflex is not the increased P_{CO_2} itself, but the concomitant increased hydrogen-ion concentration in arterial blood and brain extracellular fluid.

V. Ventilation is also stimulated, mainly via the peripheral chemoreceptors, by an increase in arterial hydrogen-ion concentration resulting from causes other than an increase in P_{CO_2}. The result of this reflex is to restore hydrogen-ion concentration toward normal by lowering P_{CO_2}.

VI. Ventilation is reflexly inhibited by an increase in arterial P_{O_2} and by a decrease in arterial P_{CO_2} or hydrogen-ion concentration.

VII. During moderate exercise, ventilation increases in exact proportion to metabolism, but the signals causing this are not known. During very strenuous exercise, ventilation increases more than metabolism.

A. The proportional increases in ventilation and metabolism during moderate exercise cause the arterial P_{O_2}, P_{CO_2}, and hydrogen-ion concentration to remain unchanged.

B. Arterial hydrogen-ion concentration increases during very strenuous exercise because of increased lactic acid production. This accounts for some of the hyperventilation seen in that situation.

VIII. Ventilation is also controlled by reflexes originating in airway receptors and by conscious intent.

HYPOXIA

I. The causes of hypoxic hypoxia are listed in Table 15-10.

II. During exposure to hypoxia, as at high altitude, oxygen supply to the tissues is maintained by the five responses listed in Table 15-11.

NONRESPIRATORY FUNCTIONS OF THE LUNGS

I. The lungs influence arterial blood concentrations of biologically active substances by removing some from systemic venous blood and adding others to systemic arterial blood.

II. The lungs also act as sieves that dissolve small clots formed in the systemic tissues.

KEY TERMS

respiration (two definitions)	intercostal muscle
respiratory system	pleural sac
pulmonary	pleura
alveoli (alveolus)	intrapleural fluid
airway	intrapleural pressure
inspiration	(P_{ip})
expiration	ventilation
pharynx	alveolar pressure (P_{alv})
larynx	atmospheric pressure
vocal cords	(P_{atm})
trachea	Boyle's law
bronchi (bronchus)	transpulmonary pressure
bronchiole	lung compliance (C_L)
conducting zone	surface tension
respiratory zone	surfactant
type I alveolar cell	lateral traction
type II alveolar cell	tidal volume
thorax	inspiratory reserve volume
diaphragm	expiratory reserve volume
	residual volume

vital capacity
minute ventilation
anatomic dead space
alveolar ventilation
alveolar dead space
physiologic dead space
respiratory quotient (RQ)
Dalton's law
partial pressure
Henry's law
hypoventilation
hyperventilation
hemoglobin
heme
globin
iron (Fe)
deoxyhemoglobin (Hb)
oxyhemoglobin (HbO$_2$)

percent hemoglobin
  saturation
oxygen-carrying capacity
oxygen-hemoglobin
  dissociation curve
2,3-diphosphoglycerate
  (DPG)
carbamino compound
carbonic anhydrase
total blood carbon dioxide
medullary inspiratory
  neuron
pulmonary stretch receptor
peripheral chemoreceptor
carotid body
aortic body
central chemoreceptor
J receptor

REVIEW QUESTIONS

1. List the functions of the respiratory system.
2. At rest, how many liters of air and blood flow through the lungs per minute?
3. Describe four functions of the conducting portion of the airways.
4. Which respiration steps occur by diffusion and which by bulk flow?
5. What are normal values for intrapleural pressure, alveolar pressure, and transpulmonary pressure at the end of an unforced expiration?
6. Between breaths at the end of an unforced expiration, in what directions are the lungs and chest wall tending to move? What prevents them from doing so?
7. State typical values for oxygen consumption, carbon dioxide production, and cardiac output at rest. How much oxygen (in milliliters per liter) is present in systemic venous and systemic arterial blood?
8. Write the equation relating air flow into or out of the lungs to atmospheric pressure, alveolar pressure, and airway resistance.
9. Describe the sequence of events that cause air to move into the lungs during inspiration and out of the lungs during expiration. Diagram the changes in intrapleural pressure and alveolar pressure.
10. What factors determine lung compliance? Which is most important?
11. How does surfactant increase lung compliance?
12. How is airway resistance influenced by airway radii?
13. List the physical factors that alter airway resistance.
14. Contrast the causes of increased airway resistance in asthma, emphysema, and chronic bronchitis.
15. What distinguishes lung capacities, as a group, from lung volumes?
16. State the formula relating minute ventilation, tidal volume, and respiratory rate. Give representative values for each at rest.

17. State the formula for calculating alveolar ventilation. What is an average value for alveolar ventilation?
18. The partial pressure of a gas is dependent upon what two factors?
19. State the alveolar partial pressures for oxygen and carbon dioxide in a normal person at rest.
20. What factors determine alveolar partial pressures?
21. What is the mechanism of gas exchange between alveoli and pulmonary capillaries? In a normal person at rest, what are the gas pressures at the end of the pulmonary capillaries, relative to those in the alveoli?
22. Why does thickening of alveolar membranes impair oxygen movement but have little effect on carbon dioxide exchange?
23. What is the major result of ventilation-perfusion inequalities throughout the lungs? Describe one homeostatic response that minimizes mismatching.
24. What generates the diffusion gradients for oxygen and carbon dioxide in the tissues?
25. In what two forms is oxygen carried in the blood? What are the normal quantities (in milliliters per liter) for each form in arterial blood?
26. Describe the structure of hemoglobin.
27. Draw an oxygen-hemoglobin dissociation curve. Put in the points that represent systemic venous and systemic arterial blood (ignore the rightward shift of the curve in systemic venous blood). What is the adaptive importance of the plateau?
28. Would breathing pure oxygen cause a large increase in oxygen transport by the blood in a normal person? In a person with a low alveolar P_{O_2}?
29. Describe the effects of increased P_{CO_2}, H^+ concentration, and temperature on the oxygen-hemoglobin dissociation curve. How are these effects adaptive for oxygen unloading in the tissues?
30. Describe the effects of increased DPG on the oxygen-hemoglobin dissociation curve. Under what conditions does an increase in DPG occur?
31. Draw figures showing the reactions carbon dioxide undergoes entering the blood in the tissue capillaries and leaving the blood in the alveoli. What fractions are contributed by dissolved carbon dioxide, bicarbonate, and carbamino?
32. What happens to most of the hydrogen ions formed in the erythrocytes from carbonic acid? What happens to blood H^+ concentration as blood flows through tissue capillaries?
33. What are the effects of P_{O_2} on carbamino formation and H^+ binding by hemoglobin?
34. In what area of the brain does automatic control of rhythmical respirations reside?
35. Describe the function of the pulmonary stretch receptors.
36. What changes stimulate the peripheral chemoreceptors? The central chemoreceptors?
37. Why does moderate anemia or carbon monoxide exposure not stimulate the peripheral chemoreceptors?
38. Is respiratory control more sensitive to small changes in arterial P_{O_2} or in arterial P_{CO_2}?
39. Describe the pathways by which increased arterial P_{CO_2} stimulates ventilation. Which pathway is more important?

40. Describe the pathway by which a change in arterial H^+ concentration independent of altered carbon dioxide influences ventilation. What is the adaptive value of this reflex?

41. What happens to arterial P_{O_2}, P_{CO_2}, and H^+ concentration during moderate and severe exercise? List other factors that may stimulate ventilation during exercise.

42. List four general causes of hypoxic hypoxia.

43. Describe two general ways in which the lungs can alter the concentrations of substances other than oxygen, carbon dioxide, and H^+ in the arterial blood.

CLINICAL TERMS

cystic fibrosis	ventilation-perfusion
pneumothorax	inequality
respiratory-distress	shunt
syndrome of the	respiratory acidosis
newborn	respiratory alkalosis
asthma	poliomyelitis
bronchodilator drugs	metabolic acidosis
chronic obstructive	metabolic alkalosis
pulmonary disease	pulmonary embolus
emphysema	dyspnea
chronic bronchitis	hypoxia
Heimlich maneuver	hypoxic hypoxia
forced expiratory volume	hypoxemia
in 1 sec (FEV_1)	anemic hypoxia
obstructive lung diseases	ischemic hypoxia
restrictive lung diseases	histotoxic hypoxia
hypoventilation	diffusion impairment
hyperventilation	hypercapnea
diffuse interstitial fibrosis	mountain sickness

THOUGHT QUESTIONS

(Answers are given in Appendix A)

1. At the end of a normal expiration, a person's lung volume is 2 L, his alveolar pressure is 0 mmHg, and his intrapleural pressure is −4 mmHg. He then inhales 800 ml, and at the end of inspiration the alveolar pressure is 0 mmHg and the intrapleural pressure is −8 mmHg. Calculate this person's lung compliance.

2. A patient is unable to produce surfactant. In order to inhale a normal tidal volume, will her intrapleural pressure have to be more or less subatmospheric during inspiration, relative to a normal person?

3. A patient is artificially ventilated by a machine during surgery at a rate of 20 breaths/min and a tidal volume of 250 ml/breath. Assuming a normal anatomic dead space of 150 ml, is this patient receiving an adequate alveolar ventilation?

4. Why must a person floating on the surface of the water and breathing through a snorkel increase his tidal volume and/or breathing frequency if alveolar ventilation is to remain normal?

5. A normal person breathing room air voluntarily increases her alveolar ventilation twofold and continues to do so until new steady-state alveolar gas pressures for oxygen and carbon dioxide are reached. Are the new values higher or lower than normal?

6. A person has an alveolar P_{O_2} of 105 mmHg and an arterial P_{O_2} of 80 mmHg. Could hypoventilation, say, due to respiratory muscle weakness, produce these values?

7. A person's alveolar membranes have become thickened enough to moderately decrease the rate at which gases diffuse across them at any given partial pressure differences. Will this person necessarily have a low arterial P_{O_2} at rest? During exercise?

8. A person is breathing 100 percent oxygen. How much will the oxygen content (in milliliters per liter of blood) of the arterial blood increase compared to when the person is breathing room air?

9. Which of the following have higher values in systemic venous blood than in systemic arterial blood: plasma P_{CO_2}, erythrocyte P_{CO_2}, plasma bicarbonate concentration, erythrocyte bicarbonate concentration, plasma hydrogen-ion concentration, erythrocyte hydrogen-ion concentration, erythrocyte carbamino concentration, erythrocyte chloride concentration, plasma chloride concentration?

10. If the spinal cord were severed where it joins the brainstem, what would happen to respiration?

11. The peripheral chemoreceptors are denervated in an experimental animal, and the animal then breathes a gas mixture containing 10 percent oxygen. What changes occur in the animal's ventilation? What changes occur when this denervated animal is given a mixture of air containing 21 percent oxygen and 5 percent carbon dioxide to breathe?

12. Patients with severe uncontrolled diabetes mellitus produce large quantities of certain organic acids. Can you predict the ventilation pattern in these patients and whether their arterial P_{O_2} and P_{CO_2} increase or decrease?

CHAPTER 16

The Kidneys and Regulation of Water and Inorganic Ions

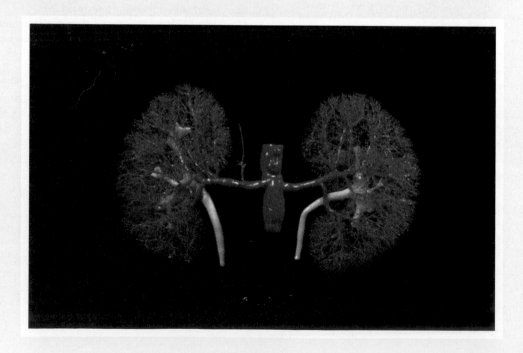

T his chapter deals with how the water and inorganic-ion composition of the internal environment are homeostatically regulated. The kidneys play the central role in these processes.

Regulation of the total-body balance of any substance can be studied in terms of the balance concept described in Chapter 7. Theoretically, a substance can appear in the body either as a result of ingestion or as a product of metabolism. On the loss side of the balance, a substance can be excreted from the body or can be metabolized. Therefore, if the quantity of any substance in the body is to be maintained at a nearly constant level over a period of time, the total amounts ingested and produced must equal the total amounts excreted and metabolized.

Reflexes that alter excretion, specifically excretion via the urine, constitute the major mechanisms that regulate the body balances of water and many of the inorganic ions that determine the properties of the extracellular fluid. The extracellular concentrations of these ions are given in Table 6-1. We will first describe how the kidneys work in general and then apply this information to how they process specific substances—sodium, water, potassium, and so on—and participate in reflexes that regulate these substances.

SECTION A
BASIC PRINCIPLES OF RENAL PHYSIOLOGY

RENAL FUNCTIONS

The adjective **renal** means "pertaining to the kidneys"; thus, for example, we refer to "renal physiology" and "renal functions."

The kidneys process the plasma portion of blood by removing substances from it and, in a few cases, by adding substances to it. In so doing they perform a variety of functions, as summarized in Table 16-1.

First, and very importantly, the kidneys play the central role in regulating the water concentration, inorganic-ion composition, and volume of the internal environment. They do so by excreting just enough water and inorganic ions to keep the amounts of these substances in the body relatively constant. For example, if you start eating a lot of salt (sodium chloride), the kidneys will in-

TABLE 16-1 FUNCTIONS OF THE KIDNEYS

1. Regulation of water and inorganic-ion balance
2. Removal of metabolic waste products from the blood and their excretion in the urine
3. Removal of foreign chemicals from the blood and their excretion in the urine
4. Gluconeogenesis
5. Secretion of hormones:
    a. Erythropoietin, which controls erythrocyte production (Chapter 14)
    b. Renin, which controls formation of angiotensin, which influences blood pressure and sodium balance (this chapter)
    c. 1,25-Dihydroxyvitamin D_3, which influences calcium balance (this chapter)

Now for the crucial point: Since inulin is filterable at the renal corpuscle but is not reabsorbed, secreted, or metabolized by the tubule, its clearance must equal the volume of plasma originally filtered, that is, C_{In} is equal to GFR.

It is important to realize that the clearance of *any* substance handled by the kidneys in the same way as inulin—filtered, but not reabsorbed, secreted, or metabolized—would equal the GFR. Unfortunately, there are no substances normally present in the plasma that meet these criteria. Because inulin must be administered, it is usually used as a research tool rather than a routine clinical measurement. For clinical purposes, the **creatinine clearance** (C_{Cr}) is commonly used to *approximate* the GFR as follows. The waste product creatinine produced by muscle is filtered at the renal corpuscle and does not undergo reabsorption. It does undergo a small amount of secretion, however, so that some plasma is cleared of its creatinine by secretion. Accordingly, the C_{Cr} overestimates the GFR but is close enough to be highly useful.

This leads to an important generalization: When the clearance of any substance is greater than the GFR, as measured by the inulin clearance, that substance must undergo tubular secretion. Look back now at our hypothetical substance X (Figure 16-7): X is filtered, and all the X that escapes filtration is secreted; no X is reabsorbed. Accordingly, all the plasma that enters the kidney per unit time is cleared of its X, and the clearance of X is therefore a measure of renal plasma flow. A substance that is handled like X is the organic anion para-aminohippurate (PAH), which is used for this purpose (unfortunately, like inulin, it must be administered intravenously).

One more example is needed: What is the clearance of glucose (C_G) in normal people? The answer is 0 mL/min. As stated earlier, no glucose is normally excreted in the urine; accordingly, the numerator of the clearance equation is 0 and C_G is 0. This leads to another important generalization: When the clearance of a filterable substance is less than the GFR, as measured by the inulin clearance, that substance must undergo reabsorption. For a substance that undergoes only partial reabsorption, sodium, for example, the clearance value will not be zero but will still be less than the inulin clearance.

The remainder of this chapter describes how the kidneys function in the homeostasis of individual substances and how renal function is coordinated with that of other organs. Before turning to these individual substances, however, we complete the general story by describing the mechanisms for eliminating urine from the body—micturition.

MICTURITION

Urine flow through the ureters to the bladder is propelled by contractions of the ureter-wall smooth muscle. The urine is stored in the bladder and intermittently ejected during urination, or **micturition.**

The bladder is a balloon-like chamber with walls of smooth muscle collectively termed the **detrusor muscle.** The contraction of the detrusor muscle squeezes on the urine in the lumen to produce urination. That part of the detrusor muscle at the base (or "neck") of the bladder where the urethra begins functions as a sphincter called the **internal urethral sphincter.** Because of its anatomy, when the detrusor muscle is *relaxed,* the sphincter is *closed.* When the detrusor muscle actively contracts, changes in the muscle's shape pull open the outlet. Beyond the internal urethral sphincter, skeletal muscle surrounds the urethra. This is the **external urethral sphincter,** the contraction of which can prevent urination even when the detrusor muscle contracts strongly.

FIGURE 16-11

Micturition reflex. As the bladder fills with urine, stimulation of the stretch receptors in the bladder walls causes reflex stimulation of the parasympathetic nerves to the bladder and inhibition of the motor nerves to the external urethral sphincter. The result is bladder contraction and sphincter relaxation. Descending input from higher centers can act on the efferent neurons to allow voluntary initiation or delay of micturition.

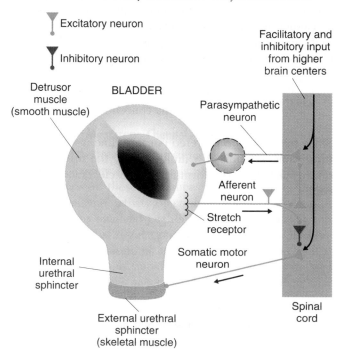

The basic micturition reflex is a local spinal reflex, but descending pathways from the brain can influence it (Figure 16-11). We describe first the isolated spinal reflex as it exists in infancy or in persons with spinal-cord damage that eliminates the contribution of the descending pathways. The bladder wall contains stretch receptors whose afferent fibers enter the spinal cord and activate the parasympathetic nerves that supply and stimulate the detrusor muscle. As the bladder fills with urine, the pressure within it increases, and this stimulates the stretch receptors. The afferent pathway from these receptors activates the parasympathetic neurons, which then cause the detrusor muscle to contract. When the bladder reaches a certain volume (300 to 400 mL), the reflexly induced contraction of the detrusor muscle becomes strong enough to pull open the internal urethral sphincter. Simultaneously, the afferent input from the stretch receptors inhibits, within the spinal cord, the somatic motor neurons that were stimulating the external urethral sphincter to contract. Both sphincters are now open and the contraction of the detrusor muscle is able to produce urination.

Voluntary prevention of micturition, learned during childhood, operates in the following way. As the bladder distends, the input from the bladder stretch receptors causes, via ascending pathways to the brain, a sense of bladder fullness and the urge to urinate. In response to this, urination can be voluntarily prevented via descending pathways that stimulate the motor nerves to the external urethral sphincter and simultaneously inhibit the parasympathetic nerves to the detrusor muscle.

In contrast, urination can be voluntarily initiated, regardless of the degree of bladder fullness, via the descending pathways.

SECTION A SUMMARY

FUNCTIONS AND STRUCTURE OF THE KIDNEYS AND URINARY SYSTEM

I. The kidneys regulate the water and ionic composition of the body, excrete waste products, excrete foreign chemicals, produce glucose during prolonged fasting, and secrete three hormones—renin, 1,25-dihydroxyvitamin D_3, and erythropoietin. The first three functions are accomplished by continuous processing of the plasma.

II. Each nephron in the kidneys consists of a renal corpuscle and a tubule.
   A. Each renal corpuscle comprises a capillary tuft, termed a glomerulus, and a Bowman's capsule, into which the tuft protrudes.
   B. The tubule extends out from Bowman's capsule and is subdivided into many segments, which can be combined for reference purposes into the proximal tubule, loop of Henle, distal convoluted tubule, and collecting duct. Beginning at the level of the collecting ducts, multiple tubules join and empty into the renal pelvis, from which urine flows through ureters to the bladder.
   C. Each glomerulus is supplied by an afferent arteriole, and an efferent arteriole leaves the glomerulus to branch into peritubular capillaries, which supply the tubule.

BASIC RENAL PROCESSES

I. The three basic renal processes are glomerular filtration, tubular reabsorption, and tubular secretion. In addition, the kidneys synthesize and/or catabolize certain substances. The excretion of a substance is equal to the amount filtered plus the amount secreted minus the amount reabsorbed.

II. Urine formation begins with glomerular filtration—approximately 180 L/day—of essentially protein-free plasma into Bowman's space.
   A. Glomerular filtrate contains all plasma substances other than proteins and substances bound to proteins in virtually the same concentrations as in plasma.
   B. Glomerular filtration is driven by the hydrostatic pressure in the glomerular capillaries and is opposed by both the hydrostatic pressure in Bowman's space and the osmotic force due to the proteins in the glomerular capillary plasma.

III. As the filtrate moves through the tubules, certain substances are reabsorbed into the peritubular capillaries.
   A. Substances to which the tubular epithelium is permeable are absorbed by diffusion because water reabsorption creates tubule-interstitium concentration gradients for them.
   B. Tubular reabsorption rates are generally very high for nutrients, ions, and water, but are lower for waste products. Reabsorption may occur by diffusion or by mediated transport.
   C. Many of the mediated-transport systems manifest transport maximums, so that when the filtered load of a substance exceeds the transport maximum, large amounts may appear in the urine.

IV. Tubular secretion (movement from the peritubular capillaries into the tubules), like glomerular filtration, is a pathway for entrance of a substance into the tubule.

V. The clearance of any substance can be calculated by dividing the mass of the substance excreted per unit time by the plasma concentration of the substance. GFR can be measured by means of the inulin clearance and estimated by means of the creatinine clearance.

MICTURITION

I. In the basic micturition reflex, bladder distention stimulates stretch receptors that trigger spinal reflexes; these reflexes lead to contraction of the detrusor muscle, mediated by parasympathetic neurons, and relaxation of the external urethral sphincter, mediated by inhibition of the motor neurons to this muscle.

II. Voluntary control is exerted via descending pathways to the parasympathetic nerves supplying the detrusor muscle and the motor nerves supplying the external urethral sphincter.

SECTION A KEY TERMS

renal
urea
uric acid
creatinine
gluconeogenesis
ureter
bladder
urethra
nephron
renal corpuscle
tubule
glomerulus
glomerular capillaries
afferent arteriole
Bowman's capsule
efferent arteriole

Bowman's space
podocyte
proximal tubule
loop of Henle
descending limb
ascending limb
distal convoluted tubule
collecting duct system
connecting tubule
cortical collecting duct
medullary collecting duct
renal pelvis
renal cortex
renal medulla
peritubular capillaries
macula densa

juxtaglomerular cells
juxtaglomerular
   apparatus (JGA)
glomerular filtration
glomerular filtrate
tubular reabsorption
tubular secretion
net glomerular filtration
   pressure
glomerular filtration rate
   (GFR)

filtered load
luminal membrane
basolateral membrane
transport maximum (*Tm*)
clearance
inulin
creatinine clearance
micturition
detrusor muscle
internal urethral sphincter
external urethral sphincter

SECTION A REVIEW QUESTIONS

1. What are the functions of the kidneys?

2. What three hormones do the kidneys secrete?

3. Fluid flows in sequence through what structures from the glomerulus to the bladder? Blood flows through what structures from the renal artery to the renal vein?

4. What are the three basic renal processes that lead to the formation of urine?

5. How does the composition of the glomerular filtrate compare with that of plasma?

6. Describe the forces that determine the magnitude of the GFR. What is a normal value for GFR?

7. Contrast the mechanisms of reabsorption for glucose and urea. Which one shows a *Tm*?

8. Diagram the sequence of events leading to micturition in infants and in adults.

SECTION B
REGULATION OF SODIUM, WATER, AND POTASSIUM BALANCE

TOTAL-BODY BALANCE OF SODIUM AND WATER

Table 16-3 summarizes total-body water balance. These are average values, which are subject to considerable normal variation. There are two sources of body water gain: (1) water produced from the oxidation of organic nutrients, and (2) water ingested in liquids and so-called solid food (a rare steak is approximately 70 percent water). There are four sites from which water is lost to the external environment: skin, respiratory passageways, gastrointestinal tract, and urinary tract. Menstrual flow constitutes a fifth potential source of water loss in women. The loss of water by evaporation from the skin and the lining of respiratory passageways is a continuous process. It is called

insensible water loss because the person is unaware of its occurrence. Additional water can be made available for evaporation from the skin by the production of sweat. Normal gastrointestinal loss of water in feces is generally quite small, but can be severe in diarrhea. Gastrointestinal loss can also be large in vomiting.

Table 16-4 is a summary of total-body balance for sodium chloride. The excretion of sodium and chloride via the skin and gastrointestinal tract is normally quite small but may increase markedly during severe sweating, vomiting, or diarrhea. Hemorrhage can also result in loss of large quantities of both salt and water.

Under normal conditions, as can be seen from Tables 16-3 and 16-4, salt and water losses exactly equal salt and water gains, and no net change in body salt and water oc-

TABLE 16-3	AVERAGE DAILY WATER GAIN AND LOSS IN ADULTS
Intake	
In liquids	1200 ml
In food	1000 ml
Metabolically produced	350 ml
Total	2550 ml
Output	
Insensible loss (skin and lungs)	900 ml
Sweat	50 ml
In feces	100 ml
Urine	1500 ml
Total	2550 ml

curs. This matching of losses and gains is primarily the result of regulation of urinary loss, which can be varied over an extremely wide range. For example, urinary water excretion can vary from approximately 0.4 L/day to 25 L/day, depending upon whether one is lost in the desert or participating in a beer-drinking contest. Similarly, some individuals ingest 20 to 25 g of sodium chloride per day, whereas a person on a low-salt diet may ingest only 50 mg. Normal kidneys can readily alter their excretion of salt over this range to match loss with gain.

We will first present the basic renal processes for sodium and water and then describe the homeostatic reflexes that influence these processes. The renal processing of chloride is usually coupled directly or indirectly to that of sodium and so we shall have little more to say about chloride even though it is the most abundant anion in the extracellular fluid.

BASIC RENAL PROCESSES FOR SODIUM AND WATER

Having low molecular weights and not being bound to protein, both sodium and water freely filter from the glomerular capillaries into Bowman's space. They both un-

TABLE 16-4	DAILY SODIUM CHLORIDE INTAKE AND LOSS
Intake	
Food	10.50 g
Output	
Sweat	0.25 g
Feces	0.25 g
Urine	10.00 g
Total	10.50 g

dergo considerable reabsorption—normally more than 99 percent (Table 16-2)—but no secretion. Most renal energy utilization goes to accomplish this enormous reabsorptive task. The bulk of sodium and water reabsorption (about two-thirds) occurs in the proximal tubule, but the major hormonal controls of reabsorption are exerted on the collecting ducts.

The mechanisms of sodium and water reabsorption can be summarized by two generalizations: (1) Sodium reabsorption is an active process occurring in all tubular segments except the descending limb of the loop of Henle; (2) water reabsorption is by diffusion and is dependent upon sodium reabsorption.

Primary Active Sodium Reabsorption

The essential feature underlying sodium reabsorption throughout the tubule is the primary active transport of sodium out of the cells and into the interstitial fluid, as illustrated for the cortical collecting duct in Figure 16-12. This transport is achieved by Na,K-ATPase pumps in the basolateral membrane of the cells. The active transport of sodium out of the cell keeps the intracellular concentration of sodium low compared to the luminal concentration, and so sodium moves "downhill" out of the lumen

FIGURE 16-12

Mechanism of sodium reabsorption in the cortical collecting duct. Movement of the reabsorbed sodium from the interstitial fluid into peritubular capillaries is shown in Figure 16-13. (As noted earlier, the *basement* membrane underlying the tubule — not to be confused with the *basolateral* membrane of each tubular cell—plays no significant role in transport and is not shown in this and subsequent figures illustrating transport.) The sizes of the letters for Na$^+$ and K$^+$ denote high and low concentrations of these ions. The fate of the potassium ions transported by the Na,K-ATPase pumps is discussed in the later section dealing with renal potassium handling.

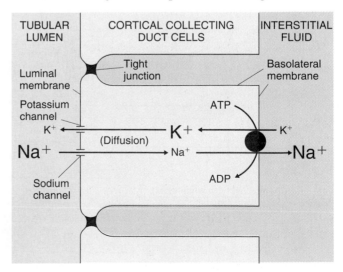

into the tubular epithelial cells. The fact that the cell interior is negatively charged relative to the lumen also contributes to the electrochemical gradient favoring this movement.

The precise mechanism of the *downhill* sodium movement across the luminal membrane into the cell varies from segment to segment of the tubule depending upon which channels and/or transport proteins are present in their luminal membranes. For example, as illustrated in Figure 16-12, the luminal entry step for sodium in the cortical collecting duct is by diffusion through sodium channels. In the proximal tubule the luminal entry step is either by cotransport with a variety of organic molecules (glucose, for example) or by countertransport with hydrogen ions (that is, the hydrogen ions move from cell to lumen as the sodium moves into the cell). In this manner, in the proximal tubule sodium reabsorption drives the reabsorption of the cotransported substances and the secretion of hydrogen ions.

While the movement of sodium downhill from lumen into cell across the *luminal membrane* varies from one segment of the tubule to another, the *basolateral membrane* step is the same in all tubular segments—the primary active transport of sodium out of the cell is via Na,K-ATPase pumps in this membrane. It is this transport process that lowers intracellular sodium concentration and so makes possible the downhill luminal entry step, whatever its mechanism.

Coupling of Water Reabsorption to Sodium Reabsorption

How does active sodium reabsorption lead to passive water reabsorption? This type of coupling was described in Chapter 6 (Figure 6-25) and is summarized again in Figure 16-13. (1) Sodium (and other solutes whose reabsorption is dependent on sodium transport, for example, glucose, amino acids, and bicarbonate) is transported from the tubular lumen to the interstitial fluid

across the epithelial cells, as described in the previous section. (2) This removal of solute lowers the osmolarity, that is, *raises* the water concentration, of the luminal fluid. It simultaneously raises the osmolarity, that is, *lowers* the water concentration, of the interstitial fluid adjacent to the epithelial cells. (3) The difference in water concentration between lumen and interstitial fluid causes net diffusion of water from the lumen across the tubular cells' plasma membranes and/or tight junctions into the interstitial fluid. (4) From there, water, sodium, and everything else in the interstitial fluid move together by bulk flow into peritubular capillaries as the final step in reabsorption.

Water movement across the tubular epithelium can occur, however, only if the epithelium is permeable to water. No matter how large its concentration gradient, water cannot cross an epithelium impermeable to it. Water permeability varies from tubular segment to segment. The water permeability of the proximal tubule is always very high, and so water molecules are reabsorbed by this segment almost as rapidly as sodium ions. As a result, the proximal tubule always reabsorbs sodium and water in the same proportions.

We will describe the water permeability of the next tubular segments—the loop of Henle and distal convoluted tubule—later. Now for the really crucial point: The water permeability of the last portions of the tubules, the collecting ducts, can be high or low because it is subject to physiological control, the only tubular segments in which water permeability is under such control.

The major determinant of this controlled permeability, and hence of water reabsorption in the collecting ducts, is a peptide hormone secreted by the posterior pituitary and known as **vasopressin**, or **antidiuretic hormone, ADH.** Vasopressin stimulates the insertion into the luminal membrane, by exocytosis, of protein mole-

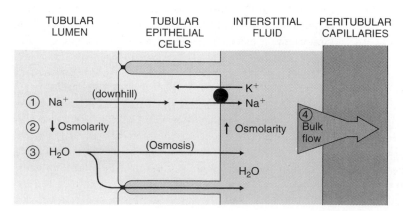

TUBULAR LUMEN   TUBULAR EPITHELIAL CELLS   INTERSTITIAL FLUID   PERITUBULAR CAPILLARIES

FIGURE 16-13

Coupling of water and sodium reabsorption. See text for explanation of numbers. The reabsorption of solutes other than sodium—for example, glucose, amino acids, and bicarbonate—also contributes to the difference in osmolarity between lumen and interstitial fluid, but the reabsorption of all these substances is ultimately dependent on direct or indirect cotransport and countertransport with sodium; therefore, they are not shown in the figure.

cules made by the collecting duct cells and known as aquaporins; these proteins function as water channels (Chapter 6). Accordingly, in the presence of a high plasma concentration of vasopressin, the water permeability of the collecting ducts is very great. Therefore, water reabsorption is maximal, and the final urine volume is small—less than 1 percent of the filtered water.

Without vasopressin, the water permeability of the collecting ducts is very low, and very little water is reabsorbed from these sites. Therefore, a large volume of water remains behind in the tubule to be excreted in the urine. This increased urine excretion resulting from low vasopressin is termed **water diuresis** (**diuresis** simply means a large urine flow from any cause). In a subsequent section we will describe the reflexes that control vasopressin secretion.

The disease ***diabetes insipidus,*** which is distinct from the other kind of diabetes mentioned earlier in this chapter (diabetes mellitus or "sugar diabetes"), illustrates what happens when the vasopressin system malfunctions. People with this disease have lost the ability to produce vasopressin, usually as a result of damage to the hypothalamus. Thus, the permeability to water of the collecting ducts is low and unchanging regardless of the state of the body fluids. Therefore a constant water diuresis is present—as much as 25 L/day.

Note that in water diuresis, there is an increased urine flow, but not an increased solute excretion. In all other cases of diuresis, termed **osmotic diuresis,** the increased urine flow is the result of a primary increase in solute excretion. For example, failure of normal sodium reabsorption causes both increased sodium excretion and increased water excretion, since, as we have seen, water reabsorption is absolutely dependent on solute reabsorption. Another example of osmotic diuresis occurs in people with uncontrolled, marked diabetes *mellitus:* In this case, the glucose that escapes reabsorption because of the huge filtered load retains water in the lumen, causing it to be excreted along with the glucose. We'll talk more about the consequences of this in Chapter 18. To summarize, any loss of solute in the urine *must* be accompanied by water loss (osmotic diuresis), but the reverse is not true, that is, water diuresis is not accompanied by equivalent solute loss.

Urine Concentration: The Countercurrent Multiplier System

Before reading this section you should review, by looking up in the glossary, several terms presented in Chapter 6—**hypoosmotic, isoosmotic,** and **hyperosmotic.**

In the section just concluded we described how the kidneys produce a small volume of urine when the plasma concentration of vasopressin is high. Under these conditions the urine is found to be concentrated (hyperosmotic) relative to plasma. This section describes the mechanisms by which this hyperosmolarity is achieved.

The ability of the kidneys to produce hyperosmotic urine is a major determinant of one's ability to survive with limited amounts of water. The human kidney can produce a maximal urinary concentration of 1400 mOsmol/L, almost five times the osmolarity of plasma, which is 290 mOsmol/L (for ease of calculation we shall round this off to 300 mOsmol/L in future discussions). The typical daily excretion of urea, sulfate, phosphate, other waste products, and ions amounts to approximately 600 mOsmol. Therefore, the minimal volume of urine water in which this mass of solute can be dissolved equals

$$\frac{600 \text{ mOsmol/day}}{1400 \text{ mOsmol/L}} = 0.444 \text{ L/day}$$

This volume of urine is known as the **obligatory water loss.** The loss of this minimal volume of urine (it would be somewhat lower if no food were available) contributes to dehydration when a person is deprived of water intake.

Urinary concentration takes place as tubular fluid flows through the medullary collecting ducts. The interstitial fluid surrounding these ducts is very hyperosmotic, and in the presence of vasopressin, water diffuses out of the ducts into the interstitial fluid and then enters the blood vessels of the medulla to be carried away. The key question is: How does the medullary interstitial fluid become hyperosmotic? The answer is: through the function of Henle's loop. Recall that Henle's loop forms a hairpin-like loop between the proximal tubule and the distal convoluted tubule. The fluid entering the loop from the proximal tubule flows down the descending limb, turns the corner, and then flows up the ascending limb. The opposing flows in the two limbs is termed a countercurrent flow, and as described next, the entire loop functions as a **countercurrent multiplier system** to create a hyperosmotic medullary interstitial fluid.

Because the proximal tubule always reabsorbs sodium and water in the same proportions, the fluid entering the desending limb of the loop from the proximal tubule has the same osmolarity as plasma—300 mOsmol/L. For the moment let's skip the descending limb since the events in it can only be understood in the context of what the *ascending* limb is doing. Along the entire length of the ascending limb, sodium and chloride are reabsorbed into the medullary interstitial fluid. In the upper (thick) portion of the ascending limb this reabsorption is achieved by transporters that actively cotransport sodium and chloride (as well as potassium); such transporters are not present in the lower (thin) portion of the ascending limb, and the reabsorption there is a passive

which is synthesized and secreted by cells in the cardiac atria. ANF acts on the tubules (several tubular segments and mechanisms are involved) to *inhibit* sodium reabsorption. It can also act on the renal blood vessels to increase GFR, which further contributes to increased sodium excretion. As would be predicted, the secretion of ANF is increased when there is an excess of sodium in the body, but the stimulus for this increased secretion is not alterations in sodium *concentration*. Rather, using the same logic (only in reverse) that applies to the control of renin and aldosterone secretion, ANF secretion increases because of the expansion of plasma volume that accompanies an increase in body sodium. The specific stimulus is increased atrial distention (Figure 16-18).

This completes our survey of the control of sodium excretion, which depends upon the control of two renal variables—the GFR and sodium reabsorption. The latter is controlled by the renin-angiotensin-aldosterone hormone system and by other factors, including atrial natriuretic factor. The reflexes that control both GFR and sodium reabsorption are essentially reflexes that regulate blood pressure, since they are most frequently initiated by changes in arterial or venous pressures.

FIGURE 16-18

Atrial natriuretic factor in the *direct* control of sodium reabsorption and hence sodium excretion.

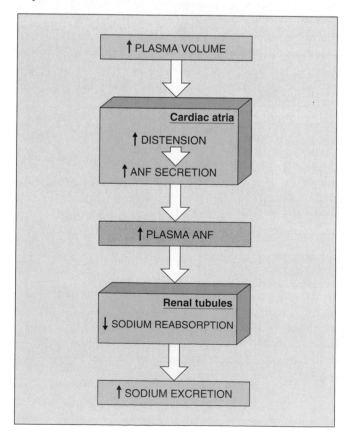

RENAL WATER REGULATION

Water excretion is the difference between the volume of water filtered (the GFR) and the volume reabsorbed. Accordingly, the baroreceptor-initiated GFR-controlling reflexes described in the previous section tend to have the same effects on water excretion as on sodium excretion. As is true for sodium, however, the major regulated determinant of water excretion is not GFR but rather the rate of water reabsorption. As we have seen, this is determined by vasopressin, and so total-body water is regulated mainly by reflexes that alter the secretion of this hormone.

As described in Chapter 10, vasopressin is produced by a discrete group of hypothalamic neurons whose axons terminate in the posterior pituitary, from which vasopressin is released into the blood. The most important of the inputs to these neurons are from baroreceptors and osmoreceptors.

Baroreceptor Control of Vasopressin Secretion

We have seen that a decreased extracellular volume, due say to diarrhea or hemorrhage, reflexly calls forth, via the renin-angiotensin system, an increased aldosterone secretion. But the decreased extracellular volume also triggers increased vasopressin secretion. This increased vasopressin increases the water permeability of the collecting ducts, more water is reabsorbed and less is excreted, and so water is retained in the body to help stabilize the extracellular volume.

This reflex is initiated by several baroreceptors in the cardiovascular system (Figure 16-19). The baroreceptors decrease their rate of firing when cardiovascular pressures decrease, as occurs when blood volume decreases. Therefore, few impulses are transmitted from the baroreceptors via afferent neurons and ascending pathways to the hypothalamus, and the result is *increased* vasopressin secretion. Conversely, increased cardiovascular pressures cause more firing by the baroreceptors, resulting in a decrease in vasopressin secretion.

In addition to its effect on water excretion, if the plasma vasopressin concentration becomes very high it, like angiotensin II, causes widespread arteriolar constriction. This helps restore arterial blood pressure toward normal (Chapter 14).

The baroreceptor reflex for vasopressin, as just described, has a relatively high threshold—that is, there must be a sizable reduction in cardiovascular pressures to trigger it. Therefore, this reflex, compared to the osmoreceptor reflex described next, generally plays a lesser role under most physiological circumstances, but it can become very important in pathological states such as hemorrhage.

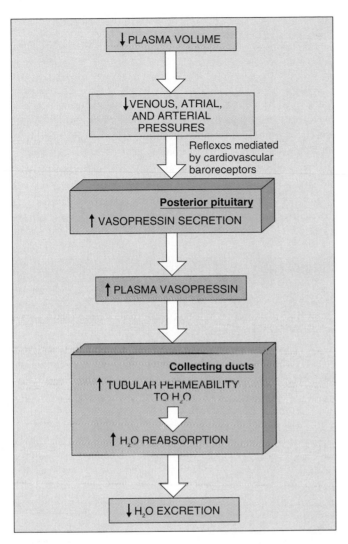

FIGURE 16-19

Baroreceptor pathway by which vasopressin secretion is increased when plasma volume is decreased. The opposite events (culminating in a decrease in vasopressin secretion) occur when plasma volume increases.

Osmoreceptor Control of Vasopressin Secretion

We have seen how changes in extracellular volume simultaneously elicit reflex changes in the excretion of *both* sodium and water. This is adaptive since the situations causing extracellular volume alterations are very often associated with loss or gain of both sodium and water in approximately proportional amounts. In contrast, we shall see now that changes in total-body water in which no change in total-body sodium occurs are compensated for by altering water excretion without altering sodium excretion.

The major change caused by water loss or gain out of proportion to sodium loss or gain is a change in the osmolarity of the body fluids. This is a key point because, under conditions due predominantly to water gain or loss,

the receptors that initiate the reflexes controlling vasopressin secretion are **osmoreceptors** in the hypothalamus, receptors responsive to changes in osmolarity (the mechanism by which they respond to changes in osmolarity is still uncertain).

As an example, take a person drinking 2 L of sugar-free soft drink, which contains little sodium or other solute, in a short time. The excess water lowers the body-fluid osmolarity (raises the water concentration), which reflexly inhibits vasopressin secretion via the hypothalamic osmoreceptors (Figure 16-20). As a result, the water

FIGURE 16-20

Osmoreceptor pathway by which vasopressin secretion is lowered and water excretion raised when excess water is ingested. The opposite events (an increase in vasopressin secretion) occur when osmolarity increases, as during water deprivation.

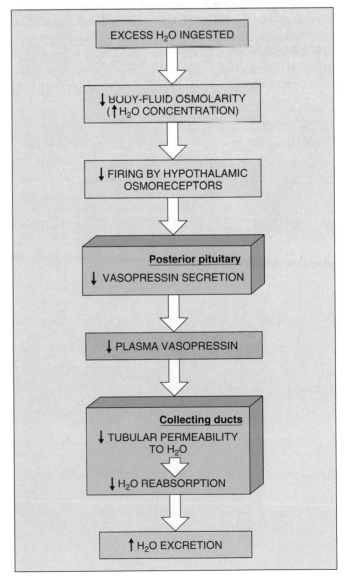

permeability of the collecting ducts becomes very low, water is not reabsorbed from these segments, and a large volume of hypoosmotic urine is excreted. In this manner, the excess water is eliminated.

At the other end of the spectrum, when the osmolarity of the body fluids increases (water concentration decreases), say, because of water deprivation, vasopressin secretion is reflexly increased via the osmoreceptors, water reabsorption by the collecting ducts is increased, and a very small volume of highly concentrated urine is excreted. By retaining relatively more water than solute, the kidneys help reduce the body-fluid osmolarity back toward normal.

We have now described two afferent pathways controlling the vasopressin-secreting hypothalamic cells, one from baroreceptors and one from osmoreceptors. To add to the complexity, these cells receive synaptic input from many other brain areas, so that vasopressin secretion, and therefore urine volume and concentration, can be altered by pain, fear, and a variety of drugs. For example, alcohol is a powerful inhibitor of vasopressin release, and this probably accounts for much of the increased urine volume produced following the ingestion of alcohol, a urine volume well in excess of the volume of beverage consumed.

A SUMMARY EXAMPLE: THE RESPONSE TO SWEATING

Figure 16-21 shows the factors that control renal sodium and water excretion in response to severe sweating. Sweat is a hypoosmotic solution containing mainly water, sodium, and chloride. Therefore, sweating causes both a decrease in extracellular volume and an increase in body-fluid osmolarity (a decrease in water concentration). The renal retention of water and sodium minimizes the deviations from normal caused by the loss of water and salt in the sweat.

THIRST AND SALT APPETITE

Now we turn to the other component of any balance—control of intake. Deficits of salt and water must eventually be compensated for by ingestion of these substances, for the kidneys cannot create new sodium ions or water, they can only minimize their excretion until ingestion replaces the losses.

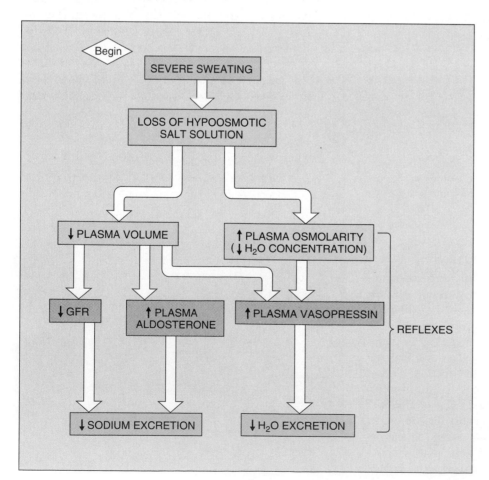

FIGURE 16-21

Pathways by which sodium and water excretion are decreased in response to severe sweating. This figure is an amalgamation of Figures 16-15, 16-17, 16-19, and the reverse of 16-20.

The subjective feeling of thirst, which drives us to obtain and ingest water, is stimulated both by a lower extracellular volume and a higher plasma osmolarity (Figure 16-22), the latter being the single most important stimulus under normal physiological conditions. Note that these are precisely the same two changes that stimulate vasopressin production, and the osmoreceptors and baroreceptors that control vasopressin secretion are identical to those for thirst. The brain centers that receive input from these receptors and mediate thirst are located in the hypothalamus, very close to those areas that produce vasopressin.

Another influencing factor is angiotensin II, which stimulates thirst by a direct effect on the brain. Thus, the renin-angiotensin system helps regulate not only sodium balance but water balance as well and constitutes one of the pathways by which thirst is stimulated when extracellular volume is decreased.

There are still other pathways controlling thirst. For example, dryness of the mouth and throat causes profound thirst, which is relieved by merely moistening them. Some kind of "metering" of water intake by the gastrointestinal tract also occurs; that is, a thirsty individual given access to water stops drinking after replacing the lost water but before most of the water has been absorbed from the gastrointestinal tract and has a chance to eliminate the stimulatory inputs to the systemic baroreceptors and osmoreceptors. How this metering occurs remains a mystery, but one function of this feedforward process is to prevent overhydration.

The analog of thirst for sodium, **salt appetite,** is an important part of sodium homeostasis in most mammals. Salt appetite consists of two components: "hedonistic" appetite and "regulatory" appetite; that is, animals "like" salt and eat it whenever they can, regardless of whether they are salt-deficient, and, in addition, their drive to obtain salt is markedly increased in the presence of bodily salt deficiency.

Human beings certainly have a strong hedonistic appetite for salt, as manifested by almost universally large intakes of salt whenever it is cheap and readily available (for example, the average American consumes 10–15 g/day despite the fact that human beings can survive quite normally on less than 0.5 g/day). However, unlike most other mammals, humans have relatively little regulatory salt appetite, at least until a bodily salt deficit becomes extremely large.

POTASSIUM REGULATION

Potassium is, as we have seen, the most abundant intracellular ion. Although only 2 percent of total-body potassium is in the extracellular fluid, the potassium concentration in this fluid is extremely important for the function of excitable tissues, notably nerve and muscle. Recall (Chapter 8) that the resting-membrane potentials of these tissues are directly related to the relative intracellular and extracellular potassium concentrations. Accordingly, either increases or decreases in extracellular potassium concentration can cause abnormal rhythms of the heart (*arrhythmias*) and abnormalities of skeletal-muscle contraction.

A normal person remains in potassium balance by daily excreting an amount of potassium in the urine equal to the amount ingested minus the amounts eliminated in the feces and sweat. Also, like sodium, potassium losses via sweat and the gastrointestinal tract are normally quite small, although vomiting or diarrhea can cause large quantities to be lost. The control of renal function is the major mechanism by which body potassium is regulated.

Renal Regulation of Potassium

Potassium is freely filterable in the renal corpuscle. Normally, the tubules reabsorb most of this filtered potassium so that very little of the filtered potassium appears in

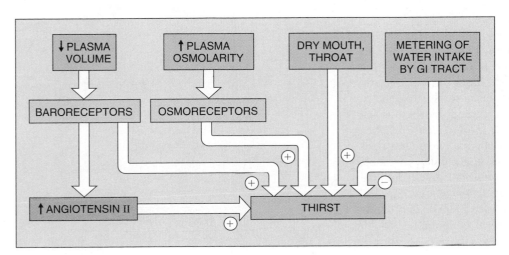

FIGURE 16-22

Inputs reflexly controlling thirst. The osmoreceptor input is the single most important stimulus under most physiological conditions. Psychosocial factors and conditioned responses are not shown.

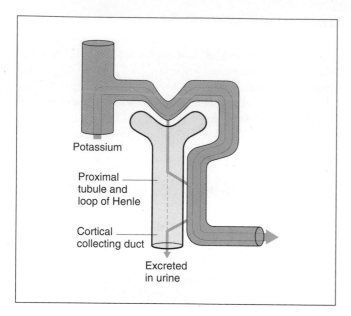

FIGURE 16-23

Simplified model of the basic renal processing of potassium under most circumstances. The cortical collecting ducts can manifest either net reabsorption or net secretion; however, except in persons who are ingesting diets very low in potassium or are otherwise potassium-depleted, net secretion occurs there, as shown in this figure.

the urine. However, the cortical collecting ducts can *secrete* potassium, and changes in potassium *excretion* are due mainly to changes in potassium *secretion* by this tubular segment (Figure 16-23).

During potassium depletion, when the homeostatic response is to minimize potassium loss, there is no potassium secretion by the cortical collecting ducts, and only the small amount of filtered potassium that escapes tubular reabsorption is excreted. In all other situations, to the small amount of potassium not reabsorbed is added a variable amount of potassium secreted by the cortical collecting ducts, an amount necessary to maintain total-body balance.

The mechanism of potassium secretion by the cortical collecting ducts was illustrated in Figure 16-12. In this tubular segment the K^+ pumped into the cell across the basolateral membrane by Na,K-ATPases diffuses into the tubular lumen through K^+ channels in the luminal membrane. Thus, the *reabsorption* of sodium by the cortical collecting duct is associated with the *secretion* of potassium by this tubular segment. Potassium secretion does not occur in other sodium-reabsorbing tubular segments because there are few potassium channels in the luminal membranes of their cells; rather, in these segments the potassium pumped into the cell by Na,K-ATPases simply diffuses back across the basolateral membrane through potassium channels located there.

What factors influence potassium secretion by the cortical collecting ducts to achieve homeostasis of bodily potassium? The single most important factor is as follows: When a high-potassium diet is ingested (Figure 16-24), plasma potassium concentration increases, though very slightly, and this drives enhanced basolateral uptake via the Na,K-ATPase pumps and hence an enhanced potassium secretion. Conversely, a low-potassium diet or a negative potassium balance, for example, from diarrhea, lowers basolateral potassium uptake; this reduces potassium secretion and excretion, thereby helping to reestablish potassium balance.

A second important factor linking potassium secretion to potassium balance is the hormone aldosterone. Besides stimulating tubular sodium reabsorption by the cortical collecting ducts, aldosterone simultaneously enhances tubular potassium secretion by this tubular segment. Recall from the discussion of aldosterone and sodium reabsorption that aldosterone stimulates synthesis of the Na,K-ATPase pumps in the basolateral membranes of the cortical collecting ducts; this is one of several mechanisms by which aldosterone stimulates potassium secretion.

The reflex by which an excess or deficit of potassium controls aldosterone production (Figure 16-24) is com-

FIGURE 16-24

Pathways by which an increased potassium intake induces greater potassium excretion.

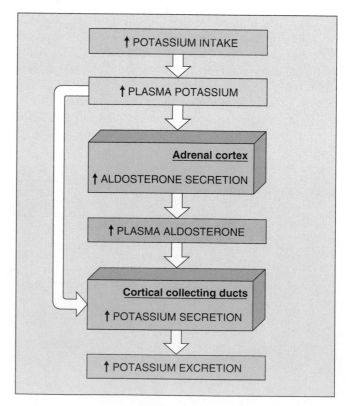

FIGURE 16-25

Summary of the control of aldosterone and its effects on sodium reabsorption and potassium secretion.

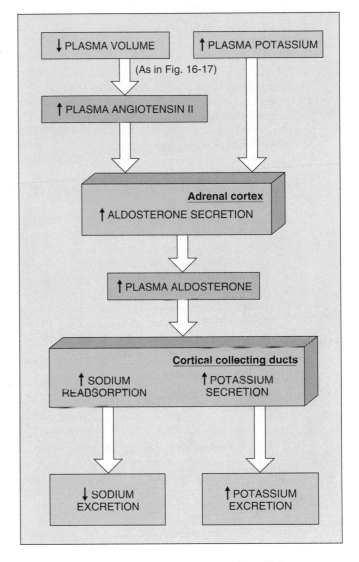

pletely different from the reflex described earlier involving the renin-angiotensin system. The aldosterone-secreting cells of the adrenal cortex are sensitive to the potassium concentration of the extracellular fluid bathing them. Thus, an increased intake of potassium leads to an increased extracellular potassium concentration, which in turn directly stimulates aldosterone production by the adrenal cortex. The resulting increased plasma aldosterone concentration increases potassium secretion and thereby eliminates the excess potassium from the body.

Conversely, a lowered extracellular potassium concentration decreases aldosterone production and thereby reduces potassium secretion. Less potassium than usual is excreted in the urine, thus helping to restore the normal extracellular concentration.

The control and major renal tubular effects of aldosterone are summarized in Figure 16-25. The fact that a single hormone regulates both sodium and potassium excretion raises the question of potential conflicts between homeostasis of the two ions. For example, if a person were sodium-deficient and therefore secreting large amounts of aldosterone, the potassium-secreting effects of this hormone would tend to cause some potassium loss even though potassium balance were normal to start with. Normally, such conflicts cause only minor imbalances because there are a variety of other counteracting controls of sodium and potassium excretion.

SECTION B SUMMARY

TOTAL-BODY BALANCE OF SODIUM AND WATER

I.   The body gains sodium and chloride by ingestion and loses them via the skin (in sweat), gastrointestinal tract, and urine.

II. The body gains water via ingestion and internal production, and it loses water via urine, the gastrointestinal tract, and evaporation from the skin and respiratory tract (as insensible loss and sweat).

III. For both water and sodium, the major homeostatic control point for maintaining stable balance is renal excretion.

BASIC RENAL PROCESSES FOR SODIUM AND WATER

I.   Sodium is freely filterable at the glomerulus, and its reabsorption is a primary active process dependent upon Na,K-ATPase pumps in the basolateral membranes of the tubular epithelium. Sodium is not secreted.

II. Sodium entry into the cell from the tubular lumen is always passive. Depending on the tubular segment it is either through channels or by cotransport or countertransport with other substances.

III. Sodium reabsorption creates an osmotic difference across the tubule, which drives water reabsorption.

IV. Water reabsorption is independent of the posterior pituitary hormone vasopressin until the collecting ducts, where it increases water-permeability. A large volume of dilute urine is produced when plasma vasopressin concentration and hence water reabsorption by the collecting ducts, is low.

V. A small volume of concentrated urine is produced by the renal countercurrent multiplier system when plasma vasopressin concentration is high.

  A. The active transport of sodium chloride by the ascending loop of Henle causes a concentration of the interstitial fluid of the medulla but a dilution of the luminal fluid.

  B. Vasopressin increases the permeability of the cortical collecting ducts to water, and so water is reabsorbed by this segment until the luminal fluid is isoosmotic to cortical interstitial fluid.

  C. The luminal fluid then enters and flows through the medullary collecting ducts, and the concentrated medullary interstitium causes water to move out of these ducts, made highly permeable to water by vasopressin. The result is concentration of the collecting duct fluid and the urine.

RENAL SODIUM REGULATION

I. Sodium excretion is the difference between the amount of sodium filtered and the amount reabsorbed.

II. GFR, and hence the filtered load of sodium, are controlled by baroreceptor reflexes. Decreased vascular pressures cause decreased baroreceptor firing and hence increased sympathetic outflow to the renal arterioles, resulting in vasoconstriction and decreased GFR. These changes are generally relatively small under most physiological conditions.

III. The major control of tubular sodium reabsorption is the adrenal cortical hormone aldosterone, which stimulates sodium reabsorption in the cortical collecting ducts.

IV. The renin-angiotensin system is one of the two major controllers of aldosterone secretion. When extracellular volume decreases, renin secretion is stimulated by three inputs: (1) stimulation of the renal sympathetic nerves to the juxtaglomerular cells by extrarenal baroreceptor reflexes; (2) pressure decreases sensed by the juxtaglomerular cells, themselves acting as intrarenal baroreceptors; and (3) a signal generated by low sodium or chloride concentration in the lumen of the macula densa.

V. Many other factors influence sodium reabsorption. One of these, atrial natriuretic factor, is secreted by cells in the atria in response to atrial distention; it inhibits sodium reabsorption and it also increases GFR.

RENAL WATER REGULATION

I. Water excretion is the difference between the amount of water filtered and the amount reabsorbed.

II. GFR regulation via the baroreceptor reflexes plays some role in regulating water excretion, but the major control is via vasopressin-mediated control of water reabsorption.

III. Vasopressin secretion by the posterior pituitary is controlled by cardiovascular baroreceptors and by osmoreceptors in the hypothalamus.

  A. Via the baroreceptor reflexes, a low extracellular volume stimulates vasopressin secretion and a high extracellular volume inhibits it.

  B. Via the osmoreceptors, a high body-fluid osmolarity stimulates vasopressin secretion and a low osmolarity inhibits it.

THIRST AND SALT APPETITE

I. Thirst is stimulated by a variety of inputs, including baroreceptors, osmoreceptors, and angiotensin II.

II. Salt appetite is not of major regulatory importance in human beings.

POTASSIUM REGULATION

I. A person remains in potassium balance by excreting an amount of potassium in the urine equal to the amount ingested minus the amounts lost in the feces and sweat.

II. Potassium is freely filterable at the renal corpuscle and undergoes both reabsorption and secretion, the latter occurring in the cortical collecting duct and being the major controlled variable determining potassium excretion.

III. When body potassium is increased, extracellular potassium concentration increases. This increase acts directly on the cortical collecting ducts to stimulate potassium secretion and also stimulates aldosterone secretion, the increased plasma aldosterone then also stimulating potassium secretion.

SECTION B KEY TERMS

insensible water loss
vasopressin (antidiuretic hormone, ADH)
water diuresis
diuresis
osmotic diuresis
hypoosmotic
isoosmotic
hyperosmotic
obligatory water loss
countercurrent multiplier system
aldosterone
renin-angiotensin system
renin
angiotensin I
angiotensinogen
angiotensin II
angiotensin converting enzyme
intrarenal baroreceptors
atrial natriuretic factor (ANF)
osmoreceptors
salt appetite

SECTION B REVIEW QUESTIONS

1. What are the sources of water gain and loss in the body? What are the sources of sodium gain and loss?

2. Describe the distribution of water and sodium between intracellular and extracellular fluids.

3. What is the relationship between body sodium and extracellular-fluid volume?

4. What is the mechanism of sodium reabsorption, and how is the reabsorption of other solutes coupled to it?

5. What is the mechanism of water reabsorption, and how is it coupled to sodium reabsorption?

6. What is the effect of vasopressin on the renal tubules, and what are the sites affected?

7. Describe the characteristics of the two limbs of the loop of Henle with regard to their transport of sodium, chloride, and water.

8. Diagram the osmolarities in the two limbs of the loop of Henle, distal convoluted tubule, cortical collecting duct, cortical interstitium, medullary collecting duct, and medullary interstitium in the presence of vasopressin. What happens to the cortical and medullary collecting duct values in the absence of vasopressin?

9. What two processes determine how much sodium is excreted per unit time?

10. Diagram the sequence of events by which a decrease in blood pressure leads to a decreased GFR.

11. List the sequence of events leading from increased renin secretion to increased aldosterone secretion.

12. What are the three inputs controlling renin secretion?

13. Diagram the sequence of events leading from decreased cardiovascular pressures or from an increased plasma osmolarity to an increased secretion of vasopressin.

14. What are the stimuli for thirst?

15. Which of the basic renal processes apply to potassium? Which of them is the controlled process and which tubular segment performs it?

16. Diagram the steps leading from increased plasma potassium to increased potassium excretion.

17. What are the two major controls of aldosterone secretion, and what are this hormone's major actions?

SECTION C
CALCIUM REGULATION

Extracellular calcium concentration normally remains relatively constant. Large deviations in either directions would cause problems. A low plasma calcium concentration increases the excitability of nerve and muscle plasma membranes, so that individuals with low plasma calcium suffer from **hypocalcemic tetany,** characterized by skeletal muscle spasms. A high plasma calcium concentration causes cardiac arrhythmias as well as depressed neuromuscular excitability. These effects reflect, in part, the ability of extracellular calcium to bind to plasma membrane proteins that function as ion channels, thereby altering membrane potentials. The binding alters the open or closed state of the channels. This effect of calcium on plasma membranes is totally distinct from its role as an intracellular excitation-contraction coupler.

EFFECTOR SITES FOR CALCIUM HOMEOSTASIS

The sections in this chapter on sodium, water, and potassium homeostasis were concerned almost entirely with the *renal* handling of these substances. In contrast, the regulation of calcium depends not only on the kidneys but also on bone and the gastrointestinal tract. The activities of the gastrointestinal tract and kidneys determine the net intake and output of calcium for the entire body and, thereby, the overall state of calcium balance. In contrast, inter-

changes of calcium between extracellular fluid and bone do not alter total body balance but, rather, the *distribution* of calcium within the body. We will first describe how the effector sites handle calcium and then discuss how they are influenced by hormones in the homeostatic control of plasma calcium concentration.

Bone

Approximately 99 percent of total-body calcium is contained in bone. Therefore, deposition of calcium in bone or its removal very importantly influences plasma calcium concentration.

Bone has functions, summarized in Table 16-5 and discussed in various chapters, other than regulating plasma calcium concentration. It is important to recognize that its

TABLE 16-5 FUNCTIONS OF BONE

1. Supports the body and imposed loads against gravity.
2. Provides the rigidity that permits locomotion.
3. Rib cage, vertebrae, and skull afford protection to the internal organs.
4. Serves as a reservoir for calcium, inorganic phosphate, and other mineral elements.
5. Bone marrow produces the blood cells.

role in maintaining normal plasma calcium concentration takes precedence over the mechanical supportive role, sometimes to the detriment of the latter.

Bone is a special connective tissue made up of several cell types surrounded by a collagen matrix, called **osteoid,** upon which are deposited minerals, particularly the crystals of calcium and phosphate known as hydroxyapatite. In some instances bones have central marrow cavities where blood cells are formed (Chapter 14). Typically, approximately one-third of a bone, by weight, is osteoid and two-thirds is mineral (the bone cells contribute negligible weight).

The three types of bone cells (marrow cells are not included in this term) are osteoblasts, osteocytes, and osteoclasts (Figure 16-26). **Osteoblasts** are the bone-forming cells. They secrete collagen to form a surrounding matrix, which then becomes calcified; just how this mineralization is brought about remains controversial. Once surrounded by calcified matrix, the osteoblasts are called **osteocytes.** The osteocytes have long cytoplasmic processes that extend throughout the bone and form tight junctions with other osteocytes. **Osteoclasts** are large multinucleated cells that break down (resorb) previously formed bone by secreting hydrogen ions, which dissolve the crystals, and hydrolytic enzymes, which digest the osteoid.

The growth of bones during childhood will be discussed in Chapter 18. What is important here is that throughout life bone is being constantly "remodeled" by the osteoblasts and osteoclasts working together. Osteoclasts resorb old bone, and then osteoblasts move into the area and lay down new matrix, which becomes calcified. This process is dependent, in part, on the stresses imposed on the bones by gravity and muscle tension, both of which stimulate osteoblastic activity. It is also influenced by many hormones, as summarized in Table 16-6, and a bewildering variety of growth factors produced locally in the bone. Of the hormones listed, only parathyroid hormone and 1,25-dihydroxyvitamin D_3 are controlled primarily by reflexes that regulate plasma calcium concentration. Nonetheless, changes in the other listed hormones have important influences on bone mass and plasma calcium concentration.

Kidneys

About 60 percent of plasma calcium is filterable at the renal corpuscle (the rest is bound to plasma protein), and most of this filtered calcium is reabsorbed. There is no tubular secretion of calcium. Accordingly, the urinary excretion of calcium is the difference between the amount filtered and the amount reabsorbed. Like that of sodium, the control of calcium excretion is exerted mainly on reabsorption; that is, reabsorption is reflexly decreased when plasma calcium concentration goes up for whatever reason, and reflexly increased when plasma calcium goes down.

FIGURE 16-26

Cross section through a small portion of bone. The light gray area is mineralized osteoid. The osteocytes have long processes that extend through small canals and connect with each other and to osteoblasts via tight junctions. (*Adapted from Goodman.*)

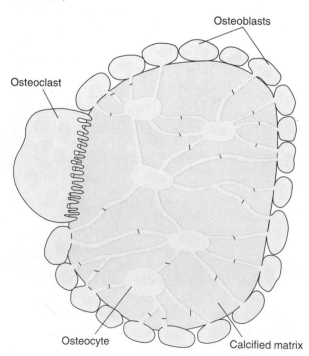

Osteoblasts

Osteoclast

Osteocyte

Calcified matrix

TABLE 16-6	REFERENCE SUMMARY OF MAJOR HORMONAL INFLUENCES ON BONE MASS*

Hormones that favor bone formation and increased bone mass
Insulin
Growth hormone
Insulin-like growth factor I (IGF-I)
Estrogen
Testosterone
1,25-Dihydroxyvitamin D_3 (influences only mineralization, not matrix)
Calcitonin

Hormones that favor increased bone resorption and decreased bone mass
Parathyroid hormone
Cortisol
Thyroid hormones (T_4 and T_3)

*The effects of all hormones except 1,25-dihydroxyvitamin D_3 are ultimately due to a hormone-induced alteration of the balance between the activities of the osteoblasts and osteoclasts. This altered balance is not always the result of a *direct* action of the hormone on these cells, however, but may result *indirectly* from some other action of the hormone. For example, large amounts of cortisol depress intestinal absorption of calcium, which in turn causes reduced plasma calcium, increased parathyroid hormone secretion, and stimulation of osteoclasts by parathyroid hormone. Also, many paracrine agents produced by the bone cells and bone-marrow connective-tissue cells influence bone formation and resorption.

In addition, as we shall see, the renal handling of phosphate plays a role in the regulation of extracellular calcium. Phosphate, too, is handled by a combination of filtration and reabsorption, the latter being hormonally controlled.

Gastrointestinal Tract

The absorption of sodium, water, and potassium from the gastrointestinal tract normally approximates 100 percent. There is some homeostatic control of these processes, but it is relatively unimportant and so we ignored it. In contrast, a considerable amount of ingested calcium is not absorbed from the intestine and simply leaves the body along with the feces. Moreover, the active transport system that moves this ion from intestinal lumen to blood is under important hormonal control. Accordingly, there can be large regulated increases or decreases in the amount of calcium absorbed. Indeed, hormonal control of this absorptive process is the major means for homeostatically regulating total-body calcium balance, more important than the control of renal calcium excretion.

HORMONAL CONTROLS

The two major hormones that homeostatically regulate plasma calcium concentration are parathyroid hormone and 1,25-dihydroxyvitamin D_3. Another hormone, calcitonin, which is secreted by the thyroid gland, has long been thought to help regulate plasma calcium, but present evidence indicates that it plays only a minor role in this regard.

Parathyroid Hormone

All three of the effector sites described above are subject, directly or indirectly, to control by a protein hormone called **parathyroid hormone,** produced by the parathyroid glands. These glands are embedded in the surface of the thyroid gland, but are distinct from it. Parathyroid hormone production is controlled by the extracellular calcium concentration acting directly on the secretory cells (via a plasma-membrane calcium receptor). *Decreased* plasma calcium concentration *stimulates* parathyroid hormone secretion, and an increased plasma calcium concentration does just the opposite.

Parathyroid hormone exerts multiple actions that increase extracellular calcium concentration, thus compensating for the decreased concentration that originally stimulated secretion of this hormone. (Figure 16-27):

1. It directly increases the resorption of bone by osteoclasts, which results in the movement of calcium (and phosphate) from bone into extracellular fluid.

2. It directly stimulates the activation of vitamin D (see below), and this latter hormone then increases intestinal absorption of calcium. Thus, the effect of parathyroid hormone on the intestinal tract is an indirect one.

3. It directly increases renal tubular calcium reabsorption, thus decreasing urinary calcium excretion.

In addition, parathyroid hormone directly reduces the tubular reabsorption of phosphate, thus raising its urinary excretion. This keeps plasma phosphate from increasing at a time when parathyroid hormone is simultaneously causing increased release of both calcium and phosphate from bone.

1,25-Dihydroxyvitamin D_3

The term **vitamin D** denotes a group of closely related compounds. One of these, called **vitamin D₃** is formed by the action of ultraviolet radiation (from sunlight, usually) on a cholesterol derivative in skin. Another form of vitamin D very similar to vitamin D_3 is ingested in food, specifically from plants.

Because of clothing and decreased outdoor living, people are often dependent upon dietary vitamin D, and for this reason it was originally classified as a vitamin. However, regardless of source, vitamin D_3 is metabolized by addition of hydroxyl groups, first in the liver and then in certain kidney tubular cells (Figure 16-28). The end result of these changes is **1,25-dihydroxyvitamin D₃** (abbreviated **1,25-$(OH)_2D_3$,** also called calcitriol), the active form of vitamin D. The major action of 1,25-$(OH)_2D_3$ is to stimulate active absorption of calcium by the intestine. Thus, the major event in vitamin D deficiency is decreased intestinal calcium absorption, resulting in decreased plasma calcium. It should be clear from this description that since 1,25-$(OH)_2D_3$ is made in the body, it is not, itself, a vitamin; instead, it fulfills the criteria for a hormone—a messenger molecule produced by one set of cells and exerting its action at other sites.

The blood concentration of 1,25-$(OH)_2D_3$ is subject to physiological control. The major control point is the second hydroxylation step, the one that occurs in the kidneys. The enzyme catalyzing this step is stimulated by parathyroid hormone. Thus, as we have seen, a low plasma calcium concentration stimulates the secretion of parathyroid hormone, which in turn enhances the production of 1,25-$(OH)_2D_3$, and both hormones contribute to restoration of the plasma calcium toward normal.

METABOLIC BONE DISEASES

Various diseases reflect abnormalities in the metabolism of bone. **Rickets** (in children) and **osteomalacia** (in adults) are conditions in which mineralization of bone matrix is

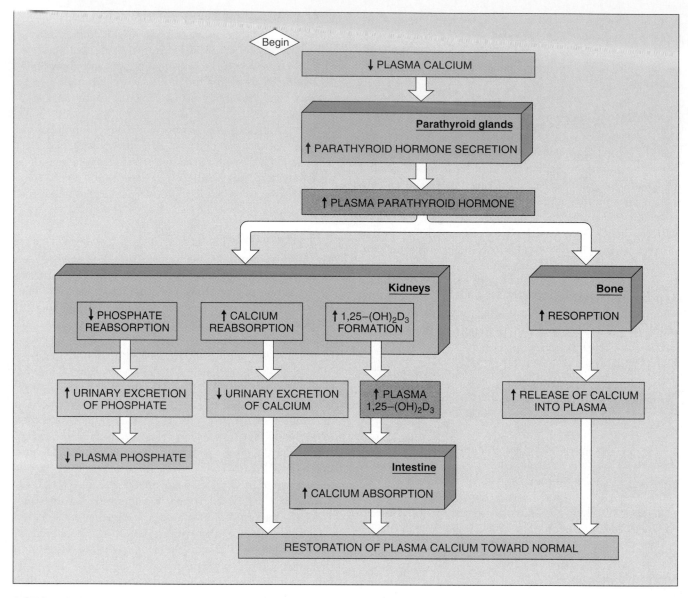

FIGURE 16-27

Reflexes by which a reduction in plasma calcium concentration is restored toward normal via the actions of parathyroid hormone. See Figure 16-28 for a description of $1,25\text{-}(OH)_2D_3$.

deficient, causing the bones to be soft and easily fractured. In addition, a child suffering from rickets typically is severely bowlegged due to the effect of weight-bearing on the legs. A major cause of these diseases is deficiency of $1,25\text{-}(OH)_2D_3$.

In contrast to rickets and osteomalacia, in *osteoporosis* both matrix and minerals are lost as a result of an imbalance between bone resorption and bone formation. The resulting decrease in bone mass and strength leads to an increased incidence of fractures. Osteoporosis can occur in people who are immobilized (disuse osteoporosis), in people who have an excessive plasma concentration of a hormone that favors bone resorption, and in people who

have a deficient plasma concentration of a hormone that favors bone formation (Table 16-6). It is most commonly seen, however, with aging. Everyone loses bone as he or she ages, but osteoporosis is much more common in elderly women than men for several reasons: Women have a smaller bone mass to begin with, and the loss that occurs with aging occurs more rapidly, particularly after menopause removes the bone-promoting influence of estrogen. Osteoporosis in postmenopausal women causes more than 1.5 million fractures each year and is responsible for thousands of deaths, usually due to complications following surgery for hip fracture.

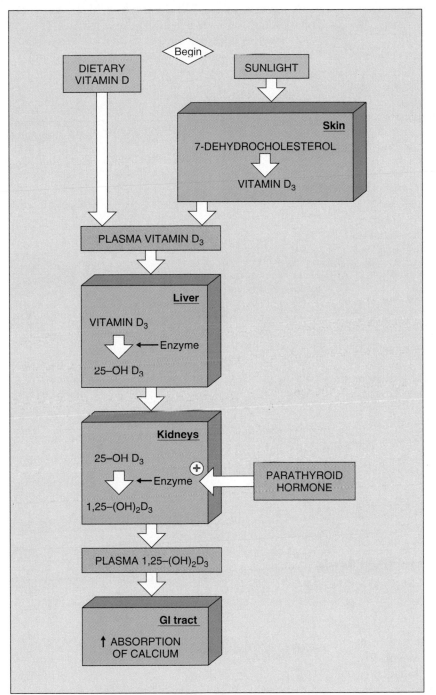

FIGURE 16-28
Metabolism of vitamin D to the active form, 1,25-(OH)$_2$D$_3$. The kidney enzyme that mediates the final step is activated by parathyroid hormone.

Prevention is the focus of attention for osteoporosis. Estrogen treatment in postmenopausal women is very effective in reducing the rate of bone loss. A regular weight-bearing exercise program (brisk walking and stair-climbing, for example) is also helpful. Adequate dietary calcium (1000 mg/day before menopause and 1200–1500 mg/day after menopause) throughout life is important to build up and maintain bone mass. Until re-

cently it was thought that there was no effective therapy once osteoporosis was established. However, several recent possibilities hold real promise as therapeutic agents. These include the hormone calcitonin, a group of drugs (called bisphosphonates) that interfere with the resorption of bone by osteoclasts, and 1,25-dihydroxyvitamin D$_3$. Estrogen is also useful not just for prevention, but for treatment as well.

EFFECTOR SITES FOR CALCIUM HOMEOSTASIS

I. The effector sites for the regulation of plasma calcium concentration are bone, the gastrointestinal tract, and the kidneys.

II. Approximately 99 percent of total-body calcium is contained in bone as minerals on a collagen matrix. Bone is constantly remodeled as a result of the interaction of osteoblasts and osteoclasts, a process which determines bone mass and provides a means for raising or lowering plasma calcium concentration.

III. Calcium is actively absorbed by the gastrointestinal tract, and this process is under hormonal control.

IV. The amount of calcium excreted in the urine is the difference between the amount filtered and the amount reabsorbed, the latter process being under hormonal control.

HORMONAL CONTROLS

I. Parathyroid hormone increases plasma calcium concentration by influencing all the effector sites.
  A. It stimulates tubular reabsorption of calcium, bone resorption with release of calcium, and formation of the hormone 1,25-dihydroxyvitamin D_3, which stimulates calcium absorption by the intestine.
  B. It also inhibits the tubular reabsorption of phosphate.

II. Vitamin D_3 is formed in the skin or ingested and then undergoes hydroxylations in the liver and kidneys, in the latter stimulated by parathyroid hormone, to the active form, 1,25-dihydroxyvitamin D_3.

osteoid
osteoblast
osteocyte
osteoclast
parathyroid hormone

vitamin D
vitamin D_3
1,25-dihydroxyvitamin D_3
$[1,25\text{-}(OH)_2D_3]$

1. List the functions of bone.

2. Describe bone remodeling.

3. Describe the handling of calcium by the kidneys and gastrointestinal tract.

4. What controls the secretion of parathyroid hormone, and what are this hormone's four major effects?

5. Describe the formation and action of 1,25-$(OH)_2D_3$. How does parathyroid hormone influence the production of this hormone?

SECTION D
HYDROGEN-ION REGULATION

Metabolic reactions are highly sensitive to the hydrogen-ion concentration of the fluid in which they occur. This sensitivity is due to the influence on enzyme function exerted by hydrogen ions, which change the shapes of proteins. Accordingly, the hydrogen-ion concentration of the extracellular fluid is closely regulated. (At this point the reader might want to review the section on hydrogen ions, acidity, and pH in Chapter 2.)

This regulation can be viewed in the same way as the balance of any other ion, that is, as the matching of gains and losses. When loss exceeds gain, the arterial plasma hydrogen-ion concentration goes down (pH goes above 7.4), and this is termed an *alkalosis.* When gain exceeds loss, the arterial plasma hydrogen-ion concentration goes up (pH goes below 7.4), and this is termed an *acidosis.*

SOURCES OF HYDROGEN-ION GAIN OR LOSS

Table 16-7 summarizes the major routes for gains and losses of hydrogen ion. First, as described in Chapter 15, a huge quantity of CO_2—about 20,000 mmol—is generated daily as the result of oxidative metabolism, and these CO_2 molecules yield hydrogen ions via the reactions:

$$CO_2 + H_2O \xrightarrow{\text{carbonic anhydrase}} H_2CO_3 \longrightarrow HCO_3^- + H^+ \tag{16-1}$$

This source does not normally constitute a net gain of hydrogen ions, however, since all the hydrogen ions generated

TABLE 16-7 SOURCES OF HYDROGEN-ION GAIN OR LOSS

Gain

1. Generation of hydrogen ions from CO_2
2. Production of nonvolatile acids from the metabolism of protein and other organic molecules
3. Gain of hydrogen ions due to loss of bicarbonate in diarrhea or other nongastric GI fluids
4. Gain of hydrogen ions due to loss of bicarbonate in the urine.

Loss

1. Loss of hydrogen ions in vomitus
2. Loss of hydrogen ions in the urine
3. Hyperventilation

via these reactions during passage of blood through the tissues are reincorporated into water when the reactions are reversed during passage of blood through the lungs (Chapter 15). Net retention of CO_2 does occur, however, in hypoventilation or respiratory disease and causes a net gain of hydrogen ions. Conversely, net loss of CO_2 occurs in hyperventilation and this causes net elimination of hydrogen ions.

The body also produces acids, both organic and inorganic, from sources other than CO_2. These are collectively termed **nonvolatile acids.** They include phosphoric acid and sulfuric acid, generated mainly by the catabolism of proteins, as well as lactic acid and several other organic acids. Dissociation of all these acids yields anions and hydrogen ions. But simultaneously the metabolism of a variety of organic anions utilizes hydrogen ions and produces bicarbonate. Thus, metabolism of "nonvolatile" solutes both generates and utilizes hydrogen ions. In the United States, where the diet is high in protein, the generation of nonvolatile acids predominants in most people, and there is an average net production of 40 to 80 mmol of hydrogen ions per day.

A third potential source of net body gain or loss of hydrogen ion is gastrointestinal secretions leaving the body. Vomitus contains a high concentration of hydrogen ions and so constitutes a source of net loss. In contrast, the other gastrointestinal secretions are alkaline; they contain very little hydrogen ion, but their concentration of bicarbonate is higher than exists in plasma. Loss of these fluids, as in diarrhea, constitutes in essence a body *gain* of hydrogen ions. This is an extremely important point: Given the mass action relationship shown in Equation 16-1, when the body loses a bicarbonate ion from the body it is the same as if the body had gained a hydrogen ion. The reason is that loss of the bicarbonate causes the reaction shown in Equation 16-1 to be driven to the right, thereby generating a hydrogen ion within the body. Similarly, when

the body gains a bicarbonate, it is the same as if the body had lost a hydrogen ion.

Finally, the kidneys constitute the fourth source of net hydrogen-ion gain or loss, that is, the kidneys can either remove hydrogen ions from the plasma or add them.

BUFFERING OF HYDROGEN IONS IN THE BODY

Any substance that can reversibly bind hydrogen ions is called a **buffer.** Between their generation in the body and their elimination, most hydrogen ions are buffered by extracellular and intracellular buffers. The normal extracellular-fluid pH of 7.4 corresponds to a hydrogen-ion concentration of only 0.00004 mmol/L (40 nanomols/L). Without buffering, the daily turnover rate of the 40 to 80 mmol of H^+ produced by our diet would cause huge changes in body fluid hydrogen-ion concentration. The general form of buffering reactions is:

$$Buffer^- + H^+ \rightleftharpoons HBuffer \qquad (16\text{-}2)$$

HBuffer is a weak acid in that it can dissociate to $Buffer^-$ plus H^+ or it can exist as the undissociated molecule (HBuffer). When H^+ concentration increases for whatever reason, the reaction is forced to the right and more H^+ is bound by $Buffer^-$ to form HBuffer. For example, when H^+ concentration is increased because of increased production of lactic acid, some of the hydrogen ions combine with the body's buffers, so the hydrogen ion concentration does not increase as much as it otherwise would have. Conversely, when H^+ concentration decreases because of the loss of hydrogen ions or the addition of alkali, Equation 16-2 proceeds to the left and H^+ is released from HBuffer. In this manner, the body buffers stabilize H^+ concentration against changes in either direction.

The major extracellular buffer is the CO_2/HCO_3^- system summarized in Equation 16-1. This system also plays some role in buffering within cells, but the major intracellular buffers are phosphates and proteins. One intracellular protein buffer is hemoglobin, as described in Chapter 15.

You must recognize that buffering does not eliminate hydrogen ions from the body or add them to the body; it only keeps them "locked-up" until balance can be restored. How balance is achieved is the subject of the rest of this section.

INTEGRATION OF HOMEOSTATIC CONTROLS

The kidneys are ultimately responsible for balancing hydrogen-ion gains and losses so as to maintain a relatively constant plasma hydrogen-ion concentration. Thus, the kidneys normally excrete the excess hydrogen ions from nonvolatile acids generated in the body from metabolism, that is, all acids other than carbonic acid. Moreover, if there is an additional net gain of hydrogen ions due to abnormally increased production of these nonvolatile acids, or to hypoventilation or respiratory malfunction, or to loss of alkaline gastrointestinal secretions, the kidneys increase their elimination of hydrogen ions from the body so as to restore balance. Alternatively, if there is a net loss of hydrogen ions from the body due to increased metabolic utilization of hydrogen ions (as in a vegetarian diet), hyperventilation, or vomiting, the kidneys replenish these hydrogen ions.

Although the kidneys are the ultimate hydrogen-ion balancers, the respiratory system also plays a very important homeostatic role. We have pointed out that hypoventilation, respiratory malfunction, and hyperventilation can *cause* a hydrogen-ion imbalance; now we emphasize that when a hydrogen-ion imbalance is due to a nonrespiratory cause, then ventilation is *reflexly altered* so as to help compensate for the imbalance. We described this phenomenon in Chapter 15 (Figure 15-32): An elevated arterial hydrogen-ion concentration stimulates ventilation, and this reflex hyperventilation causes reduces arterial P_{CO_2}, and hence, by mass action, reduced hydrogen-ion concentration. Alternatively, a decreased plasma hydrogen-ion concentration inhibits ventilation, thereby raising arterial P_{CO_2} and increasing the hydrogen-ion concentration.

Thus, the respiratory system and kidneys work together. The respiratory response to altered plasma hydrogen-ion concentration is very rapid (minutes) and keeps this concentration from changing too much until the more slowly responding kidneys (hours to days) can actually eliminate the imbalance. Of course, if the respiratory system is the actual *cause* of the hydrogen-ion imbalance, then the kidneys are the sole homeostatic responder. By the same token, malfunctioning kidneys can create a hydrogen-ion imbalance by eliminating too little or too much hydrogen ion from the body, and then the respiratory response is the only one operating.

RENAL MECHANISMS

In the previous section we wrote of the kidneys eliminating hydrogen ions from the body or replenishing them. The kidneys perform this task by altering plasma *bicarbonate* concentration. The key to understanding how altering plasma bicarbonate concentration eliminates or replenishes hydrogen ions was stated earlier: The excretion of a bicarbonate in the urine increases the plasma hydrogen-ion concentration just as if a hydrogen ion had been added to the plasma. Similarly, the addition of a bicarbonate to the plasma lowers the plasma hydrogen-ion concentration just as if a hydrogen ion had been removed from the plasma

Thus, when there is a lowering of plasma hydrogen-ion concentration (alkalosis) for whatever reason, the kidneys' homeostatic response is to excrete large quantities of bicarbonate. This raises plasma hydrogen-ion concentration back toward normal. In contrast, in response to a rise in plasma hydrogen-ion concentration (acidosis), the kidneys do not excrete bicarbonate in the urine, but instead kidney tubular cells produce *new* bicarbonate and add it to the plasma. This lowers the plasma hydrogen-ion concentration back toward normal.

Let us now look at the basic mechanisms by which bicarbonate excretion or addition of new bicarbonate to the plasma is achieved.

Bicarbonate Handling

Bicarbonate is completely filterable at the renal corpuscles and undergoes marked tubular reabsorption in various tubular segments (the proximal tubule, ascending loop of Henle, and cortical collecting ducts.) Bicarbonate can also be secreted (in the collecting ducts). Therefore:

$$HCO_3^- \text{ excretion} = HCO_3^- \text{ filtered} + HCO_3^- \text{ secreted} - HCO_3^- \text{ reabsorbed}$$

For simplicity, we will ignore the secretion of bicarbonate (because it is always quantitatively much less than tubular reabsorption) and treat bicarbonate excretion as the difference between filtration and reabsorption.

Bicarbonate reabsorption is an active process, but it is not accomplished in the conventional manner of simply having an active pump for bicarbonate ions at the luminal or basolateral membrane of the tubular cells. Instead, bicarbonate reabsorption is absolutely dependent upon the tubular secretion of hydrogen ions, which combine in the lumen with filtered bicarbonates. Figure 16-29 illustrates the sequence of events. Start this figure inside the cell with the combination of CO_2 and H_2O to form H_2CO_3, a reaction catalyzed by the enzyme carbonic anhydrase. The H_2CO_3 immediately dissociates to yield H^+ and bicarbonate (HCO_3^-). The HCO_3^- moves down its concentration gradient across the basolateral membrane into the interstitial fluid and then into the blood. Simultaneously the H^+ is secreted into the lumen, either by a primary H-ATPase pump or by Na/H countertransport, depending on the tubular segment. *But the secreted H^+ is not excreted.* Instead, it combines in the lumen with a *filtered* HCO_3^- and generates CO_2 and H_2O (both of which can diffuse into the cell and be used for another cycle of hydrogen-ion generation). The overall result is that the bicarbonate filtered from the plasma at the renal corpuscle has disappeared but its place in the plasma has been taken by the bicarbonate that was produced inside the cell, and so no

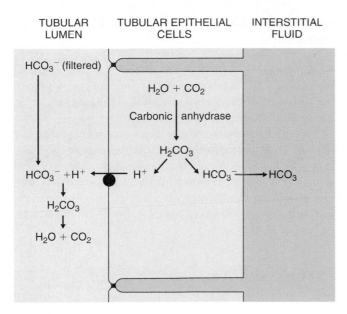

FIGURE 16-29
Reabsorption of bicarbonate. Start this figure inside the cell, with the combination of CO_2 and H_2O to form H_2CO_3. Bicarbonate reabsorption occurs in the proximal tubule, ascending limb of Henle's loop, and the collecting ducts. As shown in this figure, active H-ATPase pumps are involved in the movement of H^+ out of the cell across the luminal membrane; in several tubular segments (for example the proximal tubule) this transport step is also mediated by Na/H countertransporters.

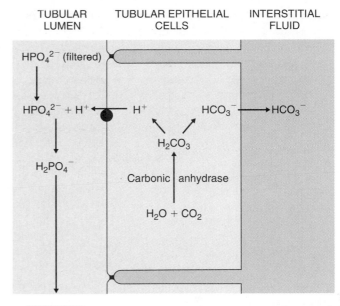

EXCRETED

FIGURE 16-30
Renal contribution of new HCO_3^- to the plasma is achieved by tubular secretion of H^+. The process of intracellular H^+ and HCO_3^- generation, with H^+ movement into the lumen and HCO_3^- into the plasma, is identical to that shown in Figure 16-29. Once in the lumen, however, the H^+ combines with filtered phosphate (HPO_4^{2-}) rather than filtered HCO_3^- and is excreted.

net change in plasma bicarbonate concentration has occurred. It may seem inaccurate to refer to this process as bicarbonate "reabsorption," since the bicarbonate that appears in the peritubular plasma is not the same bicarbonate ion that was filtered. Yet the overall result is, in effect, the same as if the filtered bicarbonate had been more conventionally reabsorbed like a sodium or potassium ion.

Addition of New Bicarbonate to the Plasma

It is essential to note in Figure 16-29 that as long as there are still significant amounts of filtered bicarbonate ions in the lumen, almost all secreted hydrogen ions will combine with them. But what happens to any secreted hydrogen ions once almost all the bicarbonate has all been reabsorbed and is no longer available in the lumen to combine with the hydrogen ions?

The answer, illustrated in Figure 16-30, is that the extra secreted hydrogen ions combine in the lumen with a filtered *nonbicarbonate* buffer, usually HPO_4^{2-}. The hydrogen ion is then excreted in the urine as part of an $H_2PO_4^-$ ion (other filtered buffers can also participate, but HPO_4^{2-} is the most important). Now for the critical point: Note in Figure 16-30 that, under these conditions, the bicarbonate generated within the tubular cell by the carbonic anhy-

drase reaction and entering the plasma constitutes a *net gain* of bicarbonate by the plasma, not merely a replacement for a filtered bicarbonate. Thus, when a secreted hydrogen ion combines in the lumen with a buffer other than bicarbonate, the overall effect is not merely one of bicarbonate conservation, as in Figure 16-29, but rather of addition to the plasma of a new bicarbonate. This raises the bicarbonate concentration of the plasma and alkalinizes it.

Significant numbers of hydrogen ions combine with filtered nonbicarbonate buffers like HPO_4^{2-} only after the filtered bicarbonate has virtually all been reabsorbed. The main reason is that there is such a large load of filtered bicarbonate—25 times more than the load of filtered nonbicarbonate buffers—competing for the secreted hydrogen ions.

There is a second mechanism by which the tubules contribute new bicarbonate to the plasma, one that involves not hydrogen-ion secretion but rather the renal production and secretion of ammonium (NH_4^+) (Figure 16-31). Tubular cells, mainly those of the proximal tubule, take up glutamine from both the glomerular filtrate and peritubular plasma and, by a series of steps, metabolize it. In the process, both NH_4^+ and bicarbonate are formed inside the cells. The NH_4^+ is actively secreted (via

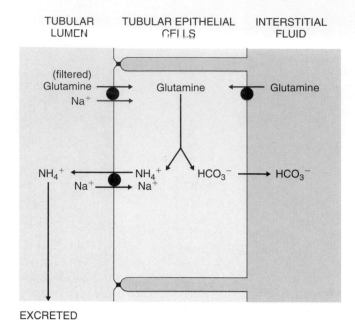

TUBULAR LUMEN   TUBULAR EPITHELIAL CELLS   INTERSTITIAL FLUID

EXCRETED

FIGURE 16-31

Renal contribution of new HCO_3^- to the plasma as achieved by renal metabolism of glutamine and excretion of ammonium (NH_4^+). Compare this figure to Figure 16-30. This process occurs mainly in the proximal tubule.

Na^+/NH_4^+ countertransport) into the lumen and excreted, while the bicarbonate moves into the peritubular capillaries and constitutes new plasma bicarbonate.

A comparison of Figures 16-30 and 16-31 demonstrates that the overall result—renal contribution of new bicarbonate to the plasma—is the same regardless of whether it is achieved by: (1) H^+ secretion and excretion on nonbicarbonate buffers such as phosphate (Figure 16-30); or (2) by glutamine metabolism with NH_4^+ excretion (Figure 16-31). It is convenient, therefore, to view the latter case as representing H^+ excretion "bound" to NH_3, just as the former case constitutes H^+ excretion bound to nonbicarbonate buffers. Thus, the amount of H^+ excreted in the urine in these two forms is a measure of the amount of new bicarbonate added to the plasma by the kidneys. Indeed, "urinary H^+ excretion" and "renal contribution of new bicarbonate to the plasma" are really two sides of the same coin and are synonomous phrases.

One last point needs to be emphasized: The last paragraph summarizes the two forms in which H^+ is excreted in the urine, but to be completely accurate there is also a third form—as free H^+. However, the amount of free H^+ is always so small that it can be ignored. For example, even the most acid of urines (4.4 is the lowest pH achievable by the tubules) contains less than 0.1 mmol of free H^+, compared to several hundred mmols of H^+ bound up in nonbicarbonate buffers and NH_4^+. This emphasizes

how important these two sources are in achieving the excretion of H^+.

Renal Responses to Acidosis and Alkalosis

We can now apply this material to the renal responses to the presence of an acidosis or alkalosis. These are summarized in Table 16-8.

Clearly, these homeostatic responses require that the rates of both hydrogen-ion secretion, glutamine metabolism, and ammonium excretion be subject to physiological control by changes in blood hydrogen-ion concentration. The specific pathways and mechanisms that bring about these rate changes are very complex, however, and are not presented here.

CLASSIFICATION OF ACIDOSIS AND ALKALOSIS

To repeat, acidosis refers to any situation in which the hydrogen-ion concentration of arterial plasma is elevated; alkalosis denotes a reduction. All such situations fit into two distinct categories (Table 16-9): (1) *respiratory acidosis* or *alkalosis;* (2) *metabolic acidosis* or *alkalosis.*

TABLE 16-8   RENAL RESPONSES TO ACIDOSIS AND ALKALOSIS

Responses to Acidosis

1. Sufficient hydrogen ions are secreted to reabsorb all the filtered bicarbonate.

2. Still more hydrogen ions are secreted, and this contributes new bicarbonate to the plasma as these hydrogen ions are excreted bound to nonbicarbonate urinary buffers such as HPO_4^{2-}

3. Tubular glutamine metabolism and ammonium excretion are enhanced, which also contributes new bicarbonate to the plasma.

Net result: Plasma bicarbonate is increased, thereby compensating for the acidosis. The urine is highly acidic (lowest attainable pH = 4.4).

Responses to Alkalosis

1. Rate of hydrogen-ion secretion is inadequate to reabsorb all the filtered bicarbonate, so that significant amounts of bicarbonate are excreted in the urine and there is little or no excretion of hydrogen ions on nonbicarbonate urinary buffers

2. Tubular glutamine metabolism and ammonium excretion are decreased so that little or no new bicarbonate is contributed to the plasma from this source.

Net result: Plasma bicarbonate concentration is decreased, thereby compensating for the alkalosis. The urine is alkaline (pH > 7.4).

TABLE 16-9   CHANGES IN THE ARTERIAL CONCENTRATIONS OF HYDROGEN ION, BICARBONATE, AND CARBON DIOXIDE IN ACID-BASE DISORDERS

Primary disorder	H^+	HCO_3^-	CO_2	Cause of HCO_3^- change	Cause of CO_2 change
Respiratory acidosis	↑	↑	↑	Renal compensation	Primary abnormality
Respiratory alkalosis	↓	↓	↓		
Metabolic acidosis	↑	↓	↓	Primary abnormality	Reflex ventilatory compensations
Metabolic alkalosis	↓	↑	↑		

As its name implies, respiratory acidosis results from altered respiration. Respiratory acidosis occurs when the respiratory system fails to eliminate carbon dioxide as fast as it is produced. Respiratory alkalosis occurs when the respiratory system eliminates carbon dioxide faster than it is produced. As described earlier, the imbalance of arterial hydrogen-ion concentrations in such cases is completely explainable in terms of mass action. Thus, the hallmark of a respiratory acidosis is an elevation in both arterial P_{CO_2} and hydrogen ion concentration; that of respiratory alkalosis is a reduction in both.

Metabolic acidosis or alkalosis includes all situations other than those in which the primary problem is respiratory. Some common causes of metabolic acidosis are excessive production of lactic acid (during severe exercise or hypoxia), or of ketone bodies (in uncontrolled diabetes mellitus or fasting, as described in Chapter 18). Metabolic acidosis can also result from excessive loss of bicarbonate, as in diarrhea. A frequent cause of metabolic alkalosis is persistent vomiting, with its associated loss of hydrogen ions as HCl from the stomach.

What is the arterial P_{CO_2} in metabolic acidosis or alkalosis? Since, by definition, metabolic acidosis and alkalosis must be due to something other than excess retention or loss of carbon dioxide, you might have predicted that arterial P_{CO_2} would be unchanged, but such is not the case. As emphasized earlier in this chapter, the elevated hydrogen-ion concentration associated with metabolic acidosis *reflexly* stimulates ventilation and lowers arterial P_{CO_2}. By mass action this helps restore the hydrogen-ion concentration toward normal. Conversely, a person with metabolic alkalosis will reflexly have ventilation inhibited. The result is a rise in arterial P_{CO_2} and, by mass action, an associated restoration of hydrogen-ion concentration toward normal.

To reiterate, the plasma P_{CO_2} changes in metabolic acidosis and alkalosis are not the *cause* of the acidosis or alkalosis but are the result of compensatory reflex responses to nonrespiratory abnormalities. Thus, in metabolic, as opposed to respiratory conditions, the arterial plasma P_{CO_2} and hydrogen-ion concentration go in opposite directions, as summarized in Table 16-9.

SECTION D SUMMARY

SOURCES OF HYDROGEN-ION GAIN OR LOSS

I.   Total-body balance of hydrogen ions is the result of both metabolic production of these ions and of net gains or losses via the respiratory system, gastrointestinal tract, and urine (Table 16-7).

II. A stable balance is achieved by regulation of urinary losses.

BUFFERING OF HYDROGEN IONS IN THE BODY

I.   Buffering is a means of minimizing changes in hydrogen-ion concentration by combining these ions reversibly with anions such as bicarbonate and intracellular proteins.

II. The major extracellular buffering system is the CO_2/HCO_3^- system, and the major intracellular buffers are proteins and phosphates.

INTEGRATION OF HOMEOSTATIC CONTROLS

I.   The kidneys and the respiratory system are the homeostatic regulators of plasma hydrogen-ion concentration.

II. The kidneys are the ultimate balancers of body hydrogen-ion balance.

III. A decrease in arterial plasma hydrogen-ion concentration causes reflex hypoventilation, which raises arterial P_{CO_2} and, hence, raises plasma hydrogen-ion concentration toward normal. An increase in plasma hydrogen-ion concentration causes reflex hyperventilation, which lowers arterial P_{CO_2} and, hence, lowers hydrogen-ion concentration toward normal.

RENAL MECHANISMS

I.   The kidneys maintain a stable plasma hydrogen-ion concentration by regulating plasma bicarbonate concentration. They can either excrete bicarbonate or contribute new bicarbonate to the blood.

II. Bicarbonate is reabsorbed when hydrogen ions, generated in the tubular cells by a process catalyzed by carbonic anhydrase, are secreted into the lumen and combine with filtered bicarbonate. The secreted hydrogen ions are not excreted in this situation.

III. In contrast, when the secreted hydrogen ions combine in the lumen with filtered phosphate or other nonbicarbonate buffer, they are excreted and the kidneys have contributed new bicarbonate to the blood.

IV. The kidneys also contribute new bicarbonate to the blood when they produce and excrete ammonium.

CLASSIFICATION OF ACIDOSIS AND ALKALOSIS

I. Acid-base disorders are categorized as respiratory or metabolic.

   A. Respiratory acidosis is due to retention of carbon dioxide, and respiratory alkalosis to excessive elimination of carbon dioxide.

   B. All other causes of acidosis or alkalosis are termed metabolic and reflect gain or loss, respectively, of hydrogen ions from a source other than carbon dioxide.

SECTION D KEY TERMS

nonvolatile acids   buffer

SECTION D REVIEW QUESTIONS

1. What are the sources of gain and loss of hydrogen ions in the body?

2. List the body's major buffer systems.

3. Describe the role of the respiratory system in the regulation of hydrogen-ion concentration.

4. How does the tubular secretion of hydrogen ions occur and how does it achieve bicarbonate reabsorption?

5. How does hydrogen-ion secretion contribute to the renal contribution of new bicarbonate to the blood. What determines whether a secreted hydrogen ion will achieve these events or will instead cause bicarbonate reabsorption?

6. How does the metabolism of glutamine by the tubular cells contribute new bicarbonate to the blood and ammonium to the urine?

7. What two quantities make up "hydrogen-ion excretion?" Why can this term be equated with "contribution of new bicarbonate to the plasma?"

8. How do the kidneys respond to the presence of an acidosis or alkalosis?

9. Classify the four types of acid-base disorders according to plasma hydrogen-ion concentration, bicarbonate concentration, and P_{CO_2}.

SECTION E
DIURETICS AND KIDNEY DISEASE

DIURETICS

Drugs used clinically to increase the volume of urine excreted are known as **diuretics.** Such agents act on the tubules to inhibit the reabsorption of sodium, along with chloride and/or bicarbonate, resulting in increased excretion of these ions. Since water reabsorption is dependent upon sodium reabsorption, water reabsorption is also reduced, resulting in increased water excretion.

A large variety of clinically useful diuretics are available and are classified according to the specific mechanisms by which they inhibit sodium reabsorption. For example, one type, called loop diuretics, acts on the ascending limb of the loop of Henle to inhibit the transport protein that mediates the first step in sodium reabsorption in this segment—cotransport of sodium and chloride into the cell across the luminal membrane.

Except for one category of diuretics, called **potassium-sparing diuretics,** all diuretics not only increase sodium excretion but also cause increased potassium excretion, an unwanted side effect. By several mechanisms, the potassium-sparing diuretics inhibit sodium reabsorption in the cortical collecting duct, and they simultaneously inhibit potassium secretion there. This explains why they do not cause increased potassium excretion.

Diuretics are among the most commonly used medications. For one thing, they are used to treat diseases characterized by renal retention of salt and water. As stated earlier, in normal individuals the mechanisms for controlling total-body sodium are so precise that sodium balance does not vary by more than a few percent despite marked changes in dietary intake or in losses due to sweating, vomiting, or diarrhea. In several types of diseases, however, sodium balance becomes deranged by the failure of the kidneys to excrete sodium normally. Sodium excretion may fall virtually to zero despite continued sodium ingestion, and the person may retain huge quantities of sodium and water, leading to abnormal expansion of the extracellular fluid and formation of *edema.* Diuretics are used to prevent or reverse this renal retention of sodium and water.

The most common example of this phenomenon is *congestive heart failure* (Chapter 14). A person with a failing heart manifests (1) a decreased GFR and (2) increased aldosterone secretion, both of which, along with other important factors, contribute to the virtual absence of sodium from the urine. The net result is extracellular volume expansion and formation of edema. The sodium-retaining responses are triggered by the lower cardiac output (a result of cardiac failure) and the resulting decrease in arterial pressure.

Another disease in which diuretics are frequently employed is hypertension (Chapter 14). It is still unclear why the decrease in body sodium and water resulting from the diuretic-induced excretion of these substances brings about arteriolar dilation and a lowering of the blood pressure.

KIDNEY DISEASE

The term kidney disease is no more specific than "car trouble," since many diseases affect the kidneys. Bacteria, allergies, congenital defects, kidney stones, tumors, and toxic chemicals are some possible sources of kidney damage. Obstruction of the urethra or a ureter may cause injury as the result of a buildup of pressure and may predispose the kidneys to bacterial infection.

Disease can attack the kidneys at any age. Experts estimate that there are at present more than 3 million undetected cases of kidney infection in the United States and that 25,000 to 75,000 Americans die of kidney failure each year.

One frequent sign of kidney disease is the appearance of protein in the urine. In normal kidneys there is a very tiny amount of protein in the glomerular filtrate because the corpuscular membranes are not completely impermeable to protein, particularly those with lower molecular weights. However, the tubular cells completely remove this filtered protein from the tubular lumen, and no protein appears in the final urine. In contrast, diseased renal corpuscles may become much more permeable to protein, and diseased tubules may lose their ability to remove filtered protein from the tubular lumen. The result is that protein will appear in the urine.

Although many diseases of the kidney are self-limited and produce no permanent damage, others progress if untreated. The end stage of progressive diseases, regardless of the nature of the damaging agent, is a shrunken, nonfunctioning kidney. Similarly, the symptoms of profound renal malfunction are relatively independent of the damaging agent and are collectively known as *uremia,* literally, "urine in the blood."

The severity of uremia depends upon how well the impaired kidneys are able to preserve the constancy of the internal environment. Assuming that the person continues to ingest a normal diet containing the usual quantities of nutrients and electrolytes, what problems arise? The key fact to keep in mind is that the kidney destruction markedly reduces the number of functioning nephrons. Accordingly, the many substances, particularly waste products, that gain entry to the tubule by filtration build up in the blood. In addition, the excretion of potassium is impaired because there are too few nephrons capable of normal tubular *secretion* of this ion. The person may also develop acidosis because the reduced number of nephrons fail to add enough new bicarbonate to the blood to compensate for the daily metabolic production of nonvolatile acids.

The remarkable fact is how large the safety factor is in renal function. In general, the kidneys are still able to perform their regulatory function quite well as long as 10 percent of the nephrons are functioning. This is because these remaining nephrons undergo alterations in function—filtration, reabsorption, and secretion—so as to compensate for the missing nephrons. For example, each remaining nephron increases its rate of potassium secretion so that the total amount of potassium excreted by the kidneys can be maintained at normal levels. The limits of regulation are restricted, however. To use potassium as our example again, if someone with severe renal disease were to go on a diet high in potassium, the remaining nephrons might not be able to secrete enough potassium to prevent potassium retention.

Other problems arise in uremia because of abnormal secretion of the hormones produced by the kidneys. Thus, decreased secretion of erythropoietin results in anemia (Chapter 15). Decreased ability to form $1,25\text{-}(OH)_2D_3$ results in deficient absorption of calcium from the gastrointestinal tract, with a resulting decrease in plasma calcium and inadequate bone calcification. Both of these hormones are now available for administration to patients with uremia.

The problem with renin, the third of the renal hormones, is rarely too little secretion but rather too much secretion by the juxtaglomerular cells of the damaged kidneys. The result is increased plasma angiotensin II concentration and the development of *renal hypertension.*

Hemodialysis, Peritoneal Dialysis, and Transplantation

As described above, failing kidneys reach a point when they can no longer excrete water and ions at rates that maintain body balances of these substances, nor can they excrete waste products as fast as they are produced. Dietary alterations can minimize these problems, for example, by lowering potassium intake and thereby reducing the amount of potassium to be excreted, but such alterations cannot eliminate the problems. The techniques used to perform the kidneys' excretory functions are hemodialysis and peritoneal dialysis. The general term dialysis means to separate substances using a membrane.

The artificial kidney is an apparatus that utilizes a process termed **hemodialysis** to remove excess substances from the blood. During hemodialysis, blood is pumped from one of the patient's arteries through tubing that is bathed by a large volume of fluid. The tubing then conducts the blood back into the patient by way of a vein. The tubing is generally made of cellophane that is highly permeable to most solutes but relatively impermeable to protein and completely impermeable to blood cells—characteristics quite similar to those of capillaries. The bath fluid is a salt solution with ionic concentrations similar to or lower than those in normal plasma. As blood flows through the tubing, the concentrations of nonprotein plasma solutes tend to reach diffusion equilibrium with those of the solutes in the bath fluid. For example, if the plasma potassium concentration of the patient is above normal, potassium diffuses out of the blood across the cellophane tubing and into the bath fluid. Similarly, waste products and excesses of other substances also diffuse into the bath and thus are eliminated from the body.

Patients with acute reversible renal failure may require hemodialysis for only days or weeks. Patients with chronic irreversible renal failure require treatment for the rest of their lives, however, unless they receive a renal transplant. Such patients undergo hemodialysis several times a week, often at home.

Another way of removing excess substances from the blood is **peritoneal dialysis,** which uses the lining of the patient's own abdominal cavity (peritoneum) as a dialysis membrane. Fluid is injected, via a needle inserted through the abdominal wall, into this cavity and allowed to remain there for hours, during which solutes diffuse into the fluid from the person's blood. The dialysis fluid is then removed by reinserting the needle and is replaced with new fluid. This procedure can be performed several times daily by a patient who is simultaneously doing normal activities.

The treatment of choice for most patients with permanent renal failure is kidney transplantation. Rejection of the transplanted kidney by the recipient's body is a potential problem with transplants, but great strides have been made in reducing the frequency of rejection (Chapter 20). Many people who might benefit from a transplant do not receive one. Presently, the major source of kidneys for transplanting is recently deceased persons, and improved public understanding should lead to many more individuals giving permission in advance to have their kidneys and other organs used following their death.

SECTION E SUMMARY

DIURETICS AND KIDNEY DISEASE

I.   Diuretics inhibit reabsorption of sodium and water, thereby enhancing the excretion of these substances. Different diuretics act on different nephron segments.

II. Many of the symptoms of uremia—general renal malfunction—are due to retention of substances because of reduced GFR and, in the case of potassium and hydrogen ion, reduced secretion. Other symptoms are due to inadequate secretion of the renal hormones.

III. Either hemodialysis or peritoneal dialysis can be used chronically to eliminate the water, ions, and waste products retained during uremia.

CHAPTER 16 CLINICAL TERMS

glucosuria
diabetes insipidus
hypertension
angiotensin converting
  enzyme (ACE) inhibitors

arrhythmias
hypocalcemic tetany
rickets
osteomalacia
osteoporosis

alkalosis
acidosis
respiratory acidosis
respiratory alkalosis
metabolic acidosis
metabolic alkalosis
diuretics

potassium-sparing diuretics
edema
congestive heart failure
uremia
renal hypertension
hemodialysis
peritoneal dialysis

CHAPTER 16 THOUGHT QUESTIONS

(Answers are in Appendix A)

1. Substance T is present in the urine. Does this prove that it is filterable at the glomerulus?

2. Substance V is not normally present in the urine. Does this prove that it is neither filtered nor secreted?

3. The concentration of glucose in plasma is 100 mg/100 ml, and the GFR is 125 ml/min. How much glucose is filtered per minute?

4. A person is found to be excreting abnormally large amounts of a particular amino acid. Just from the theoretical description of *Tm*-limited reabsorptive mechanisms in the text, list several possible causes.

5. The concentration of urea in urine is always much higher than the concentration in plasma. Does this mean that urea is secreted?

6. If a drug that blocks the reabsorption of sodium is taken, what will happen to the reabsorption of water, urea, chloride, glucose, and amino acids and to the secretion of hydrogen ions?

7. Compare the changes in GFR and renin secretion occurring in response to a moderate hemorrhage in two individuals—one taking a drug that blocks the sympathetic nerves to the kidneys and the other not taking such a drug.

8. If a person is taking a drug that completely inhibits angiotensin converting enzyme, what will happen to aldosterone secretion when the person goes on a low-sodium diet?

9. In the steady state, is the amount of sodium chloride excreted daily in the urine by a normal person ingesting 12 g of sodium chloride per day: (a) 12 g/day or (b) less than 12 g/day? Explain.

10. A young woman who has suffered a head injury seems to have recovered but is thirsty all the time. What do you think might be the cause?

11. A patient has a tumor in the adrenal cortex that continuously secretes large amounts of aldosterone. What effects does this have on the total amount of sodium and potassium in her body?

12. A person is taking a drug that inhibits the tubular secretion of hydrogen ions. What effect does this drug have on the body's balance of sodium, water, and hydrogen ion?

CHAPTER 17

The Digestion and Absorption of Food

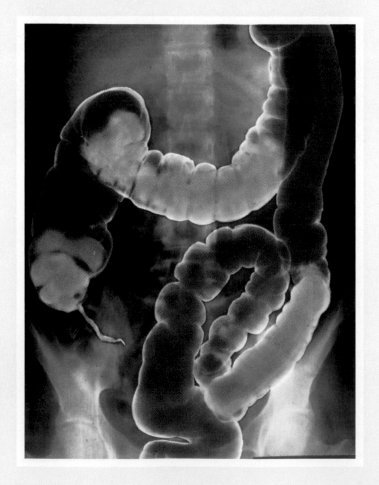

The **gastrointestinal (GI) system** (Figure 17-1) includes the **gastrointestinal tract** (mouth, pharynx, esophagus, stomach, small intestine, large intestine, and rectum) and the accessary organs (salivary glands, liver, gallbladder, and pancreas) that are not part of the tract but secrete substances into it via connecting ducts. The overall function of the gastrointestinal system is to process ingested foods into molecular forms that can be transferred, along with salts and water, from the external environment to the body's internal environment, where they can be distributed to cells by the circulatory system.

The adult gastrointestinal tract is a tube approximately 15 ft long, running through the body from mouth to anus. The lumen of the tract, like the hole in a doughnut, is continuous with the external environment, which means that its contents are technically outside the body. This fact is relevant to understanding some of the tract's properties.

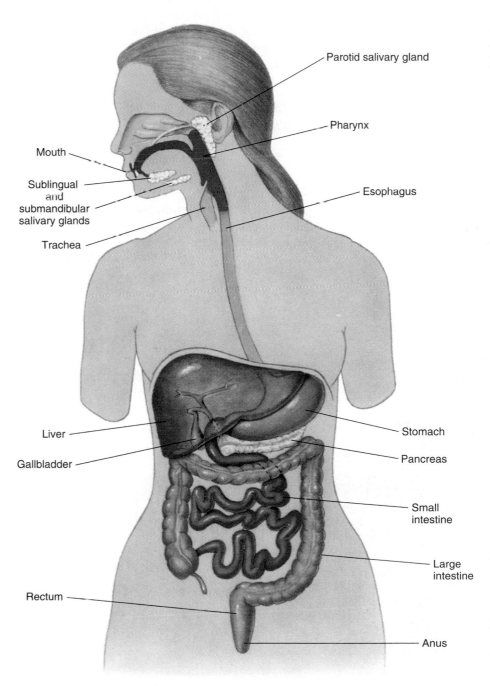

FIGURE 17-1
Anatomy of the gastrointestinal system. The liver overlies the gallbladder and a portion of the stomach, and the stomach overlies part of the pancreas.

Parotid salivary gland

Pharynx

Mouth

Sublingual and submandibular salivary glands

Esophagus

Trachea

Liver

Gallbladder

Stomach

Pancreas

Small intestine

Large intestine

Rectum

Anus

For example, the large intestine is inhabited by billions of bacteria, most of which are harmless and even beneficial in this location. However, if the same bacteria enter the blood, as may happen, for example, in the case of a ruptured appendix, they may cause a severe systemic infection.

Most food is taken into the mouth as large particles containing macromolecules, such as proteins and polysaccharides, which are unable to cross the wall of the gastrointestinal tract. Before ingested food can be absorbed, therefore, it must be dissolved and broken down into small molecules. This dissolving and breaking-down process—**digestion**—is accomplished by the action of hydrochloric acid in the stomach, bile from the liver, and a variety of digestive enzymes that are released by the system's exocrine glands. Each of these substances is released into the lumen of the GI tract by the process of **secretion.**

The molecules produced by digestion then move from the lumen of the gastrointestinal tract across a layer of epithelial cells and enter the blood or lymph. This process is called **absorption.**

While digestion, secretion, and absorption are taking place, contractions of smooth muscles in the gastrointestinal tract wall mix the luminal contents with the various secretions and move them through the tract from mouth to anus. These contractions are referred to as the **motility** of the gastrointestinal tract.

The functions of the gastrointestinal system can be described in terms of these four processes—digestion, secretion, absorption, and motility (Figure 17-2)—and the mechanisms controlling them.

The gastrointestinal system is designed to maximize absorption, and within fairly wide limits it will absorb as much of any particular substance as is ingested. With a few important exceptions, therefore, the gastrointestinal system does *not regulate* the amount of nutrients absorbed or their concentrations in the internal environment. That is primarily the function of the kidneys (Chapter 16) and a number of endocrine glands (Chapter 18).

Small amounts of certain metabolic end products are excreted via the gastrointestinal tract, primarily by way of the bile, but the elimination of most of the body's waste products is achieved by the lungs and kidneys. The material—**feces**—leaving the system at the end of the gastrointestinal tract consists almost entirely of bacteria and ingested material that was neither digested nor absorbed, that is, material that was never actually part of the internal environment.

OVERVIEW: FUNCTIONS OF THE GASTROINTESTINAL ORGANS

Figure 17-3 presents an overview of the secretions and functions of the gastrointestinal organs. The gastrointestinal tract begins with the **mouth,** and digestion starts there with chewing, which breaks up large pieces of food into smaller particles that can be swallowed without choking. **Saliva,** secreted by three pairs of **salivary glands** (Figure 17-1) located in the head, drains into the mouth through

FIGURE 17-2

Four processes carried out by the gastrointestinal tract: digestion, secretion, absorption, and motility.

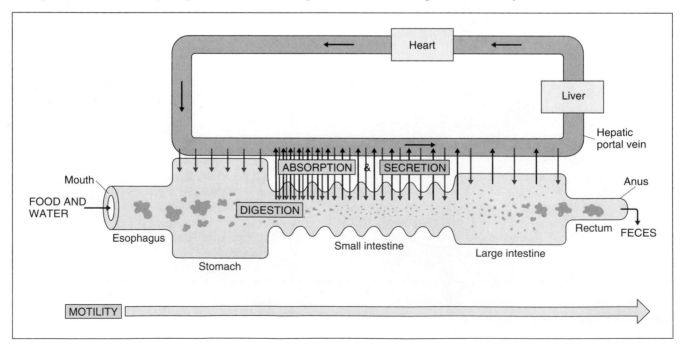

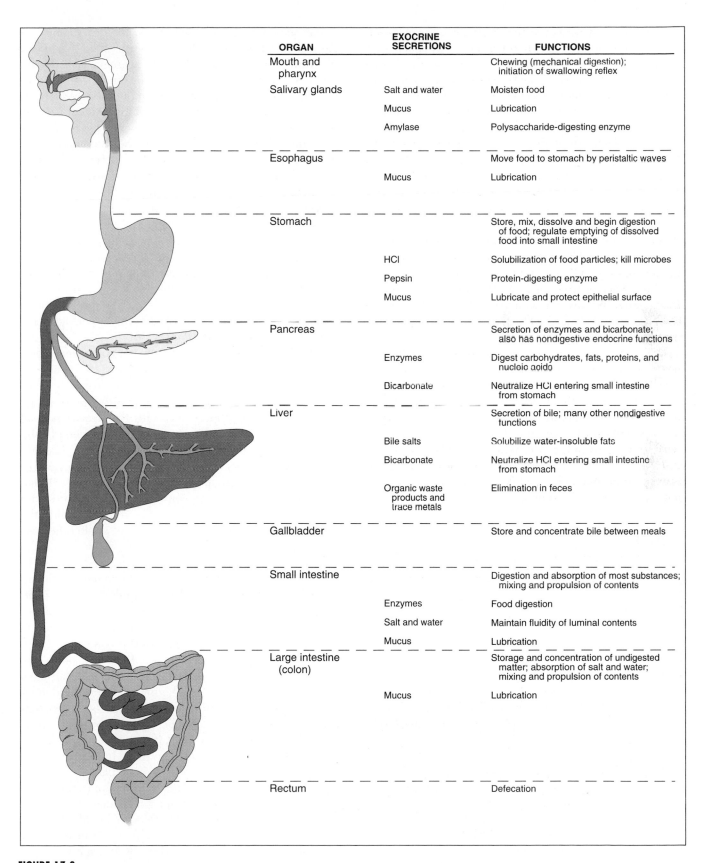

ORGAN	EXOCRINE SECRETIONS	FUNCTIONS
Mouth and pharynx		Chewing (mechanical digestion); initiation of swallowing reflex
Salivary glands	Salt and water	Moisten food
	Mucus	Lubrication
	Amylase	Polysaccharide-digesting enzyme
Esophagus		Move food to stomach by peristaltic waves
	Mucus	Lubrication
Stomach		Store, mix, dissolve and begin digestion of food; regulate emptying of dissolved food into small intestine
	HCl	Solubilization of food particles; kill microbes
	Pepsin	Protein-digesting enzyme
	Mucus	Lubricate and protect epithelial surface
Pancreas		Secretion of enzymes and bicarbonate; also has nondigestive endocrine functions
	Enzymes	Digest carbohydrates, fats, proteins, and nucleic acids
	Bicarbonate	Neutralize HCl entering small intestine from stomach
Liver		Secretion of bile; many other nondigestive functions
	Bile salts	Solubilize water-insoluble fats
	Bicarbonate	Neutralize HCl entering small intestine from stomach
	Organic waste products and trace metals	Elimination in feces
Gallbladder		Store and concentrate bile between meals
Small intestine		Digestion and absorption of most substances; mixing and propulsion of contents
	Enzymes	Food digestion
	Salt and water	Maintain fluidity of luminal contents
	Mucus	Lubrication
Large intestine (colon)		Storage and concentration of undigested matter; absorption of salt and water; mixing and propulsion of contents
	Mucus	Lubrication
Rectum		Defecation

FIGURE 17-3

Functions of the gastrointestinal organs.

a series of short ducts. Saliva, which contains mucus, moistens and lubricates the food particles before swallowing. It also contains the enzyme **amylase,** which partially digests polysaccharides. A third function of saliva is to dissolve some of the food molecules. Only in the dissolved state can these molecules react with chemoreceptors in the mouth, giving rise to the sensation of taste (Chapter 9).

The next segments of the tract, the **pharynx** and **esophagus,** contribute nothing to digestion but provide the pathway by which ingested food and drink reach the stomach. The muscles in the walls of these segments control swallowing.

The **stomach** is a saclike organ, located between the esophagus and the small intestine. Its functions are to store, dissolve, and partially digest food, and to regulate the rate at which the ingested food empties into the small intestine. The glands lining the stomach wall secrete a strong acid, **hydrochloric acid,** and several protein-digesting enzymes collectively known as **pepsin** (actually a precursor of pepsin known as pepsinogen is secreted).

The primary function of hydrochloric acid is to dissolve the particulate matter in food. The acid environment in the gastric (adjective for stomach) lumen alters the ionization of polar molecules, especially proteins, disrupting the extracellular network of connective-tissue proteins that form the structural framework of the tissues in food. The proteins and polysaccharides released by hydrochloric acid's dissolving action are partially digested in the stomach by pepsin and amylase, the latter contributed by the salivary glands. A major component of food that is not dissolved by acid is fat.

Hydrochloric acid also kills most of the bacteria that enter along with food. This process is not 100 percent effective, and some bacteria survive to take up residence and multiply in the intestinal tract, particularly the large intestine.

The digestive actions of the stomach reduce food particles to a solution known as **chyme,** which contains molecular fragments of proteins and polysaccharides and droplets of fat. None of these digestion products can cross the epithelium of the gastric wall, and thus little absorption of organic nutrients occurs in the stomach.

Digestion's final stages and most absorption occur in the next section of the tract, the **small intestine,** a tube about 1.5 inches in diameter and 9 ft in length that leads from the stomach to the large intestine. Here molecules of intact or partially digested carbohydrates, fats, and proteins are broken down by hydrolytic enzymes into monosaccharides, fatty acids, and amino acids. Some of these enzymes are on the luminal surface of the intestinal lining cells, while others are secreted by the pancreas. The products of digestion cross the epithelial cells and enter the blood and/or lymph. Vitamins, minerals, and water, which do not require enzymatic digestion, are also absorbed in the small intestine.

The small intestine is divided into three segments: An initial short segment, the **duodenum,** is followed by the **jejunum** and then by the longest segment, the **ileum.** Normally, most absorption occurs in the first quarter of the small intestine, in the duodenum and jejunum. Thus, the small intestine has a considerable functional reserve.

Two major glands—the pancreas and liver—secrete substances that flow via ducts into the duodenum. The **pancreas,** an elongated gland located behind the stomach, has both endocrine (Chapter 18) and exocrine functions, but only the latter are directly involved in gastrointestinal function and are described in this chapter. The exocrine portion of the pancreas secretes (1) digestive enzymes specific for each class of organic molecule, and (2) a fluid rich in bicarbonate ions. The high acidity of the chyme coming from the stomach would inactivate the pancreatic enzymes in the small intestine if the acid were not neutralized by the bicarbonate ions in the pancreatic fluid.

The **liver,** a large gland located in the upper right portion of the abdomen, has a variety of functions, which are described in various chapters. This is a convenient place to provide, in Table 17-1, a comprehensive reference list of these hepatic (means pertaining to the liver) functions and the chapters in which they are described. We will be concerned in this chapter only with the liver's exocrine functions that are directly related to the secretion of **bile.**

Bile contains bicarbonate ions, cholesterol, phospholipids, bile pigments, a number of organic wastes and—most important—a group of substances collectively termed **bile salts.** The bicarbonate ions, like those from the pancreas, help neutralize acid from the stomach, while the bile salts solubilize dietary fat. These fats would otherwise be insoluble in water, and their solubilization increases the rates at which they are digested and absorbed.

Bile is secreted by the liver into small ducts which join to form a single duct called the common hepatic duct. Between meals, secreted bile is stored in the **gallbladder,** a small sac underneath the liver which branches from the common hepatic duct. The gallbladder concentrates the bile by absorbing salts and water. During a meal, the smooth muscles in the gallbladder wall contract, causing a concentrated bile solution to be injected into the duodenum via the **common bile duct** (Figure 17-4), an extension of the common hepatic duct. The gallbladder can be surgically removed without impairing bile secretion by the liver or its flow into the intestinal tract. In fact, many animals that secrete bile do not have a gallbladder.

In the small intestine, monosaccharides and amino acids are absorbed by specific transporter-mediated processes in the plasma membranes of the intestinal epithelial cells, whereas fatty acids enter these cells by diffusion. Most mineral ions are actively absorbed by transporters, and water diffuses passively down osmotic gradients. Digestion and absorption have been largely completed by the middle portion of the small intestine.

TABLE 17-1   SUMMARY OF LIVER FUNCTIONS

A. Endocrine functions
1. In response to growth hormone, the liver secretes insulin-like growth factor I (IGF-I), which promotes growth by stimulating cell division in various tissues, including bone (Chapter 18).
2. Contributes to the activation of vitamin D (Chapter 16).
3. Forms triiodothyronine (T_3) from thyroxine (T_4) (Chapter 10).
4. Secretes angiotensinogen, which is acted upon by renin to form angiotensin I (Chapter 16).
5. Metabolizes hormones (Chapter 10).

B. Clotting functions
1. Produces many of the plasma clotting factors, including prothrombin and fibrinogen (Chapter 14).
2. Produces bile salts, which are essential for the gastrointestinal absorption of vitamin K, which is, in turn, needed for production of the clotting factors (Chapter 14).

C. Plasma proteins
1. Synthesizes and secretes plasma albumin (Chapter 14), acute phase proteins (Chapter 20), binding proteins for steroid hormones (Chapter 10) and trace elements (Chapter 14), lipoproteins (Chapter 18), and other proteins mentioned elsewhere in this table.

D. Exocrine (digestive) functions (Chapter 17)
1. Synthesizes and secretes bile salts, which are necessary for adequate digestion and absorption of fats.
2. Secretes into the bile a bicarbonate-rich solution of inorganic ions, which helps neutralize acid in the duodenum.

E. Organic metabolism (Chapter 18)
1. Converts plasma glucose into glycogen and triacylglycerols during absorptive period.
2. Converts plasma amino acids to fatty acids, which can be incorporated into triacylglycerols during absorptive period.
3. Synthesizes triacylglycerols and secretes them as lipoproteins during absorptive period.
4. Produces glucose from glycogen (glycogenolysis) and other sources (gluconeogenesis) during postabsorptive period and releases the glucose into the blood.
5. Converts fatty acids into ketones during fasting.
6. Produces urea, the major end product of amino acid (protein) catabolism, and releases it into the blood.

F. Cholesterol metabolism (Chapter 18)
1. Synthesizes cholesterol and releases it into the blood.
2. Secretes plasma cholesterol into the bile.
3. Converts plasma cholesterol into bile salts.

G. Excretory and degradative functions
1. Secretes bilirubin and other bile pigments into the bile (Chapter 17).
2. Excretes, via the bile, many endogenous and foreign organic molecules as well as trace metals (Chapter 20).
3. Biotransforms many endogenous and foreign organic molecules (Chapter 20).
4. Destroys old erythrocytes (Chapter 14).

The motility of the small intestine, brought about by the smooth muscles in its walls, (1) mixes the luminal contents with the various secretions, (2) brings the contents into contact with the epithelial surface where absorption takes place, and (3) slowly advances the luminal material toward the large intestine. Since most substances are absorbed in the early portion of the small intestine, only a small volume of water, salts, and undigested material is passed on to the next segment, the **large intestine.** The large intestine temporarily stores the undigested material (some of which is acted upon by bacteria) and concentrates it by absorbing salt and water. Contractions of the **rectum,** the final segment of the gastrointestinal tract, and associated sphincter muscles expel the feces—**defecation.**

The average adult consumes about 800 g of food and 1200 ml of water per day, but this is only a fraction of the material entering the lumen of the gastrointestinal tract.

An additional 7000 ml of fluid from salivary glands, gastric glands, pancreas, liver, and intestinal glands are secreted into the tract each day (Figure 17-5). Of the 8 liters of fluid entering the tract, 99 percent is absorbed; only about 100 ml is normally lost in the feces. This small amount of fluid loss represents only 4 percent of the total fluids lost by the body each day. Almost all the salts in the secreted fluids are also reabsorbed into the blood. Moreover, the secreted digestive enzymes are themselves digested, and the resulting amino acids absorbed into the blood.

This completes our overview of the gastrointestinal system. Since its major task is digestion and absorption, we begin our more detailed description with these processes. Subsequent sections of the chapter will then describe, organ by organ, regulation of the secretions and motility that produce the optimal conditions for digestion

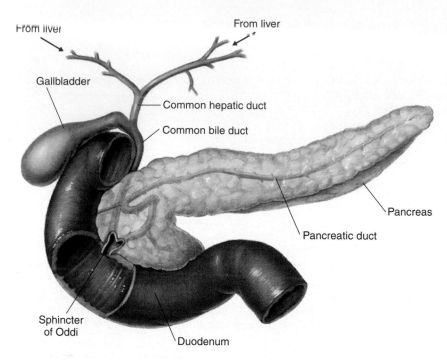

FIGURE 17-4

Bile ducts from the liver converge to form the common hepatic duct, from which branches the duct to the gallbladder. Beyond this branch the common hepatic duct becomes the common bile duct. The common bile duct and the pancreatic duct converge and empty their contents into the duodenum at the sphincter of Oddi. ✗

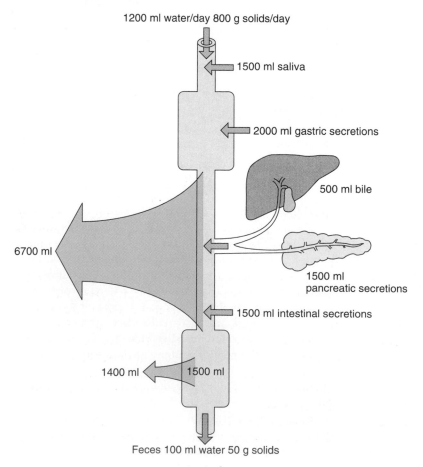

FIGURE 17-5

Average amounts of food and fluid ingested, secreted, absorbed, and excreted from the gastrointestinal tract daily.

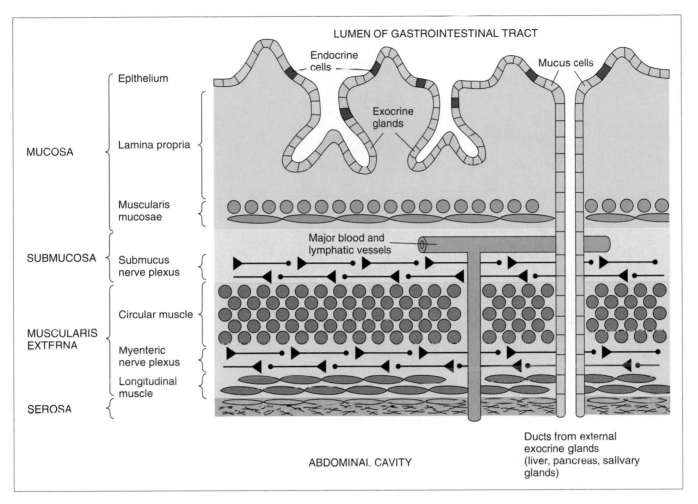

FIGURE 17-6

Structure of the gastrointestinal wall in longitudinal section. Not shown are the smaller blood vessels, neural connections between the two nerve plexuses, and neural terminations on muscles, glands, and epithelium.

and absorption. A prerequisite for all this physiology, how-ever, is a knowledge of the structure of the gastrointesti-nal tract wall.

STRUCTURE OF THE GASTROINTESTINAL TRACT WALL

From the midesophagus to the anus, the wall of the gas-trointestinal tract has the general structure illustrated in Figure 17-6. Most of the tube's luminal surface is highly convoluted, a feature that greatly increases the surface area available for absorption. From the stomach on, this surface is covered by a single layer of epithelial cells linked together along the edges of their luminal surfaces by tight junctions.

Included in this epithelial layer are exocrine cells that secrete mucus into the lumen of the tract and endocrine cells that release hormones into the blood. Invaginations of the epithelium into the underlying tissue form exocrine glands that secrete acid, enzymes, water, and ions, as well as mucus.

Just below the epithelium is a layer of connective tis-sue, the lamina propria, through which pass small blood vessels, nerve fibers, and lymphatic ducts. (These struc-tures are not shown in Figure 17-6 but are in Figure 17-7.) The lamina propria is separated from underlying tis-sues by a thin layer of smooth muscle, the muscularis mucosa. The combination of these three layers—the ep-ithelium, lamina propria, and muscularis mucosa—is called the **mucosa.**

Beneath the mucosa is a second connective tissue layer, the submucosa, containing a network of nerve cells, the

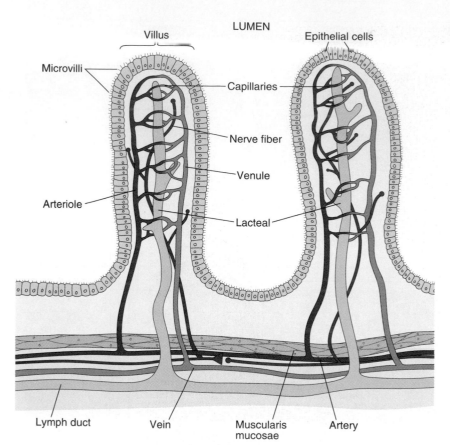

LUMEN

Villus

Microvilli

Epithelial cells

Capillaries

Nerve fiber

Venule

Arteriole

Lacteal

Lymph duct

Vein

Muscularis
mucosae

Artery

FIGURE 17-7

Structure of villi in small intestine.

submucous plexus, and blood and lymphatic vessels whose branches penetrate into both the overlying mucosa and the underlying layers of smooth muscle called the **muscularis externa.** Contractions of these muscles provide the forces for moving and mixing the gastrointestinal contents. The muscularis externa has two layers: (1) a relatively thick inner layer of **circular muscle,** whose fibers are oriented in a circular pattern around the tube such that contraction produces a narrowing of the lumen, and (2) a thinner outer layer of **longitudinal muscle,** whose contraction shortens the tube. Between these two muscle layers is a second network of nerve cells known as the **myenteric plexus.**

Finally, surrounding the outer surface of the tube is a thin layer of cells and connective tissue called the **serosa.** Thin sheets of connective tissue connect the serosa to the abdominal wall, supporting the gastrointestinal tract in the abdominal cavity.

Extending from the surface of the small intestine are fingerlike projections known as **villi** (Figure 17-7). The surface of each villus is covered with a single layer of epithelial cells whose surface membranes form small projections called **microvilli** (also known collectively as the brush border) (Figure 17-8). The combination of folded mucosa, villi, and microvilli increases the small intestine's surface area about 600-fold over that of a flat-surfaced

tube having the same length and diameter. The human small intestine's total surface area is about 300 m², the area of a tennis court.

Epithelial surfaces in the gastrointestinal tract are continuously being replaced by new epithelial cells. In the small intestine, new cells arise by mitosis from cells at the base of the villi. These cells differentiate as they migrate to the top of a villus, replacing older cells that disintegrate and are discharged into the intestinal lumen. These disintegrating cells release into the lumen their intracellular enzymes, which then contribute to the digestive process. About 17 billion epithelial cells are replaced each day, and the entire epithelium of the small intestine is replaced approximately every 5 days. It is because of this rapid cell turnover that the lining of the intestinal tract is so susceptible to damage by agents, such as radiation and anticancer drugs, that inhibit cell division.

The center of each intestinal villus is occupied both by a single blind-ended lymphatic vessel termed a **lacteal** and by a capillary network (Figure 17-7). As we will see, most of the fat absorbed in the small intestine enters the lacteals, while other absorbed nutrients enter the blood capillaries. The venous drainage from the intestinal villi, as well as from the large intestine, pancreas, and portions of the stomach does not empty directly into the vena cava but passes first, via the **hepatic portal vein,** to the liver. There it flows through a

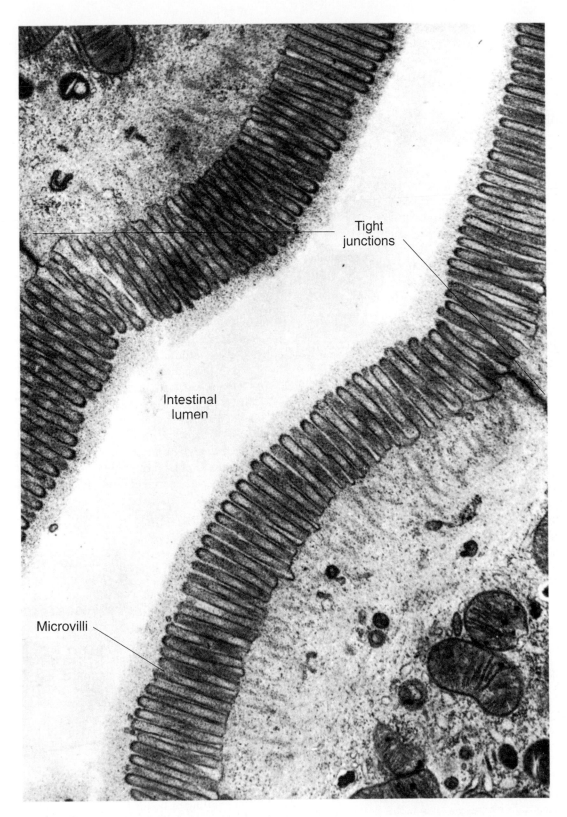

FIGURE 17-8

Microvilli on the surface of intestinal epithelial cells. *[From D. W. Fawcett, J. Histochem. Cytochem. 13: 75–91 (1965). Courtesy of Susumo Ito.]*

second capillary network before leaving the liver to return to the heart. Thus, material absorbed into the intestinal capillaries, in contrast to the lacteals, can be processed by the liver before entering the general circulation.

DIGESTION AND ABSORPTION

Carbohydrate

Carbohydrate intake per day ranges from about 250 to 800 g in a typical American diet. About two-thirds of this carbohydrate is the plant polysaccharide starch, and most of the remainder consists of the disaccharides sucrose (table sugar) and lactose (milk sugar) (Table 17-2). Only small amounts of monosaccharides are normally present in the diet. Cellulose and certain other complex polysaccharides found in vegetable matter—referred to as **fiber**—cannot be broken down by the enzymes in the small intestine and are passed on to the large intestine, where they are partially metabolized by bacteria.

Starch digestion by salivary amylase begins in the mouth and continues in the upper part of the stomach before amylase is destroyed by gastric acid. Starch digestion is completed in the small intestine by pancreatic amylase. The products produced by both amylases are the disaccharide maltose and a mixture of short, branched chains of glucose molecules. These products, along with ingested sucrose and lactose, are broken down into monosaccharides—glucose, galactose, and fructose—by enzymes located on the luminal membranes of the small-intestine epithelial cells. These monosaccharides are then transported across the intestinal epithelium into the blood. Fructose crosses the epithelium by facilitated diffusion, while glucose and galactose undergo secondary active transport coupled to sodium. Most ingested carbohydrate is digested and absorbed within the first 20 percent of the small intestine.

TABLE 17-2 CARBOHYDRATES IN FOOD

Class	Examples	Made up of
Polysaccharides	Starch	Glucose
	Cellulose	Glucose
	Glycogen	Glucose
Disaccharides	Sucrose	Glucose-fructose
	Lactose	Glucose-galactose
	Maltose	Glucose-glucose
Monosaccharides	Glucose	
	Fructose	
	Galactose	

Protein

Only 40 to 50 g of protein is required by a normal adult to supply essential amino acids and replace the amino acid nitrogen converted to urea. A typical American diet contains about 125 g of protein per day. In addition, a large amount of protein, in the form of enzymes and mucus, is secreted into the gastrointestinal tract or enters it via the disintegration of epithelial cells. Regardless of source, most of the protein in the lumen is broken down into amino acids and absorbed by the small intestine.

Proteins are broken down to peptide fragments in the stomach by pepsin, and in the small intestine by **trypsin** and **chymotrypsin**, the major proteases secreted by the pancreas. These fragments are further digested to free amino acids by **carboxypeptidase** from the pancreas and **aminopeptidase**, located on the luminal membranes of the small intestine epithelial cells. These last two enzymes split off amino acids from the carboxyl and amino ends of peptide chains, respectively.

The free amino acids then undergo secondary active transport coupled with sodium across the intestinal wall. Short chains of two or three amino acids are also absorbed by a secondary active transport that is coupled with hydrogen ions. (This is in contrast to carbohydrate absorption, in which molecules larger than monosaccharides are not absorbed.) Within the epithelial cell these di- and tripeptides are hydrolyzed to amino acids, which then leave the cell and enter the blood through a facilitated diffusion carrier in the basolateral membranes. About half of the absorbed amino acids are derived from the uptake of these short peptides. As with carbohydrates, protein digestion and absorption are largely completed in the upper portion of the small intestine.

Very small amounts of intact proteins are able to cross the intestinal epithelium and gain access to the interstitial fluid. They do so by a combination of endocytosis and exocytosis. The absorptive capacity for intact proteins is much greater in infants than in adults, and antibodies (proteins involved in the immunological defense system of the body) secreted into the mother's milk can be absorbed by the infant, providing some immunity until the infant begins to produce its own antibodies.

Fat

Fat intake ranges from about 25 to 160 g/day in a typical American diet; most is in the form of triacylglycerols. Fat digestion occurs almost entirely in the small intestine. The major digestive enzyme in this process is pancreatic **lipase**, which catalyzes the splitting of bonds linking fatty acids to the first and third carbon atoms of glycerol, producing two free fatty acids and a monoglyceride as products:

$$\text{Triacylglycerol} \xrightarrow{\text{Lipase}} \text{Monoglyceride} + 2 \text{ Fatty acids}$$

The fats in the ingested food are insoluble in water and aggregate into large lipid droplets in the upper portion of the stomach. Since pancreatic lipase is a water-soluble enzyme, its digestive action can take place only at the *surface* of a lipid droplet. Therefore, if most of the ingested fat remained in large lipid droplets, the rate of lipid digestion would be very slow. The rate of digestion is, however, substantially increased by division of the large lipid droplets into a number of much smaller droplets, each about 1 mm in diameter, thereby increasing their surface area and accessibility to lipase action. This process is known as **emulsification,** and the resulting suspension of small lipid droplets is an emulsion.

The emulsification of fat requires (1) mechanical disruption of the large fat droplets into smaller droplets, and (2) an emulsifying agent, which acts to prevent the smaller droplets from reaggregating back into large droplets. The mechanical disruption is provided by contractile activity, occurring in the lower portion of the stomach and in the small intestine, which acts to grind and mix the luminal contents. Phospholipids in food and phospholipids and bile salts secreted in the bile provide the emulsifying agents.

Phospholipids are amphipathic molecules (Chapter 2) consisting of two nonpolar fatty acid chains attached to glycerol, with a charged phosphate group located on glycerol's third carbon. Bile salts are formed from cholesterol in the liver and are also amphipathic (Figure 17-9). The nonpolar portions of the phospholipids and bile salts associated with the nonpolar interior of the lipid droplets, leaving the polar portions exposed at the water surface. There they repel other lipid droplets that are similarly coated with these emulsifying agents, thereby preventing their reaggregation into larger fat droplets (Figure 17-10).

Although digestion is speeded up by emulsification, absorption of the water-insoluble products of the lipase reaction would still be very slow if it were not for a second action of the bile salts, the formation of **micelles,** which are similar in structure to emulsion droplets but are much smaller—4 to 7 nm in diameter. Micelles consist of bile salts, fatty acids, monoglycerides, and phospholipids all clustered together with the polar ends of each molecule oriented toward the micelle's surface and the nonpolar portions forming the micelle's core. Also included in the core of the micelle are small amounts of nonpolar, fat-soluble vitamins and cholesterol.

How do micelles increase absorption? Although fatty acids and monoglycerides have an extremely low solubility

FIGURE 17-9

Structure of bile salts. (A) Chemical formula of glycocholic acid, one of several bile salts secreted by the liver (polar groups in color). (B) Three-dimensional structure of a bile salt, showing its polar and nonpolar surfaces.

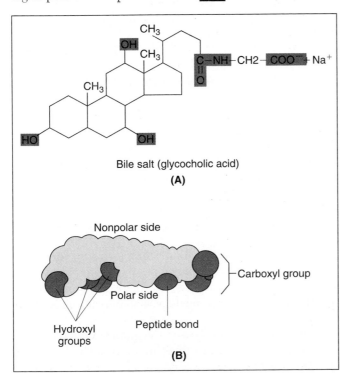

FIGURE 17-10

Emulsification of fat by bile salts and phospholipids.

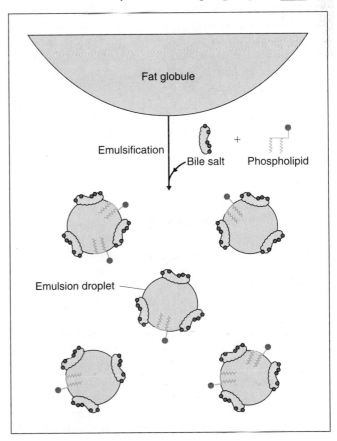

in water, a few molecules do exist in solution and are free to diffuse across the lipid portion of the luminal plasma membranes of the epithelial cells lining the small intestine.

Micelles, containing the products of fat digestion, are in equilibrium with the small concentration of fat digestion products that are free in solution. Thus, micelles are continuously breaking down and reforming. When a micelle breaks down, its contents are released into the solution and become available to diffuse across the intestinal lining. As the concentrations of free lipids falls, because of their diffusion into epithelial cells, more lipids are released into the free phase as micelles break down (Figure 17-11). Thus, the micelles provide a means of keeping most of the insoluble fat digestion products in small soluble aggregates, while at the same time replenishing the small amount of products that are free in solution and are able to diffuse into the intestinal epithelium. Note that it is not the micelle that is absorbed but rather the individual lipid molecules that are released from the micelle.

Although fatty acids and monoglycerides are the molecules that enter the epithelial cells from the intestinal lumen, it is triacyglycerol that is released on the other side of the cell into the interstitial fluid. In other words, during their passage through the epithelial cells, fatty acids and monoglycerides are resynthesized into triacylglycerols. This occurs in the agranular (smooth) endoplasmic reticulum, where the enzymes for triacylglycerol synthesis are located. Within this organelle, the resynthesized fat aggregates into small droplets coated with an amphipathic protein that performs an emulsifying function similar to that of a bile salt.

The exit of these fat droplets from the cell follows the same pathway as a secreted protein. Vesicles containing the droplet pinch off the endoplasmic reticulum, are processed through the Golgi apparatus, and eventually fuse with the plasma membrane, releasing the fat droplet into the interstitial fluid. These small, extracellular fat droplets are known as **chylomicrons.** Chylomicrons contain not only triacylglycerols but other lipids (including phospholipids, cholesterol, and fat-soluble vitamins) that have been absorbed by the same process that led to fatty acid and monoglyceride movement into cells.

The chylomicrons released from the epithelial cells pass into lacteals—lymphatic capillaries in the intestinal villi—rather than into the blood capillaries. The chylomicrons cannot enter the blood capillaries because a basement membrane (an extracellular glycoprotein layer) at the outer surface of the capillary provides a barrier to the large chylomicrons, which are about 1 μm in diameter. In contrast, the lacteals do not have basement membranes and have large slit pores between their endothelial cells through which the chylomicrons can pass into the lymph.

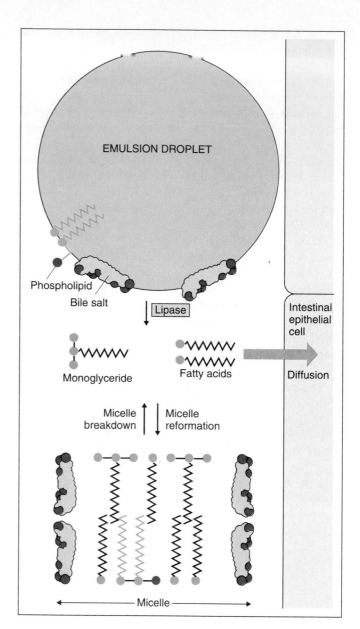

FIGURE 17-11

The products of fat digestion by lipase are held in solution in the micellar state, combined with bile salts and phospholipids. The micellar contents rapidly exchange with the free products in aqueous solution, which are able to diffuse into intestinal epithelial cells.

The lymph from the small intestine, as from everywhere else in the body, eventually empties into systemic veins. In Chapter 18 we describe how the lipids in the circulating blood chylomicrons are made available to the cells of the body.

Figure 17-12 summarizes the pathway taken by fat in moving from the intestinal lumen into the lymphatic system.

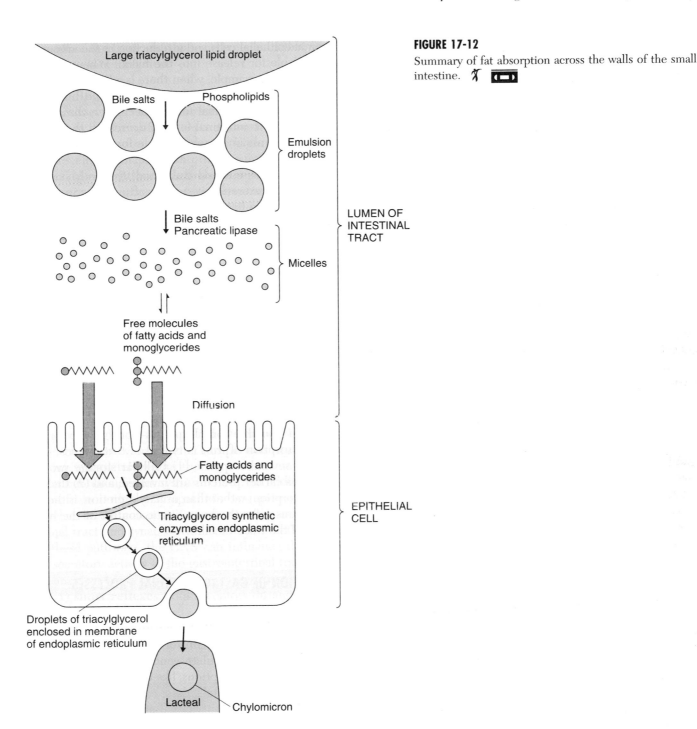

FIGURE 17-12
Summary of fat absorption across the walls of the small intestine.

Vitamins

The fat-soluble vitamins—A, D, E, and K—follow the pathway for fat absorption described in the previous section. They are solubilized in micelles; thus, any interference with the secretion of bile or the action of bile salts in the intestine decreases the absorption of the fat-soluble vitamins.

With one exception, water-soluble vitamins are absorbed by diffusion or mediated transporter. The exception, vitamin B_{12}, is a very large, charged molecule. In or-

der to be absorbed, vitamin B_{12} must first bind to a protein, known as **intrinsic factor,** produced by the acid-secreting cells in the stomach. The resulting complex then binds to specific sites on the epithelial cells in the lower portion of the ileum, where vitamin B_{12} is absorbed by endocytosis. As described in Chapter 14, vitamin B_{12} is required for erythrocyte formation, and deficiencies result in *pernicious anemia.* This form of anemia may occur when the stomach either has been removed (as, for example, to treat ulcers or gastric cancer) or fails to secrete

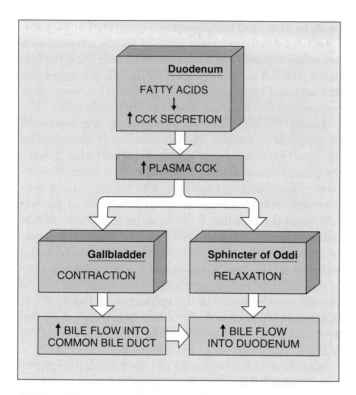

FIGURE 17-30

Regulation of bile entry into the small intestine.

intestinal epithelium has one of the highest cell renewal rates of any tissue in the body.

As stated earlier, another reason for water movement into the lumen is that the chyme entering the small intestine from the stomach may be hypertonic because of a high concentration of solutes in the meal and because digestion breaks down large molecules into many more small molecules. This hypertonicity causes the osmotic movement of water from the isotonic plasma into the intestinal lumen.

Excessive secretion of fluid by the intestinal epithelium in response to various bacterial infections can lead to diarrhea, as will be described later.

Motility. In contrast to the peristaltic waves that sweep over the stomach, the most common motion of the small intestine during a meal is a stationary contraction and relaxation of intestinal segments, with little apparent net movement toward the large intestine (Figure 17-31). Each contracting segment is only a few centimeters long, and the contraction lasts a few seconds. The chyme in the lumen of a contracting segment is forced both up and down the intestine. This rhythmical contraction and relaxation of the intestine, known as **segmentation,** produces a continuous division and subdivision of the intestinal contents, thoroughly mixing the chyme in the lumen and bringing it into contact with the intestinal wall.

These segmenting movements are initiated by electrical activity generated by pacemaker cells in the longitudinal smooth-muscle layer. Like the slow waves in the stomach, this intestinal basic electrical rhythm produces oscillations in smooth-muscle membrane potential that, if threshold is reached, trigger action potentials causing muscle contraction. The frequency of segmentation is set by the frequency of the intestinal basic electrical rhythm, but unlike the stomach, which normally has a single rhythm, the intestinal rhythm varies along the length of the intestine, each successive region having a slightly lower frequency than the one above. For example, segmentation in the duodenum occurs at a frequency of about 12 contractions/min, whereas in the last portion of the ileum the rate is only 9 contractions/min. Segmentation produces, therefore, a slow migration of the intestinal contents toward the large intestine because more chyme is forced downward, on the average, than upward.

The intensity of segmentation can be altered by hormones, the enteric nervous system, and autonomic nerves; parasympathetic activity increases the force of contraction, and sympathetic stimulation decreases it. Thus, cephalic phase stimuli, including emotional states, can alter intestinal motility. As is true for the stomach, these in-

FIGURE 17-31

Segmentation movements of the small intestine in which segments of the intestine contract and relax in a rhythmical pattern but do not undergo peristalsis. This is the pattern encountered during a meal; it mixes the luminal contents.

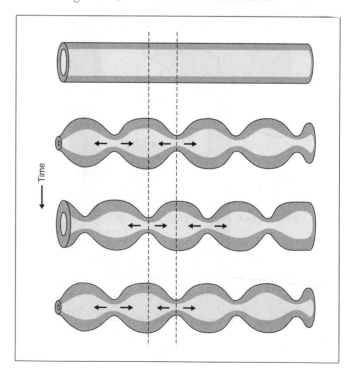

puts produce changes in the force of smooth-muscle contraction but do not significantly change the frequencies of the basic electrical rhythms.

After most of a meal has been absorbed, the segmenting contractions cease and are replaced by a pattern of peristaltic activity known as the **migrating motility complex.** Beginning in the lower portion of the stomach, repeated waves of peristaltic activity travel only a short distance (about 2 ft) along the small intestine and then die out. This short segment of peristaltic activity slowly migrates down the small intestine, taking about 2 h to reach the large intestine. By the time the migrating motility complex reaches the end of the ileum, new waves are beginning in the stomach and the process is repeated. Upon the arrival of a meal in the stomach, the migrating motility complex rapidly ceases and is replaced by segmentation.

The migrating motility complex moves into the large intestine any undigested material still remaining in the small intestine and also prevents bacteria from remaining in the small intestine long enough to grow and multiply. In diseases in which there is an aberrant migrating motility complex, bacterial overgrowth in the small intestine can become a major problem.

A rise in the plasma concentration of a candidate intestinal hormone, **motilin,** is thought to initiate the migrating motility complex, while the patterns of contractile activity during segmentation and peristalsis are coordinated by the enteric nervous system. The mechanisms of motilin action and the control of its release have not been determined.

The contractile activity in various regions of the small intestine can be altered by reflexes initiated at different points along the gastrointestinal tract. For example, segmentation intensity in the ileum increases during periods of gastric emptying, and this is known as the **gastroileal reflex.** Large distensions of the intestine, injury to the intestinal wall, and various bacterial infections in the intestine lead to a complete cessation of motility, the **intestino-intestinal reflex.**

As much as 500 ml of air may be swallowed during a meal. Most of this air travels no farther than the esophagus, from which it is eventually expelled by belching. Some of the air reaches the stomach, however, and is passed on to the intestines, where its percolation through the chyme as the intestinal contents are mixed produces gurgling sounds that are often quite loud.

Large Intestine

The large intestine is a tube 2.5 in. in diameter and about 4 ft long. Its first portion, the **cecum,** forms a blind-ended pouch from which extends the **appendix,** a small finger-like projection having no known essential function (Figure 17-32). The **colon** consists of three relatively straight segments—the ascending, transverse, and descending portions. The terminal portion of the descending colon is S shaped, forming the sigmoid colon, which empties into a relatively straight segment of the large intestine, the rectum, which ends at the anus.

Although the large intestine has a greater diameter than the small intestine, its epithelial surface area is far less, since the colon is about half as long as the small intestine, its surface is not convoluted, and its mucosa lacks villi. The secretions of the colon are scanty, lack digestive

FIGURE 17-32
The large intestine.

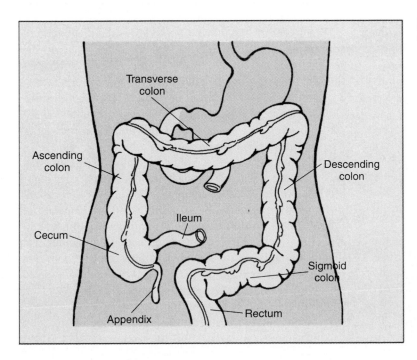

enzymes, and consist mostly of mucus and fluid containing bicarbonate and potassium ions. The primary function of the large intestine is to store and concentrate fecal material before defecation.

Chyme enters the colon through the **ileocecal sphincter.** This sphincter is normally closed, but after a meal, when the gastroileal reflex increases ileal contractions, it relaxes each time the terminal portion of the ileum contracts, allowing chyme to enter the large intestine. Distension of the colon, on the other hand, produces a reflex contraction of the sphincter, preventing fecal material from moving back into the small intestine.

About 1500 ml of chyme enters the colon from the small intestine each day. This material is derived largely from the secretions of the lower small intestine since most of the ingested food has been absorbed before reaching the large intestine. Absorption by the colon accounts for only about 4 percent of the material entering the gastrointestinal tract each day.

The primary absorptive process in the large intestine is the active transport of sodium from lumen to blood, with the accompanying osmotic absorption of water. If fecal material remains in the large intestine for a long time, almost all the water is absorbed, leaving behind hard fecal pellets. There is normally a net movement of potassium from blood into the colon lumen, and severe depletion of total body potassium can result when large volumes of fluid are excreted in the feces. There is also a net movement of bicarbonate ions into the lumen, and loss of this bicarbonate (a base) in patients with prolonged diarrhea can cause the blood to become acidic.

The large intestine also absorbs some of the products formed by the bacteria inhabiting this region. Undigested polysaccharides (fiber) are metabolized to short chain fatty acids by colonic bacteria and absorbed by passive diffusion. The bicarbonate secreted by the colon helps to neutralize the increased acidity resulting from the formation of these fatty acids. Colonic bacteria also produce small amounts of vitamins, especially vitamin K, that can be absorbed into the blood. Although this source of vitamins generally provides only a small part of the normal daily requirement, it may make a significant contribution when dietary vitamin intake is low. An individual who depends on colonic absorption of vitamins may become vitamin deficient if treated with antibiotics that inhibit other species of colonic bacteria as well as the disease-causing bacteria.

Other bacterial products include gas **(flatus),** which is a mixture of nitrogen and carbon dioxide, with small amounts of the inflammable gases hydrogen, methane, and hydrogen sulfide. Bacterial fermentation of undigested polysaccharides produces these gases in the colon at the rate of about 400 to 700 ml/day. Certain foods, beans, for example, contain large amounts of carbohydrates that cannot be digested by intestinal enzymes but are readily metabolized by bacteria in the large intestine, producing large amounts of gas.

Motility and Defecation.   Contractions of the circular smooth muscle in the colon produce a segmentation motion with a rhythm considerably slower (one every 30 min) than that in the small intestine. Because of the slow propulsion of the colonic contents, material entering the colon from the small intestine remains for about 18 to 24 h. This provides time for bacteria to grow and multiply. Three to four times a day, generally following a meal, a wave of intense contraction, known as a **mass movement,** spreads rapidly over the transverse segment of the colon toward the rectum. This usually coincides with the gastroileal reflex. Unlike a peristaltic wave, in which the smooth muscle at each point relaxes after the wave of contraction has passed, the smooth muscle of the colon remains contracted for some time after a mass movement.

The **anus,** the exit from the rectum, is normally closed by the **internal anal sphincter,** which is composed of smooth muscle, and the **external anal sphincter,** which is composed of skeletal muscle under voluntary control. The sudden distension of the walls of the rectum produced by the mass movement of fecal material into it initiates the neurally mediated **defecation reflex.**

The conscious urge to defecate, mediated by stretched mechanoreceptors, accompanies distension of the rectum. The reflex response consists of a contraction of the rectum, relaxation of the internal anal sphincter, but *contraction* of the external anal sphincter, and increased peristaltic activity in the sigmoid colon. Eventually, a pressure is reached in the rectum that triggers reflex *relaxation* of the external anal sphincter, allowing the feces to be expelled.

Brain centers can, however, via descending pathways to somatic nerves to the external anal sphincter, override the reflexive signals that eventually would relax the sphincter, thereby keeping the external sphincter closed and allowing a person to delay defecation. In this case the smooth muscle in the walls of the rectum relaxes, and the urge to defecate subsides until the next mass movement propels more feces into the rectum, increasing its volume and again initiating the defecation reflex. Voluntary control of the external anal sphincter is learned during childhood. Spinal cord damage can lead to a loss of voluntary control over defecation.

Defecation is normally assisted by a deep inspiration, followed by closure of the glottis and contraction of the abdominal and thoracic muscles, producing an increase in abdominal pressure that is transmitted to the contents of the large intestine and rectum. This maneuver also causes a rise in intrathoracic pressure, which leads to a transient rise in blood pressure followed by a fall in pressure as the venous return to the heart is decreased. The cardiovascular changes resulting from excessive strain during defecation may precipitate a stroke or heart attack, especially in constipated elderly individuals with cardiovascular disease.

PATHOPHYSIOLOGY OF THE GASTROINTESTINAL TRACT

Since the end result of gastrointestinal function is the absorption of nutrients, salts, and water, most malfunctions of this organ system affect either the nutritional state of the body or its salt and water content. The following provide a few examples of disordered gastrointestinal function.

Ulcers

Considering the high concentration of acid and pepsin secreted by the stomach, it is natural to wonder why the stomach does not digest itself. Several of the factors that protect the walls of the stomach from being digested are: (1) The surface of the mucosa is lined with cells that secrete a slightly alkaline mucus, which forms a thin layer over the luminal surface. Both the protein content of mucus and its alkalinity neutralize hydrogen ions in the immediate area of the epithelium. Thus, mucus forms a chemical barrier between the highly acid contents of the lumen and the cell surface. (2) The tight junctions between the epithelial cells lining the stomach restrict the diffusion of hydrogen ions into the underlying tissues. (3) Damaged epithelial cells are replaced every few days by new cells arising by the division of cells within the gastric pits.

Yet these protective mechanisms can prove inadequate, and erosion (***ulcers***) of the gastric surface occur. Ulcers can occur not only in the stomach but also in the lower part of the esophagus and in the duodenum. Indeed, duodenal ulcers are about 10 times more frequent than gastric ulcers, affecting about 10 percent of the U.S. population. Damage to blood vessels in the tissues underlying the ulcer may cause bleeding into the gastrointestinal lumen. On occasion, the ulcer may penetrate the entire wall, resulting in leakage of the luminal contents into the abdominal cavity.

Ulcer formation involves breaking the mucosal barrier and exposing the underlying tissue to the corrosive action of acid and pepsin, but it is not always clear what produces the initial damage to the barrier. Although acid is essential for ulcer formation, it is not necessarily the primary factor, and many patients with ulcers have normal or even subnormal rates of acid secretion.

Many factors, including genetic susceptibility, drugs, alcohol, bile salts, and an excessive secretion of acid and pepsin, may contribute to ulcer formation. The major factor, however, is the presence of a bacterium, ***Helicobacter pylori***, that is present in the stomachs of a majority of patients with ulcers or ***gastritis*** (inflammation of the stomach walls). Suppression of this bacterium with antibiotics usually leads to healing of the damaged mucosa.

Once an ulcer has formed, inhibition of acid secretion can remove the constant irritation and allow the ulcer to heal. Two classes of drugs are potent inhibitors of acid secretion. One class of inhibitors act by blocking a specific class of histamine receptors found on parietal cells, which stimulate acid secretion. These antihistamine drugs have become the most frequently prescribed drugs in America. The second class of drugs directly inhibit the H,K-ATPase pump in parietal cells. Although both classes of drugs are effective in healing ulcers, if the *Helicobacter pylori* bacteria are not removed, the ulcers tend to recur.

Despite popular notions that ulcers are due to emotional stress and despite the existence of a potential pathway (the parasympathetic nerves) for mediating stress-induced increases in acid secretion, the role of stress in producing ulcers remains unclear. Once the ulcer has been formed, however, emotional stress can aggravate it by increasing acid secretion.

Vomiting

Vomiting is the forceful expulsion of the contents of the stomach and upper intestinal tract through the mouth. Like swallowing, vomiting is a complex reflex coordinated by a region in the brainstem medulla oblongata, in this case known as the **vomiting center.** Neural input to this center from receptors in many different regions of the body can initiate the vomiting reflex. For example, excessive distension of the stomach or small intestine, various substances acting upon chemoreceptors in the intestinal wall or in the brain, increased pressure within the skull, rotating movements of the head (motion sickness), intense pain, and tactile stimuli applied to the back of the throat can all initiate vomiting.

What is the adaptive value of this reflex? Obviously, the removal of ingested toxic substances before they can be absorbed is of benefit. Moreover, the nausea that usually accompanies vomiting may have the adaptive value of conditioning the individual to avoid the future ingestion of foods containing such toxic substances. Why other types of stimuli, such as those producing motion sickness, have become linked to the vomiting center is not clear.

Vomiting is usually preceded by increased salivation, sweating, increased heart rate, pallor, and feelings of nausea. The events leading to vomiting begin with a deep inspiration, closure of the glottis, and elevation of the soft palate. The abdominal muscles then contract, raising the abdominal pressure, which is transmitted to the stomach's contents. The lower esophageal sphincter relaxes, and the high abdominal pressure forces the contents of the stomach into the esophagus. This initial sequence of events can occur repeatedly without expulsion via the mouth and is known as ***retching.*** Vomiting occurs when

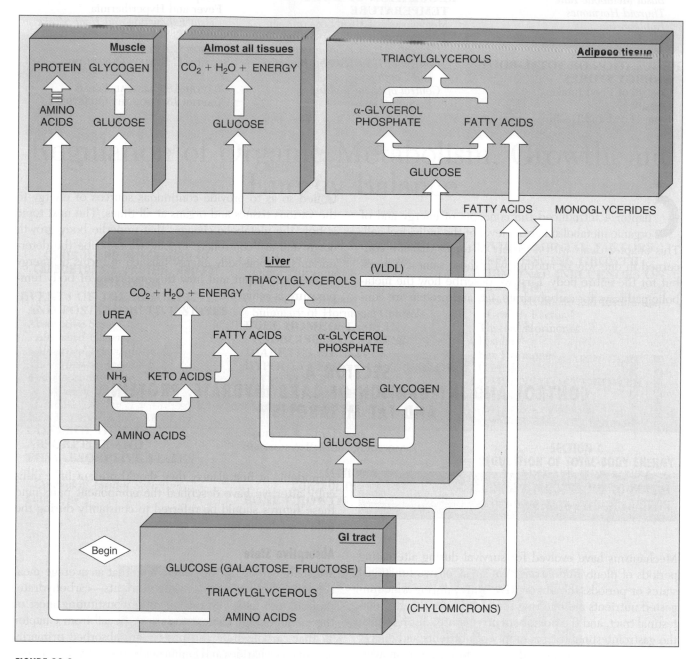

FIGURE 18-1

Major metabolic pathways of the absorptive state. The arrow from amino acids to protein in muscle is dashed to denote the fact that excess amino acids are not stored as protein (see text). All arrows between boxes denote transport of the substance via the blood. (VLDL = very low density lipoproteins)

as does glucose, we shall simply refer to absorbed carbo-hydrate as glucose.

Much of the absorbed glucose enters various body cells and is catabolized to carbon dioxide and water, providing the energy for ATP formation (Chapter 4). Indeed, and this is a key point, glucose is the body's major energy source during the absorptive state. In this regard, it should be recognized that skeletal muscle makes up the majority

of body mass and is the major consumer of metabolic fuel, even at rest.

Skeletal muscle not only catabolizes glucose during the absorptive phase, but converts some of the glucose to the polysaccharide glycogen, which is then stored in the muscle.

Adipose-tissue cells (adipocytes) also catabolize glucose for energy, but the most important fate of glucose in

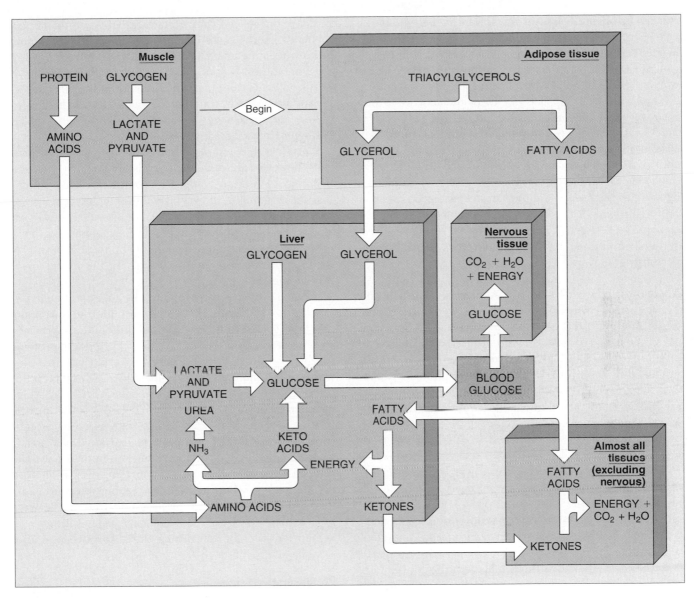

FIGURE 18-2

Major metabolic pathways of the postabsorptive state. The central focus is regulation of the blood glucose concentration. All arrows between boxes denote transport of the substance via the blood.

adipocytes during the absorptive phase is its transformation to fat (triacylglycerols). Glucose is the precursor of both α-glycerol phosphate and fatty acids, and these molecules are then linked together to form triacylglycerols.

Another large fraction of the absorbed glucose enters the liver cells. This is a very important point: During the absorptive period there is net *uptake* of glucose by the liver. It is either stored as glycogen, as in skeletal muscle, or transformed to α-glycerol phosphate and fatty acids, which are then used to synthesize triacylglycerols, as in adipose tissue. Some of the fat synthesized from glucose in the liver is stored there, but most is packaged, along with

specific proteins, into molecular aggregates of lipids and proteins called lipoproteins. These aggregates are secreted by the liver cells and enter the blood. They are called **very low density lipoproteins (VLDL)** because they contain much more fat than protein, and fat is less dense than protein. The synthesis of VLDL by liver cells occurs by processes similar to those for synthesis of chylomicrons by intestinal mucosal cells, as described in Chapter 17.

Once in the bloodstream, VLDL complexes, being quite large, do not readily penetrate capillary walls. Instead, their triacylglycerols are hydrolyzed mainly to monoglycerides (glycerol linked to one fatty acid chain) and fatty acids by the enzyme **lipoprotein lipase,** which is located on the blood-facing surface of capillary endothelial cells, especially those in adipose tissue. In adipose-tissue capillaries, the fatty acids generated by this enzyme's action diffuse across the capillary wall and into the adipocytes. There they combine with α-glycerol phosphate, supplied by glucose metabolites, to form triacylglycerols once again. Thus, most of the fatty acids in the VLDL triacylglycerol originally synthesized from glucose by the *liver* end up being stored in triacylglycerol in *adipose tissue.* The monoglycerides formed in the blood by the action of lipoprotein lipase in adipose-tissue capillaries circulate to the liver where they are metabolized.

To summarize, the major fates of glucose during the absorptive phase are utilization for energy, storage as glycogen in liver and skeletal muscle, and storage as fat in adipose tissue.

Absorbed Triacylglycerols. As described in Chapter 17, almost all absorbed chylomicrons enter the lymph, which flows into the systemic circulation. The biochemical processing of these chylomicron triacylglycerols in plasma is quite similar to that just described for VLDL produced by the liver. The fatty acids of plasma chylomicrons are released, mainly within adipose-tissue capillaries, by the action of endothelial lipoprotein lipase. The released fatty acids then enter adipocytes and combine with α-glycerol phosphate, synthesized from glucose metabolites, to form triacylglycerols.

The importance of glucose for triacylglycerol synthesis in adipocytes cannot be overemphasized. Adipocytes do not have the enzyme required for phosphorylation of glycerol, and so α-glycerol phosphate can be formed in these cells *only* from glucose metabolites and not from glycerol or any other fat metabolites.

In contrast, there are three major sources of the fatty acids found in adipose-tissue triacylglycerol: (1) glucose that enters adipose tissue and is converted to fatty acids; (2) glucose that is converted in liver to VLDL triacylglycerols, which are transported via the blood to the adipose tissue; and (3) ingested triacylglycerols transported to adipose tissue in chylomicrons. As we have seen, sources (2) and (3) require the action of lipoprotein lipase to release the fatty acids from the circulating triacylglycerols.

This description has emphasized the *storage* of ingested fat. For simplicity, we have not shown in Figure 18-1 that a fraction of the ingested fat is not stored but is oxidized during the absorptive state by various organs to provide energy. The relative amounts of carbohydrate and fat used for energy during the absorptive period depend largely on the content of the meal.

Absorbed Amino Acids. A minority of the absorbed amino acids enter liver cells. They are used to synthesize a variety of proteins, including liver enzymes and plasma proteins, or they are converted to carbohydrate-like intermediates known as **keto acids** by removal of the amino group (deamination, Chapter 4). The amino groups are used to synthesize urea, which is excreted by the kidneys. The keto acids can enter the Krebs tricarboxylic acid cycle and be catabolized to provide energy for the liver cells, or they can be converted to fatty acids, thereby participating in fat synthesis by the liver.

Most ingested amino acids are not taken up by the liver cells, however, but enter other cells (Figure 18-1), where they may be used to synthesize proteins. We have simplified the diagram by showing "nonliver" amino acid uptake only by muscle, because muscle contains by far the largest amount of body protein. It should be emphasized, however, that all cells require a constant supply of amino acids for protein synthesis and participate in the dynamics of protein metabolism.

Protein synthesis is represented by a dashed line in the muscle box in Figure 18-1 to call attention to an important fact: There is a net synthesis of protein during the absorptive period, but this basically just replaces the proteins catabolized during the postabsorptive period. In other words, excess amino acids are not *stored* as protein in the sense that glucose is stored as glycogen or that both glucose and fat are stored as fat. Rather, ingested amino acids in excess of those needed to maintain a stable protein turnover are merely converted to carbohydrate or fat. Therefore, eating large amounts of protein does not in itself cause increases in body protein. This discussion does not apply to growing children, who manifest a continuous increase in body protein, or to adults who are actively building body mass as, for example, by weight lifting.

Nutrient metabolism during the absorptive period is summarized in Table 18-1.

Postabsorptive State

As the absorptive period ends, net synthesis of glycogen, fat, and protein ceases, and net catabolism of all these

---

TABLE 18-1   SUMMARY OF NUTRIENT METABOLISM DURING THE ABSORPTIVE PERIOD

1. Energy is provided primarily by absorbed carbohydrate.
2. There is net uptake of glucose by the liver.
3. Some carbohydrate is stored as glycogen in liver and muscle, but most carbohydrate and fat in excess of that utilized for energy are stored mainly as fat in adipose tissue.
4. There is some synthesis of body proteins, but much dietary protein is utilized for energy or converted to fat.

substances begins to occur. The overall significance of these events can be understood in terms of the essential problem during the postabsorptive period: No glucose is being absorbed from the intestinal tract, yet the plasma glucose concentration must be maintained because the brain normally utilizes only glucose for energy. Too low a plasma glucose concentration can result in alterations of neural activity ranging from subtle impairment of mental function to coma and even death.

The events that maintain plasma glucose concentration fall into two categories: (1) reactions that provide sources of blood glucose, and (2) glucose sparing because of fat utilization.

Sources of Blood Glucose. The sources of blood glucose during the postabsorptive period are as follows (Figure 18-2):

1. **Glycogenolysis,** the hydrolysis of glycogen stores, occurs in the liver and skeletal muscle. In the liver, glucose is formed by this process and enters the blood. Hepatic glycogenolysis, a rapidly occurring event, is the first line of defense in maintaining plasma glucose concentration. The amount of glucose available from this source can supply the body's needs for only a few hours.

   Glycogenolysis also occurs in skeletal muscle, which contains approximately the same amount of glycogen as the liver. However, muscle, unlike liver, lacks the enzyme necessary to form glucose from the glucose 6-phosphate formed during glycogenolysis (Chapter 4). The glucose 6-phosphate undergoes glycolysis within the muscle to yield pyruvate and lactate. These substances are liberated into the blood, circulate to the liver, and are converted into glucose, which can then leave the liver cells to enter the blood. Thus, muscle glycogen contributes to the blood glucose indirectly via the liver.

2. The catabolism of triacylglycerols yields glycerol and fatty acids, a process termed **lipolysis.** The major site of lipolysis is adipose tissue, and the glycerol and fatty acids then enter the blood. The glycerol reaching the liver is converted to glucose. Thus, an important source of glucose during the postabsorptive period is the glycerol released when adipose-tissue triacylglycerol is broken down.

3. A few hours into the postabsorptive period protein becomes the major source of blood glucose. Large quantities of protein in muscle and, to a lesser extent, other tissues can be catabolized without serious cellular malfunction. There are, of course, limits to this process, and continued protein loss during a prolonged fast ultimately means functional disintegration, sickness, and death. Before this point is reached,

however, protein breakdown can supply large quantities of amino acids, particularly alanine, that enter the blood and are picked up by the liver, which converts them, via the keto acid pathway, to glucose.

In items 2 and 3 above, we described the synthesis by the liver of glucose from pyruvate, lactate, glycerol, and amino acids. Synthesis from any of these precursors is known as **gluconeogenesis,** that is, new formation of glucose. During a 24-h fast, gluconeogenesis provides approximately 180 g of glucose. (The liver is not the only organ capable of gluconeogenesis; the kidneys also perform gluconeogenesis, but mainly during a prolonged fast.)

Glucose Sparing (Fat Utilization). The 180 g of glucose per day produced by gluconeogenesis in the liver (and kidneys) during fasting supplies 720 kcal. As described later in this chapter, normal total energy expenditure for an average adult equals 1500 to 3000 kcal/day. Accordingly, gluconeogenesis cannot supply all the body's energy needs. The following essential adjustment must therefore take place during the transition from the absorptive to the postabsorptive state: Most organs and tissues markedly reduce their glucose catabolism and increase their fat utilization, the latter becoming the major energy source. This metabolic adjustment, termed **glucose sparing,** "spares" the glucose produced by the liver for use by the nervous system.

The essential step in this adjustment is lipolysis, the catabolism of adipose-tissue triacylglycerol, which liberates glycerol and fatty acids into the blood. We described lipolysis in the previous section in terms of its importance in providing *glycerol* to the liver for conversion to glucose. Now, we focus on the liberated *fatty acids*, which circulate bound to plasma albumin. [Despite this binding to protein, they are known as free fatty acids (FFA) in that they are "free" of glycerol.] The circulating fatty acids are picked up and metabolized by almost all tissues, *excluding the nervous system*. They enter the Krebs cycle by way of acetyl CoA and are catabolized to carbon dioxide and water (Chapter 4), thereby providing energy.

The liver is unique, however, in that most of the acetyl CoA it forms from fatty acids during the postabsorptive state does not enter the Krebs cycle but is processed into three compounds collectively called **ketones** (or ketone bodies). (Note that ketones are not the same as keto acids, which are metabolites of amino acids.) Ketones are released into the blood and provide an important energy source during prolonged fasting for the many tissues, *including the brain*, capable of oxidizing them via the Krebs cycle. One of the ketones is acetone, some of which is exhaled and accounts for the distinctive breath odor of individuals undergoing prolonged fasting or, as we shall see, suffering from severe untreated diabetes mellitus.

The net result of fatty acid and ketone utilization during fasting is provision of energy for the body and sparing of glucose for the brain. Moreover, as just emphasized, the brain can use ketones for an energy source, and it does so increasingly as ketones build up in the blood during the first few days of a fast. The survival value of this phenomenon is very great: When the brain reduces its glucose requirement by utilizing ketones, much less protein breakdown is required to supply amino acids for gluconeogenesis. Accordingly, the protein stores will last longer, and the ability to withstand a long fast without serious tissue disruption is enhanced.

Table 18-2 summarizes the events of the postabsorptive period. The combined effects of glycogenolysis, gluconeogenesis, and the switch to fat utilization are so efficient that, after several days of complete fasting, the plasma glucose concentration is reduced by only a few percent. After 1 month, it is decreased only 25 percent.

ENDOCRINE AND NEURAL CONTROL OF THE ABSORPTIVE AND POSTABSORPTIVE STATES

We now turn to the endocrine and neural factors that control and integrate these metabolic pathways. We shall focus primarily on the following questions, summarized in Figure 18-3: (1) What controls net anabolism of protein, glycogen, and triacylglycerol in the absorptive phase, and net catabolism in the postabsorptive phase? (2) What induces primarily glucose utilization by cells for energy during the absorptive phase, but fat utilization during the postabsorptive phase? (3) What drives net glucose uptake by the liver during the absorptive phase, but gluconeogenesis and glucose release during the postabsorptive phase?

The most important controls of these transitions from feasting to fasting, and vice versa, are two pancreatic hormones—insulin and glucagon. Also playing a role are the hormone epinephrine, from the adrenal medulla, and the sympathetic nerves to liver and adipose tissue.

Insulin and glucagon are peptides secreted by the **islets of Langerhans,** clusters of endocrine cells in the pancreas. Appropriate histological techniques reveal several distinct types of islet cells, each of which secretes a different hormone. The **beta cells** (or B cells) are the source of insulin, and the **alpha cells** (or A cells) of glucagon. (There are two other hormones—somatostatin and pancreatic polypeptide—secreted by still other islet cells, but the functions of these two pancreatic hormones in human beings are not yet fully established.)

Insulin

Insulin is the most important controller of organic metabolism. Its secretion, and hence plasma concentration, are increased during the absorptive state and decreased during the postabsorptive state. For simplicity, insulin's many actions are often divided into two broad categories: (1) *metabolic effects* on carbohydrate, lipid, and protein synthesis, and (2) *growth-promoting effects* on DNA synthesis, cell division, and cell differentiation. This section deals only with the metabolic effects; the growth-promoting effects are described later in this chapter.

The metabolic effects of insulin are exerted mainly on muscle cells (both cardiac and skeletal), adipose-tissue cells, and liver cells. The most important responses of these target cells are summarized in Figure 18-4. Compare this figure to Figure 18-1 and the left panel of Figure 18-3 and you will see that these responses to insulin are the same as the events of the absorptive-state pattern. Conversely, the effects of a reduction in plasma insulin are the same as the events of the postabsorptive pattern in Figure 18-2 and the right panel of Figure 18-3. The reasons for these correspondences is that an increased plasma concentration of insulin is the major cause of all the absorptive-state events, and a decreased plasma concentration of insulin is the major cause of all the post-absorptive events.

Like all peptide hormones, insulin induces its effects by binding to specific receptors in the plasma membrane of its target cells. This binding triggers a variety of signal transduction pathways that influence the target-cells' transport proteins and intracellular enzymes. Thus, for example, in muscle cells and adipose-tissue cells an increased insulin concentration stimulates cytoplasmic vesicles that contain glucose transporters in their membrane to fuse

---

TABLE 18-2   SUMMARY OF NUTRIENT METABOLISM DURING THE POSTABSORPTIVE PERIOD

1. Glycogen, fat, and protein syntheses are curtailed, and net breakdown occurs.

2. Glucose is formed in the liver both from the glycogen stored there and by gluconeogenesis from blood-borne lactate, pyruvate, glycerol, and amino acids. The kidneys also perform gluconeogenesis during a prolonged fast.

3. The glucose produced in the liver (and kidneys) is released into the blood, but its utilization for energy is greatly reduced in muscle and other nonneural tissues.

4. Lipolysis releases adipose-tissue fatty acids into the blood, and the oxidation of these fatty acids and of ketones produced from them by the liver provides most of the body's energy supply.

5. The brain continues to use glucose but also starts using ketones as they build up in the blood.

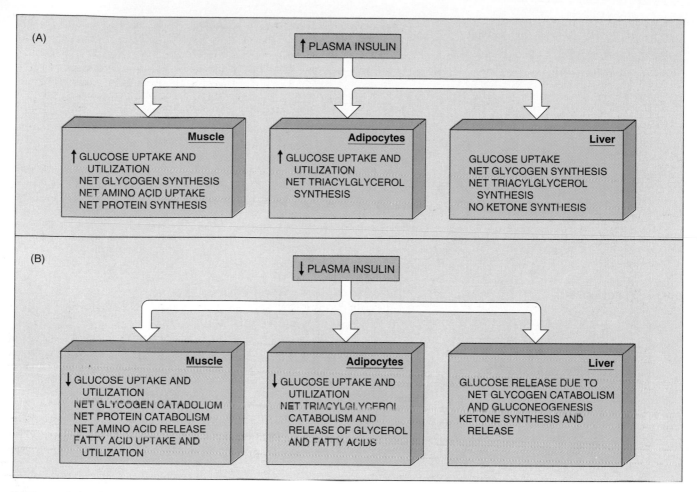

FIGURE 18-4

Summary of overall target-cell responses to (A) an increase or (B) a decrease in the plasma concentration of insulin. The responses in (A) are virtually identical to the absorptive state events of Figure 18-1 and the left panel of Figure 18-3; the responses in (B) are virtually identical to the postabsorptive state events of Figure 18-2 and the right panel of Figure 18-3. The biochemical events that underlie these responses to insulin are shown in Figure 18-6.

FIGURE 18-5

Stimulation by insulin of the translocation of glucose transporters from cytoplasmic vesicles to the plasma membrane in muscle cells and adipose-tissue cells. Note that these transporters are constantly recycled by endocytosis from the plasma membrane back into vesicles, as shown on the right side of the figure. As long as insulin levels are elevated, the entire cycle continues and the number of transporters in the plasma membrane stays high. In contrast, when insulin levels decrease, the cycle is broken, the vesicles accumulate in the cytoplasm, and the number of transporters in the plasma membrane decrease.

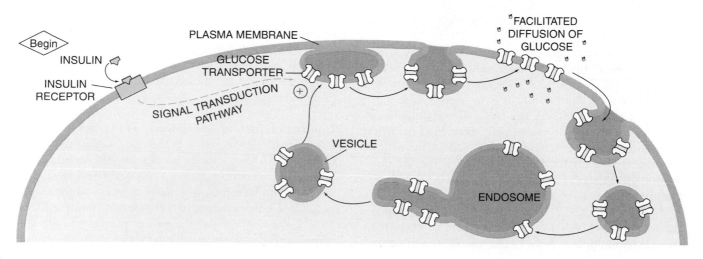

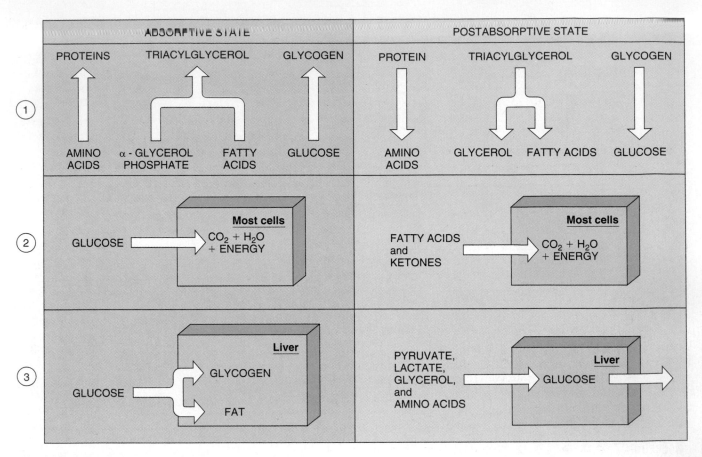

FIGURE 18-3

Summary of critical points in transition from absorptive state to postabsorptive state. The term "absorptive state" could be replaced with "actions of insulin," and the term "postabsorptive state" with "results of decreased insulin."

with the plasma membrane (Figure 18-5); the increased number of plasma-membrane glucose transporters resulting from this fusion then causes a greater rate of glucose movement from the extracellular fluid into the cells by facilitated diffusion. Recall from Chapter 6 that glucose enters virtually all cells of the body by facilitated diffusion; there are multiple subtypes of glucose transporters that mediate this process, however, and the subtype that is regulatable by insulin is found mainly in muscle cells and adipose-tissue cells.

A description of the many enzymes whose activities and/or concentrations are influenced by insulin is beyond the scope of this book, but the overall pattern is illustrated for reference in Figure 18-6. It is important here not to lose sight of the forest for the trees: The essential information (the "forest") to understand about insulin's actions is the target cells' ultimate responses, that is, the material summarized in Figure 18-4; Figure 18-6 merely shows some of the specific biochemical reactions (the "trees") that underlie these responses.

A major principle illustrated by Figure 18-6 is that, in each of its target cells, insulin brings about its ultimate responses by multiple actions. Let us take its effects on mus-

cle cells as an example. In these cells insulin favors glycogen formation and storage by (1) increasing glucose transport into the cell, (2) stimulating the key enzyme (glycogen synthase) that catalyzes the rate-limiting step in glycogen synthesis, and (3) inhibiting the key enzyme (glycogen phosphorylase) that catalyzes glycogen catabolism. Thus, insulin favors glucose transformation to and storage as glycogen in muscle through three pathways. Similarly, for protein synthesis in muscle cells, (1) insulin increases the number of active plasma-membrane transporters for amino acids, (2) stimulates the ribosomal enzymes that mediate the synthesis of protein from these amino acids, and (3) inhibits the enzymes that mediate protein catabolism.

Control of Insulin Secretion. The major controlling factor for insulin secretion is the plasma glucose concentration, which affects the beta cells directly. An increase plasma glucose concentration, as occurs after a meal, stimulates insulin secretion, whereas a decrease inhibits secretion. The feedback nature of this system is shown in Figure 18-7: Following a meal, the rise in plasma glucose concentration stimulates insulin secretion, and the insu

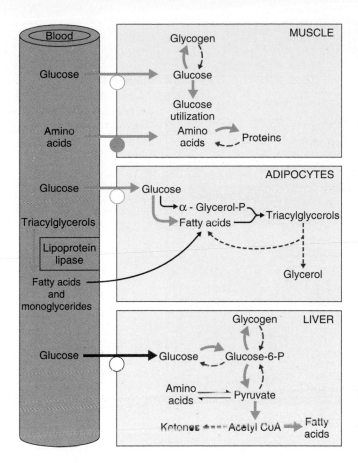

FIGURE 18-6

Reference illustration of the key biochemical events that underlie those responses of target cells to insulin summarized in Figure 18-4. Each green arrow denotes a process stimulated by insulin, whereas a dashed red arrow denotes inhibition by insulin. Except for the effects on the transport proteins for glucose and amino acids, all other effects are exerted on insulin-sensitive enzymes. The bowed arrows denote pathways whose reversibility is mediated by different enzymes (Chapter 4); such enzymes are commonly the ones influenced by insulin and other hormones. The black arrows are processes that are not *directly* stimulated by insulin but are enhanced in the presence of increased insulin as the result of mass-action.

stimulates entry of glucose into muscle and adipose tissue, as well as net uptake, rather than net output, of glucose by the liver. These effects reduce the blood concentration of glucose, thereby removing the stimulus for insulin secretion, which returns to its previous level.

In addition to plasma glucose concentration, there are numerous other insulin-secretion controls (Figure 18-8). One is the plasma concentration of certain amino acids, an elevated amino acid concentration causing enhanced insulin secretion. This is another negative feedback control:

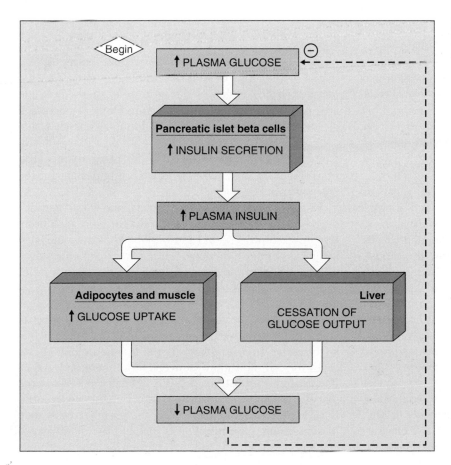

FIGURE 18-7
Negative-feedback nature of plasma glucose control over insulin secretion. ✗

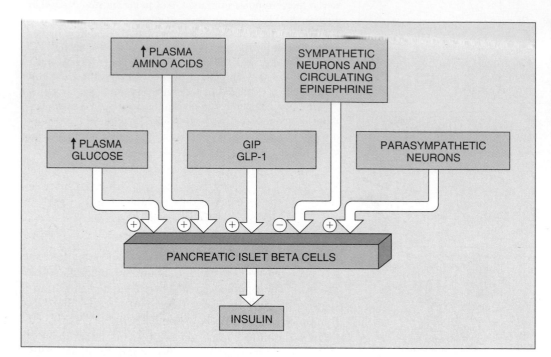

FIGURE 18-8

Major controls of insulin secretion. GIP = glucose-dependent insulinotropic peptide, and GLP-1 = glucagon-like peptide-1; both are gastrointestinal hormones.

Amino acid concentrations increase after ingestion of a protein-containing meal, and the increased plasma insulin stimulates uptake of these amino acids by muscle (and other cells as well).

There are also important hormonal controls over insulin secretion. For example, two hormones—glucose-dependent insulinotropic peptide (GIP) and glucagon-like peptide-1 (GLP-1)—secreted by the gastrointestinal tract in response to eating stimulate the release of insulin. This provides a feedforward component to glucose regulation during ingestion of a meal; thus insulin secretion rises earlier and to a greater extent than it would have if plasma glucose were the only controller.

Finally, the autonomic neurons to the islets of Langerhans also influence insulin secretion. Activation of the parasympathetic neurons, which occurs during ingestion of a meal, stimulates secretion of insulin and constitutes a second type of feedforward regulation. In contrast, activation of the sympathetic neurons to the islets or an increase in the plasma concentration of epinephrine (the hormone secreted by the adrenal medulla) inhibits insulin secretion. The significance of this relationship for the body's response to low plasma glucose (**hypoglycemia**), stress, and exercise—all situations in which sympathetic activity is increased—will be described later in this chapter.

To repeat, insulin plays the primary role in controlling the metabolic adjustments required for feasting or fasting. Other hormonal and neural factors, however, also play significant roles. They all oppose the action of insulin in one way or another and are known as **glucose-counterregulatory controls.** Of these, the most important is glucagon.

Glucagon

As noted earlier, **glucagon** is the peptide hormone produced by the alpha cells of the pancreatic islets. The major physiological effects of glucagon are all on the liver and are opposed to those of insulin (Figure 18-9): (1) increased glycogen breakdown, (2) increased gluconeogenesis, and (3) synthesis of ketones. Thus, the overall results of glucagon's effects are to increase the plasma concentrations of glucose and ketones, which are important for the postabsorptive period.

From a knowledge of these effects, one would logically suppose that glucagon secretion should increase during the postabsorptive period and prolonged fasting, and such is the case because the major stimulus for glucagon secretion is hypoglycemia. The adaptive value of such a reflex is obvious: A decreasing plasma glucose concentration induces increased release of glucagon which, by its effects on metabolism, serves to restore normal blood glucose concentration by glycogenolysis and gluconeogenesis while at the same time supplying (if the fast is prolonged) ketones for cell utilization. Conversely, an increased plasma glucose concentration inhibits glucagon's secretion, thereby helping to return the plasma glucose concentration toward normal. Thus, during the postabsorptive state plasma insulin concentration is low and plasma glucagon concentration is high, and this combined change accounts

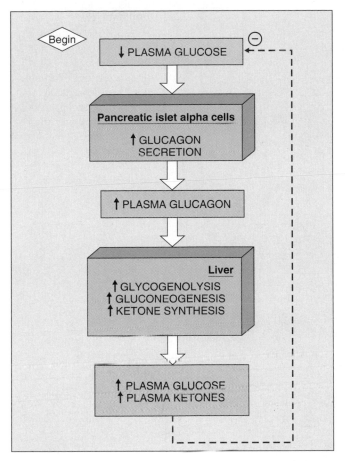

FIGURE 18-9

Negative-feedback nature of plasma glucose control over glucagon secretion.

in both the liver and skeletal muscle, (2) gluconeogenesis in the liver, and (3) lipolysis in adipocytes. Activation of the sympathetic nerves to the liver and adipose tissue elicits essentially the same responses by these organs as does circulating epinephrine.

Thus, enhanced sympathetic nervous system activity exerts effects on organic metabolism—increased plasma concentrations of glucose, glycerol, and fatty acids—that are opposite those of insulin.

As might be predicted from these effects, hypoglycemia leads reflexly to increases in both epinephrine secretion and sympathetic-nerve activity to the liver and adipose tissue. This is the same stimulus that, as described above, leads to increased secretion of glucagon, although the receptors and pathways are totally different. When the plasma glucose concentration decreases, glucose receptors in the central nervous system (and, possibly, the liver) initiate the reflexes that lead to increased activity in the sympathetic pathways to the adrenal medulla, liver, and adipose tissue. The adaptive value of the response is the same as that for the glucagon response to hypoglycemia: Blood glucose returns toward normal, and fatty acids are supplied for cell utilization.

In the compensatory response to *acute* hypoglycemia the increased activity of the sympathetic nervous system is less important than a reduced insulin concentration and an increased glucagon concentration, but nevertheless contributes. In contrast, sympathetic nervous system activity *decreases* during *prolonged* fasting or ingestion of low-caloric diets; the adaptive significance of this change is discussed later in this chapter.

Other Hormones

In addition to the three hormones already described in this section, there are many others that have various effects on organic metabolism. The secretion of all these other hormones, however, is not primarily keyed to the transitions between the absorptive and postabsorptive states. Instead, their secretion is controlled by other factors, and these hormones are for the most part involved in homeostatic processes described elsewhere in this book. Nonetheless, the effects of two of them—cortisol and growth hormone—on nutrient metabolism are important enough to warrant description here.

Cortisol. Cortisol, the major glucocorticoid produced by the adrenal cortex, plays an essential "permissive" role in the adjustments to fasting. We have described how fasting is associated with stimulation of both gluconeogenesis and lipolysis; however, neither of these critical metabolic transformations occurs to the usual degree in a person deficient in cortisol. In other words, the plasma cortisol level need not *rise* during fasting and usually does not, but the presence of even small amounts of cortisol in the blood

almost entirely for the transition from the absorptive to the postabsorptive state. Said in a different way, this shift is best explained by a rise in the glucagon:insulin ratio in the plasma.

The secretion of glucagon, like that of insulin, is controlled not only by the plasma concentration of glucose and other nutrients but also by neural and hormonal inputs to the islets. For example, the sympathetic nerves to the islets stimulate glucagon secretion—just the opposite of their effect on insulin secretion. The adaptive significance of this relationship for exercise and stress will be described subsequently.

Epinephrine and Sympathetic Nerves to Liver and Adipose Tissue

As noted earlier, epinephrine and the sympathetic nerves to the pancreatic islets inhibit insulin secretion and stimulate glucagon secretion. In addition, epinephrine also affects nutrient metabolism directly (Figure 18-10). Its major direct effects include stimulation of (1) glycogenolysis

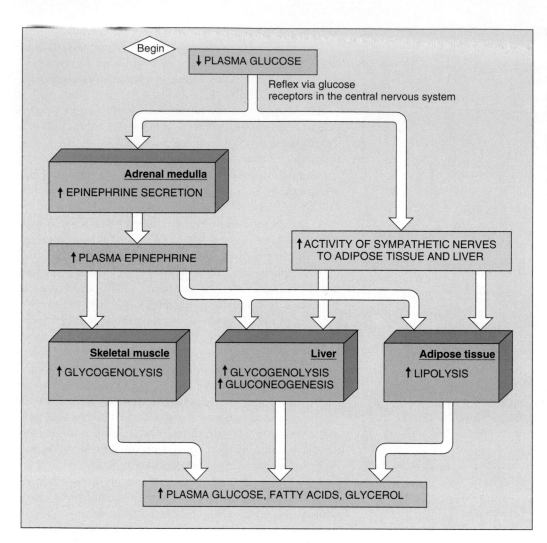

FIGURE 18-10

Participation of the sympathetic nervous system in the response to a low plasma glucose concentration (hypoglycemia). Glycogenolysis in skeletal muscle contributes to increased plasma glucose by releasing lactate and pyruvate, which are converted to glucose in the liver.

somehow maintains the concentrations of the key liver and adipose-tissue enzymes required for gluconeogenesis and lipolysis. Therefore, in response to fasting, people with a cortisol deficiency develop hypoglycemia serious enough to interfere with brain function.

Moreover, cortisol can play more than a permissive role when its plasma concentration does increase, as occurs during stress (Chapter 20). In high concentration, cortisol elicits many metabolic events ordinarily associated with fasting (Table 18-3). Clearly, here is another hormone, in addition to glucagon and epinephrine, that can exert actions opposite those of insulin. Indeed, persons with very high plasma levels of cortisol, due either to abnormally high secretion or to cortisol administration for medical reasons (Chapter 20), develop symptoms similar to those seen in individuals with insulin deficiency.

Growth Hormone.   The primary physiological effects of growth hormone are to stimulate both growth and protein anabolism. Compared to these effects, those it exerts on carbohydrate and lipid metabolism are minor. Nonetheless, as is true for cortisol, either severe deficiency or marked excess of growth hormone does produce significant abnormalities in lipid and carbohydrate metabolism. Growth hormone's effects on these nutrients are

TABLE 18-3   EFFECTS OF CORTISOL ON ORGANIC METABOLISM

1. Basal concentrations are permissive for stimulation of gluconeogenesis and lipolysis in the postabsorptive state
2. Increased plasma concentrations cause:
   a. Increased protein catabolism
   b. Increased gluconeogenesis
   c. Decreased glucose uptake by muscle cells and adipose-tissue cells.
   d. Increased triacylglycerol breakdown

Net result: Increased plasma concentrations of amino acids, glucose, and free fatty acids similar to those of cortisol and opposite those of insulin. Growth hormone (1) renders adipocytes more responsive to lipolytic stimuli, (2) increases gluconeogenesis by the liver, and (3) reduces the ability of insulin to cause glucose uptake by muscle and adipose tissue. These are often termed growth hormone's "anti-insulin effects."

Summary of Hormonal Controls

To a great extent insulin may be viewed as the "hormone of plenty." Its secretion and plasma concentration are increased during the absorptive period and decreased during postabsorption, and these changes are adequate to cause most of the metabolic changes associated with these periods. In addition, opposed in various ways to insulin's effects are the actions of four major glucose-counterregulatory controls—glucagon, epinephrine and the sympathetic nerves to the liver and adipose tissue, cortisol, and growth hormone (Table 18-4). Glucagon and the sympathetic nervous system are activated during the postabsorptive period (or in any other situation with hypoglycemia) and definitely play roles in preventing hypoglycemia, glucagon being the more important. The rates of secretion of cortisol and growth hormone are not usually coupled to the absorptive-postabsorptive pattern; nevertheless, their presence in the blood at basal concentrations is necessary for normal adjustment of lipid and carbohydrate metabolism to the postabsorptive period, and excessive amounts of either hormone cause abnormally elevated plasma glucose concentrations.

FUEL HOMEOSTASIS IN EXERCISE AND STRESS

During exercise large quantities of fuels must be mobilized to provide the energy required for muscle contraction. As described in Chapter 11, these fuels include plasma glucose and fatty acids as well as the muscle's own glycogen.

The plasma glucose used during exercise is supplied by the liver, both by breakdown of its glycogen stores and by gluconeogenesis—conversion of pyruvate, lactate, glycerol, and amino acids to glucose. The glycerol is made available to the liver by a marked increase in adipose-tissue lipolysis with a resultant release of glycerol and fatty acids into the blood, the fatty acids serving, along with glucose, as a fuel source for the exercising muscle.

What happens to blood glucose concentration during exercise? It changes very little in short-term, mild to moderate exercise and may even increase slightly with strenuous short-term activity. However, during prolonged exercise (Figure 18-11), more than 90 min, plasma glucose concentration does decrease, but usually by less than 25 percent. Clearly, glucose output by the liver increases approximately in proportion to increased glucose utilization during exercise, at least until the later stages of prolonged exercise when it begins to lag somewhat.

The metabolic profile seen in an exercising individual—increases in hepatic glucose production, triacylglycerol breakdown, and fatty acid utilization—is similar to that seen in a fasting person, and the controls are also the same: Exercise is characterized by a fall in insulin secretion and a rise in glucagon secretion (Figure 18-11), and the changes in the plasma concentrations of these two hormones are the major controls during exercise. In addition there is increased activity of the sympathetic nervous system (including increased secretion of epinephrine) and increased secretion of cortisol and growth hormone.

What triggers increased glucagon secretion and decreased insulin secretion during exercise? One signal, at least during *prolonged* exercise, is the modest decrease in plasma glucose that occurs (Figure 18-11); this is the same signal that controls the secretion of these hormones in

---

TABLE 18-4 SUMMARY OF GLUCOSE-COUNTERREGULATORY CONTROLS*

	Glucagon	Epinephrine	Cortisol	Growth hormone
Glycogenolysis	X	X		
Gluconeogenesis	X	X	X	X
Lipolysis		X	X	X
Inhibition of glucose uptake by muscle cells and adipose-tissue cells			X	X

*An X indicates that the hormone stimulates the process; no X indicates that the hormone has no major physiological effect on the process. Epinephrine stimulates glycogenolysis in both liver and skeletal muscle, whereas glucagon does so only in liver.

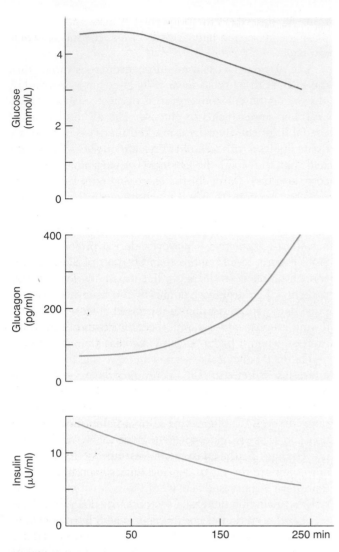

FIGURE 18-11

Plasma concentrations of glucose, glucagon, and insulin during prolonged (250 min) moderate exercise at a fixed intensity. (*Adapted from Felig and Wahren.*)

fasting. Other inputs at all intensities of exercise are increased circulating epinephrine and enhanced activity of the sympathetic neurons supplying the pancreatic islets. This sympathetic output is independent of plasma glucose concentration and is mediated by the central nervous system as part of the "preprogrammed" neural response to exercise. Thus, the increased sympathetic nervous system activity characteristic of exercise not only contributes directly to fuel mobilization by acting on the liver and adipose tissue, but contributes indirectly by inhibiting the secretion of insulin and stimulating that of glucagon.

One component of the response to exercise is quite different from the response to fasting: In exercise, glucose uptake and utilization by the muscles is increased, whereas in fasting it is markedly reduced. How is it that, during exercise, the movement, via facilitated diffusion, of glucose into muscle can remain high in the presence of reduced plasma insulin and increased plasma concentrations of cortisol and growth hormone, all of which decrease glucose uptake by skeletal muscle? By an as-yet-unidentified mechanism, muscle contraction causes migration of an intracellular store of glucose transporters to the plasma membrane.

Exercise and the postabsorptive state are not the only situations characterized by the neuroendocrine profile of decreased insulin and increased glucagon, sympathetic activity, cortisol, and growth hormone. This profile also occurs in response to a variety of nonspecific stresses, both physical and emotional. The adaptive value of these neuroendocrine responses to stress is that the resulting metabolic shifts prepare the body for exercise ("fight or flight") in the face of real or threatened injury. In addition, the amino acids liberated by catabolism of body protein stores because of decreased insulin and increased cortisol not only provide energy via gluconeogenesis but also constitute a potential source of amino acids for tissue repair should injury occur. The subject of stress and the body's responses to it are further described in Chapter 20.

DIABETES MELLITUS

The name "diabetes," meaning "syphon" or "running through," was used by the Greeks over 2000 years ago to describe the increased urinary volume excreted by people suffering from this disease. "Mellitus," meaning "sweet," distinguishes this urine from the large quantities of nonsweet ("insipid") urine produced by persons suffering from vasopressin deficiency. As described in Chapter 16, the latter disorder is known as diabetes insipidus, and the unmodified word "diabetes" is often used as a synonym for ***diabetes mellitus,*** a disease that affects nearly 15 million people in the United States.

Diabetes can be due to a deficiency of insulin or to a hyporesponsiveness to insulin, for it is not one but several diseases with different causes. Classification of these diseases rests on how much insulin the person is secreting and whether therapy requires the administration of insulin. In ***insulin-dependent diabetes mellitus (IDDM),*** or type 1 diabetes, the hormone is completely or almost completely absent from the islets of Langerhans and the plasma, and therapy with insulin is essential (this protein hormone cannot be given orally but must be injected because gastrointestinal enzymes would digest it). In ***non-insulin-dependent diabetes mellitus (NIDDM),*** or type 2 diabetes, the hormone is often present in plasma at near-normal or even above-normal levels, and therapy does not *require* administration of insulin (although such administration may be beneficial).

IDDM is less common, affecting 15 percent of diabetic patients. It is due to the total or near-total destruction of the pancreatic beta cells by the body's own white blood cells (autoimmune disease, Chapter 20). The triggering events for this autoimmune response are not yet fully established. As noted above, treatment of IDDM always involves the administration of insulin. It is likely that in the not-too-distant future transplantation of islet cells into the individual with IDDM will be possible.

Largely because of their insulin deficiency (we'll see later that glucagon also plays a role), untreated patients with IDDM always have elevated plasma glucose concentrations. This occurs both because glucose fails to enter insulin's target cells normally and because the liver continuously makes glucose—via glycogenolysis and gluconeogenesis—and releases it into the blood. Another result of the insulin deficiency is marked lipolysis with resultant elevation of plasma glycerol and fatty acids. Marked ketone formation by the liver is also present.

If extreme, these metabolic changes culminate in the acute life-threatening emergency called *diabetic ketoacidosis* (Figure 18-12). Some of the problems are due to the effects that a markedly elevated plasma glucose concentration produces on renal function. In Chapter 16, we pointed out that a normal person does not excrete glucose because all glucose filtered at the renal corpuscle is reabsorbed by the tubules. However, the elevated plasma glucose of diabetes may so increase the filtered load of glucose that the maximum tubular reabsorptive capacity is exceeded and large amounts of glucose are excreted. For the same reasons, large amounts of ketones may also appear in the urine. These urinary losses aggravate the situation by depleting the body of nutrients and leading to weight loss. Far worse, however, is the fact that these unreabsorbed solutes cause an osmotic diuresis (Chapter 16)—marked urinary excretion of sodium and water, which can lead, by the sequence of events shown in Figure 18-12, to hypotension, brain damage, and death.

The other serious abnormality in diabetic ketoacidosis is the increased plasma hydrogen-ion concentration caused by the accumulation of ketones, two of which are acids. This increased hydrogen-ion concentration causes brain dysfunction that can contribute to the development of coma and death.

Diabetic ketoacidosis is seen only in patients with untreated IDDM, that is, those with almost total inability to secrete insulin. However, 85 percent of diabetics are in the NIDDM category and never develop metabolic derangements severe enough to go into diabetic ketoacidosis. NIDDM is a disease mainly of overweight adults, typically starting in middle life. Given the earlier mention of progressive weight loss in IDDM as a symptom of dia-

betes, it may seem contradictory that most people with NIDDM are overweight. The paradox is resolved when one realizes that people with NIDDM, in contrast to those with IDDM, do not excrete enough glucose in the urine to cause weight loss.

There are several factors that combine to cause NIDDM. One major problem is target-cell hyporesponsiveness to insulin, termed *insulin resistance.* Insulin's target cells do not respond adequately to circulating insulin because of alterations either in the insulin receptors or, much more commonly, an intracellular process occurring after receptor activation. Obesity accounts for much of the insulin resistance in NIDDM, for obesity in any person—diabetic or not—induces some degree of insulin resistance, particularly in adipose-tissue cells. (One theory is that the excess adipose tissue overproduces a messenger that causes downregulation of insulin-responsive glucose transporters.) However, additional components of insulin resistance, not related to obesity and not yet understood, also usually occur with NIDDM; for example, consumption of a high-fat diet may contribute to insulin resistance independent of weight gain.

Most people with NIDDM not only have insulin resistance but also have a defect in the ability of their beta cells to secrete insulin in response to a rise in plasma glucose concentration. In other words, although insulin resistance is the primary factor inducing hyperglycemia in NIDDM, an as-yet-unidentified defect in beta-cell function prevents these cells from responding to the hyperglycemia in normal fashion.

The major therapy for obese persons with NIDDM is weight reduction, since obesity is a major cause of insulin resistance. An exercise program also is very useful, because insulin responsiveness is increased by frequent endurance-type exercise, independent of changes in body weight. (This may reflect the fact that training causes a substantial increase in the total number of plasma-membrane glucose transporters in both skeletal muscles and adipocytes.)

If plasma glucose concentration is not adequately controlled by a program of weight reduction, exercise, and dietary modification (specifically low-fat diets), then the person has traditionally been given orally active drugs, called *sulfonylureas,* that lower plasma glucose by acting on the beta cells to stimulate insulin secretion. However, several new types of drugs have recently been introduced or are undergoing clinical trials: One stops the liver from converting glycogen and protein into glucose; another slows the gastrointestinal absorption of carbohydrates; and still another, by as-yet-unidentified mechanisms, decreases insulin resistance. Finally, in some cases the use of insulin itself is warranted.

Unfortunately, people with either IDDM or NIDDM tend to develop a variety of chronic abnormalities, includ-

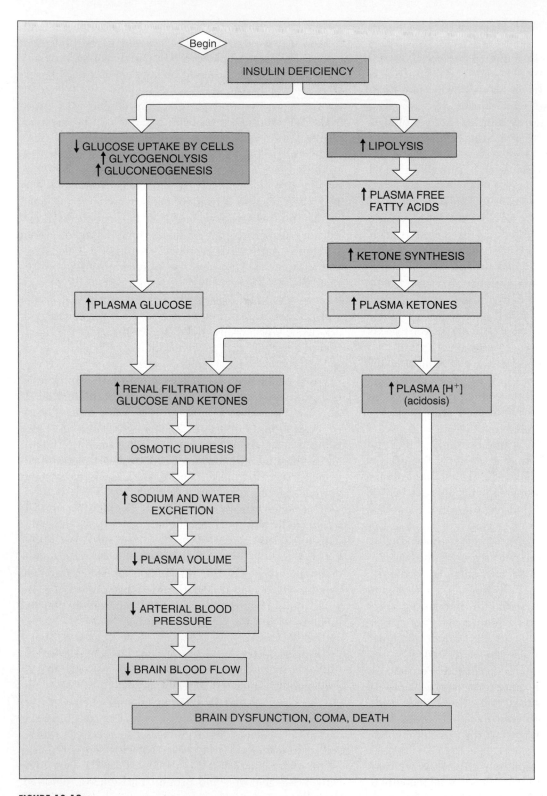

FIGURE 18-12

Diabetic ketoacidosis: Events caused by severe untreated insulin deficiency in insulin-dependent diabetes mellitus.

ing atherosclerosis, kidney failure, small-vessel and nerve disease, susceptibility to infection, and blindness. Elevated plasma glucose contributes to some, if not all, of these abnormalities by causing the intracellular accumulation of certain glucose metabolites that exert harmful effects on cells when present in high concentrations.

This discussion of diabetes has focused on insulin, but it is now clear that the hormones that elevate plasma glucose concentration may contribute to the severity of the disease. Glucagon is quite important in this regard. Since glucagon secretion is inhibited by an elevated plasma glucose level, one would expect to find a low plasma glucagon concentration in diabetic persons. However, most diabetics, particularly those with IDDM, have plasma glucagon concentrations that are either increased or unchanged (the mechanism is probably related to an influence of insulin on glucagon secretion). This absolute or relative glucagon excess contributes to the metabolic dysfunction typical of diabetes.

Finally, as we have seen, all the systems that raise plasma glucose concentration are activated during stress, which explains why stress exacerbates the symptoms of diabetes. Since diabetic ketoacidosis itself constitutes a severe stress, a positive-feedback cycle is triggered in which a marked lack of insulin induces ketoacidosis, which elicits activation of the glucose-counterregulatory systems, which worsens the ketoacidosis.

HYPOGLYCEMIA AS A CAUSE OF SYMPTOMS

As we have seen, "hypoglycemia" means a low plasma glucose concentration. Plasma glucose concentration can drop to very low values, usually during the postabsorptive state, in persons with several types of organic disorders. This is termed *fasting hypoglycemia,* and the relatively uncommon disorders responsible for it can be understood in terms of the regulation of blood glucose concentration. They include (1) an excess of insulin due to an insulin-producing tumor, a drug that stimulates insulin secretion, or the taking of too much insulin by a diabetic; and (2) a defect in one or more of the glucose-counterregulatory systems, for example, inadequate glycogenolysis and/or gluconeogenesis due to liver disease, glucagon deficiency, or cortisol deficiency.

Fasting hypoglycemia causes many symptoms. Some—increased heart rate, trembling, nervousness, sweating, and anxiety—are accounted for by activation of the sympathetic nervous system caused reflexly by the hypoglycemia. Other symptoms, such as headache, confusion, dizziness, uncoordination, and slurred speech, are direct consequences of too little glucose reaching the brain. More serious brain effects, including convulsions and coma, can occur if the plasma glucose concentration becomes low enough.

In contrast, low plasma glucose concentration has *not* been shown routinely to produce either acute or chronic symptoms of fatigue, lethargy, loss of libido, depression, or many other symptoms for which the lay press frequently holds it responsible. Despite this, a large number of persons who suffer from such symptoms, particularly several hours after eating, have been told or have assumed that the symptoms are due to so-called "functional" (or "reactive") hypoglycemia. Few of these people have ever had their blood glucose concentrations measured at the time of the symptoms, and the blood sugar is almost always within the normal range in those cases where measurements have been made. For all these reasons, most experts believe that most of the symptoms popularly ascribed to functional hypoglycemia have other causes.

REGULATION OF PLASMA CHOLESTEROL

In the previous section, we described the flow of lipids to and from adipose tissue in the form of fatty acids and triacylglycerols complexed with proteins. One very important lipid—**cholesterol**—was not mentioned earlier because it, unlike the fatty acids and triacylglycerols, serves not as a metabolic fuel but rather as a precursor for plasma membranes, bile salts, steroid hormones, and other specialized molecules. Thus, cholesterol has many important functions in the body. Unfortunately, it can also cause problems. Specifically, high plasma concentrations of cholesterol enhance the development of **atherosclerosis,** the arterial thickening that leads to heart attacks, strokes, and other forms of cardiovascular damage (Chapter 14).

A schema for cholesterol balance is illustrated in Figure 18-13. The two sources of cholesterol are dietary cholesterol and cholesterol synthesized within the body. Dietary cholesterol comes from animal sources, egg yolk being by far the richest in this lipid (a single egg contains about 250 mg of cholesterol). Not all ingested cholesterol is absorbed into the blood, however—much of it simply passes through the length of the gastrointestinal tract and is excreted in the feces.

What about cholesterol synthesis within the body? Almost all cells can synthesize some of the cholesterol required for their own plasma membranes, but most cannot do so in adequate amounts and depend upon receiving cholesterol from the blood. This is also true of the endocrine cells that produce steroid hormones from cholesterol. Thus, most cells *remove* cholesterol from the blood. In contrast, the liver and cells lining the gastrointestinal tract can produce large amounts of cholesterol, most of which *enters* the blood.

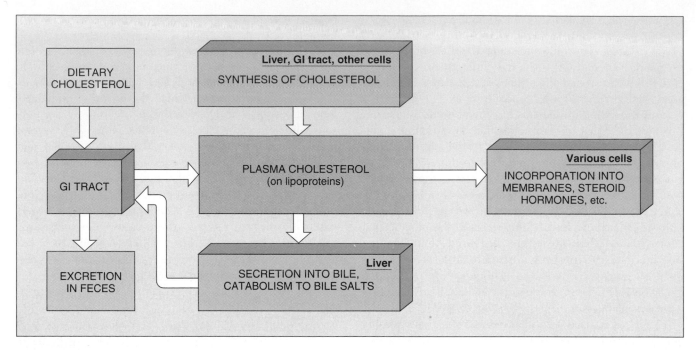

FIGURE 18-13
Cholesterol balance.

Now for the other side of cholesterol balance—the pathways, involving the liver, for net cholesterol loss from the body. First of all, some plasma cholesterol is picked up by liver cells and secreted into the bile, which carries it to the intestinal tract. Here it is treated much like ingested cholesterol, some being absorbed back into the blood and the remainder being excreted in the feces. Second, much of the cholesterol picked up by the liver cells is metabolized into bile salts (Chapter 17). After their production by the liver, these bile salts, like secreted cholesterol, flow through the bile duct into the small intestine. (As described in Chapter 17, many of these bile salts are then reclaimed by absorption back into the blood across the wall of the lower small intestine.)

The liver is clearly the center of the cholesterol universe, for it can add newly synthesized cholesterol to the blood or it can remove cholesterol from the blood, secreting it into the bile or metabolizing it to bile salts. The homeostatic control mechanisms that keep plasma cholesterol relatively constant operate on all of these hepatic processes, but the single most important response involves cholesterol production: The synthesis of cholesterol by the liver is inhibited whenever dietary cholesterol is increased. This is because cholesterol inhibits the enzyme critical for cholesterol synthesis by the liver.

Thus, as soon as the plasma cholesterol level starts rising because of increased cholesterol ingestion, hepatic synthesis is inhibited, and the plasma concentration remains close to its original value. Conversely, when dietary cholesterol is reduced and plasma cholesterol begins to fall, hepatic synthesis is stimulated (released from inhibition), and this increased production opposes any further fall. The sensitivity of this negative-feedback control of cholesterol synthesis differs greatly from person to person, but it is the major reason why, for most people, it is difficult to change plasma cholesterol very much in either direction by altering only dietary cholesterol. (A remarkable example of this is the documented case of a man who, due to a compulsion, had been eating 25 eggs a day for 15 years and yet had a relatively low plasma cholesterol; his intestinal absorption and hepatic synthesis of cholesterol were markedly below normal, and his conversion of cholesterol to bile salts was markedly elevated.)

Thus far, the relative constancy of plasma cholesterol has been emphasized. There are, however, environmental and physiological factors that can significantly alter plasma cholesterol concentrations. Perhaps the most important of these factors are the quantity and type of dietary fatty acids. Ingesting saturated fatty acids, the dominant fatty acids of animal fat (particularly high in red meats, most cheeses, and whole milk), raises plasma cholesterol. In contrast, eating either polyunsaturated fatty acids (the dominant plant fatty acids) or monounsaturated fatty acids such as those in olive or peanut oil, lowers plasma cholesterol. The various fatty acids exert their effects by altering cholesterol synthesis, excretion, and metabolism to bile salts. A variety of drugs now in common use also are capable of lowering plasma cholesterol by influencing one or more of the metabolic pathways for cholesterol—for example, inhibiting the critical enzyme for hepatic cholesterol synthesis—or by interfering with intestinal absorption of bile salts.

Based on studies of the relationship between plasma cholesterol levels and cardiovascular diseases, recent recommendations from the National Institutes of Health call a total plasma cholesterol below 200 mg/deciliter [a deciliter (dl) is 100 ml] "desirable," 200–239 mg/dl "borderline high," and 240 mg/dl or greater "high."

The story is more complicated than this, however, since not all plasma cholesterol has the same function or significance for disease. Like most other lipids, cholesterol circulates in the plasma as part of various lipoprotein complexes. These include chylomicrons (Chapter 17), VLDL (this chapter), **low-density lipoproteins (LDL),** and **high-density lipoproteins (HDL).** LDL are the main cholesterol carriers, and they deliver cholesterol to cells. LDL bind to plasma-membrane receptors specific for a protein component of the LDL, and the LDL are taken up by the cell. In contrast to LDL, HDL serve as acceptors of cholesterol from various tissues. They promote the removal of cholesterol from cells and its secretion into the bile by the liver. LDL cholesterol is often designated "bad" cholesterol since high levels of it in the plasma are associated with increased deposition of cholesterol in arterial walls and higher incidences of heart attacks; using the same criteria, HDL cholesterol has been designated "good" cholesterol. (The designation "bad" should not obscure the fact that LDL are essential for supplying cells with the cholesterol they require to synthesize cell membranes and, in the case of the gonads and adrenal glands, steroid hormones.)

The best single indicator of the likelihood of developing atherosclerotic heart disease is, therefore, not *total* plasma cholesterol but rather the *ratio* of plasma LDL-cholesterol to plasma HDL-cholesterol—the lower the ratio the lower the risk. Cigarette smoking, a known risk factor for heart attacks, lowers plasma HDL, whereas weight reduction (in overweight persons) and regular exercise increase it. Estrogen not only lowers total cholesterol and LDL but raises HDL, which explains, in part, why premenopausal women have so much less coronary artery disease than men. After menopause, the cholesterol values and coronary artery disease rates in women become similar to those in men.

Another complexity in evaluating plasma cholesterol involves a plasma substance called **lipoprotein(a),** which, when present in large quantities, accounts for much of the risk for coronary artery disease. This molecule is very similar in structure to LDL except that it contains one additional large protein dubbed apolipoprotein(a). Almost all people have some lipoprotein(a) in their blood, but its concentration varies, on a genetic basis, nearly 1,000-fold among individuals. The normal function of lipoprotein(a) and the mechanism by which an excess of it contributes to atherosclerosis remain unclear.

In summary, dietary and life-style changes, in combination with drug therapy, if necessary, can result in significant lowering of plasma cholesterol concentration and a decrease in the LDL:HDL ratio, with important consequences for reducing cardiovascular disease. For example, in a recent study of healthy men ages 45 through 64, use of a drug that inhibits hepatic cholesterol synthesis, in combination with dietary changes, significantly lowered total plasma cholesterol and LDL cholesterol, and raised HDL cholesterol; moreover, over 5 years the drug-treated group suffered 28 percent fewer fatal heart attacks than the control group given a placebo.

SECTION A SUMMARY

EVENTS OF THE ABSORPTIVE AND POSTABSORPTIVE STATES

I. During absorption, energy is provided primarily by absorbed carbohydrate, and net synthesis of glycogen, triacylglycerol, and protein occurs.

   A. Some absorbed carbohydrate not used for energy is converted to glycogen, mainly in the liver and skeletal muscle, but most is converted, in liver and adipocytes, to α-glycerol phosphate and fatty acids, which then combine to form triacylglycerol. The liver releases its triacylglycerols in very low density lipoproteins, the fatty acids of which are picked up by adipocytes.

   B. The fatty acids of some absorbed triacylglycerol are used for energy, but most are rebuilt into fat in adipose tissue.

   C. Some absorbed amino acids are converted to proteins, but excess amino acids are converted to carbohydrate and fat.

   D. There is a net uptake of glucose by the liver.

II. In the postabsorptive state, blood glucose level is maintained by a combination of glucose production by the liver and a switch from glucose utilization to fatty acid and ketone utilization by most tissues.

   A. Synthesis of glycogen, fat, and protein is curtailed, and net breakdown of these molecules occurs.

   B. The liver forms glucose by glycogenolysis of its own glycogen and by gluconeogenesis from lactate and pyruvate (from breakdown of muscle glycogen), glycerol (from adipose-tissue lipolysis), and amino acids (from protein catabolism).

C. Glycolysis is decreased, and most of the body's energy supply comes from the oxidation of fatty acids released by adipose-tissue lipolysis and of ketones produced from fatty acids by the liver.

D. The brain continues to use glucose but also starts using ketones as they build up in the blood.

ENDOCRINE AND NEURAL CONTROL OF THE ABSORPTIVE AND POSTABSORPTIVE STATES

I. The major hormones secreted by the pancreatic islets of Langerhans are insulin by the beta cells and glucagon by the alpha cells.

II. Insulin is the most important hormone controlling metabolism.

A. In muscle, it stimulates glucose uptake, glycolysis, and net synthesis of glycogen and protein; in adipose tissue, it stimulates glucose uptake and net synthesis of triacylglycerol; in liver, it inhibits gluconeogenesis and glucose release and stimulates the net synthesis of glycogen and triacylglycerols.

B. The major stimulus for insulin secretion is an increased plasma glucose concentration, but secretion is also influenced by many other factors, which are summarized in Figure 18-8.

III. Glucagon, epinephrine, cortisol, and growth hormone all exert effects on carbohydrate and lipid metabolism that are opposite, in one way or another, to those of insulin. They raise plasma concentrations of glucose, glycerol, and fatty acids.

A. Glucagon's physiological actions are all on the liver, where it stimulates glycogenolysis, gluconeogenesis, and ketone synthesis.

B. The major stimulus for glucagon secretion is hypoglycemia, but secretion is also stimulated by other inputs, including the sympathetic nerves to the islets.

C. Epinephrine released from the adrenal medulla in response to hypoglycemia stimulates glycogenolysis in the liver and muscle, gluconeogenesis in liver, and lipolysis in adipocytes. The sympathetic nerves to liver and adipose tissue exert effects similar to those of epinephrine.

D. Cortisol is permissive for gluconeogenesis and lipolysis; in higher concentrations, it stimulates gluconeogenesis and blocks glucose uptake. These last two effects are also exerted by growth hormone.

FUEL HOMEOSTASIS IN EXERCISE AND STRESS

I. During exercise, the muscles use as their energy sources plasma glucose, plasma fatty acids, and their own glycogen.

A. Glucose is produced by the liver, and fatty acids are provided by adipose-tissue lipolysis.

B. The changes in plasma insulin, glucagon, and epinephrine are similar to those that occur during the postabsorptive period and are mediated mainly by the sympathetic nervous system.

II. Stress causes hormonal changes similar to those caused by exercise.

DIABETES MELLITUS

I. Insulin-dependent diabetes is due to absolute insulin deficiency and can lead to diabetic ketoacidosis.

II. Non-insulin-dependent diabetes is usually associated with obesity and is caused by a combination of insulin resistance and a defect in beta-cell responsiveness to elevated plasma glucose concentration. Plasma insulin concentration is usually normal or elevated.

REGULATION OF PLASMA CHOLESTEROL

I. Plasma cholesterol is a precursor for the synthesis of plasma membranes, bile salts, and steroid hormones.

II. Cholesterol synthesis by the liver is controlled so as to homeostatically regulate plasma cholesterol concentration; it varies inversely with ingested cholesterol.

III. The liver also secretes cholesterol into the bile and converts it to bile salts.

IV. Plasma cholesterol is carried mainly by low-density lipoproteins, which deliver it to cells; high-density lipoproteins carry cholesterol from cells to the liver. The LDL/HDL ratio correlates with the incidence of coronary heart disease.

SECTION A KEY TERMS

absorptive state	alpha cells
postabsorptive state	insulin
very low density lipoproteins (VLDL)	hypoglycemia
lipoprotein lipase	glucose-counterregulatory controls
keto acids	glucagon
glycogenolysis	cholesterol
lipolysis	low-density lipoproteins (LDL)
gluconeogenesis	
glucose sparing	high-density lipoproteins (HDL)
ketones	
islets of Langerhans	lipoprotein(a)
beta cells	

SECTION A REVIEW QUESTIONS

1. Using a diagram, summarize the events of the absorptive period.

2. In what two organs does major glycogen storage occur?

3. How do the liver and adipose tissue metabolize glucose during the absorptive period?

4. How does adipose tissue metabolize absorbed triacyl-

glycerol, and what are the three major sources of the fatty acids in adipose tissue triacylglycerol?

5. What happens to most of the absorbed amino acids when a high-protein meal is ingested?

6. Using a diagram, summarize the events of the postabsorptive period; include the four sources of blood glucose and the pathways leading to ketone formation.

7. Distinguish between the roles of glycerol and free fatty acids during fasting.

8. List the overall responses of muscle, adipose tissue, and liver to insulin. What effects occur when plasma insulin concentration decreases?

9. List five inputs controlling insulin secretion, and state the physiological significance of each.

10. List the effects of glucagon on the liver and their consequences.

11. List two inputs controlling glucagon secretion, and state the physiological significance of each.

12. List four metabolic effects of epinephrine and the sympathetic nerves to the liver and adipose tissue, and state the net results of each.

13. List the permissive effects of cortisol and the effects that occur when plasma cortisol concentration increases.

14. List three effects of growth hormone on carbohydrate and lipid metabolism.

15. Which hormones stimulate gluconeogenesis? Glycogenolysis in liver? Glycogenolysis in skeletal muscle? Lipolysis? Blockade of glucose uptake?

16. Describe how plasma glucose, insulin, glucagon, and epinephrine levels change during exercise and stress. What causes the changes in the concentrations of the hormones?

17. Describe the metabolic disorders of severe insulin-dependent diabetes.

18. How does obesity contribute to non-insulin-dependent diabetes?

19. Hypersecretion of which hormones can induce a diabetic state?

20. Using a diagram, describe the sources of cholesterol gain and loss. Include three roles of the liver in cholesterol metabolism, and state the controls over these processes.

21. What are the effects of saturated and unsaturated fatty acids on plasma cholesterol?

22. What is the significance of the ratio of LDL cholesterol to HDL cholesterol?

SECTION B
CONTROL OF GROWTH

Growth is a complex process influenced by genetics, endocrine function, and a variety of environmental factors, including nutrition and the presence of infection. The process involves cell division and net protein synthesis throughout the body, but a person's height is determined specifically by bone growth, particularly of the vertebral column and legs.

BONE GROWTH

As described in Chapter 16, bone is a living tissue consisting of a protein (collagen) matrix upon which calcium salts, particularly calcium phosphates, are deposited. A growing long bone is divided, for descriptive purposes, into the ends, or **epiphyses,** and the remainder, the **shaft.** The portion of each epiphysis that is in contact with the shaft is a plate of actively proliferating cartilage, the **epiphyseal growth plate** (Figure 18-14). **Osteoblasts,** the bone-forming cells (Chapter 16), at the shaft edge of the epiphyseal growth plate convert the cartilaginous tissue at this edge to bone while new cartilage is simultaneously being laid down in the interior of the plate by cells called **chondrocytes.** In this manner, the epiphyseal growth plate remains intact (indeed, actually widens) and is gradually pushed away from the center of the bony shaft as the latter lengthens.

Linear growth of the shaft can continue as long as the epiphyseal growth plates exist, but ceases when the plates are themselves ultimately converted to bone as a result of hormonal influences at puberty. This is known as **epiphyseal closure** and occurs at different times in different bones. Accordingly, a person's **bone age** can be determined by x-raying the bones and determining which ones have undergone epiphyseal closure.

As shown in Figure 18-15, children manifest two periods of rapid body growth, one during the first two years of life and the second during puberty. Note in this figure that linear growth is not necessarily correlated with the rates of growth of specific organs.

The pubertal growth spurt lasts several years in both sexes, but growth during this period is greater in boys. This, plus the fact that boys grow more before puberty

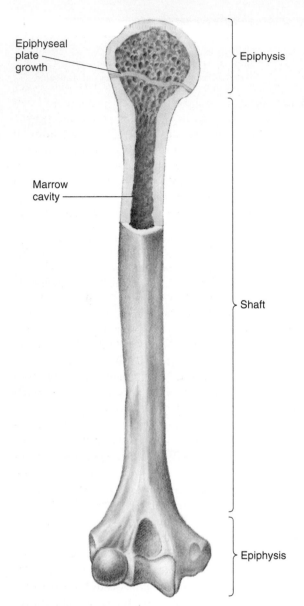

FIGURE 18-14
Anatomy of a long bone during growth.

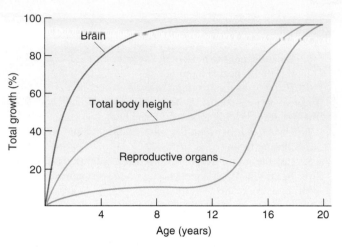

FIGURE 18-15
Relative growth in brain, total body height (a measure of long-bone and vertebral growth), and reproductive organs. Note that brain growth is nearly complete by the age 5, whereas maximal height and reproductive-organ size are not reached until the late teens.

because they begin puberty approximately two years later than girls, accounts for the differences in average height between men and women.

ENVIRONMENTAL FACTORS INFLUENCING GROWTH

Adequacy of nutrient supply and freedom from disease are the primary environmental factors influencing growth. Lack of sufficient amounts of any of the essential amino acids, essential fatty acids, vitamins, or minerals interferes with growth. Total protein and sufficient nutrients to provide energy must also be adequate. No matter how much protein is ingested, growth cannot be normal if the intake of all energy-providing nutrients is too low since the protein is simply catabolized for energy.

The growth-inhibiting effects of malnutrition can be seen at any time of development but are most profound when they occur very early in life. Thus, maternal malnutrition may cause growth retardation in the fetus. Since low birth weight is strongly associated with increased infant mortality, prenatal malnutrition causes increased numbers of prenatal and early postnatal deaths. Moreover, irreversible stunting of brain development may be caused by prenatal malnutrition. During infancy and childhood, too, malnutrition can interfere with both intellectual development and total-body growth.

Sickness can also stunt growth, but if the illness is temporary, the recovered child manifests a remarkable growth spurt ("catch-up" growth) that rapidly brings him or her up to the normal height expected for his or her age.

HORMONAL INFLUENCES ON GROWTH

The hormones most important to human growth are growth hormone, thyroid hormones, insulin, testosterone, and estrogens, all of which exert widespread effects. In addition to all these hormones, there is a huge group of peptide **growth factors** (for example, nerve growth factor), each of which is highly effective in stimulating differentiation and/or cell division of certain cell types. The general term for a chemical that stimulates cell division is a **mitogen.** There are also peptide **growth-inhibiting factors** that modulate growth by inhibiting cell division in specific tissues. Numbering more than 60 at present, these growth factors and growth-inhibiting factors are usually produced by multiple cell types rather than by discrete endocrine glands. Indeed, many of them are produced and released in the immediate vicinity of their sites of action and so are categorized as paracrine or autocrine agents.

The physiology of growth factors and growth-inhibiting factors is important not just for understanding control of normal growth but also because these factors may be involved in the development of *cancer.* Thus, some **oncogenes** (genes that are involved in causing cancer, Chapter 5) code for proteins that are identical to or very similar to growth factors, growth-factor receptors, or post-receptor components of growth factor signal transduction pathways. The problem is that these proteins have lost important regulatory constraints on their activity. For example, one oncogene codes for a version of the receptor for epidermal growth factor that is always in the activated state even in the absence of the growth factor. This activated receptor imparts a continuous growth signal to the cells containing it.

The various hormones and growth factors do not all stimulate growth at the same periods of life. For example, fetal growth is largely independent of the major hormones that stimulate growth during childhood—growth hormone, the thyroid hormones, and the sex steroids.

Growth Hormone and Insulin-like Growth Factor I

Growth hormone, secreted by the anterior pituitary, has little or no effect on fetal growth, but it is the single most important hormone for postnatal growth. Its major growth-promoting effect is stimulation (indirect, as we shall see) of cell division in its many target tissues. Thus, growth hormone promotes bone lengthening by stimulating maturation and cell division of the chondrocytes in the epiphyseal plates, thereby continuously widening the plates and providing more cartilaginous material for bone formation.

An excess of growth hormone during childhood produces **giantism,** whereas deficiency produces **dwarfism.** When excess growth hormone is secreted in adults after epiphyseal closure, it cannot lengthen the bones further, but it does produce the disfiguring bone thickening and overgrowth of other organs known as **acromegaly.**

Importantly, growth hormone exerts its cell division-stimulating (mitogenic) effect not *directly* on cells but rather *indirectly* through the mediation of a chemical messenger whose synthesis and release are induced by growth hormone. This messenger is called **insulin-like growth factor I (IGF-I)** (also known as somatomedin C). Under the influence of growth hormone, IGF-I is secreted by the liver and many other types of cells, and it is thought to act at these sites as a paracrine or autocrine agent. It remains controversial as to whether IGF-I also exerts some functions as a hormone, that is, as a blood-borne messenger. (Also controversial is the role of a second insulin-like growth factor, IGF-II.)

Current concepts of how growth hormone and IGF-I interact on the epiphyseal plates of bone are as follows: (1) Growth hormone stimulates the chondrocyte precursor cells (prechondrocytes) and/or young differentiating chondrocytes in the epiphyseal plates to differentiate into chondrocytes; (2) during this differentiation, the cells begin both to secrete IGF-I and to become responsive to IGF-I; (3) the IGF-I then acts as an autocrine or paracrine agent to stimulate the differentiating chondrocytes to undergo cell division.

It is likely that a similar interplay between growth hormone and IGF-I underlies the ability of growth hormone to stimulate growth in target tissues other than bone; that is, growth hormone stimulates precursor cells to differentiate and to secrete IGF-I, and IGF-I stimulates the newly differentiated cells to undergo cell division.

The importance of IGF-I in mediating the major growth-promoting effect of growth hormone is illustrated by the fact that dwarfism can be due not only to decreased secretion of growth hormone but also to decreased production of IGF-I or failure of the tissues to respond to IGF-I. For example, the short stature of Pygmies is probably due to a genetic mutation that impairs the ability of the cells to produce IGF-I in response to growth hormone.

Another reminder of the importance of events beyond the secretion of a hormone comes from the fact that some children with short stature have been found to have normal plasma levels of growth hormone but inability to respond normally to the hormone because of mutations in the growth hormone receptor.

The secretion and activity of IGF-I can be influenced by the nutritional status of the individual and by many hormones other than growth hormone. For example, malnutrition during childhood inhibits the production of IGF-I even though plasma growth hormone concentration is elevated and should be stimulating IGF-I secretion. To take another example, estrogen stimulates the secretion of IGF-I by cells of the uterus and ovaries.

In addition to its specific growth-promoting effect on cell division via IGF-I, growth hormone directly stimulates protein synthesis in various tissues and organs. It does this by increasing amino acid uptake by cells and the synthesis of RNA and ribosomes. All these events are essential for protein synthesis. This anabolic effect on

TABLE 18-5 MAJOR EFFECTS OF GROWTH HORMONE

1. Promotes growth: Induces precursor cells to differentiate and secrete insulin-like growth factor I (IGF-I), which stimulates cell division
2. Stimulates protein synthesis
3. Anti-insulin effects:
   a. Renders adipocytes more responsive to lipolytic stimuli
   b. Stimulates gluconeogenesis
   c. Reduces the ability of insulin to stimulate glucose uptake

protein metabolism facilitates the ability of tissues and organs to enlarge. Table 18-5 summarizes the multiple effects of growth hormone, all of which have been described in this chapter.

The control of growth hormone secretion was described in Chapter 10 (Figure 10-20). Briefly, the control system begins with two of the hypophysiotropic hormones secreted by the hypothalamus. Growth hormone secretion is stimulated by growth hormone releasing hormone (GHRH) and inhibited by somatostatin. As a result of changes in these two signals, which are virtually 180 degrees out of phase with each other (that is, one is high when the other is low), growth hormone secretion occurs in episodic bursts and manifests a striking diurnal rhythm. During most of the day, there is little or no growth hormone secreted, although bursts may be elicited by certain stimuli, including stress, hypoglycemia, and exercise. In contrast, 1 to 2 h after a person falls asleep, one or more larger, prolonged bursts of secretion may occur.

In addition to the hypothalamic controls, a variety of hormones—notably the sex hormones and thyroid hormones, as described below—influence the secretion of growth hormone. The net result of all these inputs is that the total 24-h secretion rate of growth hormone is highest during adolescence (the period of most rapid growth), next highest in children, and lowest in adults. The decreased growth hormone secretion associated with aging is responsible, in part, for the decrease in lean-body and bone mass, the expansion of adipose tissue, and the thinning of the skin that occur at that time.

The availability of large quantities of human growth hormone produced by recombinant-DNA technology has greatly facilitated the treatment of children with short stature due to deficiency of growth hormone. Controversial at present is the administration of growth hormone to short children who do not have growth-hormone deficiency, to athletes in an attempt to increase muscle mass, and to elderly persons to reverse the growth-hormone-related aging changes described in the previous paragraph.

Thyroid Hormones

The thyroid hormones (TH)—thyroxine (T_4) and triiodothyronine (T_3)—are essential for normal growth because they are required for both the synthesis of growth hormone and the growth-promoting effects of that hormone. Accordingly, infants and children with **hypothyroidism** (deficient thyroid function) manifest retarded growth due to slowed bone growth.

Quite distinct from its growth-promoting effect, TH is permissive for normal development of the central nervous system during fetal life. Inadequate production of maternal and fetal TH due to severe iodine deficiency during pregnancy is one of the world's most common preventable causes of mental retardation, termed **endemic cretinism.**

This effect on brain *development* must be distinguished from other stimulatory effects TH exerts on the nervous system throughout life, not just during infancy. A hypothyroid person exhibits sluggishness and poor mental function, and these effects are completely reversible at any time with administration of TH. Conversely, a person with **hyperthyroidism** (excessive secretion of TH) is jittery and hyperactive.

Insulin

It should not be surprising that adequate amounts of insulin are necessary for normal growth since insulin is, in all respects, an anabolic hormone. Its stimulatory effects on amino acid uptake and protein synthesis are particularly important with regard to growth.

In addition to this general anabolic effect, however, insulin exerts direct, specific growth-promoting effects on cell differentiation and cell division during fetal life (and possibly during childhood). Moreover, insulin is required for normal production of IGF-I.

Sex Hormones

As will be described in Chapter 19, sex hormone secretion (testosterone in the male and estrogen in the female) begins in earnest between the ages of 8 and 10 and progressively increases to reach a plateau over the next 5 to 10 years. A normal pubertal growth spurt, which reflects growth of the long bones and vertebrae, requires this increased production of the sex hormones. The major growth-promoting effect of the sex hormones is to stimulate the secretion of growth hormone and IGF-I.

Unlike growth hormone, however, the sex hormones ultimately *stop* bone growth by inducing epiphyseal closure. The dual effects of the sex hormones explain the pattern seen in adolescence—rapid lengthening of the bones culminating in complete cessation of growth for life.

In addition to these dual effects on bone, testosterone, but not estrogen, exerts a direct anabolic effect on protein synthesis in many nonreproductive organs and tissues of the body. This accounts, at least in part, for the increased muscle mass of men, compared with women.

This also is why synthetic testosterone-like agents termed **anabolic steroids** (or, more recently, testosterone itself) are sometimes used by athletes—both male and female—in an attempt to increase their muscle mass and strength. However, these steroids have multiple potential toxic side effects (for example, liver damage and increased risk of prostate cancer). Moreover, in females they produce masculinization.

Cortisol

Cortisol, the major hormone secreted by the adrenal cortex in response to stress, can have potent *antigrowth* effects under certain conditions. When present in high concentration, it inhibits DNA synthesis and stimulates protein catabolism in many organs, and it inhibits bone growth. Moreover, it causes bone breakdown by inhibiting osteoblasts and stimulating osteoclasts (Chapter 16). It also inhibits the secretion of growth hormone. For all these reasons, in children, the elevation in plasma cortisol that accompanies infections and other stresses is, at least in part, responsible for the retarded growth that occurs with illness.

As we shall see in Chapter 20, cortisol and very similar steroids are commonly used medically in persons with arthritis or other inflammatory disorders. A side effect of such treatment is increased protein catabolism and bone breakdown. One must carefully distinguish cortisol-type steroids (glucocorticoids) from testosterone-type steroids (anabolic steroids).

This completes our survey of the major hormones that affect growth. Their actions are summarized in Table 18-6.

COMPENSATORY GROWTH

We have dealt thus far only with growth during *childhood*. During adult life, a specific type of regenerative growth, known as **compensatory growth,** can occur in many human organs. For example, after the surgical removal of one kidney, the cells of the other kidney begin to manifest increased cell division, and the kidney ultimately grows until its total mass approaches the initial mass of the two kidneys combined. Many growth factors are known to participate in compensatory growth, but the precise signals that trigger the process are not known.

TABLE 18-6   MAJOR HORMONES INFLUENCING GROWTH

Hormone	Principal actions
Growth hormone	Major stimulus of postnatal growth: Induces precursor cells to differentiate and secrete insulin-like growth factor I (IGF-I), which stimulates cell division Stimulates protein synthesis
Insulin	Stimulates fetal growth Stimulates postnatal growth by stimulating secretion of IGF-I Stimulates protein synthesis
Thyroid hormones	Permissive for growth hormone's secretion and actions Permissive for development of the central nervous system
Testosterone	Stimulates growth at puberty, in large part by stimulating the secretion of growth hormone Causes eventual epiphyseal closure Stimulates protein synthesis
Estrogen	Stimulates the secretion of growth hormone at puberty Causes eventual epiphyseal closure
Cortisol	Inhibits growth Stimulates protein catabolism

BONE GROWTH

I. A bone lengthens as osteoblasts at the shaft edge of the epiphyseal growth plates convert cartilage to bone while new cartilage is being laid down in the plate.

II. Growth ceases when the plates are completely converted to bone.

ENVIRONMENTAL FACTORS INFLUENCING GROWTH

I. The major environmental factors influencing growth are nutrition and disease.

II. Malnutrition during in utero life and infancy may produce irreversible stunting.

HORMONAL INFLUENCES ON GROWTH

I. Growth hormone is the major stimulus of postnatal growth.
   - A. It stimulates the release of IGF-I from the liver and many other cells, and IGF-I then acts locally to stimulate cell division.
   - B. Growth hormone also acts directly on cells to stimulate protein synthesis.
   - C. Growth hormone secretion is highest during adolescence.

II. Because thyroid hormones are required for growth hormone synthesis and the growth-promoting effects of this hormone, they are essential for normal growth during childhood and adolescence. They also are permissive for brain development during infancy.

III. Insulin stimulates growth mainly during in utero life.

IV. Mainly by stimulating growth hormone secretion, testosterone and estrogen promote bone growth during adolescence, but these hormones also cause epi-physeal closure. Testosterone also stimulates protein synthesis.

V. Cortisol in a high concentration inhibits growth and stimulates protein catabolism.

epiphysis	growth factor
shaft	mitogen
epiphyseal growth plate	growth-inhibiting factor
osteoblast	insulin-like growth factor I
chondrocyte	(IGF-I)
epiphyseal closure	compensatory growth
bone age	

1. Describe the process by which bone is lengthened.
2. What are the effects of malnutrition on growth?
3. List the major hormones that control growth.
4. Describe the relationship between growth hormone and IGF-I and the roles of each in growth.
5. What are the effects of growth hormone on protein synthesis?
6. What is the status of growth hormone secretion at different stages of life?
7. State the effects of the thyroid hormones on growth and development.
8. Describe the effects of testosterone on growth, cessation of growth, and protein synthesis. Which of these effects are shared by estrogen?
9. What is the effect of cortisol on growth?

SECTION C
REGULATION OF TOTAL-BODY ENERGY BALANCE AND TEMPERATURE

BASIC CONCEPTS OF ENERGY EXPENDITURE

The breakdown of organic molecules liberates the energy locked in their molecular bonds. This is the energy cells use to perform the various forms of biological work—muscle contraction, active transport, and molecular synthesis. The first law of thermodynamics states that energy can be neither created nor destroyed, but can be converted from one form to another. Thus, internal energy liberated (ΔE) during breakdown of an organic molecule can either appear as heat (H) or be used to perform work (W).

$$\Delta E = H + W$$

During metabolism, about 60 percent of the energy released from organic molecules appears immediately as heat, and the rest is used for work. The energy used for work must first be incorporated into molecules of ATP, the subsequent breakdown of which serves as the immediate energy source for the work. It is essential to realize that the body is not a heat engine since it is totally incapable of converting heat to work, but the heat released in its chemical reactions is valuable for maintaining body temperature.

Biological work can be divided into two general categories: (1) **external work**—movement of external objects by contracting skeletal muscles; and (2) **internal work**—all other forms of work, including skeletal-muscle activity not used in moving external objects. As just stated, much of the energy liberated from nutrient catabolism appears immediately as heat. What may not be obvious is that all internal work, too, is ultimately transformed to heat except during periods of growth. For example, internal work is performed during cardiac contraction, but this energy appears ultimately as heat generated by the friction of blood flow through the blood vessels.

Thus, the total energy liberated when organic nutrients are catabolized by cells may be transformed into body heat, appear as external work, or be stored in the body in the form of organic molecules. The **total energy expenditure** of the body is therefore given by the equation

Total energy expenditure = heat produced by the body
+ external work done + energy stored

Metabolic Rate

The unit for energy is the **kilocalorie (kcal),** which is the amount of heat required to raise the temperature of one liter of water one degree celsius. (In the field of nutrition, the three terms "Calorie" with a capital C, "large calorie", and "kilocalorie" are synonyms; they are all 1000 "calories," with a small c.) Total energy expenditure per unit time is called the **metabolic rate,** which can be measured directly or indirectly. In either case, the analysis is much simpler if the person is fasting and at rest. Total energy expenditure then becomes equal to heat production since energy storage and external work are eliminated.

The most widespread method for measuring metabolic rate is indirect. One simply measures the subject's oxygen uptake per unit time by measuring total ventilation and the oxygen content of both inspired and expired air. From this value one calculates heat production based on the fundamental principle that the energy liberated by the catabolism of foods inside the body must be the same as that liberated when the foods are catabolized to the same products outside the body. We know precisely how much heat is liberated when 1 L of oxygen is utilized in the oxidation of fat, protein, and carbohydrate outside the body; this same quantity of heat (approximately 4.8 kcal/L of oxygen, regardless of the nutrient being used) must be produced when 1 L of oxygen is utilized in the body.

Determinants of Metabolic Rate

Basal Metabolic Rate. Since many factors cause the metabolic rate to vary (Table 18-7), the indirect method for evaluating it, described above, specifies certain standardized conditions, and measures what is known as the **basal metabolic rate (BMR).** The subject is at mental

TABLE 18-7 SOME FACTORS AFFECTING THE METABOLIC RATE
Age
Sex
Height, weight, and body surface area
Growth
Pregnancy, menstruation, lactation
Infection or other disease
Body temperature
Recent ingestion of food
Prolonged alteration in amount of food intake
Muscular activity
Emotional state
Sleep
Environmental temperature
Circulating levels of various hormones, especially epinephrine and thyroid hormone

and physical rest in a room at a comfortable temperature and has not eaten for at least 12 h. These conditions are arbitrarily designated "basal," even though the metabolic rate during sleep may be less than the BMR. For the following discussion, it must be emphasized that the term "BMR" can be applied to a person's metabolic rate only when the specified conditions are met; thus, a person who has recently eaten or is exercising has a metabolic rate but not a *basal* metabolic rate.

The BMR is often termed the "metabolic cost of living," and most of it is expended by the heart, liver, kidneys, and brain. The magnitude of a person's BMR is related not only to physical size but to age and sex as well. The growing child's BMR, expressed on a per weight basis, is considerably higher than the adult's, in part because the child expends a great deal of energy in net synthesis of new tissue (much of this energy remains in the body as stored energy). On the other end of the age scale, the BMR gradually decreases with advancing age. A woman's BMR is generally less than that of a man, even taking into account size differences, but increases markedly during pregnancy and lactation. Greater demands upon the body made by infection or other disease generally increase the BMR. However, when a person suffers wasting because of infection, the BMR may decrease below normal.

Thyroid Hormones. The thyroid hormones are the single most important determinant of BMR regardless of size, age, or sex. TH increases the oxygen consumption and heat production of most body tissues, a notable exception being the brain. This ability to increase BMR is termed a **calorigenic effect.** The mechanism of the TH calorigenic effect is presently uncertain.

Long term excessive TH, as in persons with hyperthyroidism, induces a host of effects secondary to the calorigenic effect. For example, the increased metabolic demands markedly increase hunger and food intake; the greater intake frequently remains inadequate to meet the metabolic needs, and net catabolism of protein and fat stores leads to loss of body weight. Also, the greater heat production activates heat-dissipating mechanisms (skin vasodilation and sweating), and the person suffers from marked intolerance to warm environments. In contrast, the hypothyroid individual complains of cold intolerance.

The calorigenic effect of TH is only one of a bewildering variety of effects exerted by these hormones. With one exception—facilitation of the activity of the sympathetic nervous system (described in Chapter 10), major functions of the thyroid hormones have all been described earlier in this chapter and are listed for reference in Table 18-8. Control of the secretion of these hormones, as well as their metabolism, is described in Chapter 10.

Epinephrine. Epinephrine is another hormone that exerts a calorigenic effect. (This effect may be related to the hormone's stimulation of glycogen and triacylglycerol catabolism, since ATP splitting and energy liberation occur in both the breakdown and subsequent resynthesis of these molecules.) Thus, when epinephrine secretion by the adrenal medulla is stimulated, the metabolic rate rises. This accounts for part of the greater heat production associated with emotional stress, although increased muscle tone also contributes.

Food-Induced Thermogenesis. The ingestion of food rapidly increases the metabolic rate by 10 to 20 percent for a few hours after eating. This effect is known as **food-induced thermogenesis.** Ingested protein produces the greatest effect, carbohydrate and fat, less. Most of the increased heat production is secondary to processing of the absorbed nutrients by the liver, not to the energy expended by the gastrointestinal tract in digestion and absorption. It is to avoid the contribution of food-induced thermogenesis that BMR tests are performed in the postabsorptive state.

To reiterate, food-induced thermogenesis is the *rapid* increase in energy expenditure in response to ingestion of a meal. As we shall see, *prolonged* alterations in food intake (either increased or decreased total calories) also have significant effects on metabolic rate but are not termed food-induced thermogenesis.

Muscle Activity. The factor that can most increase metabolic rate is altered skeletal-muscle activity. Even minimal increases in muscle contraction significantly increase metabolic rate, and strenuous exercise may raise heat production more than fifteenfold (Table 18-9). Thus, depending on the degree of physical activity, total energy expenditure may vary for a normal young adult from a value of approximately 1500 kcal/24 h to more than 7000 kcal/24 h (for a lumberjack). Changes in muscle activity also account in part for the changes in metabolic rate that occur during sleep (decreased muscle contraction), during exposure to a low environmental temperature (increased muscle contraction and shivering), and with strong emotions.

TABLE 18-8 MAJOR FUNCTIONS OF THE THYROID HORMONES (TH)

1. Required for normal maturation of the nervous system in the fetus and infant
   Deficiency: Mental retardation (cretinism)
2. Required for normal bodily growth because they facilitate the secretion of and response to growth hormone
   Deficiency: Deficient growth in children
3. Required for normal alertness and reflexes at all ages
   Deficiency: Mentally and physically slow and lethargic
   Excess: Restless, irritable, anxious, wakeful
4. Major determinant of the rate at which the body produces heat during the basal metabolic stage
   Deficiency: Low BMR, sensitivity to cold, decreased food appetite
   Excess: High BMR, sensitivity to heat, increased food appetite, increased catabolism of nutrients
5. Facilitates the activity of the sympathetic nervous system by stimulating the synthesis of one class of receptors (beta receptors) for epinephrine and norepinephrine
   Excess: Symptoms similar to those observed with activation of the sympathetic nervous system (for example, increased heart rate)

TABLE 18-9 ENERGY EXPENDITURE DURING DIFFERENT TYPES OF ACTIVITY FOR A 70-KG (154-LB) PERSON

Form of activity	Energy kcal/h
Lying still, awake	77
Sitting at rest	100
Typewriting rapidly	140
Dressing or undressing	150
Walking on level, 4.3 km/h (2.6 mi/h)	200
Bicycling on level, 9 km/h (5.5 mi/h)	304
Walking on 3 percent grade, 4.3 km/h (2.6 mi/h)	357
Sawing wood or shoveling snow	480
Jogging, 9 km/h (5.3 mi/h)	570
Rowing, 20 strokes/min	828

REGULATION OF TOTAL-BODY ENERGY STORES

The laws of thermodynamics dictate that the total energy expenditure (metabolic rate) of the body must equal the total energy intake. We have already identified the ultimate forms of energy expenditure: internal heat production, external work, and net molecular synthesis (energy storage). The source of input is the energy contained in ingested food. Therefore:

Energy from food intake = internal heat produced
+ external work + energy stored

Our equation includes no term for loss of fuel from the body via excretion of nutrients because, in normal persons, only negligible losses occur via the urine, feces, and as sloughed hair and skin. In certain diseases, however, the most important being diabetes, urinary losses of organic molecules may be quite large and would have to be included in the equation.

Let us now rearrange the equation to focus on energy storage:

Energy stored = energy from food intake
− (internal heat produced + external work)

Thus, whenever energy intake differs from the sum of internal heat produced and external work, changes in energy storage occur, that is, the total-body energy content increases or decreases (Table 18-10). Normally, as we have seen, energy storage, except in growing children, is mainly in the form of fat in adipose tissue.

It is worth emphasizing at this point that "body weight" and "total-body energy content" are not synonomous terms although there is a popular tendency to equate the two. Body weight is determined not only by the amount of fat, carbohydrate, and protein in the body, but also by the amounts of water, bone, and other minerals. For example, an individual can lose body weight quickly as the result of sweating or an unusual increase in urinary output, or can gain large amounts of weight as a result of water retention, as occurs, for example, during heart failure (Chapter 14). Moreover, even focusing only on the nutrients, a constant body weight does not mean that total-body energy content is constant. The reason is that 1 g of fat contains 9 kcal whereas 1 g of either carbohydrate or protein contains 4 kcal. Thus, for example, aging is usually associated with a gain of fat and a loss of protein; the result is that even though the person's body weight may stay constant, the total-body energy content has increased. Despite these qualifications, however, in the remainder of this chapter changes in body weight are equated with changes in total-body energy content and, more specifically, changes in body fat stores.

Body weight in adults is usually regulated around a relatively constant operating point. Theoretically this constancy can be achieved by reflexly adjusting caloric intake

TABLE 18-10 THREE POSSIBLE STATES OF ENERGY BALANCE

State	Result
1. Energy intake = (internal heat production + external work)	Body energy content remains constant (body fat content remains constant).
2. Energy intake > (internal heat production + external work)	Body energy content increases (body fat content increases).
3. Energy intake < (internal heat production + external work)	Body energy content decreases (body fat content decreases).

of energy expenditure in response to changes in body weight. It has long been assumed that regulation of caloric intake is by far the more important adjustment, and this process will be described in the next section. However, there is growing evidence that energy expenditure is also reflexly adjusted in response to changes in body weight, and we describe it first.

A typical demonstration of this process in human beings is as follows: Total daily energy expenditure was measured in nonobese subjects at their usual body weight and again after they were caused either to lose 10 percent of their body weight by underfeeding or to gain 10 percent by overfeeding. At their new body weights the overfed subjects manifested a large (15 percent) increase in both resting and nonresting energy expenditure, and the underfed subjects a similar decrease. These changes in energy expenditure were much greater than could be accounted for simply by the altered metabolic mass of the body or having to move a larger or smaller body.

The generalization that emerges from this and other similar studies is that a dietary-induced change in total-body energy stores triggers, in negative-feedback fashion, an alteration in energy expenditure that opposes the gain or loss of energy stores. This phenomenon helps explain why some dieters lose about 5 to 10 lbs of fat fairly easily and then become stuck at a plateau. It also helps explain why some very thin people have difficulty trying to gain much weight. The mechanisms that underlie the changes in energy expenditure and the pathways that trigger them are presently unclear. Another unsettled question is whether such "metabolic resistance" to changes in body weight persists indefinitely or is only a transient response to rapid changes in body weight.

Control of Food Intake

The control of food intake can be analyzed in the same way as any other biological control system. As the previous section emphasized, the variable being maintained relatively constant in this system is total body energy content, more specifically, total fat stores. Physiologists have hypothesized for more than 50 years that an essential component of such a control system must be a hormone synthesized by adipose-tissue cells themselves, and released from the cells in proportion to the amount of adipose tissue present. Recently this hormone has been identified—it is a protein called **leptin,** which acts on the hypothalamus to cause a reduction in food intake (perhaps by inhibiting the release of neuropeptide Y, a hypothalamic neurotransmitter that stimulates eating). Leptin also stimulates the metabolic rate and, therefore, probably plays an important role in the changes in energy expenditure that occur in response to overfeeding or underfeeding, as described in the previous section. Thus, as illustrated in Figure 18-16, leptin functions in a negative-feedback sys-

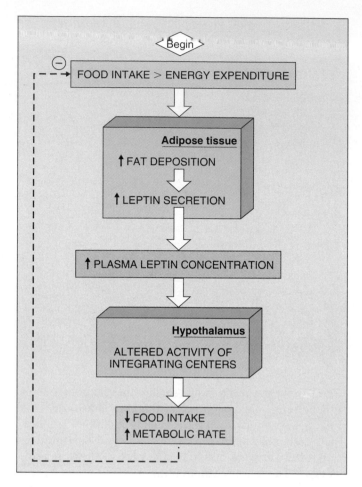

FIGURE 18-16

Role of leptin in the control of total-body energy stores.

tem to maintain total-body energy content constant by "telling" the brain how much fat is being stored.

It should be emphasized that leptin is crucial for *long-term* matching of caloric intake to energy expenditure. In addition, it is thought that various other signals act on the hypothalamus (and other brain areas) over short periods of time to regulate individual meal length and frequency (Figure 18-17). These **satiety signals** cause the person to cease feeling hungry and set the time period before hunger returns once again. For example, the rate of insulin-dependent glucose utilization by certain areas of the hypothalamus rises during eating, and this probably constitutes a satiety signal. Insulin, which increases during food absorption, may also act directly as a satiety signal. The increase in metabolic rate induced by eating tends to raise body temperature slightly, which acts as a satiety signal. Finally, there are satiety signals initiated by the presence of food within the gastrointestinal tract: These include neural signals triggered by stimulation of both stretch receptors and chemoreceptors in the stomach and duode-

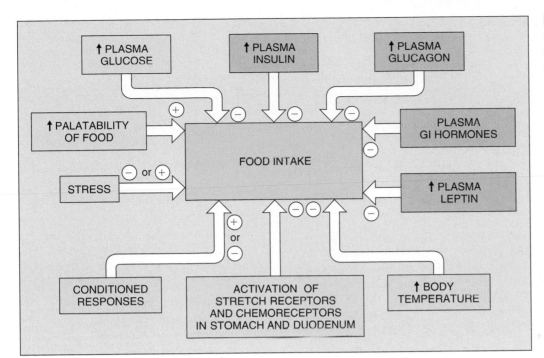

FIGURE 18-17

Inputs controlling food intake. The minus signs denote hunger suppression, and the plus signs denote hunger stimulation.

num, as well as by several of the hormones (cholecystokinin, for example) released from the stomach and duodenum during eating.

Food intake is also strongly influenced by the reinforcement, both positive and negative, of such things as smell, taste, and texture. For example, rats given a "cafeteria-style" choice of foods highly palatable to people overeat by 70 to 80 percent and soon become obese. An analogous (but reverse) experiment has shown that obese people lose weight when they must obtain all their food from a monotonous diet delivered via a tube in response to lever pressing. Thus, the behavioral concepts of reinforcement, drive, and motivation, described in Chapter 13, must be incorporated into any comprehensive theory of food-intake control.

Another significant factor that can increase food intake is stress, as demonstrated by controlled experiments in animals.

Obesity

We have been using the term obesity in this chapter rather loosely to denote a state of being overweight. However, *obesity* is formally defined as an excess of body fat resulting in a significant impairment of health from a variety of diseases, notably hypertension, atherosclerotic heart disease, and diabetes. Using this definition, a level of 20 percent or more above "desirable" body weight qualifies as obesity. But how does one determine desirable weight? The most commonly used ideal or desirable weight tables are based on mortality data and are put out by the Metropolitan Life Insurance Company (Table 18-11).

However, the presently preferred simple method for assessing degree of obesity is not the body weight per se but the **body mass index (BMI)**, which correlates quite well with the amount of body fat, as determined, for

TABLE 18-11 METROPOLITAN LIFE INSURANCE "DESIRABLE-WEIGHT" TABLE (1959), MEDIUM FRAME*

Men		Women	
Height	**Weight, lb**	**Height**	**Weight, lb**
5'2"	114–126	4'9"	94–106
5'3"	117–129	4'10"	97–109
5'4"	120–132	4'11"	100–112
5'5"	123–136	5'0"	103–115
5'6"	127–140	5'1"	106–118
5'7"	131–145	5'2"	109–122
5'8"	135–149	5'3"	112–126
5'9"	139–153	5'4"	116–131
5'10"	143–158	5'5"	120–135
5'11"	147–163	5'6"	127–140
6'0"	151–168	5'7"	128–143
6'1"	155–173	5'8"	132–147
6'2"	160–178	5'9"	136–151
6'3"	165–183	5'10"	140–155

*Height is without shoes, and weight is without clothes.

Tables issued in 1983 indicate that desirable weights are approximately 12 to 14 lb higher than those of these 1959 tables. Most experts, however, believe that the 1959 tables are more accurate predictors of health consequences.

ESSAY

FOOD AND HEALTH: WHAT SHOULD WE EAT?

For many years the major nutritional guidelines were the Recommended Dietary Allowances (RDAs) issued by the National Academy of Sciences/National Research Council. The RDAs were never meant to be recommendations for an "ideal" or "optimal" diet; rather they are the daily intakes of the various essential nutrients considered high enough to prevent specific clinical deficiencies; for example, the RDA for vitamin C is designed to prevent scurvy. The RDAs eventually came to be supplemented with recommendations to reduce dietary fat and cholesterol, largely because of the implication of these factors in coronary heart disease. In the last few years more and more dietary factors have been associated with the cause or prevention of many diseases, including not only coronary heart disease but hypertension, cancer, birth defects, osteoporosis, and a variety of other chronic diseases. These associations come mainly from animal studies, epidemiologic studies on people, and basic research concerning potential mechanisms. The problem is that the findings are often difficult to interpret and may be conflicting. To synthesize all this material in the form of simple clear recom-

SUMMARY OF NATIONAL RESEARCH COUNCIL DIETARY RECOMMENDATIONS

1. Reduce total fat intake to 30% or less of calories. Reduce saturated fatty acid intake to less than 10% of calories and the intake of cholesterol to less than 300 mg daily.
2. Every day eat five or more servings of a combination of vegetables and fruits, especially green and yellow vegetables and citrus fruits. Also, increase starches and other complex carbohydrates by eating six or more daily servings of a combination of breads, cereals, and legumes.
3. Maintain protein intake at moderate levels.
4. Balance food intake and physical activity to maintain appropriate body weight.
5. Alcohol consumption is not recommended. For those who drink alcoholic beverages, limit consumption to the equivalent of 1 ounce of pure alcohol in a single day.
6. Limit total daily intake of salt to 6 g or less.
7. Maintain adequate calcium intake.
8. Avoid taking dietary supplements in excess of the RDA in any one day.
9. Maintain an optimal intake of fluoride, particularly during the years of primary and secondary tooth formation and growth.

ESSAY

mendations to the general public is a monumental task, and all such attempts have been subjected to intense criticism. We present here two of the most commonly used recent sets of recommendations, one by the National Research Council and one—the dietary pyramid—by the U.S. Department of Agriculture. The latter, in the words of one scientist, ". . . represents a mix of well-supported findings, educated guesses, and political compromises with powerful economic interests such as the dairy and meat industries." As an example of controversy in this field, to the pyramid has been added the annotations of this scientist, whose cited article provides an excellent review and critique of the general topic.

Annotated version of the U.S. Department of Agriculture food pyramid, indicating the recommended daily servings of various food groups. The annotations in the boxes were added by Dr. Walter C. Willett to summarize issues addressed in his review. (*From Willett, W.C., Science, volume 264, pg 535, 22 April, 1994.*)

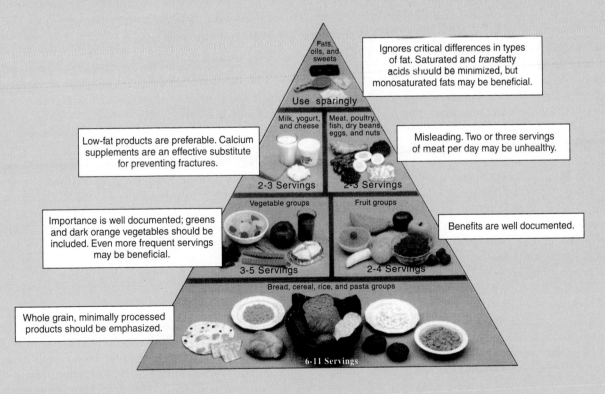

Fats, oils, and sweets

Ignores critical differences in types of fat. Saturated and *trans*fatty acids should be minimized, but monosaturated fats may be beneficial.

Use sparingly

Milk, yogurt, and cheese

Meat, poultry, fish, dry beans, eggs, and nuts

Low-fat products are preferable. Calcium supplements are an effective substitute for preventing fractures.

Misleading. Two or three servings of meat per day may be unhealthy.

2-3 Servings

2-3 Servings

Vegetable groups

Fruit groups

Importance is well documented; greens and dark orange vegetables should be included. Even more frequent servings may be beneficial.

Benefits are well documented.

3-5 Servings

2-4 Servings

Bread, cereal, rice, and pasta groups

Whole grain, minimally processed products should be emphasized.

6-11 Servings

example, by measuring skinfold thickness. BMI is calculated as weight (in kilograms) divided by the square of height (in meters). For example, a 70 kg person with a height of 180 cm would have a BMI of 21.6 ($70/1.8^2$). In 1995 the National Institutes of Health and the American Health Foundation issued guidelines that define healthy weight as a BMI below 25, based on data that show health risks increasing at BMIs greater than 25. The World Health Organization considers those with BMIs of 30 or higher to be at major risk.

Identical twins who have been separated soon after birth and raised in different households manifest strikingly similar body weights and incidences of obesity as adults; such studies have indicated that genetic factors play an important role in obesity. It has been postulated that natural selection favored the evolution in our ancestors of "thrifty genes," which boosted the ability to store fat from each feast in order to sustain people through the next fast. In today's relative surfeit in many countries of the world, such an adaptation would now be a liability.

Despite the importance of genetic factors, psychological, cultural, and social factors can also be important; for example, in the United States and other industrialized countries, obesity is much more common among members of the lowest socioeconomic group than among those of the highest group. The increasing incidence of obesity in the United States and other industrialized nations during the past few generations cannot be explained by our genes.

Much recent research has focused on possible abnormalities in the leptin system as potential causes of obesity. In mice that have severe hereditary obesity the gene—*ob*—that codes for leptin is mutated so that adipose tissue cells produce either an abnormal, inactive leptin or no leptin at all. The same is not true, however, for the vast majority of obese people: The leptin secreted by these people is normally active, and leptin concentrations in the blood are elevated, not reduced. This indicates that leptin secretion is not at fault in these people, and the obesity may be caused, at least in part, by an insensitivity of the hypothalamus to leptin. At least four other genes that, when mutated, can cause obesity have been found in rodents, but again none of these genes are mutated in obese people. This is not really surprising since most obesity researchers believe that there must be multiple genes that interact with one another and with environmental factors to influence a person's susceptibility to gain weight.

The methods and goals of treating obesity are presently undergoing extensive rethinking. An increase in body fat must be due to an excess of food intake over the metabolic rate, and low-calorie diets have long been the mainstay of therapy. However, it is now clear that such diets *alone* have limited effectiveness in obese people; over 90 percent regain all or most of the lost weight within 5 years. Another important reason for the ineffectiveness of such diets is that, as described earlier, the person's metabolic rate drops, sometimes falling low enough to prevent further weight loss on as little as 1000 calories a day. Related to this, many obese people continue to gain weight or remain in stable energy balance on a caloric intake equal to or less than the amount consumed by people of normal weight. These persons must either have less physical activity than normal or have lower basal metabolic rates. Finally, at least half of obese people—those who are more than 20 percent overweight—who try to diet down to desirable weights suffer medically, physically, and psychologically. This is what would be expected if the body were "trying" to maintain body weight (more specifically fat stores) at the higher operating point.

Such studies, taken together, indicate that crash diets are not an effective long-term method for controlling weight. Instead one should set caloric intake at a level that can be maintained for the rest of one's life; such an intake in an overweight individual should lead to a slow steady weight loss of no more than one pound per week until the body weight stabilizes at a new, lower level. Most important, any program of weight loss should include increased physical activity of the endurance type. The exercise itself utilizes calories (though depressingly few), but more importantly exercise partially offsets the tendency, described earlier, for the metabolic rate to decrease during long-term caloric restriction and weight loss. Also, the combination of exercise and caloric restriction causes the person to lose more fat and less protein than with caloric restriction alone. To restate the information of the previous two sentences in terms of control systems, exercise seems to lower the operating point around which the body regulates total-body fat stores.

Finally, as an exercise in energy balance, let us calculate how rapidly a person can expect to lose weight on a reducing diet (assuming, for simplicity, no change in energy expenditure). Suppose an individual whose steady-state metabolic rate per 24 h is 2000 kcal goes on a 1000 kcal/day diet. How much of the person's own body fat will be required to supply this additional 1000 kcal/day? Since fat contains 9 kcal/g:

$$\frac{1000 \text{ kcal/day}}{9 \text{ kcal/g}} = 111 \text{ g/day, or 777 g/week}$$

Approximately another 77 g of water is lost from the adipose tissue along with this fat (adipose tissue is 10 percent water), so that the grand total for 1 week's loss equals 854 g, or 1.8 lb. Thus, even on this rather severe diet, the person can reasonably expect to lose approximately this amount of weight per week, assuming no decrease in metabolic rate occurs. Actually, the amount of weight lost during the first week will probably be considerably greater since a large amount of *water* may be lost early in the diet, particularly when the diet contains little carbohydrate. This early loss is not really elimination of excess fat but often underlies the extravagant claims made for fad diets.

Eating Disorders: Anorexia Nervosa and Bulimia

The two major eating disorders are found almost exclusively in adolescent girls and young women. The typical person with *anorexia nervosa* becomes pathologically afraid of gaining weight and reduces her food intake so severely that she may die of starvation. It is not known whether the cause of anorexia nervosa is primarily psychological or biological. There are many other abnormalities associated with it—loss of menstrual periods, low blood pressure, low body temperature, altered secretion of many hormones. It is likely that these are simply the result of starvation although it is possible that some represent signs, along with the eating disturbances, of primary hypothalamic malfunction.

Bulimia is recurrent episodes of binge eating. It is usually associated with regular employment of self-induced vomiting, laxatives, or diuretics, as well as strict dieting, fasting, or vigorous exercise to prevent weight gain. Like individuals with anorexia nervosa, those with bulimia manifest a persistent overconcern with body weight, although they are generally within 10 percent of their ideal weight. It, too, is accompanied by a variety of physiological abnormalities, but it is unknown whether they are causal or secondary.

REGULATION OF BODY TEMPERATURE

Animals capable of maintaining their body temperatures within very narrow limits are termed **homeothermic.** The relatively constant and high body temperature frees biochemical reactions from fluctuating with the external temperature. However, the maintenance of a relatively high body temperature (37°C, in normal persons) imposes a requirement for precise regulatory mechanisms, since further large elevations of temperature cause nerve malfunction and protein denaturation. Some people suffer convulsions at a body temperature of 41°C (106°F), and 43°C is the absolute limit for survival of most people.

Several important generalizations about normal human body temperature should be stressed at the outset: (1) Oral temperature averages about 0.5°C less than rectal, which is generally used as an estimate of internal temperature; thus, not all parts of the body have the same temperature. (2) Internal temperature varies several degrees in response to activity pattern and changes in external temperature. (3) There is a characteristic circadian fluctuation of about one degree (Figure 18-18), temperature being lowest during the night and highest during the day.

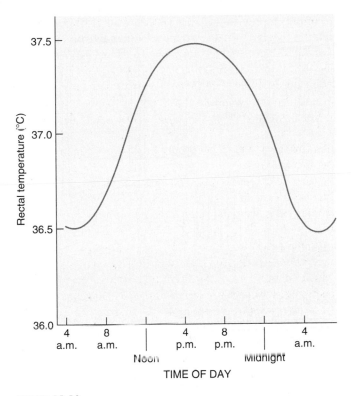

FIGURE 18-18

Circadian changes in core (measured as rectal) body temperature in normal males and in normal females in the first half of the menstrual cycle. (*Adapted from Scales et al.*)

(4) An added variation in women is a higher temperature during the last half of the menstrual cycle.

If temperature is viewed as a measure of heat "concentration," temperature regulation can be studied by our usual balance methods. The total heat content gained or lost by the body is determined by the *net difference* between heat produced and heat lost. Maintaining a constant body temperature means that, in the steady state, heat production must equal heat loss. The basic principles of heat production were described earlier in this chapter in the section on metabolic rate, and those of heat loss are described next. Then we will present the reflexes that play upon these processes specifically to regulate body temperature.

Mechanisms of Heat Loss or Gain

The surface of the body can lose heat to the external environment by radiation, conduction, and convection (Figure 18-19) and by the evaporation of water. Before defining each of these processes, however, it must be emphasized that radiation, conduction, and convection can under certain circumstances lead to heat *gain*, instead of heat loss.

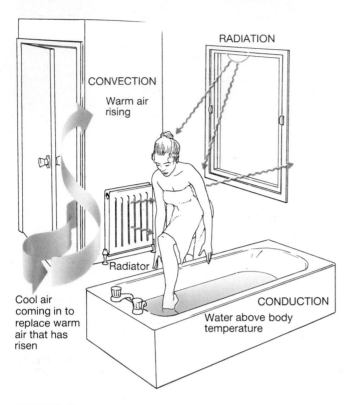

FIGURE 18-19

Mechanisms of heat transfer. In radiation, heat is transferred by electromagnetic waves. In conduction, heat moves by direct transfer of thermal energy from molecule to molecule; here, the thermal energy from the water molecules is being transferred to the molecules of the foot. Convection moves warm air away from the body to be replaced by cooler air.

Radiation is the process by which the surfaces of all objects constantly emit heat in the form of electromagnetic waves. The rate of emission is determined by the temperature of the radiating surface. Thus, if the body surface is warmer than the *average* of the various surfaces in the environment, net heat is lost from the body, the rate being directly dependent upon the temperature difference between the surfaces.

Conduction is the loss or gain of heat by transfer of thermal energy during collisions between adjacent molecules. In essence, heat is "conducted" from molecule to molecule. The body surface loses or gains heat by conduction through direct contact with cooler or warmer substances, including the air or water.

Convection is the process whereby conductive heat loss or gain is aided by movement of the air or water next to the body. For example, air next to the body is heated by conduction, moves away, and carries off the heat just taken from the body. The air that moved away is replaced by cooler air, which in turn follows the same pattern. Convection is always occurring because warm air is less dense and therefore rises, but it can be greatly facilitated by external forces such as wind or fans. Thus, convection aids conductive heat exchange by continuously maintaining a supply of cool air. Henceforth we shall also imply convection when we use the term "conduction."

Because of the great importance of air movement in aiding heat loss, attempts have been made to quantify the cooling effect of combinations of air speed and temperature. The most useful tool is the **wind-chill index,** which states the hypothetical temperature with *no* wind that would provide the same cooling effect as the actual temperature and wind velocity. For example, the wind-chill index would be −10°C if an object in a 5°C windy environment cooled as fast as it would if the temperature were actually −10°C and there were no wind at all.

Evaporation of water from the skin and membranes lining the respiratory tract is the other major process for loss of body heat. A very large amount of energy—600 kcal/L—is required to transform water from the liquid to the gaseous state. Thus, whenever water vaporizes from the body's surface, the heat required to drive the process is conducted from the surface, thereby cooling it.

Temperature-Regulating Reflexes

Temperature regulation offers a classic example of a biological control system (we used it as our example of such systems in Figure 7-2). The balance between heat production and heat loss is continuously being disturbed, either by changes in metabolic rate (exercise being the most powerful influence) or by changes in the external environment that alter heat loss or gain. The resulting changes in body temperature are detected by thermoreceptors, which initiate reflexes that change the output of various effectors so that heat production and/or loss are changed and body temperature is restored toward normal.

Figure 18-20 summarizes the specific components of these reflexes. There are two categories of thermoreceptors, one in the skin (**peripheral thermoreceptors,** described in Chapter 9) and the other (**central thermoreceptors**) in deep body structures, including the hypothalamus, spinal cord, and abdominal organs. Since it is the internal temperature, not the skin temperature, that is being maintained relatively constant, the central chemoreceptors provide the essential negative-feedback component of the reflexes. The peripheral chemoreceptors provide feedforward information, as described in Chapter 7, and also account for one's ability to identify a hot or cold area of the skin.

An area of the hypothalamus serves as the primary overall integrator of the reflexes, but other brain centers also exert some control over specific components of the reflexes.

Output from the hypothalamus and the other brain areas to the effectors is via: (1) sympathetic nerves to the

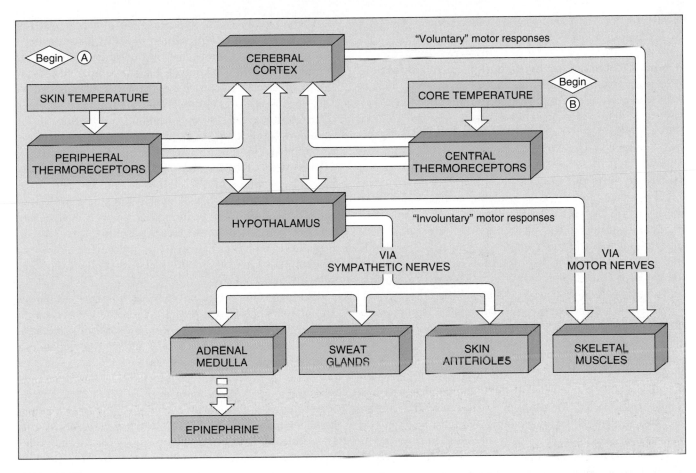

FIGURE 18-20
Summary of temperature-regulating mechanisms beginning with (A) peripheral thermoreceptors and (B) central thermoreceptors. The dashed arrow from the adrenal medulla indicates that this hormonal pathway is of minor importance in human beings. The solid arrows denote neural pathways.

sweat glands, skin arterioles, and the adrenal medulla; and (2) motor neurons to the skeletal muscles.

Control of Heat Production. Changes in muscle activity constitute the major control of heat production for temperature regulation. The first muscle changes in response to cold are a gradual and general increase in skeletal muscle contraction. This may lead to shivering, the characteristic muscle response to cold, which consists of oscillating rhythmical muscle tremors occurring at the rate of about 10 to 20 per second. During shivering, the efferent motor nerves to the skeletal muscles are controlled by descending pathways under the primary control of the hypothalamus. Because almost no external work is performed by shivering, virtually all the energy liberated by the metabolic machinery appears as internal heat and is known as **shivering thermogenesis.** People also use their muscles for voluntary heat production activities such as foot stamping and hand clapping.

Thus far, our discussion has focused primarily on the muscle response to *cold;* the opposite muscle reactions occur in response to heat. Basal muscle contraction is reflexly decreased and voluntary movement is also diminished. These attempts to reduce heat production are relatively limited, however, both because basal muscle contraction is quite low to start with and because increased body temperature acts *directly* on cells to increase metabolic rate.

Muscle contraction is not the only process controlled in temperature-regulating reflexes. In most experimental animals, chronic cold exposure induces an increase in metabolic rate, or heat production, that is not due to increased muscle activity and is termed **nonshivering thermogenesis.** Its causes are an increased adrenal secretion of epinephrine and increased sympathetic activity to adipose tissue, with some contribution by thyroid hormone as well. Nonshivering thermogenesis is quite minimal, if present at all, in adult human beings, and there is no increased secretion of thyroid hormone in response to cold. Nonshivering thermogenesis does occur in infants.

Control of Heat Loss by Radiation and Conduction.

For purposes of temperature control, it is convenient to view the body as a central core surrounded by a shell consisting of skin and subcutaneous tissue; we shall refer to this complex outer shell simply as skin. It is the temperature of the central core that is being regulated at approximately 37°C. As we shall see, the temperature of the outer surface of the skin changes markedly.

If the skin were a perfect insulator, no heat would ever be lost from the core. The temperature of the outer skin surface would equal the environmental temperature, and net conduction and radiation would be zero. The skin is not a perfect insulator, however, and so the temperature of its outer surface generally is somewhere between that of the external environment and that of the core.

The skin's effectiveness as an insulator is subject to physiological control by a change in the blood flow to it. The more blood reaching the skin from the core, the more closely the skin's temperature approaches that of the core. In effect, the blood vessels diminish the insulating capacity of the skin by carrying heat to the surface (Figure 18-21) to be lost to the external environment. These vessels are controlled largely by vasoconstrictor sympathetic nerves, the firing rate of which is increased in response to cold and decreased in response to heat. There is also a population of sympathetic neurons to the skin whose neurotransmitters (as yet unidentified) cause active *vasodilation*. Certain areas of skin participate much more than others in all these vasomotor responses, and so skin temperatures vary with location.

Three *behavioral* mechanisms for altering heat loss by radiation and conduction remain to be described: changes in surface area, in clothing, and choice of surroundings. Curling up into a ball, hunching the shoulders, and similar maneuvers in response to cold reduce the surface area exposed to the environment, thereby decreasing heat loss by radiation and conduction. In human beings, clothing is also an important component of temperature regulation, substituting for the insulating effects of feathers in birds and fur in other mammals. The outer surface of the clothes forms the true "exterior" of the body surface. The skin loses heat directly to the air space trapped by the clothes, which in turn pick up heat from the inner air layer and transfer it to the external environment. The insulating ability of clothing is determined primarily by the thickness of the trapped air layer.

Clothing is important not only at low temperatures but also at very high temperatures. When the environmental temperature is greater than body temperature, radiation and conduction favor heat *gain* rather than heat loss. People therefore insulate themselves against such temperatures by wearing clothes. The clothing, however, must be loose so as to allow adequate movement of air to permit evaporation (see below). White clothing is cooler since it reflects more radiant energy, which dark colors absorb. Loose-fitting, light-colored clothes are far more cooling than going nude during direct exposure to the sun.

The third behavioral mechanism for altering heat loss is to seek out warmer or colder surroundings, as for example by moving from a shady spot into the sunlight. Raising or lowering the thermostat of a house or turning on an air conditioner also fits this category.

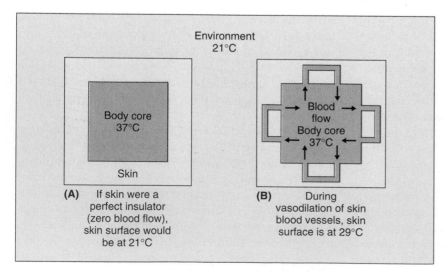

(A) If skin were a perfect insulator (zero blood flow), skin surface would be at 21°C

(B) During vasodilation of skin blood vessels, skin surface is at 29°C

FIGURE 18-21

Relationship of skin's insulating capacity to skin blood flow. (A) If skin were a perfect insulator, the temperature of its outer surface would equal that of the external environment. (B) The skin blood vessels dilate, and the increased blood flow carries heat to the body surface. In this way, the insulating capacity of the skin is reduced, and its surface temperature becomes intermediate between that of the core and that of the external environment.

Control of Heat Loss by Evaporation. Even in the absence of sweating, there is loss of water by diffusion through the skin, which is not waterproof. A similar amount is lost from the respiratory lining during expiration. These two losses are known as **insensible water loss** and amount to approximately 600 ml/day in human beings. Evaporation of this water accounts for a significant fraction of total heat loss. In contrast to this passive water loss, sweating requires the active secretion of fluid by **sweat glands** and its extrusion into ducts that carry it to the skin surface.

Production of sweat is stimulated by sympathetic nerves to the glands. (These nerves release acetylcholine rather than the usual sympathetic neurotransmitter norepinephrine.) Sweat is a dilute solution containing sodium chloride as its major solute. Sweating rates of over 4 L/h have been reported; the evaporation of 4 L of water would eliminate almost 2400 kcal from the body!

It is essential to recognize that sweat must evaporate in order to exert its cooling effect. The most important factor determining evaporation rate is the water vapor concentration of the air, that is, the relative humidity. The discomfort suffered on humid days is due to the failure of evaporation; the sweat glands continue to secrete, but the sweat simply remains on the skin or drips off.

Integration of Effector Mechanisms. Table 18-12 summarizes the effector mechanisms regulating temperature, none of which is an all-or-none response but a graded, progressive increase or decrease in activity. By altering heat loss, changes in skin blood flow alone can regulate body temperature over a range of environmental temperatures (approximately 25 to 30°C or 75 to 86°F for a nude individual) known as the **thermoneutral zone.** At temperatures lower than this, even maximal vasoconstriction cannot prevent heat loss from exceeding heat production, and the body must increase its heat production to maintain temperature. At environmental temperatures above the thermoneutral zone, even maximal vasodilation cannot eliminate heat as fast as it is produced, and another heat-loss mechanism—sweating—is therefore brought strongly into play. Indeed, at environmental temperatures above that of the body, heat is actually added to the body by radiation and conduction, and evaporation is the sole mechanism for heat loss. A person's ability to tolerate such temperatures is determined by the humidity and by the maximal sweating rate. For example, when the air is completely dry, an individual can tolerate a temperature of 130°C (255°F) for 20 min or longer, whereas very moist air at 46°C (115°F) is bearable for only a few minutes.

TABLE 18-12 SUMMARY OF EFFECTOR MECHANISMS IN TEMPERATURE REGULATION

Desired effect	Mechanism
	Stimulated by Cold (see also Figure 7-2)
Decrease heat loss	1. Vasoconstriction of skin vessels
	2. Reduction of surface area (curling up, etc.)
	3. Behavioral response (put on warmer clothes, raise thermostat setting, etc.)
Increase heat production	1. Increased muscle tone
	2. Shivering and increased voluntary activity
	3. Increased secretion of epinephrine (minimal in adults)
	4. Increased food appetite
	Stimulated by Heat
Increase heat loss	1. Vasodilation of skin vessels
	2. Sweating
	3. Behavioral response (put on cooler clothes, turn on fan, etc.)
Decrease heat production	1. Decreased muscle tone and voluntary activity
	2. Decreased secretion of epinephrine (minimal in adults)
	3. Decreased food appetite

Temperature Acclimatization

Changes in sweating onset, volume, and composition determine people's chronic adaptation to high temperatures. A person newly arrived in a hot environment has poor ability to do work; body temperature rises and severe weakness may occur. After several days, there is a great improvement in work tolerance, with little increase in body temperature, and the person is said to have acclimatized to the heat (see Chapter 7 for a discussion of the concept of acclimatization). Body temperature does not rise as much because sweating begins sooner and the volume of sweat produced is greater.

There is also an important change in the composition of the sweat, namely, a marked reduction in its sodium concentration. This adaptation, which minimizes the loss of sodium from the body via sweat, is due to increased secretion of the adrenal mineralocorticoid hormone aldosterone. The sweat-gland secretory cells produce a solution with a sodium concentration similar to that of plasma, but some of the sodium is absorbed back into the blood as the secretion flows along the sweat-gland ducts toward the skin surface. Aldosterone stimulates this absorption in a manner identical to its stimulation of sodium reabsorption in the renal tubules.

Cold acclimatization has been much less studied than heat acclimatization because of the difficulty of subjecting individuals to total-body cold stress sufficient to produce acclimatization. Moreover, groups such as Eskimos that live in cold climates generally dress very warmly and so would not develop acclimatization to the cold. The most intensive study of cold acclimatization involved Korean women who, wearing only cotton garments, used to dive for shellfish and edible seaweed in the middle of winter when the seawater temperature was as low as 10°C. The major components of their acclimatization were (1) an increase in metabolic rate, (2) an increase in the insulating ability of a given amount of skin fat, and (3) an ability to withstand a colder water temperature without shivering. The fact that these characteristics disappeared within several years after the women began using wet suits is strong evidence that these characteristics really represented acclimatization, not genetic differences.

Fever and Hyperthermia

Fever is an elevation of body temperature due to a "resetting of the thermostat" in the hypothalamus. A person with a fever still regulates body temperature in response to heat or cold but at a higher set point. The most common cause of fever is infection, but physical trauma and stress can also induce fever.

The onset of fever during infection is frequently gradual, but it is most striking when it occurs rapidly in the form of a chill. The brain thermostat is suddenly raised, the person feels cold, and marked vasoconstriction and shivering occur (Figure 18-22). The person also curls up and puts on more blankets. This combination of decreased heat loss and increased heat production serves to drive body temperature up to the new set point, where it stabilizes. It will continue to be regulated at this new value until the thermostat is reset to normal and the fever "breaks." The person then feels hot, throws off the covers, and manifests profound vasodilation and sweating.

What is the basis for the thermostat resetting? One or more chemical messengers collectively termed **endogenous pyrogen (EP)** is/are released from monocytes and macrophages in the presence of infection or other fever-producing stimuli. EP acts upon the thermoreceptors in the hypothalamus (and perhaps other brain areas), altering their rate of firing and their input to the integrating centers. The action of EP is mediated via local release of prostaglandins, which then directly alter central thermoreceptor function. Aspirin reduces fever by inhibiting prostaglandin synthesis (Chapter 7).

The term EP was coined at a time when the identity of the chemical messenger(s) was not known. At least one peptide, **interleukin 1 (IL-1),** is now known to function as an EP, but other peptides—for example, interleukin 6 (IL-6)—also released from monocytes and macrophages in response to infection and other stimuli also play a role. In addition to their effects on temperature, IL-1 and the other peptides have many other effects (described in Chapter 20) that have the common denominator of enhancing resistance to infection and promoting the healing of damaged tissue.

One would expect fever, which is such a consistent concomitant of infection, to play some important protective role, and most evidence suggests that such is the case. Increased body temperature stimulates a large number of the body's defensive responses to infection. The likelihood that fever is a beneficial response raises important questions concerning the use of aspirin and other drugs to suppress fever during infection. It must be emphasized that these questions apply to the usual modest fevers. There is no question that an extremely high fever can be harmful, particularly in its effects on the central nervous system, and must be vigorously opposed with drugs and other forms of therapy.

To reiterate, fever is an increased body temperature caused by an elevation of the thermal set point. When body temperature is elevated for any other reason, that is, when body temperature is above the set point, it is termed **hyperthermia.** The most common cause of hyperthermia in normal people is exercise; the rise in body temperature above set point is simply a physical consequence of the internal heat generated by the exercising muscles.

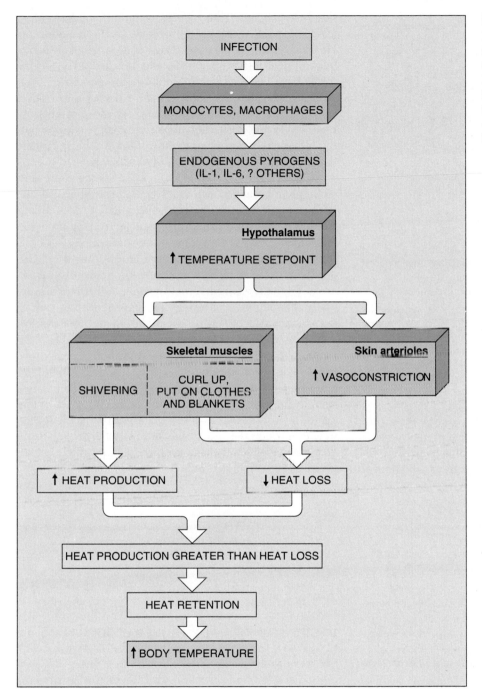

FIGURE 18-22

Pathway by which infection causes fever (IL-1 = interleukin 1, IL-6 = interleukin 6). Compare this figure to Figure 7-2: The effector responses are identical but they serve to keep body temperature *relatively constant* during exposure to cold (Figure 7-2) and *raise* body temperature during an infection.

As shown in Figure 18-23, heat production rises immediately during the initial stage of exercise and exceeds heat loss, causing heat storage in the body and a rise in the core temperature. This rise in core temperature triggers reflexes, via the central thermoreceptors, for increased heat loss; with increased skin blood flow and sweating, the discrepancy between heat production and heat loss starts to diminish but does not disappear. Therefore core temperature continues to rise. Ultimately, core temperature will be high enough to drive, via the central thermoreceptors, the heat-loss reflexes at a rate such that heat loss once again equals heat production. At this point core temperature stabilizes at this elevated value despite continued exercise.

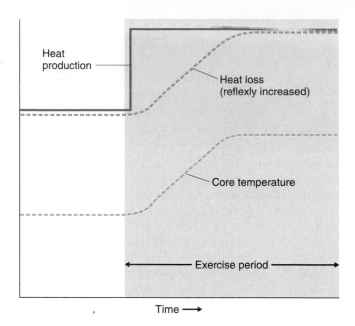

FIGURE 18-23

Thermal changes during exercise. Heat loss is reflexly increased, and when it once again equals heat production, core temperature stabilizes.

Heat Exhaustion and Heat Stroke. *Heat exhaustion* is a state of collapse, often taking the form of fainting, due to hypotension brought on by (1) depletion of plasma volume secondary to sweating, and (2) extreme dilation of skin blood vessels. Thus, decreases in both cardiac output and peripheral resistance contribute to the hypotension. Heat exhaustion occurs as a direct consequence of the activity of heat-loss mechanisms, and because these mechanisms have been so active, the body temperature is only modestly elevated. In a sense, heat exhaustion is a safety valve that, by forcing cessation of work in a hot environment when heat-loss mechanisms are overtaxed, prevents the larger rise in body temperature that would precipitate the far more serious condition of heat stroke.

In contrast to heat exhaustion, *heat stroke* represents a complete breakdown in heat-regulating systems so that body temperature keeps going up and up. It is an extremely dangerous situation characterized by collapse, delirium, seizures, or prolonged unconsciousness—all due to marked elevation of body temperature. It almost always occurs in association with exposure to or overexertion in hot and humid environments. In some individuals, particularly the elderly, heat stroke may appear with no apparent prior period of severe sweating, but in most cases, it comes on as the end stage of prolonged untreated heat exhaustion. Exactly what triggers the transition to heat stroke is not clear—impaired circulation to the brain due to dehydration is one factor—but the striking finding is that even in the face of a rapidly rising body temperature, the person fails to sweat. This sets off a positive-feedback situation in which the rising body temperature directly stimulates metabolism, that is, heat production, which further raises body temperature.

SECTION C SUMMARY

BASIC CONCEPTS OF ENERGY EXPENDITURE AND ENERGY BALANCE

  I. The energy liberated during a chemical reaction appears either as heat or work.
  II. Total energy expenditure = heat produced + external work done + energy stored.
  III. Metabolic rate is influenced by the many factors summarized in Table 18-7.
  IV. Metabolic rate is increased by the thyroid hormones and epinephrine. The other functions of the thyroid hormones are summarized in Table 18-8.
  V. Energy storage as fat can be positive or negative when the metabolic rate is less than or greater than, respectively, the energy content of ingested food.
  A. Energy storage is regulated mainly by reflex adjustment of food intake to the metabolic rate.
  B. In addition, the metabolic rate increases or decreases to some extent when food intake is chronically increased or decreased, respectively.
  VI. Food intake is controlled by leptin, secreted by adipose-tissue cells, and a variety of satiety factors, as summarized in Figure 18-17.
  VII. Obesity, the result of an imbalance between food intake and metabolic rate, increases the incidence of many diseases.

REGULATION OF BODY TEMPERATURE

  I. Core body temperature shows a circadian rhythm, being highest during the day and lowest at night.
  II. The body exchanges heat with the external environment by radiation, conduction, convection, and evaporation of water from the body surface.
  III. The hypothalamus and other brain areas contain the integrating centers for temperature-regulating reflexes, and both peripheral and central thermoreceptors participate in these reflexes.
  IV. Body temperature is regulated by altering heat production and/or heat loss so as to change total body heat content.
  A. Heat production is altered by increasing muscle tone, shivering, and voluntary activity.
  B. Heat loss by radiation, conduction, and convection

depends on the difference in temperature between the skin surface and the environment.

C. In response to cold, skin temperature is decreased by decreasing skin blood flow through reflex stimulation of the sympathetic nerves to the skin. In response to heat, skin temperature is increased by inhibiting the nerves.

D. Behavioral responses such as putting on more clothes also influence heat loss.

E. Evaporation of water occurs all the time as insensible loss from the skin and respiratory lining. Additional water for evaporation is supplied by sweat, stimulated by the sympathetic nerves to the sweat glands.

F. Increased heat production is essential for temperature regulation at environmental temperatures below the thermoneutral zone, and sweating is essential at temperatures above this zone.

V. Temperature acclimatization to heat is achieved by an earlier onset of sweating, an increased volume of sweat, and a decreased sodium concentration of the sweat.

VI. Fever is due to a resetting of the temperature set point so that heat production is increased and heat loss is decreased in order to raise body temperature to the new set point and keep it there. The stimulus is endogenous pyrogen, which is interleukin 1 and possibly other peptides as well.

VII. The hyperthermia of exercise is due to the increased heat produced by the muscles.

SECTION C KEY TERMS

external work	convection
internal work	wind-chill index
total energy expenditure	evaporation
kilocalorie (kcal)	peripheral thermoreceptor
metabolic rate	central thermoreceptor
basal metabolic rate (BMR)	shivering thermogenesis
calorigenic effect	nonshivering thermogenesis
food-induced thermogenesis	insensible water loss
leptin	sweat gland
satiety signal	thermoneutral zone
body mass index (BMI)	fever
homeothermic	endogenous pyrogen (EP)
radiation	interleukin 1 (IL-1)
conduction	hyperthermia

SECTION C REVIEW QUESTIONS

1. State the formula relating total energy expenditure, heat produced, external work, and energy storage.

2. What two hormones alter the basal metabolic rate?

3. State the equation for total-body energy balance. Describe the three possible states of balance with regard to energy storage.

4. What happens to the basal metabolic rate after a person has either lost or gained weight?

5. List five satiety signals.

6. List three beneficial effects of exercise in a weight-loss program.

7. Compare and contrast the four mechanisms for heat loss.

8. Describe the control of skin blood vessels during exposure to cold or heat.

9. With a diagram, summarize the reflex responses to heat or cold. What are the dominant mechanisms for temperature regulation in the thermoneutral zone and in temperatures below and above this range?

10. What changes are exhibited by a heat-acclimatized person?

11. Summarize the sequence of events leading to a fever and contrast this to the sequence leading to hyperthermia during exercise.

CHAPTER 18 CLINICAL TERMS

diabetes mellitus	drawfism
insulin-dependent diabetes	acrogemaly
mellitus (IDDM)	hypothyroidism
noninsulin-dependent	endemic cretinism
diabetes mellitus	hyperthyroidism
(NIDDM)	anabolic steroids
diabetic ketoacidosis	obesity
insulin resistance	anorexia nervosa
sulfonylureas	bulimia
fasting hypoglycemia	fever
atherosclerosis	hyperthermia
cancer	heat exhaustion
oncogene	heat stroke
giantism	

CHAPTER 18 THOUGHT QUESTIONS

1. What happens to the triacylglycerol concentrations in the plasma and in adipose tissue after administration of a drug that blocks the action of lipoprotein lipase?

2. A resting, unstressed person has increased plasma concentrations of free fatty acids, glycerol, amino acids, and ketones. What situations might be responsible and what additional plasma measurement would distinguish among them?

3. A normal volunteer is given an injection of insulin. The plasma concentrations of which hormones increase?

4. If the sympathetic preganglionic fibers to the adrenal medulla were cut in an animal, would this eliminate the sympathetically mediated component of increased gluconeogenesis and lipolysis during exercise? Explain.

5. A patient with insulin-dependent diabetes suffers a broken leg. Would you advise this person to increase or decrease his dosage of insulin?

In this manner, mitosis of primordial germ cells provides a supply of identical germ cells for the next stages. The timing of mitotic activity in germ cells differs greatly in females and males. In the female, mitosis of germ cells occurs exclusively during the individual's embryonic development. In the male, some mitosis occurs in the embryo to generate the population of germ cells present at birth, but mitosis really begins in earnest at puberty and usually continues throughout life.

The second stage of gametogenesis is **meiosis,** in which each resulting gamete receives only 23 chromosomes from a 46-chromosome germ cell, 1 chromosome from each homologous pair. Because a sperm and an about-to-be fertilized egg each has only 23 chromosomes, their union at fertilization results once again in a cell with a full complement of 46 chromosomes.

Let us see how meiosis works, using Figure 19-2 in which the letters are keyed to the text. Meiosis consists of two cell divisions in succession. The events preceding the first meiotic division are identical to those preceding a mitotic division, as described in Chapter 5. Recall that during the interphase preceding a *mitotic* division, the DNA of the chromosomes is replicated. Thus a resting interphase cell still has 46 chromosomes, but each chromosome consists of two identical strands of DNA, termed sister chromatids, which are joined together by a centromere (A).

As the first *meiotic* division begins, homologous chromosomes, each consisting of two identical sister chromatids, come together and line up point for point along their entire lengths. Thus, 23 four-chromatid groupings, called tetrads, are formed (B). The sister chromatids of each chromosome condense into thick rodlike structures and become highly visible (C). Then, within a tetrad, corresponding segments of homologous chromosomes come to overlap one another (D). At these points, portions of the homologous chromosomes break off and exchange with each other in a process known as **crossing-over** (E). Thus, crossing-over results in the recombination of genes on homologous chromosomes.

Following crossing over, the tetrads line up in the center of the cell (F). The orientation of each tetrad on the equator is *random,* meaning that sometimes the maternal portion points to a particular pole of the cell and sometimes the paternal portion does so. The cell then divides

FIGURE 19-2

Stages of meiosis in a generalized germ cell. For simplicity, the initial cell (A), which is in interphase, is given only 4 chromosomes rather than 46, the human number. Also, cytoplasm is shown only in (A), (H), and (I). Chromosomes from one parent are red and those from the other parent are blue. The letters are keyed to descriptions in the text. (*Adapted from Carlson.*)

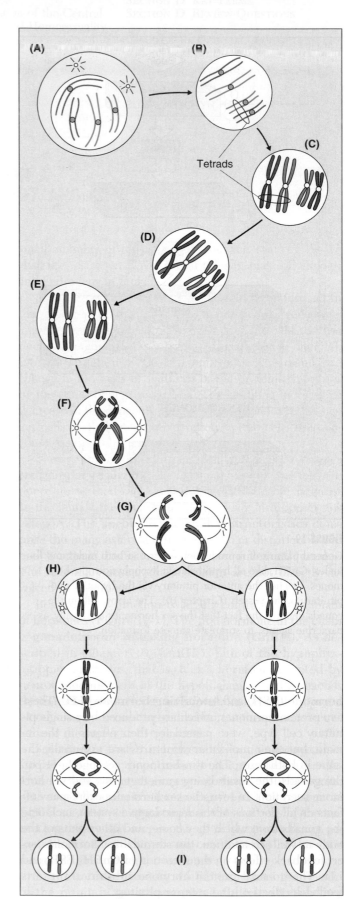

(the first division of meiosis), with the maternal chromatids of any particular tetrad going to one of the two cells resulting from the division, and the paternal chromatids going to the other (G). Because of the random orientation of the tetrads at the equator, it is extremely unlikely that all 23 maternal chromosomes will end up in one cell and all 23 paternal chromosomes in the other. Over 8 million (2^{23}) different combinations of maternal and paternal chromosomes can result during this division.

The second division of meiosis occurs without any further replication of DNA. The sister chromatids—both of which are either maternal or paternal—of each chromosome separate and move apart into the new daughter cells (H to I). The daughter cells resulting from the second meiotic division, therefore, contain 23 one-chromatid chromosomes (I).

To summarize, meiosis produces daughter cells having only 23 chromosomes, and two events during the first meiotic division contribute to the enormous genetic variability of the daughter cells: (1) crossing-over, and (2) the random distribution of maternal and paternal chromatid pairs between the two daughter cells.

SECTION A SUMMARY

The gonads have a dual function—gametogenesis and the secretion of sex hormones.

GENERAL PRINCIPLES OF GAMETOGENESIS

I. The first stage of gametogenesis is mitosis of primordial germ cells.

II. This is followed by meiosis, a sequence of two cell divisions resulting in each gamete receiving 23 chromosomes.

III. Crossing over and random distribution of maternal and paternal chromatids to the daughter cells during meiosis cause genetic variability in the gametes.

SECTION A KEY TERMS

gonad	ovary
testis	gametogenesis

gamete	secondary sexual
spermatozoa	characteristic
sperm	gonadotropin releasing
ova	hormone (GnRH)
ovum	gonadotropin
sex hormones	follicle-stimulating hormone
testosterone	(FSH)
estradiol	luteinizing hormone (LH)
progesterone	inhibin
androgen	germ cell
estrogen	meiosis
accessory reproductive	crossing-over
organ	

SECTION A REVIEW QUESTION

1. Describe the stages of gametogenesis and how meiosis results in genetic variability.

SECTION B
MALE REPRODUCTIVE PHYSIOLOGY

ANATOMY

The male reproductive system includes the two testes, the system of ducts that store and transport sperm to the exterior, the glands that empty into these ducts, and the penis. The duct system, glands, and penis constitute the male accessory reproductive organs.

The testes are suspended outside the abdomen in the **scrotum,** which is an outpouching of the abdominal wall and is divided internally into two sacs, one for each testis. During fetal development, the testes are located in the abdomen, but during the seventh month of intrauterine development, they descend into the scrotum. This descent is essential for normal sperm production during adulthood, since sperm formation requires a temperature several degrees lower than normal internal body temperature. Cooling is achieved by air circulating around the scrotum and by a heat-exchange mechanism in the blood vessels supplying the testes. In contrast to spermatogenesis, pathway, the spermatogonia would disappear, that is, would all be converted to primary spermatocytes. This does not occur because, at an early point, one of the cells of each clone "drops out" of the mitosis-differentiation cycles and reverts to being a primitive spermatogonium that, at a later time, will enter into its own full sequence of divisions. In turn, one cell of the clone it produces will do likewise, and so on. Thus, the supply of undifferentiated spermatogonia does not decrease.

Each primary spermatocyte increases markedly in size and undergoes the first meiotic division (Figure 19-6) to form two **secondary spermatocytes,** each of which contains 23 two-chromatid chromosomes. Each secondary spermatocyte, in turn undergoes the second meiotic division (Figure 19-2H to I) into **spermatids.** Thus, each primary spermatocyte, containing 46 two-chromatid chromosomes, gives rise to four spermatids, each containing 23 one-chromatid chromosomes.

The final phase of spermatogenesis is the differentiation of the spermatids into spermatozoa (sperm). This process involves extensive cell remodeling, including elongation, but no further cell divisions. The head of a sperm (Figure 19-7) consists almost entirely of the nucleus, which contains the DNA bearing the sperm's genetic information. The tip of the nucleus is covered by the **acrosome,** a protein-filled vesicle containing several enzymes that play an important role in the sperm's penetration of the egg. Most of the tail is a flagellum—a group of contractile filaments that produce whiplike movements capable of propelling the sperm at a velocity of 1 to 4 mm/min. The sperm's mitochondria form the midpiece of the sperm and provide the energy for the sperm's movement.

In any small segment of each seminiferous tubule, spermatogenesis proceeds in a regular sequence. For example, at any given time, virtually all the primary spermatocytes in one portion of the tubule are undergoing division, whereas in an adjacent segment, all the secondary spermatocytes may be dividing. The entire process, from primary spermatocyte to sperm, takes approximately 64 days. The normal human male manufactures approximately 30 million sperm per day.

Thus far, we have described spermatogenesis without regard to its orientation within the seminiferous tubules or the participation of **Sertoli cells,** the second type of cell in the seminiferous tubules with which the developing germ cells are intimately associated. As noted earlier, each seminiferous tubule is bounded by a basement membrane. Each Sertoli cell extends from the basement membrane all the way to the lumen in the center of the tubule and is joined to adjacent Sertoli cells by means of tight junctions (Figure 19-8). Thus, the Sertoli cells form an unbroken ring around the outer circumference of the seminiferous tubule, and the tight junctions divide the tubule into two compartments—a basal compartment between the base-

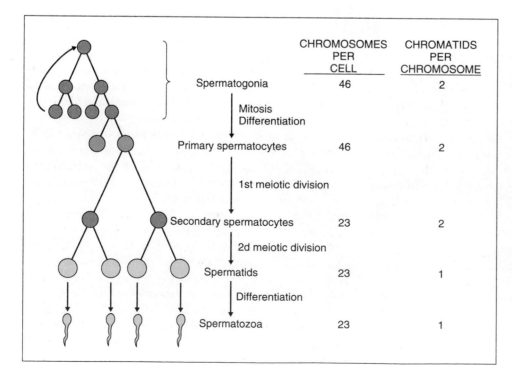

	CHROMOSOMES PER CELL	CHROMATIDS PER CHROMOSOME
Spermatogonia	46	2
Mitosis Differentiation		
Primary spermatocytes	46	2
1st meiotic division		
Secondary spermatocytes	23	2
2d meiotic division		
Spermatids	23	1
Differentiation		
Spermatozoa	23	1

FIGURE 19-6

Summary of spermatogenesis, which begins at puberty. Each spermatogonium yields, by mitosis, a clone of spermatogonia; for simplicity, the figure shows only two such cycles, with a third mitotic cycle generating two primary spermatocytes. The arrow from one of the spermatogonia back to an original spermatogonium denotes the fact that one cell of the clone does not go on to generate primary spermatocytes but reverts to being an undifferentiated spermatogonium that gives rise to a new clone. Note that each primary spermatocyte produces four spermatozoa.

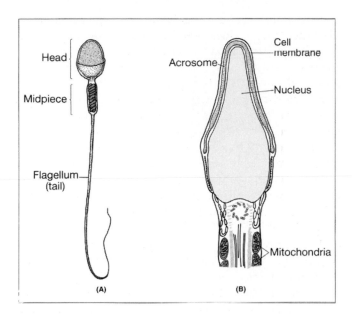

FIGURE 19-7

(A) Diagram of a human mature sperm. (B) A close-up of the head. The acrosome contains enzymes required for fertilization of the ovum.

ment membrane and the tight junctions, and a central compartment, beginning at the tight junctions and including the lumen.

This arrangement has several very important results: (1) The ring of interconnected Sertoli cells forms the **Sertoli-cell barrier** (previously termed blood-testis barrier), which prevents the movement of many chemicals from the blood into the lumen of the seminiferous tubule and thereby ensures proper conditions for germ-cell development and differentiation in the tubules; and (2) different stages of spermatogenesis take place in different compartments and, hence, in different environments.

Mitosis and differentiation of spermatogonia to yield primary spermatocytes take place entirely in the basal compartment. The primary spermatocytes then move through the tight junctions of the Sertoli cells, which open in front of them while at the same time new tight junctions form behind them, to gain entry into the central compartment. In this central compartment, the meiotic divisions of spermatogenesis occur, and the spermatids are remodeled into sperm while contained in recesses formed by invaginations of the Sertoli-cell plasma membranes. When sperm formation is completed, the Sertoli-cell cytoplasm around the sperm retracts, and the sperm are released into the lumen to be bathed by the luminal fluid.

Sertoli cells serve as the route by which nutrients reach developing germ cells, and they also secrete most

of the fluid found in the tubule lumen. This fluid has a highly characteristic ionic composition. It also contains **androgen-binding protein,** which binds the testosterone that was secreted by the Leydig cells and that has crossed the Sertoli-cell barrier to enter the tubule. This maintains a high concentration of testosterone in the lumen of the tubule (and the entire duct system).

Thus far we have emphasized how Sertoli cells influence the ionic and nutritional environment of the germ cells, but they do more than this: They act as intermediaries between the germ cells and the hormones (FSH from the anterior pituitary and testosterone from the Leydig cells) that stimulate spermatogenesis. In other words, these hormones do not act *directly* on the germ cells but rather on the Sertoli cells, which respond by secreting a variety of chemical messengers that function as paracrine agents to stimulate proliferation and differentiation of the germ cells. The traffic between the Sertoli cells and the germ cells is not just one-way because the germ cells secrete paracrine agents that, in feedback manner, help control the activities of the Sertoli cells.

In addition, the Sertoli cells secrete the protein hormone inhibin and paracrine agents that affect Leydig-cell function. The many functions of Sertoli cells, several of which remain to be described later in this chapter, are summarized in Table 19-2.

TRANSPORT OF SPERM

From the seminiferous tubules, the sperm pass through the rete testis and efferent ductules into the epididymis and thence into the vas deferens. The vas deferens and the portion of the epididymis closest to it serve as a storage reservoir for sperm, holding them until sexual arousal leads to ejaculation. Also, in the epididymis the sperm undergo a further maturation process.

Movement of the sperm as far as the epididymis results from the pressure created both by the continuous formation of fluid by the Sertoli cells back in the seminiferous tubules and by peristalsis of the tubules. The sperm themselves are nonmotile at this time. During passage through the epididymis there occurs a hundredfold concentration of the sperm by fluid absorption from the lumen of the epididymis. Therefore, as the sperm pass from the end of the epididymis into the vas deferens, they are a densely packed mass whose transport is no longer a result of fluid movement but is due to peristaltic contractions of the smooth muscle in the epididymis and vas deferens. The absence of a large quantity of fluid accounts for the fact that **vasectomy,** the surgical tying-off and removal of a segment of each vas deferens, does not cause the accumulation

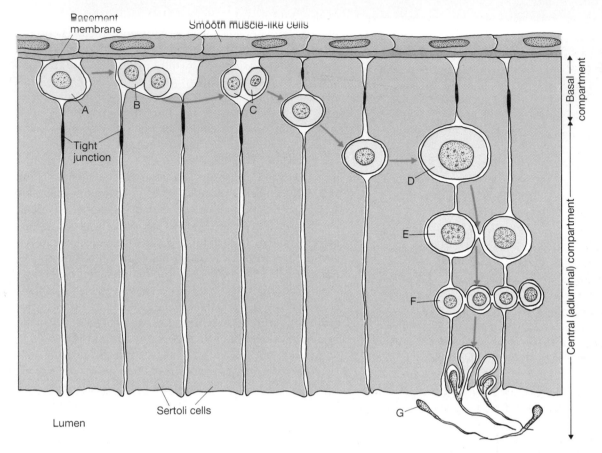

FIGURE 19-8

Relation of the Sertoli cells (green) and germ cells (yellow). The Sertoli cells form a ring (barrier) around the entire tubule. For convenience of presentation, the various stages of spermatogenesis are shown as though the germ cells move down a line of adjacent Sertoli cells; in reality, all stages beginning with any given spermatogonium take place between the same two Sertoli cells. Spermatogonia (A and B) are found only in the basal compartment (between the tight junctions of the Sertoli cells and the basement membrane of the tubule). After several mitotic cycles (A to B), the spermatogonia (B) give rise to primary spermatocytes (C). Each of the latter crosses a tight junction, enlarges (D), and divides into two secondary spermatocytes (E), which divide into spermatids (F), which in turn differentiate into spermatozoa (G). This last step involves loss of cytoplasm by the spermatids. (*Adapted from Tung.*)

TABLE 19-2 FUNCTIONS OF SERTOLI CELLS

1. Provide Sertoli-cell barrier to chemicals
2. Nourish developing sperm
3. Secrete luminal fluid, including androgen-binding protein
4. Receive stimulation by testosterone and FSH to secrete paracrine agents that stimulate sperm proliferation and differentiation
5. Secrete the protein hormone inhibin, which inhibits FSH secretion
6. Secrete paracrine agents that influence the function of Leydig cells
7. Phagocytize defective sperm
8. Secrete, during embryonic life, Müllerian inhibiting substance (MIS), which causes the primordial female duct system to regress

of much fluid behind the tie-off point. (The sperm, which are still produced in a vasectomized individual, do build up, however, and eventually dissolve, their chemical components being absorbed into the bloodstream.) Vasectomy has no effect on testosterone secretion because it does not alter the function of the Leydig cells.

The next step in sperm transport is ejaculation, usually preceded by erection, which permits entry of the penis into the vagina.

Erection

The penis's becoming rigid—**erection**—is a vascular phenomenon. The penis consists almost entirely of three cylindrical vascular compartments running its entire length. Normally the small arteries supplying the vascular compartments are constricted so that the compartments contain little blood and the penis is flaccid. During sexual excitation, the small arteries dilate, the three vascular compartments become engorged with blood at high pressure, and the penis becomes rigid. The vascular dilation is initiated by neural input to the small arteries of the penis. Moreover, as the vascular compartments expand, the veins emptying them are passively compressed, thus contributing to the engorgement. This entire process occurs rapidly, complete erection sometimes taking only 5 to 10 s.

What are the neural inputs to the small arteries of the penis? At rest, the dominant input is via sympathetic neurons; they release norepinephrine, which causes the arterial smooth muscle to contract. During erection this sympathetic input is inhibited, but much more important is the activation of nonadrenergic, noncholinergic autonomic neurons to the arteries (Figure 19-9). These neurons release **nitric oxide,** which relaxes the arterial smooth muscle.

Which receptors and afferent pathway initiate these reflexes? The primary stimulus comes from highly sensitive mechanoreceptors in the genital region, particularly in the head of the penis. The afferent fibers carrying the impulses synapse in the lower spinal cord on interneurons that control the efferent outflow.

It must be stressed, however, that higher brain centers, via descending pathways, may also exert profound stimu-latory or inhibitory effects upon the autonomic neurons to the small arteries of the penis. Thus, mechanical stimuli from areas other than the penis, as well as thoughts, emotions, sight, and odors, can induce erection in the complete absence of penile stimulation.

Impotence, the consistent inability to achieve or sustain an erection of sufficient rigidity for sexual intercourse, is a common disorder, affecting an estimated 10 million Americans. It is very much age-dependent, increasing from an incidence of 2 percent at age 40 to 25 percent at age 65. The organic causes are multiple and include damage to the efferent nerves or descending pathways, endocrine disorders, various therapeutic and "recreational" drugs (alcohol, for example), and certain diseases, particularly diabetes. Impotence can also be due to psychological factors, which are mediated by the brain and the descending pathways.

Ejaculation

The discharge of semen from the penis—**ejaculation**—is also basically a spinal reflex, the afferent pathways from penile mechanoreceptors being identical to those described for erection. When the level of stimulation is high enough, there is elicited a patterned automatic sequence

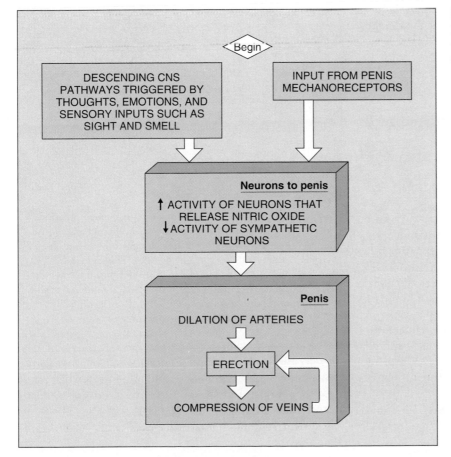

FIGURE 19-9

Reflex pathways for erection. Nitric oxide, a vasodilator, is the most important neurotransmitter to the small arteries in this reflex.

of discharge of the efferent neurons that can be divided into two phases: (1) The smooth muscles of the epididymis, vas deferens, ejaculatory ducts, prostate, and seminal vesicles contract as a result of sympathetic stimulation, emptying the sperm and glandular secretions into the urethra (**emission**); and (2) the semen (average volume 3 ml, containing 300 million sperm) is then expelled from the urethra by a series of rapid contractions of the urethral smooth muscle as well as the skeletal muscle at the base of the penis. During ejaculation, the sphincter at the base of the urinary bladder is closed so that sperm cannot enter the bladder nor can urine be expelled from it. It is worth noting that although *erection* involves *inhibition* of sympathetic nerves (to the small arteries of the penis), *ejaculation* involves *stimulation* of sympathetic nerves (to the smooth muscles of the duct system).

The rhythmical muscular contractions that occur during ejaculation are associated with intense pleasure and many systemic physiological changes, the entire event being termed an **orgasm.** Marked skeletal muscle contractions occur throughout the body, and there is a large increase in heart rate and blood pressure. This is followed by the rapid onset of muscular and psychological relaxation.

Once ejaculation has occurred, there is a latent period during which a second erection is not possible. The latent period is quite variable but may last from minutes to hours.

As is true of impotence, premature ejaculation, failure to ejaculate, or lack of a generalized orgasm at the time of ejaculation can be the result of influence by higher brain centers.

HORMONAL CONTROL OF MALE REPRODUCTIVE FUNCTIONS

Control of the Testes

Figure 19-10 summarizes the control of the testes. In a normal adult man, the GnRH-secreting neuroendocrine cells fire a brief burst of action potentials approximately every 2 h, secreting GnRH at these times. The GnRH reaching the anterior pituitary during each periodic pulse triggers the release from the anterior pituitary of both LH and FSH from the same cells, although not necessarily in equal amounts. Accordingly, systemic plasma concentrations of FSH and LH also show rhythmical episodic changes—rapid increases during the pulse followed by slow decreases over the next 90 min or so as the hormones are slowly removed from the plasma.

There is a clear separation of the actions of FSH and LH within the testes (Figure 19-10): FSH acts on the

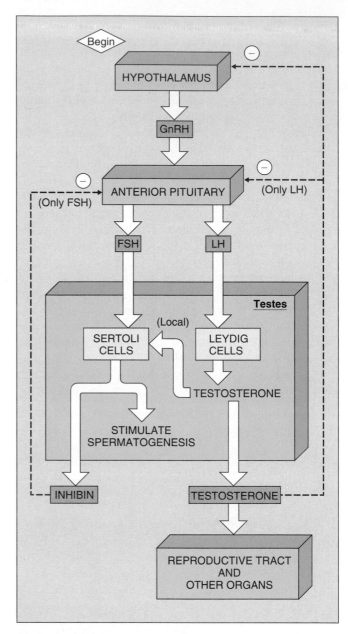

FIGURE 19-10

Summary of hormonal control of male reproductive function. GnRH reaches the anterior pituitary via the hypothalamo-pituitary portal vessels. Note that FSH acts only on the Sertoli cells, whereas LH acts only on the Leydig cells. The secretion of FSH is inhibited mainly by inhibin, a protein hormone secreted by the Sertoli cells, and the secretion of LH is inhibited mainly by testosterone, the hormone secreted by the Leydig cells. Testosterone, acting locally, is essential for spermatogenesis.

Sertoli cells to stimulate the production of paracrine agents that stimulate spermatogenesis and other Sertoli cell functions; in contrast, LH acts on the Leydig cells to stimulate testosterone secretion. (Recall that the Leydig cells lie in the small triangular interstitial spaces between the seminiferous tubules.) In addition to its many important systemic effects as a hormone the testosterone secreted by these cells also acts locally, as a paracrine agent, on spermatogenesis by moving from the interstitial spaces into the seminiferous tubules. There, testosterone enters Sertoli cells, and it is via these cells that it facilitates spermatogenesis. It must be emphasized that, despite the absence of any *direct* effect of LH on cells in the seminiferous tubules, this hormone exerts an essential *indirect* effect because the testosterone secretion stimulated by LH is required for spermatogenesis.

The last components of the hypothalamo-pituitary control of male reproduction that remain to be discussed are the negative feedbacks exerted by testicular hormones. That such inhibition exists is evidenced by the fact that **castration** (removal of the gonads) results in marked increases in the secretion of both LH and FSH. For this discussion it is important to realize that even though FSH and LH are produced by a single cell type, their secretion rates can be altered to different degrees by negative-feedback inputs.

Testosterone inhibits mainly LH secretion. It does so in two ways (Figure 19-10): (1) It acts on the hypothalamus to decrease the frequency of GnRH bursts, and the decreased amount of GnRH reaching the pituitary results in less secretion of the gonadotropins; and (2) it acts directly on the anterior pituitary to cause mainly less LH secretion in response to any given level of GnRH.

How does the presence of functioning testes reduce *FSH* secretion? The major inhibitory signal, exerted directly on the anterior pituitary, is the protein hormone inhibin secreted by the Sertoli cells (Figure 19-10). That the Sertoli cells, via inhibin, are the major source of feedback inhibition of FSH secretion makes sense since, as pointed out above, the facilitatory effect of FSH on spermatogenesis is exerted via the Sertoli cells. Thus, these cells are in all ways the link between FSH and spermatogenesis.

Despite all these complexities, one should not lose sight of the fact that the total amounts of GnRH, LH, FSH, and testosterone secreted and of sperm produced are relatively constant from day to day in the adult male. This is completely different from the large cyclical swings of activity so characteristic of the female reproductive processes.

Testosterone

In addition to its essential paracrine action within the testes on spermatogenesis and its negative-feedback effects on the hypothalamus and anterior pituitary, testosterone exerts a large number of other effects, as summarized in Table 19-3.

Accessory Reproductive Organs. The fetal differentiation, and later growth and function of the entire male duct system, glands, and penis all depend upon testosterone. Following castration in the adult, all the accessory

reproductive organs decrease in size, the glands markedly reduce their secretion rates, and the smooth-muscle activity of the ducts is diminished. Erection and ejaculation may be deficient. These defects disappear upon the administration of testosterone.

Secondary Sex Characteristics and Growth. Virtually all the male secondary sex characteristics are dependent on testosterone. For example, a male castrated before puberty does not develop a beard or either underarm or pubic hair. Other testosterone-dependent secondary sexual characteristics are deepening of the voice resulting from growth of the larynx, thick secretion of the skin oil glands (this predisposes to acne), and the masculine pattern of fat distribution. As described in Chapter 18, testosterone also stimulates bone growth, largely indirectly through its stimulation of growth hormone secretion, but ultimately shuts off bone growth by causing closure of the bones' epiphyseal plates. Also as described in Chapter 18, testosterone is an "anabolic steroid" in that it exerts a direct stimulatory effect on protein synthesis in muscle. Testosterone is necessary for expression of the genetic determinant of the common type of baldness ("male pattern") in men.

Behavior. Testosterone is essential in males for the development of sex drive at puberty. It also plays an important role in maintaining sex drive in the adult male, although men often remain sexually active, albeit usually at a reduced level, for years after castration. (This is true only if the castration occurs during adult life.)

A controversial question is whether testosterone influences other human behaviors in addition to sexual behavior; that is, are there any other inherent male-female

---

TABLE 19-3   EFFECTS OF TESTOSTERONE IN THE MALE

1. Required for initiation and maintenance of spermatogenesis (acts via Sertoli cells)
2. Decreases GnRH secretion via an action on the hypothalamus
3. Inhibits LH secretion via an action on the anterior pituitary
4. Induces differentiation of male accessory reproductive organs and maintains their function
5. Induces male secondary sex characteristics; opposes action of estrogen on breast growth
6. Stimulates protein anabolism, bone growth, and cessation of bone growth
7. Required for sex drive and may enhance aggressive behavior

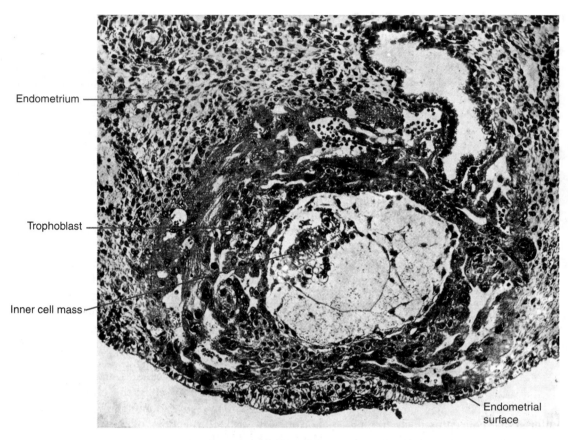

Endometrium

Trophoblast

Inner cell mass

Endometrial surface

FIGURE 19-23
Eleven-day human embryo completely embedded in the uterine lining. [*From A.T. Hertig and J. Rock, Carnegie Contrib. Embryol. 29:127 (1941).*]

dioxide, move by simple diffusion, whereas others utilize transport proteins in the plasma membranes of the epithelial cells. Still other nutrients (for example, several amino acids) and hormones are produced by the trophoblast layer of the placenta itself and added to the fetal blood. It must be emphasized that there is an *exchange* of materials between the two blood streams but no actual *commingling* of the fetal and maternal blood.

Meanwhile, a space called the **amniotic cavity** has formed between the inner cell mass and the trophoblast (Figure 19-25). The epithelial layer lining the cavity is derived from the inner cell mass and is called the **amnion,** or **amniotic sac.** It eventually fuses with the inner surface of the chorion so that only a single combined membrane surrounds the fetus. The fluid in the amniotic cavity, the **amniotic fluid,** resembles the fetal extracellular fluid, and it buffers mechanical disturbances and temperature variations.

The fetus, floating in the amniotic cavity and attached by the umbilical cord to the placenta, develops into a viable infant during the next 8 months. Note in Figure 19-25 that eventually only the amniotic sac separates the fetus from the uterine lumen.

Amniotic fluid can be sampled (**amniocentesis**) as early as the sixteenth week of pregnancy by inserting a needle into the amniotic cavity. A large number of genetic diseases can be diagnosed by the finding of certain chemicals either in the fluid or in cells suspended in the fluid. The chromosomes of these cells can also be examined for diagnosis of certain disorders as well as to determine the sex of the fetus. Another technique for fetal diagnosis is **chorionic villus sampling.** This technique, which can be performed as early as 9–12 weeks of pregnancy, involves obtaining tissue from a chorionic villus of the placenta. A third technique for fetal diagnosis is ultrasound, which provides a "picture" of the fetus without the use of x-rays. A fourth technique coming into wide use as a routine screening for various fetal abnormalities involves the obtaining only of *maternal* blood and measurement in it of several normally occurring proteins whose concentrations

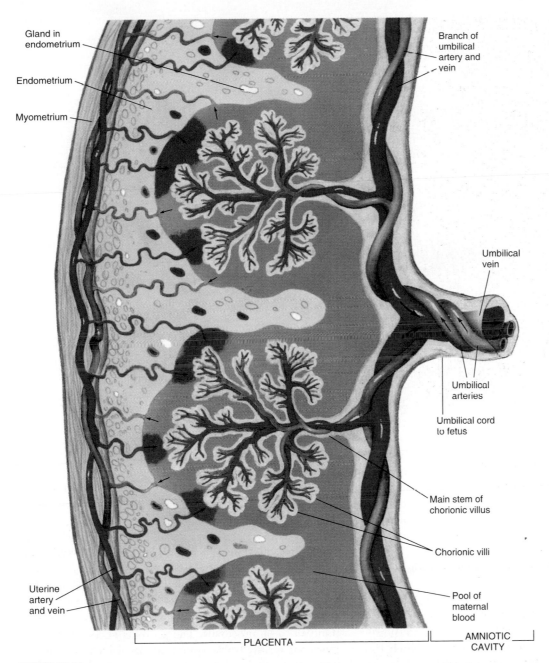

FIGURE 19-24

Interrelations of fetal and maternal tissues in formation of placenta. See Figure 19-25 for orientation of the placenta. (*From B.M. Carlson, "Patten's Foundations of Embryology," 5th ed., McGraw-Hill, New York, 1988.*)

change in the presence of these abnormalities. For example, particular changes in the concentrations of two hormones produced during pregnancy—chorionic gonadotropin and estriol—and alpha-fetoprotein (a major fetal plasma protein that crosses the placenta into the maternal blood) can identify 60 percent of cases with **Down's syndrome,** which includes mental retardation.

Maternal nutrition is crucial for the fetus. Malnutrition early in pregnancy can cause certain congenital abnormalities and/or the formation of an inadequate placenta.

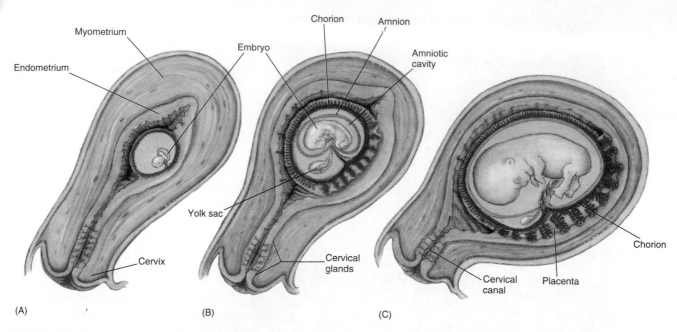

FIGURE 19-25

The uterus at (A) 3, (B) 5, and (C) 8 weeks after fertilization. Embryos and their membranes are drawn to actual size. Uterus is within actual size range. (*From B.M. Carlson, "Patten's Foundations of Embryology," 5th ed., McGraw-Hill, New York, 1988.*)

Malnutrition retards fetal growth and results in infants with higher-than normal death rates, reduced growth after birth, and an increased incidence of learning disabilities. Specific nutrients, not just total calories, are also very important, as manifested, for example, by the increased incidence of neural defects in the offspring of mothers who are deficient in the B-vitamin folate (also called folic acid and folacin). Moreover, it is likely that the mother's diet during pregnancy can have consequences for the offspring that do not show up for many years, for example, the risk of developing heart disease late in life.

The developing embryo and fetus are also subject to considerable influence by a host of other factors (noise, radiation, chemicals, viruses, and so on) to which the mother may be exposed. For example, drugs taken by the mother can reach the fetus via transport across the placenta and impair fetal growth and development. In this regard, it must be emphasized that aspirin, alcohol, and the chemicals in cigarette smoke are very potent agents, as are "street drugs" such as cocaine. Any chemical agent that can cause birth defects in the fetus is known as a ***teratogen.***

It should also be recognized that, since half of the fetal genes—those from the father—differ from those of the mother, the fetus is in essence a foreign transplant in the mother. Why the fetus is not rejected as are other transplants is still not well understood and will be discussed further in Chapter 20.

Finally, how specific tissues and organs form is presently one of the most exciting and rapidly developing areas in all of biology. Before any tissues or organs form, earlier steps must occur, steps that indicate to cells "who" they are, what tissues they should form, and where. In essence, cells must be assigned "addresses" and functions in orderly sequence. Recall that initially all the cells in the embryo undergoing cleavage are identical. How is differentiation of these totipotential cells triggered? In the past few years whole classes of genes—the *Hox genes* and *hedgehog genes,* for example—have been identified that are activated within the first few days after fertilization and guide all these processes. The proteins encoded by these genes are mainly transcription factors, which provide the signals needed to trigger proliferation and differentiation. Thus, these genes, via their encoded proteins, lay out the basic structure and organization of the body—where the head and limbs will be, and so on. Just what activates these genes in the proper sequences along the various axes of the body (top-to-bottom, right-to-left, back-to-front) remains to be determined.

Hormonal and Other Changes During Pregnancy

Throughout pregnancy, plasma concentrations of estrogen and progesterone remain high (Figure 19-26). Estrogen stimulates growth of the uterine muscle mass, which will eventually supply the contractile force needed to deliver the fetus. Progesterone inhibits uterine motility so that the

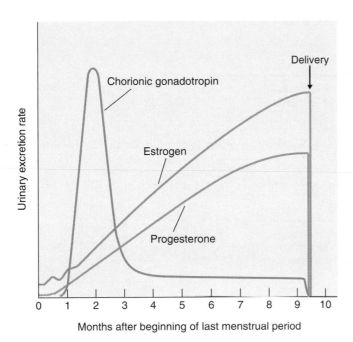

FIGURE 19-26

Urinary excretion of estrogen, progesterone, and chorionic gonadotropin during pregnancy. Urinary excretion rates are an indication of blood concentrations of these hormones.

fetus is not expelled prematurely. During approximately the first 2 months of pregnancy, almost all the estrogen and progesterone are supplied by the extremely active corpus luteum.

Recall that if pregnancy had not occurred, this gland-like structure would have degenerated within 2 weeks after its formation. The persistence of the corpus luteum during pregnancy is due to a hormone called **chorionic gonadotropin (CG),** which starts to be secreted by the trophoblast cells around the time they start their endometrial invasion. CG gains entry to the maternal circulation, and the detection of this hormone in the mother's plasma and/or urine is used as a test of pregnancy; it is positive before the next expected menstruation. This protein hormone is very similar to LH, and it not only prevents the corpus luteum from degenerating but strongly stimulates steroid secretion by it. Thus, the signal that preserves the corpus luteum comes from the conceptus, not the mother's tissues.

The secretion of CG reaches a peak 60 to 80 days after the last menstrual period (Figure 19-26). It then falls just as rapidly, so that by the end of the third month it has reached a low, but definitely detectable, level that remains relatively constant for the duration of the pregnancy. Associated with this falloff of CG secretion, the placenta begins to secrete large quantities of estrogen and progesterone. The very marked increases in plasma concentra-

tions of estrogen and progesterone during the last 6 months of pregnancy are due entirely to their secretion by the trophoblast cells of the placenta, and the corpus luteum regresses after about 3 months. Indeed, surgical removal of the ovaries during the last 7 months of pregnancy, as opposed to the first 2 months, has no effect at all upon the pregnancy.

Recall that secretion of GnRH and, hence, of LH and FSH is powerfully inhibited by high concentrations of progesterone in the presence of estrogen. Since both these steroid hormones are present in high concentrations throughout pregnancy, the secretion of the pituitary gonadotropins remains extremely low.

An important aspect of placental steroid secretion is that the placenta has the enzymes required for the synthesis of progesterone but not those for the formation of androgens, which are the precursors of estrogen. The placenta is supplied, via the fetal circulation, with these androgens, produced by an interaction between the *fetal* adrenal glands and liver. The placenta converts the androgens mainly into **estriol**, the major estrogen of pregnancy.

The trophoblast cells of the placenta produce not only CG and steroids, but inhibin and many other hormones as well, all examples of hormones produced by the conceptus and influencing the mother. Some of these (for example, thyroid-stimulating hormone) are identical to hormones normally produced by other endocrine glands, whereas some are unique. One unique hormone that is secreted in very large amounts has effects similar to those of both prolactin and growth hormone. This protein hormone, **placental lactogen** (also called chorionic somatomammotropin), may play several roles in the mother—mobilizing fat for energy and stabilizing plasma glucose at relatively high levels (growth-hormone-like effects) and facilitating development of the breasts (a prolactin-like effect). In the fetus, this hormone may exert growth-promoting effects.

In addition to all these messengers produced by the trophoblast, the endometrium also secretes a variety of hormones and growth factors important for maintaining the pregnancy.

Some of the numerous other physiological changes, hormonal and nonhormonal, in the mother during pregnancy are summarized in Table 19-9.

One comment about fluid balance during pregnancy is necessary: Approximately 5–10 percent of pregnant women retain abnormally large amounts of fluid and manifest edema, protein in the urine, and hypertension. These are the symptoms of **preeclampsia;** when convulsions also occur, the condition is termed **eclampsia.** The fetus is also affected, sometimes resulting in intrauterine growth retardation and death. The factors responsible for eclampsia are unknown, but the evidence strongly implicates altered function of the maternal blood vessels and

TABLE 19-9   MATERNAL RESPONSES TO PREGNANCY

	Response
Placenta	Secretion of estrogen, progesterone, chorionic gonadotropin, inhibin, placental lactogen, and other hormones.
Anterior pituitary	Increased secretion of prolactin and ACTH. Secretes very little FSH and LH.
Adrenal cortex	Increased secretion of aldosterone and cortisol.
Posterior pituitary	Increased secretion of vasopressin.
Parathyroids	Increased secretion of parathyroid hormone.
Kidneys	Increased secretion of renin, erythropoietin, and 1,25-dihydroxyvitamin D_3. Retention of salt and water. *Cause:* Increased aldosterone, vasopressin, and estrogen.
Breasts	Enlarge and develop mature glandular structure. *Cause:* Estrogen, progesterone, prolactin, and placental lactogen
Blood volume	Increases. *Cause:* Total erythrocyte volume is increased by erythropoietin, and plasma volume by salt and water retention.
Calcium balance	Positive. *Cause:* Increased parathyroid hormone and 1,25-dihydroxyvitamin D_3.
Body weight	Increases by average of 12.5 kg, 60 percent of which is water.
Circulation	Cardiac output increases, total peripheral resistance decreases (vasodilation in uterus, skin, breasts, GI tract, and kidneys), mean arterial pressure stays constant.
Respiration	Hyperventilation (arterial P_{CO_2} decreases).
Organic metabolism	Metabolic rate increases. Plasma glucose, gluconeogenesis, and fatty acid mobilization all increase. *Cause:* Hyporesponsiveness to insulin due to insulin antagonism by placental lactogen and cortisol.
Appetite and thirst	Increase.
Nutritional RDAs	Increase.

inadequate invasion of the endometrium by trophoblast cells, resulting in poor blood perfusion of the placenta.

Pregnancy Sickness.   The majority of women suffer from **pregnancy sickness** (popularly called morning sickness)—nausea, vomiting, changes in the perception of food palatability, and the presence of taste aversions—during the first three months (first trimester) of pregnancy. The exact cause is unknown, but high concentrations of estrogen, progesterone, and other substances secreted at this time are thought to act on the vomiting center (Chapter 17) in the brain. It has been hypothesized that pregnancy sickness is actually an adaptive (that is, beneficial) process, one that minimizes the mother's intake of potentially toxic chemicals during the first trimester, when the embryo and fetus are particularly susceptible. Certainly, the occurrence of pregnancy sickness is not associated with poor fetal outcome; if anything, the opposite is observed, supporting the hypothesis of an adaptive process.

Parturition

A normal human pregnancy lasts approximately 40 weeks. Safe survival of premature infants is now possible at about the twenty-fourth week of pregnancy, but treatment of these infants often requires heroic efforts at great costs.

Delivery of the infant is followed by the placenta, and the entire process is termed **parturition.** Parturition is produced by strong rhythmical contractions of the myometrium. Actually, weak and infrequent uterine contractions begin at approximately 30 weeks and gradually increase in both strength and frequency. During the last month, the entire uterine contents shift downward so that the baby is brought into contact with the cervix. In over 90 percent of births, the baby's head is downward and acts as the wedge to dilate the cervical canal when labor begins (Figure 19-27).

This dilation can occur because, during the last few weeks of pregnancy, the cervix becomes soft and flexible. The cervical softening (often termed "ripening") is largely the result of a breakup of collagen fibers. These biochemical changes are mediated by a variety of messengers, including estrogen and prostaglandins.

At the onset of labor or before, the amniotic sac ruptures and the amniotic fluid escapes through the vagina. When labor begins in earnest, the uterine contractions become coordinated and quite strong (although usually painless at first) and occur at approximately 10- to 15-min intervals. The contractions begin in the upper portion of the uterus and sweep downward. Such coordination is possible because the uterine smooth-muscle cells are connected to each other by gap junctions. These junctions increase markedly in number near the end of pregnancy.

As the contractions increase in intensity and frequency, the cervical canal is gradually forced open to a maximum diameter of approximately 10 cm. Until this point, the contractions have not moved the fetus out of the uterus but have served only to dilate the cervix. Now the contractions move the fetus through the cervix and vagina. At this time the mother, by bearing down to increase abdominal pressure, can help the uterine contractions to deliver the baby. The umbilical vessels and placenta are still functioning, so that the baby is not yet on its own, but within minutes of delivery both the umbilical vessels and the placental vessels completely constrict, stopping blood flow to the placenta. The entire placenta becomes separated from the underlying uterine wall, and a wave of uterine contractions delivers the placenta as the **afterbirth.**

Ordinarily, parturition proceeds automatically from beginning to end and requires no significant medical intervention. In a small percentage of cases, however, the position of the baby or some maternal defect can interfere with normal delivery. The headfirst position of the fetus is important for several reasons: (1) If the baby is not oriented headfirst, another portion of its body is in contact with the cervix and is generally a far less effective wedge. (2) Because of the head's large diameter compared with the rest of the body, if the body were to go through the cervical canal first, the canal might obstruct the passage of the head, leading to problems when the partially delivered baby attempts to breathe. (3) If the umbilical cord becomes caught between the canal wall and the baby's head or chest, mechanical compression of the umbilical vessels can result. Despite these potential problems, however, many babies who are not oriented headfirst are born normally.

What mechanisms control the events of parturition? Let us consider a set of fairly well established facts:

1. The autonomic neurons to the uterus are of little importance in parturition since anesthetizing them does not interfere with delivery.
2. The smooth-muscle cells of the myometrium have inherent rhythmicity and are capable of autonomous contractions, which are facilitated as the muscle is stretched by the growing fetus.
3. The pregnant uterus near term and during labor secretes several prostaglandins that are profound stimulators of uterine smooth-muscle contraction.
4. **Oxytocin,** one of the hormones released from the posterior pituitary, is an extremely potent uterine muscle stimulant. It not only acts directly on uterine smooth muscle but also stimulates it to synthesize the prostaglandins mentioned above. Oxytocin is reflexly released as a result of input to the hypothalamus from receptors in the uterus, particularly the cervix. During pregnancy the number of oxytocin receptors in the uterus markedly and progressively increases, mainly as a result of estrogen stimulation.
5. Throughout pregnancy progesterone exerts an essential powerful inhibitory effect upon uterine contractions by decreasing the sensitivity of the myometrium to estrogen, oxytocin, and prostaglandins. Unlike the situation in many other species, however, the rate of progesterone secretion does not decrease before or during parturition in women (until delivery of the placenta, the source of the progesterone).

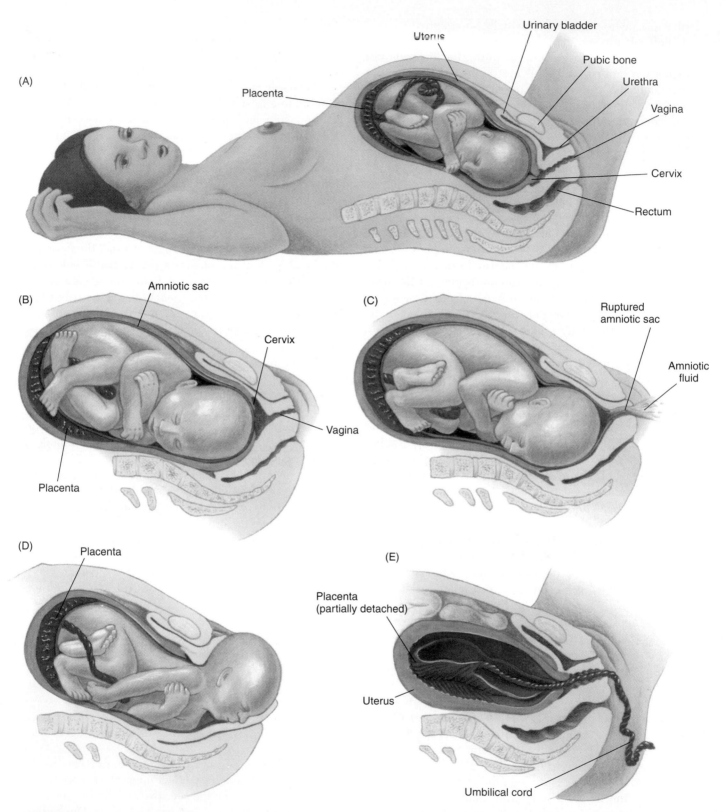

FIGURE 19-27

Stages of parturition. (A) Parturition has not yet begun. (B) The cervix is dilating. (C) The cervix is completely dilated and the fetus's head is entering the cervical canal; the amniotic sac has ruptured and the amniotic fluid escaped. (D) The fetus is moving through the vagina. (E) The placenta is coming loose from the uterine wall preparatory to its expulsion.

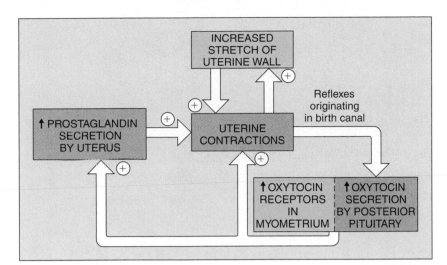

FIGURE 19-28
Factors stimulating uterine contractions during parturition. Note the positive-feedback nature of several of the inputs. The increased number of oxytocin receptors is due mainly to stimulation by estrogen.

These facts can now be put together in a unified pattern, as shown in Figure 19-28. Once started, the uterine contractions exert a positive feedback effect upon themselves via both local facilitation of inherent uterine contractions and reflex stimulation of oxytocin secretion. But precisely what the relative importance of all these factors is in *initiating* labor remains unclear.

The action of prostaglandins on parturition is the last in a series of prostaglandin effects on the female reproductive system we have described. They are summarized in Table 19-10.

Lactation

The secretion of milk by the breasts, or **mammary glands,** is termed **lactation.** The breasts contain ducts that branch all through the tissue and converge at the nipples (Figure 19-29). These ducts arise in saclike glands called **alveoli** (the same term is used to denote the lung air sacs). The breast alveoli, which are the sites of milk secretion, look like bunches of grapes with stems terminating in the ducts. The alveoli and the ducts immediately adjacent to them are surrounded by specialized contractile cells called **myoepithelial cells.**

TABLE 19-10   SOME EFFECTS OF PROSTAGLANDINS* ON THE FEMALE REPRODUCTIVE SYSTEM

Site of production	Action	Result
Late-antral follicle	Stimulate production of digestive enzymes	Rupture of follicle
Corpus luteum	May interfere with corpus luteum's hormone secretion and function	?Death of corpus luteum
Uterus	Constrict blood vessels in endometrium	Onset of menstruation
	Cause changes in endometrial blood vessels and cells early in pregnancy	Facilitate implantation
	Increase contraction of myometrium	Help initiate both menstruation and parturition
	Cause cervical "ripening"	Facilitate cervical dilation during parturition

*The term "prostaglandins" is used loosely here, as is customary in reproductive physiology, to include all the eicosanoids.

ESSAY

CLONING AN ADULT SHEEP

In February, 1997, the world was startled by the news that an adult mammal had been cloned for the first time. A normal healthy lamb, named Dolly, had been formed beginning with a single cell from the mammary gland (udder) of her genetic mother. Recall that except for the gametes, each cell in the adult body, no matter how specialized, carries in its nucleus a full complement of 46 chromosomes, 23 from each parent. Dolly received all of her 46 chromosomes from the single mammary gland cell, so she is an identical copy of her genetic mother, and she has no father.

During the previous decade scientists had come to believe that there were insurmountable practical problems in cloning a mammal from a single cell taken from an adult animal's body. The genes in such a cell (again excepting the individual gametes) carry a complete blueprint for the entire organism, but in the process of differentiation, many of the genes are "turned off," and only certain genes are expressed. It was assumed, therefore—and failures in cloning attempts had reinforced this assumption—that not all the "turned off" genes in an adult differentiated cell could be "turned on" again, an event essential for creating a new organism from that cell. The key to success for Dr. Ian Wilmut and his colleagues at the Roslin Institute in Edinburgh, Scotland, was that after removing the mammary-gland cell, these scientists cultured it for 5 days in a solution deficient in particular nutrients and growth factors. This maneuver "locked" the cell in a quiescent phase of the division cycle, which somehow made its genes more susceptible to being reprogrammed to initiate the development of a new organism when the stimulus was given.

The actual sequence of events leading to Dolly was as follows. (1) A mammary-gland cell was removed from an adult sheep (the genetic mother) and cultured, as described above. (2) A second sheep (the egg donor) was stimulated, by hormone administration, to ovulate, and her unfertilized eggs were collected; these eggs' nuclei (and thus their DNA) were then removed by microsurgery and discarded. (3) The nucleus of the mammary-gland cell was transferred into the egg cell, and this nucleus provided the egg cell with a full complement of genes from the genetic mother. (4) The egg was given an electrical stimulus (analogous to the stimulus that occurs during fertilization with a sperm), which caused it to start dividing and developing like a normal embryo. (5) At the blastocyst stage, the embryo was implanted into the uterus of still another sheep (the surrogate mother), where it continued to term and birth.

It should be emphasized that of the 277 mammary-gland-cell nuclei that were fused with enucleated eggs only 29 developed to the blastocyst stage at which they could be transplanted into surrogate mothers, and only Dolly successfully carried through to birth. Whether Dolly was simply a one-in-a-million or whether improvement of techniques will lead to much greater success rates remains unknown.

The Scottish researchers have made it clear that their interest in cloning is solely to produce animals that can be used more efficiently for particular purposes, such as the secretion of particularly valuable drugs in their milk. But it is extremely likely that if a sheep can be cloned so can a human being, and this raises profound ethical problems that will be difficult to resolve. For example, should parents be allowed to produce an identical copy of their dying child?

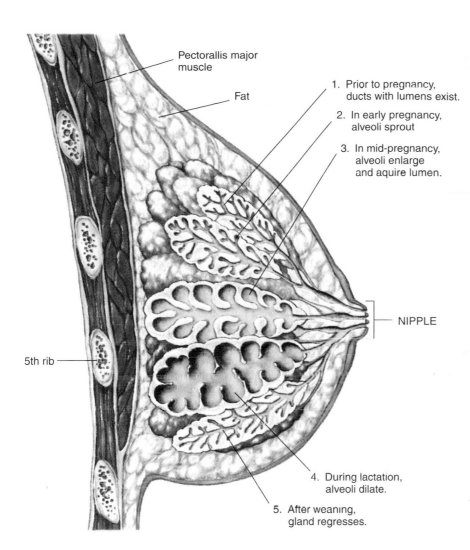

Pectorallis major muscle

Fat

1. Prior to pregnancy, ducts with lumens exist.

2. In early pregnancy, alveoli sprout

3. In mid-pregnancy, alveoli enlarge and aquire lumen.

NIPPLE

5th rib

4. During lactation, alveoli dilate.

5. After weaning, gland regresses.

FIGURE 19-29
Anatomy of the breast. The numbers refer to the sequential changes that occur over time. (*Adapted from Elias et al.*)

Before puberty, the breasts are small with little internal glandular structure. With the onset of puberty in females, the increased estrogen causes a marked enhancement of duct growth and branching but relatively little development of the alveoli, and much of the breast enlargement at this time is due to fat deposition. Progesterone secretion also commences at puberty during the luteal phase of each cycle, and this hormone contributes to breast growth by stimulating growth of alveoli.

During each menstrual cycle, the breasts undergo fluctuations in association with the changing blood concentrations of estrogen and progesterone, but these changes are small compared with the marked breast enlargement that occurs during pregnancy as a result of the stimulatory effects of high plasma concentrations of estrogen, progesterone, prolactin, and placental lactogen. This last hormone, as described earlier, is secreted by the placenta, whereas prolactin is secreted by the anterior pituitary. Under the influence of these hormones (and others not mentioned), both the ductal and the alveolar structures become fully developed.

As described in Chapter 10, the anterior pituitary cells that secrete prolactin are influenced by many hormones. They are inhibited by **dopamine,** which is secreted by the hypothalamus. They are probably stimulated by at least one **prolactin releasing factor (PRF),** also secreted by the hypothalamus (the identity of PRF is still uncertain). The dopamine and PRF secreted by the hypothalamus are hypophysiotropic hormones in that they reach the anterior pituitary by way of the hypothalamo-pituitary portal vessels. Estrogen also acts on the anterior pituitary to stimulate prolactin secretion.

Under the dominant inhibitory influence of dopamine, prolactin secretion is low before puberty. It then increases considerably at puberty in girls but not in boys, stimulated by the increased plasma estrogen concentration that occurs at this time. During pregnancy, there is a marked further increase in prolactin secretion due to stimulation by estrogen. Despite the fact that prolactin is elevated and the breasts are markedly enlarged and fully developed as pregnancy progresses, there is no secretion of milk. This is

becauoc ootrogon and progesterone, in larg⌿
tions, prevent milk production by inhibiting ⌿
action of prolactin on the breasts. Thus, altho⌿
causes an increase in the *secretion* of prola⌿
with prolactin in promoting breast growth ar⌿
tion, it, along with progesterone, is antago⌿
lactin's ability to induce milk secretion. Del⌿
the source—the placenta—of the large amo⌿
gen and progesterone and, thereby, the inhi⌿
production.

The drop in estrogen following parturi⌿
basal prolactin secretion to decrease from⌿
pregnancy levels and after several months⌿
ward prepregnancy levels even though the n⌿
ues to nurse. Superimposed upon this basal l⌿
are large secretory bursts of prolactin during⌿

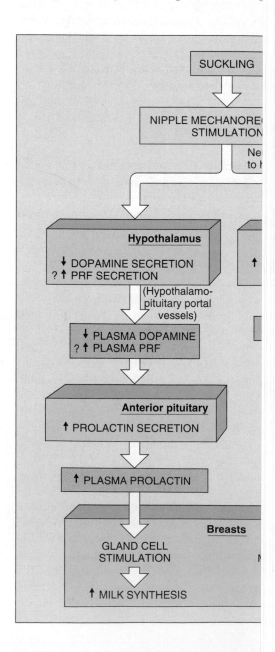

SUCKLING

NIPPLE MECHANORE(⌿
STIMULATION

Hypothalamus

↓ DOPAMINE SECRETION
? ↑ PRF SECRETION

(Hypothalamo-
pituitary portal
vessels)

↓ PLASMA DOPAMINE
? ↑ PLASMA PRF

Anterior pituitary

↑ PROLACTIN SECRETION

↑ PLASMA PROLACTIN

Breasts

GLAND CELL
STIMULATION

↑ MILK SYNTHESIS

NONSPECIFIC IMMUNE DEFENSES

Nonspecific immune defenses protect against foreign cells or matter without having to recognize their specific identities. These defenses must, of course, recognize some *general* property marking the invader as foreign; the most common such identity tags are carbohydrates that are frequent constituents of microbial cell walls. As we shall see, plasma-membrane receptors on certain immune cells, as well as at least one family of circulating proteins (called complement) are able to bind to these carbohydrates at crucial steps in nonspecific responses.

The nonspecific immune responses include defenses at the body surfaces, the response to injury known as inflammation, and a family of antiviral proteins called interferons. Another nonspecific defense—the lymphocytes called natural killer (NK) cells—is best described later in the context of the specific immune defenses that commonly mobilize them.

Defenses at Body Surfaces

The body's first lines of defense against microbes are the barriers offered by surfaces exposed to the external environment. Very few microorganisms can penetrate the intact skin, and the various skin glands and the lacrymal glands all secrete antimicrobial chemicals.

The mucus secreted by the epithelial linings of the respiratory and upper gastrointestinal tracts also contains antimicrobial chemicals, but more important, mucus is sticky. Particles that adhere to it are prevented from entering the blood. They are either swept by ciliary action up into the pharynx and then swallowed, as occurs in the upper respiratory tract, or are engulfed by macrophages in the various linings.

Other specialized surface defenses are the hairs at the entrance to the nose, the cough and sneeze reflexes (Chapter 15), and the acid secretion of the stomach and uterus, which kills microbes. Finally, a major defense against infection are the many relatively innocuous microbes normally found on the skin and other linings exposed to the external environment. Through a variety of mechanisms these microbes suppress the growth of other potentially more dangerous ones.

Inflammation

Inflammation is the body's local response to infection or injury. Regardless of cause, inflammation is relatively stereotyped since a major trigger is cell or tissue injury. The function of inflammation is to destroy or inactivate foreign invaders and/or set the stage for tissue repair. In this section we describe inflammation as it occurs in the *nonspecific* response induced by the invasion of *microbes.* Most of the same responses can be elicited by a variety of

other injuries—cold, heat, and trauma, for example. Moreover, we shall see later that inflammation is also an important component of many *specific* immune responses, in which the inflammation becomes amplified and made more effective.

The key actors in inflammation are **phagocytes.** This term denotes any cell capable of the form of endocytosis termed **phagocytosis,** whereby particulate matter is engulfed (Chapter 6). The engulfed matter is then usually destroyed inside the phagocyte. The most important phagocytes are neutrophils, monocytes, macrophages, and the macrophage-like cells.

The sequence of local events in a typical nonspecific inflammatory response to a bacterial infection, one caused, for example, by a cut with a bacteria-covered knife, is summarized in Table 20-3. The familiar manifestations of tissue injury and inflammation are local redness, swelling, heat, and pain.

The events of inflammation that underly these manifestations are induced and regulated by a huge number of chemical mediators, some of which are summarized for reference in Table 20-4 (not all of these will be described in this chapter). Any given event of inflammation, such as vasodilation, may be induced by multiple mediators, and any given mediator may induce more than one event. Based on their origins, the mediators fall into two general categories: (1) peptides (kinins, for example) generated in the infected area by the enzymatic splitting of proteins that circulate in the plasma; and (2) substances released into the extracellular fluid from cells that either already exist in the infected area (mast cells, for example) or enter it during inflammation (neutrophils, for example). The stimuli for the generation or release of the mediators are either microbial chemical substances or other chemical messengers released by tissue cells that have interacted with the microbes.

TABLE 20-3 SEQUENCE OF EVENTS IN A LOCAL INFLAMMATORY RESPONSE TO BACTERIA

1. Initial entry of bacteria into tissue
2. Vasodilation of the microcirculation in the infected area, leading to increased blood flow
3. Marked increase in protein permeability of the venules in the infected area, with resulting diffusion of protein and filtration of fluid into the interstitial fluid
4. Chemotaxis: exit of leukocytes from the venules into the interstitial fluid of the infected area
5. Destruction of bacteria in the tissue either through phagocytosis or by mechanisms not requiring prior phagocytosis
6. Tissue repair

TABLE 20-4   SOME IMPORTANT LOCAL INFLAMMATORY MEDIATORS

Mediator	Source
Kinins	Plasma proteins
Complement	Plasma proteins
Products of blood clotting	Plasma proteins
Histamine	Mast cells
Eicosanoids	Many cell types
Platelet-activating factor	Many cell types
Cytokines, including chemokines. Examples are interleukin 1, tumor necrosis factor, and interleukin 6	Monocytes, macrophages, neutrophils, lymphocytes, and several nonimmune cell types, including endothelial cells and fibroblasts
Lysosomal enzymes, nitric oxide, and other oxygen-derived substances	Neutrophils and macrophages

Let us now go step by step through the process summarized in Table 20-3, assuming that the bacterial infection in our example is localized to the tissue just beneath the skin. (If the invading bacteria enter the blood or lymph, then similar inflammatory responses would take place in any other tissue or organ invaded by the blood-borne or lymph-borne microorganisms.)

Vasodilation and Increased Permeability to Protein. A variety of chemical mediators dilate most of the microcirculation vessels in an infected and/or damaged area. The mediators also cause the local capillaries to become quite permeable to proteins by inducing the capillary endothelial cells to contract, opening spaces between them.

The adaptive value of these vascular changes is twofold: (1) The increased blood flow to the inflamed area, which accounts for the redness, increases the delivery of proteins and leukocytes, and (2) the increased capillary permeability to protein ensures that the plasma proteins that participate in inflammation—many of which are normally restrained by the intact capillary endothelium—can gain entry to the area.

As described in Chapter 14, the vasodilation and increased permeability to protein cause net filtration of plasma into the interstitial fluid and the formation of edema. This accounts for the swelling in an inflamed area, which is simply a consequence of the changes in the microcirculation and has no known adaptive value of its own.

Chemotaxis. Within 30 to 60 min of the onset of inflammation, circulating neutrophils move out of the blood across the endothelium of post-capillary venules to enter the inflamed area. This multistage process is known as **chemotaxis.** It involves a variety of protein and carbohy-

drate **adhesion molecules** on both the endothelial cell and the neutrophil, and it is regulated by messenger molecules released by cells in the area, including the endothelial cells. These messengers are collectively termed **chemoattractants** (also termed **chemotaxins** or chemotactic factors).

In the first stage, the neutrophil is loosely tethered to the endothelial cells via a particular class of adhesion molecules; this event is associated with rolling of the granulocyte along the vessel surface. In essence, this initial reversible event permits the neutrophil to be exposed to chemoattractants being released in the injured area. These chemoattractants act on the neutrophil to induce the rapid appearance of another class of adhesion molecules in its plasma membrane—molecules that bind tightly to their matching molecules in the endothelial cells. In the next stage, via still other adhesion molecules, a narrow projection of the neutrophil is inserted into the space between two endothelial cells, and the entire neutrophil squeezes through the endothelial wall and into the interstitial fluid. In this way, huge numbers of neutrophils migrate into the inflamed area and move toward the microbes.

But movement of leukocytes from the blood into the damaged area is not limited to neutrophils. Monocytes follow later, and once in the tissue they undergo anatomical and functional changes that transform them to macrophages. As we shall see later, lymphocytes also undergo chemotaxis in specific immune responses, as can basophils and eosinophils under certain conditions.

An important aspect of the multistep chemotaxis process is that it provides selectivity and flexibility for the migration of the various leukocyte types. Multiple adhesion molecules that are relatively distinct for the different leukocytes are controlled by different sets of chemoattractants. Particularly important in this regard are those

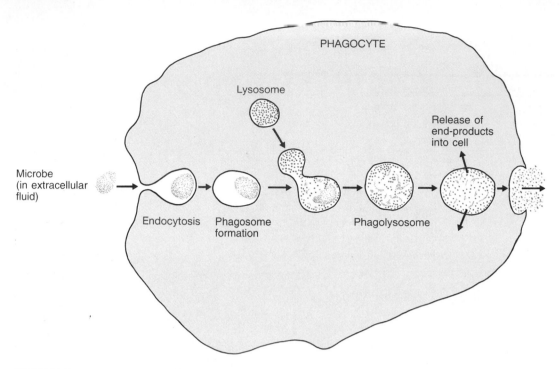

PHAGOCYTE

Lysosome

Release of
end-products
into cell

Microbe
(in extracellular
fluid)

Endocytosis   Phagosome
formation

Phagolysosome

FIGURE 20-1

Phagocytosis and intracellular digestion of a microbe. After digestion has taken place in the phagolysosome, the end products are released to the outside of the cell by exocytosis or used by the cell for its own metabolism.

cytokines that function as chemoattractants for distinct subsets of leukocytes. For example, interleukin 8 stimulates the chemotaxis of neutrophils, whereas interleukin 3 stimulates that of eosinophils. Thus, subsets of leukocytes can be stimulated to enter particular tissues at designated times during an inflammatory response, depending on the type of invader and the cytokine response it induces. The various cytokines that have chemoattractant actions are collectively referred to as **chemokines.**

Killing by Phagocytes.   The initial step in phagocytosis is contact between the surfaces of the phagocyte and microbe. One of the major triggers for phagocytosis during this contact is the interaction of phagocyte receptors with certain carbohydrates in the microbial cell walls. Contact is not itself always sufficient to trigger engulfment, however, particularly with those bacteria that are surrounded by a thick, gelatinous capsule. As we shall see, chemical factors produced by the body can bind the phagocyte tightly to the microbe and markedly enhance phagocytosis. Any substance that does this is known as an **opsonin,** from the Greek word that means "to prepare for eating."

As the phagocyte engulfs the microbe (Figure 20-1), the internal, microbe-containing sac formed in this step is called a **phagosome.** The microbe remains in the phagosome, a layer of plasma membrane separating it from the phagocyte's cytosol. The phagosome membrane then

makes contact with one of the phagocyte's lysosomes, which is filled with a variety of hydrolytic enzymes; the membranes of the two structures fuse, and the combined vesicles are now called the **phagolysosome.** Inside the phagolysosome, the microbe's macromolecules are broken down by the lysosomal enzymes. In addition, enzymes in the phagolysosome membrane produce **nitric oxide** (from the amino acid arginine, Chapter 8), as well as hydrogen peroxide and other oxygen derivatives, all of which are extremely destructive to the microbe's macromolecules.

Intracellular destruction, as described in the previous paragraph, is not the only way phagocytes can kill microbes. The phagocytes also release their antimicrobial hydrolytic enzymes and oxygen derivatives into the *extracellular* fluid, where these chemicals can destroy the microbes without prior phagocytosis. (As we shall discuss later, these chemicals can also damage normal tissue.)

The phagocytes also secrete still other substances into the extracellular fluid (Figure 20-2), some of which function as inflammatory mediators. Thus, positive feedback occurs such that when more phagocytes enter the area and encounter microbes, more inflammatory mediators are released to bring in more phagocytes.

Complement.   The family of plasma proteins known as **complement** provides another means for extracellular killing of microbes, that is killing without prior phagocy-

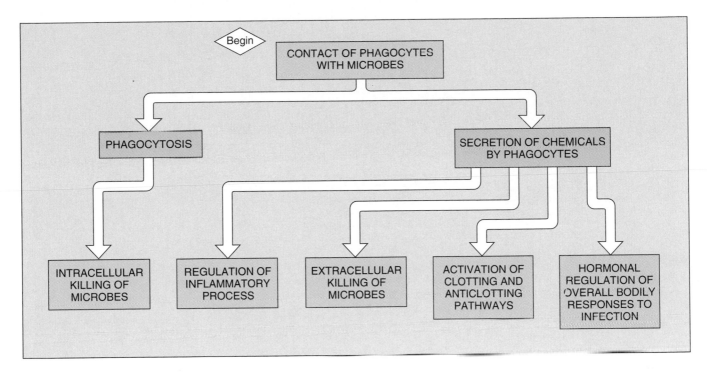

FIGURE 20-2

Role of phagocytes in nonspecific immune responses.

tosis. Certain of the complement proteins always circulate in the blood in an inactive state. Upon activation (see below) of one of the group in response to infection or damage, there occurs a cascade in which this active protein activates a second complement protein, which activates a third, and so on. In this way, multiple active complement proteins are generated in the extracellular fluid of the infected area from inactive complement molecules that have entered from the blood. Since this system consists of at least 20 distinct proteins it is extremely complex, and we shall identify the roles of only a few of the individual complement proteins.

Five of the active proteins generated in the complement cascade form a multi-unit protein, the **membrane attack complex (MAC),** which embeds itself in the microbial plasma membrane. In this manner, channels are created in the membrane, making it leaky. Water and salts enter the microbe, which disrupts the intracellular ionic environment and kills the microbe.

In addition to supplying a means for direct killing of microbes, the complement system serves other important functions in inflammation (Figure 20-3). Some of the activated complement molecules along the cascade cause, either directly or indirectly (by stimulating the release of other inflammatory mediators), vasodilation, increased capillary permeability to protein, and chemotaxis. Also, one of

the complement molecules—C3b—acts as an opsonin to attach the phagocyte to the microbe (Figure 20-4).

As we shall see later, antibodies, a class of proteins secreted by lymphocytes, are required to activate the very first protein (C1) in the full sequence known as the classical complement pathway, but lymphocytes are not involved in *nonspecific* inflammation, our present topic. How, then, is the complement sequence initiated during nonspecific inflammation? The answer is that there is an **alternate complement pathway,** one that is not antibody-dependent and bypasses C1. The alternate pathway is initiated as the result of complex interactions between carbohydrates on the surface of the microbes, inactive complement molecules beyond C1, and several other plasma proteins. The alternate pathway plugs into the classical pathway at the point of formation of C3b, the opsonin described in the previous paragraph. Not all microbes have a surface conducive to initiating the alternate pathway.

Tissue Repair.   The final stage of inflammation is tissue repair. Depending upon the tissue involved, multiplication of organ-specific cells by cell division may or may not occur during this stage. For example, liver cells multiply but neurons do not. In any case, fibroblasts (a type of connective-tissue cell) that migrate into the damaged area

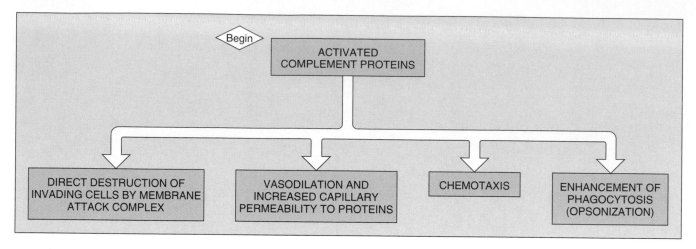

FIGURE 20-3

Functions of complement proteins. The effects on blood vessels and chemotaxis are exerted both directly by complement molecules and indirectly via other inflammatory mediators (for example, histamine) whose release the complement molecules stimulate.

divide rapidly and begin to secrete large quantities of collagen, while blood-vessel cells proliferate in the process of angiogenesis (Chapter 14). All these events are brought about by chemical mediators, particularly a group of locally produced growth factors. Finally, remodeling occurs as the healing process winds down; the final repair may be imperfect, leaving a scar.

Interferon

Interferon refers collectively to a family of cytokines that *nonspecifically* inhibit viral replication inside host cells. In response to infection by a virus, most cell types produce interferon and secrete it into the extracellular fluid. Interferon then binds to plasma-membrane receptors on the secreting cell and on other cells, whether they are infected or not (Figure 20-5). This binding triggers the synthesis of a variety of proteins by the cell. If the cell is already infected or eventually becomes infected, these proteins interfere with the ability of the viruses to replicate. It must be reemphasized that interferon is not specific. Many kinds of viruses induce interferon synthesis, and interferon in turn can inhibit the multiplication of many kinds of viruses.

We shall see in subsequent sections of this chapter that, in addition to these antiviral actions, interferon plays other important roles in host defenses.

SPECIFIC IMMUNE DEFENSES

Because of the complexity of specific immune responses it is useful to present a brief orientation before describing the various components of the response.

Overview

Lymphocytes mediate specific immune defenses. Unlike nonspecific defense mechanisms, lymphocytes must "recognize" the specific foreign matter to be attacked. Any foreign molecule that can trigger a specific immune response against itself or the cell bearing it is termed an **antigen.** Most antigens are either proteins or very large polysaccharides. The term "antigen" does not denote a specific structure in the way that anatomical terms like "microtubule" and "integal membrane protein" do. Rather it is a functional term; that is, any molecule, regardless of its lo-

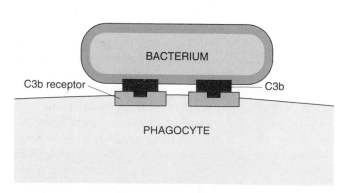

FIGURE 20-4

Function of complement C3b as an opsonin. One portion of this complement molecule binds nonspecifically to the surface of the microbe, whereas another portion binds to specific receptor sites for it on the plasma membrane of the phagocyte. The structures are not drawn to scale.

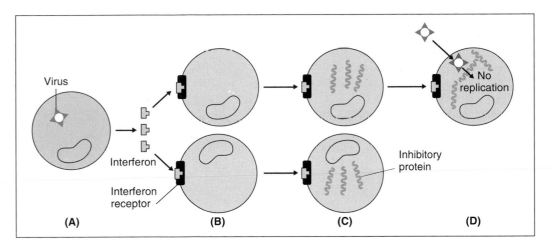

FIGURE 20-5

Role of interferon in preventing viral replication. (A) Most cell types, when infected with viruses, secrete interferon, which enters the interstitial fluid and blood and binds to interferon receptors on the secreting cells, adjacent cells, or far-removed cells (B). This induces synthesis of proteins (C) that inhibit viral replication should viruses enter the cell (D).

cation or function, that can induce a specific immune response is by definition an antigen. It is the ability of lymphocytes to distinguish one antigen from another that confers specificity upon the immune responses in which they participate.

A common example of an antigen is ragweed pollen, the antigen that causes the specific immune response we know as hay fever. In other cases the antigen is part of the surface of a cell—microbe, virus-infected body cell, tumor cell, or transplanted cell—that appears in the body. The reason for saying "appears in" rather than "enters" is that, as we shall see, body cells, themselves, produce the "foreign" molecules that act as antigens on virus-infected cells and tumor cells. These molecules are foreign in the sense that they are not present in normal cells. Also as we shall see, normal cells can be transformed in still other ways and, as a result, possess components that can serve as antigens.

A typical specific immune response can be divided into three stages: (1) antigen encounter and recognition by lymphocytes, (2) lymphocyte activation, and (3) the attack launched by the activated lymphocytes.

(1) During its development each lymphocyte synthesizes and inserts into its plasma membrane receptors that are able to bind to a specific antigen. If, at a later time, the lymphocyte ever encounters that antigen, the antigen becomes bound to the receptors. This binding is the physicochemical meaning of the word "recognize" in immunology. Accordingly, the ability of lymphocytes to distinguish one antigen from another is determined by the nature of their plasma-membrane receptors. Each lymphocyte is specific for just one type

of antigen, and it is estimated that in a typical person the lymphocyte population expresses more than 100 million distinct antigen receptors. We do not mean to imply that every lymphocyte is different from every other one; in most cases a single type of antigen receptor may be expressed by a small number of lymphocytes, termed a clone. Thus, to be more accurate, there are more than 100 million distinct small clones of lymphocytes in the body.

(2) The binding of antigen to receptor is an essential trigger for lymphocyte activation. The lymphocyte undergoes a cell division, and the two resulting daughter cells then also divide, even though only one of them still has the antigen combined with it, and so on. In other words, the original binding of antigen by a single lymphocyte specific for that antigen triggers multiple cycles of cell divisions. As a result, many lymphocytes are formed that are identical to the one that started the cycles and can recognize the antigen; this is termed clonal expansion. These lymphocytes also undergo a differentiation process, depending upon the lymphocyte type. The cell division and differentiation together are termed **lymphocyte activation.** After activation, two of the lymphocyte types—B cells and cytotoxic T cells—then function as "effector lymphocytes," which continue the responses into the attack phase. A third type of lymphocyte, called helper T cells, after activation secrete cytokines that enhance the function of B cells and cytotoxic T cells.

(3) The activated effector lymphocytes launch an attack against all antigens of the kind that initiated the immune response. Theoretically, it takes only one or two

antigen molecules to *initiate* the specific immune response that will then result in an attack on all of the other antigens of that specific kind in the body. B cells differentiate into cells termed plasma cells, which secrete antibodies into the blood, and these antibodies then recruit and guide other molecules and cells to perform the actual attack. In contrast, cytotoxic T cells directly attack and kill the cells bearing the antigens.

Lymphoid Organs and Lymphocyte Origins

Our first task is to describe the organs and tissues in which lymphocytes originate and come to reside. Then we describe the various types alluded to in the overview and summarized in Tables 20-1 and 20-2.

Lymphoid Organs. Like all leukocytes, lymphocytes circulate in the blood. At any moment, the great majority of lymphocytes are not in the blood, however, but in a group of organs and tissues collectively termed the **lymphoid organs.** These are subdivided into primary and secondary lymphoid organs.

The **primary lymphoid organs** are the bone marrow and thymus. These organs supply the secondary lymphoid organs with mature lymphocytes already programmed to perform their functions. The bone marrow and thymus are not normally sites in which lymphocytes undergo activation.

The **secondary lymphoid organs** (also termed peripheral lymphoid organs) are the lymph nodes, spleen, tonsils, and lymphocyte accumulations in the linings of the intestinal, respiratory, genital, and urinary tracts. It is in the secondary lymphoid organs that lymphocytes are activated to participate in specific immune responses.

We have stated that the bone marrow and thymus supply lymphocytes to the secondary lymphoid organs. Most of the lymphocytes in the secondary organs are not, however, cells that originated in the primary lymphoid organs. The explanation of this seeming paradox is that, once in the secondary organ, a lymphocyte can undergo cell division to produce additional identical lymphocytes, which in turn undergo cell division and so on. In other words, all lymphocytes are *descended* from ancestors that were produced in the bone marrow or thymus but may not themselves have arisen in those organs. All the progeny cells derived by cell division from a single lymphocyte constitute a lymphocyte clone.

(A distinction must be made between the "lymphoid organs" and the "lymphatic system," described in Chapter 14. The latter is a network of lymphatic vessels and the lymph nodes found along these vessels. Of all the lymphoid organs, only the lymph nodes also belong to the lymphatic system.)

There are no anatomical links, other than via the cardiovascular system, between the various lymphoid organs.

Let us look briefly at these organs, excepting the bone marrow, which was described in Chapter 14.

The **thymus** lies in the upper part of the chest. Its size varies with age, being relatively large at birth and continuing to grow until puberty, when it gradually atrophies and is replaced by fatty tissue. Before its atrophy, the thymus consists mainly of lymphocytes that will eventually migrate via the blood to the secondary lymphoid organs. It also contains endocrine cells that secrete a group of hormones that exert a still poorly understood regulatory effect on the peripheral lymphocytes of thymic origin.

Recall from Chapter 14 that the fluid flowing along the lymphatic vessels is called lymph, which is interstitial fluid that has entered the lymphatic capillaries and is being routed to the large lymphatic vessels that drain into systemic veins. During this trip, the lymph flows through **lymph nodes** scattered along the vessels. Lymph, therefore, is the route by which lymph-node cells encounter the materials that trigger their specific immune responses. Each node is a honeycomb of lymph-filled sinuses (Figure 20-6) with large clusters of lymphocytes (the lymphatic nodules) between the sinuses. There are also many macrophages and, as is the case with other secondary lymphoid organs, macrophage-like cells.

The **spleen** is the largest of the lymphoid organs and lies in the left part of the abdominal cavity between the stomach and the diaphragm. In essence, the spleen is to the circulating blood what the lymph nodes are to the lymph. Blood percolates through the vascular meshwork of the spleen's interior, where large collections of lymphocytes, macrophages, and macrophage-like cells are found. The macrophages of the spleen, in addition to interacting with lymphocytes, also phagocytize aging or dead erythrocytes.

The **tonsils** are a group of small, rounded organs in the pharynx. They are filled with lymphocytes, macrophages, and macrophage-like cells, and they have openings ("crypts") to the surface of the pharynx. Their lymphocytes respond to microbes that arrive by way of ingested food as well as inspired air. Similarly, the lymphocytes in the linings of the various tracts exposed to the external environment respond to infectious agents that penetrate the linings from the lumen of the tract.

Finally, we must describe the source of the lymphocytes in blood. Some are cells on their way from the bone marrow or thymus to the secondary organs, but the vast majority are cells that are participating in lymphocyte traffic between the secondary lymphoid organs, blood, lymph, and all the tissues of the body. Lymphocytes from the secondary lymphoid organs constantly enter the lymph and are carried, via lymphatic vessels, to the blood (Chapter 14). Simultaneously, some blood lymphocytes are pushing through the endothelium of venules all over the body to enter the interstitial fluid. From there, they move into

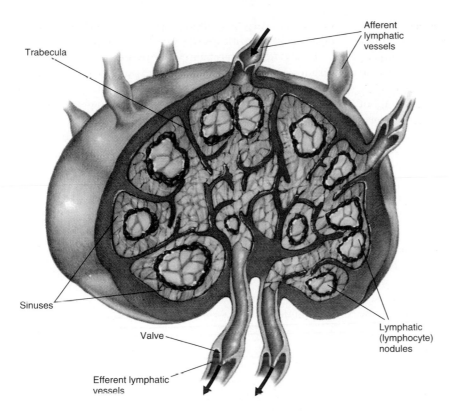

Trabecula

Afferent lymphatic vessels

Sinuses

Valve

Efferent lymphatic vessels

Lymphatic (lymphocyte) nodules

FIGURE 20-6
Anatomy of a lymph node.

lymphatic capillaries and along the lymphatic vessels to lymph nodes. They may then leave the lymphatic vessels to take up residence in the node. This recirculation is going on all the time, not just during an infection although the migration of lymphocytes into an inflamed area is greatly increased by the chemotaxis process described earlier. Lymphocyte trafficking greatly increases the likelihood that any given lymphocyte will encounter the antigen it is specifically programmed to recognize. (In contrast to the lymphocytes, granulocytes and monocytes do not recirculate; once they leave the bloodstream to enter a tissue they remain there or die.)

Lymphocyte Origins. The multiple populations and subpopulations of lymphocytes are summarized in Tables 20-1 and 20-2. The **B lymphocytes,** or simply **B cells,** mature in the bone marrow and then are carried by the blood to the secondary lymphoid organs. This overall process of maturation and migration continues throughout a person's life. All generations of lymphocytes that subsequently arise from these cells by cell division in the secondary lymphoid organs will be identical to the parent cells, that is, they will also be B cells.

In contrast to the B cells, other lymphocytes leave the bone marrow in an immature state during fetal and early neonatal life. They are carried to the thymus and mature there before moving to the secondary lymphoid organs. These cells, the **T lymphocytes** or **T cells,** constitute the

second large class of lymphocytes. Like B cells, T cells also undergo cell division in secondary lymphoid organs, the offspring being identical to the original T cells.

In addition to the B and T cells, there is a third population of lymphocytes, which are large and granular and include the **natural killer cells (NK cells).** These cells arise in the bone marrow, but their precursors and life history are still unclear. As we shall see, NK cells, unlike B and T cells, do not manifest specificity for antigens.

Functions of B Cells and T Cells

B cells, upon activation, differentiate into plasma cells, which secrete **antibodies,** proteins that travel all over the body to reach antigens identical to those that stimulated their production. The antibodies combine with these antigens and guide an attack (by phagocytes, complement, or NK cells, as we shall see) that eliminates the antigens or the cells bearing them. Antibody-mediated responses are also called humoral responses, the adjective "humoral" denoting communication by way of soluble chemical messengers, in this case, antibodies in the blood. Antibody-mediated responses have an extremely wide diversity of targets and are the major defense against bacteria, viruses, and other microbes in the extracellular fluid, and against toxic molecules (toxins).

T cells constitute a family that has several functional subsets, termed **cytotoxic T cells** and **helper T cells.** There may also be a third subset, called suppressor T cells, which have been hypothesized to inhibit the function of both B cells and cytotoxic T cells; however, the significance of these cells and even their existence are in doubt at present, and we shall have little to say about them in this chapter.

Another way to categorize T cells is not by function but rather by the presence of certain proteins, particularly those called CD4 and CD8, in their plasma membranes. Cytotoxic T cells have CD8 and so are also commonly called CD8+ cells; helper T cells have CD4 and so are also commonly called CD4+ cells.

Cytotoxic T cells are "attack" cells. They travel to the location of antigen-bearing cells, bind to them via the antigen and directly kill them, via secreted chemicals, without the intermediation of antibodies. Responses mediated by cytotoxic T cells are sometimes termed cell-mediated responses; they are directed against the body's own cells that have become cancerous or infected with viruses. (Intracellular bacteria and parasites are also targeted by

cytotoxic T cells in much the same way, but we shall not deal separately with these cases.)

It is worth emphasizing the important geographical difference in antibody-mediated responses and responses mediated by cytotoxic T cells. In the former case, the B cells remain in whatever location the recognition and activation steps occurred and send their antibodies forth, via the blood, to seek out antigens or antigen-bearing cells identical to those that triggered the response. In the responses mediated by cytotoxic T cells, the cells themselves must enter the blood and seek out the targets.

We have now assigned roles to the B cells and cytotoxic T cells. What role is performed by the helper T cells? As their name implies, these cells do not themselves function as "attack" cells but rather facilitate the activation and function of both B cells and cytotoxic T cells. Helper T cells go through the usual first two stages of the immune response in that they must first combine with antigen and then undergo activation. Once activated, they secrete cy-

FIGURE 20-7

Summary of roles of B, cytotoxic T, and helper T cells in immune responses.

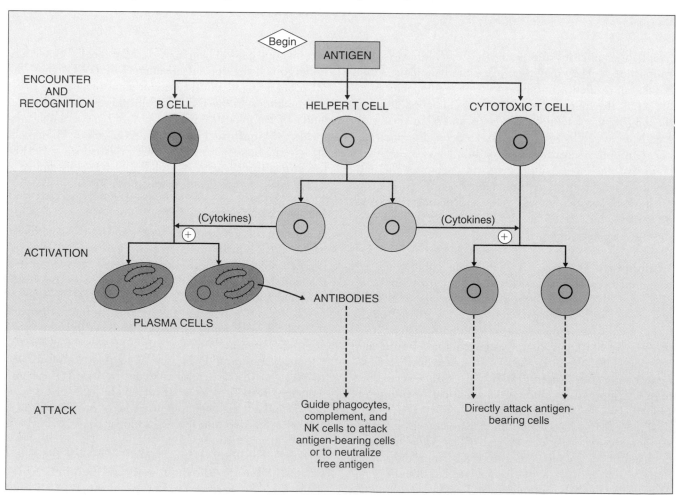

tokines that act on B cells and cytotoxic T cells that have also bound antigen. This is a very important point. With only a few exceptions, B cells and cytotoxic T cells cannot function adequately unless they are stimulated by cytokines from helper T cells.

We shall deal with helper T cells as though they were a homogeneous cell population, but in fact, there are several subtypes of helper T cells, distinguished by the different batteries of cytokines they secrete when activated. By means of these different cytokines, they "help" different sets of effector lymphocytes, as well as macrophages and NK cells. Exactly what happens during a *nonspecific* response is a major determinant of which subtype of helper T cell is preferentially recruited in the specific response; in this way, the events of nonspecific immunity play an important role in determining the events of specific immunity.

Figure 20-7 summarizes the basic interactions among B, cytotoxic T, and helper T cells presented in this section.

Lymphocyte Receptors

As stated earlier, the ability of lymphocytes to distinguish one antigen from another is determined by the lymphocytes' receptors.

B-cell Receptors. After activation, B cells proliferate and differentiate into plasma cells, which secrete antibodies. The plasma cells derived from a particular B cell can secrete only one particular antibody. Each B cell always displays on its plasma membrane copies of the particular antibody its plasma cell progeny are able to produce. This surface protein (glycoprotein, to be more accurate) acts as the receptor for the antigen specific to it.

B-cell receptors and antibodies constitute the family of proteins known as **immunoglobulins.** (The receptors are technically not antibodies since only *secreted* immunoglobulins are termed antibodies.) Each immunoglobulin molecule is composed of four interlinked polypeptide chains (Figure 20-8). The two long chains are called heavy chains, and the two short ones, light chains. There are five major classes of immunoglobulins, determined by the amino acid sequences in the heavy chains. The classes are designated by the letters A, D, E, G, and M following the symbol Ig for immunoglobulin; thus, IgA, IgD, and so on.

As illustrated in Figure 20-8, immunoglobulins have a "stem," called the **Fc** portion and comprising the lower half of the two heavy chains, and two "prongs," each containing one **antigen binding site** (or antibody combining site)—the amino acid sequences that bind antigen. The amino acid sequences of the Fc portion are identical for all immunoglobulins of a single class (IgA, IgD, and so on). In contrast, the amino acid sequences of the antigen binding sites vary from immunoglobulin to immunoglobulin in a given class. Thus, each class of antibody contains thousands, or millions, of unique immunoglobulins, each

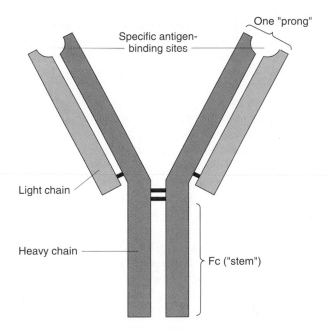

FIGURE 20-8

Immunoglobulin (antibody) structure. The Fc portions are the same for all immunoglobulins of a particular class. Each "prong" contains a single antigen binding site. The links between chains represent disulfide bonds.

capable of combining with only one specific antigen (or, in some cases, several antigens whose structures are very similar). The interaction between an antigen and antigen binding site of an immunoglobulin is analogous to the lock-and-key interactions that apply generally to the binding of ligands by proteins.

Any given B cell or clone of identical B cells produces receptors (that is, plasma membrane immunoglobulins) with unique antigen binding sites. Thus, the body has armed itself with small clones of millions of different B cells in order to ensure that receptors will exist that are specific for the vast number of different antigens the organism *might* encounter during its lifetime. The immunoglobulin that any given B cell (and all of its plasma-cell progeny) can produce was determined during the cell's maturation in the bone marrow. Diversity arises as the result of a genetic process unique to developing lymphocytes because only these cells possess the enzymes required to catalyze the process: The DNA in the gene exons that code for the immunoglobulin antigen binding site are cut into smaller segments, randomly rearranged along the gene, and then rejoined to form new exons.

One last point should be mentioned: B-cell receptors can bind antigen whether the antigen is a free molecule dissolved in the extracellular fluid or on the surface of a foreign cell. In the latter case, the B cell becomes linked to the foreign cell via the bonds between the B cell receptor and the surface antigen.

T-cell Receptors. T cells do not produce immunoglobulins, and so T-cell surface receptors for antigen are not immunoglobulins. Rather they are two-chained proteins that, like immunoglobulins, have specific regions that differ from one T cell to another. As in B cell development, multiple DNA rearrangements occur during T-cell maturation, leading to clones of millions of distinct T cells—distinct in that each cell and its offspring possess receptors of a single specificity. For T cells, this maturation occurs during their residence in the thymus.

In addition to their general structural differences, the B- and T-cell receptors differ in a much more important way: The T-cell receptor cannot combine with antigen unless the antigen is first complexed with certain of the body's own plasma-membrane proteins. The T-cell receptor combines with the entire complex of antigen and body (self) protein.

The self plasma-membrane proteins that must be complexed with the antigen in order for T-cell recognition to occur constitute a large group of proteins coded for by genes found on a single chromosome and known collectively as the **major histocompatibility complex (MHC).** The proteins are therefore called **MHC proteins.** Since no two persons other than identical twins have the same MHC genes, no two individuals have the same MHC proteins on the plasma membranes of their cells. MHC proteins are, in essence, cellular "identity tags," that is, genetic markers of biological individuality.

The MHC proteins are often termed "restriction elements" since, as we have seen, the ability of a T cell's receptor to recognize an antigen is restricted to situations in which the antigen is complexed with an MHC protein. There are two classes of MHC proteins: I and II. **Class I MHC proteins** are found on the surface of virtually all nucleated cells of a person's body, that is, virtually all cells but erythrocytes. **Class II MHC proteins** are found only on the surface of macrophages, B cells, and macrophage-like cells. The different subsets of T cells do not all have the same MHC requirements (Table 20-5): Cytotoxic T cells require antigen to be associated with class I MHC proteins, whereas helper T cells require class II MHC proteins.

At this point you might well ask: How do antigens, which are foreign, end up on the surface of the body's own cells complexed with MHC proteins? The answer is provided by the process known as **antigen presentation,** to which we now turn.

Antigen Presentation to T Cells

To repeat, T cells can bind antigen only when the antigen appears on the plasma membrane of a host cell complexed with the cell's MHC proteins. Cells bearing these complexes, therefore, function as **antigen-presenting cells (APCs).**

TABLE 20-5	MHC RESTRICTION OF THE LYMPHOCYTE RECEPTORS
Cell type	MHC restriction
B	None
Helper T	Class II, found only on macrophages, macrophage-like cells and B cells
Cytotoxic T	Class I, found on all nucleated cells of the body
NK	Presently uncertain

Presentation to Helper T Cells. Since helper T cells require class II MHC proteins and since these proteins are found only on macrophages, B cells, and macrophage-like cells, only these cells can function as APCs for helper T cells. The function of the macrophage (or macrophage-like cell) as an APC for helper T cells is easiest to visualize (Figure 20-9A) since the macrophage forms, in essence, a link between nonspecific and specific immune defenses. After a microbe or noncellular antigen has been phagocytized by a macrophage in a *nonspecific* response, it is partially broken down into smaller peptide fragments by proteolytic enzymes. The resulting fragments then bind (within endosomes) to class II MHC proteins synthesized by the cell. The fragments actually fit into a deep groove in the center of the MHC proteins. Differences in the shapes and properties of amino acid segments within the groove endow the various forms of class II MHC molecules with their affinities for different groups of peptides. The fragment-MHC complex is then transported to the cell surface where it is displayed in the plasma membrane. It is to this entire complex on the cell surface that a specific helper T cell binds.

Note that it is not the intact antigen but rather the peptide fragments, termed antigenic determinants or **epitopes,** of the antigen that are complexed to the MHC proteins and presented to the T cell. Despite this, it is customary to refer to "antigen" presentation rather than "epitope" presentation. More than one epitope can be formed from a given antigen.

How B cells process antigen and present it to helper T cells is essentially the same as the story we just described for macrophages (Figure 20-9B). An intact antigen binds to the B-cell receptor (an immunoglobulin) and this triggers receptor-mediated endocytosis, so that the antigen-receptor complex is internalized. Inside the B cell, the antigen is hydrolyzed to peptide fragments, which are complexed with class II MHC proteins. The complex is then shuttled to the cell surface where a helper T cell specific for the complex can bind to it. It must be emphasized

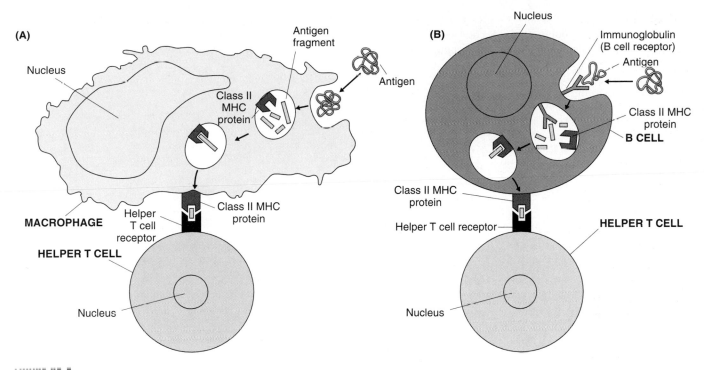

FIGURE 20-9

Sequence of events by which antigen is processed and presented to a helper T cell by (A) a macrophage or (B) a B cell. In both cases begin the figure with the antigen being internalized. (*Adapted from Gray, Setto, and Buus*)

that the ability of B cells to present antigen to helper T cells is a *second* function of B cells in response to antigenic stimulation, the other being the differentiation of the B cells into antibody-secreting plasma cells.

We have emphasized the interaction between helper T cell receptor and antigen bound to class II MHC proteins on an APC because this binding is the essential *antigen-specific* event in helper T cell activation. However, this binding by itself will not result in T cell activation. In addition, *nonspecific* interactions occur between other (nonantigenic) pairs of proteins on the surfaces of the attached helper T cell and APC, and these provide a necessary **costimulus** for T cell activation (Figure 20-10). Finally, the binding of the APC to the T cell causes the APC to secrete large amounts of the cytokine **interleukin 1 (IL-1),** which acts as a paracrine agent on the attached helper T cell to provide yet another essential stimulus for activation (Figure 20-10). Thus, the APC participates in activation of a helper T cell in three ways: antigen presentation, provision of a costimulus in the form of nonreceptor plasma-membrane proteins, and secretion of IL-1.

The activated helper T cell now secretes various cytokines that have both autocrine effects on the helper T cell and paracrine effects on B cells and cytotoxic T cells, as well as on NK cells and still other cell types; we will pick up these stories in later sections.

Presentation to Cytotoxic T Cells. Because class I MHC proteins are synthesized by virtually all nucleated cells, any such cell—not just macrophages, macrophage-like cells, or B cells—can act as an APC for a cytotoxic T cell. This distinction helps explain the major function of cytotoxic T cells—destruction of *any* of the body's own cells that have become cancerous or infected with viruses. The key point is that the antigens that complex with class I MHC proteins arise *within* body cells. They are "endogenous" antigens, foreign proteins synthesized by a body cell.

How do such antigens arise? In the case of viruses, once a virus has taken up residence inside a host cell, the viral nucleic acid causes the host cell to manufacture viral proteins, which of course are foreign proteins (Figure 20-11). A cancerous cell, as described in Chapter 4, has had one or more of its genes altered by chemicals, radiation, or other factors; the altered genes, called **oncogenes,** code for proteins that are either unique or are normally present beyond embryonic life only in very small concentrations. Such proteins can act as antigens.

In both virus-infected cells and cancerous cells some of the antigenic proteins are hydrolyzed by cytosolic enzymes into peptide fragments, complexed in the endoplasmic reticulum with the host cell's class I MHC proteins, and then shuttled by exocytosis to the plasma-membrane surface. There a cytotoxic T cell specific for the complex can bind to it.

FIGURE 20-10

Two other events are required for activation of helper T cells, in addition to the antigenic stimulus supplied by the antigen bound to a class II MHC protein on an antigen presenting cell (APC). (1) The binding of non-antigenic matching proteins in the plasma membranes of the APC and the helper T cell; this is termed a costimulus. (2) Secretion by the APC of the cytokine interleukin 1 (IL-1), which acts on the helper T cell.

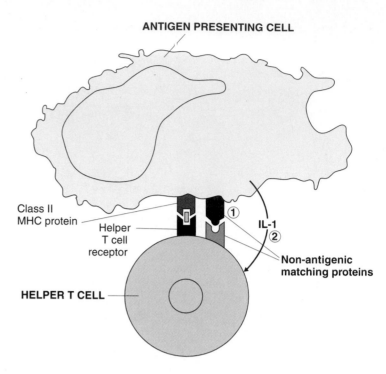

NK Cells

As noted earlier, NK cells constitute a distinct class of lymphocytes. They have several functional similarities to cytotoxic T cells: (1) Their major targets are virus-infected cells and cancer cells; and (2) they attack and kill these target cells directly, after binding to them. However, un-

like cytotoxic T cells, NK cells are not antigen specific, that is, each NK cell is able to attack virus-infected cells or cancer cells without any recognition of a specific antigen on the part of the NK cell. They have neither T cell receptors nor the immunoglobulin receptors of B cells, and the exact nature of the NK-cell surface receptors that

FIGURE 20-11

Processing and presentation of viral antigen to a cytotoxic T cell by an infected cell. The virus induces the production by the cell of viral protein, which is then hydrolyzed and the fragments complexed to the cell's class I MHC proteins in the endoplasmic reticulum. These complexes are then shuttled to the plasma membrane. (*Adapted from Gray, Sette, and Buus*)

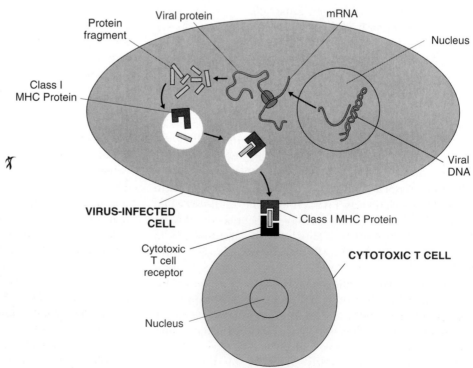

permits the cells to identify their targets is unknown (except for one case presented later). Also unclear is whether MHC proteins are involved in the activation or inhibition of NK cells.

It has been suggested that NK cells are lymphocytic "Minutemen," which can be mobilized quickly and go into action *nonspecifically* without the period of exposure and activation required by B cells and cytotoxic T cells. Why then do we deal with them in the context of *specific* immune responses? Because, as will be described subsequently, their participation in an immune response is greatly enhanced by certain antibodies or by cytokines secreted by helper T cells activated during specific immune responses.

Development of Immune Tolerance

Our basic framework for understanding specific immune responses requires consideration of one more crucial question: How does the body develop what is called **immune tolerance**—lack of immune responsiveness to *self-components*? This may seem a strange question given our definition of an antigen as a *foreign* molecule that can generate an immune response. In essence, however, we are now asking how it is that the body "knows" that its own molecules, particularly proteins, are not foreign.

Recall that the huge diversity of lymphocyte receptors is ultimately the result of multiple random DNA recombination processes. It is virtually certain, therefore, that in each person clones of lymphocytes would have emerged with receptors that could bind to that person's own proteins. The existence and functioning of such lymphocytes would, of course, be disastrous because such binding would launch an immune attack against the cells expressing these proteins; the person's own (self) proteins would act as antigens. There are at least two mechanisms, termed clonal deletion and clonal inactivation, that explain why *normally* there are no active lymphocytes that respond to self components.

First, during fetal and early postnatal life, T cells are exposed to a potpourri of self proteins in the thymus. Those T cells with receptors capable of binding self proteins are destroyed by apoptosis (Chapter 7). This process is termed **clonal deletion.**

But clonal deletion does not fully explain immune tolerance. It is very likely, for example, that some proteins in the body do not make their way to the thymus during development, and so T cells capable of binding to these proteins would not be destroyed. A second process, termed **clonal inactivation,** occurs not in the thymus but in the periphery and causes potentially self-reacting T cells to become nonresponsive.

What are the mechanisms of clonal deletion and inactivation? Recall that activation of a helper T cell requires (1) the antigen-specific stimulus (binding of the helper T cell receptor to an antigen–class II MHC protein complex on the surface of an APC) and (2) a non-specific costimulus (interaction between another protein on the APC and one on the T cell). When this costimulus is not provided, the helper T cell not only fails to become activated by antigen but dies and/or becomes inactivated forever. This is the case during early life and constitutes at least one major mechanism for clonal deletion and clonal inactivation. Precisely what accounts for the costimulus not being delivered is presently unclear. Surprisingly, this same process can also occur at any time during life, and it can fail to occur during early life under certain circumstances.

This completes our framework for understanding specific immune responses. The next two sections utilize this framework in presenting typical responses from beginning to end, fleshing out the interactions between lymphocytes and describing the attack mechanisms used by the various pathways.

Antibody-Mediated Immune Responses: Defenses Against Bacteria, Extracellular Viruses, and Toxins

A classical antibody-mediated response is one that results in the destruction of bacteria. The sequence of events, which is quite similar to the response to a virus in the extracellular fluid, is summarized in Table 20-6 and Figure 20-12, and described in the following sections.

B-Cell Recognition and Activation. This process starts the same way as for nonspecific responses, with the bacteria penetrating one of the body's linings and entering the interstitial fluid. The bacteria then enter the lymphatic system and/or bloodstream and are carried to the lymph nodes and/or the spleen, respectively. There a B cell specific for an antigen on the bacterial surface binds to the antigen via the B cell's plasma membrane immunoglobulin receptor.

In a few cases (notably bacteria with polysaccharide capsules), this binding is all that is needed to trigger B-cell activation. For the great majority of antigens, however, binding is not enough, and signals in the form of cytokines released into the interstitial fluid by helper T cells near the antigen-bound B cells are also needed.

For helper T cells to secrete cytokines they must bind to a complex of antigen and class II MHC protein on an APC. Let us assume that in this case the APC is a macrophage that has phagocytized one of the bacteria, hydrolyzed its proteins into peptide fragments, complexed them with class II MHC proteins, and displayed the complexes on its surface. A helper T cell specific for the complex then binds to it, beginning the activation of the helper T cell. Moreover, the macrophage helps this activation process in two other ways: It provides a costimulus via certain other plasma-membrane proteins, and it secretes IL-1.

TABLE 20-6 SUMMARY OF EVENTS IN ANTIBODY-MEDIATED IMMUNITY AGAINST BACTERIA

1. In peripheral lymphoid organs, bacterial antigen binds to specific receptors on the plasma membrane of B cells.

2. Simultaneously, antigen-presenting cells (for example, macrophages) (APCs) present to helper T cells processed antigen complexed to MHC Class II proteins on the APCs and secrete IL-1, which acts on the helper T cells.

3. In response, the helper T cells secrete IL-2, which activates the helper T cells to proliferate and secrete IL-2 and other cytokines. These activate the antigen-bound B cells to proliferate and differentiate into plasma cells. Some of the B cells differentiate into memory cells rather than plasma cells.

4. The plasma cells secrete antibodies specific for the antigen that initiated the response, and the antibodies circulate all over the body via the blood.

5. Antibodies combine with antigen on the surface of the bacteria anywhere in the body.

6. Presence of antibody bound to antigen facilitates phagocytosis of the bacteria by neutrophils and macrophages. It also activates the complement system, which further enhances phagocytosis and can directly kill the bacteria by the membrane attack complex. It may also induce antibody-dependent cellular cytotoxicity mediated by NK cells that bind to the antibody's Fc portion.

So far this has all been review. Now we begin new material. IL-1 stimulates the helper T cell to secrete another cytokine named **interleukin 2 (IL-2),** and to express the receptor for IL-2. This messenger, acting as an autocrine agent, then provides a proliferative stimulus to the helper T cell (Figure 20-12). The cell divides, beginning the mitotic cycles that lead to formation of a clone of activated helper T cells, and these cells then release not only IL-2 but other cytokines as well.

Certain of these cytokines provide the additional signals required to activate nearby antigen-bound B cells to proliferate and differentiate into plasma cells, which then secrete antibodies.

Thus, as shown in Figure 20-12, we are dealing with a series of protein messengers interconnecting the various cell types, the helper T cells serving as the central coordinator. The macrophage releases IL-1, which acts on the helper T cell to stimulate release of IL-2, which stimulates the helper T cell to multiply, the activated progeny then releasing still other cytokines that activate antigen-bound B cells.

Some of the B-cell progeny, however, do not differentiate into plasma cells but instead become **memory cells** (Figure 20-12), ready to respond vigorously should the antigen reappear at a future time. In other words, the proliferation triggered by a first exposure to an antigen results in a long-lasting expansion of the clone of B cells specific for that antigen.

The example we have been using employed a macrophage as the APC to helper T cells. Recall, however, that this role can also be served by B cells. The binding of the helper T cell to the antigen-bound B cell ensures maximal stimulation of the B cell by the cytokines secreted by that helper T cell and any of its progeny that remain nearby.

Antibody Secretion. Plasma cells produce thousands of antibody molecules per second before they die in a day or so. We mentioned earlier that there are five major classes of antibodies. The most abundant are the **IgG** antibodies, commonly called **gamma globulin,** and **IgM** antibodies; these two groups together provide the bulk of specific immunity against bacteria and viruses in the extracellular fluid. **IgE** antibodies mediate allergic responses and participate in defenses against multicellular parasites. **IgA** antibodies are secreted by plasma cells in the linings of the gastrointestinal, respiratory, and genitourinary tracts. IgA antibodies generally do not circulate but act locally in the linings or on their surfaces. They are also secreted by the mammary glands and hence are the major antibodies in milk. The functions of **IgD** are still unclear.

In the kind of infection we've been describing, the secreted antibodies—mostly IgG and IgM—enter the blood, which carries them from the secondary lymphoid organ in which recognition and activation occurred to all tissues and organs of the body. At sites of infection, the antibodies leave the blood (recall that nonspecific inflammation had already made capillaries leaky at these sites) and combine with the type of bacterial surface antigen that initiated the immune response (Figure 20-12). These anti-

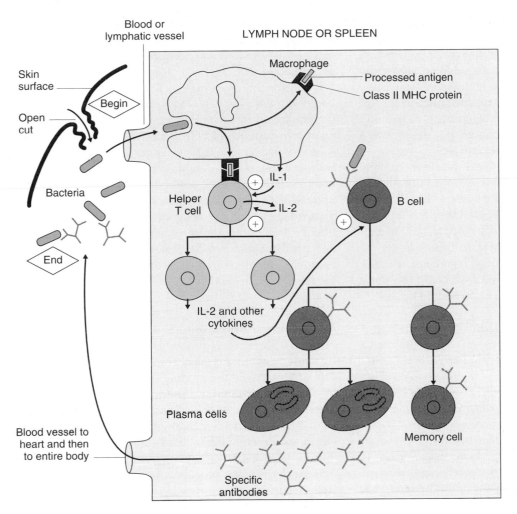

FIGURE 20-12

Summary of events by which a bacterial infection leads to antibody synthesis in peripheral lymphoid organs and the secreted antibodies travel by the blood to the site of infection where they bind to bacteria of the type that induced the response. As illustrated in Figure 20-9B, an antigen-bound B cell, rather than a macrophage, as shown in the present figure, can function as the antigen-presenting cell to the helper T cell. [For clarity, neither the intracellular processing of the antigen by the macrophage (Figure 20-9A) nor the costimulus binding of nonantigenic proteins (Figure 20-10) are shown in this figure.] ⚔

bodies then direct the attack (see below) against the bacteria to which they are now bound.

Thus, immunoglobulins play two distinct roles in immune responses: (1) During the initial recognition step, those on the surface of B cells bind to antigen brought to them; and (2) those secreted by the plasma cells (antibodies) seek out and bind to bacteria bearing the same antigens, "marking" them as the targets to be attacked.

The Attack: Effects of Antibodies. The antibodies bound to antigen on the microbial surface do not directly kill the microbe but instead link up the microbe physically to the actual killing mechanism—complement, macrophages, or NK cells. This linkage not only activates the

mechanism but ensures that the killing effects take place only in the immediate vicinity of the microbe. This protects adjacent normal structures from toxic effects of the chemicals employed by the killing mechanisms.

Activation of the complement system. As described on page 695, the complement system is activated in *nonspecific* inflammatory responses via the alternate complement pathway. In contrast, in *specific* immune responses, the presence of antibody of the IgG or IgM class bound to antigen activates the **classical complement pathway.** The first molecule in this pathway, C1, binds to the Fc portion of an antibody that has combined with antigen. This results in activation of the enzymatic portions of C1, thereby initiating the entire classical pathway. The end product of this cascade, the membrane attack complex (MAC), can kill the cells to which the antibody is bound by making their membranes leaky. In addition, as we saw in Figure 20-4, another activated complement molecule (C3b) functions as an opsonin to enhance phagocytosis of the microbe by neutrophils and macrophages, and still other activated complement molecules stimulate the steps of inflammation leading to phagocytosis.

It is important to note that C1 binds not to the unique antigen binding sites in the antibody's prongs but rather to complement binding sites in the Fc portion. Since the latter are the same in virtually all antibodies of the IgG and IgM classes, the complement molecule will bind to *any* antigen-bound antibodies belonging to these classes. In other words, there is only one set of complement molecules, and once activated, they do essentially the same thing regardless of the specific identity of the invader.

Direct enhancement of phagocytosis. Activation of complement is not the only mechanism by which antibodies enhance phagocytosis. Merely the presence of an IgG antibody attached to antigen on the microbe's surface has some enhancing effect on phagocytosis. In other words, antibodies can act directly as opsonins in addition to their indirect action via complement.

The mechanism is analogous to that for complement in that the antibody links the phagocyte to the antigen. As shown in Figure 20-13, the phagocyte has membrane receptors that bind to the Fc portion of antibodies. This linkage promotes attachment of the antigen to the phagocyte and the triggering of phagocytosis.

Antibody-dependent cellular cytotoxicity. We have seen that both a particular complement molecule (C1) and a phagocyte can bind nonspecifically to the Fc portion of antibodies. NK cells can also do this (just substitute a NK cell for the phagocyte in Figure 20-13). Thus, antibodies can link target cells to NK cells, which then kill the targets directly by secreting toxic chemicals. This is termed **antibody-dependent cellular cytotoxicity (ADCC)** because the killing (cytotoxicity) is carried out by cells (NK cells) but the process depends upon the presence of antibody. Note that it is the antibodies that confer specificity upon ADCC, just as they do on antibody-dependent complement activation and phagocytosis. As we shall see, antibodies are not the only mechanism for bringing NK cells into play, but the antibody mechanism provides the one exception, mentioned earlier, to the generalization that the mechanism by which NK cells identify their targets is unknown.

Direct neutralization of bacterial toxins and viruses. Toxins secreted by bacteria into the extracellular fluid can act as antigens to induce antibody production. The antibodies then combine with the free toxins, thus preventing interaction of the toxin with susceptible cells. Since each antibody has two binding sites for combination with antigen, clumplike chains of antibody-antigen complexes are formed (Figure 20-14), and these clumps can then be phagocytized.

A similar binding process is the major antibody-mediated mechanism for eliminating viruses in the extracellular fluid. Certain of the viral surface molecules serve as antigens, and the antibodies produced against them com-

FIGURE 20-13

Direct enhancement of phagocytosis by antibody. The antibody links the phagocyte to the bacterium.

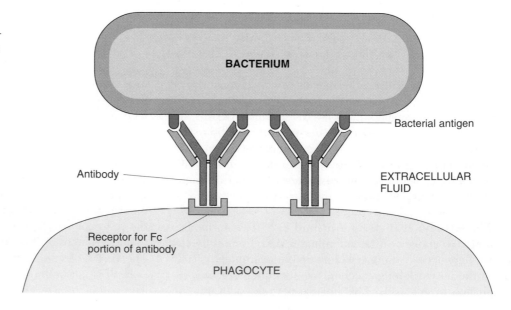

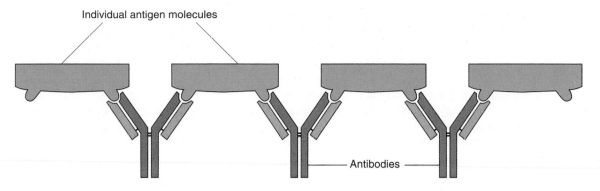

FIGURE 20-14

A chain of interlocking antigen-antibody complexes.

bine with them, preventing attachment of the virus to plasma membranes of potential host cells. This prevents the virus from entering a cell. As with bacterial toxins, chains of antibody-virus complexes are formed and can be phagocytized.

Active and Passive Humoral Immunity. We have been discussing antibody formation without regard to the course of events in time. The response of the antibody-producing machinery to invasion by a foreign antigen varies enormously, depending upon whether the machinery has previously been exposed to that antigen. Antibody production responds slowly over several days to the first contact with a microbial antigen, but any subsequent infection by the same invader elicits an immediate and marked outpouring of additional specific antibody (Figure 20-15). This response, which is mediated by the memory B cells described earlier, confers a greatly enhanced resistance toward subsequent infection with that particular microorganism. Resistance built up as a result of the body's contact with microorganisms and their toxins or other antigenic components is known as **active immunity.**

Until this century, the only way to develop active immunity was to suffer an infection, but now the injection of microbial derivatives in vaccines is used. A ***vaccine*** may consist of small quantities of living or dead microbes, small quantities of toxins, or harmless antigenic molecules derived from the microorganism or its toxin. The general principle is always the same: Exposure of the body to the agent results in the induction of the memory cells required for rapid, effective response to possible future infection by that particular organism.

A second kind of immunity, known as **passive immunity,** is simply the direct transfer of actively formed antibodies from one person (or animal) to another, the recipient thereby receiving preformed antibodies. Such transfers occur between mother and fetus since IgG is selectively transported across the placenta. Also, breast-fed children receive IgA antibodies in the mother's milk. These are important sources of protection for the infant during the first months of life, when the antibody-synthesizing capacity is relatively poor. Monoclonal antibodies produced by genetic engineering can also be used for passive immunity.

The same principle is used clinically when specific antibodies or pooled gamma globulin is given to patients exposed to or suffering from certain infections such as hepatitis. The protection afforded by this transfer of antibodies is relatively short-lived, usually lasting only a few weeks or months.

Summary. We can now summarize the interplay between nonspecific and specific humoral immune mechanisms in resisting a bacterial infection. When we encounter

FIGURE 20-15

Rate of specific antibody production following initial contact with an antigen and subsequent contact with the same antigen.

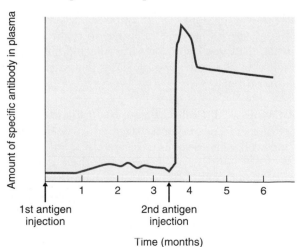

particular bacteria for the first time, *nonspecific defense* mechanisms resist their entry and, if entry is gained, attempt to eliminate them by phagocytosis and nonphagocytic killing in the inflammatory process. Simultaneously, bacterial antigens induce the specific B-cell clones to differentiate into plasma cells capable of antibody production. If the nonspecific defenses are rapidly successful, these slowly developing *specific* immune responses may never play an important role. If the nonspecific responses are only partly successful, the infection may persist long enough for significant amounts of antibody to be produced. The presence of antibody leads to both enhanced phagocytosis and direct destruction of the foreign cells, as well as to neutralization of toxins secreted by the bacteria. All subsequent encounters with that type of bacteria will be associated with the same sequence of events, with the crucial difference that the specific responses may be brought into play much sooner and with greater force; that is, the person may have active immunity against that type of bacteria.

The defenses against viruses in the extracellular fluid are similar, resulting in destruction or neutralization of the virus.

Defenses Against Virus-Infected and Cancer Cells

The previous section described how antibody-mediated immune responses constitute the major defenses against "exogenous antigens"—bacteria, viruses, and individual foreign molecules that enter the body and are encountered by the immune system in the extracellular fluid. This section now details how cytotoxic T cells, NK cells, and so-called activated macrophages (see below), all working with helper T cells, destroy the body's own cells that have become infected by viruses (or other intracellular microbes) or have been transformed into cancer cells.

What is the value of destroying virus-infected cells? Such destruction results in release of the viruses into the extracellular fluid where they can then be directly neutralized by circulating antibody, as just described. Generally, only a few host cells must be sacrificed in this way, but once viruses have had a chance to replicate and spread from cell to cell, so many virus-infected host cells may be killed by the body's own defenses that organ malfunction may occur.

Role of Cytotoxic T Cells.   Figure 20-16 summarizes a typical cytotoxic T-cell response triggered by viral infection of body cells. The response triggered by a cancer cell would be very similar. As described earlier, a virus-infected or cancer cell produces foreign proteins, "endogenous antigens," which are processed and presented on the plasma membrane of the cell complexed with MHC class I proteins. Cytotoxic T cells specific for the particular antigen can bind to the complex, but just as with B

cells, binding to antigen alone does not cause activation of the cytotoxic T cell. Cytokines from adjacent activated helper T cells are also needed.

How are the helper T cells brought into play in these cases? Figure 20-16 illustrates the most likely mechanism. Macrophages phagocytize free extracellular viruses (or, in the case of cancer, antigens released from the surface of the transformed cells) and then process and present antigen, in association with class II MHC proteins, to the helper T cells. The macrophages also provide a costimulus and secrete IL-1 (Figure 20-10), in response to which the antigen-bound helper T cell releases IL-2 and other cytokines. IL-2 then acts as an autocrine agent, as we have mentioned, to stimulate proliferation of the helper T cell.

The IL-2 also acts as a paracrine agent on the cytotoxic T cell bound to the surface of the virus-infected or cancer cell, stimulating this attack cell to proliferate. Other cytokines secreted by the activated helper T cell perform the same functions. Why is proliferation important if a cytotoxic T cell has already found and bound to its target? The answer is that there is rarely just one virus-infected cell or one cancer cell. By increasing the pool of cytotoxic T cells capable of recognizing the particular antigen, proliferation increases the likelihood that the other virus-infected or cancer cells will be encountered by the specific type of cytotoxic T cell.

There are several mechanisms of target-cell killing by cytotoxic T cells, but the best defined one is as follows. The cytotoxic T cell releases, by exocytosis, the contents of its secretory vesicles into the extracellular space between itself and the target cell to which it is bound. These vesicles contain molecules of a protein, **perforin** (also termed pore-forming protein), which is similar in structure to the proteins of the complement system's membrane attack complex. Perforin inserts into the target cell's membrane and forms channels through the membrane. In this manner, it causes the attacked cell to become leaky and die. The fact that perforin is released into an enclosed space between the tightly attached attacking cell and the target ensures that innocent bystander cells will not be killed since perforin is not at all specific.

Some cytotoxic T cells generated during proliferation following an initial antigenic stimulation do not complete their full activation at this time but remain as memory cells. Thus, active immunity exists for cytotoxic T cells just as for B cells.

Role of NK Cells and Activated Macrophages. Cytotoxic T cells are very important attack cells against virus-infected and cancer cells, but not the only ones. NK cells and "activated macrophages" also destroy such cells by secreting toxic chemicals.

In the section on antibody-dependent cellular cytotoxicity (ADCC) we pointed out that NK cells can be linked

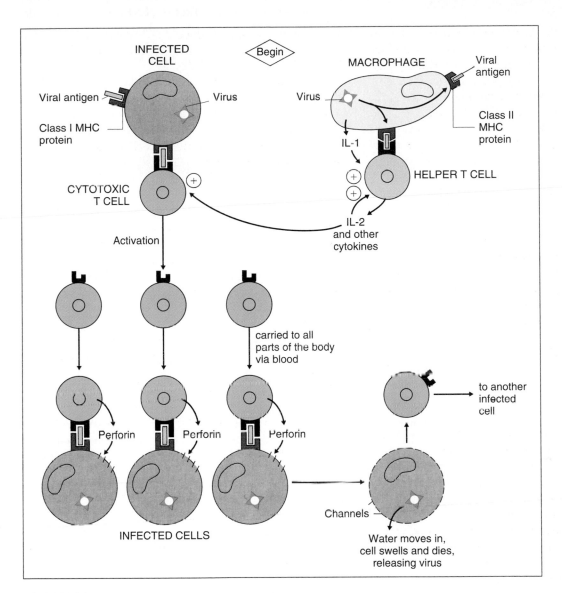

FIGURE 20-16
Summary of events in the killing of virus-infected cells by cytotoxic T cells. The released viruses can then be phagocytized. The sequence would be similar if the inducing cell were a cancer cell rather than a virus-infected cell.

to target cells by antibodies, and this certainly constitutes one potential method of bringing them into play against virus-infected or cancer cells. In most cases, however, strong antibody responses are not triggered by virus-infected or cancer cells, and the NK cell must bind *directly* to its target, without the intermediation of antibodies. As noted earlier, NK cells do not have antigen specificity; rather, they nonspecifically bind to virus-infected and cancer cells, but how they do so is not really understood.

The major signals for NK cells to proliferate and secrete their toxic chemicals are two cytokines—IL-2 and **interferon-gamma**—secreted by the helper T cells that have been activated specifically by the targets (Figure 20-16).

Thus, the attack by the NK cells is nonspecific, but a specific immune response on the part of the helper T cells is required to bring the NK cells into play. Moreover, there is a positive feedback at work here since activated NK cells can themselves secrete interferon gamma (Figure 20-17). (Whereas essentially all body cells can produce the other interferons, as described earlier, only activated helper T cells and NK cells can produce interferon-gamma.)

IL-2 and interferon-gamma act not only on NK cells but on macrophages in the vicinity to enhance their ability to kill cancer cells and cells infected with viruses and other microbes. Macrophages stimulated in this way are

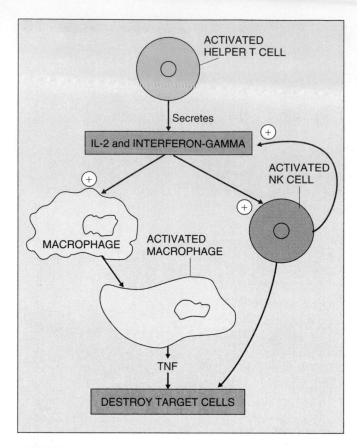

FIGURE 20-17
Role of IL-2 and interferon-gamma, secreted by activated helper T cells, in stimulating the killer ability of NK cells and macrophages.

termed **activated macrophages** (Figure 20-17). They secrete large amounts of many chemicals, particularly the cytokine **tumor necrosis factor (TNF),** that are capable of killing cells by a variety of mechanisms. (Interestingly, helper T cells themselves also secrete TNF, and so it is likely that these cells can, via TNF, kill cells on their own without the intermediation of cytotoxic T cells, NK cells, or activated macrophages.)

Table 20-7 summarizes the multiple defenses against viruses described in this chapter.

SYSTEMIC MANIFESTATIONS OF INFECTION

There are many *systemic* responses to infection, that is, responses of organs and tissues distant from the site of infection or immune response. These systemic responses are collectively known as the **acute phase response** (Figure 20-18). It is natural to think of them as part of the disease, but the fact is that most of them actually represent adaptive responses to the infection.

The single most common and striking systemic sign of infection is fever, the mechanism of which is described in Chapter 18. Present evidence suggests that fever is often beneficial, in that an increase in body temperature enhances many of the protective responses described in this chapter.

Decreases in the plasma concentrations of iron and zinc occur in response to infection and are due to changes in the uptake and/or release of these trace elements by

TABLE 20-7   SUMMARY OF HOST RESPONSES TO VIRUSES

	Main cells involved	Comment on action
Nonspecific defenses		
Anatomical barriers	Body surface linings	Physical barrier; antiviral chemicals
Inflammation	Tissue macrophages	Phagocytosis of extracellular virus
Interferon	Most cell types after viruses enter them	Interferon nonspecifically prevents viral replication inside host cells
Specific defenses		
Antibody-mediated	Plasma cells derived from B cells secrete antibodies	Antibodies neutralize virus and thus prevent viral entry into cell
		Antibodies activate complement, which leads to enhanced phagocytosis of extracellular virus
		Antibodies recruit NK cells via antibody-mediated cellular cytotoxicity
Direct cell killing	Cytotoxic T cells, NK cells, and activated macrophages	Via secreted chemicals, destroy host cell and thus induce release of virus into extracellular fluid where it can be phagocytized. Activity is stimulated by IL-2 and interferon-gamma.

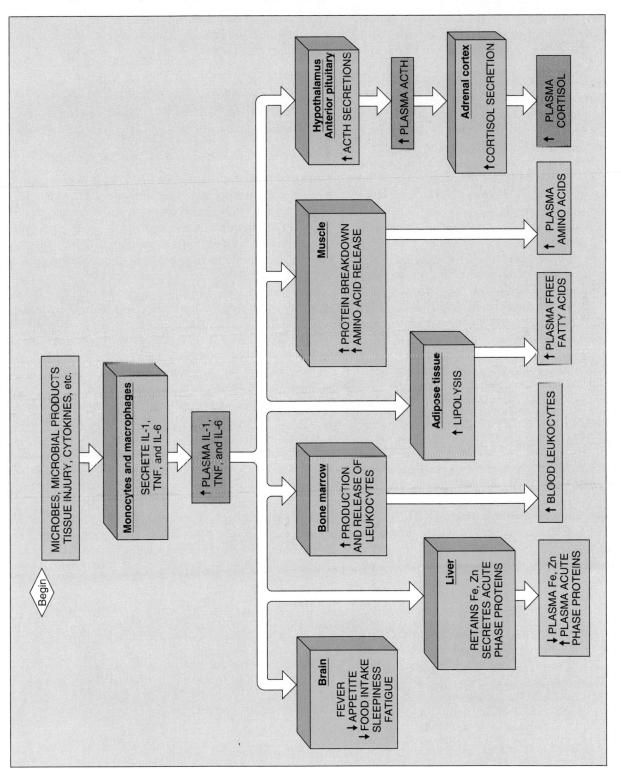

FIGURE 20-18

Systemic responses to infection or injury (the acute phase response). Other cytokines probably also participate. This figure does not include all the components of the acute phase response; for example, IL-1 and several other cytokines also stimulate secretion of insulin and glucagon. The significance of the increase in plasma cortisol is described in Section C.

liver, spleen, and other tissues. The decrease in plasma iron concentration has adaptive value since bacteria require a high concentration of iron to multiply. The role of the decrease in zinc is not known. The decrease in appetite characteristic of infection may also deprive the invading organisms of nutrients—particularly minerals—they require to proliferate.

Another adaptive response to infection is the secretion by the liver of a group of proteins known collectively as **acute phase proteins.** These proteins exert a large array of adaptive effects on the inflammatory process (including minimizing the extent of local tissue damage), immune cell function, and tissue repair.

Another response to infection, release of neutrophils and monocytes by the bone marrow, is of obvious value. There is release of amino acids from muscle, and these amino acids provide the building blocks for the synthesis of proteins required to fight the infection and for tissue repair. There is also increased release of fatty acids from adipose tissue, providing a source of energy. The secretion of many hormones is also increased in the acute phase response, notably that of cortisol by the adrenal cortex; the adaptive value of this particular hormonal response is described in Section C of this chapter.

All these responses and many others that are not listed are elicited by one or more of the cytokines released from activated macrophages and other cells. In particular, IL-1, TNF, and another cytokine—**interleukin 6 (IL-6),** all of which serve *local* roles in immune responses, also serve as *hormones* to elicit far-flung responses such as fever. This is a good place to review the various functions of macrophages in immune responses (Table 20-8).

Several other cytokines are also known to participate in the acute phase response. For example, colony stimulating factors (Chapter 14), which are secreted by macrophages, lymphocytes, endothelial cells, and fibroblasts, provide a major stimulus to the bone marrow to produce more neutrophils and monocytes.

FACTORS THAT ALTER THE BODY'S RESISTANCE TO INFECTION

There are many factors that determine the body's capacity to resist infection. We offer here only a few examples.

Protein-calorie malnutrition is, worldwide, the single greatest contributor to decreased resistance to infection, the result of impaired immune function. Deficits of specific nutrients other than protein can also lower resistance to infection.

A preexisting disease, infectious or noninfectious, can also predispose the body to infection. People with diabetes, for example, have a propensity to numerous infections, at least partially explainable on the basis of defective leukocyte function. Moreover, any injury to a tissue lowers its resistance, perhaps by altering the chemical environment or interfering with the blood supply.

Both stress and a person's state of mind can either enhance or reduce resistance to infection (and cancer). Lymphoid tissue is innervated, and the cells mediating immune defenses possess receptors for many neurotransmitters and hormones. Conversely, as we have seen, some of the cytokines released by immune cells have important effects on the brain and endocrine system. Moreover, lymphocytes secrete several of the same hormones produced by endocrine glands. Thus the immune system can alter neural and endocrine function, and in turn neural and endocrine activity modifies immune function. For example,

---

TABLE 20-8   ROLE OF MACROPHAGES IN IMMUNE RESPONSES

1. In nonspecific inflammation, phagocytize particulate matter, including microbes. Also secrete antimicrobial chemicals and protein messengers (cytokines) that function as local inflammatory mediators. The inflammatory cytokines include IL-1 and TNF.

2. Process and present antigen to helper T cells.

3. Secrete IL-1, which stimulates helper T cells to secrete IL-2 and to express the receptor for IL-2.

4. During specific immune responses, perform same killing and inflammation-inducing functions as in (1) but are more efficient because antibodies act as opsonins and because the cells are transformed into activated macrophages by IL-2 and gamma interferon, both secreted by helper T cells.

5. Secrete IL-1, TNF, and IL-6, which mediate many of the systemic responses to infection or injury.

it has been shown that the production of antibodies can be altered by psychological conditioning in much the same way as other body functions. Some possible mediators of the effects of stress on immune responses are discussed in Section C of this chapter.

The influence of physical exercise on the body's resistance to infection (and cancer) has been debated for decades. Present evidence indicates that the intensity, duration, chronicity, and psychological stress of the exercise all have important influences, both negative and positive, on a host of immune functions (for example, the level of circulating NK cells). Most experts in the field believe that, despite all these complexities, modest exercise and physical conditioning have net beneficial effects on the immune system and on host resistance.

Another factor associated with decreased immune function is sleep deprivation. For example, loss of half of a single night's sleep has been observed to reduce the activity of blood NK cells.

Resistance to infection will be impaired if one of the basic resistance mechanisms itself is deficient, as, for example, in people who have a genetic deficiency that impairs their ability to produce antibodies. These individuals experience frequent and sometimes life-threatening infections that can be prevented by regular replacement injections of gamma globulin. Another genetic defect is **combined immunodeficiency,** which is an absence of both B and T cells. If untreated, infants with this disorder usually die within their first year of life from overwhelming infections. Combined immunodeficiency can be cured by a bone-marrow transplantation, which supplies both B cells and cells that will migrate to the thymus and become T cells.

An environmentally induced decrease in the production of leukocytes is also an important cause of lowered resistance, as, for example, in patients given drugs specifically to inhibit rejection of tissue or organ transplants (see section on graft rejection below).

The most striking example of the lack of a basic resistance mechanism is the disease called **acquired immune deficiency syndrome (AIDS),** which is described in the accompanying essay.

Antibiotics

The most important of the drugs we employ in altering resistance to microbes, mainly bacteria, are **antibiotics,** such as penicillin. With few exceptions (for example, **zidovudine** in HIV infections and **acyclovir** in herpes infections), there are no antibiotics presently in common use against viruses.

Antibiotics exert a wide variety of effects, including inhibition of bacterial cell-wall synthesis, protein synthesis, and DNA replication. Antibiotics, however, must not be used indiscriminately. For one thing, they may exert toxic

effects on the *body's* cells. A second reason for judicious use is the escalating and very serious problem of drug resistance. Most large bacterial populations contain a few mutants that are resistant to the drug, and these few are capable of multiplying into large populations resistant to the effects of that particular antibiotic. Alternatively, the antibiotic can induce expression of a latent gene that confers resistance. Finally, resistance can be transferred from one microbe directly to another previously nonresistant microbe by means of DNA passed between them. A third reason for the judicious use of antibiotics is that these agents may actually contribute to a new infection by eliminating certain species of relatively harmless bacteria that ordinarily prevent growth of more dangerous ones.

HARMFUL IMMUNE RESPONSES

Until now, we have focused on the mechanisms of immune responses and their *protective* effects. In this section we shall see that immune responses can be harmful to the body.

Graft Rejection

The major obstacle to successful transplantation of tissues and organs is that the immune system recognizes the transplants, called grafts, as foreign and launches an attack against them. This is termed **graft rejection.** Although antibodies play some role, cytotoxic T cells are mainly responsible for graft rejection.

Except in the case of identical twins, the class I MHC proteins on the cells of a graft differ from the recipient's. So do the class II molecules present on the macrophages in the graft (recall that virtually all organs and tissues have macrophages). Accordingly, the MHC proteins of both classes are recognized as foreign by the recipient's T cells, and the cells bearing these proteins are destroyed by the recipient's cytotoxic T cells with the aid of helper T cells.

Some of the tools aimed at reducing graft rejection are radiation and drugs that kill actively dividing lymphocytes and thereby decrease the recipient's T-cell population. The single most effective drug, however, is **cyclosporin,** which does not kill lymphocytes but rather blocks the production of IL-2 and other cytokines by helper T cells. This eliminates a critical signal for proliferation of both the helper T cells and the cytotoxic T cells.

Adrenal corticosteroids (Chapter 10) in large doses are also used to reduce the rejection. Their possible mechanisms of action are described in Section C of this chapter.

The Placenta as a Graft. During pregnancy the fetal trophoblast cells of the placenta (Chapter 19) lie in direct contact with maternal immune cells. Since half of the

ESSAY

Acquired Immune Deficiency Syndrome (AIDS)

AIDS is caused by the ***human immunodeficiency virus* (HIV),** which incapacitates the immune system. HIV belongs to the retrovirus family, whose nucleic acid core is RNA rather than DNA. Retroviruses possess an enzyme (reverse transcriptase) that, once the virus is inside a host cell, transcribes the virus's RNA into DNA, which is then integrated into the host cell's chromosomes. Replication of the virus inside the cell causes the cell's death.

Unfortunately, the cells that HIV preferentially (but not exclusively) enters are helper T cells. HIV infects these cells because the CD4 protein on the plasma membrane of helper T cells acts as a high-affinity receptor for one of the HIV's surface proteins. Thus, the helper T cell binds the virus, making it possible for the virus to then enter the cell. The replicating virus directly kills the helper T cells and also indirectly causes their death via the body's usual immune attack, mediated in this case mainly by cytotoxic T cells, against virus-infected cells. In addition, by still poorly understood mechanisms (possibly by inducing apoptosis, Chapter 7), the HIV causes the death of many *uninfected* helper T cells. Without adequate numbers of helper T cells, neither B cells nor cytotoxic T cells can function normally. Thus the AIDS patient dies from infections and cancers that ordinarily would be readily handled by the immune system.

AIDS was first described in 1981, and it has since reached epidemic proportions. The World Health Organization estimates conservatively that, worldwide, 30 million people had been infected with HIV by the beginning of 1996. In 1995 alone, 4.7 million people were newly infected with HIV, an average of 13,000 new infections per day (about 4 percent of these occurring in the industrialized nations). Also during 1995, approximately 500,000 children were born infected with HIV. The great majority of persons presently infected with HIV have no symptoms of AIDS as yet. This is an important point: One must distinguish between the presence of the symptomatic disease—AIDS—and asymptomatic infection with HIV. (The latter is diagnosed by the presence of anti-HIV antibodies in the blood.) It is thought, however, that all infected persons will eventually develop AIDS, although at highly varying rates.

The path from HIV infection to AIDS commonly takes about 10 years. Typically, during the first five years the rapidly replicating viruses continually kill large numbers of helper T cells in lymphoid tissues, but these are replaced by new cells. Therefore, the number of helper T cells in the body stays normal (the number in blood is about 1000 cells per cubic millimeter of blood), and the person is asymptomatic. During the next 5 years, this balance is lost; the number of helper T cells, as measured in blood, drops to about half the normal level, but many people still remain asymptomatic. As the helper T cell count continues to fall, however, the symptoms of AIDS begin—infections with bacteria, viruses, fungi, and parasites. These are accompanied by systemic symptoms of weight loss, lethargy, and fever, all caused by high levels of the cytokines that induce the acute phase response. Certain cancers also occur with high frequency. Death usually ensues within two years after the onset of AIDS symptoms.

The transmission of HIV is known to occur only through (1) transfer of contaminated blood or blood products from one person to another; (2) sexual intercourse with an infected partner; or (3) transmission from an infected mother to her offspring either across the placenta during pregnancy and delivery, or via breast milk.

ESSAY

Transfusion of blood is no longer a significant source of AIDS in the United States since blood is now tested and discarded if found to be positive for the presence of antibody to HIV. There is, however, an extremely slight possibility that a blood donor was infected with HIV so recently that the antibodies have not become evident at the time of blood giving. This possibility is further minimized by careful screening of prospective donors to avoid those who might be in a high risk group.

Infection with HIV among intravenous drug users is related to exposure to contaminated blood through the practice of sharing needles.

The major reason that homosexual men are a high risk group is that anal intercourse is often associated with small tears in anal blood vessels, which allows HIV-containing semen from an infected person to pass into the blood of the partner.

HIV can definitely be transmitted during heterosexual intercourse. Such transmission occurs mainly during vaginal intercourse and can be male to female or female to male. In the United States, where most cases of HIV infection are in homosexual (and bisexual) males and intravenous drug users, only 5 to 6 percent of all HIV infections are presently attributed to heterosexual transmission. In contrast, in sub-Saharan Africa and parts of the Caribbean and Asia, most cases of HIV infection are acquired heterosexually. Some of the reasons for the high rate of heterosexual transmission in these countries are understood—for example, the seasonal migration of men in Africa in search of employment promotes casual or purchased sex and facilitates transmission to wives and girlfriends in rural villages—but many others—for example, a possible contributing role for non-HIV preexisting sexually transmitted diseases—remain unclear.

There is no evidence that HIV is transmitted through intact skin or via food, saliva, tears, sweat, feces, urine, vomit, insects, toilet seats, or swimming pools. Even prolonged close contact with persons infected with HIV does not lead to infection in the absence of the transmission routes described above. Nor is there any risk in *donating* blood.

The use of condoms markedly reduces the risk of transmission of HIV during sexual intercourse. The condom should be made of latex, and a water-based, not oil-based, lubricant, preferably one containing the anti-viral drug **noroxynol-9,** should be used with it.

There are two components to the therapeutic management of HIV-infected persons: one directed against the virus itself to delay progression of the disease, and one to prevent or treat the opportunistic infections and cancers that ultimately cause death. Until recently the major drug for the treatment of HIV infection itself was **zidovudine** (formerly called AZT), which acts by blocking the action of the enzyme (reverse transcriptase) that converts the viral RNA into the host cell's DNA. Given by itself, zidovudine or any of the other reverse-transcriptase inhibitors had only limited success, although, very importantly, zidovudine given to pregnant HIV-infected women markedly reduces the risk of viral transmission to the offspring. In contrast, beginning in 1996 studies began to document much greater effectiveness—what some experts have termed a revolution in anti-HIV therapy—from the simultaneous use of at least two drugs (and often as many as four) in combination. These combinations generally include at least one reverse-transcriptase inhibitor and a member of another class of drugs called protease inhibitors. These latter drugs block the HIV enzyme that acts late in the intracellular life cycle of the virus to cleave a large protein into structural proteins required for the assembly of infectious viral particles.

The ultimate hope for prevention of AIDS is development of a vaccine. For a variety of reasons related to the nature of the virus (there are multiple distinct subgroups) and the fact that it infects helper T cells, which are crucial for immune responses, this is not an easy task.

fetal genes are paternal all proteins coded for by these genes are foreign to the mother. Why does the mother's immune system not attack the trophoblast cells, which express such proteins, and reject the placenta? This problem is far from solved, but one critical mechanism (there are almost certainly others) is as follows: Trophoblast cells, unlike virtually all other nucleated cells, do not express the usual MHC class I proteins; instead they express a unique MHC class I protein that maternal immune cells do not recognize as foreign. How this all occurs is unknown.

Transfusion Reactions

Transfusion reactions, the illness caused when erythrocytes are destroyed during blood transfusion, are a special example of tissue rejection, one that illustrates the fact that antibodies rather than cytotoxic T cells can sometimes be the major factor in rejection. Erythrocytes do not have MHC proteins, but they do have plasma-membrane proteins and carbohydrates (the latter linked to the membrane by lipids) that can function as antigens when exposed to another person's blood. There are more than 400 erythrocyte antigens, but the ABO system of carbohydrates is the most important for transfusion reactions.

As described in the section on blood types in Chapter 14, some people have the gene that results in synthesis of the A antigen, some have the gene for the B antigen, some have both genes, and some have neither gene. (Genes, of course, do not code directly for carbohydrates; rather they code for the particular enzymes that catalyze formation of the carbohydrates.) The erythrocytes of those with neither gene are said to have O-type erythrocytes. Accordingly, the possible blood types are A, B, AB, and O (Table 20-9).

Type A individuals always have anti-B antibodies in their plasma. Similarly, type B individuals have plasma anti-A antibodies. Type AB individuals have neither anti-A nor anti-B antibody, and type O individuals have both. These antierythrocyte antibodies are called **natural antibodies.** How they arise "naturally" that is, without exposure to the appropriate antigen-bearing erythrocytes, is not presently clear.

With this information as background, we can predict what happens if a type A person were given type B blood. There are two incompatibilities: (1) The recipient's anti-B antibody causes the transfused cells to be attacked, and (2) the anti-A antibody in the transfused plasma causes the recipient's cells to be attacked. The latter is generally of little consequence, however, because the transfused antibodies become so diluted in the recipient's plasma that they are ineffective in inducing a response. It is the destruction of the transfused cells by the recipient's antibodies that produces the problems.

Similar analyses show that the following situations would result in an attack on the transfused erythrocytes: a type B person given either A or AB blood; a type A person given either type B or AB blood; a type O person given A, B, or AB blood. Type O people are, therefore, sometimes called universal donors, whereas type AB people are universal recipients. These terms are misleading, however, since besides antigens of the ABO system, there are a host of other erythrocyte antigens and plasma antibodies against them. Therefore, except in a dire emergency, the blood of donor and recipient must be tested for incompatibilities directly by the procedure called **cross-matching.** The recipient's serum is combined on a glass slide with the prospective donor's erythrocytes (a "major" cross-match), and the mixture observed for rupture (hemolysis) or clumping (agglutination) of the erythrocytes. In addition, the recipient's erythrocytes can be combined with the prospective donor's serum (a "minor" cross-match).

Another group of erythrocyte membrane antigens of medical importance is the Rh system of proteins. There are more than 40 such antigens, but the most strongly antigenic one is termed Rh_o (also called D), known commonly as the **Rh factor** because it was first studied in rhesus monkeys. Human erythrocytes either have the antigen (Rh-positive) or lack it (Rh-negative). About 85 percent of the United States population is Rh-positive.

The Rh system, unlike the "natural antibodies" of the ABO system, follows the classical immunity pattern in that

---

TABLE 20-9 HUMAN ABO BLOOD GROUPS

Blood group	Percent*	Antigen on RBC	Genetic possibilities Homozygous	Heterozygous	Antibody in blood
A	42	A	AA	AO	Anti-B
B	10	B	BB	BO	Anti-A
AB	3	A and B	—	AB	Neither anti-A nor anti-B
O	45	Neither A nor B	OO	—	Both anti-A and anti-B

*In the United States.

no one has anti-Rh antibodies unless exposed to Rh-positive cells from another person. This can occur if an Rh-negative person is subjected to multiple transfusions with Rh-positive blood, but its major occurrence involves the mother-fetus relationship. When an Rh-negative mother carries an Rh-positive fetus, some of the fetal erythrocytes may cross the placental barriers into the maternal circulation, inducing her to synthesize anti-Rh antibodies. Because this occurs mainly during separation of the placenta at delivery, a first Rh-positive pregnancy rarely offers any danger to the fetus since delivery occurs before the antibodies are made by the mother. In future pregnancies, however, these antibodies will already be present in the mother and can cross the placenta to attack and hemolyze the erythrocytes of an Rh-positive fetus. This condition, which can cause an anemia severe enough to result in death of the fetus in utero or of the newborn, is called **hemolytic disease of the newborn.** The risk increases with each Rh-positive pregnancy as the mother becomes more and more sensitized.

Fortunately, this can be prevented by giving an Rh-negative mother human gamma globulin against Rh-positive erythrocytes within 72 h after she has delivered an Rh-positive infant. These antibodies bind to the antigenic sites on any Rh-positive erythrocytes that might have entered the mother's blood during delivery and prevent them from inducing antibody synthesis by the mother. The administered antibodies are eventually catabolized.

You may be wondering whether ABO incompatibilities are also a cause of hemolytic disease of the newborn. For example, a woman with type O blood has natural antibodies to both the A and B antigens. If her fetus is type A or B, this theoretically should cause a problem. Fortunately, it usually does not, partly because the A and B antigens are not strongly expressed in fetal erythrocytes and partly because the natural antibodies, unlike the anti-Rh antibodies, are of the IgM type, which do not readily cross the placenta.

Allergy (Hypersensitivity)

Allergy or *hypersensitivity* refers to certain diseases in which immune responses to environmental antigens cause inflammation and damage to the body itself. Antigens that cause allergy are termed **allergens,** common examples of which include those in ragweed pollen and poison ivy. Most allergens themselves are relatively or completely harmless, and it is the immune responses to them that cause the damage. In essence, then, allergy is immunity gone wrong, for the response is inappropriate to the stimulus.

To develop a particular allergy, a genetically predisposed person must first be exposed to the allergen. This initial exposure causes "sensitization," and it is the subsequent exposures that elicit the damaging immune responses we recognize as the disease. The diversity of allergic responses reflects the different immunological effector pathways elicited, and the classification of allergic diseases is based on these mechanisms.

In one type of allergy, the inflammatory response is independent of antibodies. It is due to marked secretion of cytokines by sensitized helper T cells in the area. These cytokines themselves act as inflammatory mediators and also activate macrophages to secrete their potent mediators, particularly TNF. Because it takes several days to develop, this type of allergy is known as **delayed hypersensitivity.** The skin rash that appears after contact with poison ivy is an example.

In contrast to these are the various types of antibody-mediated allergic responses. One important type is termed **immune-complex hypersensitivity.** It occurs when so many antibodies (of either the IgG or IgM types) combine with free antigens that large number of antigen-antibody complexes precipitate out on the surface of endothelial cells or are trapped in capillary walls, particularly those of the renal corpuscles. These immune complexes activate complement, which then induces an inflammatory response that damages the tissues immediately surrounding the complexes. Allergy to penicillin is an example.

The most common type of antibody-mediated allergic responses, however, are those termed **immediate hypersensitivity** because they are usually very rapid in onset. They are also called **IgE-mediated hypersensitivity** because they involve IgE antibodies.

Immediate Hypersensitivity.   In immediate hypersensitivity, initial exposure to the antigen leads to some antibody synthesis and, more important, to the production of memory B cells that mediate active immunity. Upon re-exposure, the antigen elicits a more powerful antibody response. So far, none of this is unusual, but the difference is that the particular antigens that elicit immediate-hypersensitivity reactions stimulate, in genetically susceptible persons, the production of type IgE antibodies. Production of IgE requires the participation of a particular subset of helper T cells that are activated by the allergens presented by B cells; these activated cells then release cytokines that preferentially stimulate differentiation of the B cells into IgE-producing plasma cells.

Upon their release from plasma cells, IgE antibodies circulate to various parts of the body and become attached, via binding sites on their Fc portions, to connective-tissue mast cells (Figure 20-19). When subsequently an antigen enters the body and combines with the IgE bound to the mast cell, this triggers release of the mast cell's secretory vesicles, which contain many inflammatory mediators, including **histamine.** In addition to releasing these preformed inflammatory mediators stored in mast-cell vesicles, the binding of antigen to IgE triggers the mast cell

APPENDIX A
Answers To Thought Questions

Chapter 4

4-1. A drug could decrease acid secretion by (1) binding to the membrane sites that normally inhibit acid secretion, which would produce the same effect as the body's natural messengers that inhibit acid secretion; (2) binding to membrane protein that normally stimulates acid secretion but not itself triggering acid secretion, thereby preventing the body's natural messengers from binding (competition); or (3) having an allosteric effect on the binding sites, which would increase the affinity of the sites that normally bind inhibitor messengers or decrease the affinity of those sites that normally bind stimulatory messengers.

4-2. The reason for a lack of insulin effect could be either a decrease in the number of available binding sites to which insulin can bind or a decrease in the affinity of the binding sites for insulin so that less insulin is bound. A third possibility, which does not involve insulin binding, would be a defect in the way the binding site triggers a cell response once it has bound insulin.

4-3. An increase in the concentration of compound A will lead to a decrease in the concentration of compound H by the route shown below. Sequential activations and inhibitions of proteins of this general type are frequently encountered in physiological control systems.

Enzyme C

↑ [A] —Allosteric activation→ ↑ Protein kinase B activity →

↑ Activity of enzyme C

D —→ ↑ [E]

Allosteric inhibition

↓ Enzyme F activity

G —→ ↓ H

Decrease conversion of G to H. Therefore, ↓ [H]

4-4. (A) Acid secretion could be increased to 40 mmol/h by (1) increasing the concentration of compound X from $2\,pM$ to $8\,pM$, thereby increasing the number of binding sites occupied; or (2) increasing the affinity of the binding sites for compound X, thereby increasing the amount bound without changing the concentration of compound X. (B) Increasing the concentration of compound X from 18 to $28\,pM$ will not increase acid secretion because, at $18\,pM$, all the binding sites are occupied (the system is saturated), and there are no further binding sites available.

4-5. Phosphoprotein phosphatase removes the phosphate group from proteins that have been covalently modulated by a protein kinase. Without phosphoprotein phosphatase, the protein could not return to its unmodulated state and would remain in its activated state. The ability to decrease as well as increase protein activity is essential to the regulation of physiological processes.

4-6. The reactant molecules have a combined energy content of 55 + 93 = 148 kcal/mol, and the products have 62 + 87 = 149. Thus, the energy content of the products exceeds that of the reactants by 1 kcal/mol, and this amount of energy must be added to A and B to form the products C and D.

The reaction is reversible since the difference in energy content between the reactants and products is small. When the reaction reaches chemical equilibrium, there will be a slightly higher concentration of reactants than products.

4 7. The maximum rate at which the end product E can be formed is 5 molecules per second, the rate of the slowest—(rate-limiting)—reaction in the pathway.

4-8. Under normal conditions, the concentration of oxygen at the level of the mitochondria in cells, including muscle at rest, is sufficient to saturate the enzyme that combines oxygen with hydrogen to form water. The rate-limiting reactions in the electron transport chain depend on the available concentrations of ADP and P_i, which are combined to form ATP.

Thus, increasing the oxygen concentration above normal levels will not increase ATP production. If a muscle is contracting, it will break down ATP into ADP and P_i, which become the major rate-limiting substrates for increasing ATP production. With intense muscle activity, the level of oxygen may fall below saturating levels, limiting the rate of ATP production, and intensely active muscles must use anaerobic glycolysis to provide additional ATP. Under these circumstances, increasing the oxygen concentration in the blood will increase the rate of ATP production. As discussed in Chapter 14, it is not the concentration of oxygen in the blood that is increased during exercise but the rate of blood flow to a muscle, resulting in greater quantities of oxygen delivery to the tissue.

4-9. During starvation, in the absence of ingested glucose, the body's stores of glycogen are rapidly depleted. Glucose, which is the major fuel used by the brain, must now be synthesized from other types of molecules. Most of this newly formed glucose comes from the breakdown of proteins to amino acids and their conversion to glucose. To a lesser extent, the glycerol portion of fat is converted to glucose. The fatty acid portion of fat cannot be converted to glucose.

4-10. Fatty acids are broken down to acetyl coenzyme A during beta oxidation, and acetyl coenzyme A enters the Krebs cycle to be converted to carbon dioxide. Since the Krebs cycle can function only during aerobic conditions, the catabolism of fat is dependent on the presence of oxygen. In the absence of oxygen, acetyl coenzyme A cannot be converted to carbon dioxide, and the increased concentration of acetyl coenzyme A inhibits the further beta oxidation of fatty acids.

4-11. Ammonia is formed in most cells during the oxidative deamination of amino acids and then travels to the liver via the blood. The liver then detoxifies the ammonia by converting it to the nontoxic compound urea. Since the liver is the site in which ammonia is converted to urea, diseases that damage the liver can lead to an accumulation of ammonia in the blood, which is especially toxic to nerve cells. Note that it is not the liver that produces the ammonia.

Chapter 5

5-1. Nucleotide bases in DNA pair A to T and G to C. Given the base sequence of one DNA strand as:

A-G-T-G-C-A-A-G-T-C-T

(A) The complementary strand of DNA would be:

T-C-A-C-G-T-T-C-A-G-A

(B) The sequence in RNA transcribed from the first strand would be:

U-C-A-C-G-U-U-C-A-G-A

Recall that uracil U replaces thymine T in RNA.

5-2. The triplet code G-T-A in DNA will be transcribed into mRNA as C-A-U, and the anticodon in tRNA corresponding to C-A-U is G-U-A.

5-3. If the gene were only composed of the triplet exon code words, the gene would be 300 nucleotides in length since a triplet of three nucleotides codes for one amino acid. However, because of the presence of intron segments in most genes, which account for 75 to 90 percent of the nucleotides in a gene, the gene would be between 1200 and 3000 nucleotides long; moreover, there are also termination codons in the gene. Thus, the exact size of a gene cannot be determined by knowing the number of amino acids in the protein coded by the gene.

5-4. Tubulin is the protein that polymerizes to form microtubules. The tubulin monomers become linked together in a spiral that forms the walls of the hollow microtubules. Without microtubules to form the spindle apparatus, the chromosomes will not separate during mitosis. There are chemical agents in the cell that cause the polymerization of tubulin at the time of mitosis and cause it to depolymerize following cell division.

5-5. A drug that inhibits DNA replication will inhibit cell division since a duplicate set of chromosomes is necessary for this process. Since one of the characteristics of cancer cells is their ability to undergo excessive uncontrolled division, a drug that inhibits DNA replication will inhibit the multiplication of cancer cells. Unfortunately, such drugs also inhibit the division of normal cells, particularly those that divide at a high rate. These include the cells that give rise to blood cells and the epithelial lining of the gastrointestinal tract. The use of such drugs must be carefully monitored to balance the damage done to normal tissues against the inhibition of tumor growth.

Chapter 6

6-1. (A) During diffusion, the net flux always occurs from high to low concentration. Thus, it will be from 2 to 1 in A and from 1 to 2 in B. (B) At equilibrium, the concentrations of solute in the two compartments will be equal: 4 mM in case A and 31 mM in case B. (C) Both will reach diffusion equilibrium at the same rate since the difference in concentration across the membrane is the same in each case, 2 mM $[(3 - 5) = -2$, and $(32 - 30) = 2]$. The two one-way fluxes will be much larger in B than in A, but the net flux has the same magnitude in both cases, although oriented in opposite directions.

6-2. The ability of one amino acid to decrease the flux of a second amino acid across a cell membrane is an example of the competition of two molecules for the same binding site, as explained in Chapter 4. The binding site for alanine on the transport protein can also bind leucine. The higher the concentration of alanine, the greater the number of binding sites that it occupies, and the fewer available for binding leucine. Thus, less leucine will be moved into the cell.

6-3. The net transport will be out of the cell in the direction from the higher-affinity site on the intracellular surface to the lower-affinity site on the extracellular surface. More molecules will be bound to the transporter on the higher-affinity side of the membrane, and thus more will move out of the cell than into it, until the concentration in the extracellular fluid becomes large enough that the number of molecules bound to transporters at the extracellular surface is equal to the number bound at the intracellular surface.

6-4. Although ATP is not used directly in secondary active transport, it is necessary for the primary active transport of sodium out of cells. Since it is the sodium concentration gradient across the plasma membrane that provides the energy for most secondary active-transport systems, a decrease in ATP production will decrease primary active sodium transport, leading to a decrease in the sodium concentration gradient and thus to a decrease in secondary active transport.

6-5. The solution with the greatest osmolarity will have the lowest water concentration. The osmolarities are:

A. $20 + 30 + 2 \times (150) + 3 \times (10) = 380$ mOsm
B. $10 + 100 + 2 \times (20) + 3 \times (50) = 300$ mOsm
C. $100 + 200 + 2 \times (10) + 3 \times (20) = 380$ mOsm
D. $30 + 10 + 2 \times (60) + 3 \times (100) = 460$ mOsm

Thus, solution D has the lowest water concentration. (Recall that NaCl forms two ions in solution and CaCl$_2$ forms three.)

Solution B is isosmotic since it has the same osmolarity as intracellular fluid.

6-6. Initially the osmolarity of compartment 1 is $(2 \times 200) + 100 = 500$ mOsm and that of 2 is $(2 \times 100) + 300 = 500$ mOsm. The two solutions thus have the same osmolarity, and there is no difference in water concentration across the membrane. Since the membrane is permeable to urea, this substance will undergo net diffusion until it reaches the same concentration (200 mM) on the two sides of the membrane. In other words, in the steady state it will not affect the volumes of the compartments. In contrast, the higher initial NaCl concentration in compartment 1 than in compartment 2 will cause, by osmosis, the movement of water from compartment 2 to compartment 1 until the concentration of NaCl in both is 150 mM. Note that the same volume change would have occurred if there were no urea present in either compartment. It is only the concentration of nonpenetrating solutes (NaCl in this case) that determines the volume change, regardless of the concentration of any penetrating solutes that are present.

6-7. The osmolarities and nonpenetrating-solute concentrations are:

Solution	Osmolarity, mOsm	Nonpenetrating solute Concentration, mM
A	$(2 \times 150) + 100 = 400$	$2 \times 150 = 300$
B	$(2 \times 100) + 150 = 350$	$2 \times 100 = 200$
C	$(2 \times 200) + 100 = 500$	$2 \times 200 = 400$
D	$(2 \times 100) + 50 = 250$	$2 \times 100 = 200$

Only the concentration of nonpenetrating solutes (NaCl in this case) will determine the change in cell volume. Since the intracellular concentration of nonpenetrating solute is 300 mOsm, solution A will produce no change in cell volume. Solutions B and D will cause cells to swell since they have a lower concentration of nonpenetrating solute (higher water concentration) than the intracellular fluid. Solution C will cause cells to shrink because it has a higher concentration of nonpenetrating solute than the intracellular fluid.

6-8. Solution A is isotonic because it has the same concentration of nonpenetrating solutes as intracellular fluid (300 mM). Solution A is also hyperosmotic since its total osmolarity is greater than 300 mOsm, as is also true for solutions B and C. Solution B is hypotonic since its concentration of nonpenetrating solutes is less than 300 mM. Solution C is hypertonic since its concentration of nonpenetrating solutes is greater than 300 mM. Solution D is hypotonic (less than 300 mM of nonpenetrating solutes) and also hypoosmotic (having a total osmolarity of less than 300 mOsm).

6-9. Exocytosis is triggered by an increase in intracellular calcium concentration. Calcium ions are actively transported out of cells, in part by secondary countertransport coupled to the downhill entry of sodium ions on the same transporter. If the intracellular concentration of sodium ions were increased, the sodium concentration gradient across the membrane would be decreased, and this would decrease the secondary active transport of calcium out of the cell. This would lead to an increase in intracellular calcium concentration, which would trigger increased exocytosis.

Chapter 7

7-1. 4.4 mmol/L. If you answered 8 mmol/L, you ignored the fact that plasma potassium concentration is homeostatically regulated so that the doubling of input will lead to negative-feedback reflexes that oppose an equivalent increase in plasma concentration. If you answered 4 mmol/L, you ignored the fact that homeostatic control systems cannot *totally* prevent changes in the regulated variable when a perturbation occurs. Thus, there must be *some* rise in plasma potassium concentration in this situation (to serve as the error signal for the compensating reflexes), and this is consistent with the answer 4.4 mmol/L. (The actual rise would have to be experimentally determined. There is no way you could have predicted that the rise would be 10 percent. All you could predict is that it would neither double nor stay absolutely unchanged.)

7-2. No. There may in fact be one, but there is another possibility—that the altered skin blood flow in the cold represents an *acclimatization* undergone by each Eskimo during his or her lifetime as a result of performing such work repeatedly.

7-3. Patient A's drug very likely acts to block phospholipase A$_2$, whereas patient B's drug blocks lipoxygenase. (See Figure 7-6.)

7-4. The chronic loss of exposure of the heart's receptors to norepinephrine causes an up-regulation of this receptor type (that is, more receptors in the heart for norepinephrine). The drug, being an agonist of norepinephrine (that is, able to bind to norepinephrine's receptors and activate them) is now more effective since there are more receptors for it to combine with.

7-5. None. Since you are told that all six responses are mediated by the cAMP system, then blockage of any of the steps listed in the question would eliminate all six of the responses. This is because the cascade for all six responses is identical from the receptor through the formation of cAMP and activation of cAMP protein kinase. Thus, the drug must be acting at a point beyond this kinase (for example, at the level of the phosphorylated protein mediating this response).

7-6. Not in most cells, since there are other physiological mechanisms by which signals impinging on the cell can increase cytosolic calcium concentration. These include (1) second-messenger-induced release of calcium from the endoplasmic reticulum and (2) voltage-sensitive calcium channels.

Chapter 8

8-1. Little change in the resting membrane potential would occur when the pump first stops because the pump's *direct* contribution to charge separation is very small. With time, however, the membrane potential would depolarize progressively toward zero because the sodium and potassium concentration gradients, which depend on the Na,K-ATPase pumps and which give rise to the diffusion potentials that constitute most of the membrane potential, run down.

8-2. The resting potential would decrease (that is, become less negative) because the concentration gradient causing net diffusion of this positively charged ion out of the cell would be smaller. The action potential would fire more easily (that is, with smaller stimuli) because the resting potential would be closer to threshold. It would repolarize more slowly because repolarization depends on net potassium diffusion from the cell, and the concentration gradient driving this diffusion is lower. Also, the afterhyperpolarization would be smaller.

8-3. The hypothalamus was probably damaged. It plays a critical role in appetite, thirst, and sexual capacity.

8-4. The drug probably blocks cholinergic muscarinic receptors. These receptors on effector cells mediate the actions of the parasympathetic nerves. Therefore, the drug would remove the slowing effect of these nerves on the heart, allowing the heart to speed up. Blocking their effect on the salivary glands would cause the dry mouth. The drug is not blocking cholinergic nicotinic receptors because the skeletal muscles are not affected.

8-5. Since the membrane potential of the cells in question depolarizes (that is, becomes less negative) when chloride channels are blocked, one can predict that there was net chloride diffusion into the cells through these channels prior to the drug. Therefore, one can also predict that this passive inward movement was being exactly balanced by active transport of chloride out of the cells.

8-6. Without acetylcholinesterase, acetylcholine would remain bound to the receptors, and all the actions normally caused by acetylcholine would be accentuated. Thus, there would be marked narrowing of the pupils, airway constriction, stomach cramping and diarrhea, sweating, salivation, slowing of the heart, and fall in blood pressure. On the other hand, in skeletal muscles, which must repolarize after excitation in order to be excited again, there would be weakness, fatigue, and finally inability to contract. In fact, poisoning by high doses of cholinesterase inhibitors occurs because of paralysis of the muscle involved in respiration. Low doses of these compounds are used therapeutically.

8-7. These potassium channels, which open after a short delay following the initiation of an action potential, increase potassium diffusion out of the cell, hastening repolarization. They also account for the increased potassium permeability that causes the afterhyperpolarization. Therefore, the action potential would be broader (that is, longer in duration), returning to its resting level more slowly, and the afterhyperpolarization would be absent.

Chapter 9

9-1. (a) Use drugs to block transmission in the pathways that convey information about pain to the brain. For example, if substance P is the neurotransmitter at the central endings of the nociceptor afferent fibers, give a drug that blocks the substance P receptors. (b) Cut the dorsal root at the level of entry of the nociceptor fibers to prevent transmission of their action potentials into the central nervous system. (c) Give a drug that activates receptors in the descending pathways that block transmission of the incoming or ascending pain information. (d) Stimulate the neurons in these same descending pathways to increase their blocking activity (stimulation-produced analgesia or, possibly, acupuncture). (e) Cut the ascending pathways that transmit information from the nociceptor afferents. (f) Deal with the emotions, attitudes, memories, and so on, to decrease the sensitivity to the pain. (g) Stimulate nonpain, low-threshold afferent fibers to block transmission through the pain pathways (TENS). (h) Block transmission in the afferent nerve with a local anesthetic such as Novacaine or Lidocaine.

9-2. Information regarding temperature is carried via the anterolateral system to the brain. Fibers of this system cross to the opposite side of the body in the spinal cord at the level of entry of the afferent fibers (see Figure 9-18B). Damage to the left side of the spinal cord or any part of the left side of the brain that contains fibers of the pathways for temperature would interfere with awareness of a heat stimulus on the right. Thus, damage to the somatosensory cortex of the left cerebral hemisphere (that is, opposite the stimulus) would interfere with awareness of the stimulus. Injury to the spinal cord at the point at which fibers of the anterolateral system from the two halves of the spinal cord cross to the opposite side would interfere with the awareness of heat applied to either side of the body as would the unlikely event that damage occurred to relevant areas of both sides of the brain.

9-3. Vision would be restricted to the rods; therefore, it would be normal at very low levels of illumination (when the cones would not be stimulated anyway), but at higher levels of illumination clear vision of fine details would be lost, and everything would appear in shades of gray. There would be no color vision. In very bright light, there would be no vision because of bleaching of the rods' rhodopsin.

9-4. (a) The individual lacks a functioning primary visual cortex. (b) The individual lacks a functioning visual association cortex.

Chapter 10

10-1. Epinephrine falls to very low levels during rest and fails to increase during stress. The sympathetic preganglionics provide the only major control of the adrenal medulla.

10-2. The increased concentration of binding protein causes more TH to be bound, thereby lowering the plasma concentration of *free* TH. This causes less negative feedback inhibition of TSH secretion by the anterior pituitary, and the increased TSH causes the thyroid to secrete more TH until the free concentration has returned to normal. The end result is an increased *total* plasma TH—most bound to the protein—but a normal free TH. There is no hyperthyroidism because it is only the free concentration that exerts effects on TH's target cells.

10-3. Destruction of the anterior pituitary or hypothalamus. These symptoms reflect the absence of, in order, growth hormone, the gonadotropins, and ACTH (the symptom is due to the resulting decrease in cortisol secretion). One cannot tell from these data alone whether the problem is primary hyposecretion of the anterior pituitary hormones or secondary hyposecretion because the hypothalamus is not secreting hypophysiotropic hormones normally.

10-4. Vasopressin and oxytocin (that is, the major posterior pituitary hormones). The anterior pituitary hormones would not be affected because the influence of the hypothalamus on these hormones is exerted not by connecting nerves but via the hypophysiotropic hormones in the portal vascular system.

10-5. The secretion of GH increases. Somatostatin, coming from the hypothalamus, normally exerts an inhibitory effect on the secretion of this hormone.

10-6. Norepinephrine and many other neurotransmitters are released by neurons that terminate on the hypothalamic neurons that secrete the hypophysiotropic hormones. Therefore, manipulation of these neurotransmitters will alter secretion of the hypophysiotropic hormones and thereby the anterior pituitary hormones.

10-7. The high dose of the cortisol-like substance inhibits the secretion of ACTH by feedback inhibition of (1) hypothalamic corticotropin releasing hormone and (2) the response of the anterior pituitary to this hypophysiotropic hormone. The lack of ACTH causes the adrenal to atrophy and decrease its secretion of cortisol.

10-8. The hypothalamus. The low basal TSH indicates either that the pituitary is defective or that it is receiving inadequate stimulation (TRH) from the hypothalamus. If the thyroid itself were defective, basal TSH would be elevated because of less negative-feedback inhibition by TH. The TSH increase in response to TRH shows that the pituitary is capable of responding to a stimulus and so is unlikely to be defective. Therefore, the problem is that the hypothalamus is secreting too little TRH.

Chapter 11

11-1. Under resting conditions, the myosin has already bound and hydrolyzed a molecule of ATP, resulting in an energized molecule of myosin ($M^* \cdot ADP \cdot P_i$). Since ATP is necessary to detach the myosin cross bridge from actin at the end of cross bridge movement, the absence of ATP will result in rigor mortis, in which case the cross bridges become bound to actin but do not detach, leaving myosin bound to actin ($A \cdot M$).

11-2. No. The transverse tubules conduct the muscle action potential from the plasma membrane into the interior of the fiber, where it can trigger the release of calcium from the sarcoplasmic reticulum. If the transverse tubules were not attached to the plasma membrane, an action potential could not be conducted to the sarcoplasmic reticulum and there would be no release of calcium to initiate contraction.

11-3. The length-tension relationship states that the maximum tension developed by a muscle decreases at lengths below l_o. During normal shortening, as the sarcomere length becomes shorter than the optimal length, the maximum tension that can be generated decreases. With a light load, the muscle will continue to shorten until its maximal tension just equals the load. No further shortening is possible since at shorter sarcomere lengths

the tension would be less than the load. The heavier the load, the less the distance shortened before reaching the isometric state.

11-4. Maximum tension is produced when the fiber is (1) stimulated by an action potential frequency that is high enough to produce a maximal tetanic tension, and (2) at its optimum length l_o, where the thick and thin filaments have overlap sufficient to provide the greatest number of cross bridges for tension production.

11-5. Moderate tension—for example, 50 percent of maximal tension—is accomplished by recruiting sufficient numbers of motor units to produce this degree of tension. If activity is maintained at this level for prolonged periods, some of the active fibers will begin to fatigue and their contribution to the total tension will decrease. The same level of total tension can be maintained, however, by recruiting new motor units as some of the original ones fatigue. At this point, for example, one might have 50 percent of the fibers active, and 25 percent fatigued and 25 percent still unrecruited. Eventually, when all the fibers have fatigued and there are no additional motor units to recruit, the whole muscle will fatigue.

11-6. The oxidative motor units, both fast and slow, will be affected first by a decrease in blood flow since they depend on blood flow to provide both the fuel—glucose and fatty acids—and the oxygen required to metabolize the fuel. The fast-glycolytic motor units will be affected more slowly since they rely predominantly on internal stores of glycogen, which is anaerobically metabolized by glycolysis.

11-7. Two factors lead to recovery of muscle force. (1) Some new fibers can be formed by the fusion and development of undifferentiated satellite cells. This will replace some, but not all, of the fibers that were damaged. (2) Some of the restored force results from hypertrophy of the surviving fibers. Because of the loss of fibers in the accident, the remaining fibers must produce more force to move a given load. The remaining fibers undergo increased synthesis of actin and myosin, resulting in increases in fiber diameter and thus their force of contraction.

11-8. In the absence of extracellular calcium ions, skeletal muscle contracts normally in response to an action potential generated in its plasma membrane because the calcium required to trigger contraction comes entirely from the sarcoplasmic reticulum within the muscle fibers. If the motor neuron to the muscle is stimulated in a calcium-free medium, however, the muscle will not contract because the influx of calcium from the extracellular fluid into the motor nerve terminal is necessary to trigger the release of acetylcholine that in turn triggers an action potential in the muscle.

In a calcium-free solution, smooth muscles of all types would fail to respond to stimulation of the nerve supplying the muscle. However, the response to direct stimulation of the muscle's plasma membrane would depend on the type of smooth muscle. Smooth muscles that primarily depend on calcium released from the sarcoplasmic reticulum will behave like skeletal muscle and respond to direct stimulation. Smooth muscles that rely on the influx of calcium from the extracellular fluid to trigger contraction will fail to contract in response to stimulation of its plasma membrane.

11-9. The simplest model to explain the experimental observations is as follows. Upon parasympathetic nerve stimulation, a neurotransmitter is released that binds to receptors on the membranes of smooth-muscle cells and triggers contraction. The substance released, however, is not acetylcholine (ACh) for the following reason.

Action potentials in the parasympathetic nerves are essential for initiating nerve-induced contraction. When the nerves were prevented from generating action potentials by blockage of their voltage-sensitive sodium channels, there was no response to nerve stimulation. ACh is the neurotransmitter released from most, but not all, parasympathetic endings. When the muscarinic receptors for ACh were blocked, however, stimulation of the parasympathetic nerves still produced a contraction, providing evidence that some substance other than ACh is being released by the neurons and producing contraction.

Chapter 12

12-1.   None. The gamma motor neurons are important in preventing the muscle-spindle stretch receptors from going slack, but when testing this reflex, the intrafusal fibers are not flaccid. The test is performed with a bent knee, which stretches the extensor muscles in the thigh (and the intrafusal fibers within the stretch receptors). The stretch receptors are therefore responsive.

12-2.   The efferent pathway of the reflex arc (the alpha motor neurons) would not be activated, the effector cells (the extrafusal muscle fibers) would not be activated, and there would be no reflex response.

12-3.   The drawing must have excitatory synapses on the motor neurons of both ipsilateral extensor and ipsilateral flexor muscles.

12-4.   A toxin that interferes with the inhibitory synapses on motor neurons would leave unbalanced the normal excitatory input to these neurons. Thus, the otherwise normal motor neurons would fire excessively, which would result in increased muscle contraction. This is exactly what happens in lockjaw as a result of the toxin produced by the tetanus bacillus.

Chapter 13

13-1.   Dopamine is depleted in the basal ganglia of people with Parkinson's disease, and they are therapeutically given dopamine agonists, usually L-dopa. This treatment raises dopamine levels in other parts of the brain, however, where the dopamine levels were previously normal. Schizophrenia is associated with increased brain dopamine levels, and symptoms of this disease appear when dopamine levels are high. The converse therapeutic problem can occur during the treatment of schizophrenics with dopamine-lowering drugs, which sometimes causes the symptoms of Parkinson's disease to appear.

13-2.   Experiments done on anesthetized animals often involve either stimulating a brain part to observe the effects of increased neuronal activity or damaging ("lesioning") an area to observe resulting deficits. Such experiments on animals, which lack the complex language mechanisms of people, cannot help with language studies. Diseases sometimes mimic these two experimental situations, and behavioral studies of the resulting language deficits in people with aphasia, coupled with study of their brains after death, have provided a wealth of information.

Chapter 14

14-1.   No. Decreased erythrocyte volume is certainly one possible explanation, but there is a second: The person might have a normal erythrocyte volume but an abnormally increased plasma volume. Convince yourself of this by writing the hematocrit equation as: erythrocyte volume/(erythrocyte volume + plasma volume).

14-2.   A halving of tube radius. Resistance is directly proportional to blood viscosity but inversely proportional to the *fourth power* of tube radius.

14-3.   The plateau of the action potential and the contraction would be absent. You might think that contraction would persist since most calcium in excitation-contraction coupling in the heart comes from the sarcoplasmic reticulum. However, the signal for the release of this calcium is the calcium entering across the plasma membrane.

14-4.   The SA node is not functioning, and the ventricles are being driven by a pacemaker in the vicinity of the AV node.

14-5.   The person has a narrowed aortic valve. Normally, the resistance across the aortic valve is so small that there is only a tiny pressure difference between the left ventricle and the aorta during ventricular ejection. In the example given here, the large pressure difference indicates that resistance across the valve must be very high.

14-6.   This question is analogous to question 14-5 in that the large pressure difference across a valve while the valve is open indicates an abnormally narrowed valve—in this case the left AV valve.

14-7.   Decreased heart rate and contractility. These are effects mediated by the sympathetic nerves on beta-adrenergic receptors in the heart.

14-8.   120 mmHg. MAP = DP + 1/3 (SP − DP).

14-9.   The drug must have caused the arterioles in the kidneys to dilate enough to reduce their resistance by 50 percent. Blood flow to an organ is determined by mean arterial pressure and the organ's resistance to flow. Another important point can be deduced here: If mean arterial pressure has not changed even though renal resistance has dropped 50 percent, then either the resistance of some other organ or cardiac output has gone up.

14-10.   The experiment suggests that acetylcholine causes vasodilation by releasing nitric oxide or some other vasodilator from endothelial cells.

14-11.   A low plasma protein concentration. Capillary pressure is, if anything, lower than normal and so cannot be causing the edema. Another possibility is that capillary permeability to plasma proteins has increased, as occurs in burns.

14-12.   20 mmHg/L per minute. TPR = MAP/CO.

14-13.   Nothing. Cardiac output and TPR have remained unchanged, and so their product, MAP, has also remained unchanged. This question emphasizes that MAP depends on cardiac output but not on the combination of heart rate and stroke volume that produces the cardiac output.

14-14.   It increases. There are a certain number of impulses traveling up the nerves from the arterial baroreceptors. When these nerves are cut, the number of impulses reaching the medullary cardiovascular center goes to zero, just as it would physiologically if the mean arterial pressure were to decrease markedly. Accordingly, the medullary cardiovascular center responds to the absent impulses by reflexly increasing arterial pressure.

14-15. It decreases. The hemorrhage causes no immediate change in hematocrit since erythrocytes and plasma are lost in the same proportion. As interstitial fluid starts entering the capillaries, however, it expands the plasma volume and decreases hematocrit. (This is too soon for any new erythrocytes to be synthesized.)

Chapter 15

15-1. 200 ml/mmHg.

$$\text{Lung compliance} = \Delta \text{ lung volume}/\Delta\ (P_{alv} - P_{ip})$$
$$= 800 \text{ ml}/[0 - (-8)] \text{ mmHg} - [0 - (-4)] \text{ mmHg}$$
$$= 800 \text{ ml}/4 \text{ mmHg} = 200 \text{ ml/mmHg}$$

15-2. More subatmospheric than normal. A decreased surfactant level causes the lungs to be less compliant (that is, more difficult to expand). Therefore, a greater transpulmonary pressure $(P_{alv} - P_{ip})$ is required to expand them a given amount.

15-3. No.

$$\text{Alveolar ventilation} = (\text{tidal volume-dead space}) \times \text{breathing rate}$$
$$= (250 \text{ ml} - 150 \text{ ml}) \times 20 \text{ breaths/min}$$
$$= 2000 \text{ ml/min}$$

whereas normal alveolar ventilation is approximately 4000 ml/min.

15-4. The volume of the snorkel constitutes an additional dead space, and so total pulmonary ventilation must be increased if alveolar ventilation is to remain constant.

15-5. The alveolar P_{O_2} will be higher than normal, and the alveolar P_{CO_2} will be lower. If you do not understand why, review the factors that determine the alveolar gas pressures.

15-6. No. Hypoventilation reduces arterial P_{O_2} but only because it reduces alveolar P_{O_2}. That is, in hypoventilation, *both* alveolar and arterial P_{O_2} are decreased to essentially the same degree. In this problem, alveolar P_{O_2} is normal, and so the person is not hypoventilating. The low arterial P_{O_2} must therefore represent a defect that causes a discrepancy between alveolar P_{O_2} and arterial P_{O_2}. Possibilities include impaired diffusion, a shunting of blood from the right side of the heart to the left through a hole in the heart wall, and mismatching of airflow and blood flow in the alveoli.

15-7. Not at rest, if the defect is not too severe. Recall that equilibration of alveolar air and pulmonary capillary blood is normally so rapid that it occurs well before the end of the capillaries. Therefore, even though diffusion may be retarded, as in this problem, there may still be enough time for equilibration to be reached. In contrast, the time for equilibration is decreased during exercise, and failure to equilibrate is much more likely to occur, resulting in a lowered arterial P_{O_2}.

15-8. Only a few percent (specifically, from approximately 200 ml O_2/L blood to approximately 215 ml O_2/L blood). The reason the increase is so small is that almost all the oxygen in blood is carried bound to hemoglobin, and hemoglobin is almost 100 percent saturated at the arterial P_{O_2} achieved by breathing room air. The high arterial P_{O_2} achieved by breathing 100 percent oxygen does cause a directly proportional increase in the amount of oxygen *dissolved* in the blood (the additional 15 ml), but this still remains a small fraction of the total oxygen in the blood. Review the numbers given in the text.

15-9. All except plasma chloride concentration. The reasons are all given in the text.

15-10. It would cease. Respiration depends on descending input from the medulla to the nerves supplying the diaphragm and the inspiratory intercostal muscles.

15-11. The 10 percent oxygen mixture will markedly lower alveolar and thus arterial P_{O_2}, but no increase in ventilation will occur because the reflex response to hypoxia is initiated solely by the peripheral chemoreceptors. The 5 percent carbon dioxide mixture will markedly increase alveolar and arterial P_{CO_2}, and a large increase in ventilation will be elicited reflexly via the central chemoreceptors. The increase will not be as large as in a normal animal because the peripheral chemoreceptors do play a role, albeit minor, in the reflex response to elevated P_{CO_2}.

15-12. These patients have profound hyperventilation, with marked increases in both the depth and rate of ventilation. The stimulus, mainly via the peripheral chemoreceptors, is the marked increase in their arterial hydrogen-ion concentration due to the acids produced. The hyperventilation causes an increase in their arterial P_{O_2} and a decrease in their arterial P_{CO_2}.

Chapter 16

16-1. No. These are possibilities, but there is another. Substance T may be secreted by the tubules.

16-2. No. It is a possibility, but there is another. Substance V may be filtered and/or secreted, but the substance V entering the lumen via these routes may be completely reabsorbed.

16-3. 125 mg/min. The amount of any substance filtered per unit time is given by the product of the GFR and the filterable plasma concentration of the substance, in this case 125 ml/min × 100 mg/100 ml = 125 mg/min.

16-4. The plasma concentration might be so high that the T_m for the amino acid is exceeded, and so all the filtered amino acid is not reabsorbed. A second possibility is that there is a specific defect in the tubular transport for this amino acid. A third possibility is that some other amino acid is present in the plasma in high concentration and is competing for reabsorption.

16-5. No. Urea is filtered and then partially reabsorbed. The reason its concentration in the tubule is higher than in the plasma is that relatively more water is reabsorbed than urea. Therefore, the urea in the tubule becomes concentrated. Despite the fact that urea *concentration* in the urine is greater than in the plasma, the *amount excreted* is less than the filtered load (that is, net reabsorption has occurred).

16-6. They would all be decreased. The transport of all these substances is coupled, in one way or another, to that of sodium.

16-7. GFR would not go down as much, and renin secretion would not go up as much as in a person not receiving the drug. The sympathetic nerves are a major pathway for both responses during hemorrhage.

16-8. There would be little if any increase in aldosterone secretion. The major stimulus for increased aldosterone secretion is angiotensin II, but this substance is formed from angiotensin I by the action of angiotensin-converting enzyme, and so blockade of this enzyme would block the entire pathway.

16-9. (b) Urinary excretion in the steady state must be less than ingested sodium chloride by an amount equal to that lost in the sweat and feces. This is normally quite small, less than 1 g/day, so that urine excretion in this case equals approximately 11 g/day.

16-10. If the hypothalamus had been damaged, there might be inadequate secretion of ADH. This would cause loss of a large volume of urine, which would tend to dehydrate the person and make her thirsty. Of course, the area of the brain involved in thirst might have suffered damage.

16-11. Because aldosterone stimulates sodium reabsorption and potassium secretion, there will be total-body retention of sodium and loss of potassium. Interestingly, the person in this situation actually retains very little sodium because urinary sodium excretion returns to normal after a few days despite the continued presence of the high aldosterone. One hypothesized explanation for this is that GFR and atrial natriuretic factor both increase.

16-12. Sodium and water balance would become negative because of increased excretion of these substances in the urine. The person would also develop a decreased plasma bicarbonate concentration and metabolic acidosis because of increased bicarbonate excretion. The effects on acid-base status are explained by the fact that hydrogen-ion secretion—blocked by the drug—is needed both for bicarbonate reabsorption and for the excretion of hydrogen ion (contribution of new bicarbonate to the blood). The increased sodium excretion reflects the fact that much sodium reabsorption by the proximal tubule is achieved by Na/H countertransport. By blocking hydrogen-ion secretion, therefore, the drug also partially blocks sodium reabsorption. The increased water excretion occurs because the failure to reabsorb sodium and bicarbonate decreases water reabsorption (remember that water reabsorption is secondary to solute reabsorption) resulting in an osmotic diuresis.

Chapter 17

17-1. If the salivary glands fail to secrete amylase, the undigested starch that reaches the small intestine will still be digested by the amylase secreted by the pancreas. Thus, starch digestion is not significantly affected by the absence of salivary amylase.

17-2. Alcohol can be absorbed across the stomach wall, but absorption is much more rapid from the small intestine with its larger surface area. Ingestion of foods containing fat releases enterogastrones from the small intestine, and these hormones inhibit gastric emptying and thus prolong the time alcohol spends in the stomach before reaching the small intestine. This delay decreases the rate at which alcohol enters the blood. Milk, contrary to popular belief, does not "protect" the lining of the stomach from alcohol by coating it with a fatty layer. Rather, the fat content of milk decreases the rate of absorption of alcohol by decreasing the rate of gastric emptying.

17-3. Vomiting results in the loss of fluid and acid from the body. The fluid comes from the luminal contents of the stomach and duodenum, most of which was secreted by the gastric glands, pancreas, and liver and thus is derived from the blood. The cardiovascular symptoms of this patient are the result of the decrease in blood volume that accompanies vomiting.

The secretion of acid by the stomach produces an equal number of bicarbonate ions, which are released into the blood. Normally these bicarbonate ions are neutralized by hydrogen ions released into the blood by the pancreas when this organ secretes bicarbonate ions. Because gastric acid is lost during vomiting, the pancreas is not stimulated to secrete bicarbonate by the usual high-acidity signal from the duodenum, and no corresponding hydrogen ions are formed to neutralize the bicarbonate released into the blood by the stomach. As a result, the acidity of the blood decreases.

17-4. Fat can be digested and absorbed in the absence of bile salts but in greatly decreased amounts. Without adequate emulsification of fat by bile salts and phospholipids, only the fat at the surface of large lipid droplets is available to pancreatic lipase, and the rate of fat digestion is very slow. Without the formation of micelles with the aid of bile salts, the products of fat digestion become dissolved in the large lipid droplets, where they are not readily available for diffusion into the epithelial cells. In the absence of bile salts, only about 50 percent of the ingested fat is digested and absorbed. The undigested fat is passed on to the large intestine, where the bacteria there produce compounds that increase colonic motility and promote the secretion of fluid into the lumen of the large intestine, leading to diarrhea.

17-5. Damage to the lower portion of the spinal cord produces a loss of voluntary control over defecation due to disruption of the somatic nerves to the skeletal muscle of the external anal sphincter. Damage to the somatic nerves leaves the external sphincter in a continuously relaxed state. Under these conditions, defecation occurs whenever the rectum becomes distended and the defection reflex is initiated.

17-6. Vagotomy decreases the secretion of acid by the stomach. Impulses in the parasympathetic nerves directly stimulate acid secretion by the parietal cells and also cause the release of gastrin, which in turn stimulates acid secretion. Impulses in the vagus nerves are increased during both the cephalic and gastric phases of digestion. Vagotomy, by decreasing the amount of acid secreted, decreases irritation of existing ulcers, which promotes healing and decreases the probability of acid contributing to the production of new ulcers.

Chapter 18

18-1. The concentration in plasma would increase, and the amount stored in adipose tissue would decrease. Lipoprotein lipase cleaves plasma triacylglycerols, so its blockade would decrease the rate at which these molecules were cleared from plasma and would decrease the availability of the fatty acids in them for synthesis of intracellular triacylglycerols. However, this would only reduce but not eliminate such synthesis, since the adipose tissue cells could still synthesize their own fatty acids from glucose.

18-2. The person might be an insulin-dependent diabetic or might be a normal fasting person; plasma glucose would be increased in the first case but decreased in the second. Plasma insulin concentration would not be useful because it would be decreased in both cases. The fact that the person was resting and unstressed was specified because severe stress or exercise could also produce the plasma changes mentioned in the question. Plasma glucose would be increased during stress and decreased during exercise.

18-3. Glucagon, epinephrine, and growth hormone. The insulin will produce hypoglycemia, which then induces reflex increases in the secretion of all these hormones.

18-4. It might reduce it but not eliminate it. The sympathetic effects on organic metabolism during exercise are mediated not only by circulating epinephrine but also by sympathetic nerves to the liver (glycogenolysis and gluconeogenesis), to the adipose tissue (lipolysis), and to the islets (inhibition of insulin secretion and stimulation of glucagon secretion).

18-5. Increase. The stress of the accident will elicit increased activity of all the glucose-counterregulatory controls and will therefore necessitate more insulin to oppose these influences.

18-6. It will lower plasma cholesterol concentration. Bile salts are formed from cholesterol, and losses of these bile salts in the feces will be replaced by the synthesis of new ones from cholesterol. Chapter 17 describes how bile salts are normally absorbed from the small intestine so that very few of those secreted into the bile are normally lost from the body.

18-7. Plasma concentrations of HDL and LDL. It is the ratio of LDL cholesterol to HDL cholesterol that best correlates with the development of atherosclerosis (that is, HDL cholesterol is "good" cholesterol). The answer to this question would have been the same regardless of whether the person was an athlete or not, but the question was phrased this way to emphasize that people who exercise generally have increased HDL cholesterol.

18-8. In utero malnutrition. Neither growth hormone nor the thyroid hormones influence in utero growth.

18-9. Androgens stimulate growth but also cause the ultimate cessation of growth by closing the epiphyseal plates. Therefore, there might be a rapid growth spurt in response to the androgens but a subsequent premature cessation of growth. Estrogens exert similar effects.

18-10. Heat loss from the head, mainly via convection and sweating, is the major route for loss under these conditions. The rest of the body is *gaining* heat by conduction, and sweating is of no value in the rest of the body because the water cannot evaporate. Heat is also lost via the expired air (insensible loss), and some people actually begin to pant under such conditions. The rapid shallow breathing increases airflow and heat loss without causing hyperventilation.

18-11. They seek out warmer places, if available, so that their body temperature increases. That is, they use behavior to develop a fever. This is excellent evidence that the hyperthermia of infection is a fever (that is, a set-point change).

Chapter 19

19-1. Sterility due to lack of spermatogenesis would be the common symptom. The Sertoli cells are essential for spermatogenesis, and so is testosterone produced by the Leydig cells. The person with Leydig cell destruction, but not the person with Sertoli cell destruction, would also have other symptoms of testosterone deficiency.

19-2. The androgens act on the hypothalamus and anterior pituitary to inhibit the secretion of the gonadotropins. Therefore, spermatogenesis is inhibited. Importantly, even if this man were given FSH, the sterility would probably remain since the lack of LH would cause deficient testosterone secretion and *locally* produced testosterone is needed for spermatogenesis (that is, the exogenous androgen cannot do this job).

19-3. Impaired function of the seminiferous tubules, notably of the Sertoli cells. The increased plasma FSH concentration is due to the lack of negative feedback inhibition of FSH secretion by inhibin, itself secreted by the Sertoli cells. The Leydig cells seem to be functioning normally in this person, since the lack of demasculinization and the normal plasma LH indicate normal testosterone secretion.

19-4. FSH secretion. Since FSH acts on the Sertoli cells and LH acts on the Leydig cells, the answer to this question is essentially the same as that for question 1—sterility would result in either case, but the loss of LH would also cause an undesirable elimination of testosterone and its effects.

19-5. These findings are all due to testosterone deficiency. You would also expect to find that the testes and penis were small if the deficiency occurred before puberty.

19-6. They will be eliminated. The androgens act on the hypothalamus to inhibit the secretion of GnRH and on the pituitary to inhibit the response to GnRH. The result is inadequate secretion of gonadotropins and therefore inadequate stimulation of the ovaries. In addition to the loss of menstrual cycles, the woman will suffer some degree of masculinization of the secondary sex characteristics because of the combined effects of androgen excess and estrogen deficiency.

19-7. Such treatment may cause so much secretion of FSH that multiple dominant follicles are stimulated to develop simultaneously and have their eggs ovulated during the LH surge.

19-8. An increased plasma LH. The other two are due to increased plasma progesterone and so do not occur until *after* ovulation and formation of the corpus luteum.

19-9. The absence of sperm capacitation. When test-tube fertilization is performed, special techniques are used to induce capacitation.

19-10. The fetus is in difficulty. The placenta produces progesterone entirely on its own, whereas estriol secretion requires participation of the fetus, specifically, the fetal adrenal cortex.

19-11. Prostaglandin antagonists, oxytocin antagonists, and drugs that lower cytosolic calcium concentration. You may not have thought of the last category since calcium is not mentioned in this context in the text, but as in all muscle, calcium is the immediate cause of contraction in the myometrium.

19-12. This person would have normal male external genitals and testes, although the testes may not have descended fully, but would also have some degree of development of uterine tubes, a uterus, and a vagina. These internal female structures would tend to develop because no MIS was present to cause degeneration of the Müllerian duct system.

19-13. No. These two hormones are already elevated in menopause, and the problem is that the ovaries are unable to respond to them with estrogen secretion. Thus, the treatment must be with estrogen, itself.

Chapter 20

20-1. Both would be impaired because T cells would not differentiate. The absence of cytotoxic T cells would eliminate responses mediated by these cells. The absence of helper T cells would impair antibody-mediated responses because most B cells require cytokines from helper T cells to become activated.

20-2. Neutrophil deficiency would impair nonspecific inflammatory responses to bacteria. Monocyte deficiency, by causing macrophage deficiency, would impair both nonspecific inflammation and specific immune responses.

20-3. The drug might reduce but would not eliminate the action of complement, since this system destroys cells directly (via the membrane attack complex) as well as by facilitating phagocytosis.

20-4. Antibodies would bind normally to antigen but might not be able to activate complement, act as opsonins, or recruit NK cells in ADCC. The reason for these defects is that the sites to which complement C1, phagocytes, and NK cells bind are all located in the Fc portion of antibodies.

20-5. They do develop fever, although often not to the same degree as normal. They can do so because IL-1 and other cytokines secreted by macrophages cause fever, whereas the defect in AIDS is failure of helper T cell function.

20-6. This person is suffering from an autoimmune attack against the receptors. The antibodies formed then bind to the receptors and activate them just as the thyroid hormones would have.

20-7. Alcohol over time induces a high level of MES activity, which causes an administered barbiturate to be metabolized more rapidly than normal.

APPENDIX B
Glossary

A cell *see* alpha cell

absolute refractory period time during which an excitable membrane cannot generate an action potential in response to any stimulus

absorption movement of materials across an epithelial layer from body cavity or compartment toward the blood

absorptive state period during which nutrients enter bloodstream from gastrointestinal tract

accessory reproductive organ duct through which sperm or egg is transported, or a gland emptying into such a duct (in the female the breasts are usually included)

acclimatization (ah-climb-ah-tih-ZAY-shun) environmentally induced improvement in functioning of a physiological system with no change in genetic endowment

accommodation adjustment of eye for viewing various distances by changing shape of lens

acetone (ASS-ih-tone) ketone body produced from acetyl CoA during prolonged fasting or untreated severe diabetes mellitus

acetyl coenzyme A (acetyl CoA) (ASS-ih-teel koh-EN-zime A, koh-A) metabolic intermediate that trans-fers acetyl groups to Krebs cycle and various synthetic pathways

acetyl group -COCH$_3$

acetylcholine (ACh) (ass-ih-teel-KOH-leen) a neurotransmitter released by pre- and postganglionic parasympathetic neurons, preganglionic sympathetic neurons, somatic neurons, and some CNS neurons

acetylcholinesterase (ass-ih-teel-koh-lin-ES-ter-ase) enzyme that breaks down acetylcholine into acetic acid and choline

acid molecule capable of releasing a hydrogen ion; solution having an H$^+$ concentration greater than that of pure water, that is, pH less than 7; *see also* strong acid, weak acid

acidity concentration of free, unbound hydrogen ion in a solution; the higher the H$^+$ concentration, the greater the acidity

acidosis (ass-ih-DOH-sis) any situation in which arterial H$^+$ concentration is elevated above normal resting levels; *see also* metabolic acidosis, respiratory acidosis

acrosome (AK-roh-sohm) cytoplasmic vesicle containing digestive enzymes and located at head of a sperm

actin (AK-tin) globular contractile protein to which myosin cross bridges bind; located in muscle thin filaments and in microfilaments of cytoskeleton

action potential electric signal propagated by nerve and muscle cells; an all-or-none depolarization of membrane polarity; has a threshold and refractory period and is conducted without decrement

activated macrophage macrophage whose killing ability has been enhanced by cytokines, particularly IL-2 and interferon-gamma

activation *see* lymphocyte activation

activation energy energy necessary to disrupt existing chemical bonds during a chemical reaction

active hyperemia (hy-per-EE-me-ah) increased blood flow through a tissue associated with increased metabolic activity

active immunity resistance to reinfection acquired by contact with microorganisms, their toxins, or other antigenic material; *compare* passive immunity

active site region of enzyme to which substrate binds

active transport energy-requiring system that uses transporters to move ions or molecules across a membrane against an electrochemical difference; *see also* primary active transport, secondary active transport

activity *see* enzyme activity

acute (ah-KUTE) lasting a relatively short time; *compare* chronic

acute phase proteins group of proteins secreted by liver during systemic response to injury or infection

acute phase response responses of tissues and organs distant from site of infection or immune response

adaptation (evolution) a biological characteristic that favors survival in a particular environment; (neural) decrease in action-potential frequency in a neuron despite constant stimulus

adenosine diphosphate (ADP) (ah-DEN-oh-seen dy-FOS-fate) two-phosphate product of ATP breakdown

adenosine monophosphate (AMP) one-phosphate derivative of ATP

adenosine triphosphate (ATP) major molecule that transfers energy from metabolism to cell functions during its breakdown to ADP and release of P_i

adenylyl cyclase (ad-DEN-ah-lil SY-klase) enzyme that catalyzes transformation of ATP to cyclic AMP

adipocyte (ad-DIP-oh-site) cell specialized for triacylglycerol synthesis and storage; fat cell

adipose tissue (AD-ah-poze) tissue composed largely of fat-storing cells

adrenal cortex (ah-DREE-nal KOR-tex) endocrine gland that forms outer shell of each adrenal gland; secretes steroid hormones—mainly cortisol, aldosterone, and androgens; *compare* adrenal medulla

adrenal gland one of a pair of endocrine glands above each kidney; each gland consists of outer *adrenal cortex* and inner *adrenal medulla*

adrenal medulla (meh-DUL-ah) endocrine gland that forms inner core of each adrenal gland; secretes amine hormones, mainly epinephrine; *compare* adrenal cortex

adrenergic (ad-ren-ER-jik) pertaining to norepinephrine or epinephrine; compound that acts like norepinephrine or epinephrine

adrenocorticotropic hormone (ACTH) (ad-ren-oh-kor-tih-koh-TROH-pik) polypeptide hormone secreted by anterior pituitary; stimulates adrenal cortex to secrete cortisol; also called corticotropin

aerobic (air-OH-bik) in presence of oxygen

affect (AF-fect, *not* af-FECT) external expression of inner emotions

afferent (AF-er-ent) carrying toward

afferent arteriole vessel in kidney that carries blood from artery to renal corpuscle

afferent neuron neuron that carries information from receptors at its peripheral endings to CNS; cell body lies outside CNS

afferent pathway component of reflex arc that transmits information from receptor to integrating center

affinity strength with which ligand binds to its binding site

afterbirth placenta and associated membranes expelled from uterus after delivery of infant

agonist (AG-ah-nist) chemical messenger that binds to receptor and triggers cell's response; often refers to drug that mimics action of chemical normally in the body

airway tube through which air flows between external environment and lung alveoli

albumin (al-BU-min) most abundant plasma protein

aldosterone (al-doh-stir-OWN or al-DOS-stir-own) mineralocorticoid steroid hormone secreted by adrenal cortex; regulates electrolyte balance

alkaline having H^+ concentration lower than that of pure water, that is, having a pH greater than 7

alkalosis (alk-ah-LOH-sis) any situation in which arterial blood H^+ concentration is reduced below normal resting levels; *see also* metabolic alkalosis, respiratory alkalosis

all-or-none pertaining to event that occurs maximally or not at all

allele (al-EEL) a gene that differs in nucleotide sequence from other copies in the population of that same gene

allosteric modulation (al-low-STAIR-ik) control of protein binding-site properties by modulator molecules that bind to regions of the protein other than the binding site altered by them

allosteric protein protein whose binding-site characteristics are subject to allosteric modulation

alpha cell glucagon-secreting cell of pancreatic islets of Langerhans

alpha-adrenergic receptor plasma membrane receptor for epinephrine and norepinephrine that utilizes phospholipase C second messenger system or directly affects potassium and calcium channels in the membrane; also called alpha adrenoceptor; *compare* beta-adrenergic receptor

alpha-glycerol phosphate three carbon molecule that combines with fatty acids to form mono-, di-, and triacylglycerol; also called glycerol 3-phosphate

alpha motor neuron motor neuron that innervates extrafusal muscle fibers

alpha rhythm prominent 8- to 13-Hz oscillation on the electroencephalograms of awake, relaxed adults with their eyes closed

alternate complement pathway sequence for complement activation that bypasses first steps in classical pathway and is not antibody dependent

alveolar dead space (al-VEE-oh-lar) volume of inspired air that reaches alveoli but cannot undergo gas exchange with blood

alveolar pressure (P_{alv}) air pressure in pulmonary alveoli

alveolar ventilation volume of atmospheric air entering alveoli each minute

alveolus (al-VEE-oh-lus) (lungs) thin-walled, air-filled "outpocketing" from terminal air passageways in lungs; (glands) cell cluster at end of duct in secretory gland

amine hormone (ah-MEEN) hormone derived from amino acid tyrosine; includes thyroid hormones, epinephrine, norepinephrine, and dopamine

amino acid (ah-MEEN-oh) molecule containing amino group, carboxyl group, and side chain attached to a carbon atom; molecular subunit of protein

amino group $-NH_2$; ionizes to $-NH_3^+$

aminopeptidase (ah-meen-oh-PEP-tih-dase) one of a family of enzymes located in the intestinal epithelial membrane; breaks peptide bond at amino end of polypeptide

ammonia NH_3; produced during amino acid breakdown; converted in liver to urea; ionized form is ammonium

amniotic sac (am-nee-AHT-ik) membrane surrounding fetus in utero

amphipathic (am-fuh-PATH-ik) a molecule containing polar or ionized groups at one end and nonpolar groups at the other

amplitude height, how much, magnitude of change

amylase (AM-ih-lase) enzyme that partially breaks down polysaccharides

anabolism (an-NAB-oh-lizm) cellular synthesis of organic molecules

anaerobic (ah-ih-ROH-bik) in the absence of oxygen

analgesia (an-al-JEE-zee-ah) removal of pain without loss of consciousness

anatomic dead space space in respiratory tract airways whose walls do not permit gas exchange with blood

androgen (AN-dro-jen) any hormone with testosterone-like actions

anemia (ah-NEE-me-ah) reduction in total blood hemoglobin

angiogenesis (an-gee-oh-JEN-ah-sis) the development and growth of capillaries

angiotensin I small polypeptide generated in plasma by renin's action on angiotensinogen

angiotensin II hormone formed by action of angiotensin converting enzyme on angiotensin I; stimulates aldosterone secretion from adrenal cortex, vascular smooth-muscle contraction, and thirst

angiotensin converting enzyme enzyme on capillary endothelial cells that catalyzes removal of two amino acids from angiotensin I to form angiotensin II

angiotensinogen (an-gee-oh-ten-SIN-oh-jen) plasma protein precursor of angiotensin I; produced by liver

anion (AN-eye-on) negatively charged ion; *compare* cation

antagonist (muscle) muscle whose action opposes intended movement; (drug) molecule that competes with another for a receptor and binds to the receptor but does not trigger the cell's response

anterior toward or at the front

anterior pituitary anterior portion of pituitary gland; synthesizes, stores, and releases ACTH, GH, TSH, prolactin, FSH, and LH

antibody (AN-tih-bah-dee) immunoglobulin secreted by plasma cell; combines with type of antigen that stimulated its production; directs attack against antigen or cell bearing it; *see also* natural antibody

antibody-dependent cellular cytotoxicity (ADCC) killing of target cells linked to NK cells by toxic chemicals secreted by the NK cells

anticodon (an-tie-KOH-don) three-nucleotide sequence in tRNA able to base-pair with complementary codon in mRNA during protein synthesis

antidiuretic hormone (ADH) (an-ty-dy-yor-ET-ik) *see* vasopressin

antigen (AN-tih-jen) any foreign molecule that stimulates a specific immune response

antigen-presenting cell (APC) cell that presents antigen, complexed with MHC proteins on its surface, to T cells

antithrombin III (an-ty-throm-bin THREE) plasma anticlotting protein that inactivates thrombin

antrum (AN-trum) (gastric) lower portion of stomach, that is, region closest to pyloric sphincter; (ovarian) fluid-filled cavity in maturing ovarian follicle

aorta (a-OR-tah) largest artery in body; carries blood from left ventricle of heart to thorax and abdomen

aortic arch baroreceptor (a-OR-tik) *see* arterial baroreceptor

aortic body chemoreceptor chemoreceptor located near aortic arch; sensitive to arterial blood O_2 pressure and H^+ concentration

aortic valve valve between left ventricle of heart and aorta

aphasia (ah-FAY-see-ah) specific language deficit not due to mental retardation or muscular weakness

apnea (AP-nee-ah) cessation of breathing

apoptosis (a-POH-toh-sis) self-destruction of a cell programmed by intrinsic "suicide" instructions

appendix small finger-like projection from cecum of large intestine

aqueous (AH-kwee-us) watery; prepared with water

arachidonic acid (ah-rak-ah-DON-ik) polyunsaturated fatty acid precursor of eicosanoids

arrhythmia (ay-RYTH-me-ah) any

variation from normal heartbeat rhythm

arterial baroreceptor nerve endings sensitive to stretch or distortion produced by arterial blood pressure changes; located in carotid sinus or aortic arch; also called carotid sinus and aortic arch baroreceptors

arteriole (are-TEER-ee-ole) blood vessel between artery and capillary, surrounded by smooth muscle; primary site of vascular resistance

artery (ARE-ter-ee) thick-walled, elastic vessel that carries blood away from heart to arterioles

ascending limb portion of Henle's loop of renal tubule leading to distal convoluted tubule

ascending pathway neural pathway that goes to the brain; also called sensory pathway

aspartate (ah-SPAR-tate) an excitatory neurotransmitter in CNS; ionized form of the amino acid aspartic acid

association cortex *see* cortical association areas

atmospheric pressure (P_{atm}) air pressure surrounding the body (760 mmHg at sea level)

atom smallest unit of matter that has unique chemical characteristics; has no net charge; combines with other atoms to form all chemical substances

atomic nucleus dense region, consisting of protons and neutrons, at center of atom

atomic number number of protons in nucleus of atom

atomic weight value that indicates an atom's mass relative to mass of other types of atoms based on the assignment of a value of 12 to carbon atom

ATPase enzyme that breaks down ATP to ADP and inorganic phosphate; *see also* Na,K-ATPase, Ca-ATPase

atrial natriuretic factor (ANF) (nat-ry-yor-ET-ik) peptide hormone secreted by cardiac atrial cells in response to atrial distension; causes increased renal sodium excretion and vasodilation

atrioventricular (AV) node (ay-tree-oh-ven-TRIK-you-lar) region at base of right atrium near interventricular septum, containing specialized cardiac muscle cells through which electrical activity must pass to go from atria to ventricles

atrioventricular (AV) valve   valve between atrium and ventricle of heart; AV valve on right side of heart is the *tricuspid valve,* and that on left side is the *mitral valve*

atrium (AY-tree-um)   chamber of heart that receives blood from veins and passes it on to ventricle on same side of heart

atrophy (AT-roh-fee)   wasting away; decrease in size

auditory (AW-dih-tor-ee)   pertaining to sense of hearing

auditory cortex   region of cerebral cortex that receives nerve fibers from auditory pathways

autocrine agent (AW-toh-crin)   chemical messenger that is secreted into extracellular fluid and acts upon cell that secreted it

automaticity (aw-toh-mah-TISS-ih-tee)   capable of spontaneous, rhythmical self-excitation

autonomic nervous system (aw-toh-NAHM-ik)   component of efferent division of peripheral nervous system that consists of sympathetic and parasympathetic subdivisions; innervates cardiac muscle, smooth muscle, and glands; *compare* somatic nervous system

autoreceptors   receptors on a cell affected by a chemical messenger released from the same cell

autoregulation (aw-toh-reg-you-LAY-shun)   ability of an individual organ to control (self-regulate) its vascular resistance independent of neural and hormonal influence; *see also* flow autoregulation

axon (AX-ahn)   extension from neuron cell body; propagates action potentials away from cell body; also called a nerve fiber

axon terminal   end of axon; forms synaptic or neuroeffector junction with postjunctional cell

axon transport   process involving intracellular filaments by which materials are moved from one end of axon to other

B cell   (immune system) lymphocyte that, upon activation, proliferates and differentiates into antibody-secreting plasma cell; (endocrine cell) *see* beta cell

B lymphocyte   *see* B cell

B-endorphin (en-DOR-fin)   opioid peptide secreted by anterior pituitary; synthesized as part of pro-opiomelanocortin and cosecreted with ACTH

barometric pressure   *see* atmospheric pressure

baroreceptor   receptor sensitive to pressure and to rate of change in pressure; *see also* arterial baroreceptor, intrarenal baroreceptor

basal (BAY-sul)   resting level

basal ganglia   nuclei deep in cerebral hemispheres that code and relay information associated with control of body movements; specifically, caudate nucleus, globus pallidus, and putamen

basal metabolic rate (BMR)   metabolic rate when a person is at mental and physical rest but not sleeping, at comfortable temperature, and has fasted at least 12 h; also called metabolic cost of living

base   (acid-base) any molecule that can combine with H^+; (nucleotide) molecular ring of carbon and nitrogen that, with a phosphate group and a sugar, constitutes a nucleotide; *see also* purine base, pyrimidine base

basement membrane   thin layer of extracellular proteinaceous material upon which epithelial and endothelial cells sit

basic electrical rhythm   spontaneous depolarization-repolarization cycles of pacemaker cells in longitudinal smooth-muscle layer of stomach and intestines; coordinates repetitive muscular activity of GI tract

basilar membrane (BAS-ih-lar)   membrane that separates cochlear duct and scala tympani in inner ear; supports organ of Corti

basolateral membrane (bay-so-LAH-ter-al)   sides of epithelial cell facing away from lumen; also called serosal or blood side of cell

basophil (BAY-so-fill)   polymorphonuclear granulocytic leukocyte whose granules stain with basic dyes; enters tissues and becomes mast cell

beta-adrenergic receptor (BAY-ta adren-ER-jik)   plasma membrane receptor for epinephrine and norepinephrine that utilizes cAMP second messenger system; *compare* alpha-adrenergic receptor; also called beta adrenoceptor

beta cell   insulin-secreting cell in pancreatic islets of Langerhans; also called B cell

beta oxidation (ox-ih-DAY-shun)   series of reactions that generate hydrogen atoms (for oxidative phosphorylation) from breakdown of fatty acids to acetyl CoA

beta rhythm   low, fast EEG oscillations in alert, awake adults who are paying attention to (or thinking hard about) something

bicarbonate (by-KAR-bah-nate) HCO_3^-

bile   fluid secreted by liver into bile canaliculi; contains bicarbonate, bile salts, cholesterol, lecithin, bile pigments, metabolic end products, and certain trace metals

bile canaliculi (kan-al-IK-you-lee)   small ducts adjacent to liver cells into which bile is secreted

bile pigment   colored substance, derived from breakdown of heme group of hemoglobin, secreted in bile

bile salt   one of a family of steroid molecules produced from cholesterol and secreted in bile by the liver; promotes solubilization and digestion of fat in small intestine

bilirubin (bil-eh-RUE-bin)   yellow substance resulting from heme breakdown; excreted in bile as a bile pigment

binding site   region of protein to which a specific ligand binds

biogenic amine (by-oh-JEN-ik ah-MEEN)   one of family of neurotransmitters having basic formula $R-NH_2$; includes dopamine, norepinephrine, epinephrine, serotonin, and histamine

"biological clock"   neurons that function in absence of apparent external stimulation and drive body rhythms

biotransformation (by-oh-trans-for-MAY-shun)   alteration of foreign molecules by an organism's metabolic pathways

bipolar cell   retinal neuron postsynaptic to rod or cone

bladder   urinary bladder; thick-walled sac composed of smooth muscle; stores urine prior to urination

blastocyst (BLAS-toh-cyst)   particular early embryonic stage consisting of ball of developing cells surrounding central cavity

blood-brain barrier   group of anatomical barriers and transport systems in brain capillary endothelium that controls kinds of substances entering brain extracellular space from blood and their rates of entry

blood coagulation (koh-ag-you-LAY-shun) blood clotting

blood sugar glucose

blood type blood classification according to presence of A and/or B antigens or lack of them (O)

body mass index (BMI) method for assessing degree of obesity calculated as weight in kilograms divided by square of height in meters

bone marrow highly vascular, cellular substance in central cavity of some bones; site of erythrocyte, leukocyte, and platelet synthesis

Bowman's capsule blind sac at beginning of tubular component of kidney nephron

Boyle's law (boils) pressure of a fixed amount of gas in a container is inversely proportional to container's volume

bradykinin (braid-ee-KY-nin) protein formed by action of the enzyme kallikrein on precursor; dilates vessels, increases capillary permeability, and probably stimulates pain receptors

brainstem brain subdivision consisting of medulla oblongata, pons, and midbrain and located between spinal cord and forebrain

brainstem pathways descending motor pathways whose cells of origin are in the brainstem

Broca's area (BRO-kahz) region of left frontal lobe associated with speech production

bronchiole (BRON-key-ole) small airway distal to bronchus

bronchus (BRON-kus) large-diameter air passage that enters lung; located between trachea and bronchioles

buffer weak acid or base that can exist in undissociated (Hbuffer) or dissociated (H$^+$ + buffer) form

buffering reversible hydrogen-ion binding by anions when H$^+$ concentration changes; tends to minimize changes in acidity of a solution when acid is added or removed

bulbourethral gland (bul-bo-you-WREETH-ral) one of paired glands in male that secretes fluid components of semen into the urethra

bulk flow movement of fluids or gases from region or higher pressure to one of lower pressure

calmodulin (kal-MOD-you-lin) intracellular calcium-binding protein that mediates many of calcium's second-messenger functions

calorie (cal) unit of heat-energy measurement; amount of heat needed to raise temperature of 1 g of water 1°C; *compare* kilocalorie

calorigenic effect (kah-lor-ih-JEN-ik) increase in metabolic rate caused by epinephrine or thyroid hormones

cAMP *see* cyclic AMP

candidate hormone substance suspected of being a hormone but not yet proven to be one

capacitation *see* sperm capacitation

capillary smallest blood vessel type

carbamino compound (kar-bah-MEEN-oh) compound resulting from combination of carbon dioxide and protein amino groups, particularly in hemoglobin

carbohydrate substance composed of carbon, hydrogen, and oxygen according to general formula $C_n(H_2O)_n$, where n is any whole number

carbon monoxide CO; gas that reacts with hemoglobin and decreases blood oxygen-carrying capacity; possible neurotransmitter

carbonic acid (kar-BAHN-ik) H_2CO_3; an acid formed from H_2O and CO_2

carbonic anhydrase (an-HY-drase) enzyme that catalyzes the reaction $CO_2 + H_2O \rightleftharpoons H_2CO_3$

carboxyl group (kar-BOX-il) -COOH; ionizes to carboxyl ion (-COO$^-$)

carboxypeptidase (kar-box-ee-PEP-tih-dase) enzyme secreted into small intestine by exocrine pancreas as precursor, procarboxypeptidase; breaks peptide bond at carboxyl end of protein

cardiac (KAR-dee-ak) pertaining to the heart

cardiac cycle one contraction-relaxation sequence of heart

cardiac muscle heart muscle

cardiac output blood volume pumped by each ventricle per minute (not total output pumped by both ventricles)

cardiovascular center neuron cluster in brainstem medulla oblongata that serves as a major integrating center for reflexes affecting heart and blood vessels

cardiovascular system heart and blood vessels

carotid (kuh-RAH-tid) pertaining to two major arteries (carotid arteries) in neck that convey blood to head

carotid body chemoreceptor chemoreceptor near main branching of carotid artery; sensitive to blood O_2 pressure and H$^+$ concentration

carotid sinus region of internal carotid artery just above main carotid branching; location of carotid baroreceptors

carotid sinus baroreceptor *see* arterial baroreceptor

carrier *see* transporter

cascade (kas-KADE) multiplicative sequence of events in which the number of reaction products increases at one or more steps

catabolism (kuh-TAB-oh-lizm) cellular breakdown of organic molecules

catalyst (KAT ah-list) substance that accelerates chemical reactions but does not itself undergo any net chemical change during the reaction

catecholamine (kat-eh-COLE-ah-meen) dopamine, epinephrine, or norepinephrine, all of which have similar chemical structures

cation (KAT-eye-on) ion having net positive charge; *compare* anion

cecum (SEE-come) dilated pouch at beginning of large intestine into which the ileum, colon, and appendix open

cell body in cells with long extensions, the part that contains the nucleus

cell differentiation *see* differentiation

cell organelle (or-guh-NEL) membrane-bound compartment, non-membranous particle, or filament that performs specialized functions in cell

cellulose polysaccharide composed of glucose subunits; found in plant cells

center of gravity point in a body at which body mass is in perfect balance; if the body were suspended from a string attached to this point, there would be no movement

central chemoreceptor receptor in brainstem medulla oblongata that responds to H$^+$ concentration changes of brain extracellular fluid

central command fatigue muscle fatigue due to failure of appropriate regions of cerebral cortex to excite motor neurons

central nervous system (CNS) brain plus spinal cord

BUL-bar) descending pathway having its neuron cell bodies in cerebral cortex; its axons pass without synapsing to region of brainstem motor neurons

corticospinal pathway descending pathway having its neuron cell bodies in cerebral cortex; its axons pass without synapsing to region of spinal motor neurons; also called the pyramidal tract; *compare* brainstem pathway, corticobulbar pathway

corticosteroid (kor-tih-koh-STEER-oid) steroid produced by adrenal cortex or drug that resembles one

corticotropin releasing hormone (CRH) (kor-tih-koh-TROH-pin) hypophysiotropic peptide hormone that stimulates ACTH (corticotropin) secretion by anterior pituitary

cortisol (KOR-tih-sol) main glucocorticoid steroid hormone secreted by adrenal cortex; regulates various aspects of organic metabolism

cotransmitter chemical messenger released with a neurotransmitter from synapse or neuroeffector junction

cotransport form of secondary active transport in which net movement of actively transported substance and "downhill" movement of molecule supplying the energy are in the same direction; also called *symport*

countercurrent multiplier system mechanism associated with loops of Henle that creates in renal medulla a region having high interstitial-fluid osmolarity

countertransport form of secondary active transport in which net movement of actively transported molecule is in direction opposite "downhill" movement of molecule supplying the energy; also called *antiport*

covalent bond (koh-VAY-lent) chemical bond between two atoms in which each atom shares one of its electrons with the other

covalent modulation alteration of a protein's shape, and therefore its function, by the covalent binding of various chemical groups to it

cranial nerve one of 24 peripheral nerves (12 pairs) that join brainstem or forebrain with structures outside CNS

creatine phosphate (CP) (KREE-ah-tin) molecule that transfers phos-

phate and energy to ADP to generate ATP

creatinine (kree-AT-ih-nin) waste product derived from muscle creatine

creatinine clearance plasma volume from which creatinine is removed by the kidneys per unit time; approximates glomerular filtration rate

critical period time during development when a system is most readily influenced by environmental factors, sometimes irreversibly

cross bridge in muscle, myosin projection extending from thick filament and capable of exerting force on thin filament, causing the filaments to slide past each other

cross-bridge cycle sequence of events between binding of a cross bridge to actin, its release, and reattachment during muscle contraction

crossed-extensor reflex increased activation of extensor muscles contralateral to a limb flexion

crossing-over process in which segments of maternal and paternal chromosomes exchange with each other during chromosomal pairing in meiosis

crystalloid low-molecular weight solute

current movement of electric charge; in biological systems, this is achieved by ion movement

cutaneous (cue-TAY-nee-us) pertaining to skin

cyclic AMP (cAMP) cyclic 3′,5′-adenosine monophosphate; cyclic nucleotide that serves as a second messenger for many "first" chemical messengers

cyclic AMP-dependent protein kinase (KY-nase) enzyme that is activated by cyclic AMP and then phosphorylates specific proteins, thereby altering their activity; also called *protein kinase A*

cyclic endoperoxide eicosanoid formed from arachidonic acid by cyclooxygenase

cyclic GMP (cGMP) cyclic 3′,5′-guanosine monophosphate; cyclic nucleotide that acts as second messenger in some cells

cyclic GMP-dependent protein kinase (KY-nase) enzyme that is activated by cyclic GMP and then phosphorylates specific proteins,

thereby altering their activity; also called *protein kinase G*

cyclooxygenase (sy-klo-OX-ah-jen-ase) enzyme that acts on arachidonic acid and initiates production of cyclic endoperoxides, prostaglandins, and thromboxanes

cytochrome (SY-toh-krom) one of a series of enzymes that couples energy to ATP formation during oxidative phosphorylation

cytokine (SY-toh-kine) general term for protein extracellular messengers that regulate immune responses; secreted by macrophages, monocytes, lymphocytes, neutrophils, and several nonimmune cell types

cytokinesis (SY-toh-kin-EE-sis) stage of cell division during which cytoplasm divides to form two new cells

cytoplasm (SY-toh-plasm) region of cell interior outside the nucleus

cytosine (C) (SY-toh-seen) pyrimidine base in DNA and RNA

cytoskeleton cytoplasmic filamentous network associated with cell shape and movement

cytosol (SY-toh-sol) intracellular fluid that surrounds cell organelles and nucleus

cytotoxic T cell (SY-toh-TOX-ik) T lymphocyte that, upon activation by specific antigen, directly attacks a cell bearing that type of antigen and destroys it; major killer of virus-infected and cancer cells

daughter cell one of the two new cells formed when a cell divides

dead space volume of inspired air that cannot be exchanged with blood; *see also* anatomic dead space, alveolar dead space, total dead space

deamination (dee-am-in-AY-shun) removal of amino (-NH$_2$) group from a molecule

declarative memory memories of facts and events

decremental decreasing in amplitude

defecation (def-ih-KAY-shun) expulsion of feces from rectum

dendrite (DEN-drite) highly branched extension of neuron cell body; receives synaptic input from other neurons

dense body cytoplasmic structure to which thin filaments of a smooth-muscle fiber are anchored

deoxyhemoglobin (Hb, HbH) (dee-ox-see-HEE-moh-gloh-bin) hemoglobin not combined with oxygen; reduced hemoglobin

deoxyribonucleic acid (DNA) (dee-ox-see-ry-boh-noo-KLAY-ik) nucleic acid that stores and transmits genetic information; consists of double strand of nucleotide subunits that contain deoxyribose

depolarize to change membrane potential value toward zero so that cell interior becomes less negative than resting level

descending limb (of Henle's loop) segment of renal tubule into which proximal tubule drains

descending pathway neural pathways that go from the brain down to the spinal cord

desmosome (DEZ-moh-some) junction that holds two cells together; consists of plasma membranes of adjacent cells linked by fibers yet separated by a 20-nm extracellular space filled with a cementing substance

diacylglycerol (DAG) (dy-aa-syl-GLIS-er-ol) second messenger that activates protein kinase C, which then phosphorylates a large number of other proteins

dialysis (dy-AL-ih-sis) process of altering the concentration of substances in the blood by using concentration differences between plasma and a bathing solution separated by a semipermeable membrane

diaphragm (DY-ah-fram) dome-shaped skeletal muscle sheet that separates the abdominal and thoracic cavities; principal muscle of respiration

diastole (dy-ASS-toh-lee) period of cardiac cycle when ventricles are relaxing

diastolic pressure (DP) (dy-ah-STAL-ik) minimum blood pressure during cardiac cycle

diencephalon (dy-en-SEF-ah-lon) core of anterior part of brain; lies beneath cerebral hemispheres and contains *thalamus* and *hypothalamus*

differentiation (dif-fer-en-she-AY-shun) process by which unspecialized cells acquire specialized structural and functional properties

diffusion (dif-FU-shun) random movement of molecules from one location to another because of random thermal molecular motion; net diffusion always occurs from a region of higher concentration to a region of lower concentration

diffusion equilibrium state during which diffusion fluxes in opposite directions are equal, that is, the net flux equals zero

diffusion potential voltage difference created by net diffusion of ions

digestion process of breaking down large particles and high-molecular-weight substances into small molecules

dihydrotestosterone (dy-hy-droh-tes-TOS-ter-own) steroid formed by enzyme-mediated alteration of testosterone; active form of testosterone in certain of its target cells

1,25-dihydroxyvitamin D₃ (1,25-(OH₂D₃) (1-25-dy-hy-DROX-ee-vy-tah-min DEE-3) hormone that is formed by kidneys and is active form of vitamin D

2,3-diphosphoglycerate (DPG) (2-3-dy-fos-foh-GLISS-er-ate) substance produced by erythrocytes during glycolysis; binds reversibly to hemoglobin, causing it to release oxygen

disaccharide (dy-SAK-er-ide) carbohydrate molecule composed of two monosaccharides

disinhibition removal of inhibition from a neuron, thereby allowing its activity to increase

dissociation separation from

distal (DIS-tal) farther from reference point; *compare* proximal

distal convoluted tubule portion of kidney tubule between loop of Henle and collecting duct system

disulfide bond R-S-S-R

diuresis (dy-uh-REE-sis) increased urine excretion

diuretic (dy-uh-RET-ik) substance that inhibits fluid reabsorption in renal tubule, thereby increasing urine excretion

diurnal (dy-URN-al) daily; occurring in a 24-h cycle

divergence (dy-VER-gence) (neuronal) one presynaptic neuron synapsing upon many postsynaptic neurons; (of eyes) turning of eyes outward to view distant objects

dl deciliter; 0.1 L

DNA polymerase enzyme that, during DNA replication, forms new DNA strand by joining together nucleotides already base-paired with an existing DNA strand

dopamine (DOPE-ah-meen) biogenic amine (catecholamine) neurotransmitter and hormone; precursor of epinephrine and norepinephrine; *see also* prolactin inhibiting hormone

dorsal (DOR-sal) toward or at the back

dorsal root group of afferent nerve fibers that enters dorsal region of spinal cord

double bond two covalent chemical bonds formed between same two atoms; symbolized by =

double helix molecular conformation in which two strands of molecules are coiled around each other; conformation of DNA

down-regulation decrease in number of target-cell receptors for a given messenger in response to a chronic high concentration of that messenger; *compare* up-regulation

dual innervation (in-ner-VAY-shun) innervation of an organ or gland by both sympathetic and parasympathetic nerve fibers

duodenum (due-oh-DEE-num) first portion of small intestine (between stomach and jejunum)

eardrum *see* tympanic membrane

ectopic focus (ek-TOP-ik) region of heart other than SA node that assumes role of cardiac pacemaker

ectopic pregnancy implantation and development of a fetus at a site other than the uterus

edema (eh-DEE-mah) accumulation of excess fluid in interstitial space

EEG arousal transformation of EEG pattern from alpha to beta rhythm during increased levels of attention

effector (ee-FECK-tor) cell or cell collection whose change in activity constitutes the response in a control system

effector protein plasma membrane protein that serves as ion channel or enzyme in signal transduction sequence

efferent (EF-er-ent) carrying away from

efferent arteriole renal vessel that conveys blood from glomerulus to peritubular capillaries

efferent neuron neuron that carries information away from CNS

efferent pathway component of reflex arc that transmits information from integrating center to effector

egg female germ cell at any of its stages of development

eicosanoid (eye-KOH-sah-noid) general term for modified fatty acids that are products of arachidonic acid metabolism (cyclic endoperoxides, prostaglandins, thromboxanes, and leukotrienes); function as paracrine or autocrine agents

ejaculation (ee-jak-you-LAY-shun) discharge of semen from penis

ejaculatory duct (ee-JAK-you-lah-tory) continuation of vas deferens after it is joined by seminal vesicle duct; joins urethra in prostate gland

ejection fraction (EF) the ratio of stroke volume to end diastlic volume; EF = SV/EDV

electric charge particle having excess positivity or negativity

electric force force that causes charged particles to move toward regions having an opposite charge and away from regions having a like charge

electric potential (E) (or electric potential difference) *see* potential

electric signal graded potential or action potential

electric synapse (SIN-apse) synapse at which local currents resulting from electrical activity flow between two neurons through gap junctions joining them

electrocardiogram (ECG, EKG) (ee-lek-troh-KARD-ee-oh-gram) recording at skin surface of the electric currents generated by cardiac muscle action potentials

electrochemical difference force determining direction and magnitude of net charge movement; combination of electrical and chemical gradients

electrode (ee-LEK-trode) probe used to stimulate electrically, or record from, the body surface or a tissue

electroencephalogram (EEG) (eh-lek-troh-en-SEF-ah-loh-gram) recording of brain electrical activity from scalp

electrogenic pump (elec-troh-JEN-ik) active-transport system that directly separates electric charge, thereby producing a potential difference

electrolyte (ee-LEK-troh-lite) substance that dissociates into ions when in aqueous solution

electromagnetic radiation radiation composed of waves with electrical and magnetic components; includes gamma rays, x-rays, and ultraviolet, light, infrared, and radio waves

electron (ee-LEK-tron) subatomic particle that carries one unit of negative charge

embryo (EM-bree-oh) organism during early stages of development; in human beings, the first 2 months of intrauterine life

emission (ee-MISH-un) movement of male genital duct contents into urethra prior to ejaculation

emotion *see* inner emotion, emotional behavior

emotional behavior outward expression and display of inner emotions

emulsion (eh-MUL-shun) suspension of small lipid droplets

end-diastolic volume (EDV) (dy-ah-STAH-lik) amount of blood in ventricle just prior to systole

endocrine gland (EN-doh-krin) group of cells that secrete into the extracellular space hormones that then diffuse into bloodstream; also called a ductless gland

endocrine system all the body's hormone-secreting glands

endocytosis (en-doh-sy-TOH-sis) process in which plasma membrane folds into the cell, forming small pockets that pinch off to produce intracellular, membrane-bound vesicles; *see also* phagocytosis

endogenous opioid (en-DAHJ-en-us OH-pee-oid) certain neuropeptides—endorphin, dynorphin, and enkephalin

endogenous pyrogen (EP) (en-DAHJ-en-us PY-roh-jen) cytokines (including interleukin 1 and probably interleukin 6 and tumor necrosis factor) that act physiologically in the brain to cause fever

endometrium (en-doh-MEE-tree-um) glandular epithelium lining uterine cavity

endoperoxide *see* cyclic endoperoxide

endoplasmic reticulum (en-doh-PLAS-mik reh-TIK-you-lum) cell organelle that consists of interconnected network of membrane-bound branched tubules and flattened sacs; two types are distinguished *granular,* with ribosomes attached, and *agranular,* which is smooth-surfaced

endosome (EN-doh-some) intracellular vesicles and tubular elements between Golgi apparatus and plasma membrane; sorts and distributes vesicles during endo- and exocytosis

endothelium (en-doh-THEE-lee-um) thin layer of cells that lines heart cavities and blood vessels

endothelium-derived relaxing factor (EDRF) nitric oxide and possibly other substances; secreted by vascular endothelium, it relaxes vascular smooth muscle and causes arteriolar dilation

end-plate potential (EPP) depolarization of motor end plate of skeletal-muscle fiber in response to acetylcholine; initiates action potential in muscle plasma membrane

end-product inhibition inhibition of a metabolic pathway by final product's action upon allosteric site on an enzyme (usually the rate-limiting enzyme) in the pathway

end-systolic volume (ESV) (sis-TAH-lik) amount of blood remaining in ventricle after ejection

energy ability to produce change; measured by amount of work performed during a given change

enkephalin (en-KEF-ah-lin) peptide neurotransmitter at some synapses activated by opiate drugs; an endogenous opioid

enteric nervous system (en-TAIR-ik) neural network residing in and innervating walls of gastrointestinal tract

enterogastrones (en-ter-oh-GAS-trones) collective term for hormones that are released by intestinal tract and inhibit stomach activity

enterohepatic circulation (en-ter-oh-hih-PAT-ik) reabsorption of bile salts (and other substances) from intestines, passage to liver (via hepatic portal vein), and secretion back to intestines (via bile)

enterokinase (en-ter-oh-KY-nase) enzyme in luminal plasma membrane of intestinal epithelial cells; converts pancreatic trypsinogen to trypsin

entrainment (en-TRAIN-ment) adjusting biological rhythm to environmental cues

enzyme (EN-zime) protein catalyst that accelerates specific chemical reactions but does not itself undergo net chemical change during the reaction

enzyme activity rate at which enzyme converts reactant to product; may be measure of the properties of enzyme's active site as altered by allosteric or covalent modulation; affects rate of enzyme-mediated reaction

eosinophil (ee-oh-SIN-oh-fil) polymorphonuclear granulocytic leukocyte whose granules take up red dye eosin; involved in allergic responses and parasite destruction

epididymis (ep-ih-DID-eh-mus) portion of male reproductive duct system located between seminiferous tubules and vas deferens

epiglottis (ep-ih-GLOT-iss) thin cartilage flap that folds down, covering trachea, during swallowing

epinephrine (E) (ep-ih-NEF-rin) amine hormone secreted by adrenal medulla and involved in regulation of organic metabolism; a biogenic amine (catecholamine) neurotransmitter; also called *adrenaline*

epiphyseal closure (ep-ih-FIZ-ee-al) conversion of epiphyseal growth plate to bone

epiphyseal growth plate actively proliferating cartilage near bone ends; region of bone growth

epiphysis (eh-PIF-ih-sis) end of long bone

epithelial cell (ep-ih-THEE-lee-al) cell at surface of body or hollow organ; specialized to secrete or absorb ions and organic molecules; with other epithelial cells, forms an *epithelium*

epithelial transport molecule movement from one extracellular compartment across epithelial cells into a second extracellular compartment

epithelium (ep-ih-THEE-lee-um) tissue that covers all body surfaces, lines all body cavities, and forms most glands

epitope (EP-ih-tope) peptide fragment of an antigen; complexed to the MHC protein and presented to the T cell; also called an *antigenic determinant*

equilibrium (ee-quah-LIB-ree-um) no net change occurs in a system; requires no energy

equilibrium potential voltage gradient across a membrane that is equal in force but opposite in direction to concentration force affecting a given ion species

erection penis or clitoris becoming stiff due to vascular congestion

error signal steady-state difference between level of regulated variable in a control system and set point for that variable

erythrocyte (eh-RITH-roh-site) red blood cell

erythropoiesis (eh-rith-roh-poy-EE-sis) erythrocyte production

erythropoietin (eh-rith-roh-POY-ih-tin) peptide hormone secreted mainly by kidney cells; stimulates red blood cell production; one of the hematopoietic growth factors

esophagus (eh-SOF-uh-gus) portion of digestive tract that connects throat (pharynx) and stomach

essential amino acid amino acid that cannot be formed by the body at all (or at rate adequate to meet metabolic requirements) and must be obtained from diet

essential element chemical substance that must be present for normal growth and function of the body

essential nutrient substance required for normal or optimal body function but synthesized by the body either not at all or in amounts inadequate to prevent disease

estradiol (es-tra-DY-ol) steroid hormone of estrogen family; major female sex hormone

estriol (ES-tree-ol) steroid hormone of estrogen family; major estrogen secreted by placenta during pregnancy

estrogen (ES-troh-jen) group of steroid hormones that have effects similar to estradiol on female reproductive tract

euphorigen (you-FOR-ih-jen) drug that elevates mood

excitability ability to produce electric signals

excitable membrane membrane capable of producing action potentials

excitation-contraction coupling in muscle fibers, mechanism linking plasma membrane stimulation with cross-bridge force generation

excitatory postsynaptic potential (EPSP) (post-sin-NAP-tic) depolarizing graded potential in postsynaptic neuron in response to activation of excitatory synapse

excitatory synapse (SIN-apse) synapse that, when activated, increases likelihood that postsynaptic neuron will undergo action potentials or increases frequency of existing action potentials

excretion elimination of a substance from the body, mainly in urine or feces

exocrine gland (EX-oh-krin) cluster of epithelial cells specialized for secretion and having ducts that lead to an epithelial surface

exocytosis (ex-oh-sy-TOH-sis) process in which intracellular vesicle fuses with plasma membrane, the vesicle opens, and its contents are liberated into the extracellular fluid

exon (EX-on) DNA gene region containing code words for a part of the amino acid sequence of a protein

expiration (ex-pur-A-shun) movement of air out of lungs

expiratory reserve volume (ex-PY-ruh-tor-ee) volume of air that can be exhaled by maximal contraction of expiratory muscles after normal resting expiration

extension straightening a joint

extensor muscle muscle whose activity straightens a joint

external anal sphincter ring of skeletal muscle around lower end of rectum

external environment environment surrounding external surfaces of an organism

external genitalia (jen-ih-TAH-lee-ah) (female) mons pubis, labia majora and minora, clitoris, vestibule of the vagina, and vestibular glands; (male) penis and scrotum

external work movement of external objects by skeletal-muscle contraction

extracellular fluid fluid outside cell; interstitial fluid and plasma

extracellular matrix (MAY-trix) a complex consisting of a mixture of protein (and, in some cases, minerals) in which extracellular fluid is interspersed

extrafusal fiber primary muscle fiber in skeletal muscle, as opposed to modified (intrafusal) fiber in muscle spindle

extrapyramidal system *see* brainstem pathway

extrinsic (ex-TRIN-sik) coming from outside

extrinsic clotting pathway formation of fibrin clots by pathway using tissue factor on cells in interstitium; once activated, it also recruits the intrinsic clotting pathway below factor XII

facilitated diffusion (fah-SIL-ih-tay-ted) system using a transporter to move molecules from high to low concentration across a membrane; energy not required

facilitation (fah-sil-ih-TAY-shun) general depolarization of a neuron when excitatory synaptic input exceeds inhibitory input

factor XII initial factor in intrinsic pathway sequence of reactions that results in blood clotting

factor XIIIa activated plasma protein that catalyzes cross-link formation between fibrin molecules to strengthen blood clot

FAD *see* flavine adenine dinucleotide

fast fiber skeletal-muscle fiber that contains myosin having high ATPase activity

fat mobilization increased breakdown of triacylglycerols and release of glycerol and fatty acids into blood

fat-soluble vitamin a vitamin that is soluble in nonpolar solvents and insoluble in water; vitamin A, D, E, or K

fatty acid carbon chain with carboxyl group at one end through which chain can be linked to glycerol to form triacylglycerol; *see also* polyunsaturated fatty acid, saturated fatty acid, unsaturated fatty acid

Fc portion "stem" part of antibody

feces (FEE-sees) material expelled from large intestine during defecation

feedback characteristic of control systems in which output response influences input to system; *see also* negative feedback, positive feedback

feedforward aspect of some control systems that allows system to anticipate changes in a regulated variable

female internal genitalia (jen-ih-TALE-ee-ah) ovaries, uterine tubes, uterus, and vagina

ferritin (FAIR-ih-tin) iron-binding protein that stores iron in body

fertilization union of sperm and egg

fetus (FEE-tus) human beings from third month of intrauterine life until birth

fever increased body temperature due to setting of "thermostat" of temperature-regulating mechanisms at higher-than-normal level

fiber *see* nerve fiber, muscle fiber

fibrin (FY-brin) protein polymer resulting from enzymatic cleavage of fibrinogen; can turn blood into gel (clot)

fibrinogen (fy-BRIN-ah-jen) plasma protein precursor of fibrin

fibrinolytic system (fye-brin-ah-LIT-ik) cascade of plasma enzymes that breaks down clots; also called thrombolytic system

fight-or-flight response activation of sympathetic nervous system during stress

filtered load amount of any substance filtered from renal glomerular capillaries into Bowman's capsule

filtration movement of essentially protein-free plasma out across capillary walls due to a pressure gradient across the wall

first messenger extracellular chemical messenger or electric message that arrives at a cell's plasma membrane

flatus (FLAY-tus) intestinal gas expelled through anus

flavine adenine dinucleotide (FAD) coenzyme derived from the B vitamin riboflavin; transfers hydrogen from one substrate to another

flexion (FLEK-shun) bending a joint

flow autoregulation ability of individual arteries and arterioles to alter their resistance in response to changing blood pressure so that relatively constant blood flow is maintained

fluid-mosaic model (moh-ZAY-ik) cell membrane structure consists of proteins embedded in bimolecular lipid that has the physical properties of a fluid, allowing membrane proteins to move laterally within it

flux amount of a substance crossing a surface in a unit of time; *see also* net flux

folic acid (FOH-lik) vitamin of B-complex group; essential for formation of nucleotide thiamine

follicle (FOL-ih-kel) egg and its encasing follicular, granulosa, and theca cells at all stages prior to ovulation; also called *ovarian follicle*

follicle-stimulating hormone (FSH) protein hormone secreted by anterior pituitary in males and females that acts on gonads; a gonadotropin

follicular phase (fuh-LIK-you-lar) that portion of menstrual cycle during which follicle and egg develop to maturity prior to ovulation

forebrain large, anterior brain subdivision consisting of right and left cerebral hemispheres (the cerebrum) and diencephalon

fovea centralis (FOH-vee-ah) area near center of retina where cones are most concentrated; gives rise to most acute vision

Frank-Starling mechanism the relationship between stroke volume and end-diastolic volume such that stroke volume increases as end-diastolic volume increases; also called *Starling's law of the heart*

free radical atom that has an unpaired electron in its outermost orbital or molecule containing such an atom

free-running rhythm cyclical activity driven by biological clock in absence of environmental cues

frequency number of times an event occurs per unit time

frontal lobe region of anterior cerebral cortex where motor areas, Broca's speech center, and some association cortex are located

fructose (FRUK-tose) five-carbon sugar; present in sucrose (table sugar)

functional residual capacity lung volume after relaxed expiration

functional site binding site on allosteric protein that, when activated, carries out protein's physiological function; also called active site

functional unit organ subunit; all subunits of a given organ have similar structural and functional properties

fused-vesicle channel endocytotic or exocytotic vesicles that have fused to form a continuous water-filled channel through capillary endothelial cell

G protein family of regulatory proteins that reversibly bind guanosine nucleotides; plasma-membrane G proteins interact with membrane ion channels or enzymes; for the G proteins that regulate adenylyl cyclase, subscript "i" (G_i protein) denotes inhibitory action, and "s" (G$_s$ protein) denotes stimulatory action

galactose (gah-LAK-tose) six-carbon monosaccharide; present in lactose (milk sugar)

gallbladder small sac under the liver; concentrates bile and stores it between meals; contraction of gallbladder ejects bile into small intestine

gamete (GAM-eet) germ cell or reproductive cell; sperm in male and egg in female

gametogenesis (gah-mee-toh-JEN-ih-sis) gamete production

gamma-aminobutyric acid (GABA) major inhibitory neurotransmitter in CNS

gamma globulin immunoglobulin G (IgG), most abundant class of plasma antibodies

gamma motor neuron small motor neuron that controls intrafusal muscle fibers in muscle spindles

ganglion (GANG-glee-on) (pl. ganglia); generally reserved for cluster of neuron cell bodies outside CNS

ganglion cell retinal neuron that is postsynaptic to bipolar cells; axons of ganglion cells form optic nerve

gap junction protein channels linking cytoplasm of adjacent cells; allows ions and small molecules to flow between cytoplasms of the connected cells

gastric (GAS-trik) pertaining to the stomach

gastric phase (of gastrointestinal control) initiation of neural and hormonal gastrointestinal reflexes by stimulation of stomach wall

gastrin (GAS-trin) peptide hormone secreted by antral region of stomach; stimulates gastric acid secretion

gastroileal reflex (gas-troh-IL-ee-al) reflex increase in contractions of ileum during gastric emptying

gastrointestinal system (gas-troh-in-TES-tin-al) gastrointestinal tract plus salivary glands, liver, gallbladder, and pancreas

gastrointestinal tract mouth, pharynx, esophagus, stomach, and small and large intestines

gating opening or closing ion channels

gene unit of hereditary information; portion of DNA containing information required to determine a protein's amino acid sequence

gene cloning process of forming identical DNA sequences using genetic engineering techniques

genetic code three-nucleotide sequence in a gene; indicates the location of a particular amino acid in the protein specified by that gene

genome complete set of an organism's genes

germ cell cell that gives rise to male or female gametes (sperm and eggs)

gland group of epithelial cells specialized for secretion; *see also* endocrine gland, exocrine gland

glial cell (GLEE-al) nonneuronal cell in CNS; helps regulate extracellular environment of CNS; also called *neuroglia*

globin (GLOH-bin) collective term for the four polypeptide chains of the hemoglobin molecule

globulin (GLOB-you-lin) one of a family of proteins found in blood plasma

glomerular filtration (gloh-MER-you-lar) bulk flow of an essentially protein-free plasma from renal glomerular capillaries into Bowman's capsule

glomerular filtration rate (GFR) volume of fluid filtered from renal glomerular capillaries into Bowman's capsule per unit time

glomerulus (gloh-MER-you-lus) tufts of glomerular capillaries at beginning of kidney nephron

glottis opening between vocal cords through which air passes, and surrounding area

glucagon (GLOO-kah-gahn) peptide hormone secreted by alpha cells of pancreatic islets of Langerhans; leads to rise in plasma glucose

glucocorticoid (gloo-koh-KOR-tih-koid) steroid hormone produced by adrenal cortex and having major effects on nutrient metabolism

gluconeogenesis (gloo-koh-nee-oh-JEN-ih-sis) formation of glucose by the liver or kidneys from pyruvate, lactate, glycerol, or amino acids

glucose major monosaccharide in the body; a six-carbon sugar, $C_6H_{12}O_6$; also called blood sugar

glucose-counterregulatory control neural or hormonal factors that oppose insulin's actions; glucagon, epinephrine, sympathetic nerves to liver and adipose tissue, cortisol, and growth hormone

glucose-dependent insulinotropic hormone (GIP) intestinal hormone; stimulates insulin secretion in response to glucose and fat in small intestine; also called *gastric inhibitory peptide*

glucose 6-phosphate (FOS-fate) first intermediate in glycolytic pathway

glucose sparing switch from glucose to fat utilization by most cells during postabsorptive state

glutamate (GLU-tah-mate) anion formed from the amino acid glutamic acid; a major excitatory CNS neurotransmitter

glutamine (GLOO-tah-meen) glutamate having an extra NH_3

glycerol (GLISS-er-ol) three-carbon carbohydrate; forms backbone of triacylglycerol

glycine (GLY-seen) an amino acid; a neurotransmitter at some inhibitory synapses in CNS

glycocalyx (gly-koh-KAY-lix) fuzzy coating on extracellular surface of plasma membrane, consists of short, branched carbohydrate chains

glycogen (GLY-koh-jen) highly branched polysaccharide composed of glucose subunits; major carbohydrate storage form in body

glycogenolysis (gly-koh jen-NOL-ih-sis) glycogen breakdown to glucose

glycolysis (gly-KOL-ih-sis) metabolic pathway that breaks down glucose to two molecules of pyruvate (aerobically) or two molecules of lactate (anaerobically)

glycolytic fiber skeletal-muscle fiber that has a high concentration of glycolytic enzymes and large glycogen stores; white muscle fiber

glycoprotein protein containing covalently linked carbohydrates

Golgi apparatus (GOAL-gee) cell organelle consisting of flattened membranous sacs; usually near nucleus; processes newly synthesized proteins for secretion or distribution to other organelles

Golgi tendon organ tension-sensitive mechanoreceptor ending of afferent nerve fiber; wrapped around collagen bundles in tendon

gonad (GOH-nad) gamete-producing reproductive organ—testes in male and ovaries in female

gonadotropic hormone (goh-nad-oh-TROH-pik) hormone secreted by anterior pituitary that control gonadal function; FSH or LH; also called *gonadotropin*

gonadotropin releasing hormone (GnRH) hypophysiotropic hormone that stimulates LH and FSH secretion by anterior pituitary in males and females

graded potential membrane potential change of variable amplitude and duration that is conducted decrementally; has no threshold or refractory period

gradient (GRAY-dee-ent) continuous increase or decrease of a variable over distance

gram atomic mass amount of element in grams equal to the numerical value of its atomic weight

granulosa cell (gran-you-LOH-sah) cell that contributes to the layers surrounding egg and antrum in ovarian follicle; secretes estrogen, inhibin, and other messengers that influence the egg

gray matter area of brain and spinal cord that appears gray in unstained specimens and consists mainly of cell bodies and unmyelinated portions of nerve fibers

ground substance material surrounding cells and fibers in connective tissue

growth factor one of a group of peptides that is highly effective in stimulating cell division and/or differentiation of certain cell types

growth hormone (GH) peptide hormone secreted by anterior pituitary; stimulates insulin-like growth factor I release; enhances body growth by stimulating protein synthesis

growth hormone release inhibiting hormone *see* somatostatin

growth hormone releasing hormone (GHRH) hypothalamic peptide hormone that stimulates growth hormone secretion by anterior pituitary

growth-inhibiting factor one of a group of peptides that modulates growth by inhibiting cell division in specific tissues

guanine (G) (GWAH-neen) purine base of DNA and RNA

guanosine triphosphate (GTP) (GWAH-noh-seen tri-FOS-fate) energy-transporting molecule similar to ATP

except that it contains the base guanine rather than adenine

guanylyl cyclase (GUAN-ah-lil) enzyme that catalyzes transformation of GTP to cyclic GMP

H zone one of transverse bands making up striated pattern of cardiac and skeletal muscle; light region that bisects A band

habituation (hah-bit-you-A-shun) reversible decrease in response strength upon repeatedly administered stimulation

hair cell mechanoreceptor in organ of Corti and vestibular apparatus

HCO$_3^-$ bicarbonate

H$_2$CO$_3$ carbonic acid

heart rate number of heart contractions per minute

heart sound noise that results from vibrations due to closure of atrioventricular valves (first heart sound) or pulmonary and aortic valves (second heart sound)

Heimlich maneuver (HEIM-lik) forceful elevation of diaphragm produced by rescuer's fist against choking victim's abdomen; causes sudden sharp increase in alveolar pressure to expel obstructing material that is causing choking

helper T cell T cell that, via secreted cytokines, enhances the activation of B cells and cytotoxic T cells; can also activate NK cells and macrophages; binds antigen associated with MHC class II proteins

hematocrit (heh-MAT-ah-krit) percentage of total blood volume occupied by blood cells

hematopoietic growth factor (HGF) (heh-MAT-oh-poi-ET-ik) group of protein hormones and paracrine agents that stimulate proliferation and differentiation of various types of blood cells

heme (heem) iron-containing organic molecule bound to each of the four polypeptide chains of hemoglobin or to cytochromes

hemoglobin (HEE-moh-gloh-bin) protein composed of four polypeptide chains, each attached to a heme; located in erythrocytes and transports most blood oxygen

hemoglobin saturation percent of he-

moglobin molecules combined with oxygen

hemorrhage (HEM-er-age) bleeding

hemostasis (hee-moh-STAY-sis) stopping blood loss from a damaged vessel

Henle's loop *see* loop of Henle

heparin (HEP-ah-rin) anticlotting agent found on endothelial-cell surfaces; binds antithrombin III to tissues; used as an anticoagulant drug

hepatic (hih-PAT-ik) pertaining to the liver

hepatic portal vein vein that conveys blood from capillaries in the intestines and portions of the stomach and pancreas to capillaries in the liver

hertz (Hz) (hurts) cycles per second; measure used for wave frequencies

heterozygous (het-er-oh-ZY-gus) condition of having maternal and paternal copies of a gene with slightly different nucleotide sequences (alleles); *compare* homozygous

high-density lipoprotein (HDL) lipid-protein aggregate having low proportion of lipid; promotes removal of cholesterol from cells

hippocampus (hip-oh-KAM-pus) portion of limbic system associated with learning and emotions

histamine (HISS-tah-meen) inflammatory chemical messenger secreted mainly by mast cells; monoamine neurotransmitter

homeostasis (home-ee-oh-STAY-sis) relatively stable condition of extracellular fluid that results from regulatory system actions

homeostatic control system (home-ee-oh-STAT-ik) interconnected components that keep a physical or chemical parameter of internal environment relatively constant

homeothermic (home-ee-oh-THERM-ik) capable of maintaining body temperature within very narrow limits

homologous (hoh-MAHL-ah-gus) corresponding in origin, structure, and position

homozygous (hoh-moh-ZY-gus) condition of having maternal and paternal copies of a gene with identical nucleotide sequences (alleles); *compare* heterozygous

hormone chemical messenger synthesized by specific endocrine cells in response to certain stimuli and secreted into the blood, which carries it to target cells

hydrochloric acid (hy-droh-KLOR-ik) HCl; strong acid secreted into stomach lumen by parietal cells

hydrogen bond weak chemical bond between two molecules or parts of the same molecule, in which negative region of one polarized substance is electrostatically attracted to a positively charged region of polarized hydrogen atom in the other

hydrogen ion (EYE-on) H^+; single proton; H^+ concentration of a solution determines its acidity

hydrogen peroxide H_2O_2; chemical produced by phagosome and highly destructive to macromolecules

hydrolysis (hy-DRAHL-ih-sis) breaking of chemical bond with addition of elements of water (-H and -OH) to the products formed; also called hydrolytic reaction

hydrophilic (hy-droh-FIL-ik) attracted to, and easily dissolved in, water

hydrophobic (hy-droh-FOH-bik) not attracted to, and insoluble in, water

hydrostatic pressure (hy-droh-STAT-ik) pressure exerted by fluid

hydroxyl group (hy-DROX-il) -OH

hyper- increased

hypercalcemia increased plasma calcium

hypercapnea increased arterial P_{CO_2}

hyperemia (hy-per-EE-me-ah) increased blood flow; *see also* active hyperemia

hyperosmotic (hy-per-oz-MAH-tik) having total solute concentration greater than normal extracellular fluid

hyperpolarize to change membrane potential so cell interior becomes more negative than its resting state

hypersensitivity *see* allergy

hypertension chronically increased arterial blood pressure

hyperthermia increased body temperature above the setpoint

hypertonia (hy-per-TOH-nee-ah) abnormally high muscle tone

hypertonic (hy-per-TAH-nik) solutions containing a higher concentration of effectively membrane-impermeable solute particles than normal (isotonic) extracellular fluid

hypertrophy (hy-PER-troh-fee) enlargement of a tissue or organ due to increased cell size rather than increased cell number

hyperventilation increased ventilation adequate to reduce arterial P_{CO_2}

hypo- too little; below

hypoglycemia (hy-poh-gly-SEE-me-ah) low blood glucose (sugar) concentration

hypoosmotic (hy-poh-oz-MAH-tik) having total solute concentration less than that of normal extracellular fluid

hypophysiotropic hormone (hy-poh-fiz-ee-oh-TROH-pik) any hormone secreted by hypothalamus that controls secretion of an anterior pituitary hormone

hypotension low blood pressure

hypothalamic releasing hormone (hy-poh-thah-LAM-ik) *see* hypophysiotropic hormone

hypothalamus (hy-poh-THAL-ah-mus) brain region below thalamus; responsible for integration of many basic neural, endocrine, and behavioral functions, especially those concerned with regulation of internal environment

hypotonia (hy-poh-TOH-nee-ah) abnormally low muscle tone

hypotonic (hy-poh-TAH-nik) solutions containing a lower concentration of effectively nonpenetrating solute particles than normal (isotonic) extracellular fluid

hypoventilation decrease in ventilation that causes an increase in arterial P_{CO_2}

hypoxia (hy-POK-see-ah) deficiency of oxygen at tissue level

I band one of transverse bands making up repeating striations of cardiac and skeletal muscle; located between A bands of adjacent sarcomeres and bisected by Z line

IgA class of antibodies secreted by, and acting locally in, lining of gastrointestinal, respiratory, and urinary tracts

IgD class of antibodies whose function is unknown

IgE class of antibodies that mediate immediate hypersensitivity and resistance to parasites

IgG gamma globulin; most abundant class of antibodies

IgM class of antibodies that, along with IgG, provide major specific humoral immunity against bacteria and viruses

ileocecal sphincter (il-ee-oh-SEE-kal) ring of smooth muscle separating small and large intestines, that is, ileum and cecum

ileum (IL-ee-um) final, longest segment of small intestine

immune defense *see* nonspecific immune defense, specific immune defense

immune surveillance (sir-VAY-lence) recognition and destruction of cancer cells that arise in body

immunity physiological mechanisms that allow body to recognize materials as foreign or abnormal and to neutralize or eliminate them; *see also* active immunity, passive immunity

immunoglobulin (Ig) (im-mun-o-GLOB-you-lin) proteins that are antibodies and antibody-like receptors on B cells (five classes are IgG, IgA, IgD, IgM, and IgE)

implantation (im-plan-TAY-shun) event during which fertilized egg becomes embedded in uterine wall

implicit memory a memory in which one has no previous awareness of the memory and cannot necessarily remember how, when, or where the learning took place

inferior vena cava (VEE-nah KAY-vah) large vein that carries blood from lower half of body to right atrium of heart

inflammation (in-flah-MAY-shun) local response to injury or infection characterized by swelling, pain, heat, and redness

inhibin (in-HIB-in) protein hormone secreted by seminiferous-tubule Sertoli cells and ovarian granulosa cells; inhibits FSH secretion

inhibitory postsynaptic potential (IPSP) hyperpolarizing graded potential that arises in postsynaptic neuron in response to activation of inhibitory synaptic endings upon it

inhibitory synapse (SIN-apse) synapse that, when activated, decreases likelihood that postsynaptic neuron will fire an action potential (or decreases frequency of existing action potentials)

initial segment first portion of axon plus the part of the cell body where axon arises

inner ear cochlea; contains organ of Corti

inner emotion emotional feelings that are entirely within a person

innervate to supply with nerves

inorganic pertaining to substances that do not contain carbon; *compare* organic

inorganic phosphate (P_i) (FOS-fate) $H_2PO_4^-$, HPO_4^{2-}, or PO_4^{3-}

inositol trisphosphate (IP_3) (in-OS-ih-tol tris-FOS-fate) second messenger that causes release of calcium from endoplasmic reticulum into cytosol

insensible water loss water loss of which a person is unaware, that is, loss by evaporation from skin (excluding 'sweat) and respiratory-passage lining

inspiration air movement from atmosphere into lungs

inspiratory muscles muscles whose contraction contributes to inspiration

inspiratory neuron neuron in medulla oblongata that fires in synchrony with inspiration and ceases firing during expiration

inspiratory reserve volume maximal air volume that can be inspired above resting tidal volume

insulin (IN-suh-lin) peptide hormone secreted by beta cells of pancreatic islets of Langerhans; has metabolic and growth-promoting effects; stimulates glucose and amino acid uptake by most cells and stimulates protein, fat, and glycogen synthesis

insulin-like growth factor one of a group of peptides that have growth-promoting effects

insulin-like growth factor 1 (IGF-1) insulin-like growth factor that mediates mitosis-stimulating effect of growth hormone on bone and other tissues and has feedback effect on pituitary

integral membrane protein protein embedded in membrane lipid layer; may span entire membrane or be located at only one side

integrating center cells that receive one or more signals and send out appropriate response; also called an integrator

integrin (in-TEH-grin) transmembrane protein in plasma membrane; binds to specific proteins in extracellular matrix and on adjacent cells to help organize cells into tissues

intercellular cleft a narrow, water-filled space between capillary endothelial cells

intercellular fluid fluid that lies between cells; also called interstitial fluid

intercostal muscle (in-ter-KOS-tal) skeletal muscle that lies between ribs and whose contraction causes rib cage movement during breathing

interferon (in-ter-FEER-on) family of proteins that nonspecifically inhibit viral replication inside host cells; interferon-gamma also stimulates the killing ability of macrophages and NK cells

interleukin (in-ter-LOO-kin) a family of cytokines with many effects on immune responses and host defenses

interleukin 1 (IL-1) cytokine secreted by macrophages and other cells that activates helper T cells, exerts many inflammatory effects, and mediates many of the systemic, acute-phase responses, including fever

interleukin 2 (IL-2) cytokine secreted by activated helper T cells that causes antigen-activated helper T, cytotoxic T, and NK cells to proliferate; also causes activation of macrophages

interleukin 6 (IL-6) cytokine secreted by macrophages and other cells that exerts multiple effects on immune system cells, inflammation, and the acute-phase response

internal anal sphincter smooth-muscle ring around lower end of rectum

internal environment extracellular fluid (interstitial fluid and plasma)

internal genitalia (jen-ih-TAY-lee-ah) *see* female internal genitalia

internal urethral sphincter (you-REE-thrul) part of smooth muscle of urinary bladder wall that opens and closes the bladder outlet

internal work energy-requiring activities in body; *see also* work; *compare* external work

interneuron neuron whose cell body and axon lie entirely in CNS

interphase period of cell division cycle between end of one division and visible signs of beginning of next

interstitial cell stimulating hormone (ICSH) *see* lutenizing hormone

interstitial fluid extracellular fluid surrounding tissue cells; excludes plasma

interstitium (in-ter-STISH-um) interstitial space; fluid-filled space between tissue cells

interventricular septum (in-ter-ven-TRIK-you-lar) partition in heart separating right and left ventricles

intestinal phase (of gastrointestinal control) initiation of neural and hormonal gastrointestinal reflexes by stimulation of intestinal-tract walls

intestino-intestinal reflex reflex cessation of contractile activity of intestines in response to various stimuli in intestine

intracellular fluid fluid in cells; cytosol plus fluid in cell organelles, including nucleus

intrafusal fiber modified skeletal-muscle fiber in muscle spindle

intrapleural fluid (in-trah-PLUR-al) thin fluid film in thoracic cavity between pleura lining the inner wall of thoracic cage and pleura covering lungs

intrapleural pressure (P_{ip}) pressure in pleural space; also called intrathoracic pressure

intrarenal baroreceptor pressure-sensitive juxtaglomerular cells of afferent arterioles, which respond to decreased renal arterial pressure by secreting more renin

intrathoracic pressure *see* intrapleural pressure

intrinsic (in-TRIN-sik) situated entirely within a part

intrinsic clotting pathway intravascular sequence of fibrin clot formation initiated by factor XII or, more usually, by the initial thrombin generated by the extrinsic clotting pathway

intrinsic factor glycoprotein secreted by stomach epithelium and necessary for absorption of vitamin B_{12} in the ileum

intron (IN-trahn) region of noncoding base sequences in gene

inversely proportional relationship in which, as one factor increases by a given amount, the other decreases by a given amount

ion (EYE-on) atom or small molecule containing unequal number of electrons and protons and therefore carrying a net positive or negative electric charge

ionic bond (eye-ON-ik) strong electrical attraction between two oppositely charged ions

ionization (eye-on-ih-ZAY-shun) process of removing electrons from or adding them to an atom or small molecule to form an ion

ipsilateral (ip-sih-LAT-er-al) on the same side of the body

iris ringlike structure surrounding pupil of eye

irreversible reaction chemical reaction that releases large quantities of energy and results in almost all the reactant molecules being converted to product; *compare* reversible reaction

ischemia (iss-KEY-me-ah) reduced blood supply

islet of Langerhans (EYE-let of LAN-ger-hans) cluster of pancreatic endocrine cells; distinct islet cells secrete insulin, glucagon, somatostatin, and pancreatic polypeptide

isometric contraction (eye soh MET-rik) contraction of muscle under conditions in which it develops tension but does not change length

isoosmotic (eye-soh-oz-MAH-tik) having the same total solute concentration as extracellular fluid

isotonic (eye-soh-TAH-nik) containing the same number of effectively non-penetrating solute particles as normal extracellular fluid; *see also* isotonic contraction

isotonic contraction contraction of muscle under conditions in which load on the muscle remains constant but muscle shortens

isovolumetric ventricular contraction (eye-soh-vol-you-MET-rik) early phase of systole when atrioventricular and aortic valves are closed and ventricular size remains constant

isovolumetric ventricular relaxation early phase of diastole when atrioventricular and aortic valves are closed and ventricular size remains constant

J receptors receptors in the lung capillary walls or interstitium that respond to increased lung interstitial pressure

jejunum (jeh-JU-num) middle segment of small intestine

juxtaglomerular apparatus (JGA) (jux-tah-gloh-MER-you-lar) renal structure consisting of macula densa and juxtaglomerular cells; site of renin secretion and sensors for renin secretion and control of glomerular filtration rate

keto acid (KEY-toh) molecule formed from amino acid metabolism and containing carbonyl (-CO-) and carboxyl (-COOH) groups

ketone (KEY-tone) product of fatty acid metabolism that accumulates in blood during starvation and in severe untreated diabetes mellitus; acetoacetic acid, acetone, or B-hydroxybutyric acid; also called ketone body

kilocalorie (kcal) (KIL-oh-kal-ah-ree) amount of heat required to change the temperature of 1 L water by 1° C; calorie used in nutrition; also called Calorie and large calorie

kinase (KY-nase) enzyme that transfers a phosphate (usually from ATP) to another molecule

kinesthesia (kin-ess-THEE-zee-ah) sense of movement derived from movement at a joint

kininogen (ky-NIN-oh-jen) plasma protein from which kinins are generated

Krebs cycle mitochondrial metabolic pathway that utilizes fragments derived from carbohydrate, protein, and fat breakdown and produces carbon dioxide, hydrogen (for oxidative phosphorylation), and small amounts of ATP; also called tricarboxylic acid cycle or citric acid cycle

lactase (LAK-tase) small-intestine enzyme that breaks down lactose (milk sugar) into glucose and galactose

lactate ionized form of lactic acid

lactation (lak-TAY-shun) production and secretion of milk by mammary glands

lacteal (lak-TEEL) blind-ended lymph vessel in center of each intestinal villus

lactic acid (LAK-tik) three-carbon molecule formed by glycolytic pathway in absence of oxygen; dissociates to form lactate and hydrogen ions

lactose (LAK-tose) disaccharide composed of glucose and galactose; also called milk sugar

larynx (LAR-inks) part of air passageway between pharynx and trachea; contains the vocal cords

latent period (LAY-tent) period lasting several milliseconds between action potential initiation in a muscle fiber and beginning of mechanical activity

lateral position farther from the midline

lateral inhibition method of refining sensory information in afferent neurons and ascending pathways whereby fibers inhibit each other, the most active fibers causing the greatest inhibition of adjacent fibers

lateral sac enlarged region at end of each sarcoplasmic reticulum segment; adjacent to transverse tubule

law of mass action maxim that an increase in reactant concentration causes a chemical reaction to proceed in direction of product formation; the opposite occurs with decreased reactant concentration

lecithin (LESS-ih-thin) a phospholipid

lengthening contraction contraction as an external force pulls a muscle to a longer length despite opposing forces generated by the active cross bridges

lens adjustable part of eye's optical system, which helps focus object's image on retina

leukocyte (LOO-koh-site) white blood cell

leukotrienes (LOO-koh-treens) type of eicosanoid that is generated by lipoxygenase pathway and functions as inflammatory mediator

Leydig cell (LY-dig) testosterone-secreting endocrine cell that lies between seminiferous tubules of testes; also called interstitial cell

LH surge large rise in luteinizing-hormone secretion by anterior pituitary about day 14 of menstrual cycle

ligand (LY-gand) any molecule or ion that binds to protein surface by noncovalent bonds

ligand-sensitive channel membrane channel operated by the binding of specific molecules to channel proteins

limbic system (LIM-bik) interconnected brain structures in cerebrum; involved with emotions and learning

lipase (LY-pase) enzyme that hydrolyzes triacylglycerol to monoglyceride and fatty acids; *see also* lipoprotein lipase

lipid (LIP-id) molecule composed primarily of carbon and hydrogen and characterized by insolubility in water

lipid bilayer a sheet consisting of two layers of amphipathic lipids; nonprotein part of cell membrane

lipolysis (ly-POL-ih-sis) triacylglycerol breakdown

lipoprotein (lip-oh-PROH-teen) lipid aggregate partially coated by protein; involved in lipid transport in blood

lipoprotein lipase capillary endothelial enzyme that hydrolyzes triacylglycerol in lipoprotein to monoglyceride and fatty acids

lipoxygenase (ly-POX-ih-jen-ase) enzyme that acts on arachidonic acid and leads to leukotriene formation

load external force acting on muscle

local current flow movement of positive ions toward more negative membrane region and simultaneous movement of negative ions in opposite direction

local homeostatic response (home-ee-oh-STAT-ik) response acting in immediate vicinity of a stimulus, without nerves or hormones, and having net effect of counteracting stimulus

local potential small, graded potential difference between two points that is conducted decrementally and has no threshold or refractory period

locus coeruleus (sih-ROO-lee-us) brainstem nucleus that projects to many brain parts and is implicated in directed attention

long-loop negative feedback inhibition of anterior pituitary and/or hypothalamus by hormone secreted by the third endocrine gland in a sequence

long-term potentiation (LTP) process by which certain synapses undergo long-lasting increase in effectiveness when heavily used

loop of Henle (HEN-lee) hairpin-like segment of kidney nephron with *descending* and *ascending limbs;* situated between proximal and distal tubules

low-density lipoprotein (LDL) (lip-oh-PROH-teen) protein-lipid aggregate that is major carrier of plasma cholesterol to cells

lower esophageal sphincter smooth muscle of lowest 4 cm of esophagus; can close off esophageal opening into the stomach

lumen (LOO-mon) space in hollow tube or organ

luminal (LOO-min-ul) pertaining to lumen

luminal membrane portion of plasma membrane facing the lumen; also called apical or mucosal membrane

lung compliance (C_L) (come-PLY-ance) change in lung volume caused by a given change in transpulmonary pressure; the greater the lung compliance, the more stretchable the lung wall

luteal phase (LOO-tee-al) last half of menstrual cycle following ovulation; corpus luteum is active ovarian structure

luteinizing hormone (LH) (LOO-tee-en-iz-ing) peptide gonadotropic hormone secreted by anterior pituitary; rapid increase in females at mid menstrual cycle initiates ovulation; stimulates Leydig cells in males; also called interstitial cell stimulating hormone (ICSH) in male

lymph (limf) fluid in lymphatic vessels

lymph node small organ, containing lymphocytes, located along lymph vessel; a site of lymphocyte cell division and initiation of specific immune responses

lymphatic system (lim-FAT-ik) network of vessels that conveys lymph from tissues to blood and to lymph nodes along these vessels

lymphocyte (LIMF-oh-site) type of leukocyte that is responsible for specific immune defenses; B cells, T cells, and NK cells

lymphocyte activation cell division and differentiation of lymphocytes following antigen binding

lymphoid organ (LIMF-oid) lymph node, spleen, thymus, tonsil, or aggregate of lymphoid follicles; *see also* primary lymphoid organ, peripheral lymphoid organ; *compare* lymphatic system

lysosome (LY-soh-some) membrane-bound cell organelle containing digestive enzymes in a highly acid solution that break down bacteria, large molecules that have entered cell, and damaged components of cell

macrophage (MAK-roh-fayje) cell that phagocytizes foreign matter, processes it, presents antigen to lymphocytes, and secretes cytokines

(monokines) involved in inflammation, activation of lymphocytes, and systemic acute phase response to infection or injury; *see also* activated macrophage, macrophage-like cell

macrophage-like cell one of several cell types that exert functions similar to those of macrophages

macula densa (MAK-you-lah DEN-sah) specialized sensor cells of renal tubule at end of loop of Henle; component of juxtaglomerular apparatus

major histocompatibility complex (MHC) group of genes that code for major histocompatibility complex proteins, which are important for specific immune function

mammary gland milk-secreting gland in breast

mass fundamental property of an object equivalent to the amount of matter in the object

mass movement contraction of large segments of colon; propels fecal material into rectum

mast cell tissue cell that releases histamine and other chemicals involved in inflammation

maximal tubular capacity (T_m) maximal rate of mediated transport of substance across renal tubule wall

mean arterial pressure (MAP) average blood pressure during cardiac cycle; approximately diastolic pressure plus one-third pulse pressure

mechanoreceptor (meh-KAN-oh-re-sep-tor) sensory receptor that responds preferentially to mechanical stimuli such as bending, twisting, or compressing

mechanosensitive channel membrane ion channel that is opened or closed by deformation or stretch of the plasma membrane

median eminence (EM-ih-nence) region at base of hypothalamus containing capillary tufts into which hypophysiotropic hormones are secreted

mediate (MEE-dee-ate) bring about

mediated transport movement of molecules across membrane by binding to protein transporter; characterized by specificity, competition, and saturation; includes facilitated diffusion and active transport

medulla (meh-DUL-ah) innermost portion of an organ; *compare* cortex; *see* adrenal medulla and medulla oblongata

medulla oblongata (ob-long-GOT-ah) part of the brainstem closest to the spinal cord

medullary cardiovascular center *see* cardiovascular center

medullary inspiratory neuron *see* inspiratory neuron

megakaryocyte (meg-ah-KAR-ee-oh-site) large bone marrow cell that gives rise to platelets

meiosis (my-OH-sis) process of cell division leading to gamete (sperm or egg) formation; daughter cells receive only half the chromosomes present in original cell

melatonin (mel-ah-TOH-nin) candidate hormone secreted by pineal gland; suspected role in setting body's circadian rhythms

membrane structural barrier composed of lipids and proteins; provides selective barrier to molecule and ion movement and structural framework to which enzymes, fibers, and ligands are bound

membrane attack complex (MAC) group of complement proteins that form channels in microbe surface and destroy microbe

membrane potential voltage difference between inside and outside of cell

memory *see* implicit memory, declarative memory

memory cell B cell or T cell that differentiates during an initial infection and responds rapidly during subsequent exposure to same antigen

memory consolidation processes by which memory is transferred from working to long-term form

memory trace neural substrate of memory

menarche (MEN-ark-ee) onset, at puberty, of menstrual cycling in women

meninges (men-IN-jees) protective membranes that cover brain and spinal cord

menopause (MEN-ah-paws) cessation of menstrual cycling in middle age

menstrual cycle (MEN-stroo-al) cyclical rise and fall in female reproductive hormones and processes, ending in menstruation

menstruation (men-stroo-AY-shun) flow of menstrual fluid from uterus; also called menstrual period

messenger RNA (mRNA) ribonucleic acid that transfers genetic information from DNA to ribosome

metabolic acidosis (met-ah-BOL-ik ass-ih-DOH-sis) acidosis due to the build up of acids other than carbonic acid (from carbon dioxide)

metabolic alkalosis (al-kah-LOH-sis) alkalosis resulting from the removal of hydrogen ions by mechanisms other than respiratory removal of carbon dioxide

metabolic end product final molecule produced by a metabolic reaction or series of reactions

metabolic pathway sequence of enzyme-mediated chemical reactions by which molecules are synthesized and broken down in cells

metabolic rate total body energy expenditure per unit time

metabolism (meh-TAB-uhl-izm) chemical reactions that occur in a living organism

metabolite (meh-TAB-oh-lite) substance produced by metabolism

metabolize change by chemical reactions

metarteriole (MET-are-teer-ee-ole) blood vessel that directly connects arteriole and venule

methyl group -CH₃

MHC protein plasma-membrane protein coded for by a major histocompatibility complex; restricts T-cell receptor's ability to combine with antigen on cell; categorized as class I and class II

micelle (my-SEL) soluble cluster of amphipathic molecules in which molecules' polar regions line surface and nonpolar regions are oriented toward center; formed from fatty acids, 2-monoglycerides, and bile salts during fat digestion in small intestine

microbe bacterium, virus, fungus, or other parasite

microcirculation blood circulation in arterioles, capillaries, and venules

microfilament rodlike cytoplasmic actin filament that forms major component of cytoskeleton

microsomal enzyme system (MES) (my-kroh-SOM-al) enzymes, found in smooth endoplasmic reticulum of liver cells, that transform molecules into more polar, less lipid-soluble substances

microtubule tubular cytoplasmic filament composed of the protein tubulin that provides internal support for cells; allows change in cell shape and organelle movement in cell

microvilli (my-kroh-VIL-eye) small fingerlike projections from epithelial-cell surface; greatly increase surface area of cell; characteristic of epithelium lining small intestine and kidney nephrons

micturition (mik-chur-RISH-un) urination

middle-ear cavity air-filled space in temporal bone; contains three ear bones that conduct sound waves from tympanic membrane to cochlea

migrating motility complex pattern of peristaltic waves that pass over small segments of intestine after absorption of meal

milk ejection reflex process by which milk is moved from mammary gland alveoli into ducts, from which it can be sucked; due to oxytocin

milliliter (ml) (MIL-ih-lee-ter) volume equal to 0.001 L

millimol (mmol) (MIL-ih-mole) amount equal to 0.001 mol

millivolt (mV) (MIL-ih-volt) electric potential equal to 0.001 V

mineral inorganic substance, that is, without carbon; major minerals in body are calcium, phosphorus, potassium, sulfur, sodium, chloride, and magnesium

mineralocorticoid (min-er-al-oh-KORT-ih-koid) steroid hormone produced by adrenal cortex; has major effect on sodium and potassium balance; major mineralocorticoid is aldosterone

minute ventilation total ventilation per minute; equals tidal volume times respiratory rate

mitochondrion (my-toh-KON-dree-un) rod-shaped or oval cytoplasmic organelle that produces most of cell's ATP; site of Krebs cycle and oxidative phosphorylation enzymes

mitogen (MY-tuh-jen) chemical that stimulates cell division

mitosis (my-TOH-sis) process in cell division in which DNA is duplicated and an identical set of chromosomes is passed to each daughter cell

mitral valve (MY-tral) valve between left atrium and left ventricle of heart

modality (moh-DAL-ih-tee) type of sensory stimulus

modulation *see* allosteric modulation, covalent modulation

modulator molecule ligand that, by acting at an allosteric regulatory site, alters properties of other binding site on a protein and thus regulates its functional activity

mol weight of a substance in grams equal to its molecular weight; 1 mol = 6×10^{23} molecules

molarity (moh-LAR-ih-tee) number of moles of solute per liter of solution

molecular weight sum of atomic weights of all atoms in molecule

molecule chemical substance formed by linking atoms together

monoamine (mah-noh-ah-MEEN) class of neurotransmitters having the structure $R\text{-}NH_2$, where R is molecule remainder; by convention, excludes peptides and amino acids

monoamine oxidase enzyme that inactivates catecholamine neurotransmitters and serotonin

monocyte (MAH-noh-site) type of leukocyte; leaves bloodstream and is transformed into a macrophage

monoglyceride (mah-noh-GLISS-er-ide) glycerol linked to one fatty acid side chain

monosaccharide (mah-noh-SAK-er-ide) carbohydrate consisting of one sugar molecule, which generally contains five or six carbon atoms

monosynaptic reflex (mah-noh-sih-NAP-tik) reflex in which the afferent neuron directly activates the motor neuron

motilin (moh-TIL-in) candidate intestinal hormone thought to control normal GI motor activity

motivation *see* primary motivated behavior, secondary motivated behavior

motor having to do with muscles and movement

motor control hierarchy brain areas having a role in skeletal-muscle control are rank-ordered in three functional groups

motor control system CNS parts that contribute to control of skeletal-muscle movements

motor cortex strip of cerebral cortex along posterior border of frontal lobe; gives rise to many axons descending in corticospinal and multineuronal pathways; also called primary motor cortex

motor end plate specialized region of muscle-cell plasma membrane that lies directly under axon terminal of a motor neuron

motor neuron somatic efferent neuron, which innervates skeletal muscle

motor neuron pool all the motor neurons for a given muscle

motor program pattern of neural activity required to perform a certain movement

motor unit motor neuron plus the muscle fibers it innervates

mucosa (mu-KOH-sah) three layers of gastrointestinal tract wall nearest lumen, that is, *epithelium, lamina propria,* and *muscularis mucosa*

mucus highly viscous solution secreted by mucous membranes

Müllerian duct (mul-AIR-ee-an) part of embryo that, in a female, develops into reproductive system ducts, but in a male, degenerates

Müllerian inhibiting substance (MIS) protein secreted by fetal testes that causes Müllerian ducts to degenerate

multineuronal pathway pathway made up of chains of neurons functionally connected by synapses; also called multisynaptic pathway

multiunit smooth muscle smooth muscle that exhibits little, if any, propagation of electrical activity from fiber to fiber and whose contractile activity is closely coupled to its neural input

muscarinic receptor (mus-kur-IN-ik) acetylcholine receptor that responds to the mushroom poison muscarine; located on smooth muscle, cardiac muscle, some CNS neurons, and glands

muscle number of muscle fibers bound together by connective tissue

muscle fatigue decrease in muscle tension with prolonged activity

muscle fiber muscle cell

muscle-spindle stretch receptor capsule-enclosed arrangement of afferent nerve fiber endings around specialized skeletal-muscle fibers; sensitive to stretch

muscle tension force exerted by a contracting muscle on object

muscle thick filament *see* thick filament

muscle tone degree of resistance of muscle to passive stretch due to ongoing contractile activity; *see also* smooth-muscle tone

mutagen (MUTE-uh-jen) factor in the environment that increases mutation rate

mutation (mu-TAY-shun) any change in base sequence of DNA that changes genetic information

myelin (MY-uh-lin) insulating material covering axons of many neurons; consists of layers of myelin-forming-cell plasma membrane wrapped around axon

myenteric plexus (my-en-TER-ik PLEX-us) nerve cell network between circular and longitudinal muscle layers in esophagus, stomach, and intestinal walls

myo- (MY-oh) pertaining to muscle

myoblast (MY-oh-blast) embryological cell that gives rise to muscle fibers

myocardium (my-oh-KARD-ee-um) cardiac muscle, which forms heart walls

myoepithelial cell (my-oh-ep-ih-THEE-lee-al) specialized contractile cell in certain exocrine glands; contraction forces gland's secretion through ducts

myofibril (my-oh-FY-bril) bundle of thick or thin contractile filaments in cytoplasm of striated muscle; myofibrils exhibit a repeating sarcomere pattern along longitudinal axis of muscle

myogenic (my-oh-JEN-ik) originating in muscle

myoglobin (my-oh-GLOH-bin) muscle-fiber protein that binds oxygen

myometrium (my-oh-MEE-tree-um) uterine smooth muscle

myosin (MY-oh-sin) contractile protein that forms thick filaments in muscle fibers

myosin ATPase enzymatic site on globular head of myosin that catalyzes ATP breakdown to ADP and P_i, releasing the chemical energy used to produce force of muscle contraction

myosin light-chain kinase smooth muscle protein kinase; when activated by Ca-calmodulin, phosphorylates myosin light chain

Na,K-ATPase pump primary active-transport protein that splits ATP and releases energy used to transport sodium out of cell and potassium in

natural antibody antibody to erythrocyte antigens A or B; are present without prior exposure to antigen

natural killer (NK) cell type of lymphocyte that binds to virus-infected and cancer cells without specific recognition and kills then directly; participates in antibody-dependent cellular cytotoxicity

negative balance loss of substance from body exceeds gain, and total amount in body decreases; also used for physical parameters such as body temperature and energy; *compare* positive balance

negative feedback aspect of control systems in which system's response opposes the original change in the system; *compare* positive feedback

nephron (NEF-ron) functional unit of kidney; has vascular and tubular component

nerve group of many nerve fibers traveling together in peripheral nervous system

nerve cell cell in nervous system specialized to initiate, integrate, and conduct electric signals; also called *neuron*

nerve fiber axon of a neuron

nerve growth factor protein that stimulates growth and differentiation of some neurons

net amount remaining after deductions have been made; final amount

net flux difference between two one-way fluxes

neuroeffector junction "synapse" between a neuron and muscle or gland cell

neuroglia *see* glial cell

neurohormone chemical messenger that is released by a neuron and travels in bloodstream to its target cell

neuromodulator chemical messenger that acts on neurons, usually by a second-messenger system, to alter response to a neurotransmitter

neuromuscular junction synapse-like junction between an axon terminal of an efferent nerve fiber and a skeletal-muscle fiber

neuron (NUR-ahn) *see* nerve cell

neuropeptide family of more than 50 neurotransmitters composed of 2 or more amino acids; often also functions as chemical messenger in non-neural tissues

neurotransmitter chemical messenger used by neurons to communicate with each other or with effectors

neutrophil (NOO-troh-fil) polymorphonuclear granulocytic leukocyte whose granules show preference for neither eosin nor basic dyes; functions as phagocyte and releases chemicals involved in inflammation

NH_3 ammonia

NH_4^+ ammonium ion

nicotinamide adenine dinucleotide (NAD$^+$) coenzyme derived from the B vitamin niacin; transfers hydrogen from one substrate to another

nicotinic receptor (nik-oh-TIN-ik) acetylcholine receptor that responds to nicotine; primarily, receptors at motor end plate and on postganglionic autonomic neurons

nitric oxide gas molecules that function as intercellular messengers, including neurotransmitters; is endothelium-derived relaxing factor; destroys intracellular microbes

nociceptor (NOH-sih-sep-tor) sensory receptor whose stimulation causes pain

node of Ranvier (RAHN-vee-ay) space between adjacent myelin-forming cells along myelinated axon where axonal plasma membrane is exposed to extracellular fluid; also called neurofibril node

nonpolar pertaining to molecule or region of molecule containing predominantly chemical bonds in which electrons are shared equally between atoms; having few polar or ionized groups

nonspecific ascending pathway chain of synaptically connected neurons in CNS that are activated by sensory units of several different types; signals general information; *compare* specific ascending pathway

nonspecific immune defense response that nonselectively protects against foreign material without having to recognize its specific identity

norepinephrine (NE) (nor-ep-ih-NEF-rin) biogenic amine (catecholamine) neurotransmitter released at most sympathetic postganglionic endings, from adrenal medulla, and in many CNS regions

nuclear envelope double membrane surrounding cell nucleus

nuclear pore opening in nuclear envelope through which molecular messengers pass between nucleus and cytoplasm

nucleic acid (noo-KLAY-ik) nucleotide polymer in which phosphate of one nucleotide is linked to the sugar of the adjacent one; stores and transmits genetic information; includes DNA and RNA

nucleolus (noo-KLEE-oh-lus) densely staining nuclear region containing portions of DNA that code for ribosomal proteins

nucleotide (NOO-klee-oh-tide) molecular subunit of nucleic acid; purine or pyrimidine base, sugar, and phosphate

nucleus (NOO-klee-us) (pl. nuclei) (cell) large membrane-bound organelle that contains cell's DNA; (neural) cluster of neuron cell bodies in CNS

occipital lobe (ok-SIP-ih-tul) posterior region of cerebral cortex where primary visual cortex is located

Ohm's law current (I) is directly proportional to voltage (E) and inversely proportional to resistance (R) such that I = E/R

olfaction (ol-FAK-shun) sense of smell

olfactory (ol-FAK-tor-ee) pertaining to sense of smell

olfactory epithelium mucous membrane in upper part of nasal cavity containing receptors for sense of smell

olfactory nerve cranial nerve I; relays action potentials from olfactory epithelium

oligodendroglia (oh-lih-goh-den-droh-GLEE-ah) type of glial cell; responsible for myelin formation in CNS

oncogene (ON-koh-jeen) altered gene that can lead to cancer

oogenesis (oh-uh-JEN-ih-sis) gamete production in female

oogonium (oh-uh-GOH-nee-um) primitive germ cell that gives rise to primary oocyte

operating point steady-state value maintained by homeostatic control system; also called *set point*

opioid (OH-pee-oid) *see* endogenous opioid

opsin (OP-sin) protein component of photopigment

opsonin (op-SOH-nin) any substance that binds a microbe to a phagocyte and promotes phagocytosis

optic nerve cranial nerve II; relays action potentials from retina

optimal length (l_o) sarcomere length at which muscle fiber develops maximal isometric tension

organ collection of tissues joined in structural unit to serve common function

organ of Corti (KOR-tee) structure in inner ear capable of transducing sound-wave energy into action potentials

organ system organs that together serve an overall function

organelle *see* cell organelle

organic pertaining to carbon-containing substances; *compare* inorganic

orgasm (OR-gazm) inner emotions and systemic physiological changes that mark apex of sexual intercourse, usually accompanied in the male by ejaculation

orienting response behavior in response to a novel stimulus, that is, the person stops what he or she is doing, looks around, listens intently, and turns toward stimulus

osmol 1 mol of solute ions and molecules

osmolarity (oz-moh-LAR-ih-tee) total solute concentration of a solution; measure of water concentration in that the higher the solution osmolarity, the lower the water concentration

osmoreceptor (OZ-moh-ree-sep-tor) receptor that responds to changes in osmolarity of surrounding fluid

osmosis (oz-MOH-sis) net diffusion of water across a selective barrier from region of higher water concentration (lower solute concentration) to region of lower water concentration (higher solute concentration)

osmotic pressure (oz-MAH-tik) pressure that must be applied to a solution on one side of a membrane to prevent osmotic flow of water across the membrane from a compartment of pure water; a measure of the solutions' osmolarity

osteoblast (OS-tee-oh-blast) cell type responsible for laying down protein matrix of bone; called osteocyte after calcified matrix has been set down

oval window membrane covered opening between middle-ear cavity and scala vestibuli of inner ear

ovarian follicle *see* follicle

ovary (OH-vah-ree) gonad in female

ovulation (ov-you-LAY-shun) release of egg, surrounded by its zona pellucida and cumulus, from ovary

ovum (pl. ova) gamete of female; egg

oxidative (OX-ih-day-tive) using oxygen

oxidative deamination (dee-am-ih-NAY-shun) reaction in which an amino group from an amino acid is replaced by oxygen to form a keto acid

oxidative fiber muscle fiber that has numerous mitochondria and therefore a high capacity for oxidative phosphorylation; red muscle fiber

oxidative phosphorylation (fos-for-ih-LAY-shun) process by which energy derived from reaction between hydrogen and oxygen to form water is transferred to ATP during its formation

oxygen debt decrease in energy reserves during exercise that results in an increase in oxygen consumption and an increased production of ATP by oxidative phosphorylation following the exercise

oxyhemoglobin (ox-see-HEE-moh-gloh-bin) HbO_2; hemoglobin combined with oxygen

oxytocin (ox-see-TOH-sin) peptide hormone synthesized in hypothalamus and released from posterior pituitary; stimulates mammary glands to release milk and uterus to contract

P wave component of electrocardiogram reflecting atrial depolarization

pacemaker neurons that set rhythm of biological clocks independent of external cues; any nerve or muscle cell that has an inherent autorhythmicity and determines activity pattern of other cells

pacemaker potential spontaneous gradual depolarization to threshold of some nerve and muscle cells' plasma membrane

paracellular pathway the space between adjacent cells of an epithelium through which some molecules diffuse as they cross the epithelium

paracrine agent (PAR-ah-krin) chemical messenger that exerts its effects on cells near its secretion site; by convention, excludes neurotransmitters

paradoxical sleep *see* REM sleep

parasympathetic division (par-ah-sim-pah-THET-ik) portion of autonomic nervous system whose preganglionic fibers leave CNS from brainstem and sacral portion of spinal cord; most of its postganglionic fibers release acetylcholine; *compare* sympathetic division

parathyroid gland one of four parathyroid-hormone secreting glands on thyroid gland surface

parathyroid hormone (PTH) peptide hormone secreted by parathyroid glands; regulates calcium and phosphate concentrations of extracellular fluid

parietal cell (pah-RY-ih-tal) gastric gland cell that secretes hydrochloric acid and intrinsic factor

parietal lobe region of cerebral cortex containing sensory cortex and some association cortex

partial pressure (*P*) that part of total gas pressure due to molecules of one gas species; measure of concentration of a gas in a gas mixture

parturition (par-tu-RISH-un) birth; delivery of infant and placenta

passive immunity resistance to infection resulting from direct transfer of antibodies or sensitized T cells from one person (or animal) to another; *compare* active immunity

pelvic diaphragm (PEL-vik DY-ah-fram) sheet of skeletal muscle that forms floor of pelvis and help support abdominal and pelvic viscera

pepsin (PEP-sin) family of several protein-digesting enzymes formed in the stomach; breaks protein down to peptide fragments

pepsinogen (pep-SIN-ah-jen) inactive precursor of pepsin; secreted by chief cells of gastric mucosa

peptide (PEP-tide) short polypeptide chain; by convention, having less than 50 amino acids

peptide bond polar covalent chemical bond joining the amino and carboxyl groups of two amino acids; forms protein backbone

peptidergic neuron that releases peptides

percent hemoglobin saturation *see* hemoglobin saturation

perception understanding of objects and events of external world that we acquire from neural processing of sensory information

perforin protein secreted by cytotoxic T cells; forms channels in plasma membrane of target cell, which destroys it

perfusion blood flow

pericardium (per-ah-KAR-dee-um) connective-tissue sac surrounding heart

peripheral chemoreceptor carotid or aortic body; responds to changes in arterial blood P_{O_2} and H^+ concentration

peripheral nervous system nerve fibers extending from CNS

peripheral thermoreceptor cold or warm receptor in skin or certain mucous membranes

peristaltic wave (per-ih-STAL-tik) progressive wave of smooth-muscle contraction and relaxation that proceeds along wall of a tube, compressing the tube and causing its contents to move

peritoneum (per-ih-toh-NEE-um) membrane lining abdominal and pelvic cavities and covering organs there

peritubular capillary capillary closely associated with renal tubule

permeability constant (k_p) number that defines the proportionality between a flux and a concentration gradient and depends on the properties of the membrane and the diffusing molecule

permissiveness situation whereby small quantities of one hormone are required in order for a second hormone to exert its full effects

peroxisome (per-OX-ih-some) cell organelle that destroys certain toxic products of oxidative reactions

pH expression of a solution's acidity; negative logarithm to base 10 of H^+ concentration; pH decreases as acidity increases

phagocyte (FAH-go-site) any cell capable of phagocytosis

phagocytosis (fag-uh-sy-TOH-sis) engulfment of particles by a cell

pharmacological effect effect produced by much larger amounts of hormone than are normally present

pharynx (FAR-inks) throat; passage common to routes taken by food and air

phase shift a resetting of the internal clock due to altered environmental cues

phasic (FAYZ-ik) intermittent; *compare* tonic

phosphate group (FOS-fate) $-PO_4^{2-}$

phosphatidylinositol bisphosphate (PIP₂) (fos-fah-tyd-il-in-OS-ih-tol bis FOS fate) plasma membrane phospholipid that forms inositol trisphosphate and diacylglycerol when catalyzed by phospholipase C

phosphodiesterase (fos-foh-dy-ES-terase) enzyme that catalyzes cyclic AMP breakdown to AMP

phospholipase A₂ (fos-fo-LY-pase A-two) enzyme that splits arachidonic acid from plasma membrane phospholipid

phospholipase C receptor-controlled plasma-membrane enzyme that catalyzes phosphatidylinositol bisphosphate breakdown to inositol trisphosphate and diacylglycerol

phospholipid (fos-foh-LIP-id) lipid subclass similar to triacylglycerol except that a phosphate group ($-PO_4^{2-}$) and small nitrogen-containing molecule are attached to third hydroxyl group of glycerol; major component of cell membranes

phosphoprotein phosphatase (FOS-fah-tase) enzyme that removes phosphate from protein

phosphoric acid (fos-FOR-ik) acid generated during catabolism of phosphorus-containing compounds; dissociates to form inorganic phosphate and hydrogen ions

phosphorylation (fos-for-ah-LAY-shun) addition of phosphate group to an organic molecule

photopigment light-sensitive molecule altered by absorption of photic energy of certain wavelengths; consists of opsin bound to a chromophore

photoreceptor receptor sensitive to light (photic energy)

physiology (fiz-ee-OL-uh-jee) branch of biology dealing with the mechanisms by which the body functions

pinocytosis (pin-oh-sy-TOH-sis) endocytosis when the vesicle encloses extracellular fluid or specific molecules in the extracellular fluid that have bound to proteins on the extracellular surface of the plasma membrane

pitch degree of how high or low a sound is perceived

pituitary gland (pih-TOO-ih-tar-ee) endocrine gland that lies in bony pocket below hypothalamus; constitutes anterior pituitary and posterior pituitary

pituitary gonadotropin *see* gonadotropic hormone

placenta (plah-SEN-tah) interlocking fetal and maternal tissues that serve as organ of molecular exchange between fetal and maternal circulations

placental lactogen (plah-SEN-tal LAK-toh-jen) hormone that is produced by placenta and has effects similar to those of growth hormone and prolactin

plasma (PLAS-muh) liquid portion of blood; component of extracellular fluid

plasma cell cell that differentiates from activated B lymphocytes and secretes antibodies

plasma membrane membrane that forms outer surface of cell and separates cell's contents from extracellular fluid

plasma protein most are albumins, globulins, or fibrinogen

plasmin (PLAZ-min) proteolytic enzyme able to decompose fibrin and thereby to dissolve blood clots

plasminogen (plaz-MIN-oh-jen) inactive precursor of plasmin

plasminogen activator any plasma protein that activates proenzyme plasminogen

plasticity (plas-TISS-ih-tee) ability of neural tissue to change its responsiveness to stimulation because of its past history of activation

platelet (PLATE-let) cell fragment present in blood; plays several roles in blood clotting

platelet aggregation (ag-reh-GAY-shun) positive-feedback process resulting in platelets sticking together

platelet factor (PF) phospholipid exposed in membranes of aggregated platelets; important in activation of several plasma factors in clot formation

pleura (PLUR-ah) thin cellular sheet attached to thoracic cage interior (*parietal pleura*) and, folding back upon itself, is attached to lung surface (*visceral pleura*); forms two enclosed *pleural sacs* in thoracic cage

pluripotent stem cells (plur-ih-POH-tent) single population of bone marrow cells from which all blood cells are descended

polar pertaining to molecule or region of molecule containing polar covalent bonds or ionized groups; part of molecule to which electrons are drawn becomes slightly negative, and region from which electrons are drawn becomes slightly positive; molecule is soluble in water

polar body small cell resulting from unequal distribution of cytoplasm during division of primary or secondary oocyte

polar covalent bond covalent chemical bond in which two electrons are shared unequally between two atoms; atom to which the electrons are drawn becomes slightly negative, while the other atom becomes slightly positive; also called polar bond

polarized (POH-luh-rized) having two electric poles, one negative and one positive

polymer (POL-ih-mer) large molecule formed by linking together of smaller similar subunits

polymorphonuclear granulocyte (pol-ee-morf-oh-NUK-lee-er GRAN-you-loh-site) subclass of leukocytes; consisting of eosinophils, basophils, neutrophils

polypeptide (pol-ee-PEP-tide) polymer consisting of amino acid subunits joined by peptide bonds; also called peptide or protein

polysaccharide (pol-ee-SAK-er-ide) large carbohydrate formed by linking monosaccharide subunits together

polysynaptic reflex (pol-ee-sih-NAP-tik) reflex employing one or more interneurons in its reflex arc

polyunsaturated fatty acid fatty acid that contains more than one double bond

pool the readily available quantity of a substance in the body; often equals amount in extracellular fluid

portal vein vessel through which blood from several abdominal organs flows to the liver

portal vessel any blood vessel that links two capillary networks

positive balance gain of substance exceeds loss, and amount of that substance in body increases; *compare* negative balance

positive feedback aspect of control systems in which an initial disturbance sets off train of events that increases the disturbance even further; *compare* negative feedback

postabsorptive state (post-ab-SORP-tive) period during which nutrients are not being absorbed by gastrointestinal tract and energy must be supplied by body's endogenous stores

posterior toward or at the back

posterior pituitary portion of pituitary from which oxytocin and vasopressin are released

postganglionic (post-gang-glee-ON-ik) autonomic-nervous-system neuron or nerve fiber whose cell body lies in ganglion and whose axon terminals form neuroeffector junctions; conducts impulses away from ganglion toward periphery; *compare* preganglionic

postsynaptic neuron (post-sin-NAP-tik) neuron that conducts information away from a synapse

postsynaptic potential local potential that arises in postsynaptic neuron in response to activation of synapses upon it; *see also* excitatory postsynaptic potential, inhibitory postsynaptic potential

postural reflex reflex that maintains or restores upright, stable posture

potential (or potential difference) voltage difference between two points; *see also* graded potential, action potential

potentiation (poh-ten-she-AY-shun) presence of one agent enhances response to a second such that final response is greater than sum of the two individual responses

precapillary sphincter (SFINK-ter) smooth-muscle ring around capillary where it exits from thoroughfare channel or arteriole

preganglionic autonomic-nervous-system neuron or nerve fiber whose cell body lies in CNS and whose axon terminals lie in a ganglion; conducts action potentials from CNS to ganglion; *compare* postganglionic

presynaptic neuron (pre-sin-NAP-tik) neuron that conducts action potentials toward a synapse

presynaptic synapse relation between two neurons in which axon terminal of one neuron ends on axon terminal of second neuron; action potentials in first neuron affect neuro-

transmitter release from second, thereby altering effectiveness of the synapse that the second neuron makes with a third neuron

primary active transport active transport in which chemical energy is transferred directly from ATP to transporter protein

primary cortical receiving area region of cerebral cortex where specific ascending pathways end; somatosensory, visual, auditory, or taste cortex

primary lymphoid organ organs that supply secondary lymphoid organs with mature lymphocytes; bone marrow and thymus

primary motivated behavior behavior related directly to achieving homeostasis

primary motor cortex *see* motor cortex

primary oocyte (OH-uh-site) female germ cell that undergoes first meiotic division to form secondary oocyte and polar body

primary response gene (PRG) gene influenced by transcription factors generated in response to first messengers

primary spermatocyte (sper-MAT-uh-site) male germ cell derived from spermatogonia; undergoes meiotic division to form two secondary spermatocytes

primordial follicle (FAH-lik-el) *see* ovarian follicle

process long extension from neuron cell body

product molecule formed in enzyme-catalyzed chemical reaction

progestagen (proh-JES-tah-jen) progesterone-like substance

progesterone (proh-JES-ter-own) steroid hormone secreted by corpus luteum and placenta; stimulates uterine gland secretion, inhibits uterine smooth-muscle contraction, and stimulates breast growth

program related sequence of neural activity preliminary to motor act

prohormone peptide precursor from which are cleaved one or more active peptide hormones

prolactin (pro-LAK-tin) peptide hormone secreted by anterior pituitary; stimulates milk secretion by mammary glands

prolactin inhibiting hormone (PIH) dopamine, which serves as a hypophysiotropic hormone to inhibit prolactin secretion by anterior pituitary

prolactin releasing factor (PRF) a putative hypophysiotropic hormone that stimulates prolactin release from anterior pituitary

proliferative phase (pro-LIF-er-ah-tive) stage of menstrual cycle between menstruation and ovulation during which endometrium repairs itself and grows

promoter specific nucleotide sequence at beginning of gene that controls the initiation of gene transcription; determines which of the paired strands of DNA is transcribed into RNA

pro-opiomelanocortin (pro-oh-pee-oh-mel-an-oh-KOR-tin) large protein precursor for ACTH, endorphin, and several other hormones

propagation (prop-ah-GAY-shun) reproduction of self

prostacyclin eicosanoid that inhibits platelet aggregation in blood clotting; also called prostaglandin I₂ (PGI₂)

prostaglandin (pros-tah-GLAN-din) one class of a group of modified unsaturated fatty acids (eicosanoids); function mainly as paracrine or autocrine agents

prostate gland (PROS-tate) large gland encircling urethra in the male; secretes seminal fluid into urethra

protein large polymer consisting of one or more sequences of amino acid subunits joined by peptide bonds

protein binding site *see* binding site

protein C plasma protein that inhibits clotting

protein kinase (KY-nase) any enzyme that phosphorylates other proteins by transferring to them a phosphate group from ATP

protein kinase C enzyme that phosphorylates certain intracellular proteins when activated by diacylglycerol

proteolytic (proh-tee-oh-LIT-ik) breaks down protein

prothrombin (proh-THROM-bin) inactive precursor of thrombin; produced by liver and normally present in plasma

proton (PROH-tahn) positively charged subatomic particle

proximal (PROX-sih-mal) nearer; closer to reference point; *compare* distal

proximal tubule first tubular component of a nephron after Bowman's capsule; comprises *convoluted* and *straight segments*

psychoactive drug drug that affects the mind or behavior

puberty attainment of sexual maturity when conception becomes possible; as commonly used, refers to 3 to 5 years of sexual development that culminates in sexual maturity

pulmonary (PUL-mah-nar-ee) pertaining to lungs

pulmonary circulation circulation through lungs; portion of cardiovascular system between pulmonary trunk, as it leaves the right ventricle, and pulmonary veins, as they enter the left atrium

pulmonary stretch receptor afferent nerve ending lying in airway smooth muscle and activated by lung inflation

pulmonary surfactant *see* surfactant

pulmonary trunk large artery that carries blood from right ventricle of heart to lungs

pulmonary valve valve between right ventricle of heart and pulmonary trunk

pulmonary ventilation *see* total pulmonary ventilation

pulse pressure difference between systolic and diastolic arterial blood pressures

pupil opening in iris of eye through which light passes to reach retina

purine (PURE-ene) double-ring, nitrogen-containing subunit of nucleotide; adenine or guanine

Purkinje fiber (purr-KIN-jee) specialized myocardial cell that constitutes part of conducting system of heart; conveys excitation from bundle branches to ventricular muscle

pyloric sphincter (py-LOR-ik) ring of smooth muscle between stomach and small intestine

pyramidal tract *see* corticospinal pathway

pyrimidine (pi-RIM-ih-deen) single-ring, nitrogen containing subunit of nucleotide; cytosine, thymine, or uracil

pyrogen *see* endogenous pyrogen

pyruvate (PY-roo-vayt) anion formed when pyruvic acid loses a hydrogen ion

pyruvic acid (py-ROO-vik) three-carbon intermediate in glycolysis that, in absence of oxygen, forms lactic acid or, in presence of oxygen, enters Krebs cycle

QRS complex component of electrocardiogram corresponding to ventricular depolarization

R- in chemical formula, signifies remaining portion of molecule

rapid eye movement sleep *see* REM sleep

rate-limiting enzyme enzyme in metabolic pathway most easily saturated with substrate; determines rate of entire metabolic pathway

rate-limiting reaction slowest reaction in metabolic pathway; catalyzed by rate-limiting enzyme

reactant (ree-AK-tent) molecule that enters a chemical reaction; called the substrate in enzyme catalyzed reactions

reaction *see* chemical reaction

reactive hyperemia (hy-per-EE-me-ah) transient increase in blood flow following release of occlusion of blood supply

receptive field (of neuron) part of body that, if stimulated, results in activity in that neuron

receptive relaxation reflex decrease in smooth-muscle tension in walls of hollow organs in response to distension

receptor (in sensory system) specialized peripheral ending of afferent neuron, or separate cell intimately associated with it, that detects changes in some aspect of environment; (in intercellular chemical communication) specific binding site in plasma membrane or interior of target cell with which a chemical messenger combines to exert its effects

receptor activation change in receptor conformation caused by combination of messenger with receptor

receptor potential graded potential that arises in afferent-neuron ending, or a specialized cell intimately associated with it, in response to stimulation

reciprocal innervation inhibition of motor neurons activating muscles whose contraction would oppose an intended movement

recognition binding of antigen to receptor specific for that antigen on lymphocyte surface

recombinant DNA (re-KOM-bih-nent) DNA formed by joining portions of two DNA molecules previously fragmented by a restriction enzyme

recruitment activation of additional cells in response to increased stimulus strength; increasing the number of active motor units in a muscle

rectum short segment of large intestine between sigmoid colon and anus

red muscle muscle having high oxidative capacity and large amount of myoglobin

reflex (REE-flex) biological control system linking stimulus with response and mediated by a reflex arc

reflex arc neural or hormonal components that mediate a reflex; usually includes receptor, afferent pathway, integrating center, efferent pathway, and effector

reflex response final change due to action of stimulus upon reflex arc; also called effector response

refractory period (reh-FRAK-tor-ee) time during which an excitable membrane does not respond to a stimulus that normally causes response; *see also* absolute refractory period, relative refractory period

regulatory site site on protein that interacts with modulator molecule; alters functional-site properties

relative refractory period time during which excitable membrane will produce action potential but only to a stimulus of greater strength than the usual threshold strength

releasing hormone *see* hypophysiotropic hormone

REM sleep (rem) sleep state associated with small, rapid EEG oscillations, complete loss of tone in postural muscles, and dreaming; also called *rapid eye movement sleep, paradoxical sleep*

renal (REE-nal) pertaining to kidneys

renal corpuscle combination of *glomerulus* and *Bowman's capsule*

renal pelvis cavity at base of each kidney; receives urine from collecting duct system and empties it into ureter

renin (REE-nin) peptide hormone secreted by kidneys; acts as an enzyme that catalyzes splitting off of angiotensin I from angiotensinogen in plasma

replicate (REP-lih-kayt) duplicate

repolarize return transmembrane potential to its resting level

residual volume air volume remaining in lungs after maximal expiration

resistance (R) hindrance to movement through a particular substance, tube, or opening

respiration (cellular) oxygen utilization in metabolism of organic molecules; (respiratory system) oxygen and carbon dioxide exchange between organism and external environment

respiratory acidosis increased arterial H^+ concentration due to carbon dioxide retention

respiratory alkalosis decreased arterial H^+ concentration when carbon dioxide elimination from the lungs exceeds its production

respiratory pump effect on venous return of changing intrathoracic and intraabdominal pressures associated with respiration

respiratory quotient (RQ) (KWOH-shunt) ratio of carbon dioxide produced to oxygen consumed during metabolism

respiratory rate number of breaths per minute

respiratory zone portion of airways from beginning of respiratory bronchioles to alveoli; contains alveoli across which gas exchange occurs

resting membrane potential voltage difference between inside and outside of cell in absence of excitatory or inhibitory stimulation; also called resting potential

restriction element *see* MHC protein

retching strong involuntary attempt to vomit but without stomach contents passing through upper esophageal sphincter

reticular formation extensive neuron network extending through brainstem core; receives and integrates information from many afferent pathways and from other CNS regions

retina thin layer of neural tissue lining back of eyeball; contains receptors for vision

retinal (ret-in-AL) form of vitamin A that forms chromophore component of photopigment

retrograde opposite the usual course of events

reversible reaction chemical reaction in which energy release is small enough for reverse reaction to occur readily; *compare* irreversible reaction

Rh factor group of erythrocyte plasma-membrane antigens that may (Rh^+) or may not (Rh^-) be present

rhodopsin (roh-DOP-sin) photopigment in rods

ribonucleic acid (RNA) (ry-boh-noo-KLAY-ik) single-stranded nucleic acid involved in transcription of genetic information and translation of that information into protein structure; contains the sugar ribose; *see also* messenger RNA, ribosomal RNA, and transfer RNA

ribosomal RNA (rRNA) (ry-boh-SOME-al) type of RNA used in ribosome assembly; becomes part of ribosome

ribosome (RY-boh-some) cytoplasmic particle that mediates linking together of amino acids to form proteins; attached to endoplasmic reticulum as bound ribosome, or suspended in cytoplasm as free ribosome

rigor mortis (RIG-or MOR-tiss) stiffness of skeletal muscles after death due to failure of cross bridges to dissociated from actin because of the loss of ATP

RNA polymerase (POL-ih-muh-rase) enzyme that forms RNA by joining together appropriate nucleotides after they have base-paired to DNA

rod one of two receptor types for photic energy; contains the photopigment rhodopsin

saccade (sah-KADE) short, jerking eyeball movement

saliva watery solution of salts and proteins, including mucins and amylase, secreted by salivary glands

saltatory conduction propagation of action potentials along an axon such that the action potentials jump from one node of Ranvier in the myelin sheath to the next node

sarcomere (SAR-kuh-meer) repeating structural unit of myofibril; composed of thick and thin filaments; extends between two adjacent Z lines

sarcoplasmic reticulum (sar-koh-PLAZ-mik reh-TIK-you-lum) endoplasmic reticulum in muscle fiber; site of storage and release of calcium ions

satiety signal (sah-TY-ih-tee) input to food control centers that causes hunger to cease and sets time period before hunger returns

saturated fatty acid fatty acid whose carbon atoms are all linked by single covalent bonds

saturation occupation of all available binding sites by their ligand

scala tympani (SCALE-ah TIM-pah-nee) fluid-filled inner-ear compartment that receives sound waves from basilar membrane and transmits them to round window

scala vestibuli (ves-TIB-you-lee) fluid-filled inner-ear compartment that receives sound waves from oval window and transmits them to basilar membrane and cochlear duct

Schwann cell nonneural cell that forms myelin sheath in peripheral nervous system

scrotum (SKROH-tum) sac that contains testes and epididymides

second messenger intracellular substance that serves as relay from plasma membrane to intracellular biochemical machinery, where it alters some aspect of cell's function

secondary active transport active transport in which energy released during transmembrane movement of one substance from higher to lower concentration is transferred to the simultaneous movement of another substance from lower to higher concentration

secondary lymphoid organ lymph node, spleen, tonsil, or lymphocyte accumulation in gastrointestinal, respiratory, urinary, or reproductive tract; site of stimulation of lymphocyte response

secondary motivated behavior behavior not directed toward achieving homeostasis

secondary peristalsis (per-ih-STAL-sis) esophageal peristaltic waves not immediately preceded by pharyngeal phase of swallow

secondary sexual characteristics external differences between male and female not directly involved in reproduction

secondary spermatocyte (sper-MAT-toh-site) male germ cell derived from primary spermatocyte as a result of the first meiotic division

secretin (SEEK-reh-tin) peptide hormone secreted by upper small intestine; stimulates pancreas to secrete bicarbonate into small intestine

secretion (sih-KREE-shun) elaboration and release of organic molecules, ions, and water by cells in response to specific stimuli

secretory phase (SEEK-rih-tor-ee) stage of menstrual cycle following ovulation during which secretory type of endometrium develops

secretory vesicle membrane-bound vesicle produced by Golgi apparatus; contains protein to be secreted by cell

segmentation (seg-men-TAY-shun) series of stationary rhythmical contractions and relaxations of rings of intestinal smooth muscle; mixes intestinal contents

semen (SEE-men) sperm-containing fluid of male ejaculate

semicircular canal passage in temporal bone; contains sense organs for equilibrium and movement

seminal vesicle one of pair of exocrine glands in males that secrete fluid into vas deferens

seminiferous tubule (sem-ih-NIF-er-ous) tubule in testis in which sperm production occurs; lined with Sertoli cells

semipermeable membrane (sem-ee-PER-me-ah-bul) membrane permeable to some substances but not to others

sensor first component of control system; detects specific environmental changes; also called receptor

sensorimotor cortex (sen-sor-ee-MOH-tor) all areas of cerebral cortex that play a role in skeletal-muscle control

sensory information information that originates in stimulated sensory receptors

sensory pathway a group of neuron chains, each chain consisting of three or more neurons connected end-to-end by synapses; carries action potentials to those parts of the brain responsible for conscious recognition of sensory information

sensory receptor a cell or portion of a cell that contains structures or chemical molecules sensitive to changes in an energy form in the outside world or internal environment; in response to activation by this energy, the sensory receptor initiates action potentials in that cell or an adjacent one

sensory system part of nervous system that receives, conducts, or processes information that leads to perception of a stimulus

sensory unit afferent neuron plus receptors it innervates

serosa (sir-OH-sah) connective-tissue layer surrounding outer surface of stomach and intestines

serotonin (sair-oh-TONE-in) biogenic amine neurotransmitter; paracrine agent in blood platelets and digestive tract; also called 5-hydroxytryptamine, or 5-HT

Sertoli cell (sir-TOH-lee) cell intimately associated with developing germ cells in seminiferous tubule; creates blood-testis barrier, secretes fluid into seminiferous tubule, and mediates hormonal effects on tubule

serum (SEER-um) blood plasma from which fibrinogen and other clotting proteins have been removed as result of clotting

set point *see* operating point

sex chromatin (CHROM-ah-tin) nuclear mass not usually found in cells of males; condensed X chromosome

sex chromosome X or Y chromosome

sex determination genetic basis of individual's sex, XY determining male, and XX, female

sex differentiation development of male or female reproductive organs

sex hormone estrogen, progesterone, testosterone, or related hormones

shock decrease in blood flow severe enough to damage tissues and organs

short-loop negative feedback influence of hypothalamus by an anterior pituitary hormone

sigmoid colon (SIG-moid) S-shaped terminal portion of colon

signal sequence initial portion of newly synthesized protein (if protein is destined for secretion)

signal transduction pathway sequence of mechanisms that relay information from plasma-membrane receptor to cell's response mechanism; *see also* transduction

single-unit smooth muscle smooth muscle that responds to stimulation as single unit because gap junctions join fibers, allowing electrical activity to pass from cell to cell

sinoatrial (SA) node (sy-noh-AY-tree-al) region in right atrium of heart containing specialized cardiac-muscle cells that depolarize spontaneously faster than other cells in the conducting system; determines heart rate

sister chromatid (CHROM-ah-tid) one of two identical DNA threads joined together during meiosis

skeletal muscle striated muscle attached to bones or skin and responsible for skeletal movements and facial expression; controlled by somatic nervous system

skeletal-muscle pump pumping effect of contracting skeletal muscles on blood flow through underlying vessels

skeletomotor fiber *see* extrafusal fiber

sleep *see* REM sleep, slow-wave sleep

sliding-filament mechanism process of muscle contraction in which shortening occurs by thick and thin filaments sliding past each other

slow channel voltage-sensitive calcium channel in myocardial-cell plasma membrane; opens, after a short delay, upon depolarization

slow fiber muscle fiber whose myosin has low ATPase activity

slow-wave sleep sleep state associated with large, slow EEG waves and considerable postural-muscle tone but not dreaming

smooth muscle nonstriated muscle that surrounds hollow organs and tubes; controlled by autonomic nervous system, hormones, and paracrine agents; *see also* single-unit smooth muscle, multiunit smooth muscle

smooth-muscle tone smooth-muscle tension due to low-level cross-bridge activity in absence of external stimuli

sodium inactivation turning off of increased sodium permeability at action-potential peak

soft palate (PAL-et) nonbony region at back of roof of mouth

solute (SOL-yoot) substances dissolved in a liquid

solution liquid (solvent) containing dissolved substances (solutes)

solvent liquid in which substances are dissolved

somatic (soh-MAT-ik) pertaining to the body; related to body's framework or outer walls, including skin, skeletal muscle, tendons, and joints

somatic nervous system component of efferent division of peripheral nervous system; innervates skeletal muscle; *compare* autonomic nervous system

somatic receptor neural receptor in the framework or outer wall of the body that responds to mechanical stimulation of skin or hairs and underlying tissues, rotation or bending of joints, temperature changes, or painful stimuli

somatosensory cortex (suh-mat-uh-SEN-suh-ree) strip of cerebral cortex in parietal lobe in which nerve fibers transmitting somatic sensory information synapse

somatostatin (SS) (suh-mat-uh-STAT-in) hypophysiotropic hormone that inhibits growth hormone secretion by anterior pituitary; possible neurotransmitter; also found in stomach and pancreatic islets

sound wave air disturbance due to variations between regions of high air molecule density (compression) and low density (rarefaction)

spatial summation (SPAY-shul) adding together effects of simultaneous inputs to different places on a neuron to produce potential change greater than that caused by single input

specific ascending pathway chain of synaptically connected neurons in CNS, all activated by sensory units of same type

specific immune defense response that depends upon recognition of specific foreign material for reaction to it

specificity selectivity; ability of binding site to react with only one, or a limited number of, types of molecules

sperm *see* spermatozoon

sperm capacitation (kah-pas-ih-TAY-shun) process by which sperm in female reproductive tract gains ability to fertilize egg

spermatid (SPER-mah-tid) immature sperm

spermatogenesis (sper-mah-toh-JEN-ih-sis) sperm formation

spermatogonium (sper-mah-toh-GOH-nee-um) undifferentiated germ cell that gives rise to primary spermatocyte

spermatozoon (spur-ma-toh-ZOH-in) male gamete; also called sperm

sphincter (SFINK-ter) smooth-muscle ring that surrounds a tube, closing tube as muscle contracts

sphincter of Oddi (OH-dee) smooth-muscle ring surrounding common bile duct at its entrance into duodenum

sphygmomanometer (sfig-moh-mah-NOM-eh-ter) device consisting of inflatable cuff and pressure gauge for measuring arterial blood pressure

spinal nerve one of 86 peripheral nerves (43 pairs) that join spinal cord

spinal reflex reflex whose afferent and efferent components are in spinal nerves; can occur in absence of brain control

spindle fiber (muscle) *see* intrafusal fiber; (mitosis) microtubule that connects chromosome to centriole and centrioles to each other during cell division

spleen largest lymphoid organ; located between stomach and diaphragm

SRY gene gene on the Y chromosome that determines development of testes in genetic male

stable balance net loss of substance from body equals net gain, and amount of substance in body neither increases nor decreases; *compare* positive balance, negative balance

starch moderately branched plant polysaccharide composed of glucose subunits

Starling force factor that determines direction and magnitude of fluid movement across capillary wall

Starling's law of the heart *see* Frank-Starling mechanism

state of consciousness degree of mental alertness, that is, whether awake, drowsy, asleep, and so on

steady state no net change occurs; continual energy input to system is required, however, to prevent net change; *compare* equilibrium

stem cell cell that in adult body divides continuously and forms supply of cells for differentiation

stereocilia (ster-ee-oh-SIL-ee-ah) nonmotil cilia containing a score of actin filaments

steroid (STEER-oid) lipid subclass; molecule consists of four interconnected carbon rings to which polar groups may be attached

stimulus detectable change in internal or external environment

stress environmental change that must be adapted to if health and life are to be maintained; event that elicits increased cortisol secretion

stretch receptor afferent nerve ending that is depolarized by stretching; *see also* muscle-spindle stretch receptor

stretch reflex monosynaptic reflex, mediated by muscle spindle stretch receptor, in which muscle stretch causes contraction of that muscle

striated muscle (STRY-ay-ted) muscle having transverse banding pattern due to repeating sarcomere structure; *see also* skeletal and cardiac muscle

stroke volume blood volume ejected by a ventricle during one heartbeat

strong acid acid that ionizes completely to form hydrogen ions and corresponding anions when dissolved in water; *compare* weak acid

subcortical nucleus neuron cluster deep in brain; includes basal ganglia

submucosa (sub-mu-KOH-sah) connective-tissue layer under mucosa in gastrointestinal tract

submucous plexus (sub-MU-kus PLEX-us) nerve-cell network in submucosa of esophageal, stomach, and intestinal walls

substance P neuropeptide neurotransmitter released by afferent neurons in pain pathway as well as other sites

substrate (SUB-strate) reactant in enzyme-mediated reaction

substrate level phosphorylation (fosfor-ih-LAY-shun) direct transfer of phosphate group from metabolic intermediate to ADP to form ATP

subsynaptic membrane (sub-sih-NAP-tik) part of postsynaptic neuron's plasma membrane under synapse

subthreshold potential (sub-THRESH-old) depolarization less than threshold potential

subthreshold stimulus stimulus capable of depolarizing membrane but not by enough to reach threshold

sucrose (SOO-krose) disaccharide composed of glucose and fructose; also called table sugar

sulfate SO_4^{2-}

sulfhydryl group (sulf-HY-drul) -SH

sulfuric acid (sulf-YOR-ik) acid generated during catabolism of sulfur-containing compounds; dissociates to form sulfate and hydrogen ions

summation (sum-MAY-shun) increase in muscle tension or shortening in response to rapid, repetitive stimulation relative to single twitch

superior vena cava (VEE-nah KAY-vah) large vein that carries blood from upper half of body to right atrium of heart

supersensitivity increased responsiveness of target cell to given messenger due to up-regulation

suprathreshold stimulus (soo-prah-THRESH-old) any agent capable of depolarizing membrane more than its threshold potential, that is, closer to zero

surface tension attractive forces between water molecules at an air-water interface resulting in net force that acts to reduce surface area

surfactant (sir-FAK tent) detergent-like phospholipid-protein mixture produced by pulmonary type II alveolar cells; reduces surface tension of fluid film lining alveoli

sympathetic division portion of autonomic nervous system whose preganglionic fibers leave CNS at thoracic and lumbar portions of spinal cord; *compare* parasympathetic division

sympathetic trunk one of paired chains of interconnected sympathetic ganglia that lie on either side of vertebral column

sympathomimetic (sym-path-oh-mih-MET-ik) produces effects similar to those of sympathetic nervous system

synapse (SIN-apse) anatomically specialized junction between two neurons where electrical activity in one neuron influences excitability of second; *see also* chemical synapse, electric synapse, excitatory synapse, inhibitory synapse

synaptic cleft narrow extracellular space separating pre- and postsynaptic neurons at chemical synapse

synaptic potential *see* postsynaptic potential

synergistic muscle (sin-er-JIS-tik) muscle that exerts force to aid intended motion

systemic circulation (sis-TEM-ik) circulation from left ventricle through all organs except lungs and back to heart

systole (SIS-toh-lee) period of ventricular contraction

systolic pressure (SP) (sis-TAHL-ik) maximum arterial blood pressure during cardiac cycle

T cell lymphocyte derived from precursor that differentiated in thymus; *see also* cytotoxic T cell, helper T cell

T lymphocyte *see* T cell

T tubule *see* transverse tubule

T wave component of electrocardiogram corresponding to ventricular repolarization

target cell cell influenced by a certain hormone

taste bud sense organ that contains chemoreceptors for taste

tectorial membrane (tek-TOR-ee-al) structure in organ of Corti in contract with receptor-cell hairs

teleology (teel-ee-OL-oh-gee) explanation of events in terms of ultimate purpose served by them

template (TEM-plit) pattern

temporal lobe region of cerebral cortex where primary auditory cortex and Wernicke's speech center are located

temporal summation membrane potential produced as two or more inputs, occurring at different times, are added together; potential change is greater than that caused by single input

tendon (TEN-don) collagen fiber bundle that connects skeletal muscle to bone and transmits muscle contractile force to the bone

tension force; *see also* muscle tension

termination code word three-nucleotide sequence in mRNA that signifies end of protein coding sequence

testis (TES-tiss) (pl. testes) gonad in male

testosterone (test-TOS-ter-own) steroid hormone produced in interstitial cells of testes; major male sex hormone; essential for spermatogenesis and maintains growth and development of reproductive organs and secondary sexual characteristics of male

tetanus (TET-ah-nus) maintained mechanical response of muscle to high-frequency stimulation; also the disease lockjaw

tetrad (TET-rad) grouping of two homologous chromosomes, each with its sister chromatid, during meiosis

thalamus (THAL-ah-mus) subdivision of diencephalon; integrating center for sensory input on its way to cerebral cortex; also contains motor nuclei

theca (THEE-kah) cell layer that surrounds ovarian-follicle granulosa cells

thermogenesis (ther-moh-JEN-ih-sis) heat generation

thermoneutral zone temperature range over which changes in skin blood flow alone can regulate body temperature

thermoreceptor sensory receptor for temperature and temperature changes, particularly in low (cold receptor) or high (warm receptor) range

theta wave (THAY-tah) slow 4- to 8-Hz oscillation on electroencephalogram; associated with sleep

thick filament myosin filament in muscle cell

thin filament actin filament in muscle cell

thoracic cavity (thor-ASS-ik) chest cavity

thoracic wall chest wall

thorax (THOR-aks) closed body cavity between neck and diaphragm; contains lungs, heart, thymus, large vessels, and esophagus; also called the chest

threshold (THRESH-old) (or threshold potential) membrane potential to which excitable membrane must be depolarized to initiate an action potential

threshold stimulus stimulus capable of depolarizing membrane just to threshold

thrombin (THROM-bin) enzyme that catalyzes conversion of fibrinogen to fibrin; has multiple other actions in blood clotting

thrombolytic system *see* fibrinolytic system

thrombomodulin an endothelial receptor to which thrombin can bind, thereby eliminating thrombin's clot-producing effects and causing it to bind protein C

thrombosis (throm-BOH-sis) clot formation in body

thromboxane (throm-BOX-ane) a type of eicosanoid formed by the cyclooxygenase pathway; closely related to prostaglandins

thromboxane A$_2$ thromboxane that, among other effects, stimulates platelet aggregation in blood clotting

thrombus (THROM-bus) blood clot

thymine (T) (THY-meen) pyrimidine base in DNA but not RNA

thymus (THY-mus) lymphoid organ in upper part of chest; site of T-lymphocyte differentiation

thyroglobulin (thy-roh-GLOB-you-lin) large protein precursor of thyroid hormones in colloid of follicles in thyroid gland; storage form of thyroid hormones

thyroid gland endocrine gland in neck; secretes thyroid hormones and calcitonin

thyroid hormones (TH) collective term for amine hormones released from thyroid gland, that is, thyroxine (T$_4$) and triiodothyronine (T$_3$)

thyroid-stimulating hormone (TSH) glycoprotein hormone secreted by anterior pituitary; induces secretion of thyroid hormone; also called thyrotropin

thyrotropin (thy-roh-TROH-pin) *see* thyroid-stimulating hormone

thyrotropin releasing hormone (TRH) hypophysiotropic hormone that stimulates thyrotropin and prolactin secretion by anterior pituitary

thyroxine (T$_4$) (thy-ROCKS-in) tetraiodothyronine; iodine-containing amine hormone secreted by thyroid gland

tidal volume air volume entering or leaving lungs with single breath during any state of respiratory activity

tight junction cell junction in which extracellular surfaces of the plasma membrane of two adjacent cells are joined together; extends around epithelial cell and restricts molecule diffusion through space between cells

tissue aggregate of single type of specialized cell; also denotes general cellular fabric of a given organ

tissue factor extravascular plasma-membrane protein involved in initiation of clotting

tissue plasminogen activator (t-PA) plasma protein produced by endothelial cells; after binding to fibrinogen, activates the proenzyme plasminogen

tone maintained functional activity; *see also* muscle tone

tonic (TAH-nik) continuous activity; *compare* phasic

tonsil one of several small lymphoid organs in pharynx

total blood carbon dioxide sum total of dissolved carbon dioxide, bicarbonate, and carbamino-CO$_2$

total dead space alveolar and anatomic dead space; *see also* alveolar dead space, anatomic dead space

total energy expenditure sum of external work done plus heat produced plus energy stored by body

total peripheral resistance (TPR) total resistance to flow in systemic blood vessels from beginning of aorta to end of venae cavae

toxin (TOK-sin) poison

trace element mineral present in body in extremely small quantities

trachea (TRAY-key-ah) single airway connecting larynx with bronchi; windpipe

tract large, myelinated nerve-fiber bundle in CNS

transamination (trans-am-in-NAY-shun) reaction in which an amino acid amino group (-NH$_2$) is transferred to a keto acid, the keto acid thus becoming an amino acid

transcellular pathway crossing an epithelium by movement into an epithelial cell, diffusion through the cytosol of that cell, and exit across the opposite membrane

transcription formation of mRNA containing, in linear sequence of its nucleotides, the genetic information of a specific gene; first stage of protein synthesis

transcription factor one of a class of proteins that act as gene switches, regulating the transcription of a particular gene by activating or repressing the initiation process

transduction process by which stimulus energy is transformed into a response

transepithelial transport *see* epithelial transport

transfer RNA (tRNA) type of RNA; different tRNAs combine with different amino acids and with codon on mRNA specific for that amino acid, thus arranging amino acids in sequence to form specified protein

transferrin (trans-FAIR-in) iron-binding protein; carries iron in plasma

translation during protein synthesis, assembly of amino acids in correct order according to genetic instructions in mRNA; occurs on ribosomes

transmural pressure pressure difference exerted on the two sides of a wall

transport maximum (T_m) upper limit to amount of material that carrier-mediated transport can move across the renal tubule

transporter integral membrane protein that mediates passage of molecule through membrane; also called carrier

transpulmonary pressure difference between alveolar and intrapleural pressures; force that holds lungs open

transverse tubule (T tubule) tubule extending from striated-muscle plasma membrane into the fiber, passing between opposed sarcoplasmic reticulum segments; conducts muscle action potential into muscle fiber

triacylglycerol (try-ay-seel-GLISS-er-ol) subclass of lipids composed of glycerol and three fatty acids; also called fat, neutral fat, or triglyceride

tricarboxylic acid cycle *see* Krebs cycle

tricuspid valve (try-CUS-pid) valve between right atrium and right ventricle of heart

triglyceride *see* triacylglycerol

triiodothyronine (T_3) (try-eye-oh-doh-THY-roh-neen) iodine-containing amine hormone secreted by thyroid gland

triplet code three-base sequence in DNA and RNA that specifies particular amino acid

trophoblast (TROH-foh-blast) outer layer of blastocyst; gives rise to placental tissue

tropic (TROH-pik) growth promoting

tropic hormone hormone that stimulates the secretion of another hormone, and, often, growth of hormone-secreting gland

tropomyosin (troh-poh-MY-oh-sin) regulatory protein capable of re-

versibly covering binding sites on actin; associated with muscle thin filaments

troponin (troh-POH-nin) regulatory protein bound to actin and tropomyosin of striated-muscle thin filaments; site of calcium binding that initiates contractile activity

trypsin (TRIP-sin) enzyme secreted into small intestine by exocrine pancreas as precursor trypsinogen; breaks certain peptide bonds in proteins and polypeptides

trypsinogen (trip-SIN-oh-jen) inactive precursor of trypsin; secreted by exocrine pancreas

tryptophan (TRIP-toh-fan) essential amino acid; serotonin precursor

tubular reabsorption transfer of materials from kidney tubule lumen to peritubular capillaries

tubular secretion transfer of materials from peritubular capillaries to kidney tubule lumen

tumor necrosis factor (TNF) (neh-KROH-sis) cytokine that kills cells outright, stimulates inflammation, and mediates certain of the acute-phase responses

twitch mechanical response of muscle to single action potential

tympanic membrane (tim-PAN-ik) membrane stretched across end of ear canal; also called eardrum

type I alveolar cell a flat epithelial cell that with others forms a continuous layer lining the air-facing surface of the pulmonary alveoli

type II alveolar cell pulmonary cell that produces surfactant

tyrosine (TY-roh-seen) amino acid; precursor of catecholamines and thyroid hormones

tyrosine kinase protein kinase that phosphorylates tyrosine portion of proteins; may be part of plasma membrane receptor

ultrafiltrate (ul-tra-FIL-trate) essentially protein-free fluid formed from plasma as it is forced through capillary walls by pressure gradient

umbilical vessel (um-BIL-ih-kul) artery or vein transporting blood between fetus and placenta

unsaturated fatty acid fatty acid containing one or more double bonds

upper esophageal sphincter (ih-sof-ih-JEE-al SFINK-ter) skeletal-mus-

cle ring surrounding esophagus just below pharynx that, when contracted, closes entrance to esophagus

up-regulation increase in number of target-cell receptors for given messenger in response to chronic low extracellular concentration of that messenger; *see also* supersensitivity; *compare* down-regulation

uracil (U) (YOR-ah-sil) pyrimidine base; present in RNA but not DNA

urea (you-REE-ah) major nitrogenous waste product of protein breakdown and amino acid catabolism

ureter (YUR-ih-ter) tube that connects kidney to bladder

urethra (you-REE-thrah) tube that connects bladder to outside of body

uric acid (YUR-ik) waste product derived from nucleic acid catabolism

urinary bladder *see* bladder

uterine tube (YOU-ter-in) one of two tubes that carries egg from ovary to uterus; also called fallopian tube, oviduct

uterus (YOO-ter-us) hollow organ in pelvic region of females; houses fetus during pregnancy; also called womb

vagina (vah-JY-nah) canal leading from uterus to outside of body; also called birth canal

vagus nerve (VAY-gus) cranial nerve X; major parasympathetic nerve

valence (VAY-lence) number of electrons that an ion can accept or donate in a chemical reaction

van der Waals forces (walls) weak attractive forces between nonpolar regions of molecules

varicosity (vair-ih-KOS-ih-tee) swollen region of axon; contains neurotransmitter-filled vesicles; analogous to presynaptic ending

vas deferens (vas DEF-er-enz) one of paired male reproductive ducts that connect epididymis of testis to urethra; also called ductus deferens

vasectomy (vah-SEK-tuh-mee) cutting and typing off of both vas deferens, which results in sterilization of male without loss of testosterone

vasoconstriction (vays-oh-kon-STRIK-shun) decrease in blood-vessel diameter due to vascular smooth-muscle contraction

vasodilation (vays-oh-dy-LAY-shun) increase in blood-vessel diameter due to vascular smooth-muscle relaxation

vasopressin (vas-oh-PRES-sin) peptide hormone synthesized in hypothalamus and released from posterior pituitary; increases water permeability of kidneys' collecting ducts and causes vasoconstriction; also called *antidiuretic hormone* (*ADH*)

vein any vessel that returns blood to heart; but *see also* portal vein

vena cava (VEE-nah KAY-vah) (pl. venae cavae) one of two large veins that returns systemic blood to heart; *see also* superior vena cava, inferior vena cava

venous return (VR) blood volume flowing *to* heart per unit time

ventilation air exchange between atmosphere and alveoli; alveolar air flow

ventral (VEN-tral) toward or at the front of body

ventral root one of two groups of efferent fibers that leave ventral side of spinal cord

ventricle (VEN-trih-kul) cavity, as in cerebral ventricle or heart ventricle

ventricular function curve relation of the increase in stroke volume as end-diastolic volume increases, all other factors being equal

venule (VEEN-ule) small vessel that carries blood from capillary network to vein

very-low-density lipoprotein (VLDL) (lip-oh-PROH-teen) lipid-protein aggregate having high proportion of fat

vesicle (VES-ih-kul) small, membrane-bound organelle within cells

vestibular apparatus *see* vestibular system

vestibular receptor hair cell in semicircular canal, utricle, or saccule

vestibular system sense organ in temporal bone of skull; consists of three *semicircular canals,* a *utricle,* and a *saccule;* also called vestibular apparatus, sense organ of balance

villi (VIL-eye) finger-like projections from highly folded surface of small intestine, covered with single-layered epithelium

virus nuclei acid core surrounded by protein coat; lacks enzyme machinery for energy production and ribosomes for protein synthesis; thus cannot survive or reproduce except inside other cells whose biochemical apparatus it uses

viscera (VISS-er-ah) organs in thoracic and abdominal cavities

viscoelastic (viss-koh-ee-LAS-tik) having viscous and elastic properties

viscosity (viss-KOS-ih-tee) measure of friction between adjacent layers of a flowing liquid; property of fluid that makes it resist flow

visual field part of world being viewed at a given time

vital capacity maximal amount of air that can be expired, regardless of time required, following maximal inspiration

vitalism (VY-tal-ism) view that explanation of life processes requires a "life force" rather than physicochemical processes alone

vitamin organic molecule that is required in trace amounts for normal health and growth but is not manufactured in the body and must be supplied by diet; classified as *water-soluble* (vitamins C and the B complex) and *fat-soluble* (vitamins A, D, E, and K)

vocal cord one of two elastic-tissue bands stretched across laryngeal opening and caused to vibrate by air movement past them, producing sounds

volt (V) unit of measurement of electric potential between two points

voltage measure of potential of separated electric charges to do work; measure of electric force between two points

voltage-sensitive channel cell-membrane ion channel opened or closed by changes in membrane potential

vomiting center neurons in brainstem medulla oblongata that coordinate vomiting reflex

von Willebrand factor (vWF) (von-VILL-ih-brand) plasma protein secreted by endothelial cells; facilitates adherence of platelets to damaged vessel wall

vulva (VUL-vah) female external genitalia; mons pubis, labia majora and minora, clitoris, vestibule of vagina, and vestibular glands

waste product product from a metabolic reaction or series of reactions that serves no function

wavelength distance between two successive wave peaks in oscillating medium

weak acid acid whose molecules do not completely ionize to form hydrogen ions when dissolved in water; *compare* strong acid

white matter portion of CNS that appears white in unstained specimens and contains primarily myelinated nerve fibers

white muscle muscle lacking appreciable amounts of myoglobin

withdrawal reflex bending of those joints that withdraw an injured part away from a painful stimulus

Wolffian duct (WOLF-ee-an) part of embryonic duct system that, in male, remains and develops into reproductive-system ducts, but in female, degenerates

work measure of energy required to produce physical displacement of matter; *see also* external work, internal work

working memory short-term memory storage process serving as initial depository of information

X chromosome *see* sex chromosome

Y chromosome *see* sex chromosome

Z line structure running across myofibril at each end of striated-muscle sarcomere; anchors one end of thin filaments

zona pellucida (ZOH-nah peh-LOO-sih-dah) thick, clear layer separating egg from surrounding granulosa cells

zygote (ZY-goat) a newly fertilized egg

zymogen (ZY-moh-jen) enzyme precursor requiring some change to become active

APPENDIX C
English and Metric Units

	English	**Metric**
Length	1 foot = 0.305 meters 1 inch = 2.54 centimeters	1 meter = 39.37 inches 1 centimeter (cm) = 1/100 meter 1 millimeter (mm) = 1/1000 meter 1 micrometer (μ m) = 1/1000 millimeter 1 nanometer (nm) = 1/1000 micrometer
°Mass	1 pound = 433.59 grams 1 ounce = 28.3 grams	1 kilogram (kg) = 1000 grams = 2.2 pounds 1 gram (g) = 0.035 ounce 1 milligram (mg) = 1/1000 gram 1 microgram (μ g) = 1/1000 milligram 1 nanogram (ng) = 1/1000 microgram 1 picogram (pg) = 1/1000 nanogram
Volume	1 gallon = 3.785 liters 1 quart = 0.946 liters 1 pint = 0.473 liters 1 fluid ounce = 0.030 liters 1 measuring cup = 0.24 liters	1 liter = 1000 cubic centimeter = 0.264 gallons 1 liter = 1.057 quarts 1 deciliter (dl) = 1/10 liter 1 milliliter (ml) = 1/1000 liter 1 microliter (μ l) = 1/1000 milliliter

°A pound is actually a unit of force, not mass. The correct unit of mass in the English system is the slug. When we write 1 kg = 2.2 pounds, this means that one kilogram of *mass* will have a *weight* under standard conditions of gravity at the earth's surface of 2.2 pounds *force*.

APPENDIX D
Electrophysiology Equations

I. The **Nernst equation** describes the equilibrium potential for any ion species, that is, the electric potential necessary to balance a given ionic concentration gradient across a membrane so that the net passive flux of the ion is zero. The Nernst equation is

$$E = \frac{RT}{zF} \ln \frac{Co}{Ci}$$

where E = equilibrium potential for the particular ion in question

Ci = intracellular concentration of the ion

Co = extracellular concentration of the ion

z = valence of the ion (+1 for sodium and potassium, +2 for calcium, −1 for chloride)

R = gas constant [8314.9 J/(kg · mol · K)]

T = absolute temperature (temperature measured on the Kelvin scale: degrees centigrade +273)

F = Faraday (the quantity of electricity contained in 1 mol of electrons: 96,484.6 C/mol of charge)

$\ln$ = logarithm taken to the base e

II. A membrane potential depends on the intracellular and extracellular concentrations of potassium, sodium, and chloride (and other ions if they are in sufficient concentrations) and on the relative permeabilities of the membrane to these ions. The **Goldman equation** is used to calculate the value of the membrane potential when the potential is determined by more than one ion species. The Goldman equation is

$$Vm = \frac{RT}{F} \ln \frac{P_K \times Ko + P_{Na} \times Nao + P_{Cl} \times Cli}{P_K \times Ki + P_{Na} \times Nai + P_{Cl} \times Clo}$$

where Vm = membrane potential

R = gas constant [8314.9 J/(kg·mol·K)]

T = absolute temperature (temperature measured on the Kelvin scale: degrees centigrade + 273)

F = Faraday (the quantity of electricity contained in 1 mol of electrons: 96,484.6 C/mol of charge)

$\ln$ = logarithm taken to the base e

P_K, P_{Na}, and P_{Cl} = membrane permeabilities for potassium, sodium, and chloride, respectively

Ko, Nao, and Clo = extracellular concentrations of potassium, sodium, and chloride, respectively

Ki, Nai, and Cli = intracellular concentrations of potassium, sodium, and chloride, respectively

APPENDIX E
Outline of Exercise Physiology

APPENDIX F
Clinical Terms in Text

APPENDIX G
Suggested Reading

Like all scientists, physiologists report the results of their experiments in scientific journals. Approximately 1300 such journals in the life sciences publish more than 350,000 papers each year. Every physiologist must be familiar with the ever-increasing number of articles in his or her own specific field of research.

Another type of scientific writing is the review article, a summary and synthesis of the relevant research reports on a specific subject. Such reviews, which are published in many different journals, are extremely useful. Also, the American Physiological Society has a massive program of publishing comprehensive reviews in almost all fields of physiology. These reviews are gathered into a series of ongoing volumes known collectively as the "Handbook of Physiology." Another important series is the *Annual Review of Physiology*, which publishes yearly reviews of the most recent research articles in specific fields and provides the most rapid means of keeping abreast of developments in physiology.

Another type of review that is of value to the nonexpert is the *Scientific American* article. Written in a manner usually intelligible to the layperson, these articles, in addition to reviewing a topic, often present experimental data so that the reader can obtain some insight into how the experiments were done and the data interpreted. Many articles published in *Scientific American* can be obtained individually as inexpensive offprints from W.H. Freeman and Company, San Francisco, CA, 94104, which also publishes collections of *Scientific American* articles in a specific area, for example, psychobiology. Two other magazines that frequently publish excellent reviews of physiology for the nonexpert are *New England Journal of Medicine* and *Hospital Practice*. The New York Times (particularly in its Tuesday "Science" section) provides excellent coverage of physiology-related health news. Finally, in the realm of nutrition, fitness, and stress management, The University of California at Berkeley Wellness Letter, published by that university's School of Public Health, is superb; the address is P.O. Box 420148, Palm Coast, FL 32142.

Another level of scientific writing is the monograph, an extended review of a subject area broader than the review articles described above. Monographs are usually in book form and generally do not attempt to cover all the relevant literature on the topic, but to analyze and synthesize the most important data and interpretations. At the last level is the textbook, which deals with a much larger field than the monograph or review article. Its depth of coverage of any topic is accordingly much less complete and is determined by its intended audience.

We have restricted our suggested readings to textbooks and monographs because they are usually available in college and university libraries. Their bibliographies will serve as a further entry into the scientific literature. We list first a group of books that cover large areas of physiology.

BOOKS COVERING WIDE AREAS OF PHYSIOLOGY

Cell Physiology

Alberts, Bruce, et al.: "Molecular Biology of the Cell," 3rd ed, Garland, New York, 1994.

Lodish, Harvey, et al., "Molecular Cell Biology," 3rd edn., Scientific American Books, New York, 1995.

Wolfe, Stephen L.: "Molecular and Cellular Biology," Wadsworth, Belmont, CA, 1993.

Organ-System Physiology

Berne, R.M., and M.N. Levy: "Physiology," 3rd edn., Mosby, St. Louis, 1992.

Ganong, W.F.: "Review of Medical Physiology," 17th edn., Lange, Los Altos, CA, 1995.

Guyton, A.C. and J.F. Hall: "Textbook of Medical Physiology," 9th edn., Saunders, Philadelphia, 1995.

Johnson, L.R. (editor): "Essential Medical Physiology," Raven, New York, 1992.

Rhoades, R.A., and G.A. Tanner (eds.): "Medical Physiology," Little Brown, Boston, 1995.

West, John B. (ed.): "Best and Taylor's Physiological Basis of Medical Practice," 12th edn., Williams & Wilkins, Baltimore, 1991.

Anatomy

Basmajian, J.V.: "Grant's Method of Anatomy," 11th edn., Williams & Wilkins, Baltimore, 1988.

Carlson, B.M.: "Patten's Foundations of Embryology," 5th edn., McGraw-Hill, New York, 1988.

Clemente, C.D.: "Gray's Anatomy of the Human Body," 30th American edn., Lea & Febiger, Philadelphia, 1984.

Fawcett, D.W.: "Bloom and Fawcett, A Textbook of Histology," 12th edn., Chapman and Hall, New York, 1994.

Rosse, C.: "Hollinshead's Textbook of Anatomy," 5th edn., Lippincott-Raven, Philadelphia, 1997.

Woodburne, R.T.: "Essentials of Human Anatomy," 8th edn., Oxford University Press, New York, 1988.

SUGGESTIONS FOR INDIVIDUAL CHAPTERS

Chapter 1

Bernard, C.: "An Introduction to the Study of Experimental Medicine," Dover, New York, 1957 (paperback).

Brooks, C. McC., and P.F. Cranefield (eds.): "The Historical Development of Physiological Thought," Hafner, New York, 1959.

Butterfield, H.: "The Origins of Modern Science," Macmillan, New York, 1961 (paperback).

Case, R.M., and J.M. Waterhouse: "Human Physiology: Age, Stress, and the Environment," Oxford University Press, Oxford, 1994.

"The Excitement and Fascination of Science: A Collection of Autobiographical and Philosophical Essays," Annual Reviews, Palo Alto, Calif., 1966.

Fulton, J.F., and L.C. Wilson (eds.): "Selected Readings in the History of Physiology," 2d edn., Charles C Thomas, Springfield, IL, 1966.

Harvey, W.: "On the Motion of the Heart and Blood in Animals," Gateway, Henry Regnery, Chicago, 1962 (paperback).

Leake, Chauncey: "Some Founders of Physiology," American Physiological Society, Washington, D.C., 1961.

Chapter 2

Creighton, Thomas E.: "Proteins: Structures and Molecular Properties," 2d edn., Freeman, New York, 1993.

Devlin, Thomas M.: "Textbook of Biochemistry With Clinical Correlations," 3d edn., Wiley-Liss, New York, 1992.

Stryer, Lubert: "Biochemistry," 4th edn., Freeman, San Francisco, 1995.

Chapter 3

Bittar, E. Edward, and Neville Bittar (eds.): "Principles of Medical Biology—Cellular Organelles," JAI Press, Greenwich, CT, 1995.

Fawcett, D.W.: "Bloom and Fawcett, A Textbook of Histology," 12th edn., Chapman and Hall, New York, 1994.

Telford, Ira Rockwood, and Charles F. Bridgman: "Introduction to Functional Histology," 2d edn., Harper Collins, New York, 1995.

Weiss, Leon, et al. (eds.): "Cell and Tissue Biology: A Textbook of Histology," 6th edn., Urban & Schwarzenberg, Baltimore, 1988.

Chapter 4

Bittar, E. Edward and Neville Bittar (eds.): "Principles of Medical Biology—Cell Chemistry and Physiology," JAI Press, Greenwich, CT, 1995.

Creighton, Thomas E.: "Proteins: Structures and Molecular Properties," 2d edn., Freeman, New York, 1993.

Devlin, Thomas M.: "Textbook of Biochemistry With Clinical Correlations," 3d edn., Wiley-Liss, New York, 1992.

Hunt, Sara M., and James L. Groff: "Advanced Nutrition and Human Metabolism," West, St. Paul, MN, 1990.

Kinney, John M., and Hugh N. Tucker: "Energy Metabolism: Tissue Determinants and Cellular Corollaries," Raven, New York, 1992.

Voet, Donald, and Judith G. Voet: "Biochemistry," Wiley, 1990.

Chapter 5

Alberts, Bruce, et al. "Molecular Biology of the Cell," 3d edn., Garland, New York, 1994.

Gelehrter, Thomas D., and Francis S. Collins: "Principles of Medical Genetics," Williams & Wilkins, Baltimore, Maryland, 1990.

Hutchisson, Christopher, and David M. Glover (eds.): "Cell Cycle Control," IRL Press at Oxford University Press, 1995.

Lodish, Harvey, et al.: "Molecular Cell Biology," 3d edn., Scientific American Books, New York, 1995.

Mange, Arthur P., and Elaine Johnson Mange: "Genetics: Human Aspects," Sinauer, Sunderland, MA, 1990.

Old, R.W., and S.B. Primrose: "Principles of Gene Manipulation: An Introduction to Genetic Engineering," 5th edn., Blackwell Scientific, Boston, 1994.

Ruddon, Raymond W.: "Cancer Biology," 3d edn., Oxford, New York, 1995.

Steen, R. Grant: "A Conspiracy of Cells, The Basic Science of Cancer," Plenum, New York, 1993.

Steer, Clifford J., and John Hanover (eds.): "Intracellular Trafficking of Proteins," Cambridge University Press, 1991.

Stachan, Tom, and Andrew P. Read: "Human Molecular Genetics," BIOS Scientific Publishers, New York:

Wiley-Liss, 1996.

Wolffe, Alan: "Chromatin: Structure and Function," 2d edn., Academic Press, San Diego, 1995.

Watson, James D., et al.: "Recombinant DNA," 2d edn., Scientific American Books, New York, 1992.

Wolfe, Stephen L.: "Molecular and Cellular Biology," Wadsworth, Belmont, CA, 1993.

Chapter 6

Alberts, Bruce, et al.: "Molecular Biology of the Cell," 3d edn., Garland, New York, 1994.

Byrne, John H., and Stanley G. Schultz: "An Introduction to Membrane Transport and Bioelectricity," 2d edn., Raven, New York, 1994.

Lodish, Harvey, et al.: "Molecular Cell Biology," 3d edn., Scientific American Books, New York, 1995.

Petty, Howard R.: "Molecular Biology of Membranes, Structure and Function," Plenum, New York, 1993.

Schultz, Stanley G. (ed.): "Molecular Biology of Membrane Transport Disorders," Plenum, New York, 1996.

Yeagle, Philip, "The Membranes of Cells," 2d edn., Academic Press, San Diego, 1993.

Chapter 7

Arendt, J., D.S. Minors, and J.M. Waterhouse: "Biological Rhythms in Clinical Practice," Wright, London, 1990.

Bernard, C.: "An Introduction to the Study of Experimental Medicine," Dover, New York, 1957 (paperback).

Cannon, W.B.: "The Wisdom of the Body," Norton, New York, 1939 (paperback).

Dworkin, B.R.: "Learning and Physiological Regulation," University of Chicago Press, Chicago, 1994.

Goodman, H.M.: "Basic Medical Endocrinology," 2nd edn., Raven, New York, 1993.

Griffin, J.E., and S.R. Ojeda: "Textbook of Endocrine Physiology," 2nd edn., Oxford, New York, 1992.

Langley, L.L. (ed.): "Homeostasis: Origin of the Concept," Dowden, Hutchinson, and Ross, Stroudsburg, PA, 1973.

Mrosovksy, N.: "Rheostasis: The Physiology of Change," Oxford, New York, 1991.

Weintraub, B.D. (ed.): "Molecular Endocrinology: Basic Concepts and Clinical Correlations," Raven, New York, 1995.

Wilson, J.D., and D.W. Foster (eds.) "Williams Textbook of Endocrinology," 8th edn., Saunders, Philadelphia, 1991.

Chapter 8

Aidley, D.J.: "The Physiology of Excitable Cells," 3rd edn., Cambridge University Press, New York, 1989.

Byrne, J.H.: "An Introduction to Membrane Transport and Bioelectricity: Foundations of General Physiology and Electricochemical Signalling: Raven, New York, 1994.

Carpenter, M.B.: "Core Text of Neuroanatomy," 4th edn., Williams and Wilkins, Baltimore, 1991.

Cooper, J.R., F.E. Bloom, and R.H. Roth: "The Biochemical Basis of Neuropharmacology," 7th edn., Oxford, New York, 1994.

Hall, Z.W.: "An Introduction to Molecular Neurobiology, Sinauer, Sunderland, MA, 1992.

Kandel, E.R., J.H. Schwartz, and T.M. Jessell: "Principles of Neural Science," 3rd edn., Elsevier/North-Holland, New York, 1991.

Kelner, K.L., and D.E. Koshland (eds.): "Molecules to Models: Advances in Neuroscience," American Association for the Advancement of Science, Washington, DC, 1989.

Levitan, I.B., and L.K. Kaczmarek: "The Neuron: Cell and Molecular Biology," 2nd edn., Oxford, New York, 1997.

Martin, J.H.: "Neuroanatomy: Text and Atlas," 2nd edn., Prentice Hall, London, 1994.

Nauta, W.J.H., and M. Feirtag: "Fundamental Neuroanatomy," Freeman, New York, 1986.

Patton, H.D., et al. (eds.): "Textbook of Physiology: Volume 1 Excitable Cells and Neurophysiology," 21st ed, Saunders, Philadelphia, 1989.

Shepherd, G.D.: "Neurobiology," 3rd edn., Oxford, New York, 1994.

Siegel, G.J., et al.: "Basic Neurochemistry," 5th edn., Raven, New York, 1994.

Chapter 9

Aidley, D.J.: "The Physiology of

Excitable Cells," 3rd edn., Cambridge University Press, New York, 1989.

Gregory, R.L.: "Eye and Brain: The Psychology of Seeing," 4th ed, Oxford, Oxford, 1994.

Kandel, E.R., J.H. Schwartz, and T.M. Jessell: "Principles of Neural Science," 3rd edn., Elsevier/North-Holland, New York, 1991.

Patton, H.D., et al. (eds.): "Textbook of Physiology: Volume 1 Excitable cells and Neurophysiology," 21st ed, Saunders, Philadelphia, 1989.

Pickles, J.O.: "An Introduction to the Physiology of Hearing," 2nd ed, Academic Press, San Diego, 1988.

Spillman, L., and J.S. Warner (eds.): "Visual Perception: The Neurophysiological Foundations," Academic Press, San Diego, 1990.

Chapter 10

Becker, K.L. (ed.): "Principles and Practice of Endocrinology and Metabolism," 5th edn., Lippincott, Philadelphia, 1995.

Bolander, F.F.: "Molecular Endocrinology," 2nd edn., Academic Press, San Diego, 1994.

DeGroot, L.J.: "Endocrinology," 3rd ed., Saunders, Philadelphia, 1995.

Goodman, H.M.: "Basic Medical Endocrinology," 2nd edn., Raven, New York, 1993.

Greenspan, F.S., and J.D. Baxter (eds.): "Basic and Clinical Endocrinology," 4th edn., Appleton and Lange, Norwalk, 1994.

Griffin, J.E., and S.R. Ojeda: "Textbook of Endocrine Physiology," 2nd edn., Oxford, New York, 1992.

Laycock, J., and P. Wise: "Essential Endocrinology," 3rd edn., Oxford University Press, Oxford, 1996.

Porterfield, S.P.: "Endocrine Physiology," Mosby, St. Louis, 1997.

Weintraub, B.D. (ed.): "Molecular Endocrinology: Basic Concepts and Clinical Correlations," Raven, New York, 1995.

Wilson, J.D., and D.W. Foster (eds.): "Williams Textbook of Endocrinology," 8th edn., Saunders, Philadelphia, 1991.

Chapter 11

Bagshaw, Clive R.: "Muscle Contraction," 2d edn., Chapman & Hall, London, 1993.

Barany, Michael, (ed.): "Biochemistry of Smooth Muscle Contraction," Academic Press, San Diego, 1995.

Jones, D.A., and J.M. Round: "Skeletal Muscle In Health and Disease: A Textbook of Muscle Physiology," Manchester University, Manchester, England, 1990.

Keynes, R.D, and D.J. Aidley: "Nerve and Muscle," 2d edn., Cambridge, Cambridge, 1991.

McComas, Alan J.: "Skeletal Muscle, Form and Function," Human Kinetics, Champaign, IL, 1996.

Tyldesley, Barbara, and June I. Grieve: "Muscles, Nerves and Movement: Kinesiology in Daily Living," 2d edn., Blackwell Science, Cambridge, MA, 1996.

Vrbova, G., T. Gordon, and R. Jones: "nerve-Muscle Interaction," 2d edn., Chapman & Hall, London, 1995.

Walton, John: "Disorders of Voluntary Muscle," 5th edn., Churchill Livingstone, New York, 1988.

Chapter 12

Brooks, V.B.: "The Neural Basis of Motor Control," Oxford, New York, 1986.

Cordo, P., and S. Harnad: "Movement Control," Cambridge Press, Cambridge, England, 1994.

Evarts, E.V., S.P. Wise, and D. Bousfield (eds.): "The Motor System in Neurobiology," Elsevier, New York, 1985.

Kandel, E.R., J.H. Schwartz, and T.M. Jessell: "Principles of Neural Science," 3rd edn., Elsevier/North-Holland, New York, 1991.

Patton, H.D., et al. (eds): "Textbook of Physiology: Volume 1 Excitable Cells and Neurophysiology," 21st ed, Saunders, Philadelphia, 1989.

Rosenbaum, D.A.: Human Motor Control," Academic Press, San Diego, 1991.

Chapter 13

Gazzaniga, M.S.: "The Cognitive Neurosciences," MIT, Cambridge, MA, 1995.

Gregory, R.L.: "Odd Perceptions," Routledge, New York, 1986.

Kandel, E.R., J.H. Schwartz, and T.M. Jessell: "Principles of Neural Science," 3rd edn., Elsevier/North-Holland, New York, 1991.

Kaplan, H.I.: "Kaplan and Sadock's Synopsis of Psychiatry: Behavioral Sciences and Clinical Psychiatry," 7th edn., Williams & Wilkins, Baltimore, 1994.

Kelner, K.L., and D.E. Koshland (eds.): "Molecules to Models: Advances in Neuroscience," American Association for the Advancement of Science, Washington, DC, 1989.

Levitan, I.B.: "The Neuron: Cell and Molecular Biology," 2nd edn., Oxford, New York, 1997.

Martinez, J.L., Jr., and R.P. Kesner: "Learning and Memory: A Biological View," 2nd edn., Academic Press, San Diego, 1991.

Patton, H.D., et al. (eds.): "Textbook of Physiology: Volume 1 Excitable Cells and Neurophysiology," 21st ed, Saunders, Philadelphia, 1989.

Scientific American (entire issue), "Brain and Mind," September, 1992.

Chapter 14

Beck, W.S.: "Hematology," MIT Press, 5th ed, Cambridge, 1991.

Berne, R.M., and M.N. Levy: "Cardiovascular Physiology," 7th edn., Mosby, St. Louis, 1996.

High, K.A., and H.R. Roberts (eds.): "Molecular Basis of Thrombosis and Hemostasis," Marcel Dekker, New York, 1995.

Katz, A.M.: "Physiology of the Heart," Raven, New York, 1992.

Mohrman, D.E., and L.J. Heller: "Cardiovascular Physiology," 4th edn., McGraw-Hill, New York, 1996.

Rowell, L.B.: "Human Cardiovascular Control," Oxford University Press, Oxford, 1993.

Williams, W.J., et al.: "Hematology," 5th edn., McGraw-Hill, New York, 1995.

Chapter 15

Davenport, H.W.: "The ABC of Acid-Base Chemistry," 7th edn., University of Chicago Press, Chicago, 1978.

Levitzky, M.G.: "Pulmonary Physiology," 4th edn., McGraw-Hill, New York, 1995.

Mines, A.H.: "Respiratory Physiology," 3rd edn., Raven, New York, 1992.

Staub, N.C.: "Basic Respiratory Physiology," Churchill Livingstone, New York, 1991.

West, J.B.: "Respiratory Physiology: The Essentials," 5th edn., Williams &

Wilkins, Baltimore, 1994.

Crystal, R.G., and West, J.B. (eds.): "The Lung: Scientific Foundations," 2nd edn., Raven, New York, 1996.

Chapter 16

Androgue, H.J., and D.E. Wesson: "Acid-Base," Blackwell Scientific, Boston, 1994.

Brenner, B.M., and F.C. Rector (eds.): "The Kidney," 5th edn., Saunders, Philadelphia, 1995.

Davenport, H.W.: "The ABC of Acid-Base Chemistry," 7th edn., University of Chicago Press, Chicago, 1978.

Seldin, D.W., and G.H. Giebisch (eds.): "The Kidney: Physiology and Pathophysiology," 2nd edn., Raven, New York, 1992.

Koepen, B.M., and B.A. Stanton: "Renal Physiology," 2nd edn., Mosby, St. Louis, 1997.

Rose, B.D.: "Clinical Physiology of Acid-Base and Electrolyte Disorders," 4th edn., McGraw-Hill, New York, 1993.

Smith, H.W.: "From Fish to Philosopher," Anchor Books, Doubleday, Garden City, NY 1961 (paperback).

Valtin, H., and J.A. Schafer: "Renal Function," 3rd edn., Little Brown, Boston, 1995.

Vander, A.J.: "Renal Physiology," 5th edn., McGraw-Hill, New York, 1995.

Chapter 17

Arias, Irwin, N., (ed.): "The Liver: Biology and Pathobiology," 3d edn., Raven, New York, 1994.

Castell, Donald O. (ed.): "The Esophagus," 2d edn., Little Brown, Boston, 1995.

Chang, Eugene B., Michael D. Sitrin, and Dennis D. Black: "Gastrointestinal, Hepatobiliary, and Nutritional Physiology," Lippincott-Raven, Philadelphia, New York, 1996.

Granger, D.N., J.A. Barrowman, and P.R. Kvietys: "Clinical Gastrointestinal Physiology," Saunders, Philadelphia, 1985.

Gustavsson, Sven, Devinder Kumar, and David V. Graham (eds.): "The Stomach," Churchill Livingstone, New York, 1992.

Johnson, L.R. (ed.): "Gastrointestinal Physiology," 5th edn., Mosby, St. Louis, 1997.

Johnson, Leonard R., (ed.): "Physiology of the Gastrointestinal Tract," 3d edn., Raven, New York, 1994 (two volumes).

Liang, Vay, (ed.): "The Pancreas: Biology, Pathology and Disease," Raven, New York, 1993.

Chapter 18

Adolph, E.F.: "Physiology of Man in the Desert," Interscience, New York, 1947.

Becker, K.L. (ed.): "Principles and Practice of Endocrinology and Metabolism," 5th edn., Lippincott, Philadelphia, 1995.

DeGroot, L.J.: "Endocrinology," 3rd edn., Saunders, Philadelphia, 1995.

Falkner, F., and Tanner, J.M. (eds.): "Human Growth," 2nd, Plenum, New York, 1986.

Goodman, H.M.: "Basic Medical Endocrinology," 2nd edn., Raven, New York, 1993.

Greenspan, F.S.: "Basic and Clinical Endocrinology," 4th edn., Appleton and Lange, Norwalk, CT, 1994.

Griffin, J.E., and S.R. Ojeda: "Textbook of Endocrine Physiology," 2nd edn., Oxford, New York, 1992.

MacKowiak, P.A. (ed.): "Fever: Basic Mechanisms and Management," Raven, New York, 1991.

Porterfield, S.P.: "Endocrine Physiology," Mosby, St. Louis, 1997.

Vander, A.J.: "Nutrition, Stress, and Toxic Chemicals: An Approach to Environmental Health Controversies," University of Michigan Press, Ann Arbor, 1981.

Wilson, J.D., and D.W. Foster (eds.): "Williams' Textbook of Endocrinology," 7th edn., Saunders, Philadelphia, 1987.

Chapter 19

Becker, K.L. (ed.): "Principles and Practice of Endocrinology and Metabolism," 5th edn., Lippincott, Philadelphia, 1995.

Carlson, B.M.: "Patten's Foundations of Embryology," 6th edn., McGraw-Hill, New York, 1996.

Case, R.M., and J.M. Waterhouse: "Human Physiology: Age, Stress, and the Environment," Oxford University Press, Oxford, 1994.

DeGroot, L.J.: "Endocrinology," 3rd edn., Saunders, Philadelphia, 1995.

Goodman, H.M.: "Basic Medical Endocrinology," 2nd edn., Raven, New York, 1993.

Greenspan, F.S.: "Basic and Clinical Endocrinology," 4th edn., Appleton and Lange, Norwalk, CT, 1994.

Griffin, J.E., and S.R. Ojeda: "Textbook of Endocrine Physiology," Oxford, New York, 1992.

Johnson, M., and B. Everett: "Essential Reproduction," 4th edn., Blackwell Scientific Publications, Oxford, 1995.

Pollard, I.: "A Guide to Reproduction: Social Issues and Human Concerns," Cambridge University Press, New York, 1994.

Porterfield, S.P.: "Endocrine Physiology," Mosby, St. Louis, 1997.

Wilson, J.D., and D.W. Foster (eds.): "Williams Textbook of Endocrinology," 8th edn., Saunders, Philadelphia, 1991.

Yen, S.S., and Jaffee, R.B.: "Reproductive Endocrinology," 3rd edn., Saunders, Philadelphia, 1991.

Chapter 20

Asterita, M.E.: "The Physiology of Stress," Sciences Press, 1985.

Barrett, J.T.: "Medical Immunology," F.A. Davis, Philadelphia, 1991.

Case, R.M., and J.M. Waterhouse: "Human Physiology: Age, Stress, and the Environment," Oxford University Press, Oxford, 1994.

Plotnikoff, N.P., et al.: "Stress and Immunity," CRC Press, Boca Raton, FL, 1991.

Roit, I., and P.J. Delves: "Essential Immunology," 8th edn., Blackwell, St. Louis, 1995.

Roit, I.: "Immunology," 4th edn., Mosby, St. Louis, 1995.

Stites, D.P., and A.I. Terr: "Basic and Clinical Immunology," 8th edn., Lange, Norwalk, CT, 1994.

Vander, A.J.: "Nutrition, Stress, and Toxic Chemicals: An Approach to Environmental Health Controversies," University of Michigan Press, Ann Arbor, 1981.

Sapolsky, R.M.: "Why Zebras Don't Get Ulcers: A Guide to Stress, Stress-Related Disease, and Coping," Freeman, New York, 1994.

CHAPTER OPENING CREDITS

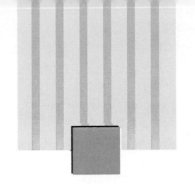

Index

Note: Page numbers in *italics* indicate illustrations. Page numbers followed by *t* indicate tables.